T0189102

Lecture Notes in Computer Science 571
Edited by G. Goos and J. Hartmanis

J. Vytopil (Ed.)

Formal Techniques in Real-Time and Fault-Tolerant Systems

Second International Symposium
Nijmegen, The Netherlands, January 8-10, 1992
Proceedings

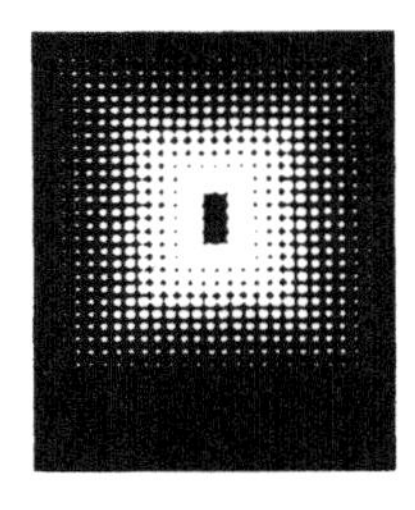

Springer-Verlag
Berlin Heidelberg New York
London Paris Tokyo
Hong Kong Barcelona
Budapest

Series Editors

Gerhard Goos
Universität Karlsruhe
Postfach 69 80
Vincenz-Priessnitz-Straße 1
W-7500 Karlsruhe, FRG

Juris Hartmanis
Department of Computer Science
Cornell University
5148 Upson Hall
Ithaca, NY 14853, USA

Volume Editor

Jan Vytopil
BSO/AT Utrecht, Kon. Wilhelminalaan 5, Postbus 8052
3503 RB Utrecht, The Netherlands; and
Catholic University of Nijmegen, Vakgroep Informatica
Toernooiveld 1, Postbus 9010, 6500 GL Nijmegen, The Netherlands

CR Subject Classification (1991): F.3, G.3, F.4.3, F.1.2, B.1.3

ISBN 3-540-55092-5 Springer-Verlag Berlin Heidelberg New York
ISBN 0-387-55092-5 Springer-Verlag New York Berlin Heidelberg

Printed in Germany

Typesetting: Camera ready by author
Printing and binding: Druckhaus Beltz, Hemsbach/Bergstr.
45/3140-543210 - Printed on acid-free paper

PREFACE

Practically every day, the media report that malfunctioning of a computer system somewhere has resulted in another tragic incident. This does not necessarily mean that the software and hardware making up such a system have not been designed with as much care as is commercially feasible. However, as the burden of controlling complicated systems is shifted onto computers, so the complexity of computer software and hardware increases. The sobering descriptions of failures of some systems has lead to the belief that there is a need for a distinct engineering discipline with its own theoretical foundations, objective design standards and supporting tools in order to develop control systems that deliver the desired functionality on time and even in presence of hardware or software failures.

The papers of this volume are devoted to considering the problems and the solutions in safety-critical system design and to examining how well the use of formal techniques for design, analysis and verification serves in relating theory to practical realities.

These papers were presented at the Symposium on Formal Techniques in Real-Time and Fault-Tolerant Systems, held at the University of Nijmegen, the Netherlands, January 8-10, 1992.

The proceedings of the previous symposium (University of Warwick, 1988) also appeared in the Lecture Notes in Computer Science, as Volume 331.

The present volume contains final versions of the selected papers. All of them underwent a careful refereeing process. I would like to thank the following people who agreed to referee the papers: Ö. Babaoglu, J.A. Bergstra, F. Cristian, S. Gerhart, N. Halbwachs, M. Joseph, A. Mok, A. Pnueli , G.M. Reed, W.P. de Roever, F.B. Schneider and D. Weber.

My special thanks go the colleagues from the organization committee, W.P. de Roever and M. Joseph, for their support and advice. The administrative work has been cheerfully and competently handled by Mirèse Willems, and Hanno Wupper has been invaluable in helping with the organization of the event.

The Symposium would not have been held but for the generous financial support of BSO Nederland.

Last but not least, I am grateful to Springer-Verlag for the quick publication.

Department of Computer Science
University of Nijmegen
November 1991

Jan Vytopil
Chairman, Programme Committee

CONTENTS

Session 8

Session 9A

Session 9B

Supplement

ISL: An Interval Logic for the Specification of Real-time Programs

Asis Goswami, Michael Bell, Mathai Joseph
Department of Computer Science *
University of Warwick

Abstract

ISL is a linear-time temporal logic for specifying properties of programs in execution intervals which are sequences of states. The end points of intervals are specified using *instances* of state predicates (or assertions) *or* time values. Abstract intervals, delimited by formulae over states in a computation, are used as the first step in constructing a timed specification. This is then transformed to incorporate timing, first by logical formulae and then using concrete time domains. Refinements are introduced to define time domains and timing properties and include refinement to programming constructs. We outline a way to specify resource limitations along with the functional and timing properties of programs. The specification method is illustrated with some examples.

1 Introduction

Real-time systems are reactive systems [HP85, Pnu86] which must satisfy timing properties. Temporal logics, both linear-time [Pnu86] and branching-time [CE81] versions, have been widely used to specify and reason about reactive systems. A linear-time temporal logic describes and verifies the properties of *state sequences* which are semantic denotations of program executions. A branching-time temporal logic, on the other hand, is used to reason about *state trees* which describe the (possible) nondeterministic behaviour of a program for a given input. Linear-time temporal logics have been used to define interval temporal logics (e.g. [SMV83]) in which intervals (in computations) and their properties can be specified and manipulated.

Several ways have been suggested to extend temporal logics to describe timed state sequences and trees. The most successful of these are the *explicit clock* and *(time-) bounded operator* approaches.

In the explicit clock approach [PH88], timing properties are described using a global time variable whose value increases from one state to the next in a computation. This has the advantage in requiring no new techniques for reasoning about the timing properties of programs. However, unrestricted use of a time variable can defeat the basic role of temporal

*Supported by Research Grant GR/F 57960 from the Science and Engineering Research Council

logics in treating time abstractly. A more elegant way is to associate time bounds with temporal operators, e.g. [Koy89], and to add proof rules to reason about the timing properties of programs [HMP91].

Comparatively little attention has been paid to the possibility of extending interval temporal logics for specifying quantitative temporal properties. As most real-time properties hold in specific intervals in a computation, a calculus of such properties can be embedded in an appropriate interval temporal. Recently, Zhao Chaochen et al [CHR91] have shown that many of the requirements of time-critical systems can be expressed and reasoned about very effectively in a calculus of durations based on an interval temporal logic.

Intervals need not always be defined using external time domains. Lamport has argued that to avoid inaccuracies in timing, only events observable within a system should be used for timing other events in the system [Lam78]. More recently, Turski [Tur88] has given examples of the possible misuse of time in program specifications. To keep the specification both abstract and general, it seems desirable that in the development of a real-time program the use of external timing events in specifications should be postponed as long as possible. However, timing can be a useful tool for obtaining inexpensive implementations (Turski agrees [Tur88]) and is sometimes is the only viable way to express the interaction of a real-time system with a physical process.

In this paper, we define an interval temporal logic ISL (*I*nterval *S*pecification *L*ogic) in which a timed temporal specification can be developed from an untimed specification by refinement. Common real-time requirements are easily expressed either using timing by system events or in terms of the times of some external time domain.

ISL is obtained from a first order linear-time temporal logic by adding three new temporal operators to deal with real-time properties. The first is the 'interval operator' which is used to describe a unique interval in a computation and the properties of the computation that hold in that interval. An interval is specified by bounds which may be internal timing events expressed as ISL formulae or points in an external time domain. Any assertion p can be said to generate a timing event each time that it is true. (Timing by such events is meaningful because a computation is described by a 'non-stuttering' sequence of states.) The next new operator defines an instance: the nth timing event generated by p is thus represented by the formula p^n, called the nth *instance* of p. Instances can also be timed by an external time domain using another new operator '@': $p^n@t$ means that p^n holds in some state of the execution if and only if t is the time of that state.

We first describe a version of ISL in which external time domains are not used and then extend it to incorporate timing. We discuss the possibility of using multiple time domains and show what time transformations would then be required. We also outline a way to specify resource limitations along with the functional and timing properties of programs. The use of the ISL logic is illustrated with some examples and it is then compared with some related logics.

2 ISL: An Interval Specification Logic

ISL is a linear-time temporal logic with an underlying assertion language. It is used for specifying the properties of programs in execution intervals which are sequences of states. An interval is bounded by *instances* of assertions and is left-closed and right-open. The properties of a program in an interval are defined by composing assertions using propositional connectives and the temporal operators $\Box$ (always) and $\Diamond$ (eventually). An *interval formula* is of the form $[l, u]r$, where l and u are, respectively, the lower and upper bounds of an interval and r is the property of the interval.

An ISL formula is built from assertions, instances of assertions, and interval formulae using propositional connectives, universal and existential quantifiers and the temporal operators $\Box$ and $\Diamond$, and $\prec$ (precedence) with the following syntactic restrictions: no temporal operator is applied over interval formulae and the precedence operator is applied only over instances of assertions (to define the ordering of interval bounds). Quantification is allowed only over variables counting instances. An ISL formula characterizes a set of program *behaviours* which represent the executions of programs.

2.1 Behaviours

Any nonempty sequence of states in which no two consecutive states are identical is called a *behaviour*. A behaviour of a command C is a sequence of its states and represents an execution of the command by the *changes* in state observed during the execution.

The state at position i in a behaviour σ is σ_i (the first state is at position 0). If σ is finite, $|\sigma|$ denotes its length and $[\sigma]$ denotes the set $[0, |\sigma| - 1]$ where, for integers i and j, $[i, j]$ is the set $\{ k \mid i \leq k \leq j \}$ of integers. If σ is an infinite sequence, $[\sigma] = \omega$, the set of nonnegative integers. For any behaviour σ, if $i, j \in [\sigma]$ such that $i < j$ then $\sigma[i, j]$ denotes the behaviour $\sigma_i, \ldots, \sigma_{j-1}$.

2.2 Meaning of ISL formulae

If σ is a behaviour, $(\sigma, j) \models p$ states that *p is true of σ at position j* (or, *σ satisfies p at j*). If $(\sigma, 0) \models p$, then we write $\sigma \models p$ (or, *σ satisfies p*). If p is satisfied by *any* behaviour, then $\models p$ (or *p is a theorem*). Theorems are "anchored" [MP89] to the initial states of behaviours, i.e., they hold only for the first positions of behaviours.

The assertion language of ISL is based on first order predicate calculus with interpreted symbols representing functions and relations on the domains of values of variables. The propositional connectives are $\neg$ (negation), $\wedge$ (conjunction), $\vee$ (disjunction), and $\rightarrow$ (implication). The symbol F represents the assertion equivalent to $r \wedge \neg r$ for any assertion r, and T represents the assertion $\neg F$. The meaning of primitive assertions and propositional connectives is defined in the usual way. In this section, we shall define only the temporal operators of ISL. For convenience, ISL symbols for propositional connectives and quantifiers are used in the metalanguage with obvious meaning.

2.2.1 Always ($\Box$)

$\Box p$ is true of behaviour σ at position j if p is true of σ at j and all positions beyond j.

$$(\sigma, j) \models \Box p \triangleq \forall k \in [\sigma] : k \geq j \rightarrow (\sigma, k) \models p$$

2.2.2 Instances

If p is an assertion and m a positive integer, p^m is the mth *instance* of p (m is called the *order* of the instance). p^m holds at position j in a behaviour σ if j is the mth position in σ at which p holds.

- $(\sigma, j) \models p^1 \triangleq (\sigma, j) \models p \wedge (\forall i \in [\sigma] : i < j \rightarrow (\sigma, j) \models \neg p)$
- For $m > 1$, $(\sigma, j) \models p^m \triangleq$
  $(\sigma, j) \models p \wedge (\exists i < j : (\sigma, j) \models p^{m-1} \wedge (\forall k \in [i+1, j-1] : (\sigma, k) \models \neg p))$

Example (*Pedestrian Lights*): The lights at a pedestrian crossing are either *red* or *green*, and requests to cross are made by pressing a *button*. If a button is pressed and the lights cannot change from red to green immediately the request is recorded (*rec*). The lights will change to allow pedestrians to cross only if there is an unserviced request. When the lights change from green to red the request is cancelled. Initially there are no requests recorded, the button is not pressed and the lights are red. A behaviour σ is:

σ	σ_0	σ_1	σ_2	σ_3	σ_4	σ_5	σ_6
rec	$(\neg rec)^1$	rec^1	rec^2	rec^3	rec^4	rec^5	$(\neg rec)^2$
button	$(\neg button)^1$	$button^1$	$(\neg button)^2$	$(\neg button)^3$	$button^2$	$(\neg button)^4$	$(\neg button)^5$
red	red^1	red^2	red^3	$(\neg red)^1$	$(\neg red)^2$	$(\neg red)^3$	red^4
green	$(\neg green)^1$	$(\neg green)^2$	$(\neg green)^3$	$green^1$	$green^2$	$green^3$	$(\neg green)^4$

The behaviour σ is defined in terms of instances of assertions. In state σ_1 the button is pressed and the request is recorded. The button is then released. At some appropiate time after σ_2 the lights change to green and whilst they are green the button is again pressed and released. Finally the lights change from green to red and the request is cancelled.

2.2.3 The interval operator

A behaviour σ satisfies the interval formula $[p^m, q^n]r$ if there are positions i, j in σ such that $i < j$, p^m holds at i, q^n holds at j and the behaviour $\sigma[i, j]$ satisfies r. If no such i and j can be found, σ satisfies $[p^m, q^n]r$. Thus[1],

$$\sigma \models [p^m, q^n]r \triangleq \forall i, j : (i < j \wedge (\sigma, i) \models p^m \wedge (\sigma, j) \models q^n) \rightarrow \sigma[i, j] \models r$$

[1] As temporal operators cannot be applied over interval formulae, it is not necessary to define $(\sigma, j) \models [p^m, q^n]r$

Example (*Pedestrian Lights*): In the behaviour σ, the second instance of *red* and the third instance of *green* hold at σ_1 and σ_5 respectively. As rec^1 holds at σ_1, $\sigma[1,5]$ satisfies *rec* - thus σ satisfies $[red^2, green^3]rec$. As *rec* holds throughout $\sigma[1,5]$, the behaviour also satisfies $[red^2, green^3]\Box rec$.

2.2.4 Derived operators

For any formula p and q, $\Diamond p$ (*eventually* p) $\triangleq \neg\Box\neg p$, and $p \Rightarrow q$ (p *entails* q) is the abbreviation of $\Box(p \rightarrow q)$. Three other operators are defined:

1. p **before** q: $p \prec q \ \triangleq\ \Diamond p \ \wedge\ (q^1 \Rightarrow \Box\neg p^1)$

   $p \prec q$ is true of a behaviour if p is eventually true and if j is the earliest position where p holds then q does not hold at or before j.

2. p **with** q: $p \approx q \ \triangleq\ \neg(p \prec q) \ \wedge\ \neg(q \prec p)$

3. p **with or before** q: $p \preceq q \ \triangleq\ (p \approx q) \vee (p \prec q)$

Example (*Pedestrian Lights*): For any behaviour, if the interval $[red^m, (\neg rec)^{m'}]$ exists and *rec* holds throughout that interval, then sometime after red^m and before $(\neg rec)^{m'}$ the lights change to green and do not change back to red until $(\neg rec)^{m'}$ holds (for all examples we take $\Rightarrow$ and $\rightarrow$ to have lower precedence than all operators other than $\forall$ and $\exists$).

$$\begin{array}{c}\forall m, m' : \ (red^m \prec (\neg rec)^{m'} \prec F) \ \wedge [red^m, (\neg rec)^{m'}]\Box rec \ \Rightarrow \\ (\exists m'' : red^{m''} \approx (\neg rec)^{m'}) \wedge [red^m, (\neg rec)^{m'}]\Diamond\Box green\end{array}$$

2.3 Reasoning about ISL specifications

Instances of assertions are used as temporal markers in behaviours and also to delimit intervals and, therefore, instances can be assumed to constitute a time domain with the ordering $\prec$ or $\preceq$. Like the ordering in commonly used time domains such as sets of integers, rationals or reals, these relations are total orders ($\preceq$ is the reflexive version of $\prec$) with least element T^1 (representing the first instant of time). A few other useful properties of $\prec$, $\approx$, $\preceq$ and instances are given below.

For convenience, we sometimes write

$$\frac{p_1, \ldots, p_n}{p}$$

to mean that $(p_1 \wedge \cdots \wedge p_n) \rightarrow p$. All interval bounds are instances of assertions.

T1. $p \prec q \ \leftrightarrow\ p^1 \prec q^1$

T2. $\Diamond p \ \leftrightarrow\ p \prec F$

T3. $\Box p \ \leftrightarrow\ \neg p \approx F$

T4. $(p \Rightarrow q) \ \rightarrow\ (q \preceq p)$

T5. $((p \vee q) \prec r) \leftrightarrow ((p \prec r) \vee (q \prec r))$

T6. $\dfrac{p \prec q,\ \ p \Rightarrow p',\ \ q' \Rightarrow q}{p' \prec q'}$

The following are some useful theorems for reasoning about instances.

T7. $p^m \Rightarrow p$

p has an instance at some state in a behaviour only if p holds there.

T8. If p is an assertion, $p \Rightarrow (\exists m : p^m)$

Any state where an assertion p holds has an instance of p. Thus any state in a behaviour has an instance of T.

T9. $[p^m, p^m]F$

p^m can hold at most once in a behaviour.

T10. $\dfrac{\Diamond p^m,\ \ m > 1}{p^{m-1} \prec p^m}$

Any instance of an assertion is preceded by all its lower order instances (if any). Thus the state at position i is characterized by the instance T^{i+1}.

T11. $(p \Rightarrow q) \leftrightarrow (\forall m : p^m \Rightarrow (\exists n \geq m : q^n)$

p entails q if and only if any instance of p corresponds to an instance of q of at least the same order.

Some properties of the interval operator are given below.

T12. $\models [l,u](r \rightarrow r') \leftrightarrow ([l,u]r \rightarrow [l,u]r')$

An ISL interval distributes over $\rightarrow$. A consequence of this is the theorem

$$[l,u]\neg r \leftrightarrow ([l,u]F \vee \neg[l,u]r)$$

T13. (a) $[l,u](r \vee s) \leftrightarrow ([l,u]r \vee [l,u]s)$
(b) $[l,u](r \wedge s) \leftrightarrow ([l,u]r \wedge [l,u]s)$
An ISL interval distributes over disjunction and conjunction.

T14. $\dfrac{[p,q]r,\ \ p' \Rightarrow p,\ \ q' \Rightarrow q}{[p',q']r}$

The properties of an interval are preserved if the interval bounds are strengthened (because the resulting interval describes a subset of the set of behaviours contained in the original interval).

T15. $\dfrac{[p,q]\Box r,\ \ p \preceq p'}{[p',q]\Box r}$

Any invariant of an interval is an invariant of all its final subintervals. Note that the hypothesis need not be strengthened by $p' \preceq q$, since if $q \prec p'$ the interval defined by the consequent is empty, and any formula is true of an empty interval.

T16. $$\frac{[p,q]\Box r, \quad r \text{ is an assertion}, \quad p \preceq p', \quad q' \Rightarrow \Diamond q}{[p',q']\Box r}$$

Note that the hypothesis does not require that q' should hold, but only that if it holds in some state, q must be true there or in some subsequent state.

T17. $$\frac{[p,q]\Box r, \quad [q,s]\Box r, \quad r \text{ is an assertion}, \quad (p \prec s \prec F) \Rightarrow \Diamond q}{[p,s]\Box r}$$

3 Timed Specifications

An ISL specification with intervals bounded by instances is timed by mapping the interval bounds into an external time domain. An instance p^m is mapped into an instant t in the time domain by adding a formula $p^m@t$, called a *timed instance.* External timing measures the distances between the states of a behaviour and hence the durations of formulae in specific intervals.

3.1 Time domains

A *time domain* is a structure $(\mathbf{T}, <, +, 0)$ where $\mathbf{T}$ is an infinite set of objects called *instants*, $<$ an irreflexive total order on $\mathbf{T}$ with least instant 0, and $+$ an 'addition' operation on $\mathbf{T}$. The minimum reflexive relation on $\mathbf{T}$ is denoted by $=$, $\leq$ is the reflexive closure of $<$, and $>$ and $\geq$ are, respectively, the converses of $\leq$ and $<$. The domains of nonnegative integers, rationals or reals with their natural ordering and addition can be used as time domains.

A subset R of $\mathbf{T}$ is said to be *finitely variable* if, for any closed interval I of $\mathbf{T}$, $R \cap I$ is finite.

3.2 Timed behaviours

A behaviour σ is *timed* by a domain $\mathbf{T}$ by extending its state to include a time variable ξ ranging over $\mathbf{T}$. The value of ξ for a state s in σ is given by $Time_\sigma(s)$. We shall omit the subscript σ if the reference is clear from the context. The set of times of the states of σ is given by $Timeset(\sigma)$.

A timed behaviour is assumed to satisfy the following conditions.

W1. $Time(\sigma_0) \ = \ 0$

The time of the initial state is 0.

W2. $\forall i \in [\sigma] : \ (i+1) \in [\sigma] \ \rightarrow \ Time(\sigma_i) < Time(\sigma_{i+1})$

Thus, time increases from one state to the next.

W3. *finite variability of states*: $Timeset(\sigma)$ is finitely variable.

3.3 Timed instances

The timed instance $p^m@t$ is defined by

$$p^m@t \triangleq (p^m \Leftrightarrow (\xi = t))$$

The notation $p@t$ will be used as an abbreviation of

$$\exists m : p^m@t$$

Example (*Pedestrian Lights*): Let the behaviour σ be timed by the time domain *Seconds* as follows:

σ	σ_0	σ_1	σ_2	σ_3	σ_4	σ_5	σ_6
Time	0	4	5	10	12	13	20

Therefore, as the button is first pressed at time 4, $button^1 @ 4$ is true, as is $button @ 4$.

The formula $(\exists m, n : T^m@t \wedge T^n@t' \wedge [T^m, T^n]r)$ is abbreviated to $\langle t, t' \rangle r$. This formula states that, if the behaviour has a subsequence such that ξ has the values t and t', respectively, at its begining and at the end, respectively, of the subsequence, The formula $\neg\langle t, t' \rangle F$ asserts that the behaviour has times t and t' such that $t < t'$. The following rules are used to relate the ordering of instances with that of time and to transform untimed intervals into intervals with time bounds.

TIME-1 $\quad (p^m@t \wedge q^n@t' \wedge \Diamond q^n) \rightarrow ((p^m \prec q^n) \leftrightarrow (t < t'))$

TIME-2 $\quad (p^m@t \wedge q^n@t' \wedge \Diamond q^n) \rightarrow ((p^m \approx q^n) \leftrightarrow (t = t'))$

TIME-3 $\quad (p^m@t \wedge q^n@t') \rightarrow ([p^m, q^n]r \leftrightarrow \langle t, t' \rangle r)$

Example (*Pedestrian Lights*): In the timed behaviour σ, the formulae $rec^1 @ 4$, $(\neg rec)^2 @ 20$ and $[rec^1, (\neg rec)^2]\Box rec$ are true, thus $\langle 4, 20 \rangle \Box rec$ is true in σ.

3.4 Durations of formulae

The *duration* $\rho(r)$ of an *assertion* r in an interval is the total length of time for which r holds in that interval. A *duration formula* is a predicate over the time domain obtained by using the timing function ρ. Interval formulae are extended to have duration formulae as conjuncts in the properties of intervals. D1 - D5 are the axioms for duration formulae.

D1. $\langle t, t' \rangle (\rho(T) = t' - t)$
D2. $\langle t, t' \rangle \Box r \Leftrightarrow \langle t, t' \rangle (\rho(r) = t' - t)$

Example (*Pedestrian Lights*): Since $\langle 4, 20 \rangle \Box rec$ is true in the timed behaviour σ, we have $\langle 4, 20 \rangle (\rho(rec) = 16)$, and using TIME-3 it can be shown that $[rec^1, (\neg rec)^2](\rho(rec) = 16)$.

D3. $\dfrac{t < t' < t'', \ \langle t, t' \rangle(\rho(r) = \delta_1), \ \langle t', t'' \rangle(\rho(r) = \delta_2)}{\langle t, t'' \rangle(\rho(r) = \delta_1 + \delta_2)}$

Example (*Pedestrian Lights*): Assume that when the lights change to green they must remain green for exactly 10 seconds, that is

$$\begin{gathered}\forall m, n : (green^m \prec red^n \prec F) \wedge [green^m, red^n]\Box green \wedge \\ (m = 1 \vee (m > 1 \wedge [green^{m-1}, green^m]\Diamond\Box red)) \wedge green^m \,@\, t \\ \Rightarrow red^n \,@\, t + 10\end{gathered}$$

In the timed behaviour σ, the lights are never green in the interval $\langle 4, 10 \rangle$, thus from TIME-3 $[rec^1, green^1](\rho(green) = 0)$. As $[green^1, (\neg rec)^2]\Box green$ is true, from the assumption above and $(\neg rec)^2 \approx red^4$ it can easily be shown that $[green^1, (\neg rec)^2](\rho(green) = 10)$.

Given these duration formulae, by applying D3, TIME-1 and TIME-3 it can be shown that $[rec^1, (\neg rec)^2](\rho(green) = 10)$ is true in the timed behaviour σ.

D4. $\langle t, t' \rangle(\rho(r) = \rho(T) - \rho(\neg r))$

Example (*Pedestrian Lights*): Given the previous result, $[rec^1, (\neg rec)^2](\rho(\neg green) = 6)$ is also true in the timed behaviour σ.

D5. $\langle t, t' \rangle((\rho(r_1 \vee r_2) = \rho(r_1) + \rho(r_2) - \rho(r_1 \wedge r_2))$

Some useful theorems of such interval formulae are given below.

T18. $\dfrac{\langle t, t' \rangle F}{\langle t, t' \rangle(\rho(T) = 0)}$

The duration of the empty interval is zero.

T19. $\dfrac{t < t'}{\langle t, t' \rangle(\rho(T) > 0)}$

The duration of a non-empty interval is greater than zero.

T20. $\langle t, t' \rangle(\rho(r) > 0) \ \Leftrightarrow \ \langle t, t' \rangle \Diamond r$

The duration of r in the interval $\langle t, t' \rangle$ is greater than zero if and only if r holds sometime in the interval.

Example (*Pedestrian Lights*): In the timed behaviour σ, the formulae $\langle 4, 20 \rangle(\rho(green) = 10)$ is true, thus $\langle 4, 20 \rangle \Diamond green$ is also true.

3.5 Transformation of time domains

The timing properties of a program which have been specified using a global domain of time must be transformed into times observable (or measurable) by local clocks when the program is implemented as a set of parallel processes. A similar transformation may also be needed if these are changes in the execution environment of programs.

A mapping f from a time domain $\mathbf{T}$ into another time domain $\mathbf{T}'$ preserves the timing properties of the underlying computation if it preserves the properties of $<$ and $+$.

a) $\forall t, t' \in \mathbf{T}: \; t < t' \;\Rightarrow\; f(t) <' f(t')$
b) $\forall t, t' \in \mathbf{T}: \; f(t + t') = f(t) +' f(t')$

We say that $\mathbf{T}'$ *embeds* $\mathbf{T}$ (or $\mathbf{T}$ *is embedded by* $\mathbf{T}'$) if such a function f exists and one of these functions is called the *standard transformation* of $\mathbf{T}$ into $\mathbf{T}'$.

A transformation must preserve the ordering of instants and also provide an interpretation of constants used for instants or durations in the specification. Durations of constant length will be preserved by the transformation f. But a constant $\tau \in \mathbf{T}$ represents an instant only in a timing formula such as $p^m@\tau$ and its transformation may be less simple. Suppose we have

$$p^m@\tau \;\wedge\; q^n@\tau' \;\wedge\; \tau < \tau'$$

In the absence of any other information, we can assume that only the ordering of τ and τ' is significant and any mapping of τ and τ' that preserves this ordering is acceptable. However, if in addition

$$[p^m, q^n](\rho(T) > t)$$

then $\tau' > \tau + t$ and a mapping g of τ and τ' must ensure that

$$g(\tau') - g(\tau) > f(t)$$

Let $\phi(\rho(T))$ be a duration formula constructed from constants in a domain $\mathbf{T}$ and the duration-valued expression $\rho(T)$, and let $\mathbf{T}'$ be any other time domain which embeds $\mathbf{T}$. Then $\phi(\rho(T))[\mathbf{T} \rightarrow \mathbf{T}']$ is a duration formula obtained from $\phi(\rho(T))$ by substituting $f(\tau)$ for all occurrences of constants $\tau \in \mathbf{T}$ and $\phi(\rho(T))[d/\rho(T)]$ is a duration formula obtained from $\phi(\rho(T))$ by substituting a duration variable d for $\rho(T)$.

Let the specification S be timed by domain $\mathbf{T}$ and let $\mathbf{T}'$ be some other time domain. If any constant in $\mathbf{T}$ occurs as durations in S, let $\mathbf{T}'$ embed $\mathbf{T}$. For a mapping $g : \mathbf{T} \rightarrow \mathbf{T}'$, let $S[\mathbf{T} \xrightarrow{g} \mathbf{T}']$ be the specification obtained from S by substituting $g(\tau)$ for all instants τ and $f(d)$ for all durations d.

The specification $S[\mathbf{T} \xrightarrow{g} \mathbf{T}']$ is a $\mathbf{T}'$-transformation of S if for all implications of S of the form

$$p^m@t_1 \;\wedge\; q^n@t_2 \;\wedge\; t_1 < t_2 \;\wedge\; [p^m, q^n]\phi(\rho(T))$$

where $\phi(\rho(T))$ is a duration formula involving only $\rho(T)$ and constants in $\mathbf{T}$, the following hold:

- $g(t_1) < g(t_2)$, and
- for the duration variable d,
  $\phi(\rho(T))[\mathbf{T} \rightarrow \mathbf{T}'][d/\rho(T)] \;\wedge\; d = g(t_2) - g(t_1)$ is satisfiable in $\mathbf{T}'$.

4 Example: Disc Head Manager

4.1 Requirements

A disc head manager schedules requests for access to a disc drive. Information on the disc is stored in *cylinders*, numbered from 0 to $C-1$, and is written and read by *heads*. The heads may access a cylinder only when they are over that cylinder.

At times, requests for disc access may arrive faster than they are serviced, and unserviced requests are queued. At any time, there are at most C requests in the queue and each request must eventually be honoured. When the heads are over a cylinder they remain there until all requests on the queue for that cylinder have been honoured. When a request is serviced it is removed from the queue. It is assumed that requests are distributed so that if there are C successive requests for sectors in cylinder i then there will be at least C successive requests for other cylinders before the next request for cylinder i.

At all times, the disc heads may move in one of two directions. When the heads are over cylinder C_i then they are moved to cylinder C_j only if there is a request for cylinder C_j, if C_j is in the direction of movement of the heads and there are no outstanding requests for C_i or for any cylinder between C_i and C_j. The direction of movement of the heads is reversed when there is no further request that can be served in that direction. When there are no outstanding requests the heads remain stationary.

The time for the heads to reach a cylinder (the 'seek' time) is linearly proportional to the distance travelled; the maximum seek time ($Tmax$) occurs when the heads are moved from cylinder 0 to cylinder $C-1$, or vice versa. Once over a cylinder, it takes a time of $Tacc$ to access the information for each request for that cylinder.

An untimed specification of a related problem, the Single Lift System, was produced by Barringer [Bar85] using propositional temporal logic. Our solution proceeds from the specification of the untimed behaviour to the behaviour with instances and then to the timed behaviour. Finally some timed properties of the system are defined.

4.2 Specification

4.2.1 State variables

Let at_i be true when the heads are stationary over cylinder C_i, and let b_i be true when the heads are *busy* serving a request on that cylinder. Thus $at_i \wedge \neg b_i$ is true when the heads are idle over cylinder C_i, and

$$\bigwedge_i ((\neg at_i \Rightarrow \neg b_i) \wedge (b_i \Rightarrow at_i)) \tag{1}$$

A request for access to the ith cylinder is denoted by req_i, and q_i denotes that a request for C_i is in the queue and the heads are not busy at C_i;

$$\bigwedge_i (q_i \Rightarrow \neg b_i) \tag{2}$$

4.2.2 Untimed behaviour

Initially, the heads are not busy at any cylinder, and no request has been made or is in the queue. The position of the heads is not known.

$$\bigwedge_i (\neg b_i \wedge \neg req_i \wedge \neg q_i) \tag{3}$$

At any time, the heads are either over one cylinder or are moving between cylinders (and are therefore considered to be over no cylinder).

$$\Box(\bigwedge_i \neg at_i \vee \bigvee_i (at_i \wedge \bigwedge_{j \neq i} \neg at_j)) \tag{4}$$

Every request for disc access must eventually be serviced.

$$\bigwedge_i (req_i \Rightarrow \Diamond b_i) \tag{5}$$

When the heads access a cylinder they may service at most C requests before becoming idle, thus the heads are continually busy over any cylinder for only a finite time.

$$\bigwedge_i (b_i \Rightarrow \Diamond \neg b_i) \tag{6}$$

4.2.3 Untimed behaviour with instances

If a request is made for cylinder C_i which cannot be serviced immediately then the request is queued. If the heads are not already busy at C_i then q_i holds until the heads access C_i.

$$\bigwedge_i \forall m \exists n : (req_i^m \preceq b_i^n \prec F) \wedge [req_i^m, b_i^n] \Box q_i \tag{7}$$

If there is no request for C_i in the queue, then q_i will remain false until a request for C_i is made when the heads are not over C_i.

$$\bigwedge_i \forall m, n : [(\neg q_i)^m, q_i^n] \Box \neg q_i \Rightarrow [(\neg q_i)^m, q_i^n] \Box \neg (req_i \wedge \neg at_i) \tag{8}$$

If the heads are busy over cylinder C_i and later access C_j without stopping at any other cylinder:

$$\textbf{from\_}C_i^m\textbf{\_to\_}C_j^n \triangleq (b_i^m \prec b_j^n \prec F) \wedge \bigwedge_{k \neq i} [b_i^m, b_j^n] \Box \neg at_k$$

If $i = j$ it is trivial to show that $n > m$. If $i \neq j$ then b_j^n is the instance when the heads become busy at C_j.

Access by the heads to cylinder C_i, then to cylinder C_j and then to cylinder C_k can thus be defined as follows:

$$C_i^l\textbf{\_via\_}C_j^m\textbf{\_to\_}C_k^n \triangleq \textbf{from\_}C_i^l\textbf{\_to\_}C_j^m \wedge \textbf{from\_}C_j^m\textbf{\_to\_}C_k^n$$

The heads leave C_i with an outstanding request for some cylinder C_k if the following formulae is true.

$$\textbf{left\_}C_i^m\textbf{\_with\_}C_k\textbf{\_unserviced} \triangleq \exists n : (b_i^m \prec (\neg at_i)^n \prec F) \wedge [b_i^m, (\neg at_i)^n](\Box at_i \wedge \Diamond \Box q_k)$$

The method (i.e. the algorithm) of access to the disc can now be expressed in a straightforward way. If at any instance the heads are busy at cylinder C_i then either a request for C_i has been made at that instance and the heads have become immediately busy, or there was an outstanding request for C_i on the queue and the heads have just become busy, or the heads became busy at some earlier instance and have been continually busy at C_i.

$$\bigwedge_i \forall m \exists n : ((req_i^n \preceq b_i^m \prec F) \wedge [req_i^n, b_i^m]\Box q_i) \vee ((n < m) \wedge [b_i^n, b_i^m]\Box b_i) \tag{9}$$

Next, the heads do not go past a cylinder which has an unserviced request and they do not remain idle over a cylinder when there is an unserviced request. If the heads consecutively service C_i and C_j, then if $i = j$ the heads remain stationary when not busy, and there is no outstanding request for any cylinder while they are idle. If $i \neq j$ there is no outstanding request for some cylinder C_k between C_i and C_j which was made before the heads left C_i.

$$\bigwedge_i \bigwedge_j \forall m, n : \textbf{from\_}C_i^m\textbf{\_to\_}C_j^n \Rightarrow \tag{10}$$

$$(i = j \Rightarrow [b_i^m, b_i^n]\Box at_i \wedge \bigwedge_{k \neq i} [b_i^m, b_i^n]\Box\neg(q_k \wedge \neg b_i)) \wedge$$

$$(i \neq j \Rightarrow \neg\exists k : \textbf{left\_}C_i^m\textbf{\_with\_}C_k\textbf{\_unserviced} \wedge min(i,j) < k < max(i,j))$$

Finally, the heads service all the requests which lie in the direction of movement before there is a change of direction. Thus if the heads service C_i, C_j and C_k consecutively, and if $i < j$ and $j > k$, then when leaving C_j there is no outstanding request for any cylinder C_h $(h > k)$. A similar constraint holds if $i > j$ and $j < k$.

$$\bigwedge_i \bigwedge_j \bigwedge_k \forall l, m, n : C_i^l\textbf{\_via\_}C_j^m\textbf{\_to\_}C_k^n \Rightarrow \tag{11}$$

$$(i < j > k \Rightarrow \neg\exists h : \textbf{left\_}C_j^m\textbf{\_with\_}C_h\textbf{\_unserviced} \wedge i < j < h) \wedge$$

$$(i > j < k \Rightarrow \neg\exists h : \textbf{left\_}C_j^m\textbf{\_with\_}C_h\textbf{\_unserviced} \wedge i > j > h)$$

4.2.4 Timed behaviour

As the seek time is linearly proportional to the distance travelled and the maximum seek time is $Tmax$, the time taken for the heads to move from cylinder i to cylinder j without stopping at any other cylinder, is the product of $(Tmax/C)$ and $|i - j|$.

$$\bigwedge_i \bigwedge_j \forall t, t', t'' : (t < t' < t'') \wedge \langle t, t' \rangle \Box at_i \wedge \Diamond at_j \wedge at_j @ t'' \wedge \bigwedge_k \langle t', t'' \rangle \Box \neg at_k \Rightarrow \tag{12}$$

$$t'' - t' = (Tmax/C) * |i - j|$$

If the heads leave cylinder C_i at time t' and move without stopping to cylinder C_j, reaching it at time t'', then $t'' - t'$ is the time it takes to move across $|i - j|$ cylinders.

From the above formulae and the statement in the requirements that each request uses $Tacc$ head time, a number of timing properties of the system can be derived.

The maximum distance the heads move without stopping is between cylinder 0 and cylinder $C-1$, therefore $Tmax$ is the maximum time the heads move without stopping.

$$\forall t, t' : \neg\langle t, t'\rangle F \wedge (\bigvee_i \langle t, t'\rangle \Box \neg at_i) \Rightarrow t' - t \leq Tmax \tag{13}$$

The maximum unbroken time that the heads are busy at a cylinder occurs when there are C consecutive requests for the same cylinder and the heads service them during one access of the cylinder. As each request takes $Tacc$, the maximum time is $C * Tacc$.

$$\forall t, t' : \neg\langle t, t'\rangle F \wedge (\bigvee_i \langle t, t'\rangle \Box b_i) \Rightarrow t' - t \leq (C * Tacc) \tag{14}$$

Therefore, the maximum duration the heads are busy in any visit to a cylinder is $C * Tacc$.

$$\bigwedge_i \forall m, n : [at_i^m, (\neg at_i)^n] \Box at_i \Rightarrow [at_i^m, (\neg at_i)^n](\rho(b_i) \leq (C * Tacc)) \tag{15}$$

The maximum time taken to service a request occurs when there is a request for C_0 (C_{C-1}) made at the moment the heads leave C_0 (C_{C-1}), and there are C other requests on the queue including at least one for C_{C-1} (C_0). In this case the heads must travel two lengths of the disc which takes time $2 * Tmax$, and they must service $C-1$ requests which takes time $(C-1) * Tacc$. Therefore,

$$\forall t, t' : \neg\langle t, t'\rangle F \wedge (t < t') \wedge \bigvee_i (req_i @ t \wedge b_i @ t' \wedge \langle t, t'\rangle \Box \neg b_i) \Rightarrow \tag{16}$$

$$t' - t \leq (((C-1) * Tacc) + (2 * Tmax))$$

If there is a request for cylinder C_i at time t and the next time the heads become busy at C_i is t', then $t' - t$ will be at most equal to the time it takes to travel two lengths of the disc and service $C-1$ requests.

4.2.5 Formal specification

The complete specification of the disc head manager can be given as a conjunction of the appropriate ISL formulae.

The initial state of the heads is specified by (3): $h\_init \triangleq \bigwedge_i(\neg b_i \wedge \neg req_i \wedge \neg q_i)$

The formulae (1), (2), (4), (5) and (6) specify the progress properties of the heads and the constraints on head position, servicing and requests:

$$h\_progress \triangleq \bigwedge_i((\neg at_i \Rightarrow \neg b_i) \wedge (b_i \Rightarrow at_i) \wedge (q_i \Rightarrow \neg b_i) \wedge (req_i \Rightarrow \Diamond b_i) \wedge (b_i \Rightarrow \Diamond \neg b_i)) \wedge \Box(\bigwedge_i \neg at_i \vee \bigvee_i(at_i \wedge \bigwedge_{j \neq i} \neg at_j))$$

The conjunction of (7), (8) and (9) specifies how requests are dealt with:

$$\begin{aligned} r\_control \triangleq \bigwedge_i \; & ((\forall m \exists n : (req_i^m \preceq b_i^n \prec F) \wedge [req_i^m, b_i^n] \Box q_i) \wedge \\ & (\forall m \exists n : ((req_i^n \preceq b_i^m \prec F) \wedge [req_i^n, b_i^m] \Box q_i) \vee ((n < m) \wedge [b_i^n, b_i^m] \Box b_i)) \wedge \\ & (\forall m, n : [(\neg q_i)^m, q_i^n] \Box \neg q_i \Rightarrow [(\neg q_i)^m, q_i^n] \Box \neg(req_i \wedge \neg at_i))) \end{aligned}$$

The formulae (10) and (11) specify the service order of the heads:

$$
\begin{aligned}
h\_order \triangleq\ & \bigwedge_i \bigwedge_j (\forall m,n : \mathbf{from\_}C_i^m\mathbf{\_to\_}C_j^n \Rightarrow \\
& (i = j \Rightarrow [b_i^m, b_i^n]\Box at_i \wedge \bigwedge_{k \neq i}[b_i^m, b_i^n]\Box\neg(q_k \wedge \neg b_i)) \wedge \\
& (i \neq j \Rightarrow \neg\exists k : \mathbf{left\_}C_i^m\mathbf{\_with\_}C_k\mathbf{\_unserviced} \ \wedge \ min(i,j) < k < max(i,j))) \\
& \wedge \\
& \bigwedge_i \bigwedge_j \bigwedge_k (\forall l,m,n : C_i^l\mathbf{\_via\_}C_j^m\mathbf{\_to\_}C_k^n \Rightarrow \\
& (i < j > k \Rightarrow \neg\exists h : \mathbf{left\_}C_j^m\mathbf{\_with\_}C_h\mathbf{\_unserviced} \ \wedge \ i < j < h) \wedge \\
& (i > j < k \Rightarrow \neg\exists h : \mathbf{left\_}C_j^m\mathbf{\_with\_}C_h\mathbf{\_unserviced} \ \wedge \ i > j > h))
\end{aligned}
$$

The speed of the head movement is specified by (12):

$$
\begin{aligned}
h\_speed \triangleq\ & \bigwedge_i \bigwedge_j \ \forall t,t',t'' :(t < t' < t'') \wedge \langle t,t'\rangle\Box at_i \wedge \Diamond at_j \wedge at_j@t'' \wedge \bigwedge_k\langle t',t''\rangle\Box\neg at_k \Rightarrow \\
& \qquad t'' - t' = (Tmax/C) * |i - j|
\end{aligned}
$$

The specification of the disc head manager is thus:

$$DHM \triangleq \ h\_init \ \wedge \ h\_progress \ \wedge \ r\_control \ \wedge \ h\_order \ \wedge \ h\_speed$$

5 Limited Resources

The execution time of a command depends on the availability of resources such as processors, memory, and communication channels [JG88]. For simplicity, let the size of a resource be represented by a positive integer and be ordered by the natural ordering $\leq$ on integers. If there are n types of resources, a *resource description* is an n-tuple of the sizes of individual resources. Assume that resource descriptions can be ordered ($\leq$) by ordering their corresponding components.

Given a command C, let $MinRes(C)$ and $MaxRes(C)$ be resource descriptions representing the minimum and the maximum amount of resources required for an execution of C. If the available resources exceed $MaxRes(C)$, the execution time of C is the same as that with $MaxRes(C)$; C cannot be executed if at least the resources in $MinRes(C)$ are not available.

If no assumptions are made in the computation model about the availability of resources, an ISL specification p can be said to have an implementation C iff p is true of any behaviour of C irrespective of the availability of resources. A necessary condition for a timed ISL specification p to have an implementation C is that C must satisfy p when the resources in $MaxRes(C)$ are available (i.e., p is true of any behaviour of C corresponding to executions with resources in $MaxRes(C)$). Suppose such an implementation is found: let this be the *maximal resource implementation*. Is it then possible to verify whether other executions of C also satisfy p? This question is relevant only if C has independent components so that when one component is waiting for its resources, another component can proceed.

For simplicity, consider a single resource – the processor. Supppose a timed ISL specification p is refined to a parallel composition of n specifications $p_1, \ldots, p_n$ and each p_i is further refined to a sequential command (let that be identified by p_i). Now define assertions $sw(p_i)$ and $fw(p_i)$ to denote the points where the command p_i 'starts waiting for processor' and 'finishes waiting for processor'. Instances of these assertions have the property that

$$\forall n : \ \Diamond sw(p_i)^n \ \rightarrow \ sw(p_i)^n \prec fw(p_i)^n \prec sw(p_i)^{n+1}$$

The intervals in which the command p_i waits for a processor are described by timing these instances. Thus if

$$\Diamond sw(p_i)^n \wedge sw(p_i)^n @ t \wedge fw(p_i)^n @ t'$$

then $\langle t, t' \rangle$ is the nth wait interval of p_i.

The set of parallel commands p_i is partitioned so that any command in a set S of the partition interleaves only with other commands in the same set. No two commands in S can execute at the same time and some command in S must execute in any wait interval of any other command in S. It is necessary to describe the control structure of an interleaving command p_i in its specification so that only legal interleavings can be specified. (The full paper shows how the location predicates [KVR83] can be used for this purpose.)

The introduction of wait intervals in a specification shifts the times of all instances which occur after that interval. Reasoning about the timing properties of interleaved computations amounts to computing the resulting times of instances. This is often very hard but can be simplified by reducing the number of control points in commands where interleaving is allowed. Restricted interleaving can also be obtained by static scheduling.

An alternative transformational method for the verification of the timing properties of limited processor executions under different interleavings (i.e. different scheduling disciplines) was given in [MJ90] and the use of the method for infinite programs with command execution times in a real interval has been described in [MJ91].

6 Comparison

ISL is an interval temporal logic for specifying the timing and functional properties of real-time programs. We have shown how time transformations and resource requirements can be specified using the logic and the full paper describes refinements to generic programming constructs.

The use of intervals in ISL has many precedents: for example, Schwartz, Melliar-Smith and Vogt's Interval Logic [SMV83], Halpern, Manna and Moszkowski's Temporal Intervals [HMM83] and the Real-Time Logic (RTL) of Mok and Jahanian [JM86] all make use of intervals. But each of these uses has served different purposes and none has focussed on their role in a *method* for developing real-time specifications.

The logic with formulae most alike those in ISL is described in [SMV83], where an interval formula has the form $[I]\alpha$: 'the next time the interval I can be constructed, the formula α will 'hold' for that interval'; if the interval I cannot be found then the interval formula is vacuously true. Interval formulae can be composed with other temporal operators to derive higher-level properties of intervals such as $[I]\Box[J]\alpha$, which states that all J intervals within interval I have property α. All intervals are derived from primitive *event intervals*. An event defined by β occurs when β becomes true, so β denotes the *interval of change* of length 2 containing the $\neg\beta$ and β states comprising the change. These primitive intervals can be combined and manipulated to form other intervals. The functions ${}_{begin}I$ and ${}_{end}I$ denote the unit interval containing the first and last states respectively of the interval term I, and the operators $\Rightarrow$ and $\Leftarrow$ are used to derive intervals from interval arguments: for example, $[x = y \Rightarrow_{begin} (y = 16)]\Box x > z$ asserts that in the interval beginning the next time that variable x is equal to y and ending at the state before y changes to the value 16, x is always greater than z. Parameterized operations are also supported by this logic.

Melliar-Smith [Mel87] has extended this interval logic to allow real-time bounds on intervals. Events can also be defined by real-time offsets by applying new operators $+$ and $-$ to an event and a duration constant to yield a new event: thus if E is an event then so are $E + 1$ *second* and $E - 1$ *day*. Bounds on the length of intervals are provided by the operators $<$ and $>$ which relate the length of an enclosing context to a duration constant. Used with an interval, they relate the length of the interval to the constant: thus, if I is an interval, $[I] < 1$ *second* is a boolean predicate on the length of that interval. These extensions to the interval logic do not affect its decidability. But as there is no simple way to refer to the unique instants when an event occurs, the formula for a specific interval can often be very long and counterintuitive and as the only reference to time is by duration constants, timed intervals can also be difficult to specify. Narayana and Aaby [NA88] present a variant of the interval logic of [SMV83] which also has capabilities for specifying qualitative and quantitative properties of real-time behaviours.

RTL [JM86] was developed to reason about the timing behaviour of systems and to analyse and describe systems that have been transformed from an event-action model. An RTL formula is based on ***constants***, ***occurrences*** and ***state predicates***, but unlike ISL it does not use temporal operators. Constants may be ***action*** constants, ***event*** constants or ***integers***. Action constants denote schedulable units of work: for example, the primitive action of reading a value. An event constant is a temporal marker and so may be a start event constant $\uparrow A$ (where A is an action), a stop event constant $\downarrow A$, a transition event constant which marks a change in a state of the system or its environment, or an external event constant which affects the system behaviour. The occurrence function '@' relates time with events. Thus $@(e, i)$ is the time of the ith occurrence of event e. It can also be used to specify durations, as in $\forall i : @(\uparrow A, i) + \delta \leq @(\downarrow A, i)$, which asserts that the end of each action A is at least a time δ after the corresponding start event $\uparrow A$. A state predicate $S[t, t']$ is the interval from the transition event which makes the state attribute S true at time t to the transition event which makes it false at time t'. A state attribute may be false in an interval, for example, $\overline{S}[t, t']$. Although RTL makes use of intervals it is not a logic for reasoning in intervals. RTL was designed so that formulae can be translated into Pressburger Arithmetic by restricting its expressiveness. It has perhaps been used more for system analysis than for specifying programs.

The interval-based temporal logic of Halpern, Manna and Moszkowski [HMM83] is an extension of linear-time temporal logic [MP81] and is used to produce executable specifications. The properties of intervals are specified by various constructs. For a formula w and an interval $\sigma_0, ..., \sigma_n$, ◈w is true if w holds in at least one subinterval of $\sigma_0, ..., \sigma_n$, and ▣ is true if the formula w holds in all subintervals of $\sigma_0, ..., \sigma_n$. Similarly the operators ◇, □ and ◇, □ reason over initial and terminal subintervals respectively. The logic also contains interval operators such as *empty*, and *beg* w and *fin* w (which test if a formula w is true in the starting state and final state of an interval). Quantitative timing properties are handled by the function *len* whose value for any interval $\sigma_0, ..., \sigma_n$ is n. This logic has found use in the specification of a variety of hardware components and can be used to prove properties such as the correctness of an implementation.

In all these logics, an interval can simply be defined as a sequence of states from a starting state to a finishing state. But the properties of an interval are not always as easily specified. The precedence operator, instances and the untimed interval formulae of ISL allow the ordering properties of a specification to be easily specified, thus providing a very simple, yet powerful method of expressing the qualitative properties of any real-time behaviour.

One reason for the development of most interval logics is to provide a simple way of specifying the qualitative and quantitative properties of real-time systems. But if the model of time is very abstract, specifying real-time behaviours and converting timing constraints from abstract to concrete time can be difficult. ISL allows any time domain which satifies the properties of Section 3.1 to be used in a specification. Thus timing functions, which can specify the duration of an interval or a predicate, and unique timed intervals, which specify the start and end times of an interval, can be introduced. This allows specific intervals and timed safety properties to be easily derived and makes it possible for an untimed specification in ISL to be adapted to express timing properties. The ability to express intervals in a very simple form, and to progressively introduce the timing properties of a real-time system allows a real-time program to be specified by refinement of a simple untimed specification to a full specification which includes both timed and untimed interval formula.

References

[Bar85] H. Barringer. Up and down the temporal way. Technical Report UMCS-85-9-3, Department of Computer Science, University of Manchester, Manchester, 1985.

[CE81] E.M. Clarke and E.A. Emerson. Design and synthesis of synchronization skeletons using branching time temporal logic. In *Lecture Notes in Computer Science 131*, pages 52–71, Heidelberg, 1981. Springer-Verlag.

[CHR91] Z. Chaochen, C.A.R. Hoare, and A.P. Ravn. A calculus of durations. Draft, 4 February, 1991. To appear in *Information Processing Letters.*

[HMM83] J. Halpern, B. Moszkowski, and Z. Manna. A hardware semantics based on temporal intervals. In *Lecture Notes in Computer Science 154*, pages 278–291. Springer-Verlag, Heidelberg, 1983.

[HMP91] T.A. Henzinger, Z. Manna, and A. Pnueli. Temporal proof methodologies for real-time systems. In *Proceedings of the 18th ACM Symposium on Principles of Programmming Languages*, 1991.

[HP85] D. Harel and A. Pnueli. On the development of reactive systems. In *Proceedings of the NATO Advanced Study Institute on Logics and Models for Verification and Specification of Concurrent Systems, NATO AFI Series F, Vol. 13*, pages 477–498, Berlin, 1985. Springer-Verlag.

[JG88] M. Joseph and A. Goswami. What's 'real' about real-time systems? In *Proceedings of the 9th IEEE Real-Time Systems Symposium*, pages 78–85, Huntsville, Alabama, 1988.

[JM86] F. Jahanian and A. Mok. Safety analysis of timing properties in real-time systems. *IEEE Transactions on Software Engineering*, 12:890–904, 1986.

[Koy89] R. Koymans. *Specifying Message Passing and Time-Critical Systems with Temporal Logic.* PhD thesis, Technical University of Eindhoven, Eindhoven, 1989.

[KVR83] R. Koymans, J. Vytopil, and W.-P. de Roever. Real-time programming and asynchronous message passing. In *Proceedings of the 2nd ACM Symposium on Principles of Distributed Computing*, pages 187–197, Montreal, 1983.

[Lam78] L. Lamport. Time, clocks, and the ordering of events in a distributed system. *Communications of the ACM*, 21(7):558–565, 1978.

[Mel87] P.M. Melliar-Smith. Extending interval logic to real time systems. In *Lecture Notes in Computer Science 398*, pages 224–242, 1987.

[MJ90] A. Moitra and M. Joseph. Implementing real-time systems by transformation. In H. Zedan, editor, *Real-time Systems: Theory and Applications*, pages 143–157. North-Holland, 1990.

[MJ91] A. Moitra and M. Joseph. Determining timing properties of infinite real-time programs. Technical Report RR172, University of Warwick, Department of Computer Science, 1991.

[MP81] Z. Manna and A. Pnueli. Verification of concurrent programs: The temporal framework. In R.S. Boyer and J.S. Moore, editors, *The Correctness Problem in Computer Science*, pages 215–273. Academic Press, London, 1981.

[MP89] Z. Manna and A. Pnueli. The anchored version of the temporal framework. In *Lecture Notes in Computer Science 354*, pages 201–284. Springer-Verlag, Heidelberg, 1989.

[NA88] K.T. Narayana and A.A. Aaby. Specification of real-time systems in real-time temporal interval logic. In *Proceedings of the 9th IEEE Real-Time Systems Symposium*, pages 86–95, Huntsville, Alabama, 1988.

[PH88] A. Pnueli and E. Harel. Applications of temporal logic to the specification of real time systems (extended abstract). In *Lecture Notes in Computer Science 331*, pages 84–98. Springer-Verlag, Heidelberg, 1988.

[Pnu86] A. Pnueli. Applications of temporal logic to the specification and verification of reactive systems: A survey of current trends. In *Lecture Notes in Computer Science 224*, pages 510–584. Springer-Verlag, Heidelberg, 1986.

[SMV83] R.L. Schwartz, P.M. Melliar-Smith, and F.H. Vogt. An interval logic for higher-level temporal reasoning. In *Proceedings of the 2nd ACM Symposium on Principles of Distributed Computing*, pages 173–186, 1983.

[Tur88] W.M. Turski. Time considered irrelevant for real-time systems. *BIT*, 28:473–486, 1988.

Appendix : A Derived Proof System for ISL

The linear-time temporal logic of [MP89] (henceforth called TL) can be used to construct a proof system for ISL. We give rules for transforming any ISL formula into an equivalent formula which does not use the interval operator. These rules, together with the proof system of TL and axioms of instances, constitute a sound and complete proof system for ISL. We define only those operators of TL that are needed in the axioms of instances.

Basic Operators:

B1. Weak previous:

$(\sigma, j) \models \oslash p \triangleq (j = 0) \vee [j > 0 \wedge (\sigma, j-1) \models p]$

B2. Weak Since:

$(\sigma, j) \models p \mathcal{S} q \triangleq \forall k : j \geq k > 0 \Rightarrow [(\sigma, k) \models p \vee [\exists i : j \geq i \geq k \wedge (\sigma, i) \models q)]]$

Derived Operators:

Strong Previous	$\ominus p \triangleq \neg \oslash \neg p$
Has always been	$\boxminus p \triangleq p \mathcal{S} F$
Strong since	$p \mathcal{S} q \triangleq (p \mathcal{S} q) \wedge \diamondminus q$

We now present the axioms which are to added to the proof system of TL [MP89].

A1. If r is an assertion, then

$\models [l, u]r \leftrightarrow ((l \Rightarrow u) \vee (l \Rightarrow \Box \neg u) \vee (l \Rightarrow r)$

A2. $\models [l, u](r \rightarrow r') \leftrightarrow ([l, u]r \rightarrow [l, u]r')$

A3. $\Box$ **elimination:** $[l, u]\Box r \leftrightarrow \forall n : (l \preceq T^n \preceq u \rightarrow [T^n, u]r)$

Since $\Diamond r \leftrightarrow \neg \Box \neg r$ these transformation rules suggest that any ISL formula can be translated to an equivalent ISL formula which does not involve the interval operator. We, therefore, need an axiomatization of instances in TL. This is given below.

- $p^1 \Leftrightarrow p \wedge \oslash \boxminus \neg p$
- $\forall m : m > 1 \Rightarrow (p^m \Leftrightarrow p \wedge \ominus(\neg p \mathcal{S} p^{m-1}))$

where $r \Leftrightarrow s$ abbreviates $(r \Rightarrow s \wedge s \Rightarrow r)$.

Duration Specifications for Shared Processors*†

Zhou Chaochen[1,2,3]
Michael R. Hansen[1]
Anders P. Ravn[1]
Hans Rischel[1]

[1] *Department of Computer Science, Technical University of Denmark*

[2] *Programming Research Group, Oxford University, England*

[3] on leave from
Institute of Software, Academia Sinica, Beijing, China

Abstract: We present a specification oriented real-time semantics for real-time programs consisting of communicating sequential processes running on a shared processor configuration. The semantics, which is given in Duration Calculus [7], separates properties of a (compiled) program from properties attributable to a scheduling strategy. This gives a clear division of concerns when a given program under a given scheduling strategy has to be proven correct wrt. hard real-time constraints.

Keywords: Duration Calculus, specifications, real-time systems, communicating systems, real-time programs, real-time semantics, scheduling.

1 Introduction

We are interested in specifying and verifying real-time properties of systems controlled by parallel programs. The systems we have in mind are occam-like [4] programs running on a transputer-like architecture, i.e. systems where several processes may share the same transputer (processor) and may share physical transputer links (channels). This requires a real-time semantics for programs $P \in \mathit{Prog}$ and we begin with a *general semantics*

$$[\![\cdot]\!]_G \in \mathit{Prog} \rightarrow \mathcal{F}$$

which associates an implementation independent meaning with each program. Assume from now on the semantics domain $\mathcal{F}$ to be logic formulas over real-time computations.

*This work is partially funded by **ProCoS** ESPRIT BRA 3104.
†This work is partially funded by the Danish Technical Research Counsil under project **RapID**.

If *Prog* includes an assignment $x := e$, the only implementation independent real-time property of this assignment is that it takes some time $t > 0$ to execute. So the inferences about time from $[\![x := e]\!]_G$ are weak; but they must be satisfied by any implementation of *Prog*.

Knowing how *Prog* is compiled, more can be said about *Prog*'s real-time properties. For instance, from the compilation of $x := e$ and from the speed of the target processor, we can calculate the time needed to execute the machine code for $x := e$. Thus, we would have a lower bound of t. Let $[\![.]\!]_{Comp}$ denote the compilation semantics of *Prog*.

If $[\![P]\!]_{Comp}$ is consistent with $[\![P]\!]_G$, then we say that $[\![P]\!]_{Comp}$ is a *refinement* of $[\![P]\!]_G$ in the sense that

$$[\![P]\!]_G \Leftarrow [\![P]\!]_{Comp}$$

In general one can have a sequence of such refinement steps, each consistently adding implementation details to the previous one.

A next possible step adds processor and channel scheduling information. From the compiler and target processor we know the time needed for e.g. $x := e$, and when several processes share a processor, a scheduler specification SCH imposes constraints on how running time is granted to the individual processes. I.e. we can add a step to the above refinement:

$$[\![P]\!]_G \Leftarrow [\![P]\!]_{Comp} \Leftarrow ([\![P]\!]_{Comp} \wedge SCH)$$

Refinement is defined by implication and that requires that $[\![.]\!]_G$, $[\![.]\!]_{Comp}$ and SCH have the same semantic domain, i.e. a single logical calculus over real-time computations. For this purpose we use the Duration Calculus [7].

A definition of $[\![.]\!]_G$ essentially abstracts from concrete time bounds in the definition of $[\![.]\!]_{Comp}$. So we focus on $[\![.]\!]_{Comp}$, giving a specification oriented real-time semantics for a Timed CSP-like language, c.f. [6], at the compilation level where the speed of the target processor is known, and where we compile into a shared processor and channel architecture. From now on $[\![.]\!]$ is used for $[\![.]\!]_{Comp}$.

The real-time semantics of a program P is a Duration Calculus formula $[\![P]\!]$ describing a set of computations, which is independent of particular scheduling policies. Processes and computations are delineated in section 2, the Duration Calculus is introduced in section 3, and the semantics of processes is given in section 4. Schedulers are specified by separate Duration Calculus formulas. We illustrate this with a number of scheduler specifications in section 5 and 6.

In section 7 we illustrate that a program P with scheduler SCH satisfies the specification $SPEC$ (see also [1]) if the formula

$$[\![P]\!] \wedge SCH \Rightarrow SPEC$$

is valid in Duration Calculus.

2 Processes

We consider systems of parallel processes communicating (signalling) synchronously via uni-directional channels. The language is kept simple so that we can concentrate on timing properties. In the discussion we sketch how to introduce program variables.

Syntax

From the basic syntactic categories: $t \in (0, \infty)$ of time claims, $c, d \in Chan$ of channels, and $p \in Pname$ of process names, we define the syntactic categories $P \in Proc$ of (sequential) process expressions, $R \in RProc$ of named recursive (sequential) processes, and $S \in Sys$ of systems of parallel processes, by the grammar:

$$\begin{aligned} P &::= p \mid \textbf{claim}\ t \mid \textbf{delay}\ t \mid c! \mid c? \mid P;P \mid \\ &\quad\ c? \rightarrow P \mathbin{[\!]} d? \rightarrow P \mid c? \rightarrow P \mathbin{[\!]} \textbf{delay}\ t \rightarrow P \\ R &::= p \stackrel{r}{=} P \\ S &::= R \mid S \parallel S \end{aligned}$$

The informal meaning is:

claim t claims time $t > 0$ of its processor. This process can be considered an abstraction of the time t needed to execute an assignment $x := e$. The value t comes from knowledge of the compilation algorithm and the speed of the target processor. This process can also simulate execution times for other internal computations such as preparation of messages.

delay t delays (deactivates) for time $t > 0$. This process is used to slow down a computation when it is required by some real-time constraint.

$c!$ sends a signal on channel c.

$c?$ receives a signal from channel c.

$P1; P2$ is sequential composition of $P1$ followed by $P2$.

$c? \rightarrow P1 \mathbin{[\!]} d? \rightarrow P2$ is a choice between $c? \rightarrow P1$ and $d? \rightarrow P2$, where the branch chosen (if any) depends on the channel scheduling.

$c? \rightarrow P1 \mathbin{[\!]} \textbf{delay}\ t \rightarrow P2$, where $t > 0$, is a choice where the branch $c? \rightarrow P1$ is taken if c is ready within time t otherwise $P2$ is taken at time t.

This notation is very close to the one in [3]; but we omit scheduling primitives from the language. The binary choice operator can be generalized to n-ary choice using for example the technique of [5]. The recursive process $p \stackrel{r}{=} P$ is a simple recursion. We assume that

p is the only process name occurring in P. The generalization to mutual recursion is well known.

A system is a parallel combination of a fixed set of recursive processes.

Restrictions on the syntax ensure that channels are uni-directional. Assume an *alphabet* $\alpha(p) \subseteq \{c! \mid c \in Chan\} \cup \{c? \mid c \in Chan\}$ associated with each (sequential) process name p such that for no channel $c : \{c!\ , c?\} \subseteq \alpha(p)$.

The *process expressions with name p*: $Proc_p$, are those processes P which do not contain other process names than p and for which P may send via channel c only if $c! \in \alpha(p)$ and P may receive from channel d only if $d? \in \alpha(p)$. Let $RProc_p$ be the set $\{p \stackrel{r}{=} P \mid P \in Proc_p\}$ of *processes with name p* (possibly recursive).

For a system of n parallel processes: $p_1 \stackrel{r}{=} P_1 \ \| \cdots \| \ p_n \stackrel{r}{=} P_n$, where $(p_i \stackrel{r}{=} P_i) \in RProc_{p_i}$, the process names must be mutually distinct and there are the obvious restrictions on their alphabets to ensure that channels are uni-directional.

Computations

We consider systems where several processes may share the same processor and the same physical channel. Such systems are described using a set $SVar$ of *state variables*. The meaning of each state variable is a function from time $[0, \infty)$ to $\{0, 1\}$, and we assume that a state variable has at most a finite number of discontinuity points in any bounded and closed time interval.

For $\{c!\ , d?\} \subseteq \alpha(p)$, the following state variables are used to describe the processes with name p:

$c!$	:	p waits to send on c
$d?$	:	p waits to receive from d
c	:	signal present on c
d	:	signal present on d
$p.run$	:	p runs on a processor
$p.rdy$	:	p is ready to run on a processor
$p.done$	:	p has terminated

The relationships between these state variables at any time instant are:

(1) $p.run \Rightarrow p.rdy$

(2) $p.rdy \Rightarrow \neg(c\xi \vee c)$ for $c\xi \in \alpha(p)$, $\xi \in \{?, !\}$

(3) $c \Rightarrow \neg(c? \vee c!)$

(4) $p.done \Rightarrow \neg(c\xi \vee c \vee p.rdy)$ for $c\xi \in \alpha(p)$

These constraints express that: (1) a process p running on its processor must be ready; (2) a ready process does not wait to communicate and does not communicate; (3) when a signal is present on channel c the communication on c is no longer waited for; and (4) if p has terminated then p is neither waiting for communication nor communicating nor ready to run.

A *computation* σ of a system is a set of state functions satisfying the constraints (1) - (4), i.e. σ has the type

$$\sigma : SVar \rightarrow ([0,\infty) \rightarrow \{0,1\})$$

3 Duration Calculus

Duration calculus is an extension to interval temporal logic where one can reason about integrals of propositional states, i.e. about their durations within a given bounded and closed interval. (From now on interval means bounded and closed interval.) The formulas of duration calculus are built from state variables such as $p.rdy, p.run$, and $c!$ above. *Composite states* are formed by propositional combinations of these state variables. If S is a composite state then $\int S$, *the duration of* S, denotes a real valued *interval function.* For a given interval $[b, e]$ of a given computation, the value of $\int S$ is

$$\int_b^e S(t)dt$$

In the semantics below we use: $\int p.run$. Thus, for a given interval $[b, e]$ of a given computation, the value of $\int p.run$ is the accumulated running time of p within this interval. We use the abbreviation ℓ for the duration $\int 1$ of a composite state which is always true, i.e. ℓ denotes the length $e - b$ of the interval $[b, e]$.

Primitive duration formulas are real arithmetic formulas of $\int S$. We shall use the abbreviations

$$\begin{array}{lcl} \lceil . \rceil & \triangleq & (\ell = 0) \\ \lceil S \rceil & \triangleq & (\int S = \ell) \ \wedge \ (\ell > 0) \end{array}$$

For a given interval $[b, e]$ of a given computation, the duration formula $\lceil S \rceil$ holds if $b < e$ and the value of $S(t)$, for $t \in [b, e]$, is 1 (except for finitely many points in $[b, e]$). So S must hold almost everywhere in this interval. For example, for a given computation and non-point interval, the duration formula $\lceil p.rdy \rceil$ holds if p is ready ($p.rdy$) almost everywhere in the interval.

Composite duration formulas are formed using the conventional logic operators with their usual meaning. A binary "chop" operator is also introduced: for duration formulas A and B, "A followed by B" (written $A \frown B$) holds for given computation and interval $[b, e]$ iff there is a m such that $b \leq m \leq e$, A holds for the initial interval $[b, m]$, and B holds for the final interval $[m, e]$.

For example, the duration formula: $(\lceil p.rdy \rceil \wedge \ell \leq 5) \frown \lceil c! \rceil$, holds for a given interval of a given computation if the interval can be partitioned into two parts where p is ready in the first part for at most 5 time units and p waits for output on c in the second part.

The modal operators $\Diamond$ (there is a sub-interval) and $\Box$ (for any sub-interval) can be defined:

$$\begin{array}{lcl} \Diamond D & \triangleq & true^\frown D^\frown true \\ \Box D & \triangleq & \neg(\Diamond\neg D) \\ \Diamond_b D & \triangleq & D^\frown true \\ \Diamond_e D & \triangleq & true^\frown D \end{array}$$

The duration formulas $\Diamond_b D$ and $\Diamond_e D$ expresses that there is a prefix interval, respectively suffix interval, for which D holds.

Any interval $[0,t]$ is called a *prefix interval* (of $[0,\infty)$). For a given computation σ and duration formula D we say that D *is true in* σ, written $\sigma \models D$, iff D holds for every prefix interval of σ. D is *valid* if it is true in any computation. (This definition of validity is new; but equivalent to that given in [2].)

4 Semantics of Processes

Let $\alpha(p) \cup \{c \in Chan \mid c\xi \in \alpha(p)\} \cup \{p.rdy, p.run, p.done\}$ be the set of state variables *induced by* a given process name p, and let $\mathcal{D}_p$ be the set of duration formulas constructed from the state variables induced by p.

Let $\alpha' \subseteq \alpha(p)$ and $\{c?, d?\} \subseteq \alpha(p)$. The following shorthand notations will be used:

$$\begin{array}{llcl} & \ell p & \triangleq & \int p.run \\ & \alpha' & \triangleq & \bigvee_{c\xi\in\alpha'} c\xi \\ & p.idle & \triangleq & \neg(\alpha(p) \vee p.rdy \vee p.done \vee \bigvee_{c\xi\in\alpha(p)} c) \\ & p.wait.c\xi & \triangleq & c\xi \wedge \neg(\alpha(p) - \{c\xi\}) \wedge \bigwedge_{d\xi\in\alpha(p)} \neg d, \quad \text{for } c\xi \in \alpha(p) \\ & p.pass.c & \triangleq & c \wedge \neg\alpha(p) \wedge \bigwedge_{d\xi\in\alpha(p), d\neq c} \neg d \\ & p.wait.c?, d? & \triangleq & c? \wedge d? \wedge \neg(\alpha(p) - \{c?, d?\}) \wedge \bigwedge_{d\xi\in\alpha(p)} \neg d \\ \text{and} & p.STOP & \triangleq & \lceil.\rceil \vee \lceil p.done \rceil \end{array}$$

We give a semantics for systems based on a continuation semantics for named processes. The continuation semantics for each (recursive) process R with name p, is a continuation function in $\mathcal{D}_p \rightarrow \mathcal{D}_p$, which for a given duration formula C (continuation) describes the set of computations for R when it is followed by C.

For a (sequential) process expression $P \in Proc_p$ (possibly) containing the name p, the continuation semantics is a function P_p in

$$(\mathcal{D}_p \rightarrow \mathcal{D}_p) \rightarrow (\mathcal{D}_p \rightarrow \mathcal{D}_p)$$

For each $Q \in (\mathcal{D}_p \rightarrow \mathcal{D}_p)$, the value of $P_p(Q)$ is the semantics of P where Q has been substituted for each occurrence of the name p.

The function P_p is inductively defined by (explanation follows after):

$$(p)_p(Q)(C) \triangleq Q(C)$$

$$(\textbf{claim } t)_p(Q)(C) \triangleq \begin{cases} \lceil . \rceil \vee (\lceil p.rdy \rceil \wedge \ell p < t) \\ \vee\, (\lceil p.rdy \rceil \wedge \Diamond_e \lceil p.run \rceil \wedge \ell p = t)^\frown C \end{cases}$$

$$(\textbf{delay } t)_p(Q)(C) \triangleq \lceil . \rceil \vee (\lceil p.idle \rceil \wedge \ell < t) \vee ((\lceil p.idle \rceil \wedge \ell = t)^\frown C)$$

$$(c\xi)_p(Q)(C) \triangleq \lceil . \rceil \vee \lceil p.wait.c\xi \rceil \vee (\lceil p.wait.c\xi \rceil^\frown \lceil p.pass.c \rceil^\frown C)$$

$$(P1; P2)_p(Q)(C) \triangleq P1_p(Q)(P2_p(Q)(C))$$

$$\begin{pmatrix} c? \rightarrow P1 \\ \| d? \rightarrow P2 \end{pmatrix}_p (Q)(C) \triangleq \begin{cases} \lceil . \rceil \vee \lceil p.wait.c?, d? \rceil \\ \vee \lceil p.wait.c?, d? \rceil^\frown \lceil p.pass.c \rceil^\frown P1_p(Q)(C) \\ \vee \lceil p.wait.c?, d? \rceil^\frown \lceil p.pass.d \rceil^\frown P2_p(Q)(C) \end{cases}$$

$$\begin{pmatrix} c? \rightarrow P1 \\ \| \textbf{delay } t \rightarrow P2 \end{pmatrix}_p (Q)(C) \triangleq \begin{cases} \lceil . \rceil \vee (\lceil p.wait.c? \rceil \wedge \ell < t) \\ \vee\, (\lceil p.wait.c? \rceil \wedge \ell < t)^\frown \lceil p.pass.c \rceil^\frown P1_p(Q)(C) \\ \vee\, (\lceil p.wait.c? \rceil \wedge \ell = t)^\frown P2_p(Q)(C) \end{cases}$$

The first line for **claim** t describes it for any interval where it has not finished execution. The second line describes a terminating **claim** t followed by C.

For **delay** t, the first two disjuncts describe it for the intervals with length less than t, and the last disjunct discribes an idle phase with length t being continued by C.

The two first disjuncts for $c\xi$ describe it for intervals where no communication on c has happened. The last disjunct describes a non-point waiting phase followed by a non-point communication phase followed by C.

Sequential composition has its standard definition with a continuation semantics and the rest of the definitions can be explained similarly as above.

The continuation semantics for (recursive) processes $p \stackrel{r}{=} P$ with name p is:

$$(p \stackrel{r}{=} P)_p(C) \triangleq \bigwedge_i (P_p^i(CHAOS)(C))$$

where the family of approximations $P_p^i \in (\mathcal{D}_p \rightarrow \mathcal{D}_p) \rightarrow (\mathcal{D}_p \rightarrow \mathcal{D}_p)$ is defined by: $P_p^0(Q) \triangleq Q$ and $P_p^{i+1}(Q) \triangleq P_p(P_p^i(Q))$, for $Q \in (\mathcal{D}_p \rightarrow \mathcal{D}_p)$. The function $CHAOS \in (\mathcal{D}_p \rightarrow \mathcal{D}_p)$ is defined by $CHAOS(C) \triangleq true$ for any $C \in \mathcal{D}_p$ (and any process name p).

For a system of n parallel processes, we denote by $\mathcal{D}$ the set of duration formulas constructed from the state variables induced by the n process names of the system. Notice that $\mathcal{D}_p \subseteq \mathcal{D}$, for each p.

The (non-continuation) semantics of named processes and systems with value in $\mathcal{D}$ is then:

$$[\![p \stackrel{r}{=} P]\!] \triangleq (p \stackrel{r}{=} P)_p(p.STOP) \quad \text{for } (p \stackrel{r}{=} P) \in RProc_p$$

$$[\![\|_{i=1}^n (p_i \stackrel{r}{=} P_i)]\!] \triangleq \bigwedge_{i=1}^n [\![p_i \stackrel{r}{=} P_i]\!] \quad \text{where } p_i \neq p_j (i \neq j)$$

5 Process scheduling

We want to distribute a set of processes statically over $n > 0$ processors. So assume a partition of the processes into n classes: $\{PS_1, \ldots, PS_n\}$. Consider from now on an arbitrary of these classes $PS = \{p_1, \ldots, p_m\}$ containing $m > 0$ different processes sharing a single processor.

There is an obvious *no conflict* requirement that rejects more than one process running on the processor at the same time:

$$\Box \bigwedge_{k \neq j} \lceil p_j.run \Rightarrow \neg p_k.run \rceil$$

A *no overhead* scheduler is specified by:

$$\Box \lceil \bigvee_j p_j.rdy \Rightarrow \bigvee_j p_j.run \rceil$$

For the special case of a *dedicated processor* — $m = 1$ — this gives $\Box \lceil p_1.rdy \Rightarrow p_1.run \rceil$. This is the maximal parallelism used for instance in [3].

The *liveness* can generally be specified by:

$$\Box((\textstyle\sum_i \int p_i.\ rdy > \delta\ ) \Rightarrow (\sum_i \ell p_i > \delta'))$$

for constants δ, δ' such that $0 \leq \delta' \leq \delta$.

We can specify an extremely fair scheduling *equal rights* for which we doubt that there is an implementation:

$$\Box \bigwedge_j ((\ell p_j \cdot \sum_i \int p_i.rdy) = (\sum_i \ell p_i) \cdot \int p_j.rdy)$$

Assume that the accumulated running and ready times are not zero. Then it says that the ratio between p_j's accumulated running time and the total accumulated running time must be the same as the ratio between p_j's accumulated ready time and the total accumulated ready time (for any j and for any interval).

Below we consider more realistic schedulers.

A *first ready first run* scheduler $FRFR_m$ is specified by:

$$\Box \bigwedge_{j \neq i} \neg(\lceil p_j.rdy\ \wedge \neg p_j.run \rceil \wedge (\lceil \neg p_i.rdy \rceil ^\frown \Diamond \lceil p_i.run \rceil))$$

There is not a p_j ready and not running when p_i becomes ready and eventually runs.

A *priority* scheduler, where p_j has higher priority than p_i if $j > i$, is specified by:

$$\Box \bigwedge_{i<j} \lceil p_j.rdy \Rightarrow \neg p_i.run \rceil$$

A $\Delta-$*fair* scheduler, with time slice $\Delta > 0$, is specified by:

$$\bigwedge_i \left(\begin{array}{l} (\lceil p_i.run \rceil ^\frown \lceil p_i.rdy \wedge \neg p_i.run \rceil) \Rightarrow \ell \geq \Delta \\ \wedge \Box ((\lceil \neg p_i.run \rceil ^\frown \lceil p_i.run \rceil ^\frown \lceil p_i.rdy \wedge \neg p_i.run \rceil) \Rightarrow \ell \geq \Delta) \\ \wedge \Box ((\lceil p_i.rdy \rceil \wedge \ell \geq 2m\Delta) \Rightarrow \Diamond \lceil p_i.run \rceil) \end{array} \right)$$

The first two conjuncts guarantee any process its slice when it is running and ready. The last conjunct guarantees a ready process to be run within an interval of $2m\Delta$, where m is the number of processes.

6 Channel Scheduling

Two kinds of channel scheduling are considered. One considers a generalized choice operator allowing a process to wait for input on channels $c_1, \ldots, c_n$. The other considers the case where several channels $d_1, \ldots, d_l$ share a "physical channel".

A *first ready choice* among $c_1, \ldots, c_n$ is specified by:

$$\Box \bigwedge_{i \neq j} ((\lceil \neg c_j \rceil \wedge (\lceil \neg c_i \rceil ^\frown \lceil c_i \rceil)) \Rightarrow \lceil c_j? \wedge c_j! \Rightarrow c_i? \wedge c_i! \rceil)$$

Notice that $c_1?, \ldots, c_n?$ occur in same choice as guards.

Suppose $d_1, \ldots, d_l$ share a "physical channel".

No conflict:

$$\Box \bigwedge_{j \neq i} \lceil d_i \Rightarrow \neg d_j \rceil$$

Upper bound on ready time $\beta > 0$:

$$\Box \bigwedge_i ((\lceil d_i? \wedge d_i! \rceil \wedge \bigwedge_j \neg \Diamond \lceil d_j \rceil) \Rightarrow \ell \leq \beta)$$

Upper bound on communication time $\delta > 0$:

$$\Box \bigwedge_i (\lceil d_i \rceil \Rightarrow \ell \leq \delta)$$

First ready first served:

$$\Box \bigwedge_{j \neq i} \neg (\lceil d_j? \wedge d_j! \rceil \wedge (\lceil \neg (d_i? \wedge d_i!) \rceil ^\frown \Diamond \lceil d_i \rceil))$$

This is similar to the first ready first run processor sharing specification.

7 Example

Consider a system with three user processes p_1, p_2, and p_3 and a database server s. Each user process p_i, inputs from channel in_i, computes for t_i time units, and deposits the result in the database system through channel d_i. The database server s serves requests from channel d_i, using t_i' time units for each request before depositing it in a storage system through the channel out_i:

$$\begin{array}{lcl} U & \triangleq & \|_{i=1}^{3} (p_i \stackrel{r}{=} in_i? \ ; \ \textbf{claim}\ t_i \ ; \ d_i! \ ; \ p_i) \\ Sys & \triangleq & U \parallel (s \stackrel{r}{=} [\!]_{i=1}^{3}\ d_i? \ \rightarrow \ \textbf{claim}\ t_i' \ ; \ out_i! \ ; \ s) \end{array}$$

where $\alpha(p_i) = \{in_i?, d_i!\}$, $i = 1, 2, 3$, and $\alpha(s) = \bigcup_{i=1,2,3}\{d_i?, out_i!\}$.

Suppose the parallel user processes U share a processor, the server s to be on its own processor, and that the communication between the two processors on the d_i's is "multiplexed" on one physical channel:

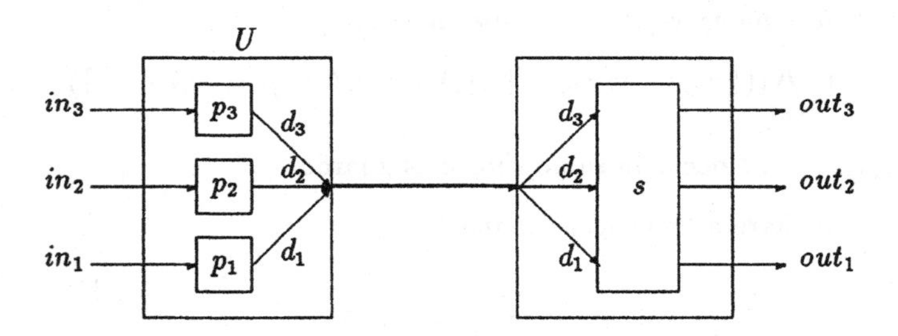

We want to guarantee an upper bound for the response time on channel out_i from a given input on channel in_i. To do so, we must make some assumptions on the environment, and on the various schedulers (where we refer to previously given specifications).

Suppose for each input channel in_i that

$$\Box\lceil in_i! \vee in_i \rceil$$

specifying a dedicated environment for in_i always being ready for sending on in_i unless a communication actually is occurring.

Similarly, suppose for each output channel out_i that it is waiting or communicating

$$\Box\lceil out_i? \vee out_i \rceil$$

Let Env be the conjunction of these assumptions about the environment.

Furthermore, assume upper bounds $\beta_{in} > 0$ and $\delta_{in} > 0$ on ready time and communication time, respectively, for each channel in_i and assume $\beta_{out} > 0$ and $\delta_{out} > 0$ to be upper bounds for ready time and communication time, respectively, for each channel out_i.

As scheduler for the three parallel processes U we choose a first ready first run $FRFR_3$ scheduler for three processes with no overhead, and as scheduler for s a no overhead scheduler is also chosen.

Concerning scheduling for d_1, d_2, and d_3 we assume the no conflict property, and $\beta_d > 0$ and $\delta_d > 0$ as upper bounds on ready time and communication time for the shared channel. Furthermore, we assume a specialized scheduling: *first ready to send first served*, on the shared channel:

$$\Box \bigwedge_{j \neq i} \neg(\lceil d_j! \rceil \wedge (\lceil \neg d_i! \rceil ^\frown \Diamond \lceil d_i \rceil))$$

Let SCH be the conjunction of all the above mentioned scheduling specifications.

Let $SPEC$ be

$$Env \Rightarrow \Box \bigwedge_i ((\Diamond_b \lceil in_i \rceil \wedge \ell > f_i) \Rightarrow \Diamond \lceil out_i \rceil)$$

where, for $i = 1, 2, 3$:

$$\begin{array}{lcl} f_i & \triangleq & time\_to\_d\_rdy + d\_com + t'_i + out\_com \\ time\_to\_d\_rdy & \triangleq & \underline{max}\{3(\underline{max}\{t'_1, t'_2, t'_3\} + out\_com), in\_com + t_1 + t_2 + t_3\} \\ d\_com & \triangleq & \beta_d + \delta_d \\ in\_com & \triangleq & \beta_{in} + \delta_{in} \\ out\_com & \triangleq & \beta_{out} + \delta_{out} \end{array}$$

One can prove that the system Sys with schedulers SCH satisfy $SPEC$, i.e. that

$$[\![Sys]\!] \wedge SCH \Rightarrow SPEC$$

is a theorem of Duration Calculus, using the following induction rule for recursive processes:

$$\begin{array}{ll} \text{if} & (X(p.STOP) \Rightarrow S) \vdash P_p(X)(p.STOP) \Rightarrow S \\ \text{then} & [\![p \stackrel{r}{=} P]\!] \Rightarrow S \end{array}$$

where P is a *guarded* process expression with name p and X stands for any function in $(\mathcal{D}_p \rightarrow \mathcal{D}_p)$.

8 Discussion

We have used Duration Calculus as semantic domain to give meaning to parallel processes signalling over uni-directional channels. The semantics was given at a level of abstraction anticipating a shared processor and channel structure — the idea being that a scheduler is specified separately as a formula in Duration Calculus which is conjuncted with the semantics of a parallel process when reasoning about its real-time properties. From this first attempt the use of Duration Calculus seems promising.

Concerning the semantics presented: The process $p \stackrel{r}{=}$ **claim** $t\ ; p$ will constantly be ready for execution and may monopolize the processor. Disliking this, the semantics of sequential composition could be changed (as in [6]) to have a minimal idle phase.

Concerning process scheduling, it must be interesting to investigate how the abstract specifications in Duration Calculus can be refined to more concrete specifications in Duration Calculus, where data structures, such as a queue of ready processes, used in implementations of schedulers are recorded in states. Each such refinement step can be proved as theorem of Duration Calculus.

This work is easy to generalize to having program variables and message passing by introducing new states to record the values of program variables and by considering operators such as $(c!e)_p$, $(c?x)_p$, and $(x := e)_p$, as functions in:

$$((V \rightarrow \mathcal{D}_p) \rightarrow (V \rightarrow \mathcal{D}_p)) \rightarrow ((V \rightarrow \mathcal{D}_p) \rightarrow (V \rightarrow \mathcal{D}_p))$$

where V is the set of functions mapping program variables to their values.

Thus, the continuation function for p is in $(V \rightarrow \mathcal{D}_p) \rightarrow (V \rightarrow \mathcal{D}_p)$ and the continuation C for the sequential composition is in $V \rightarrow \mathcal{D}_p$. However, the definition of sequential composition keeps it simple form:

$$(P1;P2)_p(Q)(C)(v) \triangleq P1_p(Q)(P2_p(Q)(C))(v)$$

Acknowledgements: The comments from C.A.R. Hoare, Z. Liu, H.H. Løvengreen, and the referees on this work are greatly appreciated.

References

[1] K.M. Hansen, A.P. Ravn, and H. Rischel, Specifying and Verifying Requirements for Critical Systems, to appear in *ACM SIGSOFT'91 Conference on software for Critical Systems*, New Orleans, Louisianna, December, 1991.

[2] M.R. Hansen and Zhou Chaochen, *A note on Completeness of the Duration Calculus*, ProCoS Techn. Rep. ID/DTH MRH 6/1, ESPRIT BRA 3104, 1991.

[3] J. Hooman, A Denotational Real-Time Semantics for Shared Processors, in *PARLE'91 Parallel Architectures and Languages Europe*, vol. II, LNCS 506, Springer-Verlag 1991, pp. 185-201.

[4] INMOS Limited, occam 2 *Reference Manual*, Prentice Hall 1988.

[5] R. Koymans, R.K. Shyamasundar, W.-P. de Roever, R. Gerth, and S. Arun-Kumar, Compositional Semantics for Real-Time Distributed Computing, *Information and Computation*, 79(3), 1988, pp. 210-256.

[6] G. Reed and A. Roscoe, Metric spaces as models for real-time concurrency, in *Proc. Workshop on Mathematical Foundations of Programming Language Semantics*, LNCS 298, Springer-Verlag 1987, pp. 331-343.

[7] Zhou Chaochen, C.A.R. Hoare, and A.P. Ravn, A Calculus of Durations, to appear in *Information Processing Letters*.

A Compositional Semantics for Fault-Tolerant Real-Time Systems

J. Coenen * J. Hooman †

Dept. of Math. and Computing Science
Eindhoven University of Technology
P.O. Box 513
5600 MB Eindhoven, The Netherlands

Abstract

Motivated by the close relation between real-time and fault-tolerance, we investigate the foundations of a formal framework to specify and verify real-time distributed systems that incorporate fault-tolerance techniques. Therefore a denotational semantics is presented to describe the real-time behaviour of distributed programs in which concurrent processes communicate by synchronous message passing. New is that in this semantics we allow the occurrence of failures, due to faults of the underlying execution mechanism, and we describe the effect of these failures on the real-time behaviour of programs. Whenever appropriate we give alternative choices for the definition of the semantics. The main idea is that making only very weak assumptions about faults and their effect upon the behaviour of a program in the semantics, any hypothesis about faults must be made explicit in the correctness proof of a program.

1 Introduction

The development of distributed systems with real-time and fault-tolerance requirements is a difficult task, which may result in complicated and opaque designs. This, and the fact that such systems are often embedded in environments where a small error can have serious consequences, calls for formal methods to specify the requirements and verify the development steps during the design process.

Unfortunately most methods that have been proposed up to the present deal either with fault-tolerance requirements, e.g. [ScSc83, Cristian85, JMS87], or with real-time requirements, e.g. [ShLa87, HoWi89, Ostroff89], but not with both simultaneously. This can be a problem, because fault tolerance is obtained by some form of redundancy. For example, a backward recovery mechanism introduces not only information redundancy

*Supported by NWO/SION Project 612-316-022: "Fault Tolerance: Paradigms, Models, Logics, Construction". E-mail: wsinjosc@win.tue.nl

†Supported by ESPRIT-BRA Project 3096: "Formal Methods and Tools for the Development of Distributed Real-Time Systems (SPEC)". E-mail: wsinjh@win.tue.nl

and modular redundancy, but also time redundancy. Hence, it is possible to obtain a higher degree of fault-tolerance by introducing more checkpoints, i.e by introducing more time redundancy. This is the main reason why program transformations that are used to transform a program into a functionally equivalent fault tolerant program, e.g. by superimposition of an agreement algorithm, may transform a real-time program into one that doesn't meet its deadlines.

The trade-off between reliability and timeliness extends to one between reliability, timeliness and functionality. An elegant way of exploiting this trade-off can be observed in graceful degrading systems. For example, if a fault occurs a system may temporary sacrifice a service in order to ensure that more important deadlines are met.

Motivated by the close relation between the reliability, timeliness and functionality of a system, we would like to reason about these properties simultaneously. Related research on the integration of these three aspects of real-time programs within one framework can be found in [HaJo89]. In that paper a probabilistic (quantitative) approach is presented, whereas we are mainly concerned with the qualitative aspects of fault-tolerance.

To illustrate how we would like to reason over fault-tolerant real-time systems, consider the Triple Modular Redundancy (TMR) system in figure 1. The TMR system consists of four components. Each of the components P_i computes for an input x on its channel c_i the same function $f(x)$ and outputs the result on its channel d_i five time units later. We consider synchronous communication over directed channels. Because communication is synchronous, a process may have to wait for its communication partner to become available.

The components P_i may use different programs (i.e. algorithms) S_i to compute $f(x)$. We will use $\langle P \Leftarrow S \rangle$ to denote that component P executes program S. The component V waits until it receives the same input at the same time on two different input channels, say d_i and d_j, and outputs this value on channel r two time units later. The informal specifications above can be formalized in, for example, a small extension of first-order predicate logic. A specification is of the form S **sat** φ, where S is a program and φ a sentence of the assertion language. The assertion language includes predicates with the following meaning

$\mathbf{fail}(P)\,\mathbf{at}\,t$	:	at time t a fault occurs in component P during the execution of its program;
$c.v\,\mathbf{at}\,t$	:	at time t the process is communicating v over channel c;
$c!v\,\mathbf{at}\,t$	:	at time t the process tries to send v over channel c.

A formula $\langle P \Leftarrow S \rangle$ **sat** φ expresses that every possible execution of the program S by component P satisfies assertion φ.

The specification of a component $\langle P_i \Leftarrow S_i \rangle$ might be $(i = 1, \ldots, n)$

$$\langle P_i \Leftarrow S_i \rangle \ \mathbf{sat}$$
$$\forall_{t' : t \leq t' \leq t+5} (\neg \mathbf{fail}(P_i)\,\mathbf{at}\,t') \rightarrow (c_i.v\,\mathbf{at}\,t \rightarrow d_i!f(v)\,\mathbf{at}\,t+5) \ .$$

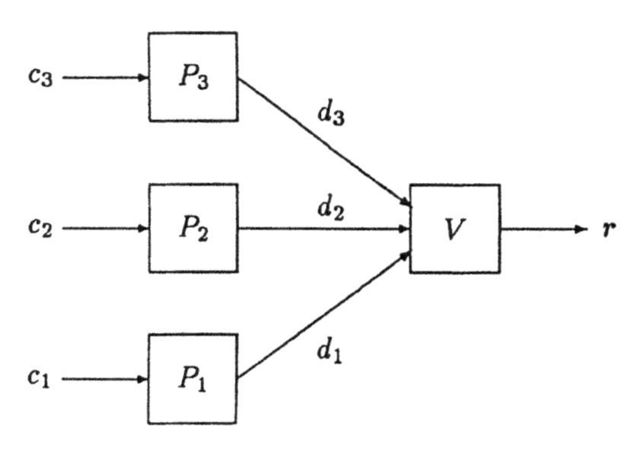

Figure 1: *TMR system*

The voter $\langle V \Leftarrow \rangle S_0$ might be specified by

$$\langle V \Leftarrow S_0 \rangle \ \mathbf{sat}$$
$$\forall_{t': t \le t' \le t+2}(\neg \mathbf{fail}(V)\ \mathbf{at}\ t') \rightarrow (d_i.u\ \mathbf{at}\ t \wedge d_j.u\ \mathbf{at}\ t \wedge i \neq j \rightarrow r!u\ \mathbf{at}\ t+2)\ .$$

The following proof rule for parallel composition ($\|$) holds under certain syntactic restrictions on the assertions

$$\frac{N_1\ \mathbf{sat}\ \varphi_1,\ N_2\ \mathbf{sat}\ \varphi_2}{N_1 \parallel N_2\ \mathbf{sat}\ \varphi_1 \wedge \varphi_2} \quad \text{(Parallel).}$$

This proof rule is the same as the one used in the proof system of [HoWi89] for real-time programs without considering the possible occurrence of faults.

If we assume that V is always ready to communicate on the channels c_1, c_2, and c_3 — this can be added formally to the specification of V — then $d_i!v\ \mathbf{at}\ t$ corresponds to $d_i.v\ \mathbf{at}\ t$. Then, by repeated application of the parallel composition rule and some predicate calculus, the following specification for the complete TMR system can be derived.

$$\langle P_1 \Leftarrow S_1 \rangle \parallel \langle P_2 \Leftarrow S_2 \rangle \parallel \langle P_3 \Leftarrow S_3 \rangle \parallel \langle V \Leftarrow S_0 \rangle\ \mathbf{sat}$$
$$(\forall_{t': t \le t' \le t+5}(\neg \mathbf{fail}(P_i)\ \mathbf{at}\ t' \wedge \neg \mathbf{fail}(P_j)\ \mathbf{at}\ t' \wedge i \neq j)$$
$$\wedge \forall_{t': t+5 \le t' \le t+7}(\neg \mathbf{fail}(V)\ \mathbf{at}\ t')) \rightarrow (c_i.v\ \mathbf{at}\ t \wedge c_j.v\ \mathbf{at}\ t \rightarrow r!f(v)\ \mathbf{at}\ t+7)$$

Since the rule for parallel composition is compositional, this can be done *without* information about the implementation of the processes S_i $(i = 0, \ldots, 3)$.

Notice that a specification typically is of the format N **sat** $(FH \rightarrow \varphi)$. The antecedent FH in the assertion is called the *fault hypothesis*. Because FH is assumed for a particular process it is called a *local* fault hypothesis, as opposed to a *global* fault hypothesis which hold for all processes. A global fault hypothesis is an axiom of the proof system, provided it is expressible in the assertion language.

A fault hypothesis characterizes faults by (c.f. [RLT78])

- Duration, i.e the time when faults occur, how long will the fault be present, etc.
- Location, i.e. the place where a fault occurs, in which processes, etc.
- Effect, i.e. the effect of the fault on the behaviour of a process, on program variables, etc.

For instance, the following fault hypothesis asserts that faults are transient

$$\textbf{fail}(P)\, \textbf{at}\, t \rightarrow \exists_{t' \geq t}(\neg \textbf{fail}(P)\, \textbf{at}\, t')\ ,$$

and another example is the following which relates the occurrence of faults in two processors (a fault P_1 will pròpagate within five time units to P_2)

$$\textbf{fail}(P_1)\, \textbf{at}\, t \rightarrow \exists_{t' : t \leq t' \leq t+5}(\textbf{fail}(P_2)\, \textbf{at}\, t')\ .$$

In this report we take a first step towards a formal method for designing real-time systems with fault-tolerance requirements. Our aim is a *compositional* proof system, i.e. is proof system in which the specification of a compound program can be inferred from the specifications of the constituent components without referring to the internal structure of these components. Compositionality is a desirable property, because it enables one to decompose a large specification of a system into smaller specifications for the subsystems. As basis for such a proof system we define a denotational (and therefore compositional) semantics, i.e. a semantics in which the semantics of a compound program is defined by the semantics of the components independently from the structure of these components.

From the discussion in the preceding paragraphs it is clear that we need semantics that simultaneously describes the following views of a system:

- Functional behaviour. The functional behaviour defines the relation between initial and final states of a program and its communication behaviour.
- Timed behaviour. For real-time systems the time at which a process terminates and the time that it communicates is of interest.
- Fault behaviour. The behaviour of a process in the presence of faults may deviate considerably from its behaviour in absence of faults. Therefore we want to distinguish the fault behaviour from the correct behaviour.

It is inevitable to make some assumptions about the fault behaviour of a process when defining a semantics. However, by making only very weak assumptions we enforce that the assumptions used when dealing with software fault-tolerance — and indeed many of the assumptions for hardware fault-tolerance — have to be made explicit by a fault hypothesis (cf. [Cristian85, BGH86, Bernstein88, Pow+88, TaWi89, CDD90]).

The remainder of this report is organized as follows. In section 2, a programming language is defined, inspired by OCCAM [OCCAM88]. We also give an informal explanation of the language constructs under the assumption that faults don't occur. Faults are taken into consideration in section 3, where we give a semantics of the language defined in section 2. Whenever appropriate we discuss alternative choices for the assumptions that are implicit in the semantics. Conclusions are present in section 4, where we also discuss some future work.

2 Programming Language

To describe real-time systems we use an OCCAM-like programming language, named RT. An RT program is a network of sequential processes that communicate over synchronous channels. Each channel is directed and connects exactly two processes. Processes can only access local variables, i.e. variables are not shared between parallel processes. Processes have unique names.

We assume that the following disjunct sets are defined:

- $(x \in)$ *VAR*, the set of program variables;
- $(e \in)$ *EXP*, the set of (integer) expressions with free occurrences of program variables only;
- $(b \in)$ *BOOL*, the set of boolean expressions with free occurrences of program variables only;
- $(c \in)$ *CHAN*, a set of channel names;
- $(P \in)$ *PID*, a set of process names.

The formal syntax of an RT program N is defined by

Statement	S	$::=$	$\mathbf{skip} \mid \mathbf{delay}\, e \mid x := e \mid c!e$
			$\mid c?x \mid S_1;\, S_2 \mid ALT \mid *ALT$
Alternative	ALT	$::=$	$[\|]_{i=1}^{n}\, b_i \rightarrow S_i]$
			$\mid [[]_{i=1}^{n}\, b_i;\, c_i?x_i \rightarrow S_i \,[]\, b_0;\, \mathbf{delay}\, e \rightarrow S_0]$
Network	N	$::=$	$\langle P \Leftarrow S \rangle \mid N_1 \parallel N_2$

If we forget about faults for the moment, and concentrate on the functional and timed behaviour of programs only, we obtain the following intended meaning for the programming language constructs above.

<u>Primitive Constructs</u>

- **skip** causes no state changes and terminates immediately. Hence, it consumes no time.
- **delay** e takes exactly $K_d + e$ time units to be executed if $e \geq 0$ and $K_d \geq 0$ time units otherwise, but has no other effect. The constant K_d is the minimal amount of time needed to execute a **delay**-statement.
- $x := e$ assigns the value of the expression e to the variable x. Its execution takes $K_a \geq 0$ time units.

- Communication takes place by synchronous message passing over directed channels. Because communication is synchronous a process may have to wait until its communication partner is ready to communicate. There are two primitives for communication:
  - The output statement $c!e$ is used to send the value of e on channel c. It causes the process to wait until the communication partner is prepared to receive a value on channel c.
  - The input statement $c?x$ is similar to the output statement, except that the process waits to receive a value on channel x. If communication takes place the received value is assigned to x.

  The actual communication itself, i.e. without the waiting period, takes exactly $K_c > 0$ time-units.

Instead of using a fixed amount of time for the execution of, for example, the assignment statement we could have chosen an interval of time or a function that assigns an amount of time to an assignment. These options, however, lead to a more difficult to understand semantics, with essentially the same properties.

Compound Constructs

- $S_1; S_2$ denotes the sequential composition of the statements S_1 and S_2. First S_1 is executed, then S_2. The total amount of time needed for execution, is the sum of the execution times of S_1 and S_2. Thus, sequential composition itself takes zero time.
- The alternative statement comes in two formats:
  - $[\|_{i=1}^{n} b_i \rightarrow S_i]$
    First the boolean expressions b_i are evaluated, which takes $K_g > 0$ time. If all the b_i evaluate to false, the statement terminates immediately after the evaluation of the guards. Otherwise, nondeterministically one of the b_i that evaluated to true is chosen and the corresponding alternative S_i is executed.
  - $[\|_{i=1}^{n} b_i; c_i?x_i \rightarrow S_i \, \| \, b_0; \textbf{delay}\, e \rightarrow S_0]$
    If all the boolean guards evaluate to false execution of this statement takes exactly $K_g > 0$ time units. Otherwise, if b_0 evaluates to false, the process waits until one of communications $c_i?x_i$ for which b_i $(i \neq 0)$ evaluated to true, is completed. After this communication, the process continues with the execution of the corresponding alternative S_i. If b_0 evaluated to true, the execution is as in the previous case, except that the process waits at most e time units for a communication. If, after evaluation of the guards, e time units have elapsed without starting a communication, the statement S_0 is executed. In this case, the process has consumed $K_g + e$ time before S_0 is executed.
- $*ALT$ denotes the iteration of an alternative statement ALT until all the boolean expressions in the guards evaluate to false. Because, the evaluation of the boolean expressions takes positive time ($K_g > 0$) only a finite number of iterations is possible in finite time.

- $\langle P \Leftarrow S \rangle$ associates the process identifier P with process S. It is not a statement that is actually executed or implemented, but it is included to enable us to reason over processes by referring to their names. Consequently, this statement consumes no time.

- $N_1 \parallel N_2$ denotes parallel composition. We assume maximal parallelism, which means that each process has its own processor. This ensures maximal progress, i.e. minimal waiting.

3 Denotational Semantics

We will define a denotational semantics that formalizes the informal description of the previous section and extends it with fault behaviour. First, we define the model and give an informal explanation. Second we define and explain the semantics of RT programs. We will distinguish between the correct behaviour of a program (defined by its normal semantics) and the behaviour of a program in the presence of faults (defined by its fault semantics).

Preliminary Definitions

The functional behaviour of a program is partially defined by the initial and final states of a program. A state $s \in STATE$ assigns to each program variable a value. Thus $STATE$ is the set of mappings $VAR \rightarrow VAL$, where VAL is the set of possible values of program variables. We use $s(e)$ to denote the value of expression e in state s, even if e is not a variable. The variant $(s|x \mapsto v)$ of a state s is defined by ($\doteq$ denotes syntactic equality):

$$(s|x \mapsto v)(y) = \begin{cases} v & , x \doteq y \\ s(y) & , \text{otherwise.} \end{cases}$$

The communication behaviour, timed behaviour and fault behaviour of a computation is described by a mapping σ over a time domain $TIME$. The time domain is dense and $t \geq 0$ for all $t \in TIME$. Furthermore, $TIME$ is linearly ordered and closed under addition and multiplication. $TIME$ includes the values of constants K_a, K_d etc. and VAL. For simplicity we assume that $TIME$ is the set of nonnegative rational numbers and that program variables are of type integer. The special symbol ∞ ($\infty \notin TIME$) denotes infinity with the usual properties.

Let Σ be the set of mappings σ of type $[0,t) \rightarrow (\mathcal{P}(CHAN \times (VAL \cup \{!,?\})) \times \mathcal{P}(PID \cup \{X\}))$, where $t \in TIME \cup \{\infty\}$. Thus for all $t \in [0,t')$, $\sigma(t)$ is a pair $(comm, fail)$ with $comm \subseteq CHAN \times (VAL \cup \{!,?\})$ and $fail \subseteq PID \cup \{X\}$. We use $\sigma(t).comm$ and $\sigma(t).fail$ to refer to respectively the first and the second field of $\sigma(t)$.

- $comm \subseteq CHAN \times (VAL \cup \{!,?\})$ defines the communication and timed behaviour. The intended meaning of $comm$ at time $t \in [0,t')$ is as follows.
  - If $(c,v) \in \sigma(t).comm$ then the value v is being communicated on channel c at time t.
  - If $(c,!) \in \sigma(t).comm$ then a process is waiting to send a value on channel c at time t.

- If $(c, ?) \in \sigma(t).comm$ then a process is waiting to receive a value on channel c at time t.

The waiting for a communication is included in the model to obtain a ***compositional*** semantics.

- $fail \subseteq PID \cup \{X\}$, $X \notin PID$. If $P \in \sigma(t).fail$ then process P is behaving according to its fault semantics. Otherwise, P is behaving correctly, i.e. according to its normal semantics. For programs S to which a name has not yet been assigned by a $\langle P \Leftarrow S \rangle$ construct, X is used as a place holder. The *fail*-field enables one to distinguish between normal behaviour (whenever $\sigma(t).fail = \emptyset$) and fault behviour (whenever $\sigma(t).fail \neq \emptyset$).

The length $|\sigma|$ of a mapping σ with domain $[0, t)$ is defined as t.

The meaning of an RT program is denoted by a set M of triples ($M \subseteq \Delta$), where Δ is the Cartesian product $STATE \times \Sigma \times STATE$. In a triple (s^0, σ, s), s^0 denotes the initial program state and s the final program state.

We define the initial part of length t of σ for $t \in [0, |\sigma|]$, notation $\sigma \downarrow t$, as

$$\begin{array}{lcl} |\sigma \downarrow t| & \hat{=} & t \\ (\sigma \downarrow t)(t') & \hat{=} & \sigma(t')\ ,\ t' \in [0, t)\ . \end{array}$$

If $t > |\sigma|$ then $\sigma \downarrow t$ is undefined. The semantics of a RT program is typically defined in two steps. First, we define the normal semantics of the program as described in the previous section, i.e. the semantics when faults do not occur. This is done by defining the interpretation function $\mathcal{M}[\![.]\!] : \mathrm{RT} \rightarrow \mathcal{P}(\Delta)$. Second, we define the interpretation function $\mathcal{M}^{\dagger}[\![.]\!] : \mathrm{RT} \rightarrow \mathcal{P}(\Delta)$ which defines the general semantics when faults are taken into account. The normal behaviour is considered to be a special case of the general behaviour, therefore for all RT programs it is guaranteed that $\mathcal{M}[\![S]\!] \subseteq \mathcal{M}^{\dagger}[\![S]\!]$, or more precisely

$$\mathcal{M}[\![S]\!] = \{(s^0, \sigma, s) \in \mathcal{M}^{\dagger}[\![S]\!] \mid \sigma(t).fail = \emptyset, \text{ for all } t \in [0, |\sigma|)\}\ .$$

The general behaviour can be partitioned into the normal behaviour and the fault behaviour that describes the behaviour if a fault occurs. This is best illustrated by the definition of the semantics of the assignment statement. First we define the normal semantics $\mathcal{M}[\![x := e]\!]$. Then we apply a function $FAIL : \mathcal{P}(\Delta) \rightarrow \mathcal{P}(\Delta)$ to $\mathcal{M}[\![x := e]\!]$, which transforms the normal behaviour into the fault behaviour. Finally we define the general semantics $\mathcal{M}^{\dagger}[\![x := e]\!]$ as the union of the normal behaviour and the fault behaviour.

Let $M \subseteq \Delta$, then $FAIL$ is defined as follows

$$\begin{array}{l} FAIL(M) \hat{=} \\ \quad \{(s^0, \sigma, s) \mid \text{ there exist } (s^0, \sigma', s') \in M \text{ and } t \in [0, \min(|\sigma|, |\sigma'|)) \\ \qquad\qquad\qquad \text{ such that } \sigma \downarrow t = \sigma' \downarrow t \text{ and for all } t' \in [t, |\sigma|) : \sigma(t').fail = \{X\}\} \end{array}$$

For a program S, $FAIL(\mathcal{M}[\![S]\!])$ defines the same behaviour as $\mathcal{M}[\![S]\!]$ up to a point in time where a fault occurs and after that the program may exhibit arbitrary behaviour. For instance it may never terminate.

Proposition 1

(*a*) $FAIL(M) = \emptyset \Leftrightarrow$ for all $(s^0, \sigma, s) \in M$: $|\sigma| = 0$.
(*b*) for all $(s^0, \sigma, s) \in FAIL(M)$ there exists a $t \in [0, |\sigma|)$ such that for all $t' \in [t, |\sigma|)$: $\sigma(t').fail \neq \emptyset$.

■

Part (*a*) of proposition 1 expresses that if, and only if, the executions in M don't consume time they cannot fail and therefore $FAIL(M)$ is empty. Part (*b*) expresses that the mappings σ of all executions in $FAIL(M)$ have a non-empty suffix — because the time domain is dense — during which the *fail*-field is continuously non-empty. As a consequence all computations in $FAIL(M)$ take time.

Skip, Delay, and Assignment

The semantics of the skip-statement is:

$$\mathcal{M}[\![\mathbf{skip}]\!] \;\hat{=}\; \{(s^0, \sigma, s^0) \mid |\sigma| = 0\}$$

Because **skip** takes no time its execution can't fail. Therefore $FAIL(\mathcal{M}[\![\mathbf{skip}]\!])$ is empty and thus the general semantics is equal to the normal semantics.

$$\begin{aligned} \mathcal{M}^\dagger[\![\mathbf{skip}]\!] &\;\hat{=}\; \mathcal{M}[\![\mathbf{skip}]\!] \cup FAIL(\mathcal{M}[\![\mathbf{skip}]\!]) \\ &= \mathcal{M}[\![\mathbf{skip}]\!] \end{aligned}$$

The definition of the semantics of the delay-statement and the assignment statement should cause no trouble after the previous discussion.

$$\begin{aligned} &\mathcal{M}[\![\mathbf{delay}\, e]\!] \;\hat{=} \\ &\quad \{(s^0, \sigma, s^0) \mid \; |\sigma| = K_d + \max(s^0(e), 0) \\ &\qquad\qquad \text{and for all } t \in [0, |\sigma|) : \sigma(t).comm = \emptyset \wedge \sigma(t).fail = \emptyset\} \end{aligned}$$

$$\mathcal{M}^\dagger[\![\mathbf{delay}\, e]\!] \;\hat{=}\; \mathcal{M}[\![\mathbf{delay}\, e]\!] \cup FAIL(\mathcal{M}[\![\mathbf{delay}\, e]\!])$$

$$\begin{aligned} &\mathcal{M}[\![x := e]\!] \;\hat{=} \\ &\quad \{(s^0, \sigma, s) \mid \; |\sigma| = K_a \wedge s = (s^0 | x \mapsto s^0(e)) \\ &\qquad\qquad \text{and for all } t \in [0, |\sigma|) : \sigma(t).comm = \emptyset \wedge \sigma(t).fail = \emptyset\} \end{aligned}$$

$$\mathcal{M}^\dagger[\![x := e]\!] \;\hat{=}\; \mathcal{M}[\![x := e]\!] \cup FAIL(\mathcal{M}[\![x := e]\!])$$

Recall from the previous section that communication is synchronous and therefore the behaviour of, for example, a send statement can be split into two parts. During the first part, the process executing the send statement waits until the communication partner is available. If the communication partner eventually is available, which is not always guaranteed, the process will continue with the second part, i.e. the communication itself. Thus a communication statement can be seen as a sequential composition of two smaller processes. Therefore, we first define sequential composition before proceeding with the communication statements.

Sequential Composition

The concatenation $\sigma_0\sigma_1$ of two mappings σ_0 and σ_1 is defined by

$$\begin{array}{lcl} |\sigma_0\sigma_1| & \hat{=} & |\sigma_0| + |\sigma_1| \\ (\sigma_0\sigma_1)(t) & \hat{=} & \begin{cases} \sigma_0(t) & \text{, if } t \in [0, |\sigma_0|); \\ \sigma_1(t - |\sigma_0|) & \text{, if } t \in [|\sigma_0|, |\sigma_0\sigma_1|). \end{cases} \end{array}$$

Sequential composition $SEQ(M_0, M_1)$ of two models $M_0, M_1 \subseteq \Delta$ is defined as follows.

$$\begin{array}{ll} \multicolumn{2}{l}{SEQ(M_0, M_1) \hat{=}} \\ & \{(s^0, \sigma_0, s) \in M_0 \mid |\sigma_0| = \infty\} \\ \cup & \{(s^0, \sigma_0\sigma_1, s) \mid \text{there exists } s' \text{ such that :} \\ & \qquad (s^0, \sigma_0, s') \in M_0 \wedge |\sigma_0| \neq \infty \wedge (s', \sigma_1, s) \in M_1\} \end{array}$$

The SEQ operator is associative, i.e.

Proposition 2

$$SEQ(SEQ(M_0, M_1), M_2) = SEQ(M_0, SEQ(M_1, M_2)) .$$

■

The normal semantics of sequential composition of two program fragments is

$$\mathcal{M}[\![S_0; S_1]\!] \quad \hat{=} \quad SEQ(\mathcal{M}[\![S_0]\!], \mathcal{M}[\![S_1]\!]) .$$

Observe that sequential composition itself doesn't consume time. Hence, faults occur in the component statements only.

A possible way to define the general semantics of sequential composition is to use the *FAIL* function as we did for delay-statement, but there are reasonable alternatives to consider.

1. Using the *FAIL* function in the same manner as in the definition of the assignment statement leads to the following definition.

$$\begin{array}{lcl} \mathcal{M}_1^\dagger[\![S_0; S_1]\!] & \hat{=} & \mathcal{M}[\![S_0; S_1]\!] \cup FAIL(\mathcal{M}[\![S_0; S_1]\!]) \\ & = & FAIL(\mathcal{M}[\![S_0]\!]) \cup SEQ(\mathcal{M}[\![S_0]\!], \mathcal{M}^\dagger[\![S_1]\!]) . \end{array}$$

   This alternative implies that once a process fails it remains failed. Note that the definition only depends on the normal semantics of the components.

2. It is also possible to assume that if a failing process terminates it will continue with the next statement:

$$\mathcal{M}_2^\dagger[\![S_0; S_1]\!] \hat{=} SEQ(\mathcal{M}_2^\dagger[\![S_0]\!], \mathcal{M}_2^\dagger[\![S_1]\!]) .$$

3. Another option is to assume absolutely nothing about the behaviour of a program once it has failed. This choice even allows the code of a program to be affected by a fault. Suppose S_0 fails and terminates some time units later. The process continues with an arbitrary behaviour, which is considered normal (because the *fail*-field of σ is empty at this time). In this case the behaviour of the program is considered to be correct, i.e. as if its code has been modified.

$$\mathcal{M}_3^\dagger[\![S_0; S_1]\!] \hat{=} SEQ(\mathcal{M}[\![S_0]\!], \mathcal{M}_3^\dagger[\![S_1]\!]) \cup SEQ(FAIL(\mathcal{M}[\![S_0]\!]), \Delta)$$

Notice that each of these definitions results in a compositional semantics, because $\mathcal{M}[\![S]\!]$ can be defined in terms of $\mathcal{M}^{\dagger}[\![S]\!]$ for all statements S in RT.

Each of the three alternatives ensures that sequential composition is associative.

Proposition 3

$$\mathcal{M}_i^{\dagger}[\![(S_0;\ S_1);\ S_3]\!] = \mathcal{M}_i^{\dagger}[\![S_0;\ (S_1;\ S_3)]\!],\ i = 1,2,3\ .$$

■

The following proposition relates the behaviors defined by these alternatives for a given program fragment S.

Proposition 4

$$\mathcal{M}_i^{\dagger}[\![S]\!] \subseteq \mathcal{M}_{i+1}^{\dagger}[\![S]\!],\ i = 1,2\ .$$

■

Although, the third alternative defines the less restrictive behaviour we prefer to use the second definition. The reason is that in case of the third alternative a process may exhibit a behaviour that is considered to be correct (i.e. the *fail*-field is empty) even if this behaviour doesn't correspond with an RT-program.

Communication

The normal semantics of the receive statement is defined as the concatenation of two models. The first model denotes the behaviour of the process while it is waiting for its communication partner ($c \in CHAN$):

$$\begin{aligned}&\mathit{WaitRec}(c) \hat{=}\\ &\quad \{(s^0, \sigma, s) \mid\ (|\sigma| < \infty \rightarrow s^0 = s)\\ &\qquad \text{and for all } t \in [0, |\sigma|) :\ \sigma(t).\mathit{comm} = \{(c, ?)\} \wedge \sigma(t).\mathit{fail} = \emptyset\}\ .\end{aligned}$$

The second model denotes the behaviour of the process while the actual communication is taking place:

$$\begin{aligned}&\mathit{CommRec}(c, x) \hat{=}\\ &\quad \{(s^0, \sigma, s) \mid\ |\sigma| = K_c \wedge \text{ there exists a } v \text{ such that } s = (s^0 | x \mapsto v)\\ &\qquad \text{and for all } t \in [0, |\sigma|) :\ \sigma(t).\mathit{comm} = \{(c, v)\} \wedge \sigma(t).\mathit{fail} = \emptyset\}\ .\end{aligned}$$

So, the complete normal behaviour of the receive statement is

$$\mathcal{M}[\![c?x]\!] \hat{=} \mathit{SEQ}(\mathit{WaitRec}(c), \mathit{CommRec}(c, x))\ .$$

For the general semantics we have similar options as in case of sequential composition. We give three reasonable alternatives.

1. The first alternative is our standard approach for the primitive constructs.

$$\mathcal{M}_1^{\dagger}[\![c?x]\!] \hat{=} \mathcal{M}[\![c?x]\!] \cup \mathit{FAIL}(\mathcal{M}[\![c?x]\!])$$

If the process fails during the waiting period and eventually terminates, it skips the communication part. Observe that while the process is still failing it may attempt to communicate because we don't want to make assumptions about the behaviour of a failing process.

2. Alternatively, it is possible to assume that if the process fails while waiting, it remains failed until communication succeeds. This models an execution mechanisms with a reliable communication channel.

$$\mathcal{M}_2^\dagger[\![c?x]\!] \hat{=} \mathcal{M}[\![c?x]\!] \cup SEQ(FAIL(WaitRec(c)), CommRec(c,x))$$

3. If one does not assume a reliable communication channel then a process that fails while waiting but does not remain failed, may thereafter attempt to communicate. Thus a successful communication is not guaranteed. The possibility of failing or not failing during the waiting period and the actual communication is modelled by $WaitRec^\dagger(c)$ and $CommRec^\dagger(c,x)$ respectively.

$$\begin{array}{lcl} WaitRec^\dagger(c) & \hat{=} & WaitRec(c) \cup FAIL(WaitRec(c)) , \\ CommRec^\dagger(c,x) & \hat{=} & CommRec(c,x) \cup FAIL(CommRec(c,x)) . \end{array}$$

The general behaviour of the receive statement is in this case

$$\mathcal{M}_3^\dagger[\![c?x]\!] \hat{=} SEQ(WaitRec^\dagger(c), CommRec^\dagger(c,x)) .$$

We prefer to use the third alternative for two reasons. One reason is that we don't want to assume a reliable communication channel. The other reason is that third alternative defines the less restrictive behaviour in case of a fault.

The send statement is defined in a similar way as the receive statement. First the behaviour of the process while it is waiting is defined. Second, the behaviour during the communication itself is defined. Finally, we define the normal behaviour as the concatenation of these behaviors.

$$\begin{array}{l} WaitSend(c) \hat{=} \\ \quad \{(s^0,\sigma,s) \mid \; (|\sigma| < \infty \rightarrow s^0 = s) \\ \qquad \text{and for all } t \in [0,|\sigma|) : \sigma(t).comm = \{(c,!)\} \wedge \sigma(t).fail = \emptyset\} \end{array}$$

$$\begin{array}{l} CommSend(c,e) \hat{=} \\ \quad \{(s^0,\sigma,s) \mid \; |\sigma| = K_c \\ \qquad \text{and for all } t \in [0,|\sigma|) : \sigma(t).comm = \{(c,s^0(e))\} \wedge \sigma(t).fail = \emptyset\} . \end{array}$$

$$\mathcal{M}[\![c!e]\!] = SEQ(WaitSend(c), CommSend(c,e))$$

For the same reasons as in case of the receive statement we define the general behaviour of the send statement by

$$\mathcal{M}^\dagger[\![c!e]\!] \hat{=} SEQ(WaitSend^\dagger(c), CommSend^\dagger(c,e)) ,$$

where $WaitSend^\dagger(c)$ and $CommSend^\dagger(c,e)$ are defined as follows.

$$\begin{array}{lcl} WaitSend^\dagger(c) & \hat{=} & WaitSend(c) \cup FAIL(WaitSend(c)) , \\ CommSend^\dagger(c,e) & \hat{=} & CommSend(c,e) \cup FAIL(CommSend(c,e)) . \end{array}$$

<u>Guarded Statements</u>

The alternative statement $ALT \doteq [\|_{i=1}^{n} b_i \rightarrow S_i]$ is is executed as follows. First the boolean guard are evaluated, and if one of the guards evaluated to true, the appropriate alternative is executed. The evaluation of the guards takes K_g time units, but has no other effect.

$$\begin{aligned}
&Guard(ALT) \hat{=} \\
&\quad \{(s^0, \sigma, s^0) \mid \quad |\sigma| = K_g \\
&\qquad\qquad \text{and for all } t \in [0, |\sigma|) : \sigma(t).comm = \emptyset \wedge \sigma(t).fail = \emptyset\}
\end{aligned}$$

If all the guards evaluated to false the remainder of the statement is skipped. Otherwise nondeterministically an appropriate alternative is chosen, and executed.

$$\begin{aligned}
&Select(ALT) \hat{=} \\
&\qquad \{(s^0, \sigma, s) \mid \text{there exists an } i \in \{1, \ldots, n\} \text{ s.t. } s^0(b_i) \wedge (s^0, \sigma, s) \in \mathcal{M}[\![S_i]\!]\} \\
&\quad \cup \quad \{(s^0, \sigma, s^0) \mid |\sigma| = 0 \wedge \bigvee_{i=1}^{n} \neg s^0(b_i)\}
\end{aligned}$$

The complete normal behaviour of the simple alternative statement is thus defined by

$$\mathcal{M}[\![ALT]\!] \hat{=} SEQ(Guard(ALT), Select(ALT)) .$$

We consider two possible definitions of the general semantics of the simple alternative statement.

1. The first possible definition is obtained by simply applying the *FAIL* function.

   $$\mathcal{M}_1^{\dagger}[\![ALT]\!] \hat{=} \mathcal{M}[\![ALT]\!] \cup FAIL(\mathcal{M}[\![ALT]\!]) .$$

   The disadvantage of this definition is that it does not discriminate between the occurrence of a fault during the evaluation of the guards and the occurrence of a fault in one of the constituent statements: both faults cause the failure of the whole alternative statement.

2. The second possibility is

   $$\begin{aligned}
   &\mathcal{M}_2^{\dagger}[\![ALT]\!] \hat{=} \\
   &\qquad \mathcal{M}[\![ALT]\!] \cup FAIL(Guard(ALT)) \\
   &\quad \cup \quad SEQ(Guard(ALT), FAIL(Select(ALT))) \\
   &\quad \cup \quad \bigcup_{i=1}^{n} SEQ(FAIL(Guard(ALT)), \mathcal{M}^{\dagger}[\![S_i]\!])
   \end{aligned}$$

   Where $\mathcal{M}^{\dagger}[\![S]\!] = \mathcal{M}_2^{\dagger}[\![S]\!]$ in case $S \doteq ALT$. This definition doesn't have the disadvantage of the previous one.

Because $\mathcal{M}_1^{\dagger}[\![ALT]\!] \subseteq \mathcal{M}_2^{\dagger}[\![ALT]\!]$ we prefer the second definition.

If $ALT \doteq [\|_{i=1}^{n} b_i;\ c_i?x_i \rightarrow S_i \| \ b_0;\ \text{delay}\ e \rightarrow S_0]$ there are three possible ways the process may continue after evaluation of the guards.

1. If all the guards are false the remainder of the *ALT* statement is skipped.
2. If one of the b_i ($i \neq 0$) is true the process waits for an input on one of the c_i for which b_i is true. If b_0 is true communication has to begin within e time units. After the input is received the process continues with the corresponding alternative.
3. If b_0 is true and the process has not received an input within e time units after the guards were evaluated it continues with the execution of S_0.

The first behaviour is defined by

$$\{(s^0, \sigma, s) \in Guard(ALT) \mid \bigwedge_{i=0}^{n} \neg s^0(b_i)\}$$

The second behaviour is defined as the concatenation of three behaviors

$$SEQ(Guard(ALT), Wait(ALT), Comm(ALT)) \ ,$$

where $Guard(ALT)$ is defined as before and $Wait(ALT)$ and $Comm(ALT)$ are defined as follows.

$$\begin{aligned} &Wait(ALT) \mathrel{\hat{=}} \\ &\quad \{(s^0, \sigma, s) \mid \ (\bigvee_{j=0}^{n} s^0(b_j)) \wedge (s^0(b_0) \rightarrow |\sigma| < \min(s^0(e), 0)) \wedge (|\sigma| < \infty \rightarrow s^0 = s) \\ &\qquad \text{and for all } t \in [0, |\sigma|) : \ \sigma(t).comm = \{(c_i, ?) \mid s^0(b_i)\}\} \end{aligned}$$

$$\begin{aligned} &Comm(ALT) \mathrel{\hat{=}} \\ &\quad \{(s^0, \sigma, s) \mid \ \text{there exists an } i \in \{1, \ldots, n\} \text{ such that} \\ &\qquad s^0(b_i) \wedge (s^0, \sigma, s) \in SEQ(CommRec(c_i, x_i), \mathcal{M}[\![S_i]\!])\} \end{aligned}$$

The third behaviour is also defined as the concatenation of three behaviors

$$SEQ(Guard(ALT), TimeOut(ALT), \mathcal{M}[\![S_0]\!]) \ ,$$

where $TimeOut(ALT)$ is defined as follows.

$$TimeOut(ALT) \mathrel{\hat{=}} \{(s^0, \sigma, s) \in Wait(ALT) \mid s^0(b) \wedge |\sigma| = \min(s^0(e), 0)\}$$

The complete normal behaviour of this *ALT* statement is the union of the three behaviors described above.

$$\begin{aligned} &\mathcal{M}[\![ALT]\!] \mathrel{\hat{=}} \\ &\qquad \{(s^0, \sigma, s) \in Guard(ALT) \mid \bigwedge_{i=0}^{n} \neg s^0(b_i)\} \\ &\quad \cup \ SEQ(Guard(ALT), Wait(ALT), Comm(ALT)) \\ &\quad \cup \ SEQ(Guard(ALT), TimeOut(ALT), \mathcal{M}[\![S_0]\!]) \end{aligned}$$

To understand the definition of the general semantics below, one must consider the places where a fault may occur while executing the *ALT* statement. We start near the end of the statement.

I Suppose a fault does not occur until the execution of one of the alternatives. Or a fault occurs while the process is communicating. If the fault behaviour is finite the process may skip the remainder of the *ALT* statement or continue with the execution of one of the alternatives which of course may also result in a fault. This possibility is captured in the following definition.

$$\begin{array}{ll} & SEQ(Guard(ALT), Wait(ALT), Comm^{\dagger}(ALT)) \\ \cup & SEQ(Guard(ALT), TimeOut(ALT), \mathcal{M}^{\dagger}[\![S_0]\!]) \end{array}$$

Where $Comm^{\dagger}(ALT)$ is defined as follows.

$$\begin{array}{l} Comm^{\dagger}(ALT) \hat{=} \\ \quad \{(s^0, \sigma, s) \mid \text{ there exists an } i \in \{1, \ldots, n\} \text{ such that} \\ \qquad s^0(b_i) \wedge (s^0, \sigma, s) \in SEQ(CommRec^{\dagger}(c_i, x_i), \mathcal{M}^{\dagger}[\![S_i]\!])\} \end{array}$$

II Suppose a fault occurs while the process is waiting to communicate. If the fault behaviour if finite the process may continue with any of the communications or alternatives for which it was waiting (i.e. those for which the guard evaluated to true). Of course each of these continuations may again lead to a fault. So we get

$$SEQ(Guard(ALT), Wait^{\dagger}(ALT)) ,$$

where $Wait^{\dagger}(ALT)$ is defined by

$$\begin{array}{l} Wait^{\dagger}(ALT) \hat{=} \\ \quad \{(s^0, \sigma, s) \mid \text{ there exist } s', \sigma_0, \text{ and } \sigma_1 \text{ such that} \\ \qquad \sigma = \sigma_0\sigma_1 \wedge (s^0, \sigma_0, s') \in FAIL(Wait(ALT)) \\ \qquad \wedge((s^0(b_0) \wedge (s', \sigma_1, s) \in \mathcal{M}^{\dagger}[\![S_0]\!]) \\ \qquad\quad \vee(\text{there exists an } i \in \{1, \ldots, n\} \text{ such that} \\ \qquad\qquad s^0(b_i) \wedge (s', \sigma_1, s) \in CommRec^{\dagger}(ALT)))\} . \end{array}$$

III Suppose the fault occurs during the evaluation of the boolean part of the guards. In this case the may wait for an arbitrary communication for an arbitrary period of time, or it may exit the alternative statement immediately. This results in the following behaviour.

$$\begin{array}{ll} & SEQ(FAIL(Guard), Wait(ALT), Comm^{\dagger}(ALT)) \\ \cup & SEQ(FAIL(Guard), TimeOut(ALT), \mathcal{M}^{\dagger}[\![S_0]\!]) \\ \cup & SEQ(FAIL(Guard), Wait^{\dagger}(ALT)) \\ \cup & \{(s^0, \sigma, s) \in FAIL(Guard) \mid \bigwedge_{i=0}^{n} \neg s(b_i)\} \end{array}$$

The general semantics of the *ALT* statement is the union of the normal semantics and the semantics given in I–III above.

Iteration

We define BB as $\bigvee_{i=1}^{n} b_i$ in case ALT is the simple alternative statement and as $\bigvee_{i=0}^{n} b_i$ otherwise. The semantics of the iteration is defined as a greatest fixed-point:

$$\begin{array}{l} \mathcal{M}[\![*ALT]\!] \hat{=} \\ \quad \nu Y.(\quad \{(s^0, \sigma, s) \mid \neg s^0(BB) \wedge (s^0, \sigma, s) \in \mathcal{M}[\![ALT]\!]\} \\ \qquad\qquad \cup\{(s^0, \sigma, s) \mid s^0(BB) \wedge (s^0, \sigma, s) \in SEQ(\mathcal{M}[\![ALT]\!], Y)\}) \end{array}$$

Because evaluation of the boolean guards takes $K_g > 0$ time greatest fixed-point exists and is not empty (cf. [Hooman91]).

We consider two possible definitions of the general semantics.

1. Using the $FAIL$ function gives the simplest definition.

   $$\mathcal{M}_1^{\dagger}[\![*ALT]\!] \hat{=} \mathcal{M}[\![*ALT]\!] \cup FAIL(\mathcal{M}[\![*ALT]\!])$$

   If a fault occurs the process will remain failed until the complete statement terminates. However, we want a definition that discriminates between, for example, a single fault in one pass of the iteration and two consecutive passes with a fault.

2. A definition that does discriminate between the above mentioned cases, and also between the place where a fault occurs is

   $$\begin{array}{l} \mathcal{M}_2^{\dagger}[\![*ALT]\!] \hat{=} \\ \quad \nu Y.(\quad \{(s^0, \sigma, s) \mid \neg s^0(BB) \wedge (s^0, \sigma, s) \in \mathcal{M}[\![ALT]\!]\} \\ \qquad\qquad \cup\{(s^0, \sigma, s) \in SEQ(\mathcal{M}^{\dagger}[\![ALT]\!], Y) \mid s^0(BB)\} \\ \qquad\qquad \cup FAIL(Guard(ALT))) \end{array}$$

   Where $\mathcal{M}^{\dagger}[\![S]\!] = \mathcal{M}_2^{\dagger}[\![S]\!]$ in case $S \doteq *ALT$. This definition allows a process to continue or exit the loop due to a failure. The existence of the greatest fixed-point follows from the fact that failing processes consume time (see proposition 1).

For the reasons mentioned above, we prefer to use the second definition.

Networks

As explained in section 2, the naming construct is not executed or implemented, but only included to facilitate reasoning over programs. Therefore it doesn't introduce new faults.

$$\begin{array}{l} \mathcal{M}^{\dagger}[\![\langle P \Leftarrow S \rangle]\!] \hat{=} \\ \quad \{(s^0, \sigma, s) \mid \quad \text{there exists } (s^0, \sigma', s) \in \mathcal{M}^{\dagger}[\![S]\!] \text{ such that } |\sigma| = |\sigma'| \\ \qquad\qquad\qquad\quad \text{and for all } t \in [0, |\sigma|) : \ \sigma(t).comm = \sigma'(t).comm \\ \qquad\qquad\qquad\quad \wedge \sigma(t).fail = \emptyset \leftrightarrow \sigma'(t).fail = \emptyset \wedge \sigma(t).fail = \{P\} \leftrightarrow \sigma'(t).fail \neq \emptyset\} \end{array}$$

The parallel composition operator doesn't consume time. Hence, it cannot introduce faults that were not already present in the component processes. We use $var(N)$ and *chan*(N) to denote the set of program variables in N and the set of channels incident with N respectively. Recall that variables are not shared and channels connect exactly two processes.

$$
\begin{array}{ll}
\mathcal{M}^{\dagger}[\![N_1 \parallel N_2]\!] \;\hat{=} & \\
\quad \{(s^0, \sigma, s) \mid & \text{there exists } (s_i^0, \sigma_i, s_i) \in \mathcal{M}^{\dagger}[\![N_i]\!] \text{ such that} \\
 & |\sigma| = \max(|\sigma_1|, |\sigma_2|) \wedge (x \in var(N_i) \rightarrow s^0(x) = s_i^0(x)) \\
 & \wedge (x \in var(N_i) \rightarrow s(x) = s_i(x)) \\
 & \wedge (x \notin var(N_1, N_2) \rightarrow s(x) = s^0(x)) \\
 & \text{and for all } t \in [0, |\sigma|),\ c \in CHAN, \text{ and } v \in VAL: \\
 & \sigma(t).comm = \sigma_1(t).comm \cup \sigma_2(t).comm \\
 & \wedge \sigma(t).fail = \sigma_1(t).fail \cup \sigma_2(t).fail \\
 & \wedge |\sigma(t).comm \cap \{(c, ?), (c, !), (c, v)\}| \leq 1 \qquad (1) \\
 & \wedge \left\{ \begin{array}{l} \text{if } c \in chan(N_1) \cap chan(N_2) \\ \text{then } (c, v) \in \sigma_1.comm \leftrightarrow (c, v) \in \sigma_2.comm \end{array} \right. \qquad (2) \\
\quad \} &
\end{array}
$$

It easily seen that parallel composition is commutative. Associativity follows from the fact that channels connect exactly two processes. Hence, the following proposition.

Proposition 5

$$
\begin{array}{lcl}
\mathcal{M}^{\dagger}[\![N_1 \parallel N_2]\!] & = & \mathcal{M}^{\dagger}[\![N_2 \parallel N_1]\!] \\
\mathcal{M}^{\dagger}[\![(N_1 \parallel N_2) \parallel N_3]\!] & = & \mathcal{M}^{\dagger}[\![N_1 \parallel (N_2 \parallel N_3)]\!]
\end{array}
$$

■

Note that (1) is the maximal progress assumption and (2) models regular communication. The assumption that (1) and (2) hold can be weakened for failing processes, by replacing them with

$$\sigma(t).fail = \emptyset \rightarrow (1) \wedge (2)\ .$$

This transformation affects commutativity nor associativity of the parallel composition operator. The weaker version has our preference.

4 Conclusions

We have taken a first step towards a formal method for specifying and verifying real-time systems in the presence of faults. A compositional semantics has been defined together with many alternative definitions. The semantics is defined such that only very weak assumptions about faults and their effect upon the behaviour of a program are made. In this way it is ensured that a proof system that takes this semantics as a basis for its soundness will include few hidden assumptions. Therefore, if one uses such a proof system to verify a real-time system, almost all assumptions about faults will have to be made explicit.

The semantics is compositional which eases the development of a compositional proof system, thereby making the verification of larger systems possible. In section 1 we discussed a small example to illustrate what a proof system might look like. Based upon the semantics defined in this report, we are currently developing a compositional proof system using a real-time version of temporal logic. Future work also includes the design of a proof system that is more like the conventional Hoare-style proof system with pre- and postconditions for sequential programs.

In our semantic definition, faults may affect any channel or local variable. For instance, a fault in a processor may affect any channel in the network, including those that are not connected to the failing processor. This is justified by our philosophy that we want to make only very few (and weak) assumptions about the effect of fault within the model itself. A first study, however, shows that it is possible to parameterize the semantics by function that restrict the set of variables and channels that might be affected by a fault during the execution of a statement.

Acknowledgment. We would like to thank the members of the NWO project "Fault Tolerance: Paradigms, Models, Logics, Construction," in particular Thijs Krol, for their remarks when this work was presented to them in the context of this project.

References

[Bernstein88] P.A. Bernstein. *Sequoia: A Fault-Tolerant Tightly Coupled Multiprocessor for Transaction Processing.* IEEE Computer pp. 37–46, February 1988.

[BGH86] J. Bartlett, J Gray & B. Horst. *Fault Tolerance in Tandem Computer Systems.* Symp. on the Evolution of Fault-Tolerant Computing, Baden, Austria, 1986.

[CDD90] F. Cristian, B. Dancey & J. Dehn. *Fault-Tolerance in the Advanced Automation System.* In "20th Annual Symp. on Fault-Tolerant Computing", 1990.

[Cristian85] F. Cristian. *A Rigorous Approach to Fault-Tolerant Programming.* IEEE Trans. on Softw. Engin. ; SE-11(1):23–31, 1985.

[HaJo89] H. Hansson & B. Jonsson. *A Framework for Reasoning About Time and Reliability.* Proc. 10th IEEE Real-Time Systems Symposium, pp. 101–111, 1989..

[Hooman91] J. Hooman. *Specification and Compositional Verification of Real-Time Systems.* LNCS 558, Springer-Verlag, 1991.

[HoWi89] J. Hooman & J. Widom. *A Temporal-Logic Based Compositional Proof System for Real-Time Message Passing.* Proc. PARLE '89 Vol. II:424–441; LNCS 366, 1989.

[JMS87] M. Joseph, A. Moitra & N. Soundararajan. *Proof Rules for Fault Tolerant Distributed Programs.* Science of Comp. Prog. ; 8:43–67, 1987.

[KLS86] N. Kronenberg, H. Levy & W. Strecker. *VAXclusters: A Closely-Coupled Distributed System.* ACM Trans. on Computer Systems, 4:130–146, 1986.

[OCCAM88] INMOS Ltd. *OCCAM 2 Reference Manual.* Prentice-Hall, 1988.

[Ostroff89] J. Ostroff. *Temporal Logic for Real-Time Systems.* Advanced Software Development Series. Research Studies Press, 1989.

[Pow+88] D. Powell, P. Verissimo, G. Bonn, F. Waeselynck & D. Seaton. *The Delta-4 Approach to Dependability in Open Distributed Computing Systems.* Proc. FTCS-18, IEEE Computer Society Press, 1988.

[RLT78] B. Randell, P.A. Lee & P.C. Treleaven. *Reliability Issues in Computing System Design.* ACM Computing Surveys, 10:123–165, 1978.

[ScSc83] R.D. Schlichting & F.B. Schneider. *Fail-stop processors: an approach to designing fault-tolerant computing systems.* ACM Trans. on Comp. Sys. ; 1(3):222–238, 1983.

[ShLa87] A.U. Shankar & S.S. Lam. *Time-Dependent Distributed Systems: Proving Safety, Liveness and Real-Time Properties.* Distributed Computing; 2:61–79, 1987.

[TaWi89] D. Taylor & G. Wilson. *Stratus.* In "Dependability of Resilient Computers", T. Anderson Ed., Blackwell Scientific Publications, 1989.

Modelling Real-Time Behavior with an Interval Time Calculus

Mats Daniels†

Department of Computer Systems
Uppsala University

Abstract

We present an extension of Milner's CCS [Mil89] with interval time. The notion of time is introduced in terms of time intervals which specify when actions are allowed to occur. We define three equivalences: *strong*, *timed weak*, and *weak* bisimulation equivalence. The strong bisimulation equivalence refines the corresponding relation in CCS by requiring strongly bisimilar processes to have the same timing behavior, the weak bisimulation equivalence is in essence the weak bisimulation equivalence of CCS, while the timed weak equivalence lays strictly between strong and weak equivalence, since it in addition to weak bisimularity also considers the timing of observable actions.

We define a refinement relation for processes specified in our calculus, which can be used to order processes according to their real-time behavior. This relation can be interpreted as "having a more precise timing specification than". For example, the refinement relation can be used to show that an implementation (which normally has more precise time requirements) meets an abstract specification in which the time requirements are specified more generously.

† Mail address: P.O. Box 520, S-751 20 Uppsala, SWEDEN
E-mail: matsd@DoCS.UU.SE

1 Introduction

The main aim of this work is to develop a calculus suitable for modelling and analysis of the real-time behavior of distributed systems. We do this by extending Milner's CCS [Mil89] with a notion of time. We use a dense time domain and introduce time in terms of time intervals which specify when actions are allowed to occur. Our motivation for introducing time intervals is that:

- It is often desirable to be vague about timing requirements in high level specifications
- There is often uncertainty in distributed systems when actions, for example a tick of a local clock, actually take place

We define three equivalences: *strong*, *timed weak*, and *weak* bisimulation equivalence. The strong bisimulation equivalence refines the corresponding relation in CCS by requiring strongly bisimilar processes to have the same timing behavior, the weak bisimulation equivalence is in essence the weak bisimulation equivalence of CCS, while the timed weak equivalence lays strictly between strong and weak equivalence, since it in addition to weak bisimilarity also considers the timing of observable actions.

It is typically the case that time requirements are less specific in high level specifications than in lower level specifications. This has motivated us to define a refinement relation between processes. The relation can be used to order processes according to their real-time behavior, i.e. a process is "more precisely specified" in terms of the timing requirements than the other. For example, the refinement relation can be used to show that a lower level specification meets the time requirements of a higher level specification.

Our timed extension of CCS differs from other CCS based timed calculi, e.g. [Han91, MT90, Wan91], by allowing actions to occur in time intervals rather than at specific points in time. Closest related to our view of time is the work on ACPρ by Baeten and Bergstra [BB89] in which time intervals are included. In parallel with the work on extending CCS with time there has been much effort devoted to extend CSP [Hoa78] with time, e.g. see [Sch90, RR86]. Other relevant work on timed calculi includes the work by Nicollin et al [NRSV90, NS90].

The syntax and operational semantics of our calculus are defined in section 2. The equivalences and the refinement relation are presented in sections 3 and 4. In section 5 we give some simple "vending machine" examples which illustrate our equivalences and the refinement relation. A brief summary of the results is given in section 6.

2 The Calculus

The syntax of our calculus is very similar to the syntax of CCS, the main difference is that time intervals are associated with actions. We will use CCSiT (CCS with interval Time) to refer to our calculus in this paper. To introduce the main ideas behind our calculus we give an informal presentation of CCSiT in section 2.1. In section 2.2 we give a definition of the syntax and the operational semantics of CCSiT.

2.1 Informal Presentation of CCSiT

The following list of properties is intended to informally introduce some basic definitions and give the intuition behind the semantics of CCSiT.

- We use a dense time domain where time is represented by real numbers.
- A time interval is associated to each action in a specification, e.g. $\mu@[0, 2]$.
- The end-points of a time interval associated to an action are relative to the occurrence of the preceding action. The interval can include or exclude the end-points.
- Actions are atomic and take no time to execute.
- When no process performs any actions time will advance. We will use *idling* to model the advance of time. Idling is synchronous, i.e. all processes idle with the same amount at the same time.
- After an action in CCS is performed a new set of actions is enabled. These, in CCS enabled actions, are denoted *time sensitive* actions in CCSiT.
- Idling the time period t leads to a shift with t of the time intervals for all time sensitive actions. For instance, idling 2 means that an interval [1, 4] shifts to [-1, 2].
- A time sensitive action is *enabled* when time 0 is in its time interval, e.g. $\alpha@[-1, 2]$ is enabled. Time 0 represents the current time position, i.e. "now".
- Synchronization, i.e. communication between processes, is as in CCS binary and synchronous. Synchronization can only occur when both participating actions are enabled.
- As in CCS, τ denotes internal (unobservable) actions. An enabled τ-action can only wait up till the upper limit of its time interval. This is a progress assumption needed to allow the specification of actions that must occur within some specified time.
- For external (observable) actions , idling may go past the upper limit of the associated interval and thus prevent the action from ever occurring, e.g. $\alpha@[-2, -1]$ is no longer enabled.

2.2 Definition of CCSiT

2.2.1 Syntax

Let Λ be a set of action symbols not containing τ or ι, and let Act = $\Lambda \cup \{\tau, \iota\}$. The action symbol τ denotes, as in CCS, an internal action and ι stands for idling. ι is not included in the syntax of CCSiT, hence it cannot be used in a specification. Let T be the time domain with values taken from the reals $\mathbb{R} \cup \{\infty, \emptyset\}$, where ∞ is the top element and $\emptyset$ is the bottom element. Let $\mu, \nu \in \Lambda \cup \{\tau\}$ denote any action except idling. Let $\alpha, \beta \in \Lambda$ denote external actions. Let L denote a subset of Λ. Let i denote an interval [l, h], (l, h], [l, h), or (l, h), where $l, h \in T$, $l \leq h$, [and] denote that the corresponding endpoint is included in the interval, and (and) denote that the endpoint is excluded. We do not allow ∞ to be included in any interval, i.e. "∞]" is an illegal construct. Intervals of the type [n, n) and (n, n) are the same as the empty interval ($\emptyset$) for any $n \in \mathbb{R}$. The endpoints will be referred to as i.l and i.h respectively. The expression "i - t" will be used to denote "i.l-t" and "i.h-t". Let P, Q, R etc. denote processes.

The syntax of CCSiT processes is defined as follows: P ::= NIL | μ@i.Q | Q+R | P|Q | P\L

Intuitively, NIL (inaction) denotes a process that cannot perform any action. μ@i.Q (prefixing) denotes a process that can perform the action μ in the time interval i, and then behave as the process Q. Q+R (choice) denotes a processes that either behaves as Q or as R, i.e. if one of them performs an action this will result in the inhibition of the other process. P|Q (parallel composition) denote two processes running in parallel, i.e. P and Q may perform actions individually (not influencing the other) or together (synchronizing on an action). Note that for simplicity we do not use complementary actions, as in CCS, to form a synchronization pair. Actions with the same name may form a synchronization pair in CCSiT. P\L (restriction) denotes that the actions listed in L cannot interact with the environment.

2.2.2 Semantics

The operational semantics of CCSiT is given in terms of a labeled transition system [Plo81]. We have $\mathcal{P}$ = (Π, Act, $\rightarrow$), where Π denotes the set of processes, Act is a set of actions, and $\rightarrow$: $\Pi \times$ Act $\times \Pi$, is the transition relation.

The passing of time is described by shifting the intervals associated to actions. This shift is in the semantic rules captured by the "idling" action, i.e. $\text{—}\iota(t)\rightarrow$ (where "t" denotes a time period during which the process does not perform any actions) the values for "t" is taken from $\mathbb{R}^+$, i.e. the positive reals.

In the definition of the semantics we will use the functions F-Next, Next, Actions, and t_limit. For a process R we define a function F_Next (R) inductively as follows:

Definition 2.1 (set of possible next actions with associated interval and following process)

$$
\begin{aligned}
\text{F\_Next (NIL)} &= \{\varnothing\} \\
\text{F\_Next } (\mu@i.P) &= \{\mu@i.P\} \\
\text{F\_Next } (P + Q) &= \text{F\_Next } (P) \cup \text{F\_Next } (Q) \\
\text{F\_Next } (P|Q) &= \text{F\_Next } (P) \cup \text{F\_Next } (Q) \cup \\
&\quad \{\tau@\{k \cap l\}.(P_i|Q_j) \mid \exists\, \alpha@k.P_i \in \text{F\_Next } (P)\ \& \\
&\quad\ \exists\, \alpha@l.Q_j \in \text{F\_Next } (Q)\ \&\ k \cap l \neq \varnothing\} \\
\text{F\_Next } (P\backslash L) &= \{\mu@i.Q \mid \exists\, \mu@i.Q \in \text{F\_Next } (P)\ \&\ \mu \notin L\}
\end{aligned}
$$

Definition 2.2 (set of possible next actions with associated interval)

$$\text{Next } (P) = \{\mu@i \mid \exists\, \mu@i.Q \in \text{F\_Next } (P) \text{ for some } Q\}$$

The purpose of Next (P) is to extract the set of actions and their corresponding time intervals. Note that some of these actions might be impossible, e.g. existence of an internal action that must occur before other actions are enabled or if the upper limit of an actions time interval is negative.

Definition 2.3 (set of next actions that is or may be enabled)

$$\text{Actions } (P) = \{\mu \mid \exists\, \mu@i \in \text{Next } (P)\ \&\ 0 < i.h\}$$

Intuitively, Actions (P) denotes the set of actions that are not "too late" already, i.e. actions with upper limits of their time intervals not less than zero.

Definition 2.4 (maximum idling allowed)

$$
\tau\text{\_limit } (P) = \begin{cases} \min\{\sup(i) \mid \tau@i \in \text{Next } (P)\ \&\ 0 < i.h\}, & \text{if } \tau \in \text{Actions } (P) \\ \infty \text{ otherwise} \end{cases}
$$

Intuitively, τ_limit (P) denotes the maximum time idling is allowed in the process. Note that an internal action that is or will become enabled is required for this upper bound to exist.

For instance,

$$\tau\text{\_limit } (\alpha@[3, 8] + \tau@[4, 7] + \tau@[4, 6]) = 6 \quad .$$

In deciding if a time point is in an interval or not we will use $\angle$ to denote < or $\leq$ depending on whether the interval includes the upper limit or not. For instance, $t \angle i.h$ represents $t \leq i.h$ if the time interval is (3, 5], and $t < i.h$ if the time interval is (3, 5)

	Rule	
Inaction	**Rule C.1:** (idling)	$\dfrac{}{NIL \xrightarrow{\iota(t)} NIL}$
Prefix	**Rule P.1:** (performing an action)	$\dfrac{}{\mu@i.P \xrightarrow{\mu} P}\ 0 \in i$
	Rule P.2: (idling for external actions)	$\dfrac{}{\alpha@i.P \xrightarrow{\iota(t)} \alpha@i\text{-}t.P}$
	Rule P.3: (idling for internal actions)	$\dfrac{}{\tau@i.P \xrightarrow{\iota(t)} \tau@i\text{-}t.P}\ t \angle i.h \text{ or } i.h < 0$
Summation	**Rule S.1:** (performing an action)	$\dfrac{P \xrightarrow{\mu} P'}{P+Q \xrightarrow{\mu} P'}$
	Rule S.1': (performing an action)	$\dfrac{Q \xrightarrow{\mu} Q'}{P+Q \xrightarrow{\mu} Q'}$
	Rule S.2: (idling)	$\dfrac{P \xrightarrow{\iota(t)} P' \quad Q \xrightarrow{\iota(t)} Q'}{P+Q \xrightarrow{\iota(t)} P'+Q'}$
Composition	**Rule PC.1:** (performing unrelated actions)	$\dfrac{P \xrightarrow{\mu} P'}{P \mid Q \xrightarrow{\mu} P' \mid Q}$
	Rule PC.1': (performing unrelated actions)	$\dfrac{Q \xrightarrow{\mu} Q'}{P \mid Q \xrightarrow{\mu} P \mid Q'}$
	Rule PC.2: (idling)	$\dfrac{P \xrightarrow{\iota(t)} P' \quad Q \xrightarrow{\iota(t)} Q'}{P \mid Q \xrightarrow{\iota(t)} P' \mid Q'}\ t < \tau\_limit\ (P \mid Q)$
	Rule PC.3: (synchronization of actions)	$\dfrac{P \xrightarrow{\alpha} P' \quad Q \xrightarrow{\alpha} Q'}{P \mid Q \xrightarrow{\tau} P' \mid Q'}$
Restriction	**Rule R.1:** (actions and idling)	$\dfrac{P \xrightarrow{\sigma} P'}{P \backslash L \xrightarrow{\sigma} P' \backslash L}\ \sigma \notin L, \text{ where } \sigma \in Act$

Table 2.1: Operational Semantics for CCSiT

Rule C.1 is the only rule for the constant (inactive) process NIL. NIL is not affected by passing of time and cannot prevent it, i.e. it can idle.

The interval in a prefix expression specifies when the corresponding action is enabled. Rule P.1 says that an action can only take place when the current time, i.e. 0, is in its associated time interval (i). The interval is shifted during idling periods ($\iota(t)$, see rules P.2 & P.3) and an interval with i.l > 0 will eventually include 0. Idling (ι) is always possible for external actions (rule P.2), but note that the action will no longer be enabled if idling pushes sup (i) below 0. Idling (ι) is possible up till the end of the interval (i) associated with an internal action, but not past this time since the progress assumption forces the internal action to occur sometime in the associated interval. Note that this means that idling can occur even when an internal action is enabled. The constraint i.h < 0 in rule P.3 is due to the legal (but non-constructive) possibility to specify a time interval with a negative upper limit. Such a specification is semantically equivalent to NIL.

Summation (+) is used to describe non-deterministic choices. Rules S.1 and S.1' are standard CCS rules. Rule S.2 show that idling can take place if both processes can idle. Remember that a process with a τ-action can only idle up to the top of this actions interval.

The operator | is used to denote parallel composition. PC.1 and PC.1' are as in CCS. These rules express that an action in one of the processes in a parallel composition can occur independently of the other process. Rule PC.2 express that idling in a parallel composition require that idling is done in both processes. is the case when both processes idle. Note that idling is not allowed past τ_limit (P|Q). The reason is that the progress assumption disallow idling past the upper limit of an enabled internal action. Note how the definition of τ_limit (P|Q) covers the case when a t-action occurs as a consequence of a synchronization between P and Q. Note in PC.3 that synchronization can only take place if the intervals of the corresponding actions overlap, i.e. the intersection $i_{\alpha\text{-in-P}} \cap i_{\alpha\text{-in-Q}}$ is not empty. The idling rule (PC.2) ensures that this synchronization cannot be idled past, i.e. the synchronization will take place latest at the end of the overlap interval.

3 Equivalences

In this section we define three equivalences: strong, timed weak, and weak bisimulation equivalence, and give some equational laws. Intuitively, the different equivalences stem from the capabilities of different observers. The observer for the strong equivalence can observe all types of actions as well as the points in time when they occur, the observer for the weak equivalence can only observe external actions and have no awareness of time, and the observer for the timed weak equivalence can in addition to observe external actions also observe the time when they occur.

3.1 Bisimulation

Two labeled transition systems are bisimular if they can simulate the behavior of each other. That is, for any action in one transition system there should be a corresponding action in the other transition system and the two systems should after having performed these actions still be able to simulate each other.

3.1.1 Strong Equivalence

Strong equivalence is built on strong bisimulation between processes. The observer is in this case very competent, since all actions of a system can be detected as well as delays between actions.

Definition 3.1 (strong equivalence)

A binary relation S is a strong bisimulation if (P, Q) $\in$ S implies, for all $\mu \in \Lambda \cup \{\tau\}$, and all t ≥ 0

(i) If P —μ→ P' then for some Q': Q —μ→ Q' and (P', Q') $\in$ S
(ii) If Q —μ→ Q' then for some P': P —μ→ P' and (P', Q') $\in$ S
(iii) If P —$\iota(t)$→ P' then for some Q': Q —$\iota(t)$→ Q' and (P', Q') $\in$ S
(iv) If Q —$\iota(t)$→ Q' then for some P': P —$\iota(t)$→ P' and (P', Q') $\in$ S

This equivalence requires that the processes must at all time instances have the same set of enabled actions, i.e. the time intervals associated with an action must also be considered when this relationship is examined. Let ~, as in [Mil89], denote the largest strong bisimulation. That ~ is a congruence in CCSiT follows from the general results on strong bisimulation by Groote and Vaandrager [GV88].

3.1.2 Weak Equivalences

We define two different weak equivalences. The first is similar to the weak equivalence as defined in [Mil89] and the other is made sensitive to the time when actions occur, i.e. a difference is detected if the processes have different timing characteristics. Let us first give some useful definitions.

Definition 3.2 (shorthand notation)

If P —ρ_1→ P_1 —ρ_2→ P_2— ....ρ_n→ P_n, then P —$\rho_1\rho_2....\rho_n$→ P_n, where $\rho_i \in$ Act

This definition is made for convenience, since it allow us to write action sequences without naming the intermediate processes.

Definition 3.3 (hiding any preceding internal actions and idling)

If P —$(\tau^*\iota(t_i)^*)^*\rho$→ P', then P $*\rho$→ P', where $\rho \in$ Act

Intuitively, this let us focus on three things: the initial process, the last action, and the resulting process, ignoring any internal actions and idling leading up to the last action. The possible existence of such preceding internal actions and idling is immaterial for definition of the weak equivalence (definition 3.5).

Definition 3.4 (external action or not)

$$\underline{\rho} = \begin{cases} \rho & \text{if } \rho \in \Lambda \\ \varepsilon & \text{otherwise, where } \varepsilon \text{ is the empty sequence} \end{cases}$$

We can now define the weak equivalence.

Definition 3.5 (weak equivalence)

A binary relation W is a weak bisimulation if (P, Q) $\in$ W implies, for all $\rho \in$ Act,

(i) If P $*\rho$→ P' then for some Q': Q $*\underline{\rho}$→ Q' and (P', Q') $\in$ W

(ii) If Q $*\rho$→ Q' then for some P': P $*\underline{\rho}$→ P' and (P', Q') $\in$ W

Intuitively, if a process can perform an external action then this should always be possible in the weakly bisimular process also. Maybe not at the same point in time, but that is not important here. Idling or performing internal actions in one process is allowed if the other process either by idling, performing internal actions, or doing nothing still is weakly bisimular with the first process. Let $\approx$ denote the largest weak bisimulation.

Definition 3.6 (transition sequence shorthand with time information)

P $*\rho[D]$→ P', if P —$\tau^*\iota(t_1)\tau^*\ldots\tau^*\iota(t_n)\tau^*\rho$→ P'

where $\rho \in$ Act and $D = \Sigma_{i\in\{1:n\}} t_i$ (+ t, if $\rho=\iota(t)$)

Intuitively, this captures one allowed value, D, of the time it takes from now to change from P to P'.

Definition 3.7 (external action or not)

$$\underline{\rho}[D] = \begin{cases} \rho[D] & \text{if } \rho \in \Lambda \\ \varepsilon[D] & \text{otherwise, where } \varepsilon \text{ is the empty sequence} \end{cases}$$

We can now define the timed weak equivalence.

Definition 3.8 (timed weak equivalence)

A binary relation T is a timed weak bisimulation if $(P, Q) \in T$ implies, for all $\rho \in$ Act and for all $t \geq 0$,

(i) If $P *\rho[t]\rightarrow P'$ then for some Q': $Q *\underline{\rho}[t]\rightarrow Q'$ and $(P', Q') \in T$

(ii) If $Q *\rho[t]\rightarrow Q'$ then for some P': $P *\underline{\rho}[t]\rightarrow P'$ and $(P', Q') \in T$

The added requirement compared to weak bisimulation is that we also consider the time when the external actions take place. Timed weak bisimular actions must be enabled in the same time interval. Let $\approx_t$ denote the largest timed weak bisimulation.

3.2 Equational Laws

Milner [Mil89] defines some basic laws for CCS, i.e. static laws for static combinators (Composition, Restriction, and Relabelling), dynamic laws for dynamic combinators (Prefix, Summation, and Constant), and the expansion law. The corresponding laws for CCSiT is presented in this section, as well as, some laws regarding behavior due to the timing of processes.

3.2.1 Basic Laws

The following laws for CCS are also valid for CCSiT. They are all sound, as can be seen from the definition of the operational semantics.

Summation	$P + Q$	$\sim$	$Q + P$	(Sum.1)
	$(P + Q) + R$	$\sim$	$P + (Q + R)$	(Sum.2)
	$P + P$	$\sim$	P	(Sum.3)
	$P + NIL$	$\sim$	P	(Sum.4)
Composition	$P \mid Q$	$\sim$	$Q \mid P$	(Com.1)
	$(P \mid Q) \mid R$	$\sim$	$P \mid (Q \mid R)$	(Com.2)
	$P \mid NIL$	$\sim$	P	(Com.3)
Restriction	$(P + Q)\backslash L$	$\sim$	$P\backslash L + Q\backslash L$	(Res.1)
	$\mu \in L$ implies: $(\mu@i.P)\backslash L$	$\sim$	NIL	(Res.2)
	$\mu \notin L$ implies: $(\mu@i.P)\backslash L$	$\sim$	$\mu@i.(P\backslash L)$	(Res.3)
	$NIL\backslash L$	$\sim$	NIL	(Res.4)

The τ-laws in [Mil89] are not applicable in general, since τ-actions are associated with a time interval. The following τ-laws are however true if we use the weak equivalence and put some restrictions on the time intervals, e.g. the upper limit of a time interval must be non-negative and must be closed at the top if the upper limit is 0. We have:

τ laws	$\tau@i.P$	$\approx$	P, where $0 \angle i.h$	(τ.1)
	$\tau@i_1.P + P$	$\approx$	$\tau@i_2.P$, where $0 \angle i_2.h$	(τ.2)
	$\alpha@i_1.\tau@i_2.P$	$\approx$	$\alpha@i_3.P$, where $0 \angle i_1.h$, $0 \angle i_2.h$, and $0 \angle i_3.h$	(τ.3)

3.2.2 Expansion Theorem

The expansion theorem defines the relation between parallel and sequential behaviors, i.e. how a parallel composition of processes can be transformed into a summation. The crux here is that idling is synchronous, i.e. for a composition (P|Q) to idle it require both P and Q to idle. Assume P = α@i.P' which in a CCS-transformation (expansion theorem) would (among other) give the term α@i.(P'|Q), where Q is not time sensitive any more (not affected by idling).

We need a way to shift time in parts of a process which normally is not time sensitive. This problem is handled with a time variable in work by Wang [Wan91]. We will here introduce the following definitions and three more semantic rules to handle this problem:

Definition 3.9 (forcing advancement of time)

F_Time (NIL, t)	=	NIL
F_Time (μ@i.P, t)	=	μ@i-t.P
F_Time (P + Q, t)	=	F_Time (P, t) + F_Time (Q, t)
F_Time (P|Q, t)	=	F_Time (P, t)|F_Time (Q, t)
F_Time (P\L, t)	=	F_Time (P, t)\L

Intuitively, F_Time (P, t) forces the time intervals for in P time sensitive actions to be shifted t towards -∞.

Rule TS.1: (an external action and idling when a time sensitive processes exist, ie P{TS})

$$\frac{}{\alpha@i.(P\{TS\}|Q) \xrightarrow{\iota(t)} \alpha@i\text{-}t.(P\{TS\}|Q')} \quad 0 \le t < \infty, \text{ where P': F\_Time (P, t)}$$

Rule TS.2: (an internal action and idling when a time sensitive processes exist, ie P{TS})

$$\frac{}{\tau@i.(P\{TS\}|Q) \xrightarrow{\iota(t)} \tau@i\text{-}t.(P'\{TS\}|Q)} \quad 0 \le t \angle \text{ i.h, where P': F\_Time (P, t)}$$

Rule TS.3: (performing an action)

$$\frac{}{\mu@i.(P\{TS\}|Q) \xrightarrow{\mu} P|Q} \quad 0 \in i$$

{TS} is used to mark processes that should be time sensitive. Note that no restriction on t is imposed here, i.e. idling might render internal actions "too old". This will not happen since that internal action will appear as a choice somewhere else in the process (another term in the sum) after use of the expansion theorem and thus put a limit on how long idling might be possible.

Theorem 3.1 (Expansion Theorem)

Let $P \equiv \Sigma_{k \in K}\, \mu_k@i_k.P_k$ and $Q \equiv \Sigma_{l \in L}\, \nu_l@i_l.Q_l$

Then $(P|Q)\backslash L \sim (\Sigma_{(k \in K,\, l \in L,\, \mu_k = \nu_l)}\, \tau@i_{k \cap l}.(P_k|Q_l)\backslash L \mid 0 \angle i_{k \cap l}.h)$
$+ (\Sigma_{k \in K}\, \mu_k@i_k.(Q\{TS\}|P_k)\backslash L \mid \mu_k \notin \{L\}\ \&\ 0 \angle i_k.h)$
$+ (\Sigma_{l \in L}\, \nu_l@i_l.(P\{TS\}|Q_l)\backslash L \mid \nu_l \notin \{L\}\ \&\ 0 \angle i_l.h)$

Where $i_{k \cap l} = i_k \cap i_l$

Note that actions listed in the restriction set (L) and/or are "too late" (their upper time limit is in the past) are removed (rather not included) in the resulting sum.

As an example, consider

$P = \alpha@[1,3].P_1 + \beta@[0,\infty).P_2$ and $Q = \alpha@[2,6].Q_1$

Applying the expansion theorem to P|Q will yield:

$P|Q \sim \tau@[2,3].(P_1|Q_1) + \alpha@[1,3].(Q\{TS\}|P_1) + \beta@[0,\infty).(Q\{TS\}|P_2) + \alpha@[2,6].(P\{TS\}|Q_1)$

where the τ-action is a consequence of an α-synchronization and the time interval comes from the intersection of the intervals [1,3] and [2,6].

3.2.3 Real-Time Aspects

Remember that the environment may delay an arbitrary amount of time before offering synchronization with an external action, and internal actions (τ-actions) must occur within the specified interval. The arbitrary delay means that an external action with an upper limit on its time interval might idle past this time, in which case the action cannot occur, i.e. it cannot participate in any synchronization. Existence of internal actions might limit the choices in a summation, e.g. an action with a time interval starting after the upper limit of an internal action will never occur, since the τ-action must occur before this action is enabled. Then we get the following law:

Equational law for summation: If $\tau\_limit(Q) < i.l$ then we have:

$$\alpha@i.P + Q \sim Q \qquad \text{(Sum.5)}$$

The semantics of CCSiT can inhibit actions due to their time interval. Actions in a process can be made impossible due to time considerations in two ways; (1) idling past the upper limit for synchronization with the environment and (2) in a restricted composition where the time intervals of the synchronizing actions (listed in the restriction set) are non-overlapping resulting in an empty interval (Ø, which is the bottom element in the time domain). The following law can be used to remove such "dead code":

Equational laws for idling: If $(\sup(i) \mid \mu@i \in Next(P)) < 0$ then we have:

$$P \sim NIL, \qquad \text{(Idl.1)}$$

4 Refinements

The previous section introduced some equivalences between processes. We will here introduce a refinement relation which relate an implementation (a more detailed description) with its specification (a more abstract specification), in the sense that the relation requires the specification to include the same visible behavior as the implementation. The relation also requires that the intervals in the specification when external actions are enabled must include the corresponding intervals in the implementation.

The timing of internal actions have to be taken into account when the intervals for external actions are decided. That is, all preceding τ-actions must be considered when defining the interval for an external action. Any τ-actions after an external action should not be accounted for in the time interval, these actions will however be considered when deciding the interval for the succeeding external actions. Consider for instance $\tau@[1,7].\alpha[1,2].\tau@[0,5].\beta@(2,4]$ where the relevant time interval for the α action is that it occur earliest at time 2 and latest at time 9. The succeeding internal action is immaterial for the time of the α action, but is important with respect to when the β action is enabled. We are interested in the interval in which an external action is enabled. In the definition of this interval we will use the following definitions:

Definition 4.1 (possible processes following a specific action)

$$
\begin{aligned}
\text{Follows\_X}(\text{NIL}, \alpha) &= \{\varnothing\} \\
\text{Follows\_X}(\beta@i.P, \alpha) &= \{\varnothing\}, \text{ where } \alpha \neq \beta \\
\text{Follows\_X}(\alpha@i.P, \alpha) &= \{P\} \\
\text{Follows\_X}(\tau@i.P, \alpha) &= \text{Follows\_X}(P, \alpha) \\
\text{Follows\_X}(P + Q, \alpha) &= \text{Follows\_X}(P, \alpha) \cup \text{Follows\_X}(Q, \alpha) \\
\text{Follows\_X}(P|Q, \alpha) &= \text{Follows\_X}(P, \alpha) \cup \text{Follows\_X}(Q, \alpha) \cup \\
&\quad \{\text{Follows\_X}(P_i|Q_j, \alpha) \mid \exists\, \mu@k.P_i \in \text{F\_Next}(P)\ \& \\
&\quad\ \exists\, \mu@l.Q_j \in \text{F\_Next}(Q)\ \&\ k \cap l \neq \varnothing\ \&\ \mu \neq \tau\} \\
\text{Follows\_X}(P\backslash L, \alpha) &= \text{Follows\_X}(P, \alpha) \text{ if } \alpha \notin L,\ \{\varnothing\} \text{ otherwise}
\end{aligned}
$$

Intuitively, Follows_X (P, α) defines the set of processes that follows α, in the cases where α is among the possible external actions process P. The definition is fairly straightforward except for the parallel composition, where we have to take into account the possible creation of new internal actions due to synchronizations in which case we have to search for succeeding α actions.

Definition 4.2 (adding an interval to a set of intervals)

$i + \bigcup_{j\in J} i_j = \bigcup_{j\in J} i_j'$, where $i_j'.l = i_j.l + i.l$ and $i_j'.h = i_j.h + i.h$

Adding an interval to a set of intervals is to adjust the lower limit of each interval in the set by the lower limit of the interval added and the same for the upper limit, i.e. the added intervals upper limit is added to the upper limit of each interval in the set. Note that a closed end of an interval can only remained so if the corresponding end of the interval added also is closed. The truth table for an and-gate where [and] represent ones and (and) represent zeroes can be used to decide the type of interval limit.

We can now define the intervals in which external actions are enabled.

Definition 4.3 (time intervals in which a specific external action is enabled

$I_{\alpha P}(NIL) = \{\varnothing\}$

$I_{\alpha P}(\beta@i.Q) = \{\varnothing\}$, where $\alpha \neq \beta$

$I_{\alpha P}(\alpha@i.Q) = \{i\}$

$I_{\alpha P}(\tau@i.Q) = \{i + I_{\alpha P}(Q)\}$, if $P \in$ Follows_X (Q, α), $\{\varnothing\}$ otherwise

$I_{\alpha P}(Q + R) = I_{\alpha P}(Q) \cup I_{\alpha P}(R)$

$I_{\alpha P}(Q|R) = I_{\alpha P}(Q) \cup I_{\alpha P}(R) \cup$
$\{k \cap l + I_{\alpha P}(Q_i|R_j) \mid \exists\, \mu@k.Q_i \in$ F_Next (Q) & $\exists\, \mu@l.R_j \in$ F_Next (R) & $k \cap l \neq \varnothing$ & $\mu \neq \tau\}$

$I_{\alpha P}(Q\backslash L) = I_{\alpha P}(Q)$ if $\alpha \notin L$, $\{\varnothing\}$ otherwise

Intuitively, $I_{\alpha P}(Q)$ defines the set of intervals in which the action α is enabled and for which the behavior after performing α is defined by P. Each interval is obtained by adding the intervals (according to definition 4.2) of all preceding internal actions that occur before α to the interval associated with α.

4.1 The Refinement Relation

Let S be a specification and R an implementation of S. Let $I_{\alpha S'}(S)$ and $I_{\alpha R'}(R)$ denote the time intervals associated with action α (and processes S' and R') in S and in R respectively.

Definition 4.2 (S is refined by R)

$S \gg R$ iff, for all $\alpha \in \Lambda$

(i) If $S *\alpha\!\rightarrow S'$ then for some R': $R *\alpha\!\rightarrow R'$ and $S' \gg R'$ and $I_{\alpha R'}(R) \subseteq I_{\alpha S'}(S)$

(ii) If $R *\alpha\!\rightarrow R'$ then for some S': $S *\alpha\!\rightarrow S'$ and $S' \gg R'$ and $I_{\alpha R'}(R) \subseteq I_{\alpha S'}(S)$

Intuitively, S » R (read: S is refined by R) if all external actions in either R or S can be matched by S and I respectively and in such a way that the time refinement relation holds for the resulting processes. $I_{\alpha R'}(R) \subseteq I_{\alpha S'}(S)$ states that all time intervals in which α can occur in the refinement must also be included in the time intervals for this action in the specification. Note that if we require the intervals to be equal ($I_{\alpha R'}(R) = I_{\alpha S'}(S)$) we get an alternative definition of timed weak bisimulation (definition 3.8).

5 Example: a vending machine

We will illustrate the calculus with a small example of a simple real-time system; a vending (or coffee) machine. In [Dan91] we describe and analyze the Alternating Bit Protocol [BSW69].

We will model a vending machine that works as follows: the thirsty user inserts a coin and gets after a slight delay a choice of selecting (and obtaining) tea or coffee, the maximum delay is different for tea and coffee, or nothing happens. Another feature is that a timer will time-out if the user waits too long to decide between coffee and tea. The following is a possible specification in CCS of the vending machine:

M = coin.(coffee.NIL + tea.NIL + τ.NIL)

The τ in this description is used to capture both malfunction and the intended time-out on too slow response. Adding time information will give us a starting point for our design of a coffee machine. Let us assume that the time-out is 30 time units, we will then get the following CCSiT description:

M1 = coin@any.(coffee@any.NIL + tea@any.NIL + τ@[0, 30].NIL)

where @any denote @[0,∞). Intuitively, a coin can be inserted at any time, whereafter either coffee or tea can be obtained at any time unless the time-out expires after 30 time units (the upper limit of the τ-action) or the machine malfunctions.

5.1 Refined Specifications

A common feature of coffee machines is that coffee and tea will not be alternatives immediately after a coin is inserted. Another thing is that we would like to distinguish between malfunction and time-out. The following is a CCSiT description with the above mentioned features:

M2 = coin@any.(coffee@[2, ∞).NIL + tea@[3, ∞).NIL + τ@any.NIL + τ@30.NIL)

There are some interesting intervals in this specification, i.e. [0, 2), [2, 3), and [3, 30]. The only action that can occur in [0, 2) is that the machine will malfunction (τ@any) and then deadlock. Two actions are enabled in the next interval, i.e. the malfunction and giving coffee. These actions are accompanied by the possibility to obtain tea in [3, 30]. This specification only allows idling up to 30 when the time-out (τ@30) is released and the process deadlocks.

We can be more specific in our refinement and actually describe the timer responsible for giving the time-out signal. The timer can be specified as:

T = start@any.T1

T1 = t_o@30.NIL

and the coffee machine would be slightly changed to, where @now means @[0, 0]:

C = coin@any.start@now.C1

C1 = coffee@[2, ∞).NIL + tea@[3, ∞).NIL + t_o@any.NIL + τ@any.NIL

5.2 Equivalence and Refinement Relations

The intention of the different descriptions of the coffee machine given above is the following: M2 and R = (C | T)\{t_o, start} should be equivalent (at least weakly) and both should be time refinements of M1. Let us examine if this is the case.

5.2.1 Equivalence

Is M2 ~ R? This requires, for instance, that (M2, R) is a pair in the strong equivalence relation. Let us find out the other pairs. M2 can synchronize on coin becoming $M2_1$, where

$M2_1$ = coffee@[2, ∞).NIL + tea@[3, ∞).NIL + τ@any.NIL + τ@30.NIL

or idle indefinitely, which does not change M2. The definition of strong bisimulation equivalence require that R can do the same, i.e. synchronize on coin and idle indefinitely. Synchronizing on coin changes R to R_1, where

R_1 = (start@now.C1 | T)\{t_o, start}

and idling does not change R. That is, the first requirement is fulfilled. Lets move on to the pair ($M2_1$, R_1). R_1 will now perform an internal action, due to synchronization on the action start, and become R_2, where

R_2 = (C1 | T1)\{t_o, start}

applying the expansion theorem and using the law P|NIL ~ P we get:

R_2 = (τ@30.NIL + coffee@[2, ∞).T1 + tea@[3, ∞).T1 + τ@any.T1)\{t_o, start}

$M2_1$ can also perform an internal action becoming NIL. (R_2 | NIL) cannot be a pair in a strong equivalence relation, i.e. M2 is not strongly equivalent to R.

Is M2 $\approx_t$ R? We can safely start where the strong equivalence failed, since the requirements for the timed weak equivalence is weaker than the requirements for strong equivalence. Starting from $M2_1$, we have that the following possibilities (using τ_limit (P) to define the upper time limit for the coffee and tea actions)

$M2_1$ ∗coffee[2, 30]→ NIL
$M2_1$ ∗tea[3, 30]→ NIL
$M2_1$ ∗τ[0, 30]→ NIL

R_1 can do

R_1 ∗τ[0, 0]→ R_2
R_1 ∗coffee[2, 30]→ NIL
R_1 ∗tea[3, 30]→ NIL
R_1 ∗τ[0, 30]→ NIL

and R_2 can do

R_2 ∗coffee[2, 30]→ NIL
R_2 ∗tea[3, 30]→ NIL
R_2 ∗τ[0, 30]→ NIL

The requirement now is that (R_1, M2) and (R_2, M2) are in the relation. This is obviously the case. (NIL, NIL) is a pair in any equivalence relation, thus we have that M2 and R are timed weakly equivalent. From this follows that they also are weakly equivalent, since the requirements are further lessened for the weak equivalence.

5.2.2 Refinement

This time we turn the attention to the most abstract of our specifications of the coffee machine, i.e. M1. **Is M1 » M2?** We have for M1 that

M1 ∗coin[0, ∞)→ $M1_1$, where $I_{coin,M1_1}(M1) = [0, \infty)$

and for M2

M2 ∗coin[0, ∞)→ $M2_1$, where $I_{coin,M2_1}(M2) = [0, \infty)$

There requirements are fulfilled for this step. Lets move on to $M1_1$, which can do

$M1_1$ ∗coffee[0, 30]→ NIL, where $I_{coffee,NIL}(M1_1) = [0, 30]$
$M1_1$ ∗tea[0, 30]→ NIL, where $I_{tea,NIL}(M1_1) = [0, 30]$
$M1_1$ ∗τ[0, 30]→ NIL, where $I_{\tau,NIL}(M1_1) = [0, 30]$

and for M2

$M2_1$ ∗coffee[2, 30]→ $M2_1$, where $I_{coffee,NIL}(M2_1) = [2, 30]$
$M2_1$ ∗tea[3, 30]→ NIL, where $I_{tea,NIL}(M2_1) = [3, 30]$
$M2_1$ ∗τ[0, 30]→ NIL, where $I_{\tau,NIL}(M2_1) = [0, 30]$

We see that all external actions can be matched by the corresponding process and that the time intervals for all external actions in M2 (the implementation) are inside the time intervals for the corresponding external action in M1. M2 is thus an implementation of M1. They are however not timed weakly equivalent, since $M1_1$, for instance, can synchronize on coffee at time 1 after accepting a coin, which $M2_1$ cannot do.

6 Conclusions

We have presented:

- CCSiT, a calculus extending CCS with time intervals.
- Three bisimulation type equivalences and a refinement relation for CCSiT processes.
- Some simple examples illustrating the calculus, the equivalences, and the refinement relation.

We are currently working on introducing recursion in the calculus and defining sound and complete axiom systems for the equivalences.

Acknowledgements

Special thanks are due to Hans Hansson and Ivan Christoff for many helpful discussions and suggestions. I would also like to thank Tharam Dillon for his help and encouragement as well as Steve Schneider and Ulla Binau for valuable comments.

References

[BB89] J.C.M. Baeten & J.A. Bergstra, 'Real Time Process Algebra', *Report P8916*, University of Amsterdam Programming Research Group, 1989

[BSW69] K.A. Bartlett, R.A. Scantlebury, P.T. Wilkinson, 'A note on reliable full-duplex transmission over half-duplex links', *Communications of the ACM*, Vol 12, No 5, pp 260-261, 1969

[Dan91] M. Daniels, 'CCS with Interval Time', *DoCS 91/32*, Department of Computer Systems, Uppsala University, 1991

[GV88] J.F. Groote & F. Vaandrager, 'Structured operational semantics and bisimulation as a congruence', *CS-R8845*, Centre for Mathematics and Computer Science, dept of Computer Science, 1988

[Han91] H. Hansson, 'Time and Probability in Formal Design of Distributed Systems', *PhD thesis DoCS 91/27*, Department of Computer Systems, Uppsala University, 1991

[Hoa78] C. Hoare, 'Communicating sequential processes', *Communications of the ACM*, Vol 21, No 8, pp 666-677, ACM, 1978

[Mil89] R. Milner, *Communication and Concurrency*, Prentice Hall, 1989

[MT90] F. Moller & C. Tofts, 'A temporal calculus of communicating systems', Proc. CONCUR '90, *LNCS 458*, pp 401-415, Springer Verlag 1990

[NRSV90] X. Nicollin, J. Richier, J. Sifakis, & J. Voiron, 'ATP: an algebra for timed processes', *Technical report RT-C16*, IMAG, Grenoble, 1990

[NS90] X. Nicollin & J. Sifakis, 'The algebra of timed processes ATP: an algebra for timed processes', *Technical report RT-C26*, IMAG, Grenoble, 1990

[Plo81] G. Plotkin, 'A structural approach to operating semantics', *FN 19, DAIMI*, Aarhus University, Denmark, 1981

[RR86] G. Reed & A. Roscoe, 'A timed model for communicating sequential processes', Proc 13th ICALP, *LNCS Vol 226*, pp 314-323, 1986

[Sch90] S. Schneider, 'Correctness and Communication in Real-Time Systems', *PhD thesis technical monograph PRG-84*, Oxford University Computing Laboratory, Oxford UK, 1989

[Wan91] Y. Wang, 'CCS + time = an interleaving model for real time systems', Proc. ICALP '91, to appear in *LNCS 510*, pp 217-228, Springer Verlag 1991

MULTICYCLES AND RTL LOGIC SATISFIABILITY

Odile MILLET
LIP, Ecole Normale Supérieure de Lyon
46, Allée d'Italie 69364 LYON Cedex 07 FRANCE
odile@frensl61.bitnet

ABSTRACT : The Jahanian-Mok semidecision procedure for safety analysis of timing properties can be expressed in RTL (Real-Time Logic). We improve this procedure by unifying variables and terms containing a Skolem function. In order to guarantee termination of the potentially infinite process, we introduce the notion of a multicycle, that represents an infinite number of cycles. When the class contains at least one positive cycle, the resolution algorithm operates through the multicycle only a finite number of times. The result is an algorithm that applies to a larger class of $\exists\forall$-formulas and operates with the same efficiency as the Jahanian-Mok procedure.

Index terms : real-time logic, real-time specification, Skolem functions, resolution, safety analysis.

I. Introduction

Among different forms used to describe timing properties in real-time systems, there is RTL Logic (Real Time Logic), proposed by F.Jahanian and A.K.Mok ([J.M.1]). As a second feature, both authors propose a semidecision procedure to relate a safety assertion A to a system specification S expressible as a subset of RTL.
The question is whether "S implies A?" or "$S\wedge(\neg A)$ is unsatisfiable?". If the answer is positive, this means that A is a theorem derivable from the specification. If there is no answer, we can try to add constraints to the system to ensure its safety (except when the negation of the safety assertion is a theorem derivable from S, in which case the answer is no and it is useless to consider such an S extension).

This work was done in collaboration with Christine Charretton and was partially supported by the PRC Mathématiques Informatique and the Esprit Basic Research Action working group Algebraic and Syntactic Methods In Computer Sciences.

For this purpose, F.Jahanian and A.Mok use a classical resolution method greatly improved by dynamically using the positive cycles of the constraints graph. These cycles are the result of the unification algorithm. To ensure termination, they prescribe that a function does not take itself as an argument ; thus infinite cycles sets are avoided. In such a set, sometimes a positive cycle is missing to finish.

We suggest an extension of their algorithm, by giving additionnal positive cycles and allowing an improvement of the performance of the original algorithm.

II. RTL Logic and Constraint graphs

Real Time Logic (RTL) is a formal language able to express timing properties of real-time systems. RTL reasons about occurrences of events :

$@(F,i)$ = time of ith occurrence of event F, denoted $f(i)$, $\in \mathbb{N}$

and the RTL formulas describing a specification and a safety assertion are represented by formulas in Presburger arithmetic with uninterpreted functions : for example, the formula $\forall x\ (f(x) + 2 \leq g(x))$ expresses that the xth occurrence of the event G occurs 2 time units at least

after the xth occurrence of the event F. In [J.M.2], the formulas are restricted to arithmetic inequalities of the form :

occurrence function $\pm$ integer constant $\leq$ occurrence function,

and the occurrence function arguments cannot be the result of interpreted functions : for example, $f(x+1)$ is not allowed.

It is well known that Presburger arithmetic with one unary function uninterpreted is undecidable. Since formulas like $f(x) + y \leq g(x)$ are not allowed, only successor function is used in $\mathbb{N}$, not an addition function. We could think that this restriction makes the theory decidable but this is not true : J.F.Pabion ([PAB]) proved that any extension of the theory of equality which has an infinite model and contains two unary uninterpreted functions is undecidable. In such a theory, even the satisfiability problem of formulas in which the only quantifier is $\forall$ is undecidable ([GUR]).

In [J.M.2], F.Jahanian and A.K.Mok transform the system specification S and the negation of the safety assertion A into clausal form and construct a graph representing the clauses so obtained (see fig 1.1). They use positive cycles detected to improve the analysis of consistency of A with respect to S, that is, the unsatisfiability of $\Sigma = S \wedge (\neg A)$. Their idea is :

– if a cycle γ is detected, the conjunction Γ of its literals is unsatisfiable. Then $\neg\Gamma$ is a valid disjunction and adding $\neg\Gamma$ to Σ does not change the eventual unsatisfiability of Σ.

– so each clause contains either only negated literals (from $\neg\Gamma$) or only unnegated literals (from Σ) and so it is easier to conclude unsatisfiability.
Since the literals corresponding to a positive cycle may belong to non unit disjunctive clauses, it is not guaranteed that the detection of a positive cycle is enough. It may be necessary to detect several positive cycles.

Their resolution algorithm is in the worst case exponential with respect to the number of positive cycles detected.

III. Multicycles

We have seen that, in the algorithm proposed for research of cycles, Jahanian and Mok forbid the unification of two terms v and v' belonging to the same cluster, if there is a Skolem function symbol occurring more than once inside of them ; for example the unification of x and $\alpha^2(Y)$ is forbidden. This is to ensure the termination of the algorithm.

Two interpretations are possible :
1) the algorithm does not recognize all cycles, so it is not complete.
2) the algorithm recognizes every cycles in a limited language : we cannot use formulas such as $...\forall x_1...\forall x_n \exists y$ ... which involve Skolem function occurrence.

III.1 Example

Here is an example where positive cycles cannot be detected although they would ensure the insatisfiability of Σ.

In a network, consider a processor which transmits a message within 4 time units of its reception. Let F be the reception event and G the emission event. Let us suppose that there is always a new message arriving at least every 2 time units.

Let A be the safety assertion : "a new reception occurs after every emission".
The system specification S is :

$\forall x \, (g(x) - 4 \leq f(x))$

$\forall x \, \exists y \, (f(x) + 2 \leq f(y))$

and the expression of A is : $\forall x \, \exists y \, (g(x) \leq f(y))$.
We want to prove that A is a theorem derivable from S.

The clausal form of $\Sigma = S \wedge$ (not A) can be written as the conjunction of the following formulas :

L : $g(x) - 4 \leq f(x)$

M : $f(x) + 2 \leq f(\alpha(x))$

N : $f(x) + 1 \leq g(Y)$
and the associated graph is represented in fig.1.1.

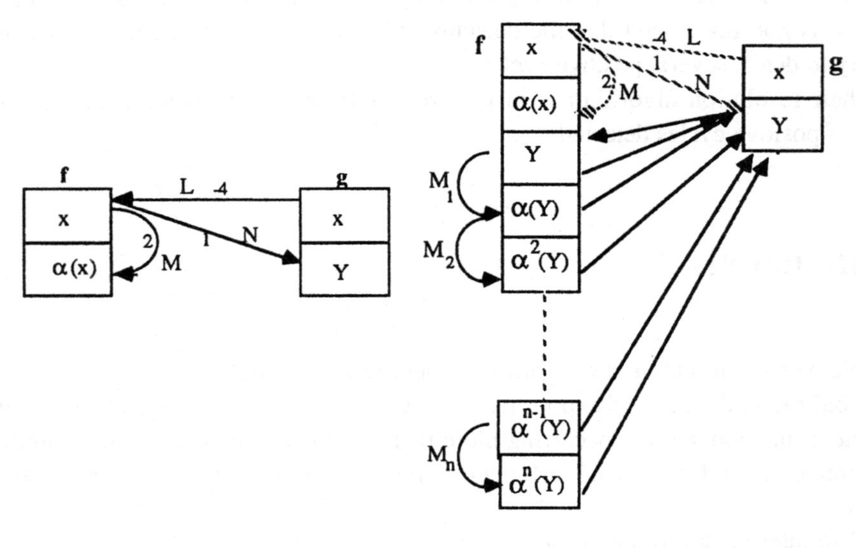

fig. 1.1 fig.1.2

The Jahanian-Mok method would detect only one cycle γ_0, LN, with weigth -3, obtained by the substitution {Y/x}, denoted L(Y) N(Y). This cycle is negative, so it will not be useful to resolve the problem.

However, if we would use $\alpha(Y),...\alpha^n(Y)$, $n \geq 0$, successively, the cycles γ_1 = L(Y) M_1 N(Y),...,γ_n = L(Y) M_1 M_2...M_n N($\alpha^n(Y)$) in which $M_k = M(\alpha^k(Y))$, occurs as shown in fig.1.2. We can write γ_n as L M^n N by dropping the argument symbols. This infinite family of cycles could not appear before although their wheight $-3+2n$ becomes positive as soon as $n \geq 2$; so they could be useful to conclude.

III.2 Example

Consider the following specification :

$\forall x \exists z (f(x) +c \leq f(z)) \vee \forall y \exists z(f(y) +d \leq f(z))$ $\quad$ c, d > 0

$\forall y (g(y) - 5 \leq f(y))$

with A : $\forall x \exists t (g(x) \leq f(t))$.

After Skolemisation, literals become :

L_α : $f(x) + c \leq f(\alpha(x))$

L_β : $g(y) + d \leq g(\beta(y))$
M : $g(x) - 5 \leq f(x)$
$\neg A = N$: $f(x) + 1 \leq g(X)$

in this case, $S = (L_\alpha \vee L_\beta) \wedge M$ and $\Sigma = (L_\alpha \vee L_\beta) \wedge M \wedge N$ and the constraint graph is represented in the fig.2.

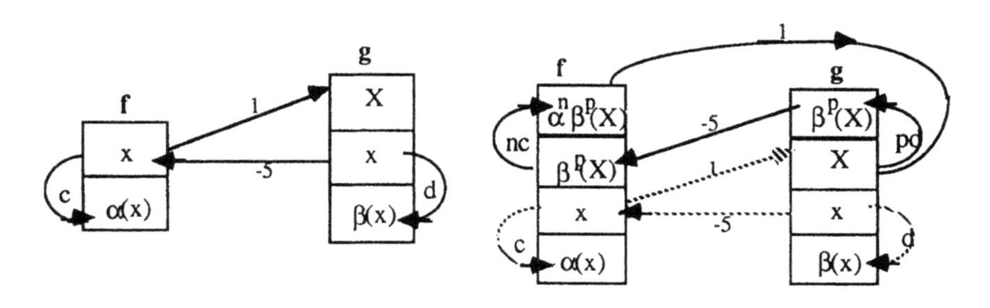

fig. 2

In this graph, if we unify x with X in the cluster g, $\beta^p(X)$ occurs and if we unify x with $\beta^p(X)$ in the cluster f, $\alpha^n(\beta^p(X))$ occurs.Then exists the cycle $\gamma_{p,n}$, in the form of $L_\beta{}^p$ M $L_\alpha{}^n$ N, and its weight is -4+pd+nc, that is an infinite family of cycles whose weight can become positive. Let us now define the formal frame work.

III.3 Definitions and notations

• The Presburger arithmetic formulas now considered will be of the form

$$f(x) \pm \textbf{integer constant} \leq g(y)$$

where x and y can be $\alpha(z)$ **with** α **unary Skolem fonction** (the unification of a variable with $\alpha(z)$ will produce $\alpha^n(z)$ with n >0).

• **A Skolem literal** will be a literal with one (or two) occurrences of Skolem functions (and a Skolem edge will be the corresponding edge). Other literals will be called **standard** literals. We point out that this paper deals only with unary Skolem functions. Our specification are characterised by the formulas

$$\exists y_1 \ldots \exists y_p \, \forall t \, \exists z_1 \ldots \exists z_n \, \forall x_1 \ldots \forall x_m \, (\ldots).$$

• To every path μ built with instances of literals $L_1, \ldots, L_p$ of the constaint graph, we associate the **word** $L_1 \ldots L_p$. A **multicycle** $\mathfrak{G}$ will be a class of cycles satisfying :

1) there exist an integer p and two sequences $u_0, \ldots, u_p$ and $v_1, \ldots, v_p$ such that :

– their terms are words built from the alphabet of the literals occurring in Σ ;

– every v_i is associated with a path whose f(x) and f(λ(x)) are endpoints, where x is a variable, λ is a function obtained by composing Skolem functions and f is an occurrence function.

2) for every element γ of $\mathfrak{C}$, there exist p positive integers $n_1,\ldots, n_p$, one of them being ≥ 1, such that γ is associated with the word $u_0 v_1^{n_1} u_1 \ldots v_p^{n_p} u_p$. Moreover for all i =1,.., p and for all n'_i greater than n_i, the cycle associated with the words $u_0 v_1^{n'_1} u_1 \ldots v_p^{n'_p} u_p$ belongs to $\mathfrak{C}$.

The cycle corresponding to the word $u_0 v_1^{n_1} u_1 \ldots v_p^{n_p} u_p$ will be denoted $\gamma_{n_1 \ldots n_p}$ and called **representative** of $\mathfrak{C}$.

The cycles of Jahanian and Mok correspond to a single word u_0.

A few remarks regarding these notations :

i) Observe that a Skolem literal can belong to a path, without producing a multicycle : for example $f(x) + c \leq g(\alpha(x))$ if there is no possibility of coming back in the cluster f. Similarly, in the case of multicycle, a standard literal can belong to one v_i. For example, if the specification produces literals Q and P_α :

$$Q \ : h(x) + e \leq f(x) \qquad P_\alpha \ : f(x) + d \leq h(\alpha(x)),$$

$v = QP_\alpha$ can be repeated.

On the other hand, a Skolem literal is necessary to produce a multicycle.

ii) In our example, the multicycle corresponds to the word LM^nN and we obtain a cycle even if n=0. It may be necessary to have n=1 to close the cycle, as in fig.3 :

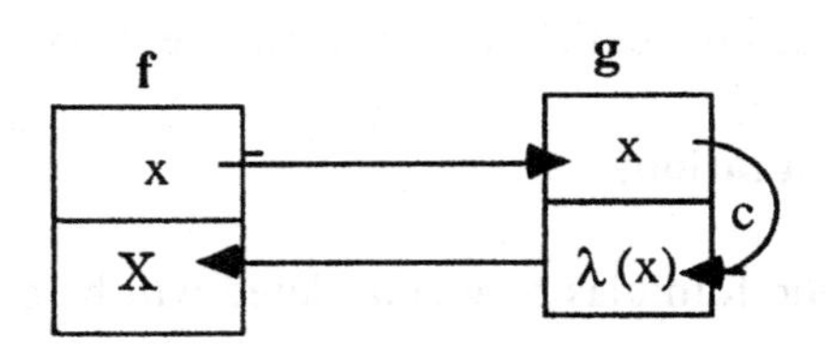

fig. 3

• The weight of $\gamma_{n_1 \ldots n_p}$ is expressed by $a+n_1c_1+\ldots+n_pc_p$ where a is the sum of the respective weights of the $u_0,\ldots,u_p$ and every c_i is the weight of v_i.

A multicycle is **positive** if it has a representative with positive weight. We have the following possibilities :

– if some c_i is positive then the multicycle is positive.

– if every c_i is ≤ 0, then, in the case of a positive multicycle, there are finitely many positive representatives (none if a ≤ 0).

The fact that the multicycle is positive or not will be deduced from the analysis of the p+1-tuple $(\mathbf{a},\mathbf{c_1},\ldots,\mathbf{c_p})$.

III.4 Multicycles as tools in the resolution algorithm. Theorem

Let $\mathfrak{G}$ be a positive multicycle, $\gamma_{n_1 \ldots n_p}$ a representative of $\mathfrak{G}$ corresponding to the word $u_0 v_1^{n_1} u_1 \ldots v_p^{n_p} u_p$.

i) If $\mathfrak{G}$ has infinitely many positive representatives, then, in the resolution tree, these positive cycles will operate exactly as the set of the p cycles $\gamma_{n_1 \ldots n_p}$ defined by the following property :

(P) there exists i with $1 \leq i \leq p$, such that $n_i = 1$ and $j \neq i$ implies $n_j = 1$ if v_j is necessary to close the cycle or $n_j = 0$ if not.

ii) If $\mathfrak{G}$ has finitely many positive representatives, then, in the resolution rule, they will operate exactly as the positive $\gamma_{n_1 \ldots n_p}$ satisfying the property (P) .

These p cycles denoted $\gamma(i)$, $1 \leq i \leq p$, will be called **significant**. They are not necessarily negative.

i) corresponds to the case "one (or more) $c_i > 0$", and ii) to the case "every $c_i \leq 0$".

Proof of **i)**

- **case p=1**

Let us suppose that v_1 is a Skolem edge M with weight c.

a) Let Σ be the set of clauses of which we want to test the insatisfiability. Let us suppose we are in a node of the resolution tree where positive standard cycles do not allow to conclude so. Let γ_n be the $\mathfrak{G}$ representative corresponding to the word $u_0 v_1^n$, with weight a+nc, and $L_1, L_2, \ldots, L_p$ the literals of u_0. If a positive γ_n makes Σ insatisfiable, then so does γ_{n_0}, where n_0 is the smallest integer $n > 0$ such that γ_{n_0} is positive.

Let us suppose that after adding disjunctions of negated literals representing standard positive cycles to Σ, we are at a node Σ' where we are unable to conclude ; let γ_n be the positive representative of $\mathfrak{G}$, formed by the set $\Gamma_n = \{L_1, \ldots, L_p, M_1, \ldots, M_n\}$ where M_i is the ith instance of M belonging to γ_n such that $\Sigma' \wedge \neg (\Gamma_n)$ are unsatisfiable. On the next level, $\Sigma' \wedge \neg L_1, \ldots, \Sigma' \wedge \neg L_p, \Sigma' \wedge \neg M_1, \ldots, \Sigma' \wedge \neg M_n$ are unsatisfiable. But as γ_{n_0} is also positive, if we put γ_{n_0} instead of γ_n, that is to say Γ_{n_0} instead of Γ_n, $\neg \Gamma_{n_0}$ is valid and we have the same result.

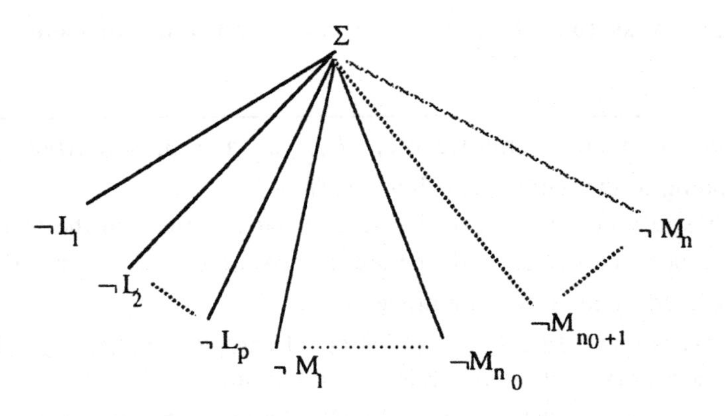

fig. 4

b) We can still simplify the result by looking at the resolution tree. If we study the branches under the nodes $\neg M_1,\ldots \neg M_{n_0}$, they differ only by the variable instanciation inside M and have the same behaviour. We can resume this behaviour on the branch under $\neg M_1$. So we can substitute γ_1 for γ_{n_0}, even if γ_1 is not positive.
The case where v_1 is a path of q edges, q>1, can be solved similarly.

• **case p>1**

This case is no more difficult to solve : if the multicycle weight is $a+n_1c_1+\ldots+n_pc_p$, so we are back in the previous case n=1, while we confirm n_i variable and we fix, for $j\neq i$, $n_j=1$ or 0 according to whether v_j is necessary to close the cycle or not.

The proof of **ii)** is analogous.

In example III.1, the negative cycle γ_1 allows us to conclude, whereas originally it was not possible to use it. In example III.2, $\gamma(1)$ and $\gamma(2)$ allow us to conclude.

IV. Detecting positive multicycles

After positive standard cycles, it is necessary to detect positive multicycles and to extract from them significant cycles.

IV.1 Construction of cycles

First, we construct paths without loops as usual. If within the same cluster v_2 and w_1 can be unified with the substitution s, for each pair of edges $<v_1, v_2>$, $<w_1,w_2>$ with weigths c and d, we add the nodes $s(v_1)$, $s(v_2)$, $s(w_1)$, $s(w_2)$ (if not already there), and the edges $<s(v_1), s(v_2)>$, $<s(w_1), s(w_2)>$ with corresponding weigths (evidently, a unification of x with $\lambda(x)$ is forbidden). But along any path, **a literal should not be instanciated more than once** to ensure termination (such a path, considered like a literals sequence, is "simple").

We obtain a cycle if the first and the last nodes can be unified.

Without Skolem literals, a classical PROLOG III program gives this result. In the case of example III.1, the program of fig.5 constructs such cycles.

```
hors_de(x,<>) ->;
hors_de(<f,x>,<<g,y>>.l) -> dif(<f,x>,<g,y>) hors_de(<f,x>,l);

chemin(<f,x>,u,<g,y>,l) -> chemin_sans_boucle(<f,x>,u,<g,y>,<<f,x>>,l')
                    reverse(l',l);

reverse(<>,<>) ->;
reverse(<x>.u,u'.<x>) -> reverse(u,u');

chemin_sans_boucle(<f,x>,u,<g,y>,l,<<g,y>>.l) -> arc(<f,x>,u,<g,y>);
chemin_sans_boucle(<f,x>,u+v,<g,y>,m,l) -> arc(<f,x>,u,<h,z>)
                                           hors_de(<h,z>,m)
                              chemin_sans_boucle(<h,z>,v,<g,y>,<<h,z>>.m,l);

circuit(<f,x>,u+v,<<f,x>>.l) -> arc(<f,x>,u,<g,y>)
                      chemin(<g,y>,v,<f,x>,l);
```

fig. 5

If we add to this program

arc(<`g`,x>,-4,<`f`,x>)->; (1)
arc(<`f`,x>,1,<`g`,`Y`>)->; (2)

> *{x=`Y`, u = -3, l = <<`f`, `Y`>, <`g`, `Y`>, <`f`, `Y`>>}*

is the answer to the query

> circuit(<`f`,x>,u,l);

The edges are standard and we are back in the case studied by Jahanian and Mok.

If we add a Skolem literal like

```
arc(<`f`,x>,2,<`f`,<`A`,x>>)->;     (3)
```

the program does not terminate and gives the infinite sequence :

```
> {x=`Y`, u = -3, l = <<`f`, `Y`>, <`g`, `Y`>, <`f`, `Y`>>}
  {x=`Y`, u = -1, l = <<`f`, `Y`>, <`f`,<α,`Y`>>,<`g`, `Y`>, <`f`, `Y`>>}
   {x=`Y`,u =1, l =<<`f`, `Y`>, <`f`,<α,`Y`>>,<`f`,<α,<α,`Y`>>>,<`g`, `Y`>, <`f`,
`Y`>>}
   ....
```

as expected.

Since we only need cycles containing single instances of literals ("simple" paths), we can transform the graph and duplicate the cluster which contains Skolem function into f_1 and f_2 to suppress the edge from f to f. So in the new graph, classical cycles are wanted cycles. The Skolem literal (3) splits into two Skolem literals :

```
arc(<`f1`,x>,2,<`f2`,<`α`,x>>)->;   (3.1)
arc(<`f2`,<`α`,x>>,1,<`g`,`Y`>)->;  (3.2)
```

Figure 6 shows the graph of example III.1 "before" and "after".

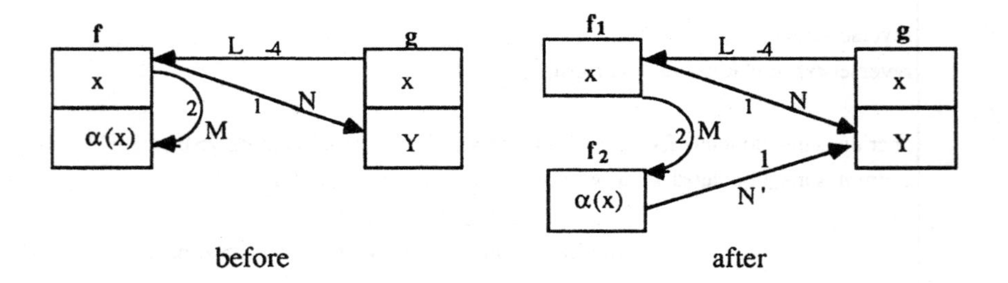

fig. 6

The answer to the same query is then :

```
> {x=`Y`, u = -3, l = <<`f1`, `Y`>, <`g`, `Y`>, <`f1`, `Y`>>}
  {x=`Y`, u = -1, l = <<`f1`, `Y``>, <`f2`,<α,`Y`>>,<`g`, `Y`>, <`f1`, `Y`>>}
>
```

and we obtain what we want.

IV.2 Determination of multicycles

Let us consider γ as such a cycle and its nodes sequence :

a) γ is not representative of a multicycle if :

– this sequence does not contain any pair $(f(t),f(\lambda(t)))$ where t is a term.

or

– there exists a pair $(f(t),f(\lambda(t)))$ belonging to this sequence and there is no path whose $f(x)$ and $f(\lambda(x))$ are endpoints, where x is a variable, of which the path joining $f(t)$ and $f(\lambda(t))$ is an instance.

In the figure below, there is no path joining $f(y)$ and $f(\alpha(y))$.

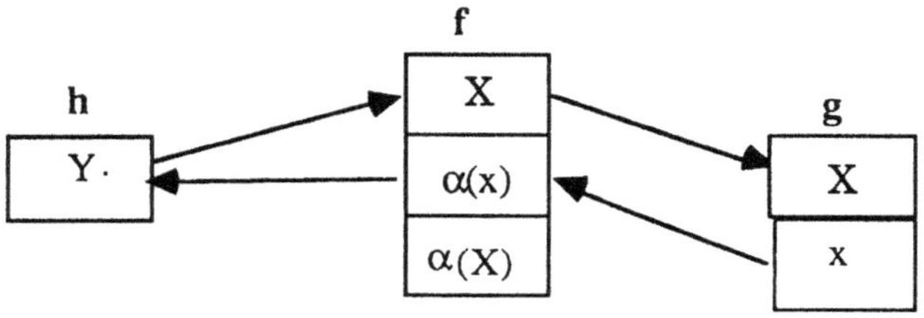

fig. 7

It is an example where a Skolem edge does not produce a multicycle.

b) Otherwise, every pair $(f_i(t), f_i(\lambda(t)))$, $1 \leq i \leq n$, corresponds to an instance of a path v_i with $f(y)$ and $f(\lambda(y))$ as endpoints and γ is representative of a multicycle.

b.1) if n=1, the multicycle weight is (a,c) where c is the weight of v and a the remaining weight. If the multicycle is positive, we use γ in the resolution algorithm for any weight of γ. In example 1, a=-3 and c=2 :

$$\gamma = g(Y) \xrightarrow{-4} \underbrace{f(Y) \xrightarrow{2} f(\alpha(Y))}_{v} \xrightarrow{1} g(Y)$$

b.2) If n >1, the relative place of the pairs must be studied. Consider two pairs $(f_i(x), f_i(\lambda(x)))$ and $(f_j(x), f_j(\mu(x)))$ that is to say v_i with weight c_i and v_j with weight c_j. Three cases are possible :

i) v_i and v_j are disjoints and the sequence is $..f_i...f_i...f_j...f_j...$

The multicycle weight is $(a, .., c_i,..., c_j,...)$ where a is the weight of the simple part. If the multicycle is positive, we use the significative cycles $\gamma(i)$ and $\gamma(j)$ for the resolution algorithm. Thus, in example III.2, we obtain the cycle

$$g(X) \xrightarrow{d} g(\beta(X)) \xrightarrow{-5} f(\beta(X)) \xrightarrow{c} f(\alpha(\beta(X))) \xrightarrow{1} g(X)$$

and according to c and d, we use the cycles $\gamma(1)$ and $\gamma(2)$ below

$$g(X) \xrightarrow{-5} f(X) \xrightarrow{c} f(\alpha(X)) \xrightarrow{1} g(X)$$

$$g(X) \xrightarrow{d} g(\beta(X)) \xrightarrow{-5} f(\beta(X)) \xrightarrow{1} g(X)$$

ii) v_j is included in v_i and the sequence is $..f_i...f_j...f_j...f_i..$
We can write $L\ v_j\ M = v_i$ and we have two multicycles. One, the weight of which is $(...,c_i,...)$, corresponding to v_i, containing $...(Lv_jM)^n...$.The other, the weight of which is $(...,c_j,...)$ corresponding to v_j, containing $...Lv_j^nM...$But the significative representative are merged in the cycle, obtained from γ, formed by Lv_jM and the simple part of γ (adding edges to close it if necessary).

On the other hand, among the cycles produced in IV.1, there exists a cycle "avoiding" v_j (or not if v_j is necessary to close). The corresponding multicycle contains $...(LM)^n...$ (or$...(Lv_jM)^n...$) and its weight is $(...,c_i\text{-}c_j,...)$ (or.$(..., c_i,...)$). The example of fig.8 shows the sequence $..f_i(x), f_j(\alpha(x)), h(\beta(\alpha(x))), f_j(\beta(\alpha(x))), f_i(\beta(\alpha(x)))$...and the sequence $...f_i(x), f_j(\alpha(x)), f_i(\alpha(x))...$

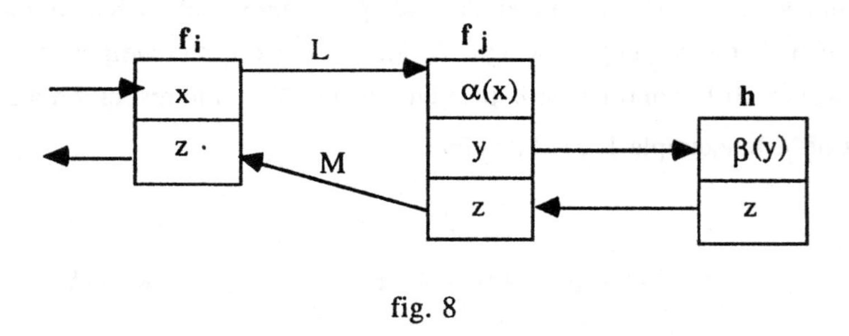

fig. 8

iii) v_i and v_j overlap and the sequence is $..f_i...f_j...f_i...f_j..$
We can write $L\ v_j = v_i\ M$. We have still two cycles : one containing $...v_i^nM...$ and the other containing $...Lv_j^n...$ as in the example below with the sequence $...f_i(x), f_j(\alpha(x)), f_i(\beta(\alpha(x))), f_j(\gamma(\beta(\alpha(x))))...$

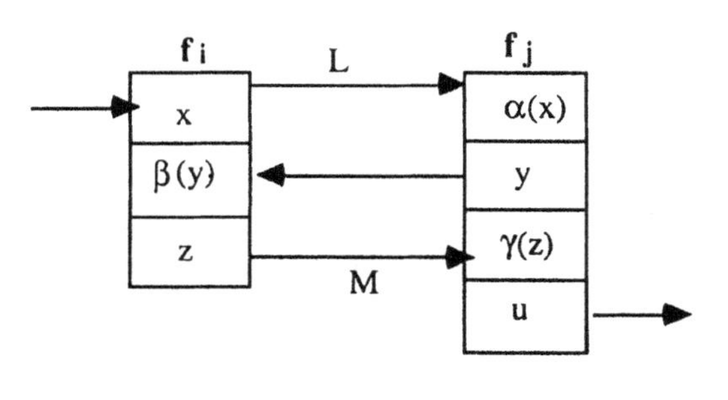

fig. 9

As in ii),the significative representatives are merged in the cycle, obtained from γ, formed by Lv_j (or v_iM) and the simple part of γ (adding edges to close if necessary).We observe that in this case, a significant cycle can be found again several times.

Thus, in the first step, we detect all cycles in which every edge is the only instance of a given literal. Two types of cycles appear :

– cycles γ of fixed size, like classical cycles (they will be traited in the same way that in the original algorithm);

– cycles γ in which a literal (at least) could produce an infinite number of instances, but does not (we only construct "simple" paths). Every such cycle is representative of a multicycle of the graph.

In the second step, we determine all parameters of every multicycle, and if a multicycle is positive, we determine which representative cycles (positive or not) are useful for the resolution algorithm.

V. Conclusion

The power of our language is currently limited in two ways :

a) not every sequence of quantifiers is allowed ;

b) only a subset of the structural properties of $\mathbb{N}$ has been exploited so far.

This paper concentrates on the study of permissible sequences of quantifiers. The introduction of the concept of multicycles allows the use of the unary Skolem functions and thus a generalization of the ideas of F. Jahanian and A.K. Mok.

By introducting this concept, we actually managed to increase the expressive power of the language. Originally, only a very rough comparison of two event occurrences was possible : for example, the formula $\exists y \forall x \ (g(y) + 3 \leq f(x))$ expresses that "every" occurrence of a event F took place at least 3 time units after one occurrence of an event G. Similarly, the formula $\forall x\ (f(x) + 2 \leq g(x))$ expresses that every occurrence of event F took place at least 2 time units before "the same" occurrence of event G. By introducing unary Skolem functions, we managed to introduce the possibility of time comparison : for example, $\forall\ x\ \exists\ y\ (f(x) + 4 \leq f(y))$ expresses that to every occurrence of event F, "another" occurrence of G can be associated that takes place at least 4 time units later.

For the moment, we can only order two events. The next step will be to try to find a way to introduce the concept of a "next" event through the f(x+1) concept.

On the other hand, we are also planning to investigate the limitation of the structural properties of $\mathbb{N}$. We hope to be able to introduce a time variable y in the formula $f(x)+y \leq g(z)$. This means that all the properties of addition in $\mathbb{N}$ will be permissible.

REFERENCES

[GUR] Y.GUREVICH, The Decision problem for standard classes, The Journal of symbolic Logic, Vol.41, N°2, 1976, pp. 460-463.

[J.M.1] F.JAHANIAN and A.K.MOK, Safety analysis of timing properties in real-time systems, IEEE Transactions on Software Engineering, Vol.SE-12, N°9, 1986, pp.890-904.

[J.M.2] F.JAHANIAN and A.K.MOK, A graph-theoretic approach for timing analysis and its implementation, IEEE Transactions on Computers. Vol.C-36, N°8, 1987, pp.961-975.

[PAB] J.-F.PABION, Communication manuscrit, 1989.

Voluntary Preemption : A Tool In The Design Of Hard Real-Time Systems

Abha Moitra *

General Electric - CR&D
Schenectady NY 12301

Abstract: A new technique for scheduling hard real-time systems is proposed. We term this technique as *voluntary preemption* as it allows the user to determine the points at which the execution of a process can be preempted. There are several advantages of this technique. First of all, the internal structure of the processes can be used in determining the points where a process may be preempted. This permits, among other things, that critical sections are not preempted as well as identifying points at which a process should be preempted. Secondly, overhead cost of context switching can be taken into account realistically without sacrificing performance. Finally, since the overhead cost of context switching depends on the point (in the execution) where a process is preempted, this technique allows the preemption points to be chosen to reduce the overhead cost of context switching. These various considerations about where a process can be preempted, can be incorporated into an automatic tool to aid in the design of real-time systems.

We describe and illustrate this technique by analyzing in detail the performance of a voluntary preemptive scheduling algorithm for scheduling a classical hard real-time problem on a single processor with context switching costs. In addition we also obtain a tight load cutoff for this algorithm. The performance of the voluntary preemptive scheduling algorithm is good as its load cutoff matches that of an optimal scheduling algorithm when context switching costs are zero. This load cutoff theorem also provides a mechanism for scheduling a multiprocessor system by partitioning the problem into independent subproblems each of which can be scheduled on a single processor.

*Part of this work done while at Odyssey Research Associates under U.S. Air Force, RADC contract F30602-86-C-0115

1 Introduction

An important class of scheduling algorithms for scheduling hard real-time system is the so called *static* scheduling algorithms. In such scheduling algorithms, a fixed priority is assigned to the sources, and whenever the scheduler has to determine the next request to be scheduled it picks the highest priority waiting request. A number of static scheduling algorithms have been studied, see for example [LL73], [Tei78] and others listed in [CSR88]. However a basic and unreasonable assumption made in them is that there is no cost (i.e. time delay) associated with context switching. While context switching costs can be taken into account when a non-preemptive scheduling policy is used, however, such policies have a poor performance in terms of the class of problems that can be feasibly scheduled. Consequently, preemptive static scheduling algorithm are normally considered. For preemptive scheduling algorithms the overhead cost of context switching can be modeled by suitably increasing the computation times of the requests. At times, this is done by adding twice the overhead cost of context switching to the computation time. However such a transformation may make feasible problems (with overhead costs) into infeasible problems. Another drawback of a pure preemptive schedule is that a process can be preempted at any arbitrary point though this problem has been addressed for processes with critical sections [Mok83], [GS88], [SRL90] and others. Preemptive scheduling by itself also precludes the possibility of using the internal structure of a process to identify the points at which a process can be preempted.

In this paper we describe a new technique, ***voluntary preemption***, that can be used in the design of scheduling algorithms for hard real-time systems. In such a technique, the structure of the processes is used in determining the points in process codes where voluntary control return instructions should be added. The driving force of the technique is that all other things being equal, the context switch is postponed as much as possible. This results in reducing the total context switching cost. Voluntary preemptive scheduling algorithms have several advantages. First of all, the internal structure of the processes can be used in determining the points where a process may be preempted. This allows, among other things, that critical sections are not preempted as well as identifying points at which a process should be preempted. Secondly, overhead cost of context switching can be taken into account realistically without sacrificing performance. Finally, reducing the overhead cost of context switching is an important goal [DJ86], and it can be reduced in two ways - reduce the number of context switches as well as the strategic placement of the context switches. Both of these types of reductions can be performed by the voluntary preemption technique.

It should be pointed out that implementations like OCCAM and Transputer scheduler also try to postpone context switches so that the switch is more efficient. Also techniques like the *period transformation* of [SLR86] use task suspension to ensure high utilization while meeting the deadline of an important long period task. Another alternative is the use kernelized monitor of [Mok83].

The technique of voluntary preemption can be used in a variety of scheduling algorithms, from static priority driven to deadline driven. The placement of voluntary control points can be automated and made an integral part of a tool to aid in the design of real-time systems.

In this paper we describe and illustrate this approach by deriving a voluntary preemptive scheduling algorithm and its tight load cutoff for scheduling a general class of hard real-time systems on a single processor. The load cutoff for the voluntary preemptive scheduling algorithm matches that of the standard optimal scheduling algorithm when context switching costs are zero. The rest of the paper is organized as follows. Section 2 discusses this classical hard real-time problem. Section 3 discusses the context switching costs. Section 4 presents the voluntary preemptive scheduling algorithm for the classical hard real-time problem with context switching costs taken into account. Section 5 analyzes this voluntary preemptive scheduling algorithm to obtain a tight load cutoff theorem. Section 6 discusses scheduling on multiprocessors. Section 7 puts this work in perspective. Section 8 sketches some possible extensions. Finally, section 9 presents the conclusions.

What should be emphasized at this point is that while most of the paper considers the effectiveness of the voluntary preemption technique for a particular class of hard real-time problems, this technique can be applied to a number of different classes of real-time problems.

2 A Class of Hard Real-Time Systems

In most of this paper we will deal with a restricted class of hard real-time systems and a class of associated scheduling algorithms. The hard real-time systems, RT, that we consider have the following properties:

- a single processor
- periodic requests
- ready time is the same as the arrival time
- no precedence relationship between requests
- the only resource needed in processing a request is the processor
- the deadline is the same as its time period

Later, we will explain why some of these restrictions are imposed and how some of them can be relaxed. In RT, a hard real-time problem can be completely specified by giving the number of processors and sources, and specifying each source S_i by (C_i, T_i) where

1. the *time period* T_i is the minimum time between successive requests from source S_i,
2. the *processing* or *computation time* C_i is the maximum time needed for processing a request from source S_i.

Throughout this paper we will assume that a hard real-time system P with n sources, $S_1, \ldots, S_n$ is denoted by $(C_1, T_1), \ldots, (C_n, T_n)$, with $T_1 \leq T_2 \leq \cdots \leq T_n$. P_i will refer to the program code as well as process corresponding to source S_i. In the worst case, these n sources can impose a *processor utilization factor* or *processor load* $U = \sum_{i=1}^{n} \frac{C_i}{T_i}$. We do not restrict the C_i and T_i values to be integers. A hard real-time system P is said to be *normalized* if all C_i are greater than or equal to one time unit.

Allowing no precedence constraints between requests, enables us to model scheduling in a large class of interesting hard real-time systems. Traditionally, scheduling problems considered were variants of jobshop scheduling problems where it was necessary to allow precedence constraints. However, most current hard real-time systems are for process control and for a large class of such applications, it is reasonable to assume that there are no precedence relationships.

In general, processes may communicate and request service from each other. However, one of our goals is to provide a quantitative analysis of a voluntary preemptive scheduling algorithm, and hence for simplicity we will not consider the internal details of the processes and make the assumption that the only resource used is the processor. However, the work described in this paper can be extended by relaxing this assumption and we will describe this as we go along.

The assumption that requests are periodic allows us to consider static scheduling algorithms, with the advantage (over dynamic scheduling algorithms) that they can be analyzed off-line, the actual run-time scheduling being simple and the overhead costs associated with scheduling small. The disadvantage is that the number and characteristics of the sources must be known (or estimated) in advance, and a slight change in the parameters of a source may require the entire schedule to be recalculated. However, for data logging and process control systems, the sources can usually be characterized appropriately. Also, it will be shown that all the changes necessary in a schedule due to changes in source parameters can be determined and effected efficiently and automatically.

The assumption that the deadline is the same as the time period is common [LL73] as it simplifies the analysis. In [Moi86] this restriction is removed, but the analysis was done with the assumption that there is no overhead cost associated with context switching.

For this restricted but well studied class of hard real-time problems, we will now describe a simple class of scheduling algorithms.

Definition 1: A priority assignment is *rate-monotonic* if it assigns priority based on the time periods, with smaller time period sources getting higher priorities.

The rate-monotonic scheduling algorithms have been well studied, right from Liu and Layland [LL73] to Sha and Goodenough [SG90]. We will restrict ourselves to ***priority-driven*** scheduling algorithms using the rate-monotonic priority assignment. Whenever the scheduler has to decide on the next request to be scheduled, the highest priority waiting request is picked. Notice that in this description we did not specify the conditions under which the scheduler chooses the next request to be processed. Traditionally, this is either as soon as a higher priority request arrives or as soon as the processing of a request is complete. The first approach is termed preemptive while the latter one is termed non-preemptive. However, a third approach, that is between these two extremes is taken in this paper. In this approach, the processes periodically relinquishes control to the scheduler. We term this voluntary preemption and the points in the process code at which control is relinquished are called *control points*. We will consider priority-driven and voluntary preemptive scheduling algorithms using the rate-monotonic priority assignment and term this class *RPV*. Any scheduling algorithm in *RPV* preprocesses the program code by adding control points.

Definition 2: A scheduling algorithm is said to be *optimal*, with respect to a class of hard real-time systems and a class of scheduling algorithms, if the scheduling algorithm is in that class of scheduling algorithms, and for any hard real-time problem in that class, the schedule it generates meets all the deadlines whenever any scheduling algorithm in that class meets all the deadlines for that problem.

The importance of rate-monotonic priority assignment follows from the following theorem.

Theorem 1: [LL73]. The rate-monotonic priority assignment is optimal in the class of fixed priority, priority-driven and preemptive scheduling algorithms for scheduling on a single processor with no overhead cost of switching processes and where the deadline is the same as the time period.

3 Context Switching Costs

In this section, we characterize the overhead costs due to context switching. The schedule generated for any arrival pattern by an *RPV* algorithm looks like

overhead; (process S_i for x_1 time units); overhead; (process S_j for x_2 time units); ...

which can be abbreviated as

overhead; $(S_i\colon x_1)$; overhead; $(S_j\colon x_2)$; ...

Notice that there are actually three components of the overhead cost: checking if a higher priority request is waiting (this can be done in the hardware without using additional time, however, our analysis will lump all the overhead costs together), writing out the process being executed and loading in a new process. At times, if no higher priority request is waiting, the overhead is simply the checking cost. For simplicity, we ignore this finer distinction of the three components of the overhead cost. The overhead cost will be denoted as Δ and will also include the cost of executing the statements that return control to the processor.

4 Voluntary Preemptive Scheduling Algorithm

We now develop a voluntary preemptive scheduling algorithm, VP, in the class RPV. This is the main result of this paper and the major issue involved is determining the placement of the control points. It is straightforward to see that the processing of a request should be delayed 'as much as possible' to minimize the overhead cost incurred. For example, if there is a feasible problem with two sources S_1 and S_2, and if S_1 is at priority 1 (1 is the highest priority) then processing of request from S_2 should be interrupted as late as possible to still allow requests from S_1 to meet their deadline.

We formulate this reasoning by determining for each source S_i the maximum amount of processing, the *slack* Sl_i, that can be allocated to a lower priority source and still have S_i meet its deadline. Sl_1 can be determined as follows. Let X amount of processing be allocated to a lower priority source (say S_i) at time $= 0$. Let a request from S_1 arrive at time $= \epsilon$; then the schedule starts out as

Δ; $(S_i\colon X)$; Δ; $(S_1\colon C_1)$;

To meet the deadline for S_1 we require that

$$T_1 \geq X + C_1 + 2 \times \Delta - \epsilon$$

and since ϵ can be arbitrarily small and we want Sl_1 to be as large as possible we get

$$Sl_1 = T_1 - C_1 - 2 \times \Delta$$

If $Sl_1 \leq 0$ then the problem is infeasible. If it turns out that $Sl_1 < C_2$ then we have to insert control points in the program code P_2. The control points are added so that the time needed for executing from any control point to the next control point is Sl_1 (except possibly for the last one). (Here we assume that the cost of executing individual statements in any program code is so small that the above condition can be met. This assumption can be removed in a straightforward manner but results in cumbersome notation.) P_2 is therefore broken into k_2 chunks of size $C_{2,1}, \ldots, C_{2,k_2}$ with $C_{2,1} = C_{2,2} = \cdots = C_{2,k_2-1} \geq C_{2,k_2}$ where $k_2 = \lceil \frac{C_2}{C_{2,1}} \rceil$ and $\sum_{j=1}^{k_2} C_{2,j} = C_2$. In fact, an alternate way of viewing the insertion of control points is to think that the source S_2 is replaced by k_2 sources, where the time period of each of these sources is T_2 and the computation time required by them is $C_{2,1}, \ldots, C_{2,k_2}$ respectively. Splitting the sources is sometimes an easier way of viewing the insertion of control points.

Ex. 1. Consider

	S_1	S_2
C	1	3
T	5	100

For simplicity assume $\Delta = 1$. Then,

$$Sl_1 = T_1 - C_1 - 2 \times \Delta = 2$$

So, it requires that control points need to be added to P_2. So the modified problem becomes

	S_1	S_2	S_3
C	1	2	1
T	5	100	100

We should point out that in Ex. 1 the 3 units of computation for the original S_2 could alternatively have been split into two chunks of 1.5 each. All that is required is that the number of splits be k_2 and that each $C_{2,i} \leq Sl_1$. However we do the splitting so that $Sl_1 = C_{2,1} = \ldots = C_{2,k_2-1} \geq C_{2,k_2}$ for notational convenience.

We should emphasize at this point that for this analysis, the placement of control points is done without taking into account the internal structure of P_2. However, if processes interact with each other, for instance there are critical sections then as a first step, control points are placed around the critical sections. Then additional control points are added if necessary to ensure that the time needed for execution from any control point to the next control point is less than or equal to $C_{2,1}$. Notice that if a critical section executes for longer than $C_{2,1}$ (and critical sections should not be preempted) then the problem is infeasible.

We now give algorithm *Split* to determine the placement of the control points. As mentioned earlier, for notational convenience we will show how the sources are to be split (rather than describe the place where the control points are to be inserted).

Algorithm *Split*
Input: Problem $P = (C_1, T_1), \ldots, (C_n, T_n)$ where $T_1 \leq T_2 \leq \cdots \leq T_n$.
Output: Problem $P' = (C_1, T_1), \ldots, (C_m, T_m)$ where $T_1 \leq T_2 \leq \cdots \leq T_m$ and $m \geq n$.

$m := n$;
$i := 1$;
while $i < m$ **do**
  $TValue_i := \{ lT_k \mid 1 \leq k \leq i-1; 1 \leq l \leq \lfloor \frac{T_i}{T_k} \rfloor \}$ /* arrival times */
  if $i = 1$
  then $Sl_1 := T_1 - C_1 - 2\Delta$;
  else
    $Sl_i :=$ maximum $X : \exists t \in TValue_i :$
    $X + \Delta + \sum_{j=1}^{i-1} \lceil \frac{t-\epsilon}{T_j} \rceil (C_j + \Delta) \leq t$ /* S_i processing starts by t */
    and $X + \Delta + \sum_{j=1}^{i-1} \lceil \frac{t-\epsilon}{T_j} \rceil (C_j + \Delta) + C_i + \Delta \leq T_i$ /* deadline met */
  fi
  if $Sl_i \leq 0$ **then stop**; /* P is infeasible */
  $Sl := \min\{ Sl_k \mid 1 \leq k \leq i \}$
  if $Sl < C_{i+1}$ /* C_{i+1} has to be split */
  then
    $k_{i+1} := \lceil \frac{C_{i+1}}{Sl} \rceil$;
    replace (C_{i+1}, T_{i+1}) **by** $(C_{i+1,1}, T_{i+1}), \ldots, (C_{i+1,k_{i+1}}, T_{i+1})$
    where $Sl = C_{i+1,1} = \cdots = C_{i+1,k_{i+1}-1} \geq C_{i+1,k_{i+1}}$
    $m := m + k_{i+1} - 1$;
  fi
  $i := i + 1$;
done

Note that in algorithm *Split* we determine the amount of processing requested at those instances of times just before a new arrival can occur. Also note that the arrival of requests from higher priority sources in the schedule will happen just after t and not exactly at time t; hence in the first condition on X we use "$\leq$" rather than "$<$".

Ex. 2. Consider

	S_1	S_2	S_3	S_4
C	1	1	1	3
T	9	9	11	100

For simplicity assume $\Delta = 1$. Now calculation will show that $Sl_1 = 6$; $Sl_2 = 4$. Consider the calculation for Sl_3. $TValue_i = \{ 9 \}$. For $t = 9$ we want to find largest X so that

$$X + \Delta + \lceil \frac{9-\epsilon}{T_1} \rceil (C_1 + \Delta) + \lceil \frac{9-\epsilon}{T_2} \rceil (C_2 + \Delta) \leq 9$$

$$\text{and } X + \Delta + \lceil \frac{9-\epsilon}{T_1} \rceil (C_1 + \Delta) + \lceil \frac{9-\epsilon}{T_2} \rceil (C_2 + \Delta) + C_3 + \Delta \leq 11$$

These two equations become

$$X + 5 \leq 9$$

$$\text{and } X + 7 \leq 11$$

So, $Sl_3 = 4$. Hence, no splitting of C_4 is required in this example. However, if $C_4 > 4$ then a splitting of C_4 would be required.

The voluntary preemptive scheduling algorithm, VP, with respect to the class of scheduling algorithms RPV and problem domain RT can now be described. The placement of the control points is determined from algorithm *Split*. The preprocessing of the program code adds statements like 'return control' to the places where a control point is to be inserted. The priority assignment is rate-monotonic and no preemption is allowed. Whenever control returns to the scheduler, the highest priority waiting task starts execution.

5 Cutoff Load for Voluntary Preemptive Scheduling Algorithm

We now show how to identify a subclass of feasible problems without considering the precise relationship between the parameters of the sources. What we will derive is a *tight cutoff load* (for feasibility); that is, it is the highest load such that a problem is feasible if its load is less than or equal to this cutoff, otherwise it may or may not be feasible. The importance of a tight cutoff load is that the processor load is a simple indicator of the difficulty of scheduling a real-time problem and it also allows quantitative performance of various scheduling algorithms to be derived and compared. It also provides a quick feasibility check that is particularly valuable if the parameters of a problem are perturbed slightly.

Theorem 2: [LL73]. The tight load cutoff for n source problems in RT under a preemptive schedule with $\Delta = 0$ is $n(2^{1/n} - 1)$.

Definition 3: A problem in RT is said to be *critical* if it can be scheduled by VP and any increase in the computation-time of any source makes it unschedulable by VP.

In order to derive the tight cutoff load theorem we first consider a subclass of problems where the algorithm *Split* does not split any computation times. This is done in theorem 3.

Theorem 3: If algorithm *Split* does not split any of the computation times for a normalized hard real-time problem Q and $U_Q \leq \frac{\ln 2}{1+\Delta}$ then Q is schedulable by VP.

Proof. Let $P \equiv (C_1, T_1), \ldots, (C_n, T_n)$ be a minimum load n source normalized problem for which algorithm *Split* does not split any computation time and it can keep the processor continously utilized for more than the time period of any of its sources. We will first investigate the structure of P and then consider the load that it imposes.

Let Seq be the sequence of requests handled for problem P that keeps the processor continously utilized for the longest possible time. We will now argue that $Seq = S_1 S_2 \ldots S_n S_1 S_2 \ldots S_n S_1$.

- Since Seq can not be extended it follows that each S_i appears at least once in Seq.
- We now argue that each S_i appears at least twice in Seq. If not then let some S_j appear just once in Seq. But this contradicts the fact that P can keep the processor continously utilized for more than T_n and Seq can not be extended.
- Let Seq_n be the smallest initial prefix of Seq that contains two occurrences of S_n. It is obvious that each S_i appears at least twice in Seq_n. We will now argue that S_1 appears exactly twice in Seq_n. This is because in processing Seq_n, each time unit allocated to S_1 has the highest 'contribution' to U_P (since T_1 is the smallest time period). Hence to minimize the load, S_1 appears twice in Seq_n. Now in Seq_n,

the second appearance of S_1 can not appear before the first appearance of S_n, because otherwise S_1 would appear more than twice in Seq_n. Consequently, $Seq = S_1S_2 \ldots S_nS_1S_2 \ldots S_nSeq_1$. Since S_1 could appear after $S_1S_2 \ldots S_n$ it follows that T_1 is such that $Seq = S_1 \ldots S_nS_1 \ldots S_nS_1Seq_2$. Now, if Seq_2 is not an empty sequence then its first element must be S_2. But in that case, a new problem satisfying the conditions of the theorem statement but with a load lower than U_P can be constructed by increasing T_2 suitably. Hence $Seq = S_1 \ldots S_nS_1 \ldots S_nS_1$, $T_1 = C_1 + \cdots + C_n + n\Delta$ and $T_i = T_{i-1} + C_{i-1} + \Delta$ for $2 \leq i \leq n$.

For notational convenience, we will write each C_i as $1 + x_i$ with $x_i \geq 0$. Hence

$$U_P = \frac{1+x_1}{n+x_1+\cdots+x_n+n\Delta} + \cdots + \frac{1+x_n}{2n-1+2x_1+\cdots+2x_{n-1}+x_n+(2n-1)\Delta}$$

Now consider a new problem $R \equiv (1+x_1+\Delta, n+x_1+\cdots x_n+n\Delta), \ldots, (1+x_n+\Delta, 2n-1+2x_1+\cdots+2x_{n-1}+x_n+(2n-1)\Delta)$. By inspection, R is a critical problem for a preemptive schedule with $\Delta = 0$. By theorem 2, we have

$$U_R = \frac{1+x_1+\Delta}{n+x_1+\cdots+x_n+n\Delta} + \cdots + \frac{1+x_n+\Delta}{2n-1+2x_1+\cdots+2x_{n-1}+x_n+(2n-1)\Delta}$$
$$= U_P + \Delta\left(\frac{1}{n+x_1+\cdots+x_n+n\Delta} + \cdots + \frac{1}{2n-1+2x_1+\cdots+2x_{n-1}+x_n+(2n-1)\Delta}\right) \geq n(2^{1/n} - 1)$$

Hence $\quad U_P + \frac{\Delta}{1+\Delta}\left(\frac{1}{n} + \cdots + \frac{1}{2n-1}\right) \geq n(2^{1/n} - 1)$

This is true for all n, hence it is true when $n \to \infty$. Now both $\frac{1}{n} + \cdots + \frac{1}{(2n-1)}$ and $n(2^{1/n} - 1)$ tend to $\ln 2$ as $n \to \infty$. Hence $U_P \to \frac{\ln 2}{1+\Delta}$ as $n \to \infty$. Notice that as the number of sources increases, the largest time period in P also increases.

We can now construct a critical problem Q from P by adding a new source $(1, T_n+1+\Delta)$. The load of Q can be made arbitrarily close to $\frac{\ln 2}{1+\Delta}$. Further this load cutoff is tight as otherwise it would contradict theorem 2 with $\Delta = 0$. □

We now use theorem 3 to derive the tight load cutoff theorem for the algorithm VP.

Theorem 4: Let P be a normalized hard real-time problem in RT. If $U_P \leq \min(\frac{1}{1+2\Delta}, \frac{\ln 2}{1+\Delta})$ then P can be scheduled by VP.

Proof. Throughout this proof we will assume that the set of problems under consideration are the normalized ones in RT that are schedulable by VP.

Let $Crit_n$ be the set of all n source critical problems, and let CU_n be the minimum load imposed by any problem in $Crit_n$. Finally let $CU = \min\ \{CU_i | i \geq 1\}$. We will first show that any problem with load less than or equal to CU is schedulable. If not, then consider some unschedulable problem R with $U_R \leq CU$. From R we can construct a critical problem R' by appropriately increasing the time periods. Now $U_{R'} < CU$, which is a contradiction.

Consider problem $Q \equiv (C_1, T_1), \ldots, (C_n, T_n)$ in $Crit_n$ and $U_Q = CU_n$. Our approach is to obtain a bound on U_Q.

Case 1. Let C_i be the lowest indexed computation time that is split by algorithm *Split*. Let the split of C_i be due to Sl_k for some $k < i$. What this means is that, if $i > 2$ then

$$\exists k < i : \forall t \in TValue_i :$$
$$C_i + \Delta + \textstyle\sum_{j=1}^{k-1} \lceil \frac{t-\epsilon}{T_j} \rceil (C_j + \Delta) > t$$
$$\text{or } C_i + \Delta + \textstyle\sum_{j=1}^{k-1} \lceil \frac{t-\epsilon}{T_j} \rceil (C_j + \Delta) + C_k + \Delta > T_k$$

and if $i = 2$ then $C_1 + \Delta + C_2 + \Delta > T_1$. Consider $t = T_k$. In that case we simply have:

$$\exists k < i : C_i + \Delta + \textstyle\sum_{j=1}^{k-1} \lceil \frac{T_k-\epsilon}{T_j} \rceil (C_j + \Delta) + C_k + \Delta > T_k$$

Let C_i be split into chunks of size $C_{i,1}$. We then have

$$\exists k < i : C_{i,1} + \Delta + \textstyle\sum_{j=1}^{k-1} \lceil \frac{T_k-\epsilon}{T_j} \rceil (C_j + \Delta) + C_k + \Delta = T_k$$
$$U_Q = \frac{C_1}{T_1} + \cdots + \frac{C_{k-1}}{T_{k-1}} + \frac{C_k}{T_k} + \cdots + \frac{C_i}{T_i} + \cdots + \frac{C_n}{T_n}$$

Our goal is to minimize the value of U_Q while keeping it a critical problem as well as the restriction imposed on T_k.

Step 1. Since the sources $S_{k+1}, \ldots, S_n$ do not play any role in the condition on T_k, we get :

$$U_Q \geq \frac{C_1}{T_1} + \cdots + \frac{C_{k-1}}{T_{k-1}} + \frac{C_k}{T_k}$$

Step 2. By the structure of U_Q and since we have a normalized problem, it follows that we can set $C_1 = \cdots = C_k = 1$ to give

$$U_Q \geq \frac{1}{T_1} + \cdots + \frac{1}{T_{k-1}} + \frac{1}{T_k}$$

where $T_k = C_{i,1} + \Delta + \sum_{j=1}^{k-1} \lceil \frac{T_k-\epsilon}{T_j} \rceil (1 + \Delta) + 1 + \Delta$

Step 3. Let $\lceil \frac{T_k-\epsilon}{T_j} \rceil = b_j$ for $1 \leq j \leq k-1$. Also, let $B = b_1 + \cdots + b_{k-1}$. That gives

$$T_k = 1 + 2\Delta + C_{i,1} + B(1 + \Delta)$$
$$\text{for } 1 \leq j \leq k-1 : T_j \leq \frac{T_k-\epsilon}{b_j}$$
$$U_Q \geq \frac{B+1}{1+2\Delta+C_{i,1}+B(1+\Delta)}$$

Now if $k = 1$ then

$$U_Q^{k=1} = \frac{1}{1+2\Delta+C_{i,1}} \leq \frac{B+1}{1+2\Delta+C_{i,1}+B(1+\Delta)}$$

Hence it follows that $U_Q \geq \frac{1}{1+2\Delta}$.

To show that this load cutoff is tight, we will show the existence of a 2 source critical problem R where algorithm *Split* splits the computation time of the second source and U_R is arbitrarily close to $\frac{1}{1+2\Delta}$. Let $R \equiv (1, 2\Delta+1+C_{2,1}), (1, (2\Delta+1+C_{2,1})/C_{2,1})$. By inspection, R is a 2 source critical problem. Note that the load imposed by R is $U_R = \frac{1+C_{2,1}}{1+2\Delta+C_{2,1}}$ and U_R decreases as $C_{2,1}$ decreases and hence U_R can be made arbitrarily close to $\frac{1}{1+2\Delta}$.

Case 2. None of the computation times are split by the algorithm *Split*. From theorem 3 it follows that $U_Q \geq \frac{\ln 2}{1+\Delta}$.

So, if $U_P \leq \min(\frac{1}{1+2\Delta}, \frac{\ln 2}{1+\Delta})$ then P can be scheduled by VP and this load cutoff is tight. □

Note that the above load cutoff requires that the computation times be greater than or equal to one time unit. Without this restriction, the load cutoff would be arbitrarily close to 0, an uninteresting and useless result. Note that such a restriction is not needed in [LL73] as $\Delta = 0$ and hence any scaling of the entire problem would still preserve its schedulability.

For most (if not all) applications, it is reasonable to assume that the computation times for handling requests is greater than or equal to the overhead cost of swapping processes. In that situation, we obtain the following corollary.

Corollary 1: Any normalized hard real-time problem P in RT where all the computation times are greater than Δ can be scheduled by VP if $U_P \leq \frac{1}{3}$.

Theorem 4 can be used in a number of ways. Whenever a hard real-time problem is considered, we can apply algorithm *Split* to determine the actual schedule and then check whether that schedule meets all the deadlines. However, for the class of normalized problems, theorem 4 provides a quick and easy (but conservative) check of schedulability.

Theorem 4 (and corollary 1) also allows us to compare the efficacy of the class of scheduling algorithms considered here with those considered elsewhere. For instance, Liu and Layland [LL73] consider the preemptive, rate-monotonic scheduling policy with zero overhead costs. They obtain a load cutoff of $\ln 2$. Our load cutoff is $\min(\frac{1}{1+2\Delta}, \frac{\ln 2}{1+\Delta})$. If $\Delta = 0$ then the two load cutoffs are the same. On the other hand, for normalized problems where the computation times are greater than or equal to the overhead cost of context switching, then our load cutoff is half of that of Liu and Layland - that is reasonable as under these stringent conditions upto half the processing could be spent in overhead.

6 Scheduling on Multiple Processors

In this section we consider briefly the more general problem of scheduling a hard real-time problem on a number of processors. For reasons of cost it is still unlikely that there are as many processors as processes. So, given a problem with n sources and m processors with $m < n$, we want to schedule the problem if possible. Now it is a well known fact [GJ79] that this problem is NP-complete even for $m = 2$. So, the problem we consider is that given a problem with n sources we determine the number of processors for which it is possible to schedule the given problem; however, this number may not be the minimum possible.

The usual approach for scheduling on multiple processors is to partition the problem into a number of independent subproblems so that each subproblem can then be scheduled on a single processor. For instance, Dhall and Liu [DL78] consider *next-fit* and *first-fit* variants of rate-monotonic priority assignment, and for both of these they give performance guarantees (in terms of the number of processors required by that approach to the minimum number of processors required) where all processors are identical. Davari and Dhall [DD86] improve the results of Dhall and Liu [DL78]. However, all these results again are on the assumption that the overhead cost of swapping processes is zero.

It is possible to devise (dynamic) schedules that would run different requests from the same source on different processors at different times. However, in keeping with our objective of making scheduling algorithms as simple as possible, we will only allow requests from each source to be processed on one processor. Our problem can now be solved in either of two ways as follows :

- try to allocate to the first processor as many sources as possible (using algorithm *Split* and checking for feasibility), and then repeat this process for the next processor,

- partition the sources so that the load imposed by each partition is less than or equal to the tight load cutoff of theorem 4, which is $\min(\frac{1}{1+2\Delta}, \frac{\ln 2}{1+\Delta})$.

7 Comparison with Other Work

In this section we compare the scheduling algorithm developed here with those existing in the literature. For our purpose, the existing work on scheduling of hard real-time systems can be classified based on the handling of the overhead cost associated with process switching.

- The overhead cost is assumed to be zero: in this case scheduling algorithms can be analyzed quantitatively. In Liu and Layland [LL73] an optimal fixed priority, priority-driven and preemptive scheduling algorithm for scheduling periodic requests on a single processor is given. For scheduling on multiple processors, the general approach is to partition the problem into a number of independent subproblems so that each subproblem can then be scheduled on a single processor. Various heuristics are employed for achieving such a partitioning, see for example Davari and Dhall [DD86]. For further references see [CSR88].

- There is no overhead cost for process switching. The work of Faulk and Parnas [FP88] falls in this category. Their work will be considered in detail below and it will turn out that they consider a different class of hard real-time problems.

- The overhead cost is taken into account. As mentioned earlier, the overhead cost of context switching can be modeled by suitable increasing the computation time of the requests. However, this leads to imprecise results. These costs are also taken into account when scheduling on distributed systems where the overhead costs are significant and cannot be ignored. However, the scheduling algorithms are not then analyzed quantitatively. An example of this approach is the work of Stankovic, Ramamritham and Cheng [SRC85] where simulation studies are performed.

We will now describe the approach taken by Faulk and Parnas [FP88] as it is related to the approach taken in this paper. Faulk and Parnas [FP88] consider issues related to synchronizing primitives of the programming languages as well as scheduling issues. The requests are assumed to be periodic but not externally activated. The programs are initially developed under the assumption that each process will run on its own processor. Then the code for each process is decomposed into a number of blocks, such that each block must be run to completion. These blocks (from all the programs) are then interleaved by the pre-run-time scheduler to satisfy all the sequencing and timing constraints. The scheduler prepares a schedule to deal with all the processing required for processing requests upto the least common multiple of the time periods of the requests involved. Externally activated events are assumed to impose low computational load and are handled by a run-time scheduler.

Hence, their work addresses a different class of problems than considered in this paper. The requests that we handle are periodic and come from the external world while in [FP88] the periodic requests correspond to internal processing. For externally activated periodic requests, an interleaving of the type proposed in [FP88] would not work unless it could be insured that the requests are *strictly* periodic, that is if a request from source S_i comes at a time t then the next request from S_i will positively arrive at $t + T_i$. If this were not the case, it is possible that the processing of a request would be started before it arrived and hence the processing may be done on an outdated state and/or with outdated information. However, their approach of decomposing into blocks is similar to our idea of inserting control points. The point of departure between these two approaches is that in our case the run-time scheduler interleaves the blocks in an appropriate manner while in their approach, the interleaving is generated off-line. Another difference is that we were able to do a quantitative analysis of our scheduling algorithm, for instance the load cutoff of theorem 4.

8 Various Extensions

In this section we briefly sketch how some of the restrictions imposed in section 2 on the model of real-time problems as well as the class of scheduling algorithms considered can be relaxed. This will also indicate how the technique of voluntary preemption can be used in the design of other scheduling algorithms.

For any real-time problem P let the deadline for source S_i be D_i. We now consider how to relax the assumption that $D_i = T_i$. We can relax it to $D_i \leq T_i$ as follows. Let *deadline based* priority assignment be the priority assignment where smaller deadlines get higher priority. From [Tei78] it follows that the deadline-based priority assignment is optimal in the class of fixed priority, priority-driven and preemptive scheduling algorithms for scheduling on a single processor with no overhead cost of switching processes and where for each source the deadline is less than or equal to its time period.

So, when $D_i \leq T_i$, the deadline based priority assignment should be chosen. The class of scheduling algorithms to be considered should be, DPV, for priority driven and voluntary preemptive scheduling algorithms using the deadline based priority assignment. A voluntary preemptive scheduling algorithm in this class would be developed in a manner similar to the development of VP. Note that we can construct infeasible problems with arbitrary low load (e.g. a two source problem where $C_i = D_i = 1$ and $T_i = x$, $i = 1, 2$, would not be schedulable and its load would be $\frac{2}{x}$). Hence we would only obtain a trivial cutoff load. It should be pointed out that a *deadline driven* scheduling algorithm (where the request with the earliest deadline is chosen for processing) is a *dynamic* scheduling algorithm, while any scheduling algorithm in DPV would be a static scheduling algorithm. We expect that even the restriction that $D_i \leq T_i$ can be removed by taking an approach similar to that in [Moi86].

A restriction that we have placed on the class of scheduling algorithms is that only a particular fixed priority assignment is considered. It would be interesting to do away with this restriction, that is, both the priority assignment as well as the placement of control points should be determined by the scheduling algorithm. Consider Ex. 3 below (where we assume that $\Delta = 0$ for simplicity).

Ex. 3. Consider

	S_1	S_2	S_3
C	1.1	.9	.1
T	2	3	5

Now Ex. 3 can not be scheduled by VP, but assigning priorities 1, 3, 2 to S_1, S_2, S_3 respectively and not splitting any computation meets all the deadlines.

9 Conclusions

We have demonstrated the effectiveness of the voluntary preemption technique by deriving a new scheduling algorithm for a classical hard real-time problem. However, the voluntary preemption technique is a general purpose technique that can be applied to a number of other real-time classes. For instance, the processes need not be independent, the internal structure of the processes can be taken into account, the scheduling can be on a multi-processor, etc.

There are several advantages of using voluntary preemption. The main advantage is that it allows the internal structure of processes to be taken into account. Secondly, the scheduling algorithms so designed can take into account the overhead cost of switching processes; this is done realistically and without sacrificing performance. This approach can also be used to schedule on a multi-processor system. Finally, voluntary preemptive scheduling also allows reducing the overhead cost of context switching by reducing the number of preemptions and/or the strategic placement of the preemption points.

References

[CSR88] S. Cheng, J. A. Stankovic, and K. Ramamritham. Scheduling algorithms for hard real-time systems - a brief survey. In J. A. Stankovic and K. Ramamritham, editors, *Tutorial on Hard Real-Time Systems*, pages 150–173. Computer Society Press, 1988.

[DD86] S. Davari and S. K. Dhall. An on line algorithm for real-time task allocation. In *IEEE Real-Time Systems Symposium*, 1986.

[DJ86] M. D. Donner and D. H. Jameson. A real-time juggling robot. Technical Report RC 12111, IBM T. J. Watson Research Center, Yorktown Heights, NY 10598, 1986.

[DL78] S. K. Dhall and C. L. Liu. On a real-time scheduling problem. *Operations Research*, 26:127–140, 1978.

[FP88] S. R. Faulk and D. L. Parnas. On synchronization in hard-real-time systems. *Commun. of the Association for Computing Machinery*, 31:274–287, 1988.

[GJ79] M. R. Garey and D. S. Johnson. *Computers and Intractability : A Guide to the Theory of NP-Completeness.* W. H. Freeman and Co., 1979.

[GS88] J. B. Goodenough and L. Sha. The priority ceiling protocol : A method for minimizing the blocking of high priority Ada tasks. In *Proceedings of the 2nd ACM International Workshop on Real-Time Ada Issues*, 1988.

[LL73] C. L. Liu and J. W. Layland. Scheduling algorithms for multiprogramming in a hard-real-time environment. *Journal of the Association for Computing Machinery*, 20:46–61, 1973.

[Moi86] A. Moitra. Scheduling of hard real-time systems. In K. V. Nori, editor, *Foundations of Software Technology and Theoretical Computer Science*, pages 362–381, December 1986. in LNCS, vol. 241, Springer-Verlag.

[Mok83] A. K. Mok. Fundamental design problems of distributed systems for the hard-real-time environment. Ph.D. Thesis, MIT, 1983.

[SG90] L. Sha and J. B. Goodenough. Real-time scheduling theory and Ada. *COMPUTER*, 23(4):53–62, 1990.

[SLR86] L. Sha, J. P. Lehoczky, and R. Rajkumar. Solutions for some practical problems in prioritized preemptive scheduling. In *IEEE Real-Time Systems Symposium*, pages 181–191, 1986.

[SRC85] J. A. Stankovic, K. Ramamritham, and S. Cheng. Evaluation of a flexible task scheduling algorithm for distributed hard real-time systems. *IEEE Transactions on Computers, C-34*, pages 1130–1143, 1985.

[SRL90] L. Sha, R. Rajkumar, and J. P. Lehoczky. Priority inheritance protocols : An approach to real-time synchronization. *IEEE Transactions on Computers, C-39*, pages 1175–1185, 1990.

[Tei78] T. Teixeira. Static priority interrupt scheduling. In *7th Texas Conf. on Computing Systems*, pages 5–13:5–18, 1978.

Observing Task Preemption in Ada 9X

K.T.Narayana[1]
ORA Corporation
(Formerly Odyssey Research Associates, Inc.)
301A Harris B Dates Drive
Ithaca, NY14850

October 21, 1991

Abstract

Ada 9X mapping proposals revise Ada83 and introduce several novel concepts for programming real-time systems. Dynamic priorities of tasks and a family of priority models are critical elements in this revision. Because of dynamic priorities, observing task preemption becomes necessary. When processors have local clocks, this becomes an important technical and correctness issue. In this paper, we consider a small rendezvous based subset derived from Ada 9X mapping proposals and give limited parallelism semantic models for the subset language. From an unconstrained semantic model, we develop a simple priority model, and the simple $priority^{+83}$ model. The simple $priority^{+83}$ model is a submodel of the core priority model of Ada 9X mapping proposals.

1 Introduction

In priority driven real-time systems, task switching is generally governed by a rule that a high priority task must always execute in preference to a low priority one. If a high priority task becomes ready, then in order to dispatch that task onto a processor an executing task on that processor must be preempted. Hardware features allow this only at processor designated interruption points. These interruption points represent consistent processor states. For example, in some multi-processor architectures, tasks can only be preempted on the processor clock boundaries and in others on instruction boundaries (whatever the number of cycles that an instruction takes). Thus if a high priority task becomes ready at some time, then at that time it may not be able to execute on a processor of the multi-processor system. The task may have to wait so that an interruption point on a processor is reached. This phenomenon has to be accounted for in a semantic model for real-time concurrency which allows preemption. Thus observing task preemption becomes a technical and correctness issue, because without observing that processors cannot be interrupted, it is not possible to construct a semantic model which is consistent with the operational behavior of the system. Just as architectural issues have to be taken into account in the real-time model, so also any language defined uninterruptable actions.

We approach this technical issue by defining a limited parallelism semantic model for a 'toy' language derived from Ada 9X mapping proposals and based in part on its priority

[1] The author is partially funded under the contract number F8635-90-C-0308 from the Eglin Airforce Base. The Office of Naval Research under the grant number N00014-89-J-1171 funded the author's work on real-time system specification and verification. Any findings and opinions contained in this paper are those of the author and not of the Department of Defense.

model. Ada 9X [2] mapping proposals revise Ada83[1] and introduce several new features like protected records, dynamic priorities, etc., for programming embedded systems. The primitives are supported by a complex web of priority models and a core model of it gives the least deterministic computations of the program. In defining the notion of priority, Ada 9X introduces two basic concepts that of *base* and *active* priorities of a task. From a semantic point of view only the base priorities of tasks are observable. The active priorities of tasks may be inferred from the observables of the program and several rules for computing the active priorities of tasks are given in [2]. These informal rules account for the synchronization state of tasks—for example if a task is in rendezvous, then the active priority of the caller task contributes to the active priority of the acceptor task, etc. Tasks are almost always(!) scheduled on the basis that a low active priority task may not execute in preference to a runnable high active active priority task.

Henceforth in this paper whenever we refer to Ada 9X, we mean the mapping proposals contained in [2]. For reasons of exposition and to clarify why observables must be sufficiently discriminating, we develop the semantic models in stages with each of them addressing a particular concern. With the limited parallelism semantic model of Shade and Narayana[13] as the basis, we will consider a sublanguage L of Ada 9X consisting of rendezvous, dynamic priorities and delay commands. L does not even have conditional or repetitive commands, and thus is a 'toy' language for demonstrating the semantic issues. But the issues addressed are applicable to a wide range of languages. As in [13] we choose an execution model of a multi-processor system in which the number of processors is less than the number of tasks in the program, each processor runs at possibly different speed—called its clock cycle—from the others, and in which each processor can only be interrupted on its cycle boundary—an assumption that is realistic and practical.

In the first step, we develop for the sublanguage L the meaning function for parallel composition which yields nondeterministic schedules[13] and in which task switching between processors honors cycle boundaries, and no priority model is imposed. Local clocks are handled in a simple way. Every observable which designates an executable action is potentially made on any processor. Because access to shared memory can only be resolved in global time, for the duration of the clock cycle, observables recording shared memory access are translated into arbitrary waiting for access, followed by an access, and then a sequence of compensating internal actions. An observable designating an internal action is translated by replicating it over the global duration of the clock cycle. A central element in the semantics is that the observer is forgetful about the duration of the local clocks when obtaining parallel composition.

In the second step, we consider a *simple priority model* in which the base priorities alone determine the set of tasks that can execute at any time. While this model is only applicable for a very restricted subset of Ada 9X, the exposition contained here illustrates the technical issues in a simple way. In Ada 9X the base priority of a task is an atomic (with respect to updates) shared variable. A system function `setpriority` assigns the given value as the base priority of the given task. Since we model shared variables by requiring atomic update, we can take for reasons of simplicity the set priority call as an assignment to a distinguished shared variable denoting the base priority of the task. In the simple priority model, we need to observe tasks differently from the nondeterministic model. When a task is suspended for any reason, we need to record the observable relative to (the base priority of) the task. Further it is not sufficient to translate an internal action by replicating it over the clock duration. When translating an internal action observed on the local clock, we need to distinguish its initiation and its continuation over the duration of the clock cycle. Similarly any compensating internal actions for translating observables recording shared memory access must be recorded as a continuation of internal steps. Without these changes to observations, we cannot enforce the requirement that a processor cannot be preempted any time during its clock cycle and assure that task preemption occurs on cycle boundary.

In the third step, we give the *simple priority*$^{+83}$ *model.* This model brings together dynamic priorities and the Ada83 requirements on the priority of a task executing while in rendezvous. The simple *priority*$^{+83}$ model is distinct from the simple priority model —semantically they are incomparable— and requires the notion of the active priorities of tasks. Here the observables must be more discriminating than the previous step. In particular we also need to observe the tasks relative to their base priorities when they are executing as acceptors in a rendezvous. In order to infer active priorities, we need to construct for each time point the set of all *rendezvous dependence traces.* The simple *priority*$^{+83}$ model is a submodel of the core priority model of Ada 9X.

We will only be giving the most nondeterministic semantic models in each of the category. For example, when tasks have the same base (or active) priority, we will be assuming that they are arbitrarily interleaved. Ada 9X prescribes (sometimes mandates) a suite of policies to achieve determinism —`run-until blocked`, FIFO, round-robin etc.

Organization of the paper

The paper is organized as follows. Section 2 discusses the execution environment and models. It also presents the syntax and an informal operational semantics of the language L. Section 3 gives some preliminaries and notations. In section 4 we develop the limited parallelism semantic model for rendezvous. In section 5, we introduce the simple priority model. In section 6 we give the simple *priority*$^{+83}$ model. Section 7 closes with a discussion.

2 Environments and Execution Models

We consider a physical environment of a MIMD (multiple-instruction, multiple-data) *shared-memory machine* with one or more physical processors running concurrently. We will always assume that there are ρ processors, numbered from 1 to ρ. The value of ρ is fixed for a given program, but need not be the same for all programs. Each physical processor has its own local memory, and in addition there is a single shared memory which can be accessed by any processor.

We assume that there is a conceptual global clock which acts as a standard for measurement. Each physical processor has its own conceptual local clock which runs at a fixed speed relative to the global clock. Each "tick" of a local clock takes a discrete non-zero amount of global time, which is the *speed* of the processor or the duration of the processor cycle. The function θ maps processors to their speeds. For example, if $\theta(4) = 3$ and $\theta(1) = 2$, then processor 4 runs 50% more slowly than processor 1. For simplicity we consider a CREW model of shared memory access. We assume that all memory accesses will be resolved within a finite period of time. δ denotes the maximum amount of global time that a process must wait before it is granted access to a variable. The function δ is a parameter of the model, and is obviously dependent on ρ.

Domains

$$\begin{aligned}
c \in \mathcal{C} &\equiv \text{constants}\\
u \in \mathcal{V}_L &\equiv \text{local variables}\\
\mathcal{V}_E &\equiv \text{explicit shared variables}\\
\mathcal{V}_T &\equiv \text{implicit task specific shared variables}\\
y \in \mathcal{V}_S = \mathcal{V}_E \cup \mathcal{V}_T &\equiv \text{shared variables}\\
v \in \mathcal{V} = \mathcal{V}_L \cup \mathcal{V}_S &\equiv \text{variables}
\end{aligned}$$

Language Syntax

$$\begin{aligned}
\Pi &::= \|_{i=1}^{n} t_i :: S_i\\
S &::= v := E \mid S; S \mid p \mid \mathbf{delay}\ E \mid \mathbf{accept}\ p\ \mathbf{do}\ S\ \mathbf{end}\\
E &::= (E \otimes E) \mid (\ominus E) \mid v \mid c
\end{aligned}$$

where $\Pi \in$ programs, $E \in$ expressions, $S \in$ statements, $p \in$ entries, and $t \in$ tasks.

Figure 1: The Language L

We consider the limited parallelism model [13] which is a compromise between interleaving and maximal parallelism[8] models. ρ may be less than the number of tasks. Maximal throughput is enforced, with the added constraint that processors may not go unused if there is a task waiting for execution. Since there may not be enough processors to go around, task *scheduling* is required to make good use of the available processors. We will make minimal assumptions about the scheduler. However we do require that there be a delay of at least one global time unit before a task is switched from one processor to another.[2] This is to prevent pointless task swapping.

2.1 The Language L

We will define a language L which captures features sufficient to model dynamic priorities, and rendezvous. The abstract syntax of L is shown in figure 1.

A program Π is the parallel composition of one or more sequential tasks. We do not consider nested parallelism, in which a task may contain parallel subtasks. A task consists of one or more statements $S_1; S_2; \ldots; S_k$, where ";" denotes sequential composition. The assignment statement $v := E$ has the usual operational meaning: the expression E is fully evaluated and the resulting value is assigned to the variable v. Expressions and variables can take on any values from the domain $\mathcal{C}$; we assume that all expressions and assignments are type-correct. We use $\otimes$ and $\ominus$ to denote arbitrary binary and unary operations over $\mathcal{C}$, respectively. All shared variable accesses are atomic. To avoid having to deal with blocks and scope, we assume that the local variables of distinct tasks are disjoint. The `delay` E statement suspends for at least e units of time and necessarily executes for n units of time prior to that, where e is the value of the expression E and n is the amount of time required to evaluate E. An entry call p causes the task to suspend for a rendezvous with the acceptor. When an acceptor is ready, the rendezvous occurs.

[2]This restriction can be easily dropped from our model and is there for reasons of realism.

The task continues to be suspended during the rendezvous and becomes enabled only after the rendezvous is complete. An accept statement **accept** p **do** S **end** causes the task to suspend itself for a caller on the entry p. When a caller is available, the task (then on called the acceptor) enters rendezvous immediately by executing the statement S. The rendezvous is said to be completed when execution of S is completed and the suspended caller is signaled of the end of the rendezvous.

3 Preliminaries

In a linear-history style semantics [3], the semantic domain $\mathcal{D}$ consists of non-empty, prefix-closed sets of state-history pairs. A state-history pair is written $\langle s, h\rangle$, where s is the *state* and h is the *history*. The pair $\langle s, h\rangle$ denotes a computation: intuitively, s is the state at the end of the computation and h is the sequence of actions which led to s.

The state domain is denoted by $\mathcal{S}$. It can be partitioned into a set $\mathcal{S}_P$ of *proper* states, and a set $\mathcal{S}_I$ of *improper* states. Usually, a proper state is a "snapshot" of program memory which records the values of the variables at a given instant of time. For simple languages, a proper state may simply indicate successful termination. The set of improper states $\mathcal{S}_I$ will always contain $\bot$, which denotes an incomplete computation. Naturally, the precise choice of $\mathcal{S}$ depends on the language being modeled.

An observation is a bag of observables. Typical *observables* include waiting for synchronization, accessing a shared variable, and executing a rendezvous. The set of all observables is denoted by O. Since a program may contain two or more tasks running in parallel, it is possible that two different tasks may be simultaneously performing actions which are denoted by the same observable. Therefore, an observation is modeled by a *bag* (or *multiset*).

Intuitively, a bag is a "set" in which there may be multiple copies of an element. Formally, for any set X, a *bag* over X is a function $b: X \rightarrow \mathcal{N}$ that maps elements of X to natural numbers; $b(x)$ represents the number of occurrences of x in b. We can extend the standard set operators to bags as follows:

$$\begin{aligned}
x \in b &\equiv b(x) > 0\\
b_1 \subseteq b_2 &\equiv \forall x (x \in b_1 \Rightarrow b_1(x) \leq b_2(x))\\
b_1 \cup b_2 &= \{(x,n) \mid b_1(x) + b_2(x) = n \wedge n > 0\}\\
b_1 \cap b_2 &= \{(x,n) \mid \min\{b_1(x), b_2(x)\} = n \wedge n > 0\}\\
b_1 - b_2 &= \{(x,n) \mid x \in b_1 \wedge n = b_1(x) - b_2(x) \wedge n > 0\}.
\end{aligned}$$

$\mathcal{B}(X)$ denotes the set of all bags over X. A finite bag is denoted by $\{\!|x_1, \ldots, x_n|\!\}$, where the x_i need not be distinct. We will not require infinite bags in this paper. For reasons of conciseness, we sometimes need to coerce a set into a bag. We allow this type coercion by the definition: $b\ op\ a = b\ op\ \{\!|c \mid c \in a|\!\}$, and $a\ op\ b = \{\!|c \mid c \in a|\!\}\ op\ b$, where a is a set of observables, and op any operator defined on bags.

A history is a finite sequence of observations. The ith element of a history contains the observable actions at time unit i. The context of its use determines whether the time unit is a local clock or a global clock. A history element is a member of $\mathcal{O} = \mathcal{B}(O)$, and the history domain $\mathcal{H}$ equals $\mathcal{P}(\mathcal{O}^*)$. A history is a finite sequence of the form $o_1 o_2 \ldots o_n$, where $o_i \in \mathcal{O}$. The empty history is denoted by ϵ. Let h, h' be histories and

H be a non-empty set of histories. $|h|$ denotes the length of h, and hh' denotes the usual concatenation of h and h'. If $1 \leq i \leq |h|$ then $h[i]$ denotes the ith element of h, otherwise $h[i] = \{\!|\,|\!\}$. Some additional operations on histories are defined below.

- $h^n = \textbf{if } n \leq 0 \textbf{ then } \epsilon \textbf{ else } hh^{n-1}$.
- $h[i..j] = \textbf{if } 1 \leq i \leq |h| \wedge i \leq j \textbf{ then } h[i] \ldots h[j] \textbf{ else } \epsilon$.
- $h[i..] = h[i..|h|]$.
- $H[i] = \bigcup_{h \in H} h[i]$.
- $h \prec h' \iff |h| < |h'| \wedge h = h'[1..|h|]$.
- $Cmb(h_1, \ldots, h_n) = H[1] \ldots H[\max\{|h| \mid h \in H\}]$, where $H = \{h_1, \ldots, h_n\}$.

A set $X \subseteq \mathcal{S} \times \mathcal{H}$ is *prefix closed* if $\langle s, h \rangle \in X$ and $h' \prec h$ imply $\langle \perp, h' \rangle \in X$. The *prefix closure* of X is denoted $Cls(X)$ and is given by

$$Cls(X) = X \cup \{\langle \perp, h' \rangle \mid \langle s, h \rangle \in X \wedge h' \prec h\}.$$

The semantic domain $\mathcal{D}$ consists of sets of prefix-closed state-history pairs. Formally,

$$\mathcal{D} = \{Cls(X) \mid X \subseteq (\mathcal{S} \times \mathcal{H}) \wedge X \neq \emptyset\}.$$

$\mathcal{D}$ can be partially ordered by subset inclusion, where the minimal element is $\{\langle \perp, \epsilon \rangle\}$. For $D \in \mathcal{D}$, the natural projections are given by

$$\begin{aligned} \pi_{\mathcal{S}}(D) &= \{s \mid \exists h. \langle s, h \rangle \in D\} \\ \pi_{\mathcal{H}}(D) &= \{h \mid \exists s. \langle s, h \rangle \in D\}. \end{aligned}$$

For every valid program Π of a given language, we define a function $\mathcal{M}[\![\Pi]\!]: \mathcal{S}_P \rightarrow \mathcal{D}$ such that $\mathcal{M}[\![\Pi]\!]s$ denotes the behavior of Π starting from the initial proper state s. $\mathcal{M}[\![S_1; S_2]\!]s$ (where ";" denotes sequential composition) should be the functional "composition" of $\mathcal{M}[\![S_2]\!]$ and $\mathcal{M}[\![S_1]\!]s$. Since such functions cannot be composed in the usual sense because their domain and range are disjoint, we introduce the following definitions. Let f be a function from $\mathcal{S}_P$ to $\mathcal{D}$. Then $\tilde{f}: \mathcal{S} \rightarrow \mathcal{D}$ and $\hat{f}: \mathcal{D} \rightarrow \mathcal{D}$ are given by

$$\begin{aligned} \tilde{f}(s) &= \textbf{if } s \in \mathcal{S}_P \textbf{ then } f(s) \textbf{ else } Cls\{\langle s, \epsilon \rangle\} \\ \hat{f}(D) &= \{\langle s', hh' \rangle \mid \langle s, h \rangle \in D \wedge \langle s', h' \rangle \in \tilde{f}(s)\}. \end{aligned}$$

We can now write $\mathcal{M}[\![S_1; S_2]\!]s = \widehat{\mathcal{M}[\![S_2]\!]}(\mathcal{M}[\![S_1]\!]s)$.

A state-history pair $\langle s, h \rangle$, where $s \in \mathcal{S}_P$, denotes a successful terminating computation with history h. The pair $\langle \perp, h \rangle$ denotes an incomplete computation which is either an approximation of a terminating computation $\langle s, hh' \rangle$, or an element of an infinite sequence of incomplete computations. Thus, if $D \in \mathcal{D}$ is the meaning of a program Π, Π terminates if there exists $\langle s, h \rangle \in D$ such that $s \in \mathcal{S}_P$, and diverges if D contains an infinite set $\{\langle \perp, h_i \rangle \mid i \geq 0\}$ such that $h_i \prec h_{i+1}$. Note that more than one of these possibilities may hold for a nondeterministic program.

4 Formal Semantics of L

It order to simplify the discussion, we will define the semantics incrementally by considering sublanguages of L. The modeling of these sublanguages captures the important semantic issues associated with L. Of primary importance are the consistency checks which need to be imposed at parallel composition. The structure of these checks and the difficulties associated with their imposition provide us an insight into the complexity of the correctness problem for real-time computing using Ada 9X.

4.1 Shared Variables: The Language L_A

We consider a small language $L_A \subseteq L$ which contains sequential composition, parallel composition, and assignment. For now we only consider the maximal parallelism model. In a maximal parallelism model, the notion of priorities has no effect. The main purpose of this subsection is to illustrate how the shared memory is modeled. This subsection is mostly extracted from Shade and Narayana[13]. The language L_A is defined as follows:

$$\begin{array}{rcl} \Pi & ::= & \|_{i=1}^{n} S_i \\ S & ::= & v := E \mid S;S \\ E & ::= & (E \otimes E) \mid (\ominus E) \mid v \mid c \end{array}$$

$\mathcal{S}_P$, the proper states, is the set of all partial functions $s: \mathcal{V} \rightarrow \mathcal{C}$. $s[c/v]$ denotes the state s' which agrees with s except that $s'(v) = c$. For $V \subseteq \mathcal{V}$, $s|V$ denotes the restriction of s to the variables in V. As observables we take the set

$$\mathcal{O} = \{A_{y \rightarrow c}, A_{c \rightarrow y}, I, W_y \mid c \in \mathcal{C} \wedge y \in \mathcal{V}_S\}.$$

I denotes a local internal action of a task, a computation which does not involve the shared memory. W_y indicates that a task has requested (or is currently requesting) access to shared variable y but has not yet been granted access. $A_{y \rightarrow c}$ indicates that a task is reading the value c from shared variable y, and $A_{c \rightarrow y}$ indicates that c is being written to y. The following definitions will be useful, where $o \in \mathcal{O}$:

$$\begin{array}{rcl} Reads(y, o) & = & \sum_{c \in \mathcal{C}} o(A_{y \rightarrow c}) \\ Writes(y, o) & = & \sum_{c \in \mathcal{C}} o(A_{c \rightarrow y}) \\ Accesses(y, o) & = & Reads(y, o) + Writes(y, o). \end{array}$$

$Reads(y, o)$ gives the number of reads on the variable y in the observation o. Similarly $Writes(y, o)$ gives the number of writes to y in o. Then the number of accesses to y in the observation o is the sum of the number of reads and the number of writes to y in o.

The semantics of an expression E is defined as a function $\mathcal{E}[\![E]\!]: \mathcal{S}_P \rightarrow \mathcal{C} \times \mathcal{H}$ as shown below. The presence of (c, h) in $\mathcal{E}[\![E]\!]s$ indicates that one possible evaluation of E starting from the proper state s produces the history h and yields the value c. Note that if an expression E contains a reference to a shared variable, it is impossible to locally determine the value of E, because other tasks may modify the shared variable while the evaluation of E is taking place. For example, before evaluating $(u + 3) * y$ the variable y may have the value 0, but while the subexpression $(u + 3)$ is evaluated another task might overwrite y with some non-zero value. The solution is to assume that a shared variable can have *any* value, and then reject all incorrect assumptions at the time of parallel composition, when all activity involving the shared variable is known. The formal definitions are as follows:

$$\begin{aligned}
\mathcal{E}[\![(E_1 \otimes E_2)]\!]s &= \{(c_1 \otimes c_2, h_1 h_2 \{\!|I|\!\}) \mid (c_1, h_1) \in \mathcal{E}[\![E_1]\!]s \wedge (c_2, h_2) \in \mathcal{E}[\![E_2]\!]s\} \\
\mathcal{E}[\![(\ominus E)]\!]s &= \{(\ominus c, h\{\!|I|\!\}) \mid (c, h) \in \mathcal{E}[\![E]\!]s\} \\
\mathcal{E}[\![c]\!]s &= \{(c, \{\!|I|\!\})\} \\
\mathcal{E}[\![u]\!]s &= \{(s(u), \{\!|I|\!\})\} \\
\mathcal{E}[\![y]\!]s &= \{(c, \{\!|W_y|\!\}^i \{\!|A_{y \to c}|\!\}) \mid 0 \le i \le \delta \wedge c \in \mathcal{C}\}.
\end{aligned}$$

The denotation of $(\ominus E)$ indicates that first the expression E is evaluated, where h is the sequence of steps performed and c is the resulting value; the application of $\ominus$ to c requires one internal action, and the result is $\ominus c$. The other definitions are similar.

The semantics of the primitive commands are given by

$$\begin{aligned}
\mathcal{M}[\![u := E]\!]s &= \mathit{Cls}\{\langle s[c/u], h\{\!|I|\!\}\rangle \mid (c, h) \in \mathcal{E}[\![E]\!]s\} \\
\mathcal{M}[\![y := E]\!]s &= \mathit{Cls}\{\langle s, h\{\!|W_y|\!\}^i \{\!|A_{c \to y}|\!\}\rangle \mid (c, h) \in \mathcal{E}[\![E]\!]s \wedge 0 \le i \le \delta\} \\
\mathcal{M}[\![S_1; S_2]\!]s &= \widehat{\mathcal{M}[\![S_2]\!]}(\mathcal{M}[\![S_1]\!]s).
\end{aligned}$$

The internal action in the history of $\mathcal{M}[\![u := E]\!]s$ represents the actual assignment of the value c to u. Note that the state s is updated to reflect the new value of u, but no such update is performed in the assignment $y := E$. There are two reasons for this. First, as indicated above, the global state may be changed at any time by another task. Second, a complete record of all shared-variable accesses is kept in the histories.

The meaning of a program $\|_{i=1}^{n} t_i :: S_i$ is obtained by *merging* the denotations of the S_i. Intuitively, each task must make assumptions about the value of a shared variable at a given time instant, since the variable may have been changed by another task. Thus, it is possible that different tasks may make contradictory assumptions. Merging consists of detecting such inconsistencies and combining consistent state-history pairs from each task. Let $\mathcal{V}_L(S_i)$ denote the set of all local variables accessed by S_i, and for $D \in \mathcal{D}$, let h_D denote the history $\mathit{Cmb}(\pi_{\mathcal{H}}(D))$. Then

$$\mathcal{M}[\![\, \|_{i=1}^{n} t_i :: S_i]\!]s = \{\mathit{Merge}(s, \bigcup_{i=1}^{n} \{\langle s_i | \mathcal{V}_L(S_i), h_i\rangle\}) \mid \langle s_i, h_i\rangle \in \mathcal{M}[\![S_i]\!]s\},$$

where

$$\begin{aligned}
\mathit{Merge}(s, D) &= \textbf{if } \mathit{Comp}(D) \wedge \mathit{Cons}_A(s, D) \textbf{ then } \langle \mathit{SM}_A(s, D), h_D\rangle \textbf{ else } \langle \bot, \epsilon\rangle \\
\mathit{Comp}(D) &= \forall \langle s, h\rangle \in D (s = \bot \Rightarrow |h| = |h_D|) \\
\mathit{Cons}_A(s, D) &= \mathit{Mem}(h_D) \wedge \mathit{NoWait}(h_D) \wedge \mathit{Verify}(s, h_D)
\end{aligned}$$

The arguments to the *Merge* function are the initial state s, and a set D which contains one state-history pair from each task. *Comp* is the *compatibility* predicate, which enforces the technical requirement that histories whose corresponding state is $\bot$ are not merged with longer histories. (Intuitively, it would make no sense to merge a long completed history with a short incomplete one.) *Cons* is the *consistency* predicate; consistent histories are merged by simply combining them element-wise into a single history. SM_A is the state merge function. *Cons* and SM_A are defined below.

To guarantee consistency, three properties must be satisfied. First, the value read from a shared variable must be the same as the value most recently written to it. This is ensured by the *Verify* predicate, where

$$Verify(s,h) = \forall i(A_{y\to c} \in h[i] \Rightarrow LastWr(y,s,h[1..i-1],c))$$

and the predicate $LastWr(y,s,h,c)$ is true iff

$$(\sum_i Writes(y,h[i]) = 0 \wedge s(y) = c) \vee \exists j(A_{c\to y} \in h[j] \wedge \sum_{k=j+1}^{|h|} Writes(y,h[k]) = 0)$$

LastWr says that if y had the value c at time i, then the last value written to y *before* time i was the value c.

Second, there must be no memory conflicts, and finally, there must be no unnecessary waiting to access a shared variable. These conditions are enforced by the *Mem* and *NoWait* predicates, respectively:

$$\begin{aligned} Mem(h) &= \forall i,y(Accesses(y,h[i]) > 1 \Rightarrow Writes(y,h[i]) = 0) \\ NoWait(h) &= \forall i,y(W_y \in h[i] \Rightarrow Writes(y,h[i]) > 0) \end{aligned}$$

These definitions are for the CREW model. *Mem* says that at every time instant i, if more than one task is accessing the shared variable y, then they are all read accesses. *NoWait* ensures that if a task is waiting to access y, it is because some other task is currently writing to y.

Merging states is performed as follows. First, if any of the states is $\perp$, then the combined state is $\perp$ since at least one of the tasks has not completed its computation. Now assume that all states are proper. Note that proper states are restricted to their local components at parallel composition, and the local variables of each task are disjoint. Changes to the global state are recorded only in the histories. It therefore suffices to form the union of the local states and compute the global state using the initial state s and any changes recorded in the associated histories:

$$\begin{aligned} SM_A(s,D) &= \textbf{if } \perp \in \pi_S(D) \textbf{ then } \perp \textbf{ else } Global(s,h_D) \cup \bigcup \pi_S(D) \\ Global(s,h) &= \{(y,c) \mid y \in \text{dom}(s) \wedge LastWr(y,s,h,c)\}. \end{aligned}$$

4.2 Rendezvous and Limited Parallelism: The Language L_S

The sublanguage L_S contains parallel and sequential composition, delay and rendezvous commands, and assignment statements. Unlike in the previous section, we consider here the limited parallelism model, so there is a fixed number ρ of processors whose speeds are defined by θ. This language illustrates how limited parallelism and rendezvous are all handled. The syntax of L_S is as follows:

$$\begin{aligned} P &::= \|_{i=1}^{n}\, t_i :: S_i \\ S &::= \texttt{delay } c \mid x := e \mid S;S \mid \texttt{accept } p \texttt{ do } S \texttt{ end} \mid p \end{aligned}$$

We take the semantic categories to be the same as earlier. The set O of observables is given by the following: $A_{y\to c}$, $A_{c\to y}$ W_y, and I denote the same as in the previous section, and

Z	::	delay action of a task
$R_{\leftarrow p}$	::	suspended for rendezvous on entry p
$J_{\leftarrow p}$	::	suspended in rendezvous on entry p
$L_{\leftarrow p}$	::	leaving rendezvous on entry p
$R_{\rightarrow p}$	::	suspended as acceptor for entry p
$J_{\rightarrow p}$	::	in rendezvous as acceptor of entry p
$L_{\rightarrow p}$	::	leaving rendezvous as acceptor of entry p
Q	::	waiting for a processor
U_i	::	processor i allocated to the task

The nonexecutable observables are the elements of the set

$$\widehat{O} = \{Z, R_{\leftarrow p}, R_{\rightarrow p}, J_{\leftarrow p}, L_{\leftarrow p}, Q \mid p \in \mathit{entries}\}$$

These observables record the suspension of tasks. Since suspension is a nonexecutable activity not involving a processor, these observations are always made against a conceptual global clock. The executable observables are elements of the set

$$O_E = \{W_y, A_{c\to y}, A_{y\to c}, I \mid c \in \mathcal{C}, y \in \mathcal{V}_S\}$$

and these observations are made against local clock. These actions are always observed when a task is executing on a processor. The observables $J_{\rightarrow p}$ and $L_{\rightarrow p}$ are auxiliary and are always associated with another observable. Its association context determines whether the observation is made against a global clock, or a local clock. For example, if an internal action I is simultaneously observed with respect to $J_{\rightarrow p}$, then it means that a task performs an internal action while in rendezvous as an acceptor of an entry call on p. Such an observation is made against the local clock because I is observed against the local clock. Similarly if a Q action is associated with $J_{\rightarrow p}$, then it means that a task is suspended for a processor while in rendezvous as an acceptor of entry call p. Such an observation is made against the global clock because Q is observed against the global clock. The observable Q is brought about by the limited parallelism environment. On the other hand the observables in $\widehat{O}_L$

$$\widehat{O}_L = \{Z, R_{\leftarrow p}, R_{\rightarrow p}, J_{\leftarrow p}, L_{\leftarrow p} \mid p \in \mathit{entries}\}$$

are language specific nonexecutable activities of tasks. This distinction is essential for prohibiting unnecessary suspension—when the suspension is caused by explicit program actions. Whether a task is unnecessarily suspended for a processor is resolved by imposing maximal thoughput from the system.

We give the a priori semantics of statements not considered earlier in L_A. The a priori semantics of sequential composition remains the same in spite of limited parallelism and is repeated below. Here we only consider delay commands with constant expressions.

$$\begin{aligned}
\mathcal{M}[\![\texttt{delay}\ c]\!]s &= \mathit{Cls}\{\langle s, \{\!|Z|\!\}^c\rangle\} \\
\mathcal{M}[\![p]\!]s &= \mathit{Cls}\{\langle s, \{\!|R_{\leftarrow p}|\!\}^i\{\!|J_{\leftarrow p}|\!\}^j\{\!|L_{\leftarrow p}|\!\}\rangle \mid i \ge 0 \wedge j \ge 1\} \\
\mathcal{M}[\![S_1; S_2]\!]s &= \widehat{\mathcal{M}[\![S_2]\!]}(\mathcal{M}[\![S_1]\!]s) \\
\mathcal{M}[\![\texttt{accept}\ p\ \texttt{do}\ S\ \texttt{end}]\!]s &= \mathit{Cls}\{\langle s', \{\!|R_{\rightarrow p}|\!\}^i \mathit{Cmb}(h, \{\!|J_{\rightarrow p}|\!\}^{|h|}\{\!|I, L_{\rightarrow p}|\!\}\rangle \mid \\
&\qquad i \ge 0 \wedge \langle s', h\rangle \in \mathcal{M}[\![S]\!]s\}
\end{aligned}$$

Prefix closure ensures that the set $\{(\perp, \{R_{\leftarrow p}\}^i) \mid i \geq 0\}$ is contained in the denotation of p. This models the case when the task is suspended forever to rendezvous. Similar remarks apply to the denotation of **accept** p **do** S **end**. In $\mathcal{M}[\![$**accept** p **do** S **end**$]\!]s$, the task is suspended for an unspecified duration and then begins execution of S in rendezvous with the caller of p. We add $J_{\rightarrow p}$ to all elements of h to indicate that the execution of S is simultaneously observed as rendezvous. The final element of of a rendezvous is recorded as an internal action which is simultaneously observed as signaling the end of rendezvous. Signaling the end of rendezvous as a separate action is necessary to prohibit simultaneous completion of rendezvous when rendezvous of individual tasks are nested. Further it has the consequence of capturing the synchronization aspect associated with the end of the rendezvous without which sequential composition of two entry calls cannot be adequately modeled(!). An entry call on p causes the task to be suspended for an arbitrary amount of time and remains suspended during the rendezvous. However during rendezvous the observation $J_{\rightarrow p}$ distinguishes the suspension of the caller.

The basic strategy for computing $\mathcal{M}[\![\ ||_{i=1}^{n} t_i :: S_i]\!]s$ is as follows. Let $D_i = \mathcal{M}[\![S_i]\!]s$. Figure out all of the different ways that the histories in D_i could be executed in a limited parallelism environment with ρ processors, taking the processor speeds into account, but ignoring the other tasks. Let this *scheduled* meaning be D_i'. Merge all of the D_i' as usual and then reject inconsistent histories while simultaneously enforcing maximum throughput. The resulting histories yield pure nondeterministic schedules and are consistent with respect to the scheduling model adopted in [13].

Since $\mathcal{M}[\![S_i]\!]s$ does not impose limited parallelism, we have to explicitly impose the intervention of a scheduler to construct the a priori meaning $\mathcal{M}[\![t_i :: S_i]\!]s$ of a task. We do this by scheduling the observed histories of the task.

Scheduling consists of three distinct phases. Let h be a history to be scheduled.

Phase 1: Interleave Q elements. Since there are fewer processors than processes (in general), it is possible that actions of h may be delayed due to the lack of an available processor. This is modeled by interleaving arbitrary length sequences of Q elements with the actions of h. However there are some restrictions. Q's cannot be inserted between elements denoting a nonpreemptable sequence of actions and between a sequence of actions (or at the beginning of an action) which already records that the task is suspended. These cases are as follows.

1. The W_y elements denoting waiting for shared variable access denote nonpreemptable actions and holds the processor.

2. The Z elements denote that a task is suspended under program control by the use of a delay command. We cannot model the notion of the earliest successful guard in an alternative (or select) command if we interleave Q elements in that sequence.

3. The set of observables $\{R_{\leftarrow p}, R_{\rightarrow p}, J_{\leftarrow p}, L_{\leftarrow p}\}$ all denote suspended actions. We cannot prohibit unnecessary suspension if we interleave Q elements between sequences of them.

This phase will produce a set H_1 of histories. Certain semantic ramifications are introduced as a result of the decisions made in regard to interleaving of Q elements.

Phase 2: Add U_i elements. Each executable action must be executed on *some* processor. There is no local way to decide which processor is used. So for each history $h_1 \in H_1$, and for each executable action in h_1, generate ρ different histories, each corresponding to the assumption that the action was executed on processor i. To record this assumption, add the element U_i to the action. Again, there is a restriction: contiguous sequences of executable actions must all run on the same processor, since a context-switch must be preceded by at least one Q action.

Phase 3: Expand the executable actions. The duration of processor cycles must be taken into account. If an action is executed on processor i, there must be some way to indicate that execution of that action takes $\theta(i)$ units of global time. The idea is to expand the action into an appropriate subhistory of length $\theta(i)$. How should the expansion be performed? For purely local actions, such as the I denoting an internal action, repetition is sufficient. For example, if $\theta(3) = 4$, the expansion of $\{\!|U_3, I|\!\}$ would be $\{\!|U_3, I|\!\}^4$. However, some actions are measured against the global clock. For example, the exact moment that a memory access occurs must be measured in global time, and could therefore occur in the middle of a local processor cycle. The action must be recorded in global time, so the expansion consists of an appropriately chosen prefix denoting waiting and a suffix of internal actions that compensate for the unexpended clock duration. These are defined below in the *Exp* function. Before that we introduce some useful notations.

Definition 1 *Given a bag of observables o, Rn(o) gives the bag of all observables of the form $J_{\rightarrow p}$ contained in o. More formally,*

$$Rn(o) = \{\!| J_{\rightarrow p} \mid p \in entries \wedge J_{\rightarrow p} \in o |\!\}$$

Definition 2 *Given two bags of observables o and x, we say that an observable α transitions from o to x if and only if there exists an $\bar{\alpha}$ such that $\alpha \in o$ is of the form $R_{\leftarrow p}$, and $\bar{\alpha} \in x$ is of the form $J_{\leftarrow p}$, or $\alpha \in o$ is of the form $J_{\leftarrow p}$, and $\bar{\alpha} \in x$ is of the form $L_{\leftarrow p}$. We write the same as $Trans(\alpha, o, x)$.*

If $Trans(\alpha, o, x)$ holds, then it tells us that the task has moved from a state of being suspended for rendezvous to rendezvous on an entry, or from the state of being in rendezvous on an entry to the end of that rendezvous.

Definition 3 *Given two bags of observables o and x, we say that an observable $\alpha \in \widehat{O}_L$ continues from o to x if and only if $\alpha \in o$ and $\alpha \in x$. We write the same as $Cont(\alpha, o, x)$.*

If $Cont(\alpha, o, x)$ holds for two consecutive observations o and x, then the task continues either: the delay, the suspension for rendezvous via an entry call, the suspension while in rendezvous during an entry call, or the suspension for rendezvous at at the accept of an entry.

We now define a function *LgMap* which maps an executable observation made on processor p to a set of sequences of observations on the global time scale.[3]

Definition 4

$$\begin{array}{lcl}
LgMap(I, p) & = & \{\{\!|I, U_p|\!\}^{\theta(p)}\} \\
LgMap(W_y, p) & = & \{\{\!|W_y, U_p|\!\}^{\theta(p)}\} \\
LgMap(A_{y\rightarrow c}, p) & = & \{\{\!|W_y, U_p|\!\}^i \{\!|A_{y\rightarrow c}, U_p|\!\}\{\!|I, U_p|\!\}^j \mid i + j + 1 = \theta(p)\} \\
LgMap(A_{c\rightarrow y}, p) & = & \{\{\!|W_y, U_p|\!\}^i \{\!|A_{c\rightarrow y}, U_p|\!\}\{\!|I, U_p|\!\}^j \mid i + j + 1 = \theta(p)\}
\end{array}$$

[3] We note that p is used both for designating a processor and also for an entry. The explicit context of their use resolves the ambiguity for us.

Scheduling is formally handled by the recursive function *Sch*, which takes three arguments: h, the history to be scheduled; o, the previous action; and p, the processor on which o was executed.

$$Sch(h,o,p) = \begin{cases} \{\epsilon\} & \text{if } h=\epsilon \\ \bigcup_{\substack{1\le p'\le \rho \\ h_2 \in Sch(h[2..],h[1],p') \\ h_1\in Exp(h[1],p')}} Sch_{12}(h[1],h_1,h_2,o,p,p') & \text{otherwise.} \end{cases}$$

where

$$Sch_{12}(x,h_1,h_2,o,p,p') = \bigcup \left\{ \begin{array}{l} \{h_1h_2\} \\ \qquad \text{if } o\cap O_E = \{\!|\,|\!\} \Rightarrow p=p' \\ \{\{\!|Q|\!\}^i h_1h_2 \mid i\ge 1\} \\ \qquad \text{if } Intr(x,o)\wedge \neg Inrendz(x,o) \\ \{(\{\!|Q|\!\}\cup Rn(x))^i h_1h_2 \mid i\ge 1\} \\ \qquad \text{if } Intr(x,o)\wedge Inrendz(x,o) \end{array} \right\}$$

and

$$\begin{array}{rcl} Intr(x,o) & = & \neg(W_y\in o \vee Trans(\alpha,o,x)\vee Cont(\alpha,o,x)) \\ Inrendz(x,o) & = & (J_{\rightarrow q}\in o\wedge J_{\rightarrow q}\in x)\vee(J_{\rightarrow q}\in o\wedge L_{\rightarrow q}\in x) \\ Exp(x,p) & = & \left\{ \begin{array}{ll} \{Cmb(\tau, Rn(x)^{\theta(p)}) \mid \tau\in LgMap(\alpha,p)\} \\ \qquad \text{if } \alpha\in O_E\cap x\wedge L_{\rightarrow q}\notin x \\ \{Cmb(\tau, Rn(x)^{\theta(p)}, \{\!|J_{\rightarrow q}|\!\}^{\theta(p)-1}\{\!|L_{\rightarrow p}|\!\}) \mid \tau\in LgMap(\alpha,p)\} \\ \qquad \text{if } \alpha\in O_E\cap x\wedge L_{\rightarrow q}\in x \\ x \qquad \text{otherwise.} \end{array}\right. \end{array}$$

Intr defines those actions which can be interrupted. These exclude the sequence of initiated waiting actions for access to shared memory and the access thereof, and the sequence of language specific suspended actions. *Inrendz* records the occurrence of two consecutive actions in which a rendezvous $J_{\rightarrow q}$ at entry q is simultaneously observed.

The clauses in the *Sch* function can be explained as follows. The first clause says that if the history h to be scheduled is empty, then there is nothing to schedule. The second clause is a bit involved. In order to schedule a history h, we the first element of the history appropriately, and in the context of the processor allocated for that schedule the rest of the history. The predicate $h_1 \in Exp(h[1],p')$ captures a history from the set generated by the *LgMap* function for the first component of h. The predicate $h_2 \in Sch(h[2..],h[1],p')$ captures a history from the scheduling done for the rest of the sequence h. These schedules are formed by assuming arbitrary processor allocation. The function Sch_{12} then performs the appropriate scheduling of the first element —for which processor p' was allocated— in the context of the allocation of processor p for the previous action o.

The Sch_{12} function returns a set of histories. Essentially it appropriately interleaves the Q elements and rejects any inconsistent assignment of processors. The clauses in the Sch_{12} function can be explained as follows. The first clause gives the task behavior when the scheduler does not intervene at the current observation or intervenes to execute the suspended task. The second clause records that, while not in rendezvous, the task can unconditionally wait on the scheduler provided its current action can be interrupted. The third clause records that while in rendezvous the task can unconditionally wait on the scheduler, denoted by $(\{\!|Q|\!\} \cup Rn(x))^i$, when its current action can be interrupted.

Exp performs the necessary expansion of global actions.

Definition 5 *A history h is* **legal** *if for all* $1 \leq i \leq |h|$ *exactly one of the following is true: either 1)* $h[i] \cap \hat{O} \neq \{\!|\,|\!\}$, *or 2)* $U_p \in h[i]$ *for some unique p,* $1 \leq p \leq \rho$.

Lemma 6 *Let* $\langle s', h\rangle = \mathcal{M}[\![S]\!]s$. *If* $h' \in Sch(h, o, p)$, *then h' is legal.*

For a history to be legàl, no processor must be allocated to observations involving suspension, and that every observation involving an executable observable must have a processor assignment. The lemma says that the scheduling function guarantees that every history processed by it shall be legal.

Formally, parallel composition is handled in much the same way as it was in the previous section. The only major change is the use of the scheduling function *Sch* to construct the meaning function $\mathcal{M}[\![t_i :: S_i]\!]s$. We also assume that for each S_i, $Init(S_i)$ is the processor on which task $t_i :: S_i$ will initially run. It is possible that $Init(S_i) = Init(S_j)$ for $i \neq j$, in which case tasks $t_i :: S_i$ and $t_j :: S_j$ will compete for the processor. However, we insist that *Init* fully utilize the available processors, *i.e.* it cannot map two tasks to the same processor unless there is at least one task mapped to every processor. The function *Init* can be viewed as a part of the program (although it depends on ρ). *Init* was not required in the previous section since all processors were identical.

$$\mathcal{M}[\![t_i : S_i]\!] = Cls\{\langle s_i, h_i\rangle \mid \langle s_i, h\rangle \in \mathcal{M}[\![S_i]\!]s \wedge h_i \in Sch(h, \{\!|\,|\!\}, Init(S_i))\}$$

$$\mathcal{M}[\![\ \|_{i=1}^{n} t_i :: S_i]\!]s \quad = \quad Cls\{Merge(s, \bigcup_{i=1}^{n}\{\langle s_i \mid \mathcal{V}_L, h_i\rangle\}) \mid \langle s_i, h_i\rangle \in \mathcal{M}[\![t_i :: S_i]\!]s))\},$$

where

$$\begin{aligned}
Merge(s, D) &= \textbf{if } Comp(D) \wedge Cons(s, D) \textbf{ then } \langle SM(s, D), h_D\rangle \textbf{ else } \langle \bot, \epsilon\rangle \\
Comp(D) &= \forall \langle s, h\rangle \in D(s = \bot \Rightarrow |h| = |h_D|) \\
Cons(s, D) &= Cons_A(s, D) \wedge MaxTPut(D) \wedge Rendezvous(h_D) \wedge Sync(h_D) \\
SM(s, D) &= SM_A(D)
\end{aligned}$$

The compatibility predicate *Comp* is unchanged. The consistency predicate *Cons* plays the same role as before, although the details have changed.

$$\begin{aligned}
MaxTPut(h) &= \forall i((Q \in h[i] \Rightarrow Usage(h[i]) = \rho) \wedge \forall j(h[i](U_j) \leq 1)) \\
Usage(o) &= \textstyle\sum_{j=1}^{\rho} h[i](U_j) \\
Rendezvous(h) &= \forall i, p((R_{\leftarrow p} \in h[i] \Rightarrow h[i](R_{\rightarrow p}) = 0) \wedge \\
&\qquad (J_{\leftarrow p} \in h[i] \Leftrightarrow J_{\rightarrow p} \in h[i]) \wedge (h[i](J_{\leftarrow p}) \leq 1)) \\
Sync(h) &= \forall i, p(L_{\leftarrow p} \in h[i] \Leftrightarrow L_{\rightarrow p} \in h[i]) \wedge (h[i](L_{\leftarrow p}) \leq 1).
\end{aligned}$$

To guarantee consistency, three additional properties must be satisfied. First, processors must execute at most one task at a time, and tasks must not wait for processors unless all processors are in use. *MaxTPut* ensures that no processor remains idle unless necessary and it is independent of any language specific observable. *Rendezvous* ensures that at most one caller is in rendezvous with the acceptor, prohibits unnecessary waiting for rendezvous, and ensures that the caller is suspended for the duration of the rendezvous and only that. *Sync* ensures a two party synchronization of the acceptor and the caller at the end of the rendezvous.

Theorem 7 *The limited parallelism model guarantees that task switching occurs only at processor cycle boundaries.*
Proof. Omitted (See [10]). □

Theorem 7 signifies the following. A nondeterministic scheduler behaves like an *oracle* performing task switching only at processor cycle boundaries. At each processor cycle boundary, provided that the processor is not executing an uninterruptable action, the nondeterministic scheduler will schedule on that processor any of the runnable tasks. Uninterruptability of actions plays only a local role and that the parallel composition is free of any such notions.

5 A Simple Priority Model

The parallel composition of the previous section interleaves computations subject to maximal thoughput. The scheduling function *Sch* and the *MaxTPut* predicate together ensure that each processor cycle time is honored without interruption. The semantic model does not impose priorities —though task priorities are regarded as distinct shared variables from $\mathcal{V}_T$.

In this section, we will impose a simple priority model. The task specific shared variables $y_i \in \mathcal{V}_T$ denote the base priorities of tasks. We will require that a low priority task may not execute in preference to a runnable higher priority task, unless that task's action(s) cannot be interrupted. Actions that cannot be interrupted arise from two fronts. First access to shared memory —including waiting for such access— cannot be interrupted. Second an executing task cannot be interrupted during a processor cycle. That means any task preemption must occur only on processor cycle boundaries. Operationally then a high priority task must necessarily wait to be scheduled if it becomes runnable during the processor cycle of a low priority task. The waiting must terminate at the cycle boundary of the executing low priority task.

To accommodate the limited parallelism semantics, the a priori semantics of tasks forced task suspension for a processor with the introduction of Q elements. In the simple priority model, the a priori semantics of tasks cannot make any *unilateral* assumption about a task's suspension or otherwise for a processor. This is because of the asynchronous nature of dynamic priorities —task base priority is a shared variable that can be assigned by any task. At each time point the base priority of a task will only be known after parallel composition. Thus the simple priority model can only be imposed on top of a limited parallelism semantics of the previous section. But unfortunately the parallel composition above is cast on the global time scale and needs slight adjustments —but semantically significant— before a simple priority model can be imposed.

Adjustments to the limited parallelism semantics above are necessary for the following reasons: (1) The observables of the previous section are not discriminating enough. Any association that a task's a priori semantics can establish with the task specific shared variables from $\mathcal{V}_T$ is lost in the parallel composition. (2) The $LgMap$ function replicates internal actions I for the global duration of the processor's local clock cycle, and appends every shared memory access with a tail sequence of I elements (so that the processor cycle is expended). As a result on the global time scale of the parallel composition, we cannot infer the cycle boundaries of processors. (3) The observables of the previous section are not sufficient to infer the base priorities of executing actions that cannot be interrupted. This becomes relevant for imposing that any executing action that can be interrupted will have a base priority equal or higher than a task suspended for a processor.

We can rectify these concerns as follows: (1) Making the nonexecutable observables $\widehat{O}$ more discriminatory and task specific —actually specific to base priority variables $y_i \in \mathcal{V}_T$— will be sufficient to discriminate the base priorities of nonexecuting tasks from the executing ones. Therefore we introduce a revised set of nonexecutable observables

$$\widehat{O}_S = \{Z^{y_i}, R^{y_i}_{\leftarrow p}, R^{y_i}_{\rightarrow p}, Q^{y_i}, J^{y_i}_{\leftarrow p}, L^{y_i}_{\leftarrow p} \mid p \in entries \wedge y_i \in \mathcal{V}_T\}$$

and similarly the language specific nonexecutable observables will be

$$\widehat{O}_{S_L} = \{Z^{y_i}, R^{y_i}_{\leftarrow p}, R^{y_i}_{\rightarrow p}, J^{y_i}_{\leftarrow p}, L^{y_i}_{\leftarrow p} \mid p \in entries \wedge y_i \in \mathcal{V}_T\}$$

(2) On the global scale, distinguishing the start of an internal action and its continuation over a processor cycle is necessary. Thus the observable I denotes the start of the initial action, and a new companion observable $\vec{I}$ designates the continuation of I. In $LgMap$ function instead of replicating I elements, we distinguish the start of an internal action with the element I and its continuation over the processor cycle with the new element $\vec{I}$. Similarly instead of appending a sequence of I elements in $LgMap$ function, we append a sequence of $\vec{I}$ elements. (3) In our framework, the observables W_y and $\vec{I}$ are sufficient to identify actions which cannot be interrupted. Thus we require that W_y and $\vec{I}$ be observed relative to the base priority variables of the tasks.

With this change, the meaning function $\mathcal{M}[\![t_i :: S_i]\!]s$ of tasks requires revision. This is accomplished as follows: We first define a *Rename* function.

$$Rename(h, y_i) = \begin{cases} \epsilon & \text{if } h = \epsilon \\ Renamebag(h[1], y_i).Rename(h[2..], y_i) & \text{otherwise.} \end{cases}$$

$$\begin{aligned} Renamebag(B, y_i) = \; & \{\!|\alpha \mid \alpha \in B \wedge \alpha \in (O_E - \{W_y\})|\!\} \cup \\ & \{\!|\beta^{y_i} \mid \beta \in B \wedge (\beta \in \widehat{O} \vee \beta = W_y)|\!\} \end{aligned}$$

Next we revise $LgMap$ function to $LgMap_R$. This makes the distinction between the start of the internal action and its continuation over the processor cycle. Similarly all compensating internal actions are recorded as continuing actions over the processor cycle.

Definition 8

$$\begin{aligned} LgMap_R(I, p, y_i) &= \{\{\!|I, U_p|\!\}\{\!|\vec{I}^{y_i}, U_p|\!\}^{\theta(p)-1}\} \\ LgMap_R(W^{y_i}_y, p, -) &= \{\{\!|W^{y_i}_y, U_p|\!\}^{\theta(p)}\} \\ LgMap_R(A_{y\rightarrow c}, p, y_i) &= \{\{\!|W^{y_i}_y, U_p|\!\}^i\{\!|A_{y\rightarrow c}, U_p|\!\}\{\!|\vec{I}^{y_i}, U_p|\!\}^j \mid i+j+1 = \theta(p)\} \\ LgMap_R(A_{c\rightarrow y}, p, y_i) &= \{\{\!|W^{y_i}_y, U_p|\!\}^i\{\!|A_{c\rightarrow y}, U_p|\!\}\{\!|\vec{I}^{y_i}, U_p|\!\}^j \mid i+j+1 = \theta(p)\} \end{aligned}$$

where p denotes a processor.

Note that $LgMap_R$ takes an additional argument which is the base priority variable of the task, and generates appropriately the continuation actions $\vec{I}$ relative to the base priority of the task. The function Sch is then revised to Sch_R. Sch_R is the same as Sch except that in place of $LgMap$ it uses $LgMap_R$ with the appropriately chosen set of observables . The revised meaning $\mathcal{M}_R[\![t_i :: S_i]\!]s$ of $\mathcal{M}[\![t_i :: S_i]\!]s$ will then be:

$$\mathcal{M}_R[\![t_i : S_i]\!]s = Cls\{\langle s_i, h_i\rangle \mid \langle s_i, h\rangle \in \mathcal{M}[\![S_i]\!]s \wedge h_i \in Sch_R(Rename(h, y_i), \{\!|\,|\!\}, Init(S_i))\}$$

We revise the definition of a legal history by taking $\widehat{O}_S$ in place of $\widehat{O}$. This revised definition is implied henceforth.

Some changes occur in the meaning function for parallel composition. First we need to take $\mathcal{M}_R[\![t_i :: S_i]\!]s$ in place of $\mathcal{M}[\![t_i :: S_i]\!]s$ in the parallel composition. Second we need to take into account renamed observables in the verification conditions. All other things will remain the same except the predicate $Cons$. In $Cons(s, D)$, the definitions of $Cons_A(s, D)$, $MaxTPut(h_D)$, $Rendezvous(h_D)$, and $Sync(h_D)$ are exactly same as above except that the superscripts on observables from $\widehat{O}_S$ are ignored. Similarly the superscripts on W_y are also ignored. With this adjustment the meaning function for parallel composition shall remain the same as in the previous section except that task specific superscripts appear on the nonexecutable observables and on $\vec{I}$ and W_y. Call the meaning function at this stage as $\mathcal{M}_R[\![\, \|_{i=1}^{n} t_i :: S_i]\!]s$.

We now define a predicate $Dyn(s, h)$ for imposing the simple priority model.

Definition 9 *Given a history h and an instant* $i > 1$, $UnInt(i, h)$

$$UnInt(i, h) = \{\!|y_j \mid \vec{I}^{y_j} \in h[i]|\!\} \cup \{\!|y_j \mid W_y^{y_j} \in h[i-1]|\!\}$$

gives the bag of the base priority variables of uninterruptable *observables at i.*

If $W_y^{y_j} \in h[i-1]$, then the task with that base priority variable y_j cannot be interrupted at the instant i —either it continues to wait or accesses the shared memory at the instant i.

Relative to an initial state s, and a history h, classify, based on attributes, the bag of base priorities of tasks at instant i as follows.

$Base(i, s, h)$	::	All tasks.
$Base_{SP}(i, s, h)$	::	Suspended tasks.
$Base_Q(i, s, h)$	::	Tasks suspended for a processor.
$Base_E(i, s, h)$	::	Executing Tasks.
$Base_{E_l}(i, s, h)$	::	Executing lower priority tasks
$Base_U(i, s, h)$	::	Uninterruptable tasks.

Definition 10

$$\begin{aligned}
Base(i, s, h) &= \{\!|c \mid LastWr(y_j, s, h[1..i], c) \wedge y_j \in \mathcal{V}_T|\!\}\\
Base_{SP}(i, s, h) &= \{\!|c \mid LastWr(y_j, s, h[1..i], c) \wedge \alpha^{y_j} \in h[i] \cap \widehat{O}_S|\!\}\\
Base_Q(i, s, h) &= \{\!|c \mid LastWr(y_j, s, h[1..i], c) \wedge Q^{y_j} \in h[i]|\!\}\\
Base_E(i, s, h) &= Base(i, s, h) - Base_{SP}(i, s, h)\\
Base_{E_l}(i, s, h) &= \{\!|c \mid c \in Base_E(i, s, h) \wedge c < max(Base_Q(i, s, h))|\!\}\\
Base_U(i, s, h) &= \{\!|c \mid LastWr(y_j, s, h[1..i], c) \wedge y_j \in UnInt(i, h)|\!\}
\end{aligned}$$

where y_j is the shared variable designating the base priority of task t_j. Recall that $LastWr(y_j, s, h[1..i], c)$ holds if the applicable value of the shared variable y_j at the instant i is c.

Definition 11 *The predicate $Dyn(s,h)$ is given by*

$$Dyn(s,h) \quad = \quad Base_{E_t}(1,s,h) = \{\!\!\}\ \wedge \forall i > 1(Base_{E_t}(i,s,h) \subseteq Base_U(i,s,h))$$

$Dyn(s,h)$ is true if and only if initially no low priority task executes in preference to a high priority task, and at any other instant all the low priority tasks executing in preference to a high priority task must be *uninterruptable*.

Now we can impose the simple priority model on top of the adjusted parallel composition of the previous section.

$$\mathcal{M}_S[\![\, \|_{i=1}^n t_i :: S_i]\!] s = \{\langle s', h\rangle \mid \langle s', h\rangle \in \mathcal{M}_R[\![\, \|_{i=1}^n t_i :: S_i]\!] s \wedge Dyn(s,h)\}$$

Theorem 12

$$\mathcal{M}_S^{\natural}[\![\, \|_{i=1}^n t_i :: S_i]\!] s \subseteq \mathcal{M}[\![\, \|_{i=1}^n t_i :: S_i]\!] s$$

where $\mathcal{M}_S^{\natural}$ drops the task specific shared variables and the top vectors on $\vec{I}$ from the histories in $\mathcal{M}_S$.

Theorem 13 *The parallel composition under the simple priority model guarantees that task preemption occurs on processor cycle boundaries.*

Theorem 12 has the consequence that any property which is provable under the nondeterministic scheduler holds when the scheduler observes the simple priority model. By theorem 13, task switching occurs on processor cycle boundaries. It means that processors are always left in a consistent state. When task preemption occurs, uninterruptable actions are not only locally significant but also play a global role with $Base_U(i,s,h)$ in parallel composition.

6 A Simple Priority^{+83} Model

In the above simple priority model, the priority at which a task executes is determined directly by the base priority of that task. In Ada 9X revision, upward compatibility of the language with Ada 83 is maintained— Ada 83 imposes additional requirements on the priority with which an acceptor task executes in a rendezvous with a caller task.

This kind of priority model is distinct from the simple priority model—semantically they are incomparable—and hence the name *a simple priority^{+83} model.* To impose the model, we need an auxiliary concept known as the *active priority* of a task. Informally the active priority of a task is its base priority if the task is not an acceptor in rendezvous; otherwise it is the maximum of the active priority of the caller task (with respect to which the rendezvous is taking place) and of itself. We will later give a formal definition of the active priority of a task that will also take into account rendezvous nesting to arbitrary depth.

In defining the simple priority model, it was sufficient to observe the nonexecutable alphabet, W_y, and $\vec{I}$ relative to the task specific base priority variables. This is not sufficient for imposing the simple *priority^{+83}* model because it does not distinguish an acceptor task's execution priority while in rendezvous. We rectify this by making the auxiliary observables $\{J_{\rightarrow p}, L_{\rightarrow p}\}$ task specific. That is we define a new set of auxiliary observables

$$\{J^{y_i}_{\rightarrow p}, L^{y_i}_{\rightarrow p} \mid y_i \in \mathcal{V}_T \wedge p \in entries\}$$

Accordingly, the meaning function $\mathcal{M}_R[t_i :: S_i]s$ of tasks should be revised such that histories in $\mathcal{M}_R[\![t_i :: S_i]\!]s$ will reflect that for all $p \in entries$, $J^{y_i}_{\rightarrow p}$ and $L^{y_i}_{\rightarrow p}$ appear in place of $J_{\rightarrow p}$ and $L_{\rightarrow p}$ respectively. We will assume this revised meaning function when we refer to $\mathcal{M}_R[t_i :: S_i]s$.

The scheduling rules will now say that a low active priority task may not execute in preference to a high active priority task unless that task is engaged in an *uninterruptable* action.

To construct the parallel composition, we start from a meaning function $\mathcal{M}_R[\![\ \|_{i=1}^{n} t_i :: S_i]s$ which takes the extended observables into account. On top of that we impose the predicate $Dyn_{83}(s, h)$ to cater to the simple $priority^{+83}$ model.

The predicate Dyn_{83} makes use of the formally defined notion of active priorities of tasks and is constructed as follows.

Definition 14 *Given a history h and an instant i, $1 \leq i \leq |h|$, we define an irreflexive* depends *relation, notated $\longrightarrow$, at $h[i]$ as follows:*

If $J^{y_i}_{\rightarrow p} \in h[i]$, then

$$\begin{array}{lll} J^{y_i}_{\rightarrow p} \longrightarrow J^{y_j}_{\rightarrow q} & \textit{if } J^{y_j}_{\rightarrow q}, J^{y_j}_{\leftarrow p} \in h[i], \\ J^{y_i}_{\rightarrow p} \longrightarrow J^{y_j}_{\leftarrow p} & \textit{otherwise.} \end{array}$$

If $L^{y_i}_{\rightarrow p} \in h[i]$, then

$$\begin{array}{lll} J^{y_i}_{\rightarrow p} \longrightarrow L^{y_j}_{\rightarrow q} & \textit{if } J^{y_j}_{\rightarrow q}, L^{y_j}_{\leftarrow p} \in h[i], \\ L^{y_i}_{\rightarrow p} \longrightarrow L^{y_j}_{\leftarrow p} & \textit{otherwise.} \end{array}$$

Example. Assume that at some time instant task t_1 is in rendezvous with the task t_2 which in turn is in rendezvous with task t_3. Then informally $t_2 \longrightarrow t_3$ and $t_1 \longrightarrow t_2$ holds.

Definition 15 *For any history h and an instant $1 \leq i \leq |h|$, the longest derivation $\xrightarrow{+}$ of the dependence relation $\longrightarrow$ is called a* rendezvous dependence trace($rdt(i, h)$) *of h at i.*

Example. From the previous example, it is clear that $t_1 \xrightarrow{+} t_3$ holds.
We note that for a given history h and an instant i, there can be several rendezvous dependence traces at $h[i]$.

Definition 16 $Rdt(i, h) = \{\tau \mid \tau \text{ is a } rdt(i, h)\}$

$Rdt(i, h)$ gives the set of all rendezvous dependence traces of h at i and every member of it is always finite.

Definition 17 *Let $\tau \in Rdt(i, h)$ be of the form $\alpha^{y_k} \longrightarrow \ldots \longrightarrow \alpha^{y_n}$. Then $Head(\tau) = \alpha^{y_k}$.*

Given a rendezvous dependence trace τ, $Head(\tau)$ gives the first element in the trace τ.

Definition 18 *Given a history h and an instant i,* $1 \leq i \leq |h|$, *define the set of all base priority variables of a rendezvous dependence trace* τ *at i, notated* $Pvar(i, h, \tau)$, *as* $Pvar(i, h, \tau) = \{y_1, y_2 \ldots y_n\}$ *where* $\tau = \alpha_n \xrightarrow{+} \alpha_1$, *and* $y_j \in \mathcal{V}_T (1 \leq j \leq n)$ *are the base priority variables on the observables* $\alpha_j (1 \leq j \leq n)$.

We can now define the active priority of an acceptor task which is in rendezvous as the maximum of the base priorities of tasks which constitute the rendezvous dependence trace of that task.

Definition 19 *Relative to an initial state s, a history h, an* $i \geq 1$, *and* $\tau \in Rdt(i, h)$, *the active priority of trace* τ, *notated* $ActP(i, s, h, \tau)$,

$$ActP(i, s, h, \tau) = max\{c_j \mid LastWr(y_j, s, h[1..i], c_j) \wedge y_j \in Pvar(i, h, \tau)\}$$

$ActP(i, s, h, \tau)$ gives the applicable active priority of the observable at the head of τ.

Relative to an initial state s, and a history h, classify based on attributes the bag of active priorities of tasks at instant i as follows.

$Active(i, s, h)$	::	All tasks.
$Active_{SP}(i, s, h)$	::	Suspended tasks.
$Active_Q(i, s, h)$	::	Tasks suspended for a processor.
$Active_E(i, s, h)$	::	Executing Tasks.
$Active_{E_l}(i, s, h)$	::	Executing low active priority tasks
$Active_U(i, s, h)$	::	Uninterruptable tasks.

Before we define the active priorities, we will introduce some short hand notation for specific bags of interest.

Definition 20

$$A_1 = \{\!| c \mid c = ActP(i, s, h, \tau) \wedge \tau \in Rdt(i, h) |\!\}$$

gives the bag of active priorities of tasks which are in rendezvous.

$$A_2 = \{\!| c \mid LastWr(y_j, s, h[1..i], c) \wedge y_j \notin \bigcup_{\tau \in Rdt(i,h)} Pvar(i, h, \tau) |\!\}$$

gives the bag of active priorities of tasks which are not in rendezvous.

$$A_3 = \{\!| c \mid c = ActP(i, s, h, \tau) \wedge Head(\tau) = \alpha^{y_j} \wedge \beta^{y_j} \in h[i] \cap \widehat{O}_S \wedge \tau \in Rdt(i, h) |\!\}$$

gives the bag of active priorities of tasks in rendezvous but are suspended.

$$A_4 = \{\!| c \mid LastWr(y_j, s, h[1..i], c) \wedge y_j \notin \bigcup_{\tau \in Rdt(i,h)} Pvar(i, h, \tau) \wedge \beta^{y_j} \in h[i] \cap \widehat{O}_S |\!\}$$

gives the bag of active priorities of tasks not in rendezvous but are suspended.

$$A_5 = \{\!| c \mid c = ActP(i, s, h, \tau) \wedge Head(\tau) = \alpha^{y_j} \wedge Q^{y_j} \in h[i] \wedge \tau \in Rdt(i, h) |\!\}$$

gives the bag of active priorities of tasks in rendezvous but are suspended for a processor.

$$A_6 = \{\!| c \mid LastWr(y_j, s, h[1..i], c) \wedge y_j \notin \bigcup_{\tau \in Rdt(i,h)} Pvar(i, h, \tau) \wedge Q^{y_j} \in h[i] |\!\}$$

gives the bag of active priorities of tasks not in rendezvous but are suspended for a processor.

$$A_7 = \{\!|c \mid c = ActP(i,s,h,\tau) \wedge \tau \in Rdt(i,h) \wedge Head(\tau) = \alpha^{y_j} \wedge y_j \in UnInt(i,h)|\!\}$$

gives the bag of active priorities of tasks which are in rendezvous and cannot be interrupted.

$$A_8 = \{\!|c \mid LastWr(y_j, s, h[1..i], c) \wedge y_j \in UnInt(i,h) \wedge y_j \notin \bigcup_{\tau \in Rdt(i,h)} Pvar(i,h,\tau)|\!\}$$

gives the bag of active priorities of tasks which are not in rendezvous and cannot be interrupted.

Definition 21

$$\begin{array}{lcl} Active(i,s,h) & = & A_1 \cup A_2 \\ Active_{SP}(i,s,h) & = & A_3 \cup A_4 \\ Active_Q(i,s,h) & = & A_5 \cup A_6 \\ Active_E(i,s,h) & = & Active(i,s,h) - Active_{SP}(i,s,h) \\ Active_{E_\ell}(i,s,h) & = & \{\!|c \mid c \in Active_E(i,s,h) \wedge c < max(Active_Q(i,s,h))|\!\} \\ Active_U(i,s,h) & = & A_7 \cup A_8 \end{array}$$

Definition 22 *The predicate $Dyn_{83}(s,h)$ is given by*

$$Dyn_{83}(s,h) \quad = \quad Active_{E_\ell}(1,s,h) = \{\!|\,|\!\} \wedge \forall i > 1 (Active_{E_\ell}(i,s,h) \subseteq Active_U(i,s,h))$$

$Dyn_{83}(s,h)$ is true if and only if initially no low active priority task executes in preference to a high priority task, and at any other instant all the low active priority tasks executing in preference to a high active priority task must be *uninterruptable* at that instant.

Now we can impose the simple priority model on top of the adjusted parallel composition of the previous section.

$$\mathcal{M}_{83}[\![\, \|_{i=1}^n t_i :: S_i]\!] s = \{\langle s', h\rangle \mid \langle s', h\rangle \in \mathcal{M}_R[\![\, \|_{i=1}^n t_i :: S_i]\!] s \wedge Dyn_{83}(s,h)\}$$

Theorem 23 $\quad \mathcal{M}^{\natural}_{83}[\![\, \|_{i=1}^n t_i :: S_i]\!] s \subseteq \mathcal{M}[\![\, \|_{i=1}^n t_i :: S_i]\!] s$

Theorem 24 *The parallel composition under the simple priority^{+83} model guarantees that task preemption occurs on processor cycle boundaries.*

Theorem 25 *If $\rho = n$, then no Q elements will appear in $\mathcal{M}[\![\, \|_{i=1}^n t_i :: S_i]\!] s$.*

Theorems 23 and 24 signify the same things as those of theorems 12 and 13 respectively. The difference is in the semantic model. Theorem 25 implies that no executing task shall be preempted. However it does not imply that a task shall execute only on its initially assigned processor. It is possible that a task may be suspended as a result of language specific actions and when it becomes executable thereafter a different processor may be assigned to that task from that prior to suspension. On the other hand if the processors are also synchronous, then our limited parallelism model reduces to maximal parallelism when the U_i elements are ignored.

7 Discussion

Semantic models for real-time concurrency, like the maximal parallelism models of [7] and the limited parallelism model [13, 12], introduce the time-metric on the operational basis that processors do not idle. The operational basis leads to the notion of prohibition of unnecessary waiting of tasks for synchronization. In limited parallelism models, maximal throughput enforces that tasks do not unnecessarily wait on the scheduler. This paper brings out that tasks have to necessarily wait (for a bounded duration) on the scheduler because processors cannot be preempted during a clock cycle—an aspect relevant when modeling task preemption and dynamic priorities[4]. Our main result is that in the presence of local clocks, it is necessary to distinguishingly observe processor actions as initiating actions and their respective continuation over the processor cycle. Even in the case of multi-processor systems in which processors can only be interrupted on instruction boundaries, observing task preemption requires observing instruction boundaries, unless all the instructions are of equal duration. For such architectures similar techniques as developed here are applicable. We further should note that on a uniprocessor system, one need not globally observe the clock or instruction boundaries as the case may be. Apart from that the observables must still be discriminating enough to impose the priority models.

A **run-until-blocked**(rub) scheduling paradigm can be imposed on top of the simple priority model and the simple *priority*$^{+83}$ model. We can then talk of a hierarchy of models as follows.

$$\mathcal{M}[\![\, \|_{i=1}^{n}\, t_i :: S_i]\!]s \supseteq \begin{array}{l} \mathcal{M}^{\natural}_{S}[\![\, \|_{i=1}^{n}\, t_i :: S_i]\!]s \supseteq \mathcal{M}^{\natural}_{S_{rub}}[\![\, \|_{i=1}^{n}\, t_i :: S_i]\!]s \\ \mathcal{M}^{\natural}_{83}[\![\, \|_{i=1}^{n}\, t_i :: S_i]\!]s \supseteq \mathcal{M}^{\natural}_{83_{rub}}[\![\, \|_{i=1}^{n}\, t_i :: S_i]\!]s \end{array}$$

where the subscript *rub* designates the run-until-blocked model of the appropriate kind.

Reasoning about priorities in the simple priority model has the same complexity as that of reasoning about shared variables. However in practice shared variables are always enclosed in atomic actions which reduces the complexity. On the other hand, no discipline is prescribed for manipulating task specific base priority variables. In the simple *priority*$^{+83}$ model, we have to reason about the set of rendezvous dependent traces for each time point of the program. This necessarily complicates the design of proof systems for Ada's rendezvous, unlike those considered in [4].

Relation to Other Work

Hooman[6] in a contemporary work introduces the denotational semantics of a small language fragment addressing dynamic priorities. In this work tasks can set their own priorities only and thus the primitives used for setting priorities are not fully asynchronous. In addition, the environment considered is that of a uniprocessor system. That is tasks can only execute on the processor on which they are assigned unlike our execution environment. We must note however that in its essentials Hooman's model will be a very restricted version of our model when $\rho = 1$. See Roscoe and Reed[11], [9], [12], and [5] for other approaches and models for real-time concurrency.

[4] When tasks obey static priority schemes and the language specifies that task waiting at synchronization or delay points be regarded as suspension, task preemption can arise.

Acknowledgements

The author wishes to gratefully acknowledge the criticisms and comments on the drafts of this paper by Eric Shade, David Guaspari, Fred Schneider, Wolf Polak, and Mark Saaltnik.

References

[1] "The Programming Language Ada Reference Manual," *Lecture Notes in Comput. Sci.* **155**, Springer-Verlag, Berlin (1983).

[2] "Ada 9X Mapping Document: Mapping Specification," *Draft Ada 9X Project Report*, Office of the Undersecretary of Defense Acquisition, Washington DC (Aug 1991).

[3] Francez, N., Lehman, D., and Pnueli, A.: "A Linear-history Semantics for Languages for Distributed Programming," *Theoret. Comput. Sci.* **32**, 25—46 (1984).

[4] Gerth,R.T., de Reover, W.P.: "A Proof System for Concurrent Ada Programs," *Science of Computer Programming,*, pp. 159-204 (1984).

[5] Henzinger,T.A.: "Temporal Specification and Verification of Real-Time Systems," *Ph. D Thesis*, Stanford University (1991).

[6] Hooman,J.:"A Denotational Real-Time Semantics for Shared Processors," *Lecture Notes in Comp. Sci.*, **506**, pp. 184—201 (1991).

[7] Huizing, C., Gerth R., de Roever, W.P.: "Full Abstraction of a Real-Time Denotational Semantics for an OCCAM-like Language," *Proc. 14th ACM Symposium on Principles of Programming Languages*, 223—238 (1987).

[8] Koymans, R., Shyamasundar, R.K., de Roever, W.P., Gerth, R., and Arun-Kumar, S.: "Compositional Denotational Semantics for Real-time Distributed Computing," *Information and Control* **79** (3), 210—256 (1988).

[9] Milner, R.: "Calculi for Synchrony and Asynchrony," *Theoretical Computer Science* **25**, 267—310 (1983).

[10] Narayana,K.T.: "Observing Task Preemption in Ada 9X," *ORA Research Report* (June 1991).

[11] Reed, G.M., and Roscoe, A.W.: "A Timed Model for Communicating Sequential Processes," *Lecture Notes in Comput. Sci.* **226**, 314—323 (1986).

[12] Shade, E.: "Concepts and Models for Real-Time Concurrency Under Limited Parallelism," *Ph.D Thesis*, Department of Computer Science, Whitmore Laboratory, The Pennsylvania State University, University Park, Pa16802 (June 1991).

[13] Shade, E., Narayana, K.T.: "Real-Time Semantics for Shared-variable Concurrency," *Information and Computation*, In Press (1991(2?)).

REAL-TIME SCHEDULING BY QUEUE AUTOMATA[1]

L.Breveglieri* A.Cherubini** S.Crespi-Reghizzi*
Politecnico di Milano, Piazza Leonardo, 32 - Milano, Italy 20133
*Dipt. Elettronica - **Dipt. Matematica

Abstract

Real-time schedulers are modelled by finite-state transition systems using FIFO queues as auxiliary memory. The intuitive notion of hard real time is related to the definition of quasi-real-time behaviour of an automaton. Then, starting with simple schedulers for independent tasks, a modular approach to the design of schedulers (FIFO, static priority based, preemptive, dynamic priorities) is presented. This is based on recent results on recognition power and closure properties of quasi-real-time queue automata w.r.t. intersection, shuffle and reverse homomorphism. The treatment of readers-writers schedulers is compared with recent proposals based on intersections of context-free languages. Possible developments are in the conclusion.

1. Introduction

In a multi-programming computer system, concurrent activities (tasks or processes) are serialized for execution by a scheduler [Cooling 1990]. External events causing interrupt signals, as well as internal events (e.g. termination of a program) are the sources of requests for computing services. Various policies can be applied to select which request to serve in case of conflict: FIFO and priority driven are the simplest; more complex policies take into account other parameters of the request, such as deadlines, service already obtained, etc. Priorities of requests can be fixed (static) or dynamic.
The scheduler, a central part of an operating system kernel, provides data-structures for representing the requests, the states of tasks, as well as the state of the computer, and implements the scheduling algorithms in accordance with the selected policies. Simple schedulers are included in the so called Real-Time kernels (e.g. VRTX, PSOS2, Real-time-craft), small system packages widely used for programming embedded systems on microcomputers. More complex schedulers have been developed for the larger operating systems. Concurrent programming languages, e.g. Ada, include also a scheduler in their run-time support.
The decision logic of RT schedulers is usually described informally, a situation not fully satisfactory which prevents exact comparisons of different policies. The data-structures used for representing the states of tasks are lists of task descriptors of some kind, in the simplest cases FIFO lists (queues). Therefore it is appealing to model a scheduler by means of an automaton taking as input the incoming events, and making a state transition based on the state, represented by one or more queues. Relying on recent results on queue automata (Brandenburg [1988], Cherubini et al. [1991]), we investigate their use for the simplest scheduling problems. A method is then suggested for modularly building more complex schedulers, starting with the basic ones, composing them by set operations (shuffle and intersection), and adding precedence constraints formulated as regular expressions. The

[1]This work was supported by Italian MURST 60% and by ESPRIT-BRA project ASMICS.

schedulers thus obtained are correct by construction and preserve the real-time properties of the building blocks.

The paper is organized as follows. In Sect.2 we define the problem to be modelled and introduce the scheduling events and actions. We propose a representation of schedulers as automata recognizing traces of events, and their conversion into automata with output (transducers) enforcing scheduling decisions. Then we consider the property of hard real time [Manacher 1967, Liu 1973], i.e. the requirement that the time to serve a request be bounded, and its relation to the Quasi-Real-Time property of automata. An automaton operates in QRT if it never performs unboundedly many transitions without consuming an input symbol. In Sect.3 we recall the basic definitions of queue automata and list relevant properties. In Sect.4 we present a basic scheduler as a queue automaton; then we show that more complex schedulers (with priorities, preemption, readers-writers) can be systematically derived by exploiting compositional and closure properties of quasi-real-time queue automata. We prove that k-priority scheduling in quasi-real-time requires no less than k queues. Through the readers-writers example we compare this work with the approach of Hemmendinger [1990] for modelling concurrent events by means of context-free grammars. Our last case considers dynamic priorities, a feature that requires, in addition to the queues, the use of finite-state information. In the conclusion we briefly examine the potential of our approach for designing practical schedulers.

2. Modelling scheduling activities

In order to build an automata-based model of scheduling we categorize the set of events which are input to the scheduler and its reactions. For a recent survey of scheduling models refer to [Herrtwich 1990]. The scheduler, roughly speaking, waits for arrivals of processes (or tasks or clients) requesting a service, schedules their start times, dispatches for execution, monitors termination and expells the finished processes. We assume that there are $N \geq 1$ classes of *processes* or *services* S_i, requiring distinct services. Several processes in the same class may simultaneously exist in the system.

A first class of events are the *external events*, which arise in the equipment controlled by the computer or inside the computer and are notified to the scheduler. The following external events are considered:

- *Arrival of external requests of a service*: $a_1, a_2, ..., a_N$ (later on we shall consider priorities of requests).
- *Termination (end) of service*: $e_1, e_2, ..., e_N$

The *set of events* is denoted $E_N = \{ a_i, e_i \mid 1 \leq i \leq N \}$.

We use the term *action* to refer to the decisions enforced by the scheduler. *Independent actions* do not depend on other actions but just on events[2]. The following independent actions will be considered:

- *Beginning of service*: $b_1, b_2, ..., b_N$
- *Departure of a client*: $d_1, d_2, ..., d_N$
- *Change of priority* (to be considered later): c and $\hat{c}$

2. This is true for the simple scheduling policies of this paper.

The *dependent actions*, which depend on other actions and possibly on events, are:

- *Suspension of service*: $s_1, s_2, ..., s_N$
- *Resumption of service*: $r_1, r_2, ..., r_N$

The *set of actions* (without considering priorities) is denoted $A_N = \{ b_i, d_i, s_i, r_i \mid 1 \leq i \leq N \}$. The total alphabet of the scheduler is $\Sigma_N = E_N \cup A_N$.
Moreover the scheduler can undertake *internal actions*, from time to time, which are necessary for readjusting the scheduler state and queues, but do not produce immediately visible consequences. Internal or invisible actions are also called ε-actions.
A larger set of events and actions then the ones considered in this paper could be introduced for modeling more complex situations (e.g. lock and release of a shared resource).

Schedulers as automata

Given the alphabet Σ_N (the subscript will be later dropped) of events and actions, a string over Σ is a *trace* or scheduling history (we do not consider infinite traces); a set of traces $T \subseteq \Sigma^*$ is a *trace language* and can be taken to represent the behaviour of a scheduler. The problem henceforth is to define devices to handle such languages.
Two descriptions of schedulers are possible, by recognizers and by transducers. A *recognizer* is the acceptor of a trace language $T \subseteq \Sigma^*$. This stresses the temporal order of events and actions. A *transducer* defines a one-to-many mapping from events to actions $\psi : E^* \rightarrow 2A^*$. Performance reasons require ψ to be a function (otherwise the scheduler would be non-deterministic). A transducer is a more realistic model than a recognizer, since it describes the algorithm used for dispatching, suspending and resuming requests. We suggest in the conclusion that the transducer model could be actually exploited for designing and implementing customizable real-time schedulers.
A transducer can be easily derived from a recognizer by moving the action symbols from the input to the output domain. Let $T \subseteq \Sigma^*$ be a trace language, and Π_E and Π_A be the natural projections of the total alphabet Σ onto E and A respectively. Then the transduction ψ corresponding to T is defined: $\psi(u) = \{ v \mid \exists\ w \in T : u = \Pi_E(w) \wedge v = \Pi_A(w) \}$. If the transducer is deterministic ψ is a function: clearly determinism of the recognizer is necessary but not sufficient for the transducer to be deterministic. In addition, for any u, the output $\psi(u)$ must be single valued. Conversion from a transducer to a recognizer is also straightforward; a one-to-one mapping from transducer to recognizer for a restricted class of recognizers is later defined.

Hard-Real-Time (HRT) and Quasi-Real-Time (QRT) characterization of scheduling models

Here we discuss how the intuitive notion of hard-real-time (Manacher [1967]) is related to the definition of quasi-real-time behaviour of an automaton, still without referring to specific automata models. The usual definition of HRT states that the time between an arrival a_i and the corresponding termination of service e_i must be bounded. Depending on the arrival rate and distribution of service times, a given computer may or may not be able to serve in HRT a given request. Since in our model time is not considered, but only events, we cannot deal in full with HRT but only provide necessary conditions for HRT to be possible.

Consider the time interval between an arrival a_i and the corresponding termination e_i. If the time interval has to be bounded, the scheduler has to execute boundedly many steps or transitions between the two events. For a scheduler represented as a transducer a transition can read an external event in the set $E_N=\{ a_i, e_i \mid 1 \le i \le N \}$, or nothing for an internal ε-move. Therefore a necessary condition for HRT operation is that the transducer does not perform unbounded runs of internal steps. This is precisely the automata- theoretical property known as Quasi-Real-Time behaviour. A machine (recognizer or transducer) operates in QRT if there exists a constant $k \ge 1$ such that in any k successive transition steps the machine reads at least one input symbol. A QRT machine is thus guaranteed to complete the computation in ($k \times$ *length of input*) steps. When $k=1$, the machine operates in real-time. QRT behaviour is a special case of linear-time complexity. We shall study precisely the scheduling problems which can be handled in QRT by various models of automata. Notice however that QRT operation of the automaton is necessary but not sufficient to ensure hard-real-time response to external events. This is because the QRT property bounds the runs of invisible actions but not the runs of other events (such as arrivals and terminations) that could cause unbounded delay in service.

3. Queue automata

The theory of Queue (or FIFO) Automata (QA) was not developed until recently by Brandenburg [1988] and Cherubini et al. [1991], after the pioneering works of Vollmar [1970] and Vauquelin [1980]. This in contrast to LIFO automata, usually called push-down automata (PDA), which are the fundamental model to design compilers and interpreters of programming languages. Here we recall the basic definitions and relevant properties of deterministic multi-queue automata. For a more complete and rigorous presentation the reader is referred to Cherubini et al. [1991]. A one-queue automaton (1-QA) is the same as a PDA except that its memory tape is managed according to a FIFO instead than a LIFO discipline. A QA can have $n \ge 1$ queues. A n-queues recognizer A is defined:

$A = (\Sigma, Q, q_0, Q_f, \Gamma, Z_0, \delta)$ where Σ is the input or terminal alphabet, Q is the set of internal states, $q_0 \in Q$ is the initial state, $Q_f \subseteq Q$ is the set of final states, Γ is the queue alphabet, $Z_0 \in \Gamma$ is the initial symbol of the queue, and δ is the transition function

$\delta: (\Sigma \cup \{\varepsilon\}) \times Q \times (\Gamma \times D)^n \rightarrow (Q \times (\Gamma^*)^n)$ is a partial function and $D=\{S, H\}$.

Note that recognition is by final state. A configuration is defined:

$c = \langle aw, q, \alpha_1, \alpha_2, ..., \alpha_n \rangle \quad w \in \Sigma^*, \ a \in \Sigma \cup \{\varepsilon\}, \quad q \in Q, \quad \alpha_i \in \Gamma^*, \ 1 \le i \le n$

where a is the current input character, w is the remainder of the input, q is the present state, and the α_i's are the contents of the n queues. Given a move:

$\delta(a, q, (A_1, D_1), ..., (A_n, D_n)) = (q', \beta_1, ..., \beta_n) \qquad A_i \in \Gamma, \ D_i \in D, q' \in Q, \ \beta_i \in \Gamma^*, \ 1 \le i \le n$

and the configuration $c' = \langle w, q', \gamma_1, \gamma_2, ..., \gamma_n \rangle$, where $\gamma_i \in \Gamma^*, 1 \le i \le n$, the transition $c \vdash c'$ implies that:

$\forall i, 1 \le i \le n: \ \alpha_i = A_i \alpha'_i$

$\forall i, 1 \le i \le n: \ (D_i = H \Rightarrow \gamma_i = A_i \alpha'_i \beta_i) \wedge (D_i = S \Rightarrow \gamma_i = \alpha'_i \beta_i)$

In state q the automaton checks the input character a and the tape symbols A_i on the fronts of the queues. If these symbols match a move possible in that state, the automaton switches to the next state q', advances to the next character on the input, deletes (S for 'shifts') or holds (H) the symbols on the front of each queue and enqueues n strings at the rears of the queues. The machine stops when the move is not defined or any queue is empty. Recognition is defined by final state (and occasionally by empty queue).

In Cherubini et al [1991] the family of n-queue QRT automata is studied. Obviously a QA with no queues is a finite state machine.The following result is relevant:

The family of languages accepted by n-queue deterministic QRT automata is strictly contained in the family of languages accepted by (n+1)-queue deterministic QRT automata, for any $n \geq 1$.[3]

A *n*-queue *transducer* is defined:

$T = \{E, Q, \Gamma, A, q_0, Q_f, \delta, Z_0\}$ $q_0 \in Q$, $Q_f \subseteq Q$

$\delta_T: (E \cup \{\varepsilon\}) \times Q \times (\Gamma \times D)^n \rightarrow (Q \times (A \cup \{\varepsilon\}) \times (\Gamma^*)^n)$ is a partial function and $D = \{S, H\}$.

The definition is derived from the recognizer in the usual way: in a move the transducer does the same as the recognizer and additionally outputs a character from $A \cup \{\varepsilon\}$. Note that a transducer reads external events from the input and writes strings of actions on the output.

A systematic transformation from transducer to recognizer is presented in Fig.1. In practice, the recognizer works in two steps. First it reads a without modifying the queues; then it reads b (the symbol emitted by the transducer) acting on the queues as the original transducer. If the underlying recognizer (with input alphabet E) is deterministic, the derived recognizer is so too. We are only interested in recognizers which are derived from transducers as shown in Fig.1. Hence also the correspondence from recognizers to

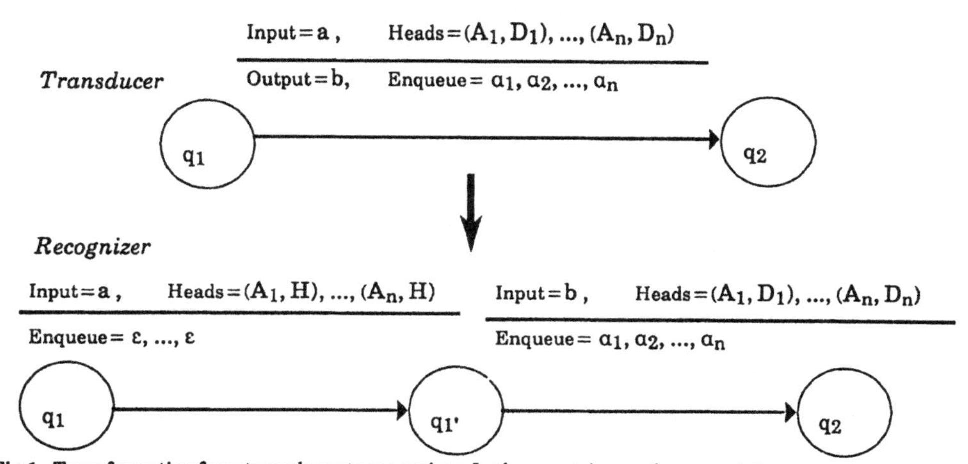

Fig.1 - Transformation from transducer to recognizer. In the recognizer $q_{1'}$ is a new state.

transducers is a one-to-one mapping. Clearly a transducer is QRT if it derives from a recognizer with the same properties, and conversely.

The family of n-queue QRT languages has several closure properties listed in Fig.2, to be later used to build complex schedulers from simple ones, which are proved in Cherubini et al

3. Our definition slightly differs from Cherubini et al [1991], since we do not allow repeated reading of the same input character. The previous results as well as the relevant closure properties still hold.

[1991] or immediately derivable. We use the following parameters: m, n, r: the minimum

Property	QRT
$L_1 \cup L_2$	$r \leq \max(m,n)$ $s \leq 2 \times \max(p,q)$
$L_1 \cap L_2$	$r \leq \max(m,n)$ $s \leq 2 \times \max(p,q)$
$L_1 \cap R$ R regular	$r \leq m$ $s \leq p$
ϕ^{-1} reverse homomorphism, with ϕ alphabetic hom.	$r = m$ $s = p$
$L_1 // L_2$ shuffle with disjoint alphabets	$r \leq m + n$ $s \leq p + q$
$L_1 L_2$ catenation	$r \leq m + n$ $s = \max(p,q)$

Fig. 2 - Closure properties of n-queue QRT languages.

number of queues required by languages L_1, L_2 and by the result; p, q, s: the QRT constants of L_1, L_2 and of the result.

Finally we discuss the role of finite state information for queue automata. Clearly the current state can be stored in an extra queue, which always contains exactly one element, so that no computational power is lost by considering one-state multi-queue automata. For such a machine recognition must of course be defined by empty queue(s), or by a stated final configuration. We say that for an automaton states are superfluous if there exists an equivalent monostatic automaton, with the *same* number of queues (recognizing by empty queue or by a stated final configuration).[4]

4. Queue schedulers and their composition

The main purpose of this section is to model a scheduler by means of deterministic queue automata or transducers for some simple queueing and service disciplines. First the automata are presented for some elementary cases, then we use results on the composition of automata to treat more complex situations, and to prove minimality of the solutions.

Next we consider service of independent jobs. We recall that an arrival of a request of service S_i is denoted a_i, the initiation by b_i, the end by e_i, and the departure of the customer by d_i.

Case 1) FIFO service of independent tasks without priorities

The following alphabets will be used:

Input events alphabet $E = \{ a_i, e_i \mid 1 \leq i \leq N \}$

Output actions alphabet $A = \{ b_i, d_i \mid 1 \leq i \leq N \}$

The problem is specified by the following conditions:

a. Requests of service S_i, $1 \leq i \leq N$, arrive, are stored and serviced in the order of arrival (FIFO policy).

b. One request at a time can be serviced (mutual exclusion).

c. As soon as the server is idle (i.e. the current service is ended and departs) it begins to serve the next pending request (immediate service).

d. Initially the server is idle.

e. Eventually all requests must be serviced.

f. As soon as a customer ends it departs (immediate departure).

4. In the course of this study we have found that for any fixed number of queues, the family of languages recognized by automata with at most k internal states form an infinite hierarchy with respect to k.

A trace of events meeting the above conditions is:

$a_3\, b_3\, a_1\, a_2\, e_3\, d_3\, b_1\, e_1\, d_1\, b_2\, a_4\, e_2\, d_2\, b_4\, e_4\, d_4$

On the contrary the following traces violate some conditions:

$a_1\, b_1\, a_2\, a_3\, e_1\, d_1\, b_3\, e_3\, d_3\, b_2\, e_2\, d_2$ violates *a*. $a_1\, b_1\, a_2\, b_2\, e_1\, d_1\, e_2\, d_2$ violates *b*.
$a_1\, a_2\, b_1\, e_1\, d_1\, b_2\, e_2\, d_2$ violates *c*. $a_1\, e_2\, d_2\, b_1\, e_1\, d_1$ violates *d*.
$a_1\, b_1\, a_2\, e_1\, d_1$ violates *e*. $a_1\, b_1\, a_2\, e_1\, b_2\, d_1\, e_2\, d_2$ violates *f*.

The transducer T_1 corresponding to the deterministic queue automaton recognizing the set of legal traces L_{1a} is shown in Fig.3. Each transition arrow is labeled by an expression:

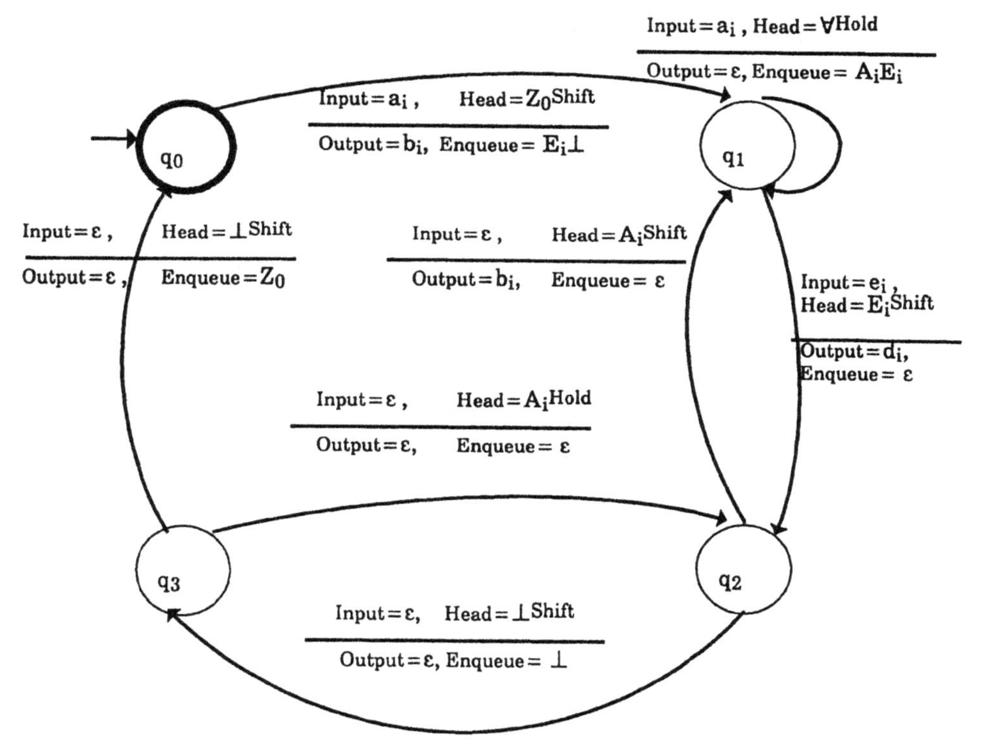

Fig.3 - State-transition-action diagram of queue automaton *T1* (transducer) for trace language L_{1a}: FIFO service of independent tasks without priorities (immediate service, no suspension). The symbol ∀ stands for any element of Γ.

$$\frac{x,\ (Y,D)}{z,\ \omega}$$

where x is a character of E or ε (as customary ε denotes the null string), Y is a character of the memory alphabet $\Gamma = \{A_i, E_i \mid 1 \le i \le N\}$ or ε , z is a string in A^*, ω is a string in Γ^* and $D \in \{S, H\} = \{\text{Shift, Hold}\}$. The meaning of this label is the following. Let T_1 be in a state q_i , and consider the arc $(q_i\, ,\, q_j)$ labelled by $x,(Y,D)/_{z,\omega}$; the transition has the following effect: reading x from the input, Y from the head of the queue, the transducer moves to state q_j, outputs z and enqueues ω; it dequeues Y if D=S. The initial state is q_0, which is also the final state. Initially the queue contains the starting symbol Z_0. Briefly, in state q_0 the trasducer is idle and without pending requests, so it receives and begins to service a request or terminates. When it begins a request (by emitting b_i), it also enqueues the two symbols $E_i\bot$; E_i predicts the future occurrence of the corresponding end e_i, while the special symbol $\bot$ indicates that the queue contained just Z_0 when this transition occurred. In state q_1 the machine is servicing a request and receives and stores further requests or ends the service moving to state q_2. When the trasducer is in the state q_2 it is idle, so if no request is pending it moves to q_3, then it terminates by going to q_0 if the input is ended, otherwise it restarts. If a request is pending i.e. A_i is in front of the queue, the trasducer begins its service returning to state q_1. The special simbol $\bot$ is used in Γ to avoid empty queue and state q_3 is introduced to control if there is a pending request when the special symbol $\bot$ is on the head of the queue. The instantaneous configuration of the automaton after processing, for instance, the input string $a_3\ a_1\ a_2\ e_3$ is : state$=q_2$, queue$=\bot A_1E_1A_2E_2$ and output $= b_3d_3$. Notice that the machine is deterministic and operates in QRT but not in HRT because unboundedly many requests may be pending when a new request arrives. Note also that some states are superfluous for L_{1a}, but are used for clarity sake.

In order to simplify the presentation of queue automata, we shall drop from now on the initial symbol Z_0 and allow instead explicit testing of emptiness of the queue(s). It is clear that the theoretical results recalled in Sect.3 continue to apply, since the simplified automata can be replaced by equivalent ones complying with the formal definition. A simplified version of T_1 is shown in Fig.3a, where states have been reduced. By changing the

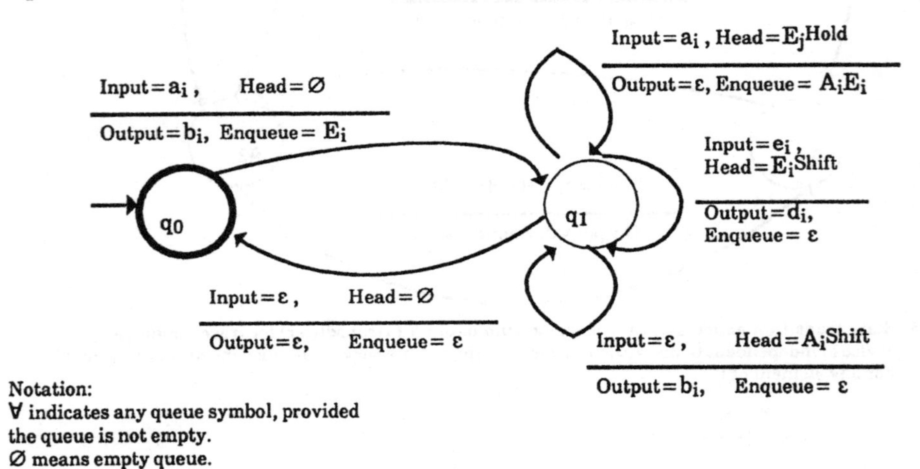

Fig.3a - 2-states version of queue automaton T_1 (transducer) for trace language L_{1a}: FIFO service of independent tasks without priorities (immediate service, no suspension). No less than 2 states are possible as recognition is by final state.

recognition mode internal states could be altogether suppressed. In the coming examples of cases 2, 3 and 4, states could be similarly reduced to two or suppressed, but we prefer not to do so because readability would somewhat deteriorate. On the contrary we shall see that case 5, dynamic priorities, makes essential use of internal states.

Case 2) FIFO service of independent tasks with static priorities, no preemption

First we prove that no 1-queue transducer exists which performs this scheduling job in QRT. Then we prove that it suffices a transducer with k queues to perform the job in QRT, where k is the number of priority levels. In what follows $S^h{}_i$ and $S^k{}_j$ denote two services with priorities h and k respectively, meaning that if $h<k$ the request of service h is less urgent than the request of k. For service $S^h{}_i$ arrivals, initiations, ends and departures are denoted by $a^h{}_i$, $b^h{}_i$, $e^h{}_i$ and $d^h{}_i$. Similarly for $S^k{}_j$.

The problem can be specified by supplementing conditions *a.* to *f.* of Case 1 with the next one:

g. no request of priority level i $(1 \le i \le k)$ can begin service if a request of level $j>i$ is awaiting service.

For simplicity henceforth we consider two levels of priority, but the results are not affected by this assumption.

Statement 1. No deterministic 1-queue automaton can recognize the language of event traces of case 2 (priority scheduler) in QRT.

Hint of proof: The proof is similar to the one of Th.4.1 of Cherubini et al. [1991]. Pending requests must be stored in the queue as an unbounded string of task descriptors (denoted by $A1_h$ and $A2_i$ for requests $a^1{}_h$ and $a^2{}_i$). Upon termination of a service, the machine must scan through the queue to find the first descriptor of a prioritarian request, i.e. the first $A2_i$ (if any). But this could be preceded in the queue by unboundedly many requests $A1_h$ of lower priority. Hence the machine must perform an unbounded number of moves without reading any input since cond. *c.* forbids further reading of requests when the server is idle.

Of course it is possible to construct a deterministic non-QRT scheduler with only one queue for case 2. The machine circulates the queue to find the first prioritarian request (if any). In order to recognize when the whole queue has been circulated, a special character $ is enqueued at the beginning of the circulation phase. We remark that a deterministic 1-queue automaton not constrained to QRT operation is a very powerful computational device, as it can simulate any Turing machine [Cherubini et al. 1991]. For real-time applications it is necessary to operate in QRT, that is, to be sure that scheduling decisions are taken within a constant time bound. But we recall that QRT does not guarantee HRT.

Now we prove that there exists a deterministic QRT two queues transducer for Case 2 scheduling, with two priority levels. Let $\{S^1{}_i \mid 1 \le i \le N_1)\}$ and $\{S^2{}_i \mid 1 \le i \le N_2\}$ be the sets of services of priority 1 and 2. Let L_2 be the language of event traces of case 2 and let $L^1{}_2$ and $L^2{}_2$ be the projections of L_2 onto the alphabets corresponding to services of priority 1 and 2 respectively. The latter two are languages similar to case 1, except that they do not satisfy condition *c.* (immediate service); this happens because the projection can erase a *begin* symbol, giving rise to an adjacency which violates condition *c*. This situation is illustrated by the trace in L_2:

$\ldots a^1{}_5\, d^1{}_3\, b^2{}_j\, a^1{}_4\, e^2{}_j\, b^1{}_5 \ldots$

which after the projection onto the priority 1 alphabet becomes

$\ldots a^1{}_5\, d^1{}_3\, \varepsilon\, a^1{}_4\, \varepsilon\, b^1{}_5 \ldots = \ldots a^1{}_5\, d^1{}_3\; a^1{}_4\; b^1{}_5 \ldots$

a valid trace of $L^1{}_2$, which violates *c* because after $d^1{}_3$ the server is idle but event $b^1{}_5$ is delayed. We denote L_{1b} a language such as $L^1{}_2$ and $L^2{}_2$. It is recognized by a QRT 1-queue automaton A_{1b}, shown in Fig.4. Notice that the original language L_{1a} (of Fig.3) can be

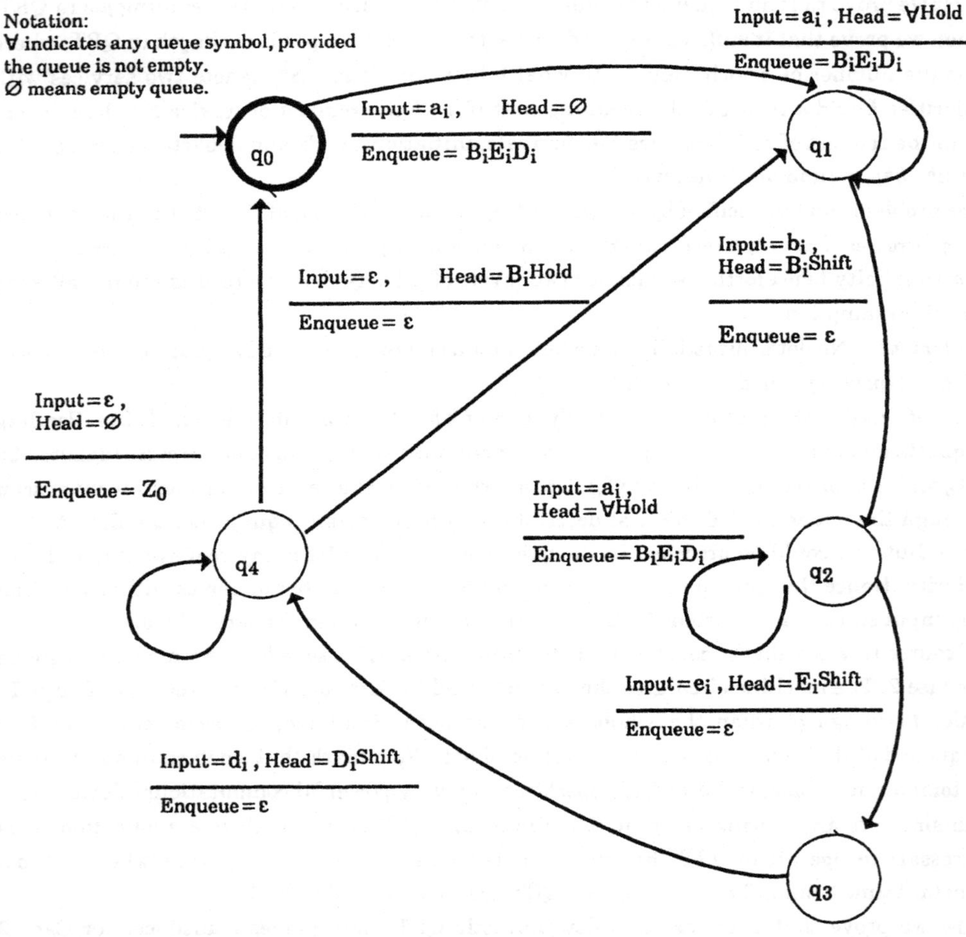

Fig.4 - Queue recognizer A_{1b} for language L_{1b}: FIFO service of independent tasks without priorities (non-immediate service, no suspension).

reobtained by intersecting L_{1b} with the regular language:

$R = \neg[\{x\, d_i\, a_j\, a_l\, y \mid x,y \in \Sigma^*\}$ when a_j arrives the server is idle, because a departure d_i just occurred, but before it starts servicing a_j a new customer a_l arrives.

$\cup\{a_j\, a_l\, x \mid x \in \Sigma^*\}$ before the first requests a_j starts a new request a_l arrives.

$\cup\{x\, a_j\, y\, d_i\, a_l\, z \mid x,z \in \Sigma^*, y \in (\Sigma-\{b_j\})^*\}]$ after d_i departs the server is idle, but initiation of the pending request a_j is delayed by the arrival of a_l

which forbids delayed service of any request a_j.

Next we construct the recognizer of the language of case 2, L_2, by composition and refinement. An operation combining two languages X and Y is the *shuffle product*, a traditional but not frequent operation in formal language theory, defined as follows:

$X \,//\, Y = \{x_1y_1x_2y_2 \ldots x_ny_n \mid n \geq 1, x = x_1x_2 \ldots x_n \in X, y = y_1\, y_2 \ldots y_n \in Y, \forall i \in 1\ldots n: x_i \in \Sigma^*, y_i \in \Sigma^*\}$

The shuffle product $L^1{}_2 \,//\, L^2{}_2$ clearly satisfies conditions *a.*, *d.* and *e.*: for each class 1 and 2, service is FIFO, initially the server is idle and eventually all requests are ended and depart. Now the constraints which must be added to $L^1{}_2 \,//\, L^2{}_2$ for a trace to be in L_2 are expressed by the regular languages in Fig.5. Thus $L_2 = (L^1{}_2 \,//\, L^2{}_2) \cap R_1 \cap R_2 \cap R_3 \cap R_4$ and from the

Constraint	Regular language and conditions satisfied
b	$R_1 = \neg\{y\, b^i{}_h\, x\, b^j{}_k\, z \mid x \in (\Sigma - \{e^i{}_h\})^*,\ y,z \in \Sigma^*;\ 1 \le h \le N_i,\ 1 \le k \le N_j\}$ $(L^1{}_2 \,//\, L^2{}_2) \cap R_1$ meets condition a, d, e, b.
g	$R_2 = \neg\{y\, a^2{}_h\, x\, b^1{}_k\, z \mid x \in (\Sigma - \{b^2{}_h\})^*,\ y,z \in \Sigma^*;\ 1 \le h \le N_2,\ 1 \le k \le N_1\}$ $(L^1{}_2 \,//\, L^2{}_2) \cap R_1 \cap R_2$ meets condition a, d, e, b, g.
f	$R_3 = \neg \bigcup_{\forall a \in \Sigma - \{d^i{}_h\}} \{y\, e^i{}_h\, a\, z \mid y,z \in \Sigma^*;\ 1 \le h \le N_i\}$ $(L^1{}_2 \,//\, L^2{}_2) \cap R_1 \cap R_2 \cap R_3$ meets condition a, d, e, b, g, f.
c	$R_4 = \neg[\bigcup_{\forall a \in \Sigma - \{b^1{}_i, b^2{}_j \mid 1 \le i \le N_1, 1 \le j \le N_2\}}$ $\{y\, a^i{}_h\, x\, d^j{}_k\, a\, z \mid x \in (\Sigma - \{b^i{}_h\})^*,\ y,z \in \Sigma^*;\ 1 \le h \le N_i,\ 1 \le k \le N_j\} \cup$ $\{x\, a^i{}_r\, a^l{}_s\, y \mid x = \varepsilon \text{ or}, x = x' d^i{}_k,\ x', y \in \Sigma^*,\ 1 \le r \le N_i,\ 1 \le k \le N_i,\ 1 \le s \le N_l\}]$ $(L^1{}_2 \,//\, L^2{}_2) \cap R_1 \cap R_2 \cap R_3 \cap R_4 = L_2$ meets condition a, d, e, b, g, f, c.

Fig.5 - Regular constraints of case 2 for imposing priorities onto $L^1{}_2 \,//\, L^2{}_2$.

closure properties of QA's (v.s. Fig.2) L_2 can be recognized by a QRT automaton with two queues[5].

Case 3) FIFO service of independent tasks with static priorities, and preemption

A more complex scheduling policy, intended to ensure rapid and predictable response to urgent events, allows a higher priority request to preempt the lower priority one under service. The suspended task is placed back into its priority queue. We assume that a request originating from a suspended task is undistinguisheable from an external one. For preemption two more symbols are introduced: $s^h{}_i$ for suspension, and $r^h{}_i$ for resumption of service $S^h{}_i$. The problem can be specified as follows. Requests are grouped by priority as in case 2. Case 3) is specified by conditions {a,b,c,d,e,f,h,k} where *h* and *k* are:

h. no request of priority level i can begin or continue service if a request of level $j > i$ is awaiting service.
k. a request suspended at a certain instant is treated as a request arriving at that instant.

Consider for simplicity two priorities. Let L_3 be the language of event traces of Case 3, and let $L^1{}_3$ and $L^2{}_3$ be the projections of L_3 onto the alphabets corresponding to requests of services of priority 1 and 2, respectively. $L^2{}_3$ is a language corresponding to a scheduling policy of case 1 and is recognized by a deterministic QRT 1-queue automaton, as in Fig.3. $L^1{}_3$ differs from L_{1b} (Fig.4), because a service can be suspended and resumed: notice that this scheduling policy time-shares its services among the requests. For brevity we do not present a 1-queue QRT recognizer for $L^1{}_3$. From these components, we construct the recognizer of the shuffle product $L^1{}_3 // L^2{}_3$, which satisfies condition *d.*, *e.* and *k.*

The remaining constraints to be met by any trace in L_3 are expressed by the regular

5. The shuffle of two 1-queue languages is recognized by a sort of Cartesian product machine which keeps the two original queues separate and inspects them alternatively.

languages defined in Fig.6. Thus $L_3 = (L^1_3 // L^2_3) \cap R_3 \cap R_5 \cap R_6 \cap R_7 \cap R_8 \cap R_9$ and from the

Constraint	Regular language and conditions satisfied
no word contains a substring where an *end* is immediately followed by a character other than the corresponding *departure*	$R_3 = \neg \bigcup_{\forall a \in \Sigma - \{d^i_h\}} \{y\, e^i_h\, a\, z \mid y,z \in \Sigma^*; 1 \le h \le N_i,\}$ (as in Fig.5) $(L^1_3 // L^2_3) \cap R_3$ meets condition d,e,f,k
no word contains a substring such that a service begins or is resumed before the actual service ends or is suspended (condition b.)	$R_5 = \neg [\, \{y\, a\, x\, b^j_k\, z \mid a \in \{b^1_h, r^1_h\}, x \in (\Sigma_1 - \{s^1_h, e^1_h\})^*,$ $y,z \in \Sigma^*, 1 \le k \le N_j,\ 1 \le h \le N_1\} \cup \{y\, b^2_h\, x\, a\, z \mid a \in \{b^i_k, r^1_j\},$ $x \in (\Sigma - \{e^2_h\})^*, y,z \in \Sigma^*, 1 \le h \le N_2, 1 \le k \le N_i, 1 \le j \le N_1 \}]$ $(L^1_3 // L^2_3) \cap R_3 \cap R_5$ meets conditions b,d,e,f,k
no word contains a substring such that a request of lower priority is not preempted by a higherpriority one	$R_6 = \neg \{x\, a\, y\, a^2_j\, b\, z \mid$ $a \in \{b^1_i, r^1_i \mid 1 \le i \le N_1\}, b \in (\Sigma - \{s^1_i, e^1_i\}), y \in (\Sigma - \{s^1_i, e^1_i\})^*, x,z \in \Sigma^*\}$
no word contains a substring such that a request of lower priority starts or resumes service when a higher priority request is waiting	$R_7 = \neg \{x\, a^2_h\, b\, y \mid b \in \{b^1_i, r^1_i\}, 1 \le i \le N_1, x,y \in \Sigma^*\}$
no word contains a substring such that a lower priority service is suspended by an equally prioritarian one (i.e. lower priority requests are serviced in FIFO order if no higher priority request arrives)	$R_8 = \neg \{x\, a\, s^1_k\, y \mid a \in (\Sigma - \{a^2_h\}, 1 \le h \le N_2), x,y \in \Sigma^*, 1 \le k \le N_1\}$
no word contains a substring such that when the server is idle and a request is pending the corresponding service is not started or resumed immediately	$R_9 = \neg [\{x\, a^i_h\, y\, c\, a\, z \mid x,z \in \Sigma^*, y \in (\Sigma - \{e^i_h\})^*, c \in \{d^j_k, s^1_h\}$ $a \in (\Sigma - \{b^1_l, r^1_l, b^2_s, r^2_s\}, 1 \le l \le N_1, 1 \le s \le N_2, 1 \le k \le Nj)\} \cup$ $\{x\, a^i_l\, y\, s^1_k\, a\, z \mid x,z \in \Sigma^*, y \in (\Sigma - \{e^i_l\})^*, a \in (\Sigma - \{b^1_i, r^1_i, b^2_j, r^2_j\}),$ $1 \le i \le N_1, 1 \le j \le N_2, 1 \le l \le N_i\} \cup$ $\{a^i_h\, b\, x \mid b \in (\Sigma - \{b^i_h\}, x \in \Sigma^*, 1 \le h \le N_i\}]$ $(L^1_3 // L^2_3) \cap R_3 \cap R_5 \cap R_6 \cap R_7 \cap R_8 \cap R_9 = L_3$

Fig.6 - Regular constraints of case 3 for refining the shuffle $L^1_3 // L^2_3$.

closure properties of QA's L_3 can be recognized by a QRT automaton with two queues, which is obtained by a Cartesian product of the states and a disjoint union of the queues.

With an argument similar to the one used for Statement 1 we can prove:

Statement 2.

No 1-queue automaton exists to solve Case 3 (preemptive priority scheduling) in QRT.

Case 4: readers-writers with writers precedence, no preemption

Scheduling concurrent accesses to a shared resource is a typical important duty of operating

systems. The many existing versions of the readers-writers problem [Courtois 1971] share the basic assumption that readers are allowed concurrent access, but a writer is mutually exclusive of other writers or readers accesses. The versions differ w.r.t. the scheduling policy. In our formulation the problem can be specified as follows. There are two sorts of requests: reading requests $rr_1, rr_2, \dots$; writing requests $rw_1, rw_2, \dots$ .Beginning and ending for reader i and writer j are resp. denoted: br_i, bw_j and er_i , ew_j .
Scheduling must satisfy the following conditions:

a. reader requests begin service in order of arrival;
b. any number of readers can be serviced at any instant;
c. writer requests begin service in order of arrival;
d. when a writer is under service no other request can be serviced;
e. writers take precedence over readers, i.e. when a writer is waiting, no pending reader can start;
f. eventually all requests must be serviced;
g. writers do not preempt readers.

Since this problem includes two levels of priority, by a reasoning similar to the proof of Stat.1 it is easy to show that no QRT 1-queue automaton can solve Case 4.
Let L_4 be the language of valid sequences of events w.r.t. this scheduling policy. The projection L_{w4} of L_4 onto the alphabet $\{rw_i, bw_i, ew_i \mid i \in W\}$, W set of writers, is a language similar to L_{1b} of Case 2. We can construct a recognizer for scheduling writers by replacing a_i with rw_i, e_i with ew_i, b_i with bw_i and d_i with ε in the recognizer represented in Fig.4.
Let L be the projection of L_4 onto the alphabet $\{rr_i, br_i, er \mid i \in R\}$, R set of readers; there exists a deterministic QRT 2-queue automaton A recognizing L (one of the queues is actually a counter). Roughly speaking, upon reading rr_i from the input, A writes A_i onto the first queue, then it matches br_i with A_i writing E on the second queue and finally it erases E if an *er* occurs. If all matchings are true the word is in L.
The shuffle product $L_{w4} \,//\, L$ satisfies conditions *a.*, *b.*, *c.* and *f.* Now consider the constraints which must be met by any trace in L_4, expressed by the languages represented in Fig. 7.

Constraint	Language and condition satisfied
d	$R_1 = \neg\{x\, bw_i\, y \mid x \in \Sigma^*, y \in (\Sigma - \{ew_i\})^* \{br_j, bw_h\} \Sigma^*\}$ $(L_{w4} \,//\, L) \cap R_1$ meets conditions a,b,c,f and d.
e	$R_2 = \neg\{x\, rw_i\, y \mid x \in \Sigma^*, y \in (\Sigma - \{bw_i\})^* br_j\, \Sigma^*\}$ $(L_{w4} \,//\, L) \cap R_1 \cap R_2$ meets conditions a,b,c,f,d and e.
g	$C_3 = \neg\{x\, bw_i\, y \mid y \in \Sigma^*, x \in \psi^{-1}(L^\circ)$ where $L^\circ = \{w \in \{br_i, rr_i, er\}^* \mid \sum_{i \in R}(\text{the no. of } br_i \text{ in } w) = (\text{the no. of } er \text{ in } w)\}\}$ $(L_{w4} \,//\, L) \cap R_1 \cap R_2 \cap C_3 = L_4$ meets conditions a,b,c,f ,d, e and g.

Fig.7 - Constraints of case 4 (not all regular).The first two languages R_1 and R_2 are regular and the third one C_3 is a 1-queue language, the catenation of two terms x and ($bw_i\, y$). The term x is the reverse homomorphic image of a language recognizable by a deterministic QRT 1-queue automaton, hence again a 1-queue QRT language. The term $bw_i\, y$ is a regular language.

Thus $L_4 = (L_{w4} \,//\, L) \cap R_1 \cap R_2 \cap C_3$ and from the closure properties of QA's (Fig.2) L_4 can be recognized by a QRT automaton with four queues, which is obtained by construction (similar to cases 2 and 3). Hence its correctness is easily proved. This solution is not minimal:

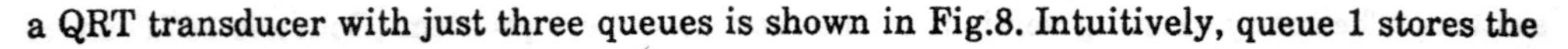
a QRT transducer with just three queues is shown in Fig.8. Intuitively, queue 1 stores the

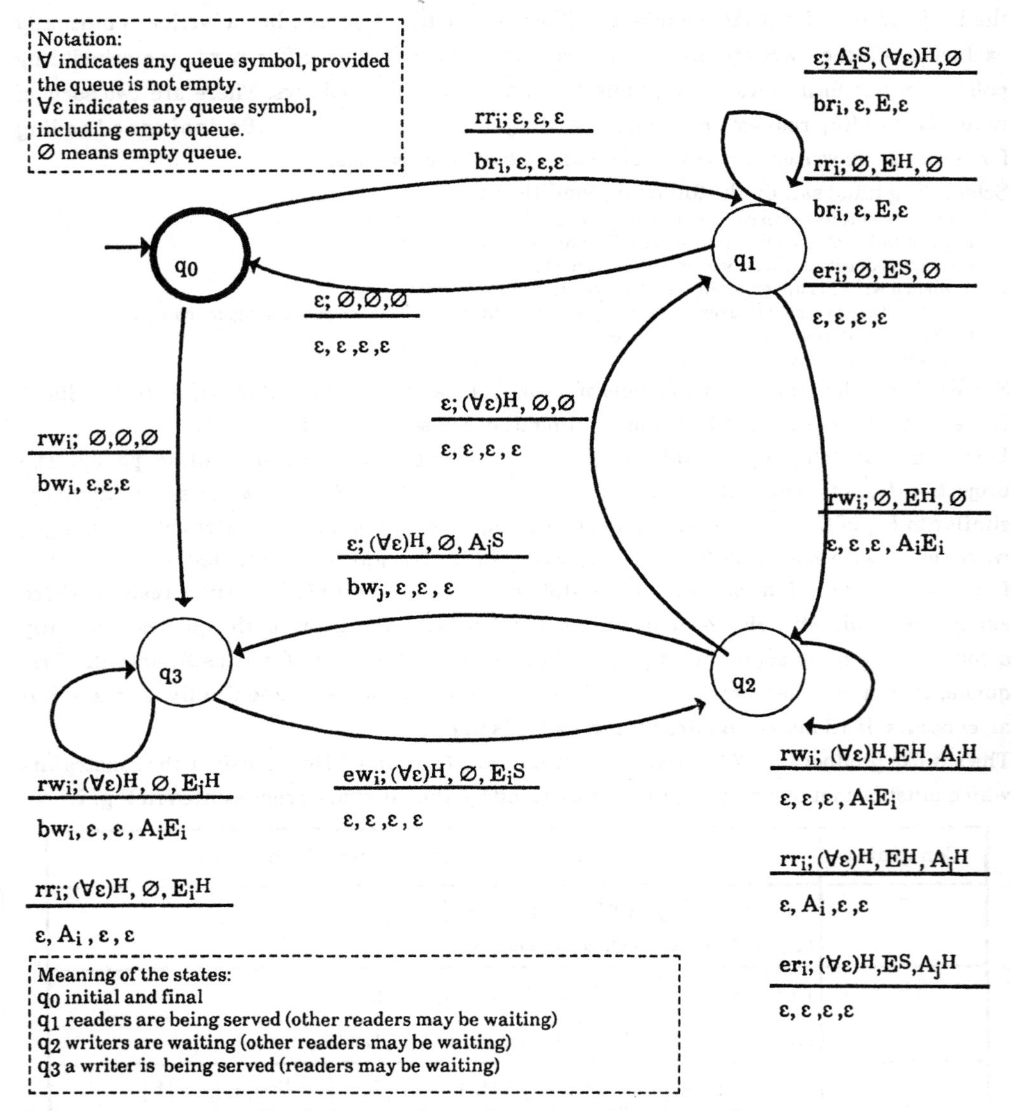

Fig.8 - 3-queues transducer T_{rw} for language L_4 : readers-writers with writer preference.

pending readers requests, queue 2 contains the number of currently active readers and queue 3 stores the pending writer requests. As noticed before, the number of states could be reduced to two. We do not know of a solution with only two queues.

The reader-writers problem is treated by Hemmendinger [1990] by means of intersections of context-free languages, which corresponds to multi-push-down automata. We observe that his solution does not consider different requests to be distinct; because of the PDA model, processing of requests is implicitly LIFO, which is unnatural for schedulers. Notice that if reading (and writing) requests are considered undistinguisheable, with our model the queues are used as counters, hence Hemmendinger 's solution is reobtained.

Case 5: dynamic priorities

In some schedulers priorities of requests are not constant but change in time; this policy is frequently followed in order to meet deadlines. We consider the situation where priorities can be dynamically adjusted for classes of clients, as it may occur in a computing center when certain classes of jobs are raised in priority. As before we consider for simplicity only two priority levels, with 2 more urgent than 1. We assume that the external command $\hat{c}$ cancels priorities, and conversely c (re)introduces priorities. For simplicity we do not consider preemption.

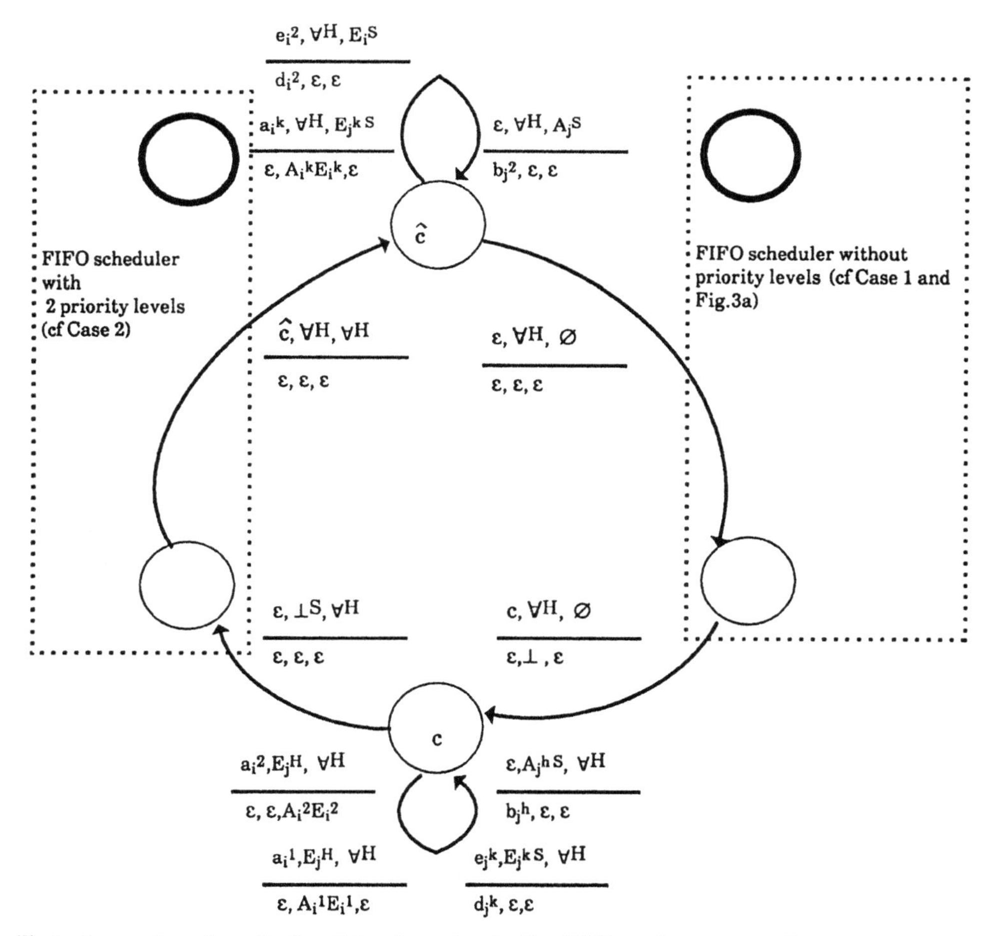

Fig.9 - 2-queue transducer for Case 5: two dynamic priorities, FIFO service, no preemption.

The problem is specified by the following conditions:

1. Initially the scheduler is in the no-priority mode and behaves as Case 1 (FIFO scheduling). Thus all pending requests have the same priority.

2. In this mode c causes the scheduler to switch to the priority mode (same as Case 2). Event $\hat{c}$ has no effect.

3. In the priority mode, event $\hat{c}$ forces the scheduler back into the no-priority mode.

The following supplementary conditions (which are partly arbitrary and motivated by simplicity) specify the effect of mode changes on the pending high-priority requests. The scheduler works initially in the non-priority mode (denoted $1=2$), when a c is received it changes to the priority mode (denoted $1<2$), terminating the pending requests without priority in FIFO order of arrival: that is the pending requests are not shadowed by the incoming prioritarian ones. Arrival of $\hat{c}$ switches the scheduler back to policy ($1=2$), terminating the services already scheduled with higher priority, in their order of arrival. Fig. 9 presents the scheduler as the combination of the machines of Case 2 (FIFO scheduler with two static priority levels) and Case 1 (FIFO scheduler without priority for two classes of requests labelled 1 and 2). Suitable transitions connect the two machines and are entered when c or $\hat{c}$ arrive. It is not difficult to see that this scheduler requires at least 4 states:

i. initial and terminal state (recognition by empty queues);

ii. stationary operation as no-priority or priority scheduler;

iii. two transition states for switching from ($1<2$) to ($1=2$) and conversely.

Simplification of this solution is possible, by merging the initial and final states of the two automata. As in previous cases, preemption or other queueing disciplines could be added by intersection, shuffle and homomorphism.

5. Conclusion

We have applied recent development in queue automata theory to formally describe simple real-time scheduling algorithms, with the aim of designing schedulers which can be proved correct. This new approach is accurate, natural (since FIFO queues are the essential data-structure of practical schedulers) and modular. Modularity allows specification of more complex scheduling disciplines by composition and refinement of simpler ones. This is possible because quasi-real-time multi-queue automata [Cherubini et al. 1991] have several closure properties, notably with respect to intersection, shuffle product and complementation. We have proved lower bounds on the number of queues necessary to model priority based schedulers by quasi-real-time automata and shown actual automata as solutions. The latter have been obtained in most cases by modular composition using products of automata and are sometimes non-optimal.

In our opinion this approach is promising for the actual design of real-time event-driven kernels, for embedded microprocessors and for run-time support of concurrent programming languages, such as Ada. A similar proposal, using less convenient definitions based on intersections of context-free languages (push-down automata), was made by Hemmendinger [1990]. It is clear that queues are more suitable than push-down stacks to model the FIFO and round-robin disciplines of schedulers.

In order to use our method for designing schedulers, several practical aspects need however to be examined, including interrupt handling and inter-process dependencies. A data structure for efficiently performing queue operations has to be selected .

FIFO queues are the primary, but not the unique data-structure of real-time kernels. In particular LIFO structures are also needed for interrupt processing and for (remote) procedure stacking. Combination of stacks and queues in the quasi-real-time model seems attractive for the purpose of addressing the requirements of real applications. A formal model for multi-queue-stack automata, presented in Cherubini et al. [1990] and in Breveglieri et al. [1990], allows for any number of queues and stack heads operating on a single memory tape.
Finally we observe that timing aspects are currently not treated by our model. But this omission is common to the event-driven schedulers of current commercial real-time multi-tasking kernels, which usually do not provided deadline-driven scheduling, because it is computationally too expensive.

Aknowledgment
We thank Claudio Citrini and Dino Mandrioli for their comments.

4. References

Brandenburg F.J. "On the intersections of stacks and queues", *Theor. Comp. Sc.*, 58, 69-80, 1988.
Breveglieri L. et al. "Stacks, queues and their languages", Rept.N°. 90053, Dipt. Elettronica, Politecnico di Milano, 1990.
Cherubini A., et al. "Breadth and depth grammars and dequeue automata", *Foundations of Comp. Sc.*, 1, 3, 1990, 219-232.
Cherubini A. et al. "Quasi-real-time FIFO automata, breadth-first grammars and their relations", *Theor. Comp. Sc.*, 1991.
Cooling J.E. *Software design for real-time systems*, Chapman and Hall, London 1990.
Courtois P.J., Heyman R. and Parnas D.I. "Concurrent control with readers and writers", *Comm.ACM*, 14, 1971, 667-668.
Hemmendinger D. "Specifying Ada server tasks with executable formal grammars", *IEEE Trans.SE*, 16, 7, Jul. 1990, 741-754.
Herrtwich R.G. "An introduction to real-time schedulers", Rept. N°90-035, Int. Comp. Sc. Inst., July, 1990.
Liu C.L. and Layland J.W. "Scheduling algorithms for multiprogramming in a hard-real-time environment", *Journ. ACM*, 20, 1, Jan. 1973, 46-61.
Manacher G.K. "Production and stabilization of real-time task schedules", *Journ. ACM*, 14, 3, Jul. 1967, 439-465.
Vauquelin B. and Franchi Zannettacci P. "Automates à file", *Theor. Comp. Sc.*, 11, 221-225, 1980.
Vollmar R. " Ueber einem Automaten mit Pufferspeicherung", *Computer, 5*, 50-70, 1970.

Broadcast Communication for Real-time Processes

Jim Davies Dave Jackson Steve Schneider

Oxford University Computing Laboratory
Programming Research Group
11 Keble Road, Oxford
OX1 3QD

Abstract

Timed Communicating Sequential Processes (Timed CSP) is a mathematical approach to the design and analysis of timed concurrent systems. In the existing treatment of concurrency, a Timed CSP process and its environment must cooperate on all observable actions. In a description of a real-time process, it is sometimes convenient to include observable events that are not synchronisations. This paper extends Timed CSP to include signals: broadcast events which do not require the cooperation of the environment.

1 Introduction

In established process algebras, agents or processes are characterised by the occurrence and availability of observable events. In Timed CSP, these events are considered to be synchronisations with the environment of the process. A process cannot observe an event unless its cooperation is required for the event to occur. This provides for a simple and elegant treatment of concurrency, and leads to a compositional semantics.

However, such a view of communication proves inadequate if we wish to model broadcasting behaviour, in which a process may perform output events without the cooperation of its environment. Applications that exhibit such behaviour include satellite communication systems, packet radio systems, and the lowest level of many local area network protocols. This does not fit comfortably within a synchronous model of communication. In the existing models of Timed CSP, any process which offers to perform an event can always be prevented from doing so.

Further, when describing the behaviour of a real-time process, we may wish to include observable events that are not synchronisations. These signals may make it easier to describe and analyse certain aspects of behaviour, providing useful reference points in a history of the system. For example, an audible bell might be considered to be part of the user interface to a telephone network, even though the bell may ring without the cooperation of the user.

In this paper, we show that the language of Timed CSP may be extended to provide a simple treatment of broadcast communication. We introduce signals as distinguished events whose time of occurrence is determined only by the originating process. We present a semantic model for the extended language, and demonstrate that it is consistent with the existing Timed Failures model [Reed 88]. The material is illustrated with a simple example of broadcast communication: a satellite packet broadcasting protocol. The paper concludes with a discussion of alternative approaches.

2 Timed CSP with Signals

Hoare (1985) uses the word *process* to denote the behaviour pattern of an object, viewed through the occurrence and availability of certain *events*—atomic communications between an object and its environment. In previous studies, every event has been drawn from a universal alphabet of synchronisations Σ. Every event from Σ requires the participation of all observers. The addition of timing information [Reed & Roscoe 86] makes it easier to add signals—output events which occur as soon as they are offered—to our universal alphabet.

We add an alphabet of signals $\widehat{\Sigma}$ to our language: any event $\widehat{a}$ from $\widehat{\Sigma}$ will be hatted to distinguish it from the corresponding synchronisation a. The set of all communication events becomes

$$\widetilde{\Sigma} \triangleq \widehat{\Sigma} \cup \Sigma$$

Events from $\widehat{\Sigma}$ correspond to output events that may occur without the participation of the environment. The syntax of Timed CSP with signal events is a process algebra $TCSP_{sig}$, generated as follows:

$$\begin{aligned} P \; ::= \; & STOP \mid SKIP \mid WAIT\, t \mid a \xrightarrow{t} P \mid \widehat{a} \xrightarrow{t} P \mid P \,;\, P \mid \\ & P \,\Box\, P \mid a : A \xrightarrow{t_a} P_a \mid P \sqcap P \mid P \overset{t}{\rhd} P \mid \\ & P \;{}_A\|_B\; P \mid P \,|||\, P \mid P \setminus A \mid f(P) \mid \mu X \circ F(X) \end{aligned}$$

This is a modified form of the syntax presented in [Reed 88], with the addition of signal events and a timeout operator.

The term *STOP* represents a broken process which will never engage in external communication. Without an explicit treatment of internal activity, we cannot distinguish between the undesirable phenomena of deadlock and divergence. The term *SKIP* represents a process which does nothing except terminate immediately. As in untimed CSP, the special event $\checkmark$ is used to indicate that a process has terminated.

The term *WAIT t* is a delayed form of *SKIP*. It represents a process which does nothing except terminate successfully after time t. In Timed CSP, time values are drawn from the non-negative real numbers; we place no lower bound on the interval between consecutive events. This allows us to model asynchronous processes in a satisfactory fashion, without artificial constraints upon the times at which independent events may be observed.

The event prefix operator $\xrightarrow{t}$ allows us to introduce communication events into the behaviour pattern of a process. The process $a \xrightarrow{t} P$ is initially prepared to engage in synchronisation event a. If this event is performed, there is a delay of time t before control is passed to process P. In contrast, the process $\widehat{a} \xrightarrow{t} P$ will perform event $\widehat{a}$ immediately—signal events do not require the cooperation of the environment—and control is passed to process P at time t.

As an example, consider the simple process

$$a \xrightarrow{2} \widehat{b} \xrightarrow{1} STOP$$

which is initially prepared to engage in synchronisation a. If this synchronisation occurs, then the process will broadcast the signal $\widehat{b}$ exactly two time units later. One time unit after the signal has been broadcast, the process begins to behave as $STOP$.

The sequential composition operator provides a means of transferring control after a process has terminated. In the construct $P ; Q$, control is passed from P to Q if and when P performs the termination event $\checkmark$. This event is not visible to the environment, and occurs as soon as it becomes available. This is consistent with our assumption of *maximal parallelism*: if sufficient resources are available then every communication should occur as soon as all participants are ready.

Timed CSP provides two forms of choice: external and internal. An external choice $P \,\Box\, Q$ may be resolved by the environment of the process. If the environment is prepared to cooperate with process P, but not with process Q, then the choice behaves as P: the choice is resolved by the first communication. However, signal events do not require the cooperation of the environment; in our extended language, an external choice may be resolved by the transmission of a signal. In contrast, the environment has no influence over an internal choice $P \sqcap Q$: the outcome of such a choice is nondeterministic.

As an example of the interaction of choice and signals, consider the following external choice process:

$$(a \xrightarrow{2} SKIP) \quad \Box \quad (WAIT\ 1 ; \widehat{b} \xrightarrow{1} STOP)$$

This process is initially prepared to engage in event a, represented by the left-hand component of the choice. If this synchronisation occurs, then the process will terminate successfully 2 time units later. However, if a has not been performed by time 1, the right-hand component is able to perform a signal $\widehat{b}$. This communication then resolves the choice, and the process stops one time unit later.

A restricted form of indexed external choice may be used to model value-passing communication. If x is a value drawn from type T, the set of values that may be passed on synchronisation channel c, then the prefix choice

$$c?x : T \xrightarrow{t_x} P_x$$

denotes a process which ready for input on channel c. The environment of the process dictates which value x will arrive. In a similar way, we use the expression $c!v$ to denote the output of value v on channel c. No choice construct is required in this case; the value transmitted is determined by the sending process.

The timeout operator $P \overset{t}{\triangleright} Q$ transfers control from process P to process Q if no communications occur before time t. As above, this time may be any non-negative real number. If an attempt at communication involving process P is made at time t precisely, then the outcome will be nondeterministic. If either of the components should terminate, then the entire timeout construct terminates immediately.

In Timed CSP, the parallel combination of two processes P and Q is parametrised by two sets of events. In the construct

$$P \ {}_{A}\|_{B} \ Q$$

process P may perform only those synchronisations in A, process Q may perform only those synchronisations in B, and the two processes must cooperate on synchronisations drawn from the intersection of A and B. Synchronisations that are in neither A nor B are proscribed.

We intend that signals should be propagated through a parallel combination, and that available synchronisations are *triggered* by the corresponding signal events: if a signal $\widehat{a}$ is observed, then any process waiting to perform synchronisation a will be allowed to proceed. In the parallel combination above, any signal performed by P will be propagated to Q and to the environment, provided that it is declared in set A. The two sets A and B allow us to specify the connectivity of the parallel combination.

As an example, consider the parallel combination:

$$a \xrightarrow{1} b \xrightarrow{1} STOP \ {}_{\{a,b\}}\|_{\{a,\widehat{b}\}} \ a \xrightarrow{2} \widehat{b} \xrightarrow{2} STOP$$

This process is initially prepared to perform synchronisation a. If this event occurs, the right-hand side will broadcast signal $\widehat{b}$ two time units later. The inclusion of $\widehat{b}$ in the corresponding event set means that this signal will be propagated. Meanwhile, the left-hand side becomes ready to engage in synchronisation b one time unit after a is observed. This synchronisation will be triggered by the corresponding signal.

In the synchronous parallel combination described above, the inclusion of event b in both event sets, as either a signal or synchronisation, means that every occurrence of synchronisation b requires the cooperation of both components. The asynchronous parallel operator provides a much simpler form of parallel combination. In the process $P \mid\mid\mid Q$, both components evolve concurrently without interacting.

The hiding operator provides a mechanism for abstraction in Timed CSP. The process $P \setminus A$ behaves as P except that signals and synchronisations in A are concealed from the environment. Concealed synchronisations no longer require the cooperation of the environment, and so occur as soon as P is ready to perform them.

The relabelled process $f(P)$ has a similar control structure to P, with observable events renamed according to function f. We must insist that f does not map signal events to synchronisations, nor synchronisation events to signals. Finally, we use the expression $\mu X \circ F(X)$ to denote a process that behaves as $F(X)$, with X representing a recursive invocation of the process. This process will satisfy the equation $X = F(X)$.

Later in this paper, we will show how each of the operators described above can be given a formal semantics using the denotational model presented in the next section.

3 The Timed Signals Model

We will now formalise our intuitions about broadcast communication by providing a semantic model for the language $TCSP_{sig}$. The model is based upon the denotational models for Timed CSP devised by Reed and Roscoe; in particular, it may be seen as an extension of the Timed Failures model presented in [Reed 88].

Every observable event in the history of a process is labelled with a time of occurrence, where the domain of time values is taken to be the non-negative real numbers $[0, \infty)$. All observations are recorded with reference to an imaginary global clock, but this clock cannot be accessed by any part of the system being modelled. If system clocks are required, they may be modelled as Timed CSP processes.

In our denotational model, each piece of process algebra will be associated with a set of observations. Each observation is represented by a triple $(s, \aleph, t)$. The first component is a timed trace s: a record of observable events performed and the times at which they were observed. The second is a timed refusal $\aleph$: a record of synchronisations refused, and the times at which they were refused. The third is a time value t, representing the end of the current observation.

The addition of broadcast events to our computational model means that timing information is required to ensure the consistency of the timed trace information. If an observation extends to a time t, then all signals scheduled to occur before that time must be included in the trace. The notion of timed refusal is left unchanged from Reed's model: by their nature, signal events may not be refused.

Timed Observations

A timed event is a pair (t, e) where t is a time value drawn from the non-negative real numbers, and e is a communication event. The domain of time values is defined to be

$$TIME \triangleq [0, \infty)$$

and the set of all communication events is given by

$$\tilde{\Sigma} \triangleq \Sigma \cup \hat{\Sigma}$$

where Σ is the set of all possible synchronisations, and $\hat{\Sigma}$ is the set of all signals. The set of all timed events is thus

$$T\tilde{\Sigma} \triangleq TIME \times \tilde{\Sigma}$$

A timed trace is a chronologically ordered finite sequence of timed events; the set of all timed traces is given by:

$$T\tilde{\Sigma}^{*}_{\leqslant} \triangleq \{s \in \text{seq}\, T\tilde{\Sigma} \mid (t, e) \text{ precedes } (t', e') \text{ in } s \Rightarrow t \leqslant t'\}$$

The presence of a timed event (t, e) in a timed trace will correspond to the observation of the signal or synchronisation e at time t.

A timed refusal set is a set of timed synchronisations, formed as a finite union of refusal tokens. Each token is a Cartesian product set of the form $I \times A$, where I is a half-open time interval and A is a set of events. If we take $TINT$ to be the set of all such intervals, and $RTOK$ to be the set of all refusal tokens, then the set of all timed refusal sets is given by

$$RSET \triangleq \{\bigcup C \mid C \subseteq_{\text{fin}} RTOK\}$$

where

$$\begin{aligned} TINT &\triangleq \{[b, e) \mid 0 \leqslant b < e < \infty\} \\ RTOK &\triangleq \{I \times A \mid I \in TINT \wedge A \in \mathbb{P}\Sigma\} \end{aligned}$$

The presence of a timed event (t, a) in refusal set $\aleph$ corresponds to the refusal of the process to engage in synchronisation a at time t.

The set of possible observations is given by $\widetilde{TF}$, where

$$\widetilde{TF} \triangleq T\widetilde{\Sigma}^*_{\leqslant} \times RSET \times TIME$$

Each observation is a triple, consisting of a timed trace from $T\widetilde{\Sigma}^*_{\leqslant}$, a timed refusal set from $RSET$, and a time value.

Notation

To give a semantics to our language, and to simplify the process of reasoning about it, we define a variety of simple operators on timed traces, timed refusals, and timed observations. We inherit the following notation from [Hoare 85]:

$\frown$	concatenation of traces	**in**	contiguous subsequence
$\langle\rangle$	the empty trace	$\#$	length of a sequence

The predicate s_1 **in** s_2 holds precisely when trace s_1 is a contiguous subsequence of s_2. If s is a timed trace, then $\#(s)$ returns the number of timed events in that trace.

The *begin* and *end* operators return the times of occurrence of the first and last event in a timed trace, respectively:

$$\begin{aligned} begin(\langle\rangle) &\triangleq \infty & end(\langle\rangle) &\triangleq 0 \\ begin(\langle(t, a)\rangle \frown s) &\triangleq t & end(s \frown \langle(t, a)\rangle) &\triangleq t \end{aligned}$$

The values chosen for the empty trace are the most convenient for the subsequent mathematics: the possibility of a trace being empty will not require special consideration in our specifications and proofs. We will apply these operators, with similar interpretations, to timed refusals and to timed (trace, refusal) pairs:

$$\begin{aligned} begin(\aleph) &\triangleq inf(\{t \mid \exists\, a \bullet (t, a) \in \aleph\}) \\ begin(\{\}) &\triangleq \infty \\ end(\aleph) &\triangleq sup(\{t \mid \exists\, a \bullet (t, a) \in \aleph\}) \\ end(\{\}) &\triangleq 0 \end{aligned}$$

$$\begin{aligned} begin(s, \aleph) &\triangleq min\{begin(s), begin(\aleph)\} \\ end(s, \aleph) &\triangleq max\{end(s), end(\aleph)\} \end{aligned}$$

We define the *during* ($\uparrow$) operator on timed traces and refusals, returning the part of the trace or refusal that lies in a specified time interval:

$$\begin{aligned} \langle\rangle \uparrow I &\mathrel{\widehat{=}} \langle\rangle \\ (\langle(t,a)\rangle^\frown s) \uparrow I &\mathrel{\widehat{=}} \langle(t,a)\rangle^\frown(s \uparrow I) \quad \text{if } t \in I \\ &\phantom{\mathrel{\widehat{=}}} (s \uparrow I) \quad \text{otherwise} \\ \aleph \uparrow I &\mathrel{\widehat{=}} \aleph \cap (I \times \Sigma) \end{aligned}$$

where I is a set of real numbers. In the case that $I = \{t\}$ for some time t, we may omit the set brackets.

In many cases, we are interested only the behaviour of a process up until a particular time t. It will prove convenient to define a *before* ($\upharpoonright$) operator as an abbreviation:

$$\begin{aligned} s \upharpoonright t &\mathrel{\widehat{=}} s \uparrow [0,t] \\ \aleph \upharpoonright t &\mathrel{\widehat{=}} \aleph \uparrow [0,t) \end{aligned}$$

The definition of *before* on refusal sets differs from that on timed traces. For traces, $s \upharpoonright t$ includes events at t; in the case of refusals, such events are excluded. This choice of definitions is the most convenient for timed failures specifications.

We define alphabet operators on traces and refusals, yielding the set of synchronisations or signals present:

$$\begin{aligned} \sigma(s) &\mathrel{\widehat{=}} \{a \in \Sigma \mid \exists\, t \bullet \langle(t,a)\rangle \text{ in } s\} \\ \hat{\sigma}(s) &\mathrel{\widehat{=}} \{\hat{a} \in \hat{\Sigma} \mid \exists\, t \bullet \langle(t,\hat{a})\rangle \text{ in } s\} \\ \sigma(\aleph) &\mathrel{\widehat{=}} \{a \in \Sigma \mid \exists\, t \bullet (t,a) \in \aleph\} \end{aligned}$$

and an operator that shifts timed traces and refusals through time:

$$\begin{aligned} \langle\rangle + t &\mathrel{\widehat{=}} \langle\rangle \\ (\langle(t_1,a)\rangle^\frown s) + t &\mathrel{\widehat{=}} (\langle(t_1+t,a)\rangle^\frown(s+t)) \uparrow [0,\infty) \\ \aleph + t &\mathrel{\widehat{=}} \{(t_1+t,a) \mid (t_1,a) \in \aleph\} \uparrow [0,\infty) \end{aligned}$$

It proves economical to extend this operator to timed observations:

$$(s,\aleph,t_1) + t \mathrel{\widehat{=}} (s+t, \aleph+t, t_1+t)$$

The Timed Signals Model

We will associate each process P with a set of possible observations. Recalling that the set of all timed observations is given by $T\widetilde{F}$, this means mapping each process algebra term to an element of $TS_{\widetilde{F}}$, where

$$TS_{\widetilde{F}} \mathrel{\widehat{=}} \mathbb{P}\, T\widetilde{F}$$

We define the Timed Signals model $TM_{\widetilde{F}}$ to be those elements of $TS_{\widetilde{F}}$ which satisfy a certain set of healthiness conditions. These conditions are expressed as axioms of the semantic model.

If S is to be an element of $TM_{\tilde{F}}$, then

1. $(\langle\rangle, \{\}, 0) \in S$

2. $(s, \aleph, t) \in S \Rightarrow t \geqslant end(s, \aleph)$

3. $(s, \aleph, t) \in S \wedge t' \geqslant t \Rightarrow \exists\, s' \bullet \sigma(s') = \{\} \wedge (s \frown (s' + t), \aleph, t') \in S$

4. $(s \frown w, \aleph, t) \in S \wedge end(s) \leqslant t' \leqslant min\{t, begin(w)\} \Rightarrow (s, \aleph \restriction t', t') \in S$

5. $(s, \aleph, t) \in S \wedge \aleph' \in RSET \wedge \aleph' \subseteq \aleph \Rightarrow (s, \aleph', t) \in S$

6. $(s, \aleph, t) \in S \Rightarrow \exists\, \aleph' \in RSET \bullet \aleph \subseteq \aleph' \wedge (s, \aleph', t) \in S \wedge$
$$\forall\, t' : TIME\, ; a : \Sigma \bullet (t' \leqslant t \wedge (t', a) \notin \aleph') \Rightarrow (s \restriction t' \frown \langle (t', a) \rangle, \aleph' \restriction t', t') \in S$$

7. $\forall\, t : TIME \bullet \exists\, n(t) \in \mathbb{N} \bullet (s, \aleph, t) \in S \Rightarrow \#(s) \leqslant n(t)$

The first axiom insists that every process admits at least one observation—the empty trace and the empty refusal, observed up until time 0. The second axiom insists that no information is recorded after the observation is supposed to have ended: in the statement of the axiom, neither the trace s nor the refusal $\aleph$ may extend beyond time t.

The third axiom states that any observation may be extended into the future; to do this, it may be necessary to add a series of signal events to the current trace. The fourth axiom states that any observation of a process must give rise to another observation of the same process if truncated, and the fifth asserts that if a process may refuse the entirety of a refusal set $\aleph$, then it may refuse any subset of $\aleph$.

The remaining pair of axioms correspond to our assumption of *finite variability*: no process may undergo infinitely many changes of state in a finite time. The sixth states that for any observation $(s, \aleph, t)$ there will always be a complete refusal set $\aleph'$ that captures all of the refusal information for the current trace. Given any time t', every timed synchronisation (t', a) not in $\aleph'$ is a possible extension of $s \restriction t'$. As $\aleph'$ is a refusal set, it must be a *finite* union of refusal tokens, and hence represents only finitely many changes of state.

The final axiom places a similar condition upon traces. For any process S, we can exhibit a function n that places a bound upon the number of events observed before a given time. If trace s ends at or before time t, then the length of s must be no greater than $n(t)$. This bounded speed condition leads to constraints upon the application of infinitary operators such as infinite choice and indexed nondeterminism.

To give a semantics to recursive process definitions, we will define a metric upon the space of timed observation sets. For any $S \in TS_{\tilde{F}}$ and $t \in TIME$, consider

$$S(t) \;\triangleq\; \{(s, \aleph, t') \in S \mid t' \leqslant t\}$$

which yields the set of observations from S that end at or before time t.

Our distance metric upon $TS_{\tilde{F}}$ is then given by

$$\tilde{d}(S,T) \;\hat{=}\; inf(\{2^{-t} \mid S(t) = T(t)\} \cup \{1\})$$

The distance between two processes is determined by the first time at which they exhibit dissimilar behaviours.

In the following sections, we will define a semantic function $\mathcal{F}_S$ by recursion upon the terms in our process algebra $TCSP_{sig}$

$$\mathcal{F}_S \;\in\; TCSP_{sig} \rightarrow TS_{\tilde{F}}$$

We will write $\mathcal{F}_S[\![P]\!]$ to denote the semantics of term P.

4 Sequential Processes

Atoms

The only trace of the broken process $STOP$ is the empty trace $\langle\rangle$, and any timed event may be refused:

$$\mathcal{F}_S[\![STOP]\!] \;\hat{=}\; \{(s, \aleph, t) \mid s = \langle\rangle \wedge t \geqslant end(\aleph)\}$$

For consistency, we insist that all refusals are recorded before the observation ends.

The terms $SKIP$ and $WAIT\ t$ are closely related. $SKIP$ denotes a process which is ready to terminate immediately, while $WAIT\ t$ denotes a process which becomes ready to terminate at time t. In any reasonable semantic model, we would expect the following equivalence to hold:

$$SKIP \;\equiv\; WAIT\ 0$$

Indeed, we will use this equivalence to define the semantics of $SKIP$.

The only event which may appear in a trace of the delay process $WAIT\ t$ is the special synchronisation $\checkmark$. This event is made available at time t, and remains available until it occurs. If no events have been observed, then $\checkmark$ is available from time t until the end of the current observation; this corresponds to its exclusion from the refusal set $\aleph$ during the interval $[t, t')$. Any synchronisation may be refused before time t.

$$\begin{aligned}\mathcal{F}_S[\![WAIT\ t]\!] \;\hat{=}\; & \{(\langle\rangle, \aleph, t') \mid \checkmark \notin \sigma(\aleph \uparrow [t, t')) \wedge t' \geqslant end(\aleph)\} \\ & \cup \\ & \{(\langle(t'', \checkmark)\rangle, \aleph, t') \mid t'' \geqslant t \wedge t' \geqslant max\{t'', end(\aleph)\} \\ & \qquad \wedge \checkmark \notin \sigma(\aleph \uparrow [t, t''))\}\end{aligned}$$

If $\checkmark$ is observed at some time t'', then it must have been available since time t.

Prefix

The semantics of the event prefix operator depends upon whether the event in question is a synchronisation or a signal. If the event is a synchronisation, then it is made available immediately, and remains available until it occurs:

$$
\begin{aligned}
\mathcal{F}_S[\![a \xrightarrow{t} P]\!] \;\; \widehat{=} \;\; & \{(\langle\rangle, \aleph, t') \mid a \notin \sigma(\aleph) \wedge t' \geqslant end(\aleph)\} \\
& \cup \\
& \{(\langle(t'', a)\rangle^\frown s, \aleph, t') \mid \; a \notin \sigma(\aleph \upharpoonright t'') \wedge \\
& \qquad begin(s) \geqslant t'' + t \wedge \\
& \qquad t' \geqslant max\{t'', end(s, \aleph)\} \wedge \\
& \qquad (s, \aleph, t') - (t'' + t) \in \mathcal{F}_S[\![P]\!] \}
\end{aligned}
$$

If the event is a signal, then it should occur immediately.

$$
\begin{aligned}
\mathcal{F}_S[\![\hat{a} \xrightarrow{t} P]\!] \;\; \widehat{=} \;\; & \{(\langle\rangle, \{\}, 0)\} \\
& \cup \\
& \{(\langle(0, \hat{a})\rangle^\frown s, \aleph, t') \mid \; t' \geqslant end(s, \aleph) \wedge \\
& \qquad begin(s) \geqslant t \wedge \\
& \qquad (s, \aleph, t') - t \in \mathcal{F}_S[\![P]\!] \}
\end{aligned}
$$

In either case, once the event has been observed, control is passed to process P after a delay of time t. During this delay, no events may be observed, and any synchronisation may be refused. The subsequent behaviour, if shifted to start at time 0, will be a possible behaviour of P.

Sequential Composition

In the sequential composition $P\,;Q$ control is transferred from P to Q as soon as P offers to synchronise upon the termination event $\checkmark$. If control has not been transferred, then we observe a non-terminating behaviour of P: a behaviour in which $\checkmark$ has not been made available. In this case, $\checkmark$ must be absent from the trace, and may be added to the refusal set at any time.

$$
\begin{aligned}
\mathcal{F}_S[\![P\,; Q]\!] \;\; \widehat{=} \;\; & \{(s, \aleph, t') \mid \checkmark \notin \sigma(s) \wedge (s, \aleph \cup ([0, t') \times \{\checkmark\}), t') \in \mathcal{F}_S[\![P]\!] \} \\
& \cup \\
& \{(s^\frown w, \aleph, t') \mid \checkmark \notin \sigma(s) \wedge (w, \aleph, t') - t \in \mathcal{F}_S[\![Q]\!] \\
& \qquad \wedge \\
& \qquad (s^\frown\langle(t, \checkmark)\rangle, \aleph \upharpoonright t \cup ([0, t) \times \{\checkmark\}), t) \in \mathcal{F}_S[\![P]\!] \}
\end{aligned}
$$

If control is transferred at time t, then the subsequent behaviour, if shifted to start at time 0, will be a possible behaviour of Q. The behaviour observed before time t must correspond to a terminating behaviour of P; it may be followed immediately by an occurrence of $\checkmark$. In the presence of the sequential composition operator, the termination event $\checkmark$ occurs as soon as it is made available, and is concealed from the environment.

Choice

An external choice $P \Box Q$ is resolved by the first observable event that occurs. If no events have been observed, then any event refused must be refused by both components:

$$\begin{aligned}
\mathcal{F}_S[\![P \Box Q]\!] \;\hat{=}\;& \{(\langle\rangle, \aleph, t') \mid (\langle\rangle, \aleph, t') \in \mathcal{F}_S[\![P]\!] \cap \mathcal{F}_S[\![Q]\!]\} \\
& \cup \\
& \{(s, \aleph, t') \mid s \neq \langle\rangle \wedge (s, \aleph, t') \in \mathcal{F}_S[\![P]\!] \cup \mathcal{F}_S[\![Q]\!] \\
& \qquad \wedge \\
& \qquad (\langle\rangle, \aleph \restriction begin(s), begin(s)) \in \mathcal{F}_S[\![P]\!] \cap \mathcal{F}_S[\![Q]\!]\}
\end{aligned}$$

If an event have been observed, then the current observation must be a possible observation of either P or Q. Again, any refusal information observed before the beginning of the trace must be common to both processes.

The prefix choice operator may be used to offer the environment an infinite choice of inputs to a process. Signals correspond to output events, so there is no reason to include signals in a prefix choice construct.

$$\begin{aligned}
\mathcal{F}_S[\![a : A \xrightarrow{t_a} P_a]\!] \;\hat{=}\;& \{(\langle\rangle, \aleph, t') \mid A \cap \sigma(\aleph) = \{\}\} \\
& \cup \\
& \{(\langle(t, a)\rangle ^\frown s, \aleph, t') \mid a \in A \wedge t \geqslant 0 \;\wedge \\
& \qquad\qquad begin(s) \geqslant t + t_a \;\wedge \\
& \qquad\qquad A \cap \sigma(\aleph \restriction t) = \{\} \;\wedge \\
& \qquad\qquad (s, \aleph, t') - (t + t_a) \in \mathcal{F}_S[\![P_a]\!]\}
\end{aligned}$$

We assume that the set A contains only synchronisation events, and that the set of alternatives $\{P_a \mid a \in A\}$ is *uniformly bounded*—there must exist a bounding function $n : TIME \rightarrow \mathbb{N}$ such that

$$\forall a : A \,;\, t : TIME \;\bullet\; (s, \aleph, t') \in \mathcal{F}_S[\![P_a]\!] \wedge end(s) \leqslant t \Rightarrow \#(s \restriction t) \leqslant n(t)$$

This assumption is necessary to ensure compliance with the bounded speed assumption of our model, as expressed by axiom 7 above.

An observation of the internal choice process $P \sqcap Q$ may be an observation of either of the component processes:

$$\mathcal{F}_S[\![P \sqcap Q]\!] \;\hat{=}\; \mathcal{F}_S[\![P]\!] \cup \mathcal{F}_S[\![Q]\!]$$

This semantic equation may be extended to give a meaning to indexed nondeterministic choice, provided that the set of alternatives is uniformly bounded.

In the timeout process $P \overset{t}{\triangleright} Q$, control is passed from P to Q if no events are observed before time t. Any observation of P in which the first observable event occurs at or before time t is a possible observation of this process.

$$\begin{aligned}
\mathcal{F}_S[\![P \overset{t}{\triangleright} Q]\!] \;\hat{=}\;& \{(s, \aleph, t') \mid begin(s) \leqslant t \wedge (s, \aleph, t') \in \mathcal{F}_S[\![P]\!]\} \\
& \cup \\
& \{(s, \aleph, t') \mid begin(s) \geqslant t \wedge (\langle\rangle, \aleph \restriction t, t) \in \mathcal{F}_T[\![P]\!] \\
& \qquad \wedge \\
& \qquad (s, \aleph, t') - t \in \mathcal{F}_S[\![Q]\!]\}
\end{aligned}$$

If control has been transferred, we know that any event refused before time t will be refused by process P, and that the subsequent behaviour must correspond to an observation of process Q.

Relabelling

The process $f(P)$ behaves as P, but with the observable events renamed according to relabelling function f. We do not allow signal events to be relabelled as synchronisations, or *vice versa*. If f is such that

$$\forall a : \Sigma ; \hat{a} : \hat{\Sigma} \bullet f(a) \in \Sigma \wedge f(\hat{a}) \in \hat{\Sigma}$$

then the semantics of $f(P)$ is given by

$$\mathcal{F}_S[\![f(P)]\!] \mathrel{\widehat{=}} \{(f(s), \aleph, t) \mid (s, f^{-1}(\aleph), t) \in \mathcal{F}_S[\![P]\!]\}$$

In the above equation, the expression $f^{-1}(\aleph)$ denotes the set

$$\{(t, a) \mid (t, f(a)) \in \aleph\}$$

This is the inverse image of refusal set $\aleph$ under function f.

Abstraction

In the process $P \setminus A$, every event in set A is concealed from the environment. Our maximal parallelism assumption means that any synchronisation will occur as soon as all of the processes involved are ready to cooperate. A hidden synchronisation does not require the cooperation of any other process, and so occur as soon as P is ready. A hidden signal event is simply not propagated to the environment.

If s represents a trace in which every instance of a is performed as soon as it becomes available, then any additional offers of this synchronisation will be refused. The observation $(s, \aleph, t')$ corresponds to a behaviour of P in which every instance of a occurs as soon as possible iff

$$(s, \aleph \cup ([0, t') \times \{a\}), t')$$

is also an observation of P.

This leads to the following semantic equation for the hiding operator:

$$\mathcal{F}_S[\![P \setminus A]\!] \mathrel{\widehat{=}} \{(s \setminus A, \aleph, t') \mid (s, \aleph \cup ([0, t') \times (A \cap \Sigma)), t') \in \mathcal{F}_S[\![P]\!]\}$$

where the hiding operator on traces is defined by

$$\begin{aligned} \langle\rangle \setminus A &\mathrel{\widehat{=}} \langle\rangle \\ (\langle(t, e)\rangle \frown s) \setminus A &\mathrel{\widehat{=}} s \setminus A && \text{if } e \in A \\ & \phantom{\mathrel{\widehat{=}}} \langle(t, e)\rangle \frown (s \setminus A) && \text{otherwise} \end{aligned}$$

In the above definition, e denotes any communication event.

Recursion

The expression $\mu X \circ F(X)$ denotes a process that behaves as $F(X)$, with X representing a recursive invocation of the process. The semantics of this expression is given by

$$\mathcal{F}_S[\![\mu X \circ F(X)]\!] \triangleq \text{the unique fixed point of the mapping } M_F \text{ on } TM_{\widetilde{F}} \text{ corresponding to syntactic function } F$$

Our semantic model is a complete metric space under the metric $\widetilde{d}$ defined in section 3. From the results presented in [Davies 91], we may infer that the mapping M_F has a unique fixed point when every occurrence of X in $F(X)$ is guarded by a strictly positive time delay. For example, if we define

$$F(X) \triangleq (\widehat{a} \xrightarrow{1} X) \Box (WAIT\ 2\ ; X)$$
$$G(X) \triangleq STOP \sqcap X$$

then $\mu X \circ F(X)$ has a well-defined semantics, whereas $\mu X \circ G(X)$ does not.

5 Signals and Concurrency

We intend that signals should be propagated through a parallel combination, and that available synchronisations should be *triggered* by the corresponding signal events: if a signal $\widehat{a}$ is observed, then any process waiting to perform synchronisation a will be allowed to proceed. Observable synchronisations require the participation of the environment, so if a signal forms part of the interface between two processes then the corresponding synchronisation must be concealed.

In the parallel combination $P\ {}_A\|_B\ Q$, the sets A and B determine which synchronisations may be performed by processes P and Q, respectively. By adding signals to these sets, we may also determine which signals are propagated. It would be possible to conceal only those synchronisations which occur at the same time as the corresponding signal events, allowing a process to signal and synchronise upon the same event. However, we obtain a more intuitive semantics for concurrency if we proscribe this dynamic reconfiguration of input and output channels.

As an example, consider the following choices for A and B:

$$A \triangleq \{a, b, \widehat{c}\} \qquad B \triangleq \{\widehat{a}, b, \widehat{c}\}$$

In this case, either component may broadcast signal $\widehat{c}$ to the environment, and Q may broadcast $\widehat{a}$. If Q broadcasts $\widehat{a}$, then P may perform synchronisation a, but only the signal will be propagated to the environment. As before, both components must cooperate upon any synchronisation in $A \cap B$.

For convenience, we define an operator *sync* upon timed traces and sets of timed events. For any $s \in T\widetilde{\Sigma}^*_{\leqslant}$ or $A \subseteq T\widetilde{\Sigma}$,

$$sync(\langle\rangle) \triangleq \langle\rangle$$
$$sync(\langle(t,a)\rangle^\frown s) \triangleq \langle(t,a)\rangle^\frown sync(s)$$
$$sync(\langle(t,\widehat{a})\rangle^\frown s) \triangleq \langle(t,a)\rangle^\frown sync(s)$$
$$sync(A) \triangleq \{a \in \Sigma \mid a \in A \vee \widehat{a} \in A\}$$

This operator returns the set of synchronisation events involved, as either synchronisations or signals. We also define an inclusion relation on timed traces:

$$s_1 \subseteq s_2 \;\Leftrightarrow\; \forall t, a \bullet \langle(t,a)\rangle \text{ in } s_1 \Rightarrow \langle(t,a)\rangle \text{ in } s_2$$

We say that a trace s_1 is included in trace s_2 if and only if each timed event in s_1 is also present in s_2. We will require a restriction operator for traces and refusals:

$$\begin{array}{rcll} \langle\rangle \restriction A & \hat{=} & \langle\rangle & \\ (\langle(t,e)\rangle^\frown s) \restriction A & \hat{=} & \langle(t,e)\rangle^\frown(s \restriction A) & \text{if } e \in A \\ & & s \restriction A & \text{otherwise} \\ \aleph \restriction A & \hat{=} & \{(t,a) \mid a \in A \wedge (t,a) \in \aleph\} & \end{array}$$

This operator removes any events not in A from the trace or refusal.

To derive the semantic equation for the parallel combination $P \;{}_A\|_B\; Q$, suppose that the traces performed by components P and Q are s_P and s_Q respectively. Our assumption that each component may treat each event as a signal or synchronisation, but not both, corresponds to the following restriction upon sets A and B:

$$\forall S \in \{A, B\} \;\bullet\; \{a \mid a \in \Sigma \wedge a \in S \wedge \hat{a} \in S\} \;=\; \{\}$$

Any synchronisation common to both sets must be performed by both components:

$$s \restriction \Sigma \restriction (A \cap B) \;=\; s_P \restriction \Sigma \restriction (A \cap B) \;=\; s_Q \restriction \Sigma \restriction (A \cap B)$$

If s is a trace of the parallel combination arising from traces s_P and s_Q, then all three traces must agree when restricted to synchronisations from the set $A \cap B$.

If a synchronisation a is declared in A, and the corresponding signal $\hat{a}$ is declared in B, then component P may perform a only when Q broadcasts $\hat{a}$. This corresponds to the following condition upon traces:

$$\begin{array}{rcl} s_P \restriction (A \cap sync(B \cap \hat{\Sigma})) & \subseteq & sync(s_Q \restriction \hat{\Sigma} \restriction B) \\ s_Q \restriction (B \cap sync(A \cap \hat{\Sigma})) & \subseteq & sync(s_P \restriction \hat{\Sigma} \restriction A) \end{array}$$

A timed synchronisation (t, a) may appear in s_P only if $(t, \hat{a})$ appears in s_Q and is propagated. This is true whenever (t, a) appears in the trace $sync(s_Q \restriction \hat{\Sigma} \restriction B)$. A similar condition applies for synchronisations of Q.

A synchronisation that is exclusive to one component will be observed if it is performed by that component, and not hidden by a corresponding signal:

$$\begin{array}{rcl} s \restriction \Sigma \restriction (A - B) & = & s_P \restriction (\Sigma - sync(B \cap \hat{\Sigma})) \restriction (A - B) \\ s \restriction \Sigma \restriction (B - A) & = & s_Q \restriction (\Sigma - sync(A \cap \hat{\Sigma})) \restriction (B - A) \end{array}$$

The above conditions completely determine the synchronisations that may be observed of the parallel combination. Signal events are observed iff they are declared in A or B and performed by the corresponding process:

$$s \restriction \hat{\Sigma} \;\in\; s_P \restriction \hat{\Sigma} \restriction A \,|||\, s_Q \restriction \hat{\Sigma} \restriction B$$

where the set of possible interleavings of s_1 and s_2 is defined by

$$s_1 \mid\mid\mid s_2 \;\widehat{=}\; \{s \mid s \in T\widetilde{\Sigma}^*_{\leqslant} \wedge s \text{ permutes } s_1 \frown s_2 \wedge s_1 \subseteq s \wedge s_2 \subseteq s\}$$

Observe that simultaneous events from different components may occur in any order in an interleaved trace.

Each component may perform signals that are not declared in A and B, but these events will not be propagated. If we define

$$C \;\widehat{=}\; A \cap \widehat{\Sigma} \qquad D \;\widehat{=}\; B \cap \widehat{\Sigma}$$

to abbreviate the sets of signal declared in A and B respectively, then we may characterise the set of traces that may arise from component traces s_P and s_Q:

$$\begin{aligned} s \in s_P \;{}_A\widetilde{\|}_B\; s_Q \;\Leftrightarrow\; & s \upharpoonright \Sigma \upharpoonright (A \cap B) = s_P \upharpoonright \Sigma \upharpoonright (A \cap B) = s_Q \upharpoonright \Sigma \upharpoonright (A \cap B) \;\wedge \\ & s \upharpoonright \Sigma \upharpoonright (A - B) = s_P \upharpoonright (\Sigma - sync(D)) \;\wedge \\ & s \upharpoonright \Sigma \upharpoonright (B - A) = s_Q \upharpoonright (\Sigma - sync(C)) \;\wedge \\ & s_P \upharpoonright (A \cap sync(D)) \subseteq sync(s_Q) \;\wedge \\ & s_Q \upharpoonright (B \cap sync(C)) \subseteq sync(s_P) \;\wedge \\ & s \upharpoonright \widehat{\Sigma} \in s_P \upharpoonright C \mid\mid\mid s_Q \upharpoonright D \;\;\wedge\; s \upharpoonright \Sigma = s \upharpoonright \Sigma \upharpoonright (A \cup B) \end{aligned}$$

If P performs a signal $\widehat{a}$ at time t, and $\widehat{a}$ is declared in set A, then this signal will be propagated. If the corresponding synchronisation a is declared in B, then the occurrence of $\widehat{a}$ should trigger as many copies of a as process Q is ready to perform.

By our assumption of maximal parallelism, every instance of a available at time t has been performed if and only if the timed event (t, a) would be refused if offered. The observation $(s_Q, \aleph_Q, t')$ corresponds to a behaviour of Q in which every available a is triggered at time t iff we may add (t, a) to the refusal set. If s_P and s_Q are the traces performed by components P and Q, then we insist that for any time t

$$\begin{aligned} sync(\sigma(s_Q \upharpoonright D \uparrow t)) &\subseteq \sigma(\aleph'_P \upharpoonright A \uparrow t) \\ sync(\sigma(s_P \upharpoonright C \uparrow t)) &\subseteq \sigma(\aleph'_Q \upharpoonright B \uparrow t) \end{aligned}$$

The synchronisations corresponding to the signals performed by Q at time t must be refused by component P at time t, if they are contained in set A. A similar condition applies to signals performed by P.

Each observation $(s, \aleph, t)$ of the parallel combination will arise from a consistent pair of observations, one from each component:

$$(s_P, \aleph_P \cup \aleph'_P, t) \in \mathcal{F}_S\llbracket P \rrbracket \;\;\wedge\;\; (s_Q, \aleph_Q \cup \aleph'_Q, t) \in \mathcal{F}_S\llbracket Q \rrbracket$$

Any synchronisation from A that is refused by component P must be refused by the parallel combination. A similar condition applies to events from B.

$$\aleph_P \subseteq \aleph \upharpoonright A \;\;\wedge\;\; \aleph_Q \subseteq \aleph \upharpoonright B$$

We need consider only those refusals of P and Q which correspond to events declared in A and B: any other synchronisation may be refused by the parallel combination.

Any event in $A \cup B$ which is refused by the parallel combination must be refused by at least one component, or concealed by the declaration of the corresponding signal:

$$\aleph \downharpoonright (A \cup B) \setminus sync(C \cup D) \quad = \quad (\aleph_P \setminus sync(D)) \cup (\aleph_Q \setminus sync(C))$$

Recall that an event $\widehat{a}$ is declared as a signal if and only if the corresponding synchronisation a is an element of $sync(C)$ or $sync(D)$.

Combining these conditions, we obtain the semantic equation for synchronising parallel combination in the signals model:

$$\begin{aligned} \mathcal{F}_S[\![P \;{}_A\|_B\; Q]\!] \;\; \widehat{=} \;\; \{(s, \aleph, t) \mid \exists\, s_P, \aleph_P, \aleph'_P, s_Q, \aleph_Q, \aleph'_Q \bullet \forall t' \bullet \\ s \in s_P \;{}_A\widetilde{\|}_B\; s_Q \;\wedge\; \aleph \in \aleph_P \;{}_A\widetilde{\|}_B\; \aleph_Q \;\wedge \\ sync(\sigma(s_Q \downharpoonright D \uparrow t')) \subseteq \sigma(\aleph'_P \downharpoonright A \uparrow t') \;\wedge \\ sync(\sigma(s_P \downharpoonright C \uparrow t')) \subseteq \sigma(\aleph'_Q \downharpoonright B \uparrow t') \;\wedge \\ (s_P, \aleph_P \cup \aleph'_P, t) \in \mathcal{F}_S[\![P]\!] \;\wedge \\ (s_Q, \aleph_Q \cup \aleph'_Q, t) \in \mathcal{F}_S[\![Q]\!] \;\} \end{aligned}$$

where

$$\begin{aligned} \aleph \in \aleph_P \;{}_A\widetilde{\|}_B\; \aleph_Q \;\Leftrightarrow\; & \aleph_P \subseteq \aleph \downharpoonright A \;\wedge\; \aleph_Q \subseteq \aleph \downharpoonright B \;\wedge \\ & \aleph \downharpoonright (A \cup B) \setminus sync(C \cup D) = (\aleph_P \setminus sync(D)) \cup (\aleph_Q \setminus sync(C)) \end{aligned}$$

The semantics of asynchronous parallel combination is considerably simpler:

$$\begin{aligned} \mathcal{F}_S[\![P \;|||\; Q]\!] \;\; \widehat{=} \;\; \{(s, \aleph, t) \mid \exists\, s_P, s_Q \bullet s \in s_P \;|||\; s_Q \wedge (s_P, \aleph, t) \in \mathcal{F}_S[\![P]\!] \\ \wedge (s_Q, \aleph, t) \in \mathcal{F}_S[\![Q]\!]\} \end{aligned}$$

The two components may perform and refuse synchronisations independently. Signal events are propagated to the environment, but do not trigger synchronisations in the other component.

6 Consistency

The Timed Signals model is an extension of Reed's Timed Failures model TM_F. If a process P does not involve signal events, then the semantics of P in $TM_{\widetilde{F}}$ will be equivalent to the semantics of P in TM_F. In the latter, processes are represented by sets of timed failures: timed (trace, refusal) pairs.

Theorem If we use ι to denote the natural injection mapping

$$\iota\, S \;\; \widehat{=} \;\; \{(s, \aleph, t) \mid t \geqslant end(s, \aleph) \wedge (s, \aleph) \in S\}$$

then we may establish that

$$\mathcal{F}_S[\![P]\!] \;\; = \;\; \iota\,(\mathcal{F}_T[\![P]\!])$$

for any process P constructed entirely without signal events. $\heartsuit$

Proof It is easy to show that the defining axioms of TM_F, given in [Reed 88], are consistent with those of $TM_{\tilde{F}}$, in the sense that

$$S \in TM_F \;\Rightarrow\; \iota(S) \in TM_{\tilde{F}}$$

If a set satisfies the axioms of the Timed Failures model, then the image of S satisfies the axioms of the Timed Signals model. The proof may be completed by structural induction upon the syntax of $TCSP_{sig}$. □

7 Example

As an application of the Signals model, we consider a Timed CSP description of a satellite packet broadcasting protocol, similar to the slotted ALOHA protocol described in [Tanenbaum 81]. In this protocol, the satellite acts as a repeater, rebroadcasting messages received from independent ground stations. The satellite also broadcasts a clock signal, dividing time into discrete intervals, or *slots*. A ground station which is ready to transmit must wait for the next slot before broadcasting. If only one station transmits during a particular slot, then the satellite will receive the message intact, and rebroadcast it to all ground stations.

We may model the satellite as the parallel combination of two processes:

$$SATELLITE \;\hat{=}\; CLOCK \;{}_{\{\widehat{pip}\}}\|_{\{pip,up,down\}}\; REFLECT$$

Messages are received on channel *up*, and rebroadcast on channel *down*. The clock process broadcasts a $\widehat{pip}$ signal every slot time t_s

$$CLOCK \;\hat{=}\; \widehat{pip} \xrightarrow{t_s} CLOCK$$

and the rebroadcasting process is triggered by this event

$$REFLECT \;\hat{=}\; pip \xrightarrow{t_1} (up?m \xrightarrow{0} R_m) \overset{t_2}{\triangleright} REFLECT$$

Shortly after a *pip* occurs, the rebroadcasting process is ready to receive any message m on channel *up*. If one message arrives, then the process behaves as R_m, where

$$R_m \;\hat{=}\; (\Box_{n \neq m}\, up.n \xrightarrow{t_3} RELAY_g) \overset{t_2}{\triangleright} RELAY_m$$

This process watches for another message until time t_2 has elapsed. To ensure correct operation of the protocol, any messages must be received within time $t_1 + t_2$ of the beginning of a slot. This leads to a constraint upon the time taken for a ground station to begin transmission following the arrival of a clock signal, modelled by delay t_6 below. If another message $n \neq m$ arrives before time t_2, then a garbled message g will result. Otherwise, the first message m will be rebroadcast:

$$RELAY_m \;\hat{=}\; (pip \xrightarrow{t_4} \widehat{down}.m \xrightarrow{0} STOP) \;|||\; REFLECT$$

The retransmission of a message m is modelled by a spawned process which does nothing except broadcast m on the channel *down* after the end of current slot.

Each ground station k is the asynchronous combination of two processes, a transmitter and a receiver:

$$STATION_k \;\widehat{=}\; TRANS_k \;|||\; REC_k$$

The transmitter process is initially prepared to accept a message for transmission. Once a message m has been accepted, it is held until the next *epip* event occurs, and then broadcast on channel *k.send*. After a delay of t_r, the round trip time to the satellite, the transmitter is prepared to receive a signal on channel *rec*. If the same message m is received, we may conclude that the transmission was successful, and the transmitter returns to its original state.

$$\begin{aligned} TRANS_k &\;\widehat{=}\; k.in?m \xrightarrow{t_5} HOLD_k(m) \\ HOLD_k(m) &\;\widehat{=}\; epip \xrightarrow{t_6} k.\widehat{send}.m \xrightarrow{t_r} SENT_k(m) \\ SENT_k(m) &\;\widehat{=}\; rec?n \xrightarrow{t_7} \textbf{if } n = m \textbf{ then } TRANS_k \textbf{ else } BACK\,;HOLD_k(m) \end{aligned}$$

If the expected message does not arrive, then the process backs off for a random period—modelled by the process *BACK*. It then returns to the holding state and retransmits the message.

The receiver process is prepared to receive any signal on channel *rec*. Incoming messages are tested for validity and destination, and any uncorrupted message intended for the current station is offered as output on channel *k.out*. This offer will be withdrawn if output does not occur within time t_9.

$$\begin{aligned} REC_k \;\widehat{=}\; rec?n \xrightarrow{t_8} \textbf{if } valid(n) \wedge dest(n) = k \;&\textbf{then } k.out!n \overset{t_9}{\rhd} REC_k \\ &\textbf{else } REC_k \end{aligned}$$

In any case, the receiver will return to its initial state within time $t_8 + t_9$.

In the above description, the event *epip* corresponds to the arrival of the clock signal at ground level. To model the relationship between these signals, we require a process which behaves as a broadcast medium. The generic process defined by

$$M(l, r, d, c) \;\widehat{=}\; l?x \xrightarrow{0} ((WAIT\ d\,;\widehat{r}!x \xrightarrow{0} STOP) \;|||\; (WAIT\ c\,; M(l, r, d, c)))$$

may be used to represent a medium which accepts a message x on channel l and broadcasts it on channel r after a delay of d. This medium is capable of accepting a new message every c time units: this delay models the limited bandwidth of the medium.

Our example will require three media. The transmission of signals from the satellite is modelled by the processes *PIP* and *DOWN*, and the transmission of messages to the satellite is modelled by the process *UP*:

$$\begin{aligned} PIP &\;\widehat{=}\; M(pip, epip, delay, cycle) \\ DOWN &\;\widehat{=}\; M(down, rec, delay, cycle) \\ UP &\;\widehat{=}\; |||_{k:K} M(k.send, up, delay, cycle) \end{aligned}$$

The *UP* process is the asynchronous combination of several media, one for each ground station. In an implementation of the protocol, the stations share a single transmission frequency; this is modelled by the shared output channel *up* and the satellite's inability to receive two messages in a single slot without corruption.

The transmission media and collection of ground stations are modelled as follows:

$$MEDIA \;\hat{=}\; UP \,|||\, DOWN \,|||\, PIP \qquad EARTH \;\hat{=}\; |||_{k:K}\, STATION_k$$

and the complete protocol *ALOHA* is the parallel combination of the satellite, the media, and the ground stations:

$$ALOHA \;\hat{=}\; SATELLITE \;{}_{\substack{\widehat{pip} \\ \widehat{down},\, up}}\|_{\substack{pip \\ down,\, \widehat{up}}}\; MEDIA \;{}_{\substack{\widehat{epip} \\ K.send,\, \widehat{rec}}}\|_{\substack{epip \\ K.\widehat{send},\, rec \\ K.in,\, K.out}}\; EARTH$$

8 Discussion

In this paper, we have proposed a language which includes both broadcasting and synchronous communication. The language is an extension of the CSP notation presented in [Hoare 85], and the semantic model is consistent with the timed models for CSP, so we may draw upon the existing body of work in the area, including behavioural specifications, proof systems, and timewise refinement. The addition of signal events—outputs which do not require cooperation of the environment—allows us to model broadcast communication in a clear and intuitive fashion, without sacrificing the power of the CSP synchronisation.

Our representation of broadcasting behaviour rests upon the assumption—axiom three of the model—that any process behaviour may be observed up until any finite time. The semantic equations insist that time may not pass unless scheduled signal events are recorded. The principle of deferring the passage of time until an event has occurred may be found in other timed models: TCCS [Moller & Tofts 90], ATP [Nicollin *et al.* 90], and Timed ACP [Baeten & Bergstra 91] all permit such an interpretation. However, these languages include constructs which can block the occurrence of signal events. The use of restriction in the first two models may result in a temporal deadlock which must be eliminated from any implementation. In the third, restricting a signal causes the transmitter to deadlock, although it can allow time to pass.

Further, if we wish to interpret communication events as signal events in the above models, then we require other constructs to model synchronisation. The TCCS delay, the ATP timeout, and the Timed ACP integration are all possible candidates. Other timed models, [Hennessy & Regan 90] and [Wang 90] for example, support only synchronised communication.

Several theoretical models exist which allow instantaneous communication between more than two processes, but few have managed to combine this with a treatment of broadcast communication. Prasad's Calculus of Broadcasting Systems [Prasad 91] provides a model in which every output event is a signal, and is received instantaneously by any listening process. However, it is not clear how this model may be extended to include timing information. It seems inevitable that any attempt to include synchronisation and broadcasting communication in a single model will introduce significant complexity. By extending the theory of Timed CSP, we have been able to combine both forms of communication in a uniform theory of timed concurrency.

Acknowledgements

The authors would like to thank Mike Reed and Bill Roscoe for their helpful advice and inspiration. We are also very grateful to the referees for their detailed analysis of the paper. We have not been able to address all of their suggestions here; the others will be addressed in a reworked and expanded version of the paper.

Jim Davies would like to thank the Royal Signals and Radar Establishment—and in particular Steve Giess, Colin O'Halloran, and Tim White—for their continued support. Dave Jackson is an employee of Rolls-Royce plc, and Steve Schneider is supported by the ESPRIT SPEC project, BRA 3096.

References

[Baeten & Bergstra 91]

J.C.M. Baeten and J.A. Bergstra, *Real Time Process Algebra,* Formal Aspects of Computing 3:142–188, 1991.

[Davies 91]

J. Davies, *Specification and Proof in Real-time Systems,* Programming Research Group Technical Monograph PRG–93, Oxford University 1991.

[Hennessy & Regan 90]

M. Hennessy and T. Regan, *A Temporal Process Algebra,* Technical Report 2–90, University of Sussex 1990.

[Hoare 85]

C.A.R. Hoare, *Communicating Sequential Processes,* Prentice-Hall 1985.

[Moller & Tofts 90]

F. Moller and C. Tofts, *A Temporal Calculus of Communicating Systems*, Proceedings of CONCUR 90, Springer LNCS 458 (1990).

[Nicollin *et al.* 90]

X. Nicollin, J.-L. Richier, J. Sifakis and J. Voiron, ATP: an Algebra for Timed Processes, Proceedings of the IFIP Working Conference on Programming Concepts and Methods, 1990.

[Prasad 91]

K.V.S. Prasad, *A Calculus of Broadcasting Systems*, Proceedings of TAPSOFT '91, Springer LNCS 493 (1991).

[Reed & Roscoe 86]

G.M. Reed and A.W. Roscoe, *A Timed Model for Communicating Sequential Processes,* Proceedings of ICALP '86, Springer LNCS 226 (1986); Theoretical Computer Science 58 (1988).

[Reed 88]
G.M. Reed, *A Uniform Mathematical Theory for Real-time Distributed Computing,* Oxford University D.Phil thesis 1988.

[Tanenbaum 81]
A.S. Tanenbaum, *Computer Networks,* Prentice-Hall International 1981.

[Wang 90]
Wang Yi, *Real-time Behaviour of Asynchronous Agents,* Proceedings of CONCUR 90, Springer LNCS 458 (1990).

Analysis of Timeliness Requirements in Safety–Critical Systems

Rogério de Lemos, Amer Saeed and Tom Anderson
Computing Laboratory
University of Newcastle upon Tyne, NE1 7RU, UK

Abstract

Requirements analysis plays a vital role in the development of safety–critical systems since any faults in the requirements specification will corrupt the subsequent stages of system development. Experience in safety–critical systems has shown that faults in the requirements can and do cause accidents. This paper presents a general framework for the analysis of timeliness requirements in safety–critical systems. The analysis is performed in two distinct phases; for each phase we propose different formalisms and time structures. The specification of the timing constraints is based on an event/action model. To illustrate the proposed approach an example based on a train set crossing is presented.

Keywords: safety–critical systems, requirements analysis, timeliness requirements, formal models, time modelling.

1. Introduction

A safety–critical system is a system for which there exists at least one failure that can be adjudged to cause a catastrophe (e.g. loss of life). A major motivation for the work presented in this paper is to extend the framework for the requirements analysis of safety–critical systems, previously introduced in /Saeed 91/, to allow analysis of timeliness requirements. The aim of the framework is to locate and remove faults related to timing issues introduced during the requirements analysis in the development of software for safety–critical systems. (Although "safety" is an attribute of the system rather than just software, in this paper attention is restricted to problems related to "software safety".)

The approach to be followed for the analysis of the requirements is based on a clear separation of the mission and the safety requirements. The mission requirements focus on what the system is supposed to achieve in terms of function, timeliness and some dependability requirements – namely the attributes of reliability, availability and security. On the other hand, the safety requirements focus on the elimination and control of hazards, and the limitation of damage in the case of an accident; thus they are related to the safety attribute of dependability /Laprie 90/. In the proposed framework we are concerned with the timeliness requirements that are related to the safety attribute of dependability.

As a general structure, safety–critical systems (in particular process control systems) are usually partitioned into three distinct components: the *operator*, the *controller* and the *physical process* (or plant). The *environment* is that part of the rest of the world which may affect, or be affected by, the system. This structure is further decomposed to reflect the decision to separate the mission and safety issues: mission and safety operators, and mission and safety controllers. The safety operator and the safety controller are the components which must ensure that the system does not enter a hazardous state.

The rest of the paper is organized as follows. The next section discusses the behaviour of real–time systems and presents a general model of time that can be specialized to represent either dense or discrete time. Section 3 describes the framework for requirements analysis of safety–critical systems. In section 4, we discuss how the model of time presented in section 2 is integrated in the framework. In section 5, a case study based on a train set crossing is discussed in accordance with the time modelling adopted for the framework. Finally, section 6 presents some concluding remarks.

2. Behaviour of Real–Time Systems

Two classes of temporal relationships are distinguished for real–time systems, in accordance with the causal relationships between events /Koymans 88/:

> *qualitative* temporal relationships are only concerned with the ordering of events in time, which means that there is no explicit referring to the occurrence of an event;
>
> *quantitative* temporal relationships, referred to in this paper as *timing constraints*, are concerned with occurrences of events where the time reference is explicitly mentioned.

Timing constraints can be further subdivided into *timeliness requirements* and *performance requirements*; in this paper we are only concerned with timeliness requirements. These are intervals of time which dictate the response time of the system to stimuli (either physical or logical). The timeliness requirements concerning a sequence of events can be specified in one of the following forms:

> an *absolute* time representation, associated with one of the events in the sequence;
>
> a *relative* time representation, stipulating the elapsed time between the occurrence of any two events in a sequence of events.

2.1. A General Model of Time

In this section we consider a time structure based on traditional point structures, consisting of points in time ordered by a relation of precedence (“earlier”, “before”) /van Benthem 90/. We then present *intervals* in terms of the basic time structure and a *containment* relation between time structures.

Time Structure

We define a time structure as the triple $\mathcal{T} = \langle T, <, \Delta \rangle$, where $T (\subseteq \mathbf{R})$ is a non-empty set of *time points*, $<$ the *precedence* relation, and Δ ($\in \mathbf{R}_+ \cup \{0\}$) the *granularity* (distance between any two adjacent time points). The following properties must hold:

Transitivity
For any three time points t_1, t_2, t_3 in T if t_1 precedes t_2 and t_2 precedes t_3 then t_1 precedes t_3.
$\forall t_1, t_2, t_3 \in T: t_1 < t_2 \wedge t_2 < t_3 \Rightarrow t_1 < t_3$.

Irreflexive
A time point in T cannot precede itself.
$\forall t \in T: \neg(t < t)$

Linearity
For any two time points in T, either one precedes the other or they are the same point.
$\forall t_1, t_2 \in T: t_1 < t_2 \vee t_2 < t_1 \vee t_1 = t_2$.

Δ–density
For any two time points t_1, t_2 in T, if and only if t_1 precedes t_2 by more than Δ then there exists a time point in *T* which lies between t_1, t_2.
$\forall t_1, t_2 \in T: t_1 < t_2 - \Delta \Leftrightarrow \exists t \in T: t_1 < t < t_2$.

Both dense and discrete time structures can be defined as special cases of *Δ–density*; specifically $\Delta = 0$ characterizes a dense time structure and $\Delta > 0$ a discrete time structure with granularity Δ. In the case $\Delta = 1$, the time set T is normally a subset of the integers.

Time Intervals

We define two sorts of intervals for a time structure $\mathcal{T}$:

Convex Intervals
A convex interval is a subset of T that represents an uninterrupted stretch of time.
$Int \subseteq T$ is a *convex interval* iff $\forall t_1, t_2 \in Int: \forall t_3 \in T: (t_1 < t_3 < t_2 \Rightarrow t_3 \in Int)$.

Interval Set
CI(T) is the set of all convex intervals contained within T.

Relationships Between Time Structures

Containment
We say that a time structure $\mathcal{T}_2$ contains another time structure $\mathcal{T}_1$ (denoted by $\mathcal{T}_2$ **con** $\mathcal{T}_1$) iff the granularity of $\mathcal{T}_1$ is greater than the granularity of $\mathcal{T}_2$ and the set of time points of $\mathcal{T}_1$ is a strict subset of the set of time points of $\mathcal{T}_2$.

$\mathcal{T}_2 \ \mathbf{con} \ \mathcal{T}_1 \Leftrightarrow^{def} \Delta(\mathcal{T}_1) > \Delta(\mathcal{T}_2) \wedge T(\mathcal{T}_1) \subset T(\mathcal{T}_2)$.

In the rest of this section we use the term interval as shorthand for closed convex intervals.

2.2. An Event/Action Model

From the general model of time defined in the previous section we introduce the concept of a "timeline", a line marked with the time points of T, for which consecutive time points are separated by a distance Δ.

In order to capture temporal ordering we introduce an *event/action* model based on a timeline, as defined in /MacEwen 88, PDCS 90/.

> An *event* is a temporal marker of no duration, represented as a cut in the timeline (i.e. an element of T); the time period between two events is a bounded interval (i.e. an element of CI(T)); a *duration* is a measure of the length of a bounded interval (i.e. the distance between the events that mark the boundaries of the interval).
>
> An *action* is a basic unit of activity which implies a duration in its execution; it is manifested in the system by its associated set of events: a *start event* marks the initiation of an action, and a *finish event* marks the completion of an action. The period of the execution of an action is represented as a bounded interval.

Remark: if time structure $\mathcal{T}_2$ contains time structure $\mathcal{T}_1$, an event observed in $\mathcal{T}_1$ at time point t may be an action when observed in $\mathcal{T}_2$ that occurs during a convex interval that is a subset of the interval $[t+\Delta(\mathcal{T}_1), t-\Delta(\mathcal{T}_2)]$.

A relative time representation of the timeliness requirements of a sequence of events can impose one of three basic types of timing relations /Dasarathy 85/:

> *minimum* – no less than t time units must elapse between the occurrence of two events;
>
> *maximum* – no more than t time units must elapse between the occurrence of two events;
>
> *durational* – exactly t time units must elapse between two events.

2.3. Specification of Timeliness Requirements

Here we show how timeliness requirements will be expressed at different levels of abstraction in the analysis of the safety requirements.

At the highest level of abstraction we introduce an "event–occurrence function" E(t), which maps the timeline into the occurrence of events. The timing constraints imposed on the occurrence of one or more events may also be specified in order to consider time uncertainties, which may be expressed through the timing relations previously introduced. Assuming T to be the set of all points in the timeline and t^e the time point at which the event occurs, the event–occurrence function is formalised as follows:

$$E(t): T \to \{0,e\}$$
$$E(t) = 0 \qquad \forall t: t \neq t^e$$
$$E(t) = e \qquad \forall t: t = t^e.$$

At a lower level of abstraction in the specification of the timeliness requirements, the execution of an action can be better represented by means of a "utility function" U(t), or value function /Jensen 85/, which represents the usefulness (versus time) of an item of information to be consumed/produced by the action under consideration. In this paper we are concerned with the class of utility functions which are typically found in safety–critical systems – *discrete* or *critical utility functions*: for such functions the usefulness of the service to be delivered by the execution of an action can only assume maximum and minimum values. For the definition of a utility function four main time attributes are established:

earliest starting time (t^{EST}) – the earliest time point at which the execution of the action can be started;

latest starting time (t^{LST}) – the latest time point at which the execution of the action can be started;

earliest finishing time (t^{EFT}) – the earliest time point at which the execution of the action can be finished;

latest finishing time (t^{LFT}) – the latest time point at which the execution of the action can be finished.

As far as critical utility functions are concerned, it is usual to refer to t^{EST} as "delay" and t^{LFT} as "deadline". These time attributes can be expressed by step utility functions, respectively positive and negative step functions. A critical utility function can be formalised as follows:

$$
\begin{aligned}
&U(t): T \rightarrow \{0,u\} \\
&\qquad U(t)=0 \qquad\qquad t < t^{EST} \\
&\qquad U(t)=u \qquad\qquad t^{EST} \leq t \leq t^{LFT} \\
&\qquad U(t)=0 \qquad\qquad t^{LFT} < t.
\end{aligned}
$$

At the lowest level of abstraction, the actual execution of an action can be represented by a "action–execution function" A(t), which represents the time interval in which a resource is actually used, for example at the level of the controller this function might represent the execution of a process (task). For the definition of an action–execution function the following time attributes are established from the timing relations, previously presented:

start time (t^{sta}) – is the time point at which the start event of an action occurs;

finish time (t^{fin}) – is the time point at which the finish event of an action occurs. Due to possible uncertainties associated with the execution of an action we may have to specify an interval which contains the finish time; this interval is defined by the following time points:

earliest finish time (t^{Efin}) – the earliest time point at which the execution of the action can be finished;

latest finish time (t^{Lfin}) – the latest time point at which the execution of the action can be finished;

execution time (t^{ext}) – is the duration of the interval during which the action is executed.

An action–execution function may be formalised as follows:

$$A(t): T \rightarrow \{0,a\}$$
$$A(t)=0 \qquad t < t^{sta}$$
$$A(t)=a \qquad t^{sta} \leq t \leq t^{fin}$$
$$A(t)=0 \qquad t^{fin} > t.$$

The action–execution function is used to define the timeliness requirements of a utility function whenever the duration of the action is known a priori; if no information is available some estimates must be made. The value of t^{LST} and t^{EFT}, defined for an utility function, may be used to simulate the detection of errors in the execution of an action; hence, the value of t^{LST} should be related to the maximum duration in the execution of the action, and the t^{EFT} to its minimum duration.

In the following, we present an example, illustrated in figure 1, which shows how the functions defined above are interrelated in the specification of the timeliness requirements at different levels of abstraction. The initial step in the specification of the timeliness requirements for a particular activity is to consider the time points associated with the occurrence of events in the execution of an action or sequence of actions; this includes possible time uncertainties which are shown in the figure as timing relations imposed on the event t^e. The utility function is defined in terms of the time points established before, eliminating some of the time uncertainties of the previous step. Finally, the actual execution of the action is represented by the function A(t), and must be performed within the timing constraints imposed by the utility function.

If an action has two or more utility functions imposed on it then, for the execution of that action to be useful, it must comply with all the imposed utility functions. Below we present a lemma which allows us to define a utility function which represents the superposition of more than one utility function on an action.

Lemma 1
When two or more utility functions, representing the usefulness in executing the same action, are superimposed, the resulting utility function imposes the following timing constraints at the start time and finish time of an action:

$$t^{sta} \in \{t \mid \max(t_1^{EST}, ..., t_n^{EST}) \leq t \leq \min(t_1^{LST}, ..., t_n^{LST})\};$$
$$t^{fin} \in \{t \mid \max(t_1^{EFT}, ..., t_n^{EFT}) \leq t \leq \min(t_1^{LFT}, ..., t_n^{LFT})\}.$$

Proof. Immediate.

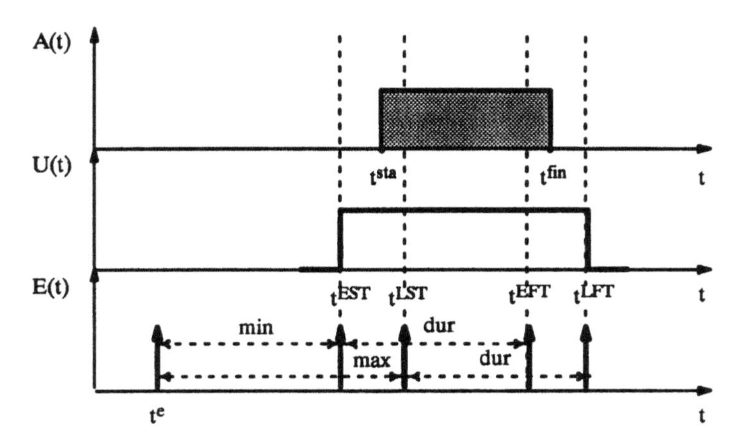

Figure 1. Abstraction Levels of Timeliness Requirements

An example is the situation where the same exception handler is used to treat two or more exceptions which impose different timing constraints upon the execution of the handler. In practical terms this could represent the occurrence of two different alarms which are treated by the same action.

3. Framework for the Requirements Analysis of Safety-Critical Systems

A general framework for the requirements analysis of safety-critical systems /Saeed 91/, which provides a systematic approach to the production of the requirements specification of the safety controller, is shown in figure 2. The basic aim behind our framework for the requirements analysis of safety-critical systems is to subdivide the whole problem into smaller domains where the analysis of the requirements can be simplified, thereby leading to more accurate requirements specifications. This is obtained by, first, splitting the mission from safety requirements, and second, subdividing the analysis of the safety requirements into a sequence of phases.

- **Conceptual Analysis** – the objective of this phase is to produce an initial, informal statement of the aim and purpose of the system and to determine what is meant by safety for the system. As a product of the Conceptual Analysis we obtain the Safety Requirements, enumerating the potential *accidents* (an accident is an unintended event or sequence of events that cause death, injury, environmental or material damage), and the *hazards* (a hazard is a system condition that can lead to an accident under certain environmental circumstances) related to these accidents. It is the identification of the hazards of a system, by performing a Preliminary Hazard Analysis, that allows us to make a distinction between the mission and safety issues of the system. For each of the identified hazards an estimate of the associated risk must also be determined.
- **Safety Requirements Analysis** – the objective of this phase is the identification of the real world properties relevant to the safety requirements: physical laws and rules of

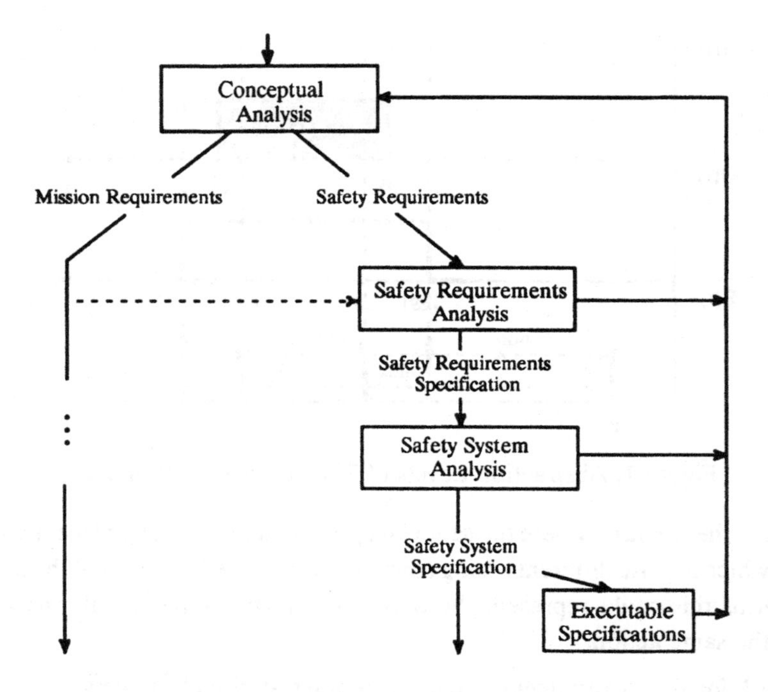

Figure 2. A Framework for Requirements Analysis.

operation. As a product of this real world analysis, we obtain the Safety Requirements Specification, containing the *safety constraints* (a safety constraint is a condition imposed on the system that is the negation of a hazard, modified to incorporate safety margins) and the *safety strategies* (a safety strategy is a way of maintaining a safety constraint, and is defined as a set of conditions over the physical process).

- **Safety System Analysis** – the objective of this phase is the identification of the interface between the safety controller and the physical process, and the specification of the system behaviour that must be observed at the identified interface. Also in this phase a top level organization of the system is realized in terms of the properties of the system components, and the effects of the possible failures of these components. This phase leads to the production of the Safety System Specification, containing the *safety controller strategies* (a safety controller strategy is a refinement of a safety strategy incorporating some of the system components, such as the sensors and actuators, and their relationship with the real world).

- **Executable Specifications** – the objective of this phase is to construct a low-cost prototype of the system, which will enable the user of the system to check if the Safety System Specification does indeed capture the intentions of the user, and if it is consistent with the mission requirements. There are other ways in which errors can be detected in the Safety System Specification, such as general correctness criteria which must be satisfied for all systems of a certain type /Jaffe 91/.

The hierarchical structure which captures the relationship of the specifications produced at each phase of the analysis of the safety requirements is called the Safety Specification Hierarchy. Because of the differing characteristics of each phase, instead of seeking a single formalism, our approach is to use different formalisms for each phase. This has the advantage of allowing us to select formalisms in accordance with the properties that should be expressed at each phase of development. In /Saeed 91/ we proposed the utilization of a logical formalism, such as Timed History Logic - THL /Saeed 90/, for the Safety Requirements Analysis, and a net formalism, such as Predicate–Transition nets - PrT nets /Genrich 87/, for the Safety System Analysis. The Conceptual Analysis is by its nature informally recorded, and for the executable specifications we suggest simulation of the PrT net model already constructed during the Safety System Analysis.

4. The Analysis of the Timeliness Requirements in the Framework

The basic approach suggested in this paper, of dividing the analysis of the timeliness requirements into distinct phases, and performing the analysis at each phase with the appropriate time model and formalism, has not previously been investigated. What has usually been presented is the utilization of a single formalism such as Temporal Logic /Gorski 86/, Petri nets /Leveson 87/, or THL /Saeed 90/. However, there are two approaches in the literature which use different formalisms for the analysis of the timeliness requirements /Ostroff 87, Jahanian 86/. These papers do not address the issue of which sort of time structures are most appropriate during the different stages of requirements analysis, and how different time structures can be accommodated in a single framework - themes which are central to this paper. In /Ostroff 87/, ESMs are used to model "plant–controller processes"; from the paths of these ESMs, trajectories are obtained which can be used to provide a formal operational semantics. The specification of plant behaviour is then given by RTTL formulae over these trajectories, and verified by demonstrating that the trajectories defined by the ESMs do indeed comply with RTTL formulae. In /Jahanian 86/, the specification of the system is realized in terms of RTL and Modecharts; Modecharts produce a decision procedure for classes of properties expressed as RTL formulas. The system properties are verified using Computation Graphs, obtained from the Modecharts, to check if the corresponding RTL formulas comply with the Modecharts.

4.1. Safety Requirements Analysis

We propose the use of a dense time structure (i.e. $\Delta = 0$) during the Safety Requirements Analysis. Several formalisms use a dense time structure; these include extensions to Temporal Logic /Pnueli 88/, CSP /Reed 86/ and CCS /Milner 83/. Some justification for selecting a dense time structure for the Safety Requirements Analysis is presented below.

Physical laws. Physical laws govern the properties of the environment in which a system works, and impinge on the behaviour of the system. To perform a thorough analysis of system behaviour these laws must be expressed and their impact on the system analysed. Many of these physical laws are defined in terms of physical time, e.g. the distance travelled by a train for a duration of time is given by an integral over the velocity of the train. Hence, an appropriate formalism to express these laws should have a dense time structure.

Minimum separation. The physical process of a control system consists of many concurrent interacting processes; for such a system it is not possible to give a minimum separation between events in the different processes. The lack of minimum separation makes it difficult to define a *granularity* for a discrete time structure, hence a dense representation of time is more suitable for the real world analysis. In fact the *granularity* of a suitable time structure is a lower level concept since it depends on limitations of the controller to observe and manipulate the real world.

4.1.1. Proposed Time Base

We define the real world time structure as the triple $\mathcal{T}_{RW} = \langle \mathbf{R}, <, 0 \rangle$, where **R** is the set of non–negative real numbers. This is a well defined time structure; the *transitivity*, *irreflexive*, *linearity* and *Δ–density* properties follow from the properties of **R**. Since the granularity is zero, $\mathcal{T}_{RW}$ is a *dense* time structure, meaning that between any two time points there is an intermediate time point.
$\forall t_1, t_2 \in \mathbf{R}: t_1 < t_2 \Rightarrow \exists t \in \mathbf{R}: t_1 < t < t_2.$

4.1.2. Timed History Logic

THL is a formalism based on the time structure $\mathcal{T}_{RW}$, and consists of three main concepts: *histories*, *relations* and *modes*. Here we present an overview of histories and relations; a more detailed description of the model is given elsewhere /Saeed 90/. In the following we use the term interval for closed convex interval.

Over any interval the following functions are defined: *start point* s(Int) – the earliest time point in Int; *end point* e(Int) – the latest time point in Int; and *interval set* SI(Int) – the set of all intervals contained within Int. The system lifetime (ST) is an interval which represents the operational lifetime of the system.

For a system with n state variable we have the state vector: $Sv = \langle p_1, ..., p_n \rangle$. The set of possible values of a state variable (say, p_i) is defined by its variable range (Vp_i). The state space of a system (Γ) is defined as the cross product of the variable ranges. A history H of a system is a function of the form $H: ST \rightarrow \Gamma$. The set of all "possible" histories of a system is defined as the universal history set (ΓH). For a history H the sequence of values taken by a state variable p_i is denoted by the function $H.p_i: ST \rightarrow Vp_i$.

Invariant relations are used to express relationships over the state variables which hold at every time point within T; these are formulated as *system predicates*.

A *system predicate* is a predicate built using n free value variables $p_1, ..., p_n$ of types $Vp_1, ..., Vp_n$. No other free variables may be used. A tuple of values $V = \langle x_1, ..., x_n \rangle$, where x_i is of type Vp_i, satisfies a system predicate SysPred if and only if substitution of each x_i for p_i within SysPred evaluates to **true**. This is denoted by: V **sat** SysPred.

A system predicate SysPred is an invariant relation for a history H if and only if: $\langle H.p_1(t), ..., H.p_n(t) \rangle$ for all $t \in ST$. This will be abbreviated as H **sat** SysPred.

History relations are used to express relationships over the state variables which hold during every interval included within ST; these are formulated as *history predicates.*

A *history predicate* is a predicate built using two free time variables T_0, T_1 and n free function variables $p_1, ..., p_n$ (where p_i is a function of class Cp_i). No other free variables may be used. A history H satisfies a history predicate HistPred for an interval Int if and only if the expression resulting from substituting: i) s(Int) for T_0, ii) e(Int) for T_1 and iii) $H.p_i$ for p_i for all i, evaluates to **true**. This is denoted by: H **sat** HistPred@Int.

A history predicate HistPred is a history relation for a history H if and only if H **sat** HistPred@Int for all Int $\in$ SI(ST). This will be abbreviated as H **sat** HistPred.

4.2. Safety System Analysis

For the analysis of timeliness requirements during the Safety System Analysis we propose the use of a discrete time structure (i.e. $\Delta > 0$). The choice of using a discrete time structure seems natural, in the sense that it is the most convenient and expressive model of time for discrete computation. The timeliness analysis to be performed at this phase is essentially related to the safety requirements of the safety controller. Some concepts related to discrete time structures, presented in the following, are based on the work reported in /PDCS 90/.

For a discrete time structure, the timeline is a sequence of discrete points. A time grid is a set of equidistant points on the timeline, and to any grid we associate a granularity. A timeline may be represented by a physical clock, with which we associate exactly one time grid. We define a reference clock which has a sufficiently fine granularity to enable the measurement (in principle) of the duration between ticks of any other clock. As far as the occurrence of events are concerned on a discrete time structure, it is not possible to observe an event itself, only its consequence(s) may be observed. The reference clock can either unambiguously order any two events, or define the events to be simultaneous.

4.2.1. Proposed Time Base

We define the safety controller time structure as the triple $\mathcal{T}_{SC} = \langle \mathbf{N}, <, 1 \rangle$, where **N** is the set of non-negative integers. This is a well defined time structure, the *transitivity*, *irreflexive*, *linearity* and *Δ-density* properties follow from the properties of **N**. Since the granularity is greater than zero, $\mathcal{T}_{SC}$ is a *discrete* time structure, meaning that between two adjacent time

points, no intermediate point exists.
$\forall t_1, t_2 \in N: t_1 < t_2 \Rightarrow t_1+1 \leq t_2$.

4.2.2. Predicate-Transition Nets

In the following we present the semantics of PrT nets /Genrich 87/, a form of high-level Petri net, which will be the formalism used in the phase of Safety System Analysis. Petri nets are mainly used for the modelling and analysis of discrete-event systems which are concurrent, asynchronous, and non-deterministic. The use of PrT nets, instead of (Timed) Petri nets /Leveson 87/, adds to the modelling power of the latter the formal treatment of *individuals* (i.e. the notion of token identity) and their changing properties and relations.

Below, we present an informal definition of PrT nets; a formal definition is given elsewhere /Genrich 87/.

Let S, T, F be finite sets. The triple N = (S, T, F) is called a *directed net* iff the following conditions hold: $S \cap T = \emptyset, S \cup T \neq \emptyset, F \subseteq (S \times T) \cup (T \times S)$, and domain (F) $\cup$ codomain (F) $= S \cup T$.

For a given net N = (S, T, F), S is the set of *places* of N, T is the set of *transitions* of N, and F is the *flow* relation containing the *arcs* of N. For $x \in S \cup T$, $I(x) = \{y \in S \cup T \mid (y, x) \in F\}$ is called the preset of x, and $O(x) = \{y \in S \cup T \mid (x, y) \in F\}$ is called the postset of x.

A PrT net consists of the following constituents:

(1) a directed net (S, T, F), where S is the set of predicates and T is the set of transitions;

(2) predicates are variable relations amongst individuals ("first-order" places);

(3) the transitions are schemes of elementary changes of markings representing the actions carried out by the system;

(4) an arc label specifies a variable extension of a predicate to which the arc is connected;

(5) a marking is a mapping that assigns to each predicate formal sums of n-tuples of individual symbols, also called tuples.

The graphical representation of a PrT net is obtained by representing a predicate by a circle, a transition by a box, an element of $F \cap (S \times T)$ by a directed arc from a circle to a box, and an element of $F \cap (T \times S)$ by a directed arc from a box to a circle.

The incorporation of time in PrT nets will follow what has been proposed for ER nets /Ghezzi 91/. To each tuple we associate a timestamp which represents the time when the tuple was produced, which depends on the relational expression associated to the transitions. However, three properties of monotonicity must be satisfied in order to deal with the notion of time: first, the timestamp of a tuple produced by the firing of a transition cannot precede the value of the timestamp in any tuple removed by the firing, second, the firing time of a transition cannot precede the time of the previous firing of the same transition, and third, for any firing sequence, the times of the firings should be monotonically nondecreasing with respect to their occurrence in the sequence.

The basic timing relations associated to a PrT net transition, that we will be assuming, are expressed as an interval [Tmin, Tmax] which represents, respectively, the minimum and maximum time delay that must elapse between the enabling of the transition and its firing time. Other timing relations which are not directly related to the timestamp of a tuple may also be employed.

We assume a Weak Time Semantics (WTS) rather than a Strong Time Semantics (STS),which means that a transition does not have to fire when its maximum firing time has been reached; if it does fire it does so within the time set specified by the time condition. The utilization of WTS will permit the representation of the event/action model, previously presented, in terms of PrT nets, as shown in figure 3.

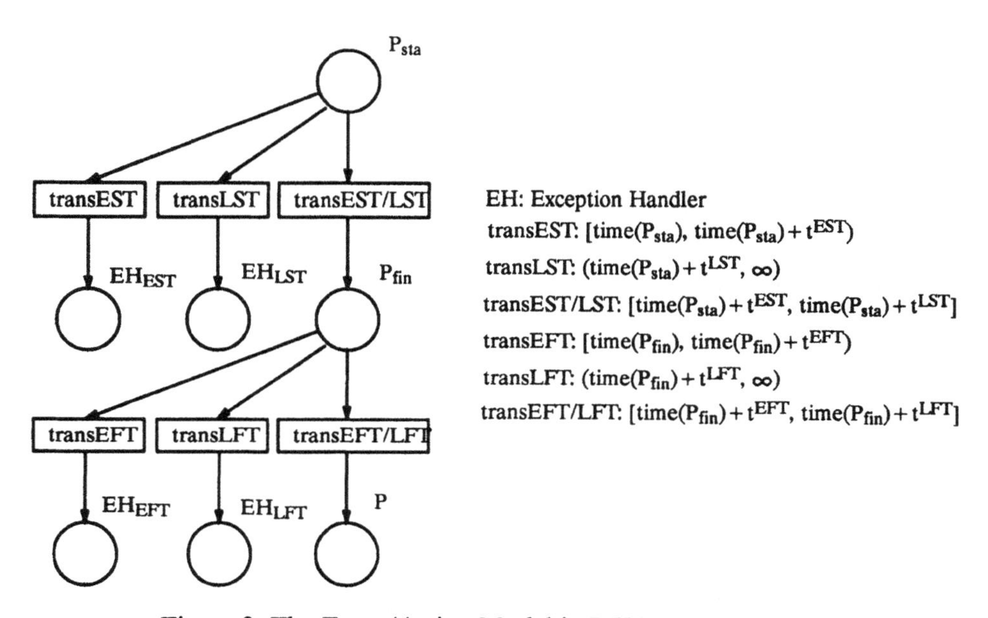

Figure 3. The Event/Action Model in PrT Nets

5. The Train Set Crossing

With the aim of exemplifying the modelling of time in the proposed framework, a train set crossing was selected as a case study. The train set crossing described below raises safety–critical issues that are similar to those found at the traditional level crossing (i.e. road–rail). In /de Lemos 92/ we presented the analysis of the safety requirements; here we demonstrate how the analysis of the timeliness requirements is performed. The formalism employed for the Safety Requirements Analysis is THL /Saeed 90/, and the formalism employed for the Safety System Analysis is PrT nets /Genrich 87/.

Conceptual Analysis

The physical process consists of two track circuits Cp and Cs, and two types of trains – primary (Trp) and secondary (Trs). The circuits are divided into sections and there are two separate crossing sections. A crossing section is that part of the track which consists of the sections (one from each circuit) at which the two circuits intersect. Trains of type Trp travel around circuit Cp and trains of type Trs travel around circuit Cs; both types of train travel in one direction only. The longest train is shorter than the smallest section. The primary trains always take priority over the secondary trains at the crossing sections. Specifically, a primary train must not be made to wait for a secondary train at a crossing section. We consider only one crossing section CC. Several accidents are associated with the system, but we only consider one potential accident: trains of different type collide. The associated hazard is that some part of a primary train and a secondary train are in crossing section CC at the same time. The circuits Cp, Cs and the crossing section CC are illustrated in figure 4.

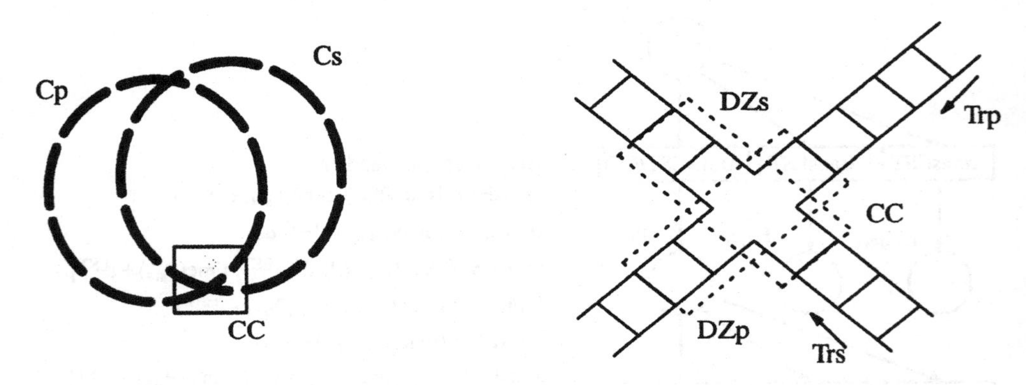

Figure 4. The train set circuits and the crossing section

General Model

The type of circuit is denoted by $c \in L$, $L = \{p, s\}$, the trains which run on Cc are denoted by $x, y \in Trc = \{1, ..., Ntc\}$, the sections of Cc are denoted by $i, j \in Sc$, $Sc = \{0, ..., Nsc\}$, and the velocity of a train is denoted by $u, v \in V$, $V = \{v \in \mathbf{R} \mid 0 \leq v \leq Vmax\}$. Addition $\oplus$ and subtraction $\ominus$ on circuit section numbers are performed modulo the number of sections of the circuit. The danger zone (DZ) on circuit Cc is defined as: $DZc = \{CCc, CCc\oplus 1\}$, where CCc is the number of the section of Cc that is part of CC. The danger zones DZp and DZs are illustrated in figure 4. We define two functions dmax(c, i, j) and dmin(c, i, j) which give the maximum and minimum distance that a train in section i must travel to enter section j on circuit c. That is, dmax(c, i, j) gives the distance to section j when a train is at the start of section i, and dmin(c, i, j) the distance when a train is at the end of section i. The behaviour of the system is captured by three state variables Ptrain, Rtrain and Vtrain (a continuous variable) described below.

No.	Name	Range	Comments
p_1	Ptrain	$Sp^{Ntp} \times Ss^{Nts}$	The position of each train expressed as a section number, that is, the section containing the front of a train.
p_2	Rtrain	$Rp^{Ntp} \times Rs^{Nts}$	The reservation sets of the trains, where $Rc = \mathcal{P}(Sc)$.
p_3	Vtrain	$V^{(Ntp+Nts)}$	The velocity of the trains.

Ptrain(c, x) denotes the state variable for the position of train x on circuit Cc, Rtrain(c, x) the reservation set of train x on circuit Cc, and Vtrain(c, x) the velocity of train x on circuit Cc. ΓH is the set of all functions $H: T \rightarrow Sp^{Ntp} \times Ss^{Nts} \times Rp^{Ntp} \times Rs^{Nts} \times V^{(Ntp+Nts)}$. A history from ΓH satisfies a requirements specification, if and only if the system (resp. history) predicates that describe it are invariant (resp. history) relations for that history.

5.1. Safety Requirements Analysis

The analysis performed in /de Lemos 92/ concentrated on the qualitative temporal relationships, here we are concerned with introducing the dimension of time and performing a quantitative temporal analysis.

Physical Law

In the following we present the law (as a history predicate) which captures the relationship between intervals, the velocity of trains, and the positions of trains.

Hr_1 – For any train, if during any interval $[T_0, T_1]$ the train travels more than the maximum distance between the section it occupies at T_0 and section *i*, then the train must occupy section *i* at some time point in $[T_0, T_1]$; and if the train travels less than the minimum distance between the section it occupies at T_0 and section *i*, then the train is never in section *i* during $[T_0, T_1]$.

$\forall c \in L: \forall x \in Trc: \forall i \in Sc:$

$$\int_{T_0}^{T_1} Vtrain(c, x).dt \geq dmax(c, Ptrain(c, x)(T_0), i) \Rightarrow \exists t \in [T_0, T_1]: Ptrain(c, x)(t) = i \quad \wedge$$

$$\int_{T_0}^{T_1} Vtrain(c, x).dt < dmin(c, Ptrain(c, x)(T_0), i) \Rightarrow \forall t \in [T_0, T_1]: Ptrain(c, x)(t) \neq i.$$

It is vital that the above law is confirmed to be a property of the system, since during the subsequent analysis this law is treated as an axiom of the model. More precisely, henceforth we consider only those histories of ΓH for which Hr_1 is a history relation.

As a result of a qualitative analysis of the crossing section, the following safety constraint and safety strategy were derived.

Safety Constraint

Either the front of no primary train is in the danger zone DZp or the front of no secondary train is in the danger zone DZs, formalized as a system predicate:

$\exists c \in L$: $\forall x \in Trc$: $Ptrain(c, x) \notin DZc$.

Safety Strategy (without priority constraint)

The safety strategy is based on a reservation scheme. The two essential rules which impose the safety strategy are as follows.

ssa. *Reservation Constraint*
For any train the current section (i.e. the position of the train) and the section behind the current section must always be reserved:

$\forall c \in L$: $\forall x \in Trc$: $\{Ptrain(c, x) \ominus 1, Ptrain(c, x)\} \subseteq Rtrain(c, x)$.

ssb. *Exclusion Constraint*
Section CCp and section CCs cannot both be reserved at the same time:

$\exists c \in L$: $(\forall x \in Trc$: $CCc \notin Rtrain(c, x))$.

Although the safety strategy for the crossing section is sufficient to prevent collisions involving primary and secondary trains, it ignores the priority of primary trains over secondary trains. Here we consider how the introduction of time can be used to analyse the priority constraint.

We recall the priority constraint, "a primary train must not be made to wait for a secondary train at a crossing section". By inspection of rule *ssb*, this can be stated in terms of the reservation scheme as: the section CCs cannot be reserved by a secondary train, if any primary train will need to reserve section CCp before the secondary train releases section CCs. From rule *ssa* we can infer that a secondary train can release section CCs when it enters section $CCs \oplus 2$, and a primary train must reserve section CCp when it enters section CCp. Hence a secondary train cannot reserve section CCs if a primary train may enter section CCp before the secondary train will enter section $CCs \oplus 2$. To analyse this situation in the event/action model, we describe the utility functions $U_{Trs}(t)$ and $U_{Trp}(t)$ illustrated by the graphs in figure 5, these graphs depict the case when a secondary train does not stop when approaching DZs and the case when a secondary train must stop.

Events:
- EN, marks the time when a secondary train starts to approach DZs.

Actions:
- APc, a train of type Trc approaching the danger zone DZc, after a secondary train enters $CCS \ominus 1$.
- CZc, a train of type Trc crossing danger zone DZc.

For simplicity, we assume that the velocity of a moving train x on circuit Cc near to the crossing section, remains constant and at the value given by the function V(c, x) or is zero.

In other words we rather crudely assume instantaneous acceleration and deceleration of the trains.

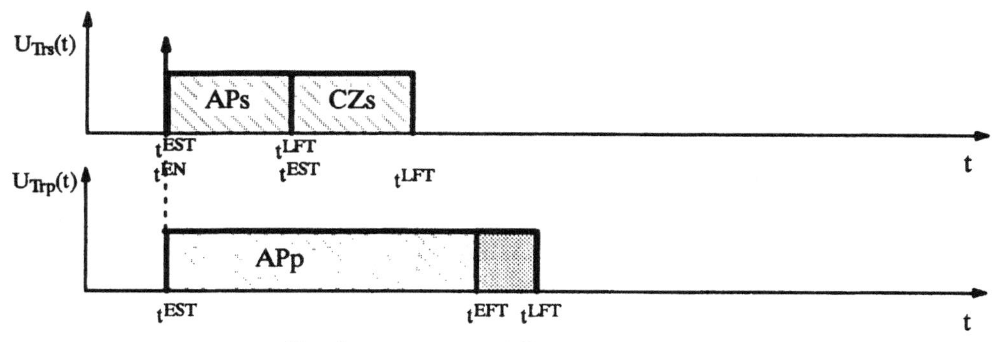

a. Trs does not stop while approaching DZs

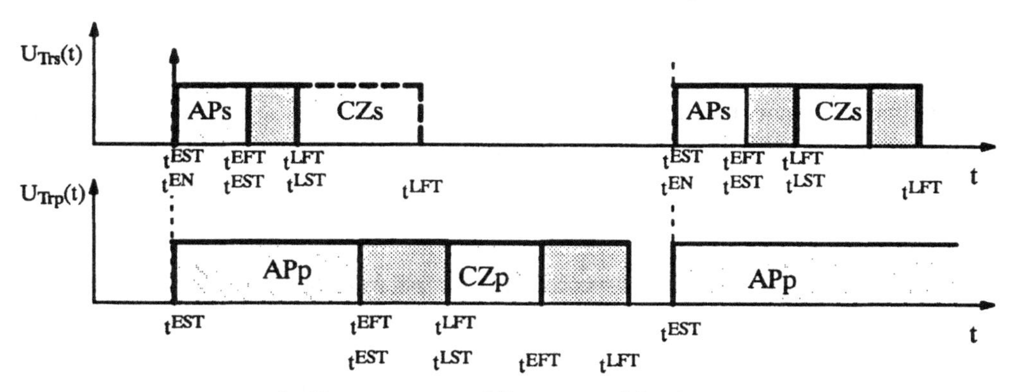

b. Trs must stop while approaching DZs

Figure 5. Timeliness Requirements for Priority Constraint

The duration of the utility function of action *ACc* for train *x* on circuit *Cc* is abbreviated to δ_x^{ACc}. For example, δ_x^{APs} denotes the duration of action *APs* for train *x* on circuit *Cs*.

From Hr_1 and the constant velocity assumption we can infer the following bounds on the duration of action APs:

$$\delta_x^{APs}|_{min} = 0 \text{ and } \delta_x^{APs}|_{max} = dmax(s, CCs\ominus 1, CCs)/V(s, x).$$

The duration for the action APs has an upper and a lower bound since the action may be aborted: the minimum distance is zero since $CCs\ominus 1$ immediately precedes DZs, and the maximum duration is for the case when a train crosses the entire section $CCs\ominus 1$.

Bounds on the duration for action APp that starts at time point t^{EN}, can be derived from Hr_1:

$$\delta_y^{APp}(t^{EN})|_{min} = dmin(p, Ptrain(p, y)(t^{EN}), CCp)/V(p, y) \text{ and}$$

$$\delta_y^{APp}(t^{EN})|_{max} = dmax(p, Ptrain(p, y)(t^{EN}), CCp)/V(p, y).$$

The duration for the action APp has an upper and a lower bound since the controller knows that the train is somewhere in section Ptrain(p, y), but does not know exactly where.

Similarly, we can infer the following bounds on the duration of action CZc:

$$\delta_x^{CZc} = dmax(c, CCc, CCc\oplus 2)/V(c, x).$$

There are no uncertainties, for the action CZs, since a train always passes a fixed distant at a constant velocity.

If we consider the behaviour of a secondary train x that is approaching the danger zone DZs with respect to a primary train y, we observe that train x may cross the danger zone DZs if it will cross DZs before train y reaches DZp, which can be expressed in terms of the durations of the actions:

$$\delta_x^{APs}|_{max} + \delta_x^{CZs} < \delta_y^{APp}|_{min}.$$

Now, consider the behaviour of a secondary train x that is approaching the danger zone DZs with respect to a sequence of primary trains approaching the danger zone DZp. We observe that train x may start to cross danger zone DZs only if it will cross DZs before the "nearest primary train" (i.e. the primary train which must pass the smallest number of sections to enter DZp) enters DZp. This can be expressed in terms of the durations of the actions: $\forall y \in Trp$:

$$\delta_x^{APs}|_{max} + \delta_x^{CZs} < \delta_y^{APp}(t^{EN})|_{min}.$$

We specify the priority constraint as a history predicate that describes the conditions in which a secondary train cannot reserve the section CCs; we say that a history satisfies the priority constraint if the history predicate that describes it is a history relation. To complete the safety strategy for the crossing section we must add the priority constraint (rule *ssc*) to the rules *ssa* and *ssb*.

ssc. *Priority Constraint*

For any interval $[T_0, T_1]$, for any secondary train x, if CCs is not reserved by x at T_0 and during $[T_0, T_1]$ the maximum duration for all parts of train x to pass the danger zone DZs from section CCs⊖1 (when it is moving) is at least the minimum duration for a primary train y to approach the danger zone DZp or the current section is not CCs⊖1, then the section CCs cannot be reserved by secondary train x during $[T_0, T_1]$.

$\forall x \in Trs$: [$CCs \notin Rtrain(s, x)(T_0) \wedge$

$\forall t \in [T_0, T_1]: \exists y \in Trp: \delta_x^{APs}|_{max} + \delta_x^{CZs} \geq \delta_y^{APp}(t)|_{min} \vee Ptrain(s, x) \neq CCs\ominus 1$

$\Rightarrow \forall t \in [T_0, T_1]: CCs \notin Rtrain(s, x)(t)$].

5.2. Safety System Analysis

In the previous phase of the requirements analysis, the event/action model was employed with the aim to understand the temporal behaviour of the physical process, in terms of quantitative relationships, and to identify those timing constraints, or "thresholds", which will be part of the safety strategy. In this phase of the analysis, we investigate how the safety strategy is to be implemented by the safety controller as *safety controller strategy*; or alternatively, the safety strategies must be mapped onto a set of sensors and actuators. The event/action model, at this stage, considers the timing uncertainties related to the components of the safety controller, and those that arise from the sensors and actuators.

Safety Controller Strategy

The safety controller strategy must maintain the three rules of safety strategy: rules *ssa* and *ssb* which say that any part of a primary and a secondary train cannot be at their respective danger zones, at the same time, and rule *ssc* which says that a primary train must not be made to wait for a secondary train at a crossing section. The PrT net model of the physical process and the safety controller which implements the rules mentioned above is shown in figure 6, to specify the timing properties we use: $\delta s = \delta_x^{APs}|_{max} + \delta_x^{CZs}$ and $\delta p = \delta_y^{APp}(t)|_{min}$. From the net model we noticed that, a primary train is never made to wait for a secondary train if the time uncertainties of key actions remain within known bounds: minimum time for a primary train to approach the crossing section, the maximum time for a secondary train to approach the crossing section, and the time it takes for a secondary train to cross the danger zone. Whenever the time uncertainties of one of the key actions are no longer in the known bounds the access to the crossing section is imposed by the mutual exclusion technique, which ignores the priority contraint.

The predicates of the PrT net model of the crossing section are the following:

S0cx, S1cx, S2cx, S3cx – a section of the circuit c is occupied by train x;
CCcx – the crossing section of circuit c is occupied by train x;
ICPcxj – train x in circuit c is allowed to enter section j (= 4, the crossing section);
IPCcxjδ_c – train x in circuit c has entered section j;
APPpxj – train x in circuit p is in section j =3, (i.e. x is about to enter DZ);
APSsxjδ_s – train x in circuit s is in section j =3, (i.e. x is about to enter DZ), and the actions approach and crossing danger zone will take δ_s;
ZDPpxj –train x in circuit p has access to section j (= 4, the crossing section);
ZDSsxj –train x in circuit s has access to section j (= 4, the crossing section);
ME – either primary or secondary trains allowed to enter the crossing section
LDZpxj – the primary train has left the danger zone;
PPpxiδ_p – train x in circuit p is in section j and the approach action will take δ_p.

Further refinements in the PrT net model of the safety controller will consist of introducing the timing constraints imposed on the exchange of information between the physical process and the safety controller. We consider, for instance, the introduction of sensors and actuators, and the estimated execution time of the safety controller actions that are part of the control loop.

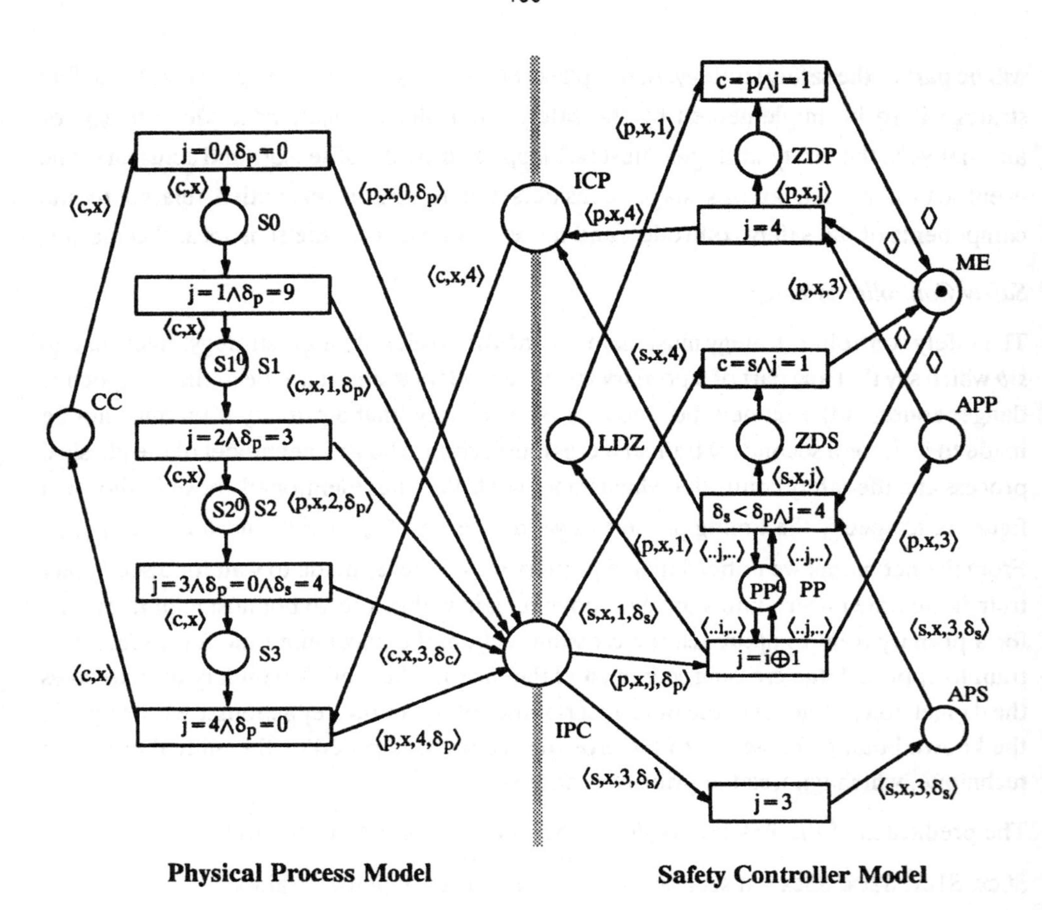

$S1^0 = \{\langle p,1,1\rangle; \langle s,1,1\rangle\}$; $S2^0 = \{\langle p,2,2\rangle; \langle s,2,2\rangle\}$; $PP^0 = \{\langle p,1,1\rangle; \langle p,2,2\rangle\}$

Figure 6. The PrT net Model of the Crossing Section

6. Conclusions

Timeliness requirements play a crucial role in the requirements of many safety-critical systems, hence an approach to the requirements analysis of such systems must provide some means to perform a quantitative analysis of the temporal relationships that exist in the system. The adoption of our framework for requirements analysis facilitates this quantitative analysis, by enabling it to be performed in clearly defined stages. For the analysis of timeliness requirements within the framework we propose the use of different time structures, specifically, a dense time structure during the Safety Requirements Analysis and a discrete time structure during the Safety System Analysis.

A problem that may arise when using different time structures is a loss of information as the analysis proceeds. In the approach presented here, we attempt to overcome this problem by providing an event/action model for the timing analysis, which can be used at the different stages. This model enables a link to be established between the timing constraints specified during the different stages. The practical utility of this approach was illustrated by performing the analysis of the timeliness requirements of a train set crossing section.

Acknowledgements

The authors would like to acknowledge the financial support of BAe (DCSC), CAPES/Brazil, and the ESPRIT Basic Research Action PDCS.

References

/Dasarathy 85/ B. Dasarathy. "Timing Constraints of Real–Time Systems: Constructs for Expressing them, Methods of Validating them". *IEEE Transactions on Software Engineering* Vol. SE–11(1). January, 1985. pp 80–86.

/de Lemos 92/ R. de Lemos, A. Saeed, T. Anderson. "A Train set as a Case Study for the Requirements Analysis of Safety–Critical Systems". *The Computer Journal*. February 1992 (to appear).

/Genrich 87/ H. Genrich. "Predicate/Transition Nets". *Petri Nets: Central Models and their Properties*. Eds: W. Brauer, W. Reisig, G. Rozemberg. Lectures Notes in Computer Science Vol. 254. 1987. pp 206–247.

/Ghezzi 91/ C. Ghezzi, D. Mandrioli, S. Morasca, M. Pezzè. "A Unified High–Level Petri Net Formalism for Time–Critical Systems". *IEEE Transactions on Software Engineering* Vol. SE–17(2). February, 1991. pp 160–172.

/Gorski 86/ J. Gorski. "Design for Safety using Temporal Logic". *SAFECOMP'86*. Sarlat, France. October, 1986. pp 149–155.

/Jaffe 91/ M. S. Jaffe, N. G. Leveson, M. P. E. Hiemdahl, B. E. Melhart. "Software Requirements Analysis for Real–Time Process–Control Systems". *IEEE Transactions on Software Engineering*, Vol SE–17 (3). March 1991. pp 241–258.

/Jahanian 88/ F. Jahanian, D. A. Stuart. "A Method for Verifying Properties of Modechart Specifications". *Proceedings of the Real–Time Systems Symposium 1988*. Huntsville, AL. December, 1988. pp 12–21.

/Jensen 85/ E. Jensen, D. Locke, H. Tokuda. "A Time–Driven Scheduling Model for Real–Time Operating Systems". *Proceedings of the Real–Time Systems Symposium 1985*. San Diego, CA. December, 1985. pp 112–122.

/Koymans 88/ R. Koymans, R. Kuiper, E. Zijlstra. "Paradigms for Real–Time Systems". *Proceedings of the Symposium in Formal Techniques in Real–Time and Fault–Tolerant Systems*. LNCS 331. Springer–Verlag. M. Joseph (Ed.). Warwick, UK. September, 1988. pp 159–174.

/Laprie 90/ J.C. Laprie. "Dependability: Basic Concepts and Associated Terminology". *ESPRIT PDCS Report No 31*. 1990.

/Laprie 91/ J.-C. Laprie, B. Littlewood. "Quantitative Assessement of Safety-Critical Software: Why and How?". *Probabilistic Safety Assessment and Management Conference*. Beverly Hills, CA. February, 1991.

/Leveson 87/ N. G. Leveson, J. Stolzy. "Safety Analysis Using Petri Nets". *IEEE Transactions on Software Engineering* Vol. SE-13(3). March, 1987. pp 386-397.

/Leveson 91/ N. G. Leveson. "Software Safety in Embedded Computer Systems". *Communications of the ACM*, Vol 34 (2). February, 1991. pp 34-46.

/MacEwen 88/ G. MacEwen, D. Skillicorn. "Using High-Order Logic for Modular Specifications of Real-Time Distributed Systems". *Proceedings of the Symposium in Formal Techniques in Real-Time and Fault-Tolerant Systems.* LNCS 331. M. Joseph (Ed.). Warwick, UK. September, 1988. pp 36-66.

/Milner 83/ R. Milner, "Calculi for Synchrony and Asynchrony". *Theoretical Computer Science* Vol. 25. 1983. pp 267-310.

/Ostroff 87/ J. S. Ostroff, W. M. Wonham. "Modelling, Specifying and Verifying Real-Time Embedded Computer Systems". *Proceedings of the Real-Time Systems Symposium 1987*. San Jose, CA. December 1987. pp 124-132.

/PDCS 90/ "Real-Time Systems (Specific Closed Workshop)". *ESPRIT PDCS Workshop Report W6*. London, UK. September, 1990.

/Pnueli 88/ A. Pnueli, E. Harel, "Applications of Temporal Logic to the Specification of Real Time Systems". *Proceedings of the Symposium in Formal Techniques in Real-Time and Fault-Tolerant Systems.* LNCS 331. Springer-Verlag. M. Joseph (Ed.). Warwick, UK. September, 1988. pp. 84- 97.

/Reed 86/ G. M. Reed, A.W. Roscoe, "A timed model for communicating sequential processes". *Proceedings of 13th International Colloquium on Automata, Languages and Programming*. LNCS 226. Springer-Verlag. Laurent Kott (Ed.). Rennes, France. July, 1986. pp 314-323.

/Saeed 90/ A. Saeed, T. Anderson, M. Koutny. "A Formal Model for Safety-Critical Computing Systems". *SAFECOMP'90*. London, UK. October, 1990. pp 1-6.

/Saeed 91/ A. Saeed, R. de Lemos, T. Anderson. "The Role of Formal Methods in the Requirements Analysis of Safety-Critical Systems: a Train Set Example". *Proceedings of the 21st Symposium on Fault-Tolerant Computing*. Montreal, Canada. June, 1991. pp 478-485.

/van Benthem 90/ J. van Benthem. "*The Logic of Time*". Kluwer Academic Publishers. 1991.

Verification of a Reliable Net Protocol

Victor Yodaiken
Krithi Ramamritham *
Computer Science Dept.
University of Massachusetts
Amherst, MA 01003

Abstract

We specify and prove correctness of a real-world fault-tolerance algorithm. The algorithm, developed by Chang and Maxemchuk [CM84], guarantees delivery of broadcast messages over a broadcast medium (e.g., an ethernet) in the presence of faults that may cause messages to be lost or only partially delivered. Instead of describing the operation of the algorithm in pseudo-code, as the authors of the algorithm have done, we generate a precise mathematical specification which is amenable to reasonably simple proof techniques. The formal method that we use in this paper is based on modal (state dependent) functions called the *modal primitive recursive* (m.p.r.) functions. Our analysis clarifies the workings of the algorithm by discarding the complex program scaffolding that obscures the original exposition.

1 Introduction

Suppose that we have an ethernet-style broadcast network connecting a collection of *source* and *destination* sites. If we assume that the network may occasionally lose messages, but that spurious or erroneous messages are filtered out at a lower level, we may want to implement a protocol to compensate for the imperfect nature of the connection. The obvious protocol requires each destination site to acknowledge each message that it receives from a source site. The obvious disadvantage of this *positive acknowledgement* strategy is that it imposes a very high cost. Message loss on well designed networks will be occasional, but the

*This research was funded in part by the Office of Naval Research under contract N00014-85-K-0398 and N00014-92-J-1048.

the overhead costs of the positive acknowledgment protocol remain high even when the network is error free. The Chang/Maxemchuk protocol [CM84] takes advantage of a clever observation about broadcast networks to provide for reliable message transfer, while minimizing overhead. In this paper, we will specify and formally verify the correctness (safety) of the Chang/Maxemchuk protocol. We hope to show that a formal treatment can actually clarify the workings of a somewhat slipperly algorithm by highlighting exactly which properties of the protocol and which assumptions about the environment are necessary for the protocol to function correctly.

The formal method. The formal method that we use in this paper is based on modal (state dependent) functions called the *modal primitive recursive* (m.p.r.) functions. Very little of the machinery of formal logic is present in this method. Instead, we have endeavored to remain within the familiar world of functions, sets, and relations. In order to accommodate the dynamic nature of discrete systems where the contents of a register or the truth of a proposition can change with the state of the system, we allow for functions that may depend on the effect of system inputs. For example, the function $\mathsf{Xmitted}(m)$ can be defined to have value 0 (false) in the initial state, and so that an input $\mathsf{xmit}.x$ will cause the value of $\mathsf{Xmitted}(m)$ to become 1 (true). The m.p.r. functions are obtained from the primitive recursive functions [Pet67, Goo71] by adding a *single* new function composition rule to those of substitution and primitive recursion. Methods for describing composition and concurrency can be derived from the resulting mathematical system, and require no further extensions. The underlying semantics of m.p.r. functions derives from Moore machines (finite state machines with output) [Moo64, HU79] and products of Moore machines [Gec86]. For any state dependent m.p.r. function $\mathsf{f}(x)$ and any fixed argument n the expression $\mathsf{f}(n)$ corresponds to a unique (minimal) Moore machine. Conversely, every Moore machine, including those which contain concurrent sub-machines, can be precisely specified by some m.p.r. expression. By remaining within the context of m.p.r. functions and using general properties of the functions as well as standard rules of arithmetic and algebra, we can avoid actually constructing the very large-scale state machines needed to model a computer network. Details of the m.p.r. formal semantics can be found in [Yod91b, Yod90] and [Yod91a].

In the next section we specify the workings of the protocol in a "high-level" manner by specifying seven abstract properties. These properties arise naturally from an informal description of the protocol and our expectations of the behavior of a broadcast network. We formally state these properties, and then show that when all seven properties hold, the protocol is *safe* — a source of a broadcast message will conclude that the message has been received by all destinations only if all destination sites have received the message. The analysis of section two has been specifically developed to omit as much implementation detail as possible, in order to expose the underlying structure of the protocol. In section three, we develop a more concrete treatment in which such data structures as queues and message buffers are utilized. This, more concrete, treatment provides some of the details of the local algorithms that must be followed at each site. The destination sites must maintain queues of arriving acknowledgment and broadcast messages, and counters which keep track of the next expected message sequence numbers. The source sites must maintain information about which acknowledgements have been seen, and must buffer transmitted messages until the protocol can assure their safe delivery. We complete the specification and verification of the protocol by connecting the local algorithms to form a model of a network and by showing that the resulting system satisfies the seven properties developed in section two. As a result, we can conclude that the local algorithms and the specific network interconnection we describe, satisfy the safety requirement of the protocol. The process can be elaborated to cover more details of the implementation (e.g., to delimit queue lengths and to verify the correctness of error recovery), and to cover issues such as liveness and real-time requirements, but we will not do so here.

This paper serves four purposes. First, we wish to illustrate the practical expressive power of a formal method which rests on an extremely elementary mathematical basis — one which we believe is significantly more tractable than those underlying the process algebras [Mil79, Hoa85], the temporal logics [Pnu85] or the net based formalisms (e.g., [VGR87]). Second, the correctness of a quite interesting and practical fault tolerance algorithm is verified, and, it is hoped, the workings of the algorithm are clarified. Third, we illustrate a specific approach to algorithm verification in which the abstract properties of the algorithm are first formalized as high-level properties, the properties are shown to be sufficient to ensure a desired system condition, and then an implementation is shown to actually implement the high-level properties. The final purpose of this paper is to provide a meaningful benchmark for comparing various formal approaches to fault-tolerance and real-time. The state of the art in the field is no longer served by purely synthetic or trivial examples. In order to evaluate the practical expressive range of proposed methods, we require more substantial examples, and examples that confront real-world constraints. The Chang/Maxemchuk algorithm is short enough to be presented in a reasonable space and is amenable to intuitive understanding, yet it is manifestly non-trivial. We look forward to seeing this protocol become part of a suite of standard examples which can be used to expose the weaknesses and strengths of competing formal approaches.

2 The algorithm

Chang and Maxemchuk's algorithm depends on the observation that in a broadcast network, the acknowledgment messages can also be broadcast. Thus, if only one of the destinations sends the source an acknowledgment, all the other destinations can "listen in" on the conversation. We arrange the destinations in a cycle, and cause each destination to take a turn as the *tokensite* responsible for acknowledging the next message from a source. We constrain destinations so that a site will only act as tokensite for message m if it has not missed any of the previous messages that have been acknowledged. When tokensite responsibility has been rotated through the entire set of destinations, it must be the case that the first message acknowledged has been received by all destinations. Failures cause additional message traffic, but at a reasonable level. The theorem that we want to prove in this paper states that the algorithm is *safe*. That is, when a source site concludes that a broadcast has completed, all the destination sites must actually have received the message. Our strategy will be to formalize the algorithm in steps, at each step we state only what we need to advance the proof to the next step. In this way we will clarify our understanding of exactly what makes the algorithm work.

In order to prove the correctness of the algorithm or even to state the algorithm, we need to develop some model of a network. On the other hand, we want to leave the network as loosely specified as possible, both in order to increase generality, and to allow us to get to the interesting part of the algorithm without a lengthy pre-amble. To begin with, therefore, our model is rather sketchy. We start with a set $\mathcal{I}$ of site identifiers, a set $\mathcal{M}$ of messages, and two subsets of $\mathcal{I}$: the destination sites $\mathcal{D} \subset \mathcal{I}$ and and the source sites $\mathcal{S} \subset \mathcal{I}$. Note that $\mathcal{D}$ and $\mathcal{S}$ may overlap. Each message $m \in \mathcal{M}$ should be labeled by a type. For now let's limit ourselves to 2 types, BCAST (reliable broadcast) and ACK (acknowledgment). A message with type BCAST should be labeled with a sequence number, and the source identifier. A message with type ACK should be labeled with the identifier of the source of the message being acknowledged, the sequence number of the message being acknowledged, and a *timestamp* (essentially a sequence number for acknowledgments). We expect to have a map $\mathsf{tokensite}: \mathsf{Timestamps} \rightarrow \mathcal{D}$ so that $\mathsf{tokensite}(t)$ is the identifier of the destination site responsible for sending acknowledgments with timestamp t. That is, if $\mathsf{ts}(m)$ is the timestamp of m, then if m is an acknowledgment message $\mathsf{responsible}(m) = \mathsf{tokensite}(\mathsf{ts}(m))$ identifies the destination site that is responsible for generating m. Let's formalize these descriptions as follows:

$\mathcal{M}$	set of messages
$\mathcal{I}$	set of site identifiers
$\mathcal{D} \subset \mathcal{I}$	destination site identifiers
$\mathcal{S} \subset \mathcal{I}$	source site identifiers
$\mathsf{Types} = \{\mathsf{BCAST}, \mathsf{ACK}\} \bigcup \mathsf{Others}$	
$\mathcal{T} = 0 \leqslant \mathsf{t} \leqslant \mathsf{maxts}$	consecutive timestamps
$\mathcal{Q} = 0 \leqslant \mathsf{q} \leqslant \mathsf{maxseq}$	consecutive sequence numbers
$\mathsf{tokensite} : \mathcal{T} \rightarrow \mathcal{D}$	responsibility map
For $\mathsf{m} \in \mathcal{M}$	
$\mathsf{type} : \mathcal{M} \rightarrow \mathsf{Types}$	message type
$\mathsf{source} : \mathcal{M} \rightarrow \mathcal{S}$	source of data
$\mathsf{seq} : \mathcal{M} \rightarrow \mathcal{Q}$	broadcast sequence number
$\mathsf{ts} : \mathcal{M} \rightarrow \mathcal{T}$	the message timestamp
$\mathsf{responsible} : \mathcal{M} \rightarrow \mathcal{D}$	The site responsible for an ack
$\mathsf{responsible}(\mathsf{m}) \stackrel{\text{def}}{=} \mathsf{tokensite}(\mathsf{ts}(\mathsf{m}))$	

Messages on the broadcast network

Simplifying assumption. We assume that sequence numbers and timestamps never need to be rolled over. If timestamps and sequence numbers are represented by 48 bit long bit sequences, and if a broadcast is generated every millisecond for each group, it will take approximately 100 years before we run out of timestamps. In fact, the algorithm will work perfectly well when timestamps need to be recycled, but the proof is slightly more complex. The simplifying assumption allows us to use only the most elementary information about state: we do not have to show that by the time a sequence number is re-used, previous messages with the same number have been flushed from the network.

The source can conclude that a message has been successfully broadcast if the following conditions have been met.

1. *If the source has sent a broadcast message* b*;*

2. *And if the source has received an acknowledgment message* a *which acknowledges* b*;*

3. *And if the source has received an acknowledgment message* a' *for any message (the acknowledgment may acknowledge a message sent by another*

source), where the timestamp of a' *is greater than or equal to* $|\mathcal{D}|$ *(the number of elements of* $\mathcal{D}$*) plus the timestamp of* a*;*

Let's formalize these conditions. We first define a function to test whether or not one message is an acknowledgment for a second.

$$\mathsf{acks}(b, a) \stackrel{\text{def}}{=} (\mathsf{source}(a) = \mathsf{source}(b) \bigwedge \mathsf{seq}(a) = \mathsf{seq}(b))$$

Let $\mathsf{Sent}(s, m)$ and $\mathsf{Received}(s, m)$ be state dependent boolean functions that indicate whether or not the named site has sent or received the named message. We will provide a more concrete definition of these two functions later, but we need only an informal understanding of their intended use for now. To simplify notation, we let a, a', a'' denote messages with type ACK, and let b, b', b'' denote messages of type BCAST. This notational device will permit us to avoid repetitive checks on the message type. Thus, $(\exists a)\mathsf{Received}(s, a)$ is true iff $(\exists m)\mathsf{Received}(s, m) \bigwedge \mathsf{type}(m) = \mathsf{ACK}$.

Now we can make the safety theorem precise.

Theorem 2.1

$$\left(\begin{array}{l} \mathsf{Sent}(s, b) \\ \bigwedge \mathsf{source}(b) = s \\ \bigwedge (\exists a, a') \left(\begin{array}{l} \mathsf{acks}(b, a) \\ \bigwedge \mathsf{Received}(s, a) \\ \bigwedge \mathsf{Received}(s, a') \\ \bigwedge \mathsf{ts}(a') \geqslant \mathsf{ts}(a) + |\mathcal{D}| \end{array}\right) \end{array}\right) \rightarrow (\forall s' \neq s \in \mathcal{D})\mathsf{Received}(s', b)$$

We specify the algorithm by listing seven properties which must be satisfied by the network in order for the algorithm to work. We then show that the theorem can be deduced from these properties. The goal here is to assume as little as possible about the global properties of the network, and to try to rely only on local properties of individual sites. The first property is, in fact, the only global assumption that we require — we assume that the network does not deliver spurious messages. The second property involves the workings of the map between timestamps and token sites. We assume that any sequence of $|\mathcal{D}|$ consecutive timestamps must cover the entire set of destinations. These assumptions seem modest enough. The other properties are all local. That is, these properties can be guaranteed by individual sites, no matter how other sites or the network behave.

To state the first property we need to be able to assert that in the initial or start state, no site has received any message, and also that no site can receive a message until some other site has sent the message. To formalize the first of these assertions, we use the modal boolean function StartState() which is true only in the initial system state, and stays false thereafter. The assertion $\mathrm{StartState}() \rightarrow \mathrm{Received}(s, m) = \mathit{false}$ then is true if in the start (initial) state, no site has received a message. We also use the m.p.r function modifier *after* which allows us to examine future values of m.p.r. functions. When we speak of the value of a m.p.r function f(x), we usually are concerned with its ***current value.*** If y is an input symbol, then $\mathit{after}(\langle y\rangle, f(x))$ denotes the value of f(x), not in the current state, but in the state reached by forcing the system to accept input y. The expression $\mathit{after}(\langle y\rangle, \mathrm{Received}(s, m))$ then is a boolean expression that is true iff a single input y takes us to a state where Received(s, m) is true. The assertion $\mathit{after}(\langle y\rangle, \mathrm{Received}(s, m)) \rightarrow (\exists s' \neq s \in \mathcal{I})\mathrm{Sent}(s', m)$ is then true if and only if $\mathrm{Sent}(s', m)$ must be true in the current state in order to permit Received(s, m) to be true in the next state. [1]

1. *Network safety:* Messages cannot be received unless they have been sent, and once received or sent, a message cannot be "un-received" or "un-sent".

   A. $\mathrm{StartState}() \rightarrow \mathrm{Received}(s, m) = \mathit{false}$
   B. $\mathrm{StartState}() \rightarrow \mathrm{Sent}(s, m) = \mathit{false}$
   C. $\mathit{after}(\langle x\rangle, \mathrm{Received}(s, m)) \rightarrow (\exists s' \neq s)\mathrm{Sent}(s', m)$
   D. $\mathrm{Received}(s, m) \rightarrow \mathit{after}(\langle x\rangle, \mathrm{Received}(s, m))$
   E. $\mathrm{Sent}(s, m) \rightarrow \mathit{after}(\langle x\rangle, \mathrm{Sent}(s, m))$

2. *Tokensite coverage:* Any sequence of $|\mathcal{D}|$ timestamps should map to the entire set of destinations, i.e.:

$$t \leqslant \mathrm{maxts} - |\mathcal{D}| \rightarrow \{\mathrm{tokensite}(t+i) : 0 \leqslant i < |\mathcal{D}|\} = \mathcal{D}.$$

3. *Source safety:* A broadcast message b with source s can be sent by site s′ iff s = s′ or s′ previously received b. The second condition allows rebroadcast.

$$\mathrm{Sent}(s, b) \rightarrow (\mathrm{source}(b) = s \bigvee \mathrm{Received}(s, b))$$

[1] Those familiar with temporal logic will note that *after* and the temporal operator $\bigcirc$ are related. One might translate the m.p.r assertion $\mathit{after}(\langle y\rangle, P) \rightarrow Q$ into the temporal logic assertion $(\bigcirc P) \rightarrow Q$, but this means, "if P must be true in the next state, then Q must be true in the current state," quite different from "if a y input makes P true, then Q must currently be true," which is the informal meaning of the m.p.r assertion. The more accurate temporal logic translation would be $\neg Q \rightarrow \bigcirc\neg P$. But this is also not quite equivalent because $\bigcirc$ means "after one time unit" or "after one program statement", in contrast to the m.p.r. assertion "after the next state change." A true equivalence can only be obtained by adding history variables to the temporal logic: a reasonably involved translation ensues.

4. *Sequence uniqueness:* A source cannot send two distinct broadcast messages with the same sequence number.

$$\left(\begin{array}{l}\mathsf{Sent}(s,b)\\ \bigwedge \mathsf{Sent}(s,b')\\ \bigwedge s = \mathsf{source}(b) = \mathsf{source}(b')\\ \bigwedge \mathsf{seq}(b) = \mathsf{seq}(b')\end{array}\right) \rightarrow b = b'$$

5. *Destination safety:* An acknowledgment message a can be sent by site s iff $s = \mathsf{responsible}(a)$ or s previously received a. The second condition allows rebroadcast.

$$\mathsf{Sent}(s,a) \rightarrow (\mathsf{responsible}(a) = s \bigvee \mathsf{Received}(s,a))$$

6. *Timestamp uniqueness:* A tokensite cannot send two distinct acknowledgements with the same timestamp.

$$\left(\begin{array}{l}\mathsf{Sent}(s,a)\\ \bigwedge \mathsf{Sent}(s,a')\\ \bigwedge \mathsf{ts}(m) = \mathsf{ts}(m')\\ \bigwedge \mathsf{responsible}(a) = s\end{array}\right) \rightarrow a = a'$$

7. *Tokensite safety:* A destination cannot send an acknowledgement message a unless it has received a broadcast matching the acknowledgment, and unless for every timestamp t less than $\mathsf{ts}(a)$ the destination has seen both an acknowledgment and a matching broadcast.

$$\mathsf{Sent}(s,a) \rightarrow (\forall t \leqslant \mathsf{ts}(a))(\exists a', b) \left(\begin{array}{l}\mathsf{acks}(b,a')\\ \bigwedge \left(\begin{array}{l}(\mathsf{source}(b) = s \bigwedge \mathsf{Sent}(s,b))\\ \bigvee \mathsf{Received}(s,b)\end{array}\right)\\ \bigwedge \left(\begin{array}{l}(\mathsf{responsible}(a') = s \bigwedge \mathsf{Sent}(s,a'))\\ \bigvee \mathsf{Received}(s,a')\end{array}\right)\end{array}\right)$$

We can show that the properties listed above guarantee two key derived properties. First, we can show that if a site has received a broadcast message b, then it must be the case that the site $\mathsf{source}(b)$ has previously sent b. Similarly, we can show that if a site has received an acknowledgment message a, it must be the case that $\mathsf{responsible}(a)$ previously sent a.

Lemma 2.2 Source origin: *A broadcast message* b *must be sent by* $\mathsf{source}(b)$ *before it can be received by any site.*

$$\mathsf{Received}(s,b) \rightarrow \mathsf{Sent}(\mathsf{source}(b),b)$$

The proof of this lemma is based on something called induction on inputs. We say that we have proved a property P by induction on inputs if we can prove $\mathsf{StartState}() \to \mathsf{P}$ for the base case, and $\mathsf{P} \to \mathsf{after}(\langle y \rangle, \mathsf{P})$ for the induction step. If we have proved both of these cases, then it must follow that P is true in every state. The base step ensures that in the initial state of the system P, will be true, and the induction step ensures that each input that drives the system to a new state must keep P true.

Proof. by induction on input

{*Base case* }

1. $\mathsf{StartState}()$

$\Rightarrow$ 2. $\mathsf{Received}(s, b) = \mathsf{false}$

$\Rightarrow$ 3. $(\mathsf{Received}(s, b) \to$ Anything $)$

$\Rightarrow$ 4. $(\mathsf{Received}(s, b) \to \mathsf{Sent}(\mathsf{source}(b), b))$

{*Induction step* }

1. $\mathsf{after}(\langle x \rangle, \mathsf{Received}(s, b))$

$\Rightarrow$ {*by network safety* }

2. $(\exists s')\mathsf{Sent}(s', b)$

$\Rightarrow$ {*by source safety* }

3. $(\mathsf{Received}(s', b) \bigvee s' = \mathsf{source}(b))$

$\Rightarrow$ {*induction hypothesis* }

4. $(\mathsf{Sent}(\mathsf{source}(b), b) \bigvee s' = \mathsf{source}(b))$

$\Rightarrow$ {*from 2 and second clause of 4* }

5. $(\mathsf{Sent}(\mathsf{source}(b), b) \bigvee \mathsf{Sent}(\mathsf{source}(b), b)$

$\Rightarrow$ 6. $\mathsf{Sent}(\mathsf{source}(b), b)$

$\Rightarrow$ {*network safety* }

7. $\mathsf{after}(\langle x \rangle, \mathsf{Sent}(\mathsf{source}(b), b)$

Lemma 2.3 Timestamp origin: *An acknowledgement message* a *must be sent by the site* $\mathsf{responsible}(a)$ *before it can be received by or rebroadcast by any other site.*

$$\mathsf{Received}(s, a) \to \mathsf{Sent}(\mathsf{responsible}(a), a)$$

Proof *Slight variation on the above proof*

There are now two simple lemmas that we need to show that if a site has received a broadcast message b, and if the source of b has sent a message b' so that $\mathsf{seq}(b) = \mathsf{seq}(b')$, then $b = b'$. A similar lemma about acknowledgments demonstrates that there are never two distinct acknowledgment messages with the same timestamp floating around in the system.

Lemma 2.4

$$\text{Received}(s,b)\bigwedge \text{Sent}(\text{source}(b),b') \ \bigwedge \text{seq}(b)=\text{seq}(b')\bigwedge \text{source}(b')=\text{source}(b) \rightarrow b=b'$$

Proof.

1. $\text{Received}(s,b)$

$\Rightarrow$ {by lemma 2.2 }

2. $\text{Sent}(\text{source}(b),b)$

{hypothesis }

3. $\text{Sent}(\text{source}(b),b')\bigwedge \text{seq}(b')=\text{seq}(b)\bigwedge \text{source}(b')=\text{source}(b)$

$\Rightarrow$ {Sequence uniqueness, 2, and 3 }

4. $b=b'$

Lemma 2.5

$$\text{Received}(s,a)\bigwedge \text{Sent}(\text{responsible}(a),a')\bigwedge \text{ts}(a)=\text{ts}(a') \rightarrow a=a'$$

Proof.

1. $\text{Received}(s,a)$

$\Rightarrow$ {by lemma 2.3 }

2. $\text{Sent}(\text{responsible}(s),a)$

{hypothesis }

3. $\text{Sent}(\text{responsible}(a),a')\bigwedge \text{ts}(a')=\text{ts}(a)$

$\Rightarrow$ {timestamp uniqueness, 2, and 3 }

4. $a=a'$

We have a final lemma to prove before we prove the main theorem. We use the tokensite safety property and the lemmas we have just proven to show that a destination that has sent an acknowledgment with a timestamp at least as great as that of the acknowledgment message that acknowledges some broadcast message b, must have received b. Once this lemma is proved, the main theorem is quite easy.

Lemma 2.6

$$\begin{aligned}&(\text{Sent}(\text{source}(b),b)\bigwedge \text{Received}(\text{source}(b),a')\bigwedge \text{Acks}(b,a'))\\ &\bigwedge \text{Sent}(s,a)\\ &\bigwedge \text{ts}(a)\geqslant \text{ts}(a')\\ &\rightarrow \text{Received}(s,b)\bigvee s=\text{source}(b)\end{aligned}$$

Proof.

1. $Sent(s,a) \bigwedge ts(a) \geqslant ts(a')$

$\Rightarrow$ *{by tokensite safety}*

2. $(\exists a'') \; ts(a'') = ts(a') \bigwedge (Received(s,a'') \bigvee Sent(s,a''))$

$\Rightarrow$ *{by lemma 2.3}*

3. $Sent(responsible(a''), a'')$

{premise}

4. $Received(source(b), a') \bigwedge Acks(b, a')$

$\Rightarrow$ *{by lemma 2.3}*

5. $Sent(responsible(a'), a')$

$\Rightarrow$ *{by 2,3, 5,lemma 2.5}*

6. $a' = a''$

$\Rightarrow$ *{1,2,6, tokensite safety}*

7. $(\exists b') \; Acks(b', a') \bigwedge (Received(s, b') \bigvee (Sent(s, b') \bigwedge source(b') = s)$

$\Rightarrow$ {*def of* aks(b, a), 10 *and 7*}

8. $source(a') = source(b')$

$\Rightarrow$ 9. $source(b') = source(b)$

$\Rightarrow$ 10. *{9, lemma 2.2}*

11. $Sent(source(b), b')$

$\Rightarrow$ 12. $b = b'$

$\Rightarrow$ 13. $Received(s, b) \bigvee source(b) = s$

Now we can prove theorem 1.

	{Theorem premises}
1	$Sent(s,b) \bigwedge source(b) = s$
2.	$acks(b,a) \bigwedge Received(s,a)$
3.	$Received(s,a') \bigwedge ts(a') \geqslant ts(a) + \lvert\mathcal{D}\rvert$
$\Rightarrow$	*{by lemma 2.3, 3}*
4.	$Sent(tokensite(ts(a'),a')$
$\Rightarrow$	*{by tokensite safety}*
5.	$(\forall t \leqslant ts(a'))(\exists a'')\ ts(a'') = t \bigwedge Received(tokensite(ts(a'),a'') \bigvee Sent(tokensite(ts(a'),a'')$
$\Rightarrow$	*{by lemma 2.3}*
6.	$(\forall t \leqslant ts(a'))(\exists a'') ts(a'') = t \bigwedge Sent(tokensite(ts(a''),a')$
$\Rightarrow$	*{by arithmetic}*
7.	$(\forall ts(a'') - \lvert\mathcal{D}\rvert \leqslant t \leqslant ts(a'))(\exists a'')\ ts(a'') = t \bigwedge Sent(tokensite(ts(a''),a')$
$\Rightarrow$	*{tokensite coverage}*
8.	$(\forall d \in \mathcal{D})(\exists a'') ts(a'') \geqslant ts(a) \bigwedge Sent(d,a'')$
$\Rightarrow$	*{by lemma 2.6}*
12.	$(\forall d \in \mathcal{D}) Received(d,b) \bigvee d = source(b)$

3 Verification of the implementation

Let us turn our attention from the, rather ethereal, high-level considerations of the previous section to a lower level of detail where we must be concerned with data structures such as message queues and counters. We will now attempt to formalize the detailed assumptions and algorithms presented in the article in which Chang and Maxemchuk describe their protocol [CM84]. There are two parts to the more detailed presentation of the algorithm: local and global. The local algorithms control the operation of individual sites. The global specification defines how sites interact. We first develop a collection of state dependent functions that capture the local algorithms and then show that these algorithms satisfy local safety properties analogous to the properties 3-7 of the previous section. Finally, we develop a function that encapsulates and connects the "local" functions. This composite function captures the behavior of the network, and we show that the local safety properties and the global properties of the network together implement all seven of the abstract safety properties.. Thus, the network satisfies theorem 1.

3.1 The local algorithm

We suppose that a site reacts to inputs of the form receive.m and send.m with $m \in \mathcal{M}$. The functions SrcCtl(s) and DestCtl(s) capture the algorithms for reliable sources and for reliable destinations, respectively. The source algorithm is trivial: when a broadcast message b is sent, the source caches b until it receives an acknowledgment, and then the message plus its acknowledgment are placed on a queue. The source also keeps track of the highest timestamp number on any incoming acknowledgment message.

$$
\begin{array}{l}
\mathsf{StartState}() \bigvee s \notin \mathcal{S} \rightarrow \mathsf{SrcCtl}(s) = (\mathsf{nullm}, \langle\rangle, 0, 0) \\
s \in \mathcal{S} \bigwedge \mathsf{SrcCtl}(s) = (m, q, n, t) \rightarrow \\
\quad \mathsf{after}(\langle x\rangle, \mathsf{SrcCtl}(s)) = \\
\quad \begin{cases}
(b, q, n+1, t) & \textbf{if } x = \mathsf{send}.b \\
 & \textbf{and } m = \mathsf{nullm}; \\
 & \textbf{and } \mathsf{source}(b) = s \\
 & \textbf{and } \mathsf{seq}(b) = n; \\
(\mathsf{nullm}, \langle(m, a)\rangle q, n, \max\{t, ts(a)\}) & \textbf{if } x = \mathsf{receive}.a \\
 & \textbf{and } \mathsf{acks}(m, a); \\
(m, q, n, \max\{t, ts(a)\}) & \textbf{else if } x = \mathsf{receive}.a; \\
(m, q, n, t) & \textbf{otherwise} .
\end{cases}
\end{array}
$$

So, $\mathsf{SrcCtl}(s) = (m, q, n, t)$ where m is the current message being transmitted, q is the queue of acknowledged messages, n is the sequence number of the next broadcast message, and t is the largest timestamp seen on a message. We define SafeSent(s, b) to be true if b is in the queue of acked messages, and if:

$$
\mathsf{SafeSent}(s, b) \stackrel{\text{def}}{=} \begin{cases}
\mathit{true} & \textbf{if } \mathsf{SrcCtl}(s) = (m, q, n, t) \\
 & \textbf{and } (\exists a)(b, a) \in q \bigwedge ts(a) + |\mathcal{D}| \leqslant t; \\
\mathit{false} & \textbf{otherwise} .
\end{cases}
$$

The destination algorithm is a little more complex. Here is a slightly edited quote from Chang and Maxemchuk that sketches out part of the destination algorithm.

> Each site i maintains the following information:
>
> - $\mathsf{Nseq}_i[s]$ the number of the next broadcast message it expects from site s;
> - nts_i the next timestamp it expects to receive (page 256).
>
> [...] The receivers store broadcast messages in a queue Q_B and process the acknowledgments in the order they are received. At site

> i, $ACK(ts, B(s,n))$ is only processed when $nts_i = ts$ and $B(s,n)$ ...is in Q_B. When acknowledgement ts is processed, nts_i is incremented and if a message has been acknowledged, the next message from the source is changed to $Nseq_i[s] = n+1$. If $ts < nts_i$, this acknowledgment has been previously processed, and is not processed again. If $ts > nts_i$, acknowledgment messages have been lost. Before processing acknowledgement ts, the missing acknowledgements are requested and processed. If $B(s,n)$, the broadcast message being acknowledged, is not in Q_B it must be obtained before the acknowledgment is processed. ...Any acknowledgements that are received while waiting for a retransmitted message are stored in a queue Q_C and are processed in the order they arrive. (page 257).

There are quite a few details here. We can begin by focusing on the key data structures: the counters $nseq$ and nts, and the message queues Q_C (the queue of acknowledgments yet to be processed) and Q_B the queue of broadcast messages waiting to be acknowledged. Although not mentioned here, we also need a queue Q_T of pairs (a, b) to record *processed* broadcast messages plus the acknowledgment messages that acknowledges them. The queue Q_B is really just used as a buffer, order does not matter, so let $b \in Q_B$ be true if any one of the elements of Q_B matches b, and let $Q_B \setminus b$ denote the result of removing all b elements from Q_B. The counter $nseq$ is really a list of counters. For $s \in \mathcal{S}$, write $nseq[s]$ to denote the element of $nseq$ that gives us the number of the next expected sequence number from site s.

The receiver can *process* an acknowledgement a, only when: a is at the head of Q_C, $nts = ts(a)$, and there is a message in Q_B that a acknowledges. We define a recursive function to process the waiting messages in a recursive manner, calling itself until it has no more acknowledgments to process.

$$\mathrm{process}_s((Q_C, Q_B, Q_T, nts, nseq)) \stackrel{\text{def}}{=} \begin{cases} \mathrm{process}_s((Q'_C, Q'_B, Q'_T, nts', nseq')) & \textbf{if } Q_C = q\langle a\rangle; \\ & \textbf{and } (\exists b \in Q_B)\mathrm{Acks}(b, a) \\ & \textbf{and } nts = ts(a) \\ & \textbf{and } nts' = nts + 1 \\ & \textbf{and } nseq'[source(b)] = seq(b) \\ & \textbf{and } nseq'[s \neq source(b)] = nseq[s] \\ & \textbf{and } Q'_C = q \\ & \textbf{and } Q'_B = Q_B \setminus b \\ & \textbf{and } Q'_T = \langle (a, b)\rangle Q_T; \\ \mathrm{process}_s((\langle a\rangle, Q_B, Q_T, nts, nseq)) & \textbf{if } Q_C = \langle\rangle \\ & \textbf{and } tokensite(nts) = s \\ & \textbf{and } ts(a) = nts \\ & \textbf{and } (\exists b \in Q_B) acks(b, a) \\ & \textbf{and } seq(b) = nseq[source(b)]; \\ (Q_A, Q_C, Q_T, nts, nseq) & \textbf{otherwise}\ . \end{cases}$$

A site s processes $(Q_C, Q_B, Q_T, nts, nseq)$ by, recursively, removing the ACK message at the head of Q_C and the message which it acknowledges from Q_B, appending the two messages to the queue Q_T, and updating the counters. If the head of Q_C does not have a timestamp equal to nts, or if there is no message in Q_B which is acknowledged by that message, then we can do nothing. If $tokensite(nts) = s$, then this site is the tokensite, and it should generate an acknowledgment for some waiting source message. If this site is the tokensite, but $Q_C \neq \langle\rangle$, then we must have already generated the acknowledgment. Note that nts must be incremented after we generate an acknowledgment.

The function process tells us nothing about the state dependent behavior of the destination sites. We formalize some of this behavior with the function $\mathrm{DestCtl}(s)$.

$$\mathrm{StartState}() \bigvee s \notin \mathcal{D} \rightarrow \mathrm{DestCtl}(s) = (\langle\rangle, \langle\rangle, \langle\rangle, (0^{|\mathcal{S}|}), 0)$$

$$\mathrm{DestCtl}(s) = (Q_A, Q_C, Q_T, nseq, nts) \rightarrow$$

$$\mathrm{after}(\langle x\rangle, \mathrm{DestCtl}(s)) \stackrel{\text{def}}{=} \begin{cases} \mathrm{process}_s((\langle a\rangle Q_C, Q_B, Q_T, nseq, nts)) & \textbf{if } x = receive.a \\ & \textbf{and } ts(a) \geqslant nts \\ & \textbf{and } a \notin Q_C \\ \mathrm{process}_s((Q_C, \langle b\rangle Q_B, Q_T, nseq, nts)) & \textbf{if } x = receive.b \\ & \textbf{and } seq(b) \geqslant nseq[source(b)] \\ & \textbf{and } b \notin Q_B; \\ (Q_C, Q_B, Q_T, nseq, nts) & \textbf{otherwise}\ . \end{cases}$$

Note that if $\mathsf{ts}(a) \leqslant \mathsf{nts}$, or if $\mathsf{seq}(b) \leqslant \mathsf{nseq}[\mathsf{source}(b)]$ we ignore these messages — they have already been seen.

We now tie the control functions to together with a function $\mathsf{Site}(s)$. The output of $\mathsf{Site}(s)$ will either be idle or (send, m) for some $m \in M$ or error. Output (send, m) is a request to the network to transmit message m. We don't really care about messages that are not type ACK or BCAST, but we do want to make sure that ACK messages have been previously processed by $\mathsf{DestCtl}(s)$ and that BCAST messages either are ok with $\mathsf{SrcCtl}(s)$ or are retransmissions of messages processed by $\mathsf{DestCtl}(s)$. If the site is forced to send a message against its will, e.g., when $\mathsf{Site}(s) \neq (\mathsf{send}, m)$, we make the site go to an error state, and remain there. Note that the site can fail for other reasons. But $\mathsf{Site}(s) = \mathsf{error}$ indicates only that the site has been forced to erroneously transmit a message.

Specification of the site behavior

$$
\begin{aligned}
&\mathsf{Site}(s) = (\mathsf{send}, b) \rightarrow \mathsf{source}(b) = s \\
&\qquad\qquad\qquad\qquad\qquad \bigwedge \mathsf{SrcCtl}(s) = (m, q, n, t) \\
&\qquad\qquad\qquad\qquad\qquad \bigwedge\ b = m \\
&\qquad\qquad\qquad\qquad\qquad\quad \bigvee (\exists a)(b, a) \in q \\
&\qquad\qquad\qquad\qquad\qquad\quad \bigvee m = \mathsf{nullm} \bigwedge n = \mathsf{seq}(b) \\
&\qquad\qquad\qquad \bigvee\ \mathsf{DestCtl}(s) = (Q_A, Q_C, Q_T, \mathsf{nseq}, \mathsf{nts}) \\
&\qquad\qquad\qquad\quad \bigwedge (\exists a)(b, a) \in Q_T \\
&\mathsf{Site}(s) = (\mathsf{send}, a) \rightarrow \mathsf{DestCtl}(s) = (Q_A, Q_C, Q_T, \mathsf{nseq}, \mathsf{nts}) \\
&\qquad\qquad\qquad\quad \bigwedge (\exists b)(b, a) \in Q_T \\
&\mathsf{StartState}() \rightarrow \mathsf{Site}(s) \neq \mathsf{error} \\
&\mathsf{after}(\langle x \rangle, \mathsf{Site}(s)) = \mathsf{error} <+> (x = \mathsf{send}.m \bigwedge \mathsf{Site}(s) \neq (\mathsf{send}, m) \\
&\qquad\qquad\qquad\qquad\qquad \bigvee \mathsf{Site}(s) = \mathsf{error}
\end{aligned}
$$

3.2 Local conditions and the safety assumptions

Let us now define local analogs of Received and Sent, and show that the sites must obey local versions of network safety properties 3-7.

$$\mathsf{StartState}() \rightarrow \mathsf{LocalSent}(s, m) = \mathsf{false}$$

$$\mathsf{after}(\langle x \rangle, \mathsf{LocalSent}(s, m)) = \begin{cases} \mathsf{true} & \textbf{if } x = \mathsf{send}.m \\ \mathsf{LocalSent}(s, m) & \textbf{otherwise}\ . \end{cases}$$

$$\mathsf{StartState}() \rightarrow \mathsf{LocalReceived}(s, m) = \mathsf{false}$$

$$\mathsf{after}(\langle x \rangle, \mathsf{LocalReceived}(s, m)) = \begin{cases} \mathsf{true} & \textbf{if } x = \mathsf{receive}.m \\ \mathsf{LocalReceived}(s, m) & \textbf{otherwise}\ . \end{cases}$$

Lemma 3.1 *Local source safety.*

$$\text{LocalSent}(b) \bigwedge \text{Site}(s) \neq \text{error} \rightarrow (\text{source}(b) = s \bigvee \text{LocalReceived}(b))$$

Proof.

{by input induction }
$\text{StartState}() \rightarrow \text{LocalSent}(b) = \text{false}$

{induction step }

1. $\text{after}(\langle x \rangle, \text{LocalSent}(b)) = \text{true}$

$\Rightarrow$ *{definition of LocalSent }*

2. $\text{LocalSent}(b) \bigvee x = \text{send}.b$

$\Rightarrow$ *{induction hypothesis }*

3. $(\text{source}(b) \bigvee \text{LocalReceived}(b)) \bigvee x = \text{send}.b$

$\Rightarrow$ *{the source of* b *will not change and }*
{by definition, LocalReceived *cannot become false once true }*

4. $\text{after}(\langle x \rangle, (\text{source}(b) = s \bigvee \text{LocalReceived}(b)) \bigvee x = \text{send}.b$

$\Rightarrow$ *{introduce an abbreviation }*

5. $\text{LemmaTrue} \bigvee x = \text{send}.b$

$\Rightarrow$ *{specification of site behavior }*

6. $\text{LemmaTrue} \bigvee \text{Site}(s) = (\text{send}, b)$

$\Rightarrow$ *{specification of site behavior }*

7. $\text{LemmaTrue} \bigvee \ (\text{source}(b) = s \bigvee \ \text{DestCtl}(s) = (Q_C, Q_B, Q_T, \text{nseq}, \text{nts}) \bigwedge (\exists a)(b, a) \in Q_T$

$\Rightarrow$ *{boolean calculus }*

8. $\text{LemmaTrue} \bigvee \ \text{DestCtl}(s) = (Q_C, Q_B, Q_T, \text{nseq}, \text{nts}) \bigwedge (\exists a)(b, a) \in Q_T$

$\Rightarrow$ *{definition of* DestCtl *}*

9. $\text{LemmaTrue} \bigvee \text{LocalReceived}(b)$

$\Rightarrow$ *{Again,* LocalReceived *cannot become false in the next state }*

10. LemmaTrue

End Proof

We will assert without proving the correctness of the following.

Lemma 3.2 Local sequence uniqueness

$$\begin{pmatrix} \text{Site}(s) \neq \text{error} \\ \bigwedge \text{LocalSent}(b) \\ \bigwedge \text{LocalSent}(b') \\ \bigwedge s = \text{source}(b) = \text{source}(b') \\ \bigwedge \text{seq}(b) = \text{seq}(b') \end{pmatrix} \rightarrow b = b'$$

Local Destination safety:

$$\mathsf{Site}(s) \neq \mathsf{error} \bigwedge \mathsf{LocalSent}(a) \rightarrow (\mathsf{responsible}(a) = s \bigvee \mathsf{LocalReceived}(a))$$

Local Timestamp uniqueness:

$$\begin{pmatrix} \mathsf{Site}(s) \neq \mathsf{error} \\ \bigwedge \mathsf{LocalSent}(a) \\ \bigwedge \mathsf{LocalSent}(a') \\ \bigwedge \mathsf{ts}(m) = \mathsf{ts}(m') \\ \bigwedge \mathsf{responsible}(a) = s \end{pmatrix} \rightarrow a = a'$$

Local Tokensite safety:

$$\mathsf{LocalSent}(a) \\ \rightarrow (\forall t \leqslant \mathsf{ts}(a))(\exists a', b) \begin{pmatrix} \mathsf{acks}(b, a') \\ \bigwedge \begin{pmatrix} (\mathsf{source}(b) = s \bigwedge \mathsf{LocalSent}(b)) \\ \bigvee \mathsf{LocalReceived}(b) \end{pmatrix} \\ \bigwedge \begin{pmatrix} (\mathsf{responsible}(a') = s \bigwedge \mathsf{LocalSent}(a')) \\ \bigvee \mathsf{LocalReceived}(a') \end{pmatrix} \end{pmatrix}$$

3.3 Modeling the interconnection

A state dependent m.p.r. function can be considered as a finite state machine, accepting input from the environment and generating an output that is the value of the function. We describe composite systems by composing m.p.r. functions in such a way that the input symbols provided to the composed functions depends on the output (values) of the other functions. That is, we *encapsulate* m.p.r. functions so that their input comes through a filter and not directly from the environment. The formal semantics of encapsulation is derived from that of the feedback automata product [Yod91b, Gec86]. The practical import of encapsulation is that we can connect specifications of arbitrary systems using an interconnection semantics of our choice — there is no fixed paradigm of communication or synchronization embedded in the formal method itself. In this case we will sketch the definition of a function $\mathsf{Net}(x)$ so that for $s \in \mathcal{I}$, $\mathsf{Net}(s) \stackrel{\text{def}}{=} \mathsf{Encap}(\mathsf{Site}(s), \mathsf{filter}(\mathsf{input}, s))$. Let $\mathsf{Components} = \mathcal{I} \bigcup \{\mathsf{cable}\}$ be a set of identifiers for the component subsystems of the network. The filter function generates input for $\mathsf{Site}(s)$ from system input and from the the current values of $\mathsf{Net}(c) : c \in \mathsf{Components}$. Thus, when the network gets input symbol x, the input sequence given by $\mathsf{filter}(x, s)$ is generated for a each encapsulated $\mathsf{Site}(s)$.

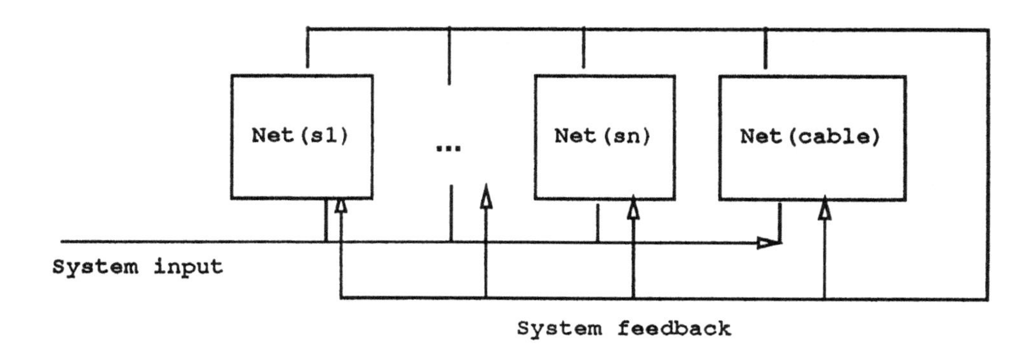

A view of the network

Informally, filter(input, cable) will generate input which indicates to the cable state machine which of the sites are attempting to transmit messages. Similarly, filter(input, s) will generate input $\langle receive.m \rangle$ if Net(cable) has a value indicating that the cable is delivering m to site s, and filter(input, s) will generate $\langle send.m \rangle$ when the cable indicates that it will accept a transmission from site s.

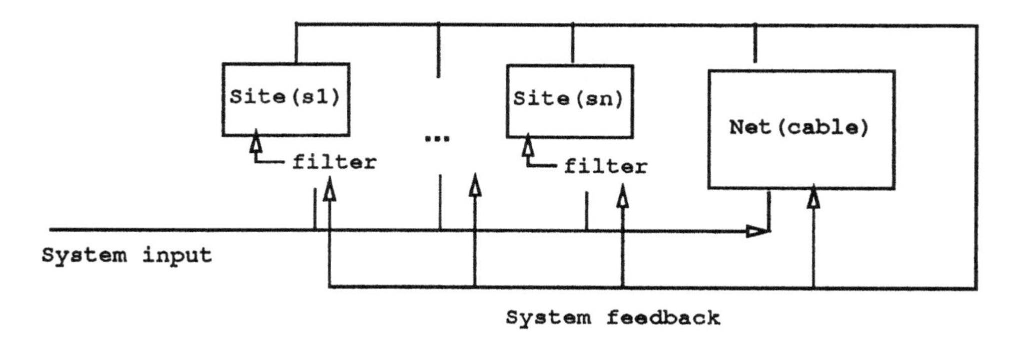

A second view of the network (showing the filter).

Note that we haven't said anything about the input alphabet of the network: it may be very simple, containing only the single input symbol nextstep, or it may contain symbols representing the passage of real-time, symbols representing commands that should be passed on to the sites, or symbols representing failures of the cable. At this level of abstraction we don't need to know anything about this alphabet. All we need to know is that the network obeys two simple safety properties. The first property makes sure that sites only send messages that they intend to send.

Composite Net Safety I.

$$\text{filter}(x, s) = u\langle send.m \rangle v \rightarrow (u = \langle\rangle \bigwedge \text{Net}(s) = (send, m))$$

Recall that Net(s) is an encapsulated copy of Site(s). So Net(s) = (send, m) when the filtered input provided to Site(s) causes it to reach a state where it has value (send, m). Thus, the network will only cause Site(s) to input send.m when s is attempting to transmit. We make $u = \langle\rangle$ in order to prevent the network from committing itself to sending a message too early. For example, if $\text{filter}(x, s) = \langle z, \text{send.m}\rangle$ then the z input might cause Site(s) to take on a value not equal to (send, m).

We can now conclude that the sites are never forced to transmit erroneous messages.

Lemma 3.3

$$\text{Net}(s) \neq \text{error}$$

Proof

	{By input induction }
1.	$\text{StartState}() \rightarrow \text{Net}(s) \neq \text{error}$
2.	$\text{filter}(x, s) = u\langle\text{send..m}\rangle w$
$\Rightarrow$	$\text{filter}(x, s) = \langle\text{send.m}\rangle w$
$\Rightarrow$	$\text{Net}(s) = (\text{send}, m)$
$\Rightarrow$	$\text{after}(x, \text{Net}(s)) \neq \text{error}$

Now we want to make sure that sites only receive messages that have previously been sent. We need a function in order to define this property, and we will call this function Sent. We will have to show, later, that Sent as defined here, obeys the network safety property of the previous section.

Definition 3.4

$$\text{StartState}() \rightarrow \text{Sent}(s, m) = \text{false}$$
$$\text{after}(\langle x\rangle,\ \text{Sent}(s, m)) = \begin{cases} \text{true} & \textbf{if } (\exists u, v)\text{feed}(x, s) = u\langle\text{send.m}\rangle v; \\ \text{Sent}(s, m) & \textbf{otherwise} . \end{cases}$$

The second composite network safety property is now straightforward.

Composite Net Safety II.

$$\text{filter}(x, s) = u\langle\text{receive.m}\rangle v \rightarrow (\exists s' \neq s)\text{Sent}(m, s')$$

Let's now make Received more precise.

Definition 3.5

$$\text{StartState}() \rightarrow \text{Received}(s, m) = \text{false}$$
$$\text{after}(\langle x\rangle,\ \text{Received}(s, m)) = \begin{cases} \text{true} & \textbf{if } (\exists u, v)\text{filter}(x, s) = u\langle\text{receive.m}\rangle v; \\ \text{Received}(s, m) & \textbf{otherwise} . \end{cases}$$

Lemma 3.6 *Composite net safety II and the definitions of* Sent *and* Received *imply the 5 parts of the network safety property.*

Proof

{*By definition* }

A. StartState() → Received(s, m) = false

B. StartState() → Sent(s, m) = false

{*easy induction on inputs* }

C. after(⟨x⟩, Received(s, m)) = true → $(\exists s' \neq s)$Sent(s′, m)

{*By definition* }

D. Received(s, m) → after(⟨x⟩, Received(s, m))

E. Sent(s, m) → after(⟨x⟩, Sent(s, m))

Finally, we want to relate Received and Sent to LocalRecieved and LocalSent. We write in(Net(s), f(x)) to ask for the value of f(x) within the context of Net(s) : that is we make the input to f(x) come through filter(y, s). Thus, in(Net(s), LocalReceived(m)) tells us whether or not the site encapsulated within Net(s) saw input receive.m [2]. We can now fix Received(s, m) $\stackrel{\text{def}}{=}$ in(Net(s), LocalReceived(m)) and Sent(s, m) $\stackrel{\text{def}}{=}$ in(Net(s), LocalSent(m)). Since the local safety properties depend only on the behavior of the Site functions, it must be the case that for each local safety property P, in(Net(s), P) must hold. That is, since Site(s) ensures that, for example, *Local destination safety* must hold, and since in(Net(s), P) is evaluated by testing to see if the encapsulated Site(s) obeys P, we can conclude that in(Net(s), *Local Destination Safety*) must be true. All that remains is to verify that $(\forall s)$ in(Net(s), LocalP) implies P — i.e., if the local safety property holds in every encapsulated site, then the original safety property must hold for the network. This proof is a trivial consequence of how we have defined Sent and Received. Thus, the implementation does satisfy all 7 safety properties, and the algorithm as specified here must be safe.

4 Conclusion

In this paper we have provided a hierarchical specification and proof of correctness (safety) of the Chang/Maxemchuk distributed fault-tolerant message protocol. An extension of the analysis carried out here can treat error recovery, liveness, and real-time performance. We have attempted to cover four issues:

[2] The functional in is a map from m.p.r. functions to m.p.r. functions. Essentially, in(*Net*(*s*), *f*(*x*)) rewrites the definition of a *Net* to get a new function *Net′*, that also encapsulates *f*(*x*) in such a way as to make sure that the inputs provided to *f*(*x*) are identical to the inputs provided to the encapsulated *Site*(*s*).

- The use of the m.p.r. functions, a formal method based on finite state machines and primitive recursive functions, to verify distributed algorithms;
- The correctness and intuitive basis of the protocol itself;
- An approach to specification of a distributed algorithm that postpones the issues of communication between modules until both the required high-level correctness properties of the system, and the detailed operation of the local algorithms have been covered;
- The use of a substantive illustrative example for a formal method, in place of such examples as concurrent programs which compute **skip**, fifo queues, dining philosophers, or highly stylized versions of plant control problems.

References

[CM84] J.M. Chang and N.F. Maxemchuk. Reliable broadcast protocols. *ACM Trans. Computer Systems*, 2(3), august 1984.

[Gec86] Ferenc Gecseg. *Products of Automata.* Monographs in Theoretical Computer Science. Springer Verlag, 1986.

[Goo71] R. L. Goodstein. *Development of Mathematical Logic.* Logic Press, London, 1971.

[Hoa85] C. A. R. Hoare. *Communicating Sequential Processes.* Prentice-Hall, 1985.

[HU79] John E. Hopcroft and Jeffrey D. Ullman. *Introduction to Automata Theory, Languages, and Computation.* Addison-Welsey, Reading MA, 1979.

[Mil79] R. Milner. *A Calculus of Communicating Systems*, volume 92 of *Lecture Notes in Computer Science.* Springer Verlag, 1979.

[Moo64] E.F. Moore, editor. *Sequential Machines: Selected Papers.* Addison-Welsey, Reading MA, 1964.

[Pet67] Rozsa Peter. *Recursive functions.* Academic Press, 1967.

[Pnu85] A. Pnueli. Applications of temporal logic to the specification and verification of reactive systems: a survey of curent trends. In J.W. de Bakker, editor, *Current Trends in Concurrency*, volume 224 of *Lecture Notes in Computer Science.* Springer-Verlag, 1985.

[VGR87] K. Voss, H.J. Genrich, and G Rozenberg, editors. *Concurrency and Nets: Advances in Petri Nets.* Springer-Verlag, 1987.

[Yod90] Victor Yodaiken. *A Modal Arithmetic for Reasoning About Multi-Level Systems of Finite State Machines.* PhD thesis, University of Massachusetts (Amherst), 1990.

[Yod91a] Victor Yodaiken. The algebraic feedback product of automata. In *Papers from the DIMACS Workshop on Computer Aided Verification*, AMS-DIMACS Series. American Mathematical Society, 1991.

[Yod91b] Victor Yodaiken. Modal functions for concise representation of finite automata. *Information Processing Letters*, to appear 1991.

Mechanical Verification of a Generalized Protocol for Byzantine Fault Tolerant Clock Synchronization

Natarajan Shankar*
Computer Science Laboratory
SRI International
Menlo Park, CA 94025 USA
shankar@csl.sri.com
Phone: (415)859-5272

October 23, 1991

Abstract

Schneider [Sch87] generalizes a number of protocols for Byzantine fault-tolerant clock synchronization and presents a uniform proof for their correctness. We present a mechanical verification of Schneider's protocol leading to several significant clarifications and revisions. The verification was carried out with the EHDM system [RvHO91] developed at the SRI Computer Science Laboratory. The mechanically checked proofs include the verification that the *egocentric mean* function used in Lamport and Melliar-Smith's Interactive Convergence Algorithm [LMS85] satisfies the requirements of Schneider's protocol. Our mechanical verification raises a number of issues regarding the verification of fault-tolerant, distributed, real-time protocols that are germane to the design of a special-purpose logic for such problems.

1 Introduction

Synchronizing clocks in the presence of faults is a classic problem in distributed computing. Even the most accurate clocks do drift at significant rates, both with respect to a time standard and relative to each other. In order for independent processors to exhibit cooperative behavior, it is often required that their local clocks be synchronized. Such synchrony is the basis for distributed algorithms that use timeouts, time stamps, and rounds of message passing. Synchronization is also assumed when the same computation is executed on multiple, independent processors in order to mask processor failures.

*This work was supported by NASA Contract NAS1-18226. John Rushby, Friedrich von Henke, Fred Schneider, and Rick Butler provided considerable guidance and encouragement. I also thank Paul Miner (NASA Langley Research Center) and the referees for their comments and clarifications.

Synchronizing clocks in the presence of faults is a difficult problem. Maintaining synchrony by periodically broadcasting a global clock has the drawback of creating a single point of failure. The basic way to achieve fault-tolerant synchronization is for each processor to periodically execute a protocol that involves exchanging clock values with the other processors, computing a consensus reading from these values, and appropriately adjusting its local clock to reflect the consensus. The difficulty is that processors can fail in arbitrary, unpredictable ways and can upset the consensus by communicating one clock value to one processor and a different clock value to another. An algorithm that can cope with such failures is said to be *Byzantine* fault-tolerant [LSP82]. There are a number of known algorithms for Byzantine fault-tolerant clock synchronization. These algorithms themselves are fairly simple to describe, but the reasoning required to establish their correctness is extremely delicate. The difficulty stems from having to simultaneously deal with relative and absolute clock drifts, processor failures, reading errors, and the complicated arithmetic that is involved. Correctly modelling the behavior of clocks can itself be a difficult problem under these conditions.

Schneider [Sch87] presents a clock synchronization scheme (abbreviated here as SCS) that captures the mathematics behind a number of individual synchronization algorithms. His scheme alleviates some of the complexity of reasoning about these protocols. Schneider regards each processor as maintaining a local clock by periodically adjusting its value to one computed by a *convergence function* applied to the readings of all of the clocks. Schneider places certain natural conditions on the behavior of suitable convergence functions and shows that these conditions are sufficient for demonstrating that at any time, the readings of two nonfaulty clocks are always within a fixed bound of each other. The convergence functions used by individual protocols can then be shown to meet Schneider's conditions.

The generality of Schneider's formulation made it an appropriate candidate for mechanical verification. Our verification employed the EHDM specification/verification environment developed at the Computer Science Laboratory of SRI International. The verification provides a rigorous formalization of the behavior of clocks in the presence of Byzantine faults, and careful and tight derivations of the conditions needed to achieve synchronization. The use of EHDM led to the clarification of a number of details from Schneider's original presentation. For instance, Schneider employs a monotonicity condition on convergence functions that was found to be inessential for the proof. The monotonicity condition actually fails for several protocols (see Section 5). The mechanized proof clears up some minor inaccuracies in Schneider's derivation of several of the inequality constraints on the various quantities. The powerful decision procedures for linear equalities and inequalities provided by EHDM were extremely useful for this proof.

There are some other related efforts aimed at mechanically verifying distributed protocols. Rushby and von Henke [RvH91] have used EHDM to check the proofs of Lamport and Melliar-Smith's interactive convergence clock synchronization algorithm (ICA) [LMS85]. This verification followed the original presentation of ICA in modelling local clocks as mapping clock time to real time. Our verification formalizes clocks as mapping real time to clock time. We compare these approaches in Section 6. Rushby [Rus91] uses EHDM to model and verify fault masking and recovery in a synchronous N-plex system for fault-tolerance. Bevier and Young [BY90] have used the Boyer-Moore theorem prover to verify the Oral Messages algorithm for Byzantine agreement [LSP82] assuming that the processors are already synchronized.

Schneider's results appear in a technical report and have not had the benefit of widespread scrutiny so it is not surprising that errors were discovered during verification. On the other hand, the interactive convergence algorithm appears in a widely read journal paper [LMS85] and the verification attempt by Rushby and von Henke [RvH91] discovered flaws in the proof that had, till then, safely survived the social process.

The above machine-assisted proofs of distributed protocols were formalized in general-purpose logics. EHDM, for example, uses a simply typed higher-order logic. Despite the mechanical assistance and the relative abstractness of the SCS protocol, our verification still required a significant amount of effort. It is an interesting challenge to devise a logic that is more specialized to the task of describing fault-tolerant distributed protocols and proving their correctness.

In this paper, we describe one outcome of our mechanized verification, namely, a precise description of Schneider's clock synchronization scheme. We also examine the issues relating to the formal description and verification of such protocols. In Section 2 we discuss the problem of Byzantine fault-tolerant clock synchronization. Section 3 is a careful outline of the SCS protocol as refined by the mechanized verification. Section 4 is a brief sketch of the proof that was mechanically checked. Section 5 illustrates how the egocentric mean function of the ICA protocol satisfies Schneider's conditions. Some observations on the proof are presented in Section 6. The Appendices present the informal proof and some highlights of the mechanized verification. An expanded description of the proof is available as a technical report [Sha91].

2 Byzantine Fault-tolerant Clock Synchronization

In any implementation of synchronized clocks, each processor has a *physical* clock that is typically a crystal clock. Such a physical clock drifts away from the fixed standard time ("real time") at a rate that can be bounded. By periodically applying an adjustment to the reading of the physical clock, each processor also maintains a *logical,* or *virtual,* clock. The adjustment is computed by a protocol involving the exchange of clock readings by the various processors. The primary requirement that any algorithm for clock synchronization must satisfy is that at any instant, the absolute difference, or the *skew*, between two virtual clock readings should be within some fixed, acceptable bound.

Processor failure adds a significant dimension of complexity to the problem of clock synchronization. As an illustration of the difficulty of synchronizing clocks in the presence of Byzantine failures, consider the case of three clocks a, b, and c, where only c is faulty, and nonfaulty clocks can gain or lose up to one minute during an hour. This means that two nonfaulty clocks could drift apart by two minutes over an hour since one clock can gain a minute while the other loses a minute. Suppose that the goal is to keep the nonfaulty clocks synchronized to within three minutes, where clock a gains a minute each hour and b loses a minute each hour. The clocks start synchronized at 12 noon. At 1pm, clock a reads 1:01pm, clock b reads 12:59pm, and clock c has failed. The clocks exchange their readings, and c maliciously communicates its reading as 1:03pm to a and as 12:57pm to b. At this point, a natural way for a clock to resynchronize would be to reset itself to the average of the acceptable clock readings, namely those readings that are within three minutes of its own reading. Then a resets itself to 1:01pm and b resets itself to 12:59pm, so that they remain two minutes apart. Continuing thus, at 2pm, clock a reads 2:02pm whereas clock b reads 1:58pm. The clocks a and b are now four minutes apart, thereby exceeding the acceptable bound on the skew between nonfaulty clocks.

The above scenario illustrates one of the early clock synchronization protocols capable of tolerating Byzantine processor failures: the Interactive Convergence Algorithm (ICA) of Lamport and Melliar-Smith [LMS85]. ICA tolerates up to F failures for N processors where $3F < N$, so that in the above case, at least four clocks are needed to tolerate a single failure. In ICA, a processor p resynchronizes for the i'th time when its clock reads iR. Processor p then reads the difference between the other clock readings and its own clock reading. By ignoring clock readings that differ from its own by more than a certain fixed value, a processor p computes the *egocentric mean* of the remaining clock readings as the required resynchronized clock reading. A number of functions can be used in place of egocentric mean function in order to compute the correction. For example, the fault-tolerant mean function takes the average of the clock readings that remain after the top F and the bottom F readings are discarded [LL84].

The following section describes the formalization of the SCS protocol arising from our verification.

3 Schneider's Schema for Clock Synchronization

The SCS protocol is described below in careful detail so that it can be compared with Schneider's original presentation [Sch87]. Section 3.1 describes how the logical clock is computed from the physical clock using the convergence function. Section 3.2 describes Schneider's conditions on the behavior of clocks and on suitable convergence functions. Note that we only deal with the case when the clocks are resynchronized with an instantaneous adjustment, whereas Schneider also deals with the situation when the adjustment is applied in a continuous manner. Since certain "clock ticks" might be lost or repeated in an instantaneous resynchronization, no critical events can be scheduled for these clock ticks. Another point to note is that both real time and clock readings range over real numbers in the formalization below, but the machine verification has also been carried out with natural number values for the time parameters.

3.1 Defining Clocks

The physical and logical clocks are presented as functions from real time (as given by some external standard) to clock readings. This real time thus forms the frame of reference and is often referred to simply as "time." The variable t ranges over this real time. Values ranging over real time are written in lower case and clock times are in upper case. Synchronization takes place in *rounds*. The time at which processor p adjusts its clock following the i'th round of synchronization is represented by t_p^i. The starting time t_p^0 is taken to be zero.

$PC_p(t)$ is the reading of p's physical clock at real time t, and $VC_p(t)$ is p's virtual or logical clock reading. The virtual clock reading at time t_p^i is computed by applying an adjustment adj_p^i to the physical clock reading $PC_p(t_p^i)$. In its i'th *interval* of operation, *i.e.*, when $t_p^i \le t < t_p^{i+1}$, the virtual clock reading, $VC_p(t)$ is given by $PC_p(t) + adj_p^i$. The virtual clock in the interval between t_p^i and t_p^{i+1} is modelled by an abstraction called the interval clock whose value at time t is $IC_p^i(t)$.

At round 0, the adjustment adj_p^0 is taken to be 0 so that for $t < t_p^1$, the reading $VC_p(t)$ is just $PC_p(t)$. For $i > 0$, we let Θ_p^i be an array of clock readings so that $\Theta_p^i(q)$ is p's reading of $IC_q^{i-1}(t_p^i)$, *i.e.*, q's $(i-1)$'th interval clock reading at time t_p^i, since p is really estimating q's clock reading without taking into account the adjustment adj_q^i. The corrected value of the clock at time t_p^i, namely $VC_p(t_p^i)$, is computed by a *convergence function*, $cfn(p, \Theta_p^i)$.[1]

The above description leads to following definitions where i ranges over the natural numbers and $t > 0$.

$$adj_p^{i+1} = cfn(p, \Theta_p^{i+1}) - PC_p(t_p^{i+1}) \quad (3.1)$$

$$adj_p^0 = 0 \quad (3.2)$$

$$IC_p^i(t) = PC_p(t) + adj_p^i \quad (3.3)$$

$$VC_p(t) = IC_p^i(t), \text{ for } t_p^i \le t < t_p^{i+1} \quad (3.4)$$

It is easy to derive the following from Definitions (3.1), (3.3), and (3.4).

$$VC_p(t_p^{i+1}) = IC_p^{i+1}(t_p^{i+1}) = cfn(p, \Theta_p^{i+1}) \quad (3.5)$$

$$IC_p^{i+1}(t) = cfn(p, \Theta_p^{i+1}) + PC_p(t) - PC_p(t_p^{i+1}) \quad (3.6)$$

In the next section, we enumerate the constraints on these quantities when p is a *nonfaulty* processor. The constraints on the behavior of the convergence function are particularly significant. The main result obtained from these constraints and the above definitions is a bound δ on the skew between the logical clocks of two correct processors p and q.

Theorem 3.1 (bounded skew) *For any two clocks p and q that are nonfaulty at time t,*

$$|VC_p(t) - VC_q(t)| \le \delta \quad (3.7)$$

The proof of Theorem 3.1 is outlined in Section 4.

3.2 Clock conditions

In formalizing the laws constraining the behavior of individual clocks, we must ensure that no assumptions are made regarding the faulty clocks since we are dealing with Byzantine failures. These laws which are conditions on the behavior of clocks are enumerated below. Individual protocols and clock implementations are expected to satisfy these conditions. Note that N is the total number of processors, and F is the maximum number of faulty clocks that the algorithm is expected to tolerate. To start, the following condition asserts that the nonfaulty clocks are synchronized to within the quantity δ_S at time 0.

[1] In the EHDM formalization, the array of observed clock readings Θ_p^i, is actually represented as a function from clocks to readings. Since Θ_p^i is a function, *cfn* is a higher-order function.

Condition 1 (initial skew) *For nonfaulty processors p and q*

$$|PC_p(0) - PC_q(0)| \leq \delta_S \tag{3.8}$$

The nonfaulty physical clocks must keep good enough time so that they do not drift away from real time by a rate greater than ρ.

Condition 2 (bounded drift) *If clock p is nonfaulty at time s, $s \geq t$, then*

$$(1-\rho)(s-t) \leq PC_p(s) - PC_p(t) \leq (1+\rho)(s-t) \tag{3.9}$$

A useful corollary to *bounded drift* is that two physical clocks p and q that are not faulty[2] at time s, for $s \geq t$, can drift further apart over the interval $s-t$ by $2\rho(s-t)$, since both p and q can drift by $\rho(s-t)$ with respect to real time, but in opposite directions.

$$|PC_p(s) - PC_q(s)| \leq |PC_p(t) - PC_q(t)| + 2\rho(s-t) \tag{3.10}$$

Each protocol has some mechanism for triggering the resynchronization of the clocks. Schneider postulates the existence of a global synchronization signal, t^i_G, which occurs at a period bounded from above and below. Our description dispenses with the notion of a global synchronization signal and bounds the period between the local synchronization signals and the range within which these signals occur.

Condition 3 (bounded interval) *For nonfaulty clock p*

$$0 < r_{min} \leq t^{i+1}_p - t^i_p \leq r_{max} \tag{3.11}$$

Condition 4 (bounded delay) *For nonfaulty clocks p and q*[3]

$$|t^i_q - t^i_p| \leq \beta \tag{3.12}$$

Condition 5 (initial synchronization) *For nonfaulty clock p*

$$t^0_p = 0 \tag{3.13}$$

The following condition ensures that there is no overlap between synchronization periods, *i.e.*, by the time any nonfaulty processor is ready to synchronize for the $(i+1)$'th time, all nonfaulty processors have already synchronized for the i'th time.

Condition 6 (nonoverlap)

$$\beta \leq r_{min} \tag{3.14}$$

[2]In the mechanized verification, great pains are taken to indicate the times at which the clocks are required to be nonfaulty. The informal discussion here makes the simplifying assumption that clocks are either faulty or nonfaulty, and often disregards the time at which clocks are asserted as being nonfaulty.

[3]Rushby [private communication] observes that this condition is a somewhat stringent one since for most protocols, it entails the assumption that the clocks are already synchronized. This is an important observation since it implies that the demonstration that a particular protocol meets this condition will have to be carried out by induction over the number of synchronization rounds.

An important corollary of the *bounded interval* and *bounded delay* conditions is that for any two nonfaulty clocks p and q,

$$0 \leq t_p^{i+1} - t_q^i \leq r_{max} + \beta. \tag{3.15}$$

For a nonfaulty clock p, the value $\Theta_p^{i+1}(q)$ represents p's observation of q's i'th clock reading at time t_p^{i+1}, *i.e.*, it is p's estimate of $IC_q^i(t_p^{i+1})$. The error in this reading is assumed to be bounded by Λ.

Condition 7 (reading error) *For nonfaulty clocks p and q,*

$$|IC_q^i(t_p^{i+1}) - \Theta_p^{i+1}(q)| \leq \Lambda \tag{3.16}$$

Condition 8 (bounded faults) *At any time t, at most F processors are faulty. If p is nonfaulty at time t, then it is nonfaulty at any time s prior to t.*

The conditions below are mathematical constraints placed on the convergence function, e.g., clocks, drifts, and failures, do not play any role in the statements. The isolation of these constraints makes it possible to demonstrate that the egocentric mean function of ICA satisfies the conditions below independent of the context of its use. The condition of *translation invariance* indicates that adding x to the value of the convergence function should be the same as adding x to each clock reading instead.

Condition 9 (translation invariance) *For any function θ mapping clocks to clock values,*

$$cfn(p, (\lambda n\colon \theta(n) + x)) = cfn(p, \theta) + x \tag{3.17}$$

The next condition of *precision enhancement* facilitates a comparison between values of the convergence function based on the range of values of some subset C of the clock readings. C is to be intuitively interpreted as the subset of nonfaulty processors. *Precision enhancement* captures the convergence behavior of the convergence function by asserting that the "closer" two arrays of clock readings γ and θ are to each other, the closer are the results of the convergence function applied to γ and θ, respectively. In the statement of *precision enhancement*, the above intuitive interpretation of C as the subset of nonfaulty clocks is permissible by the *bounded faults* condition. The "closeness" of γ and θ is formalized by asserting that all the readings in γ and θ, respectively, of clocks in C lie in an interval of width y, and that the corresponding readings of γ and θ of any clock in C are no more than x apart. The function $\pi(x, y)$ captures the closeness of the values that result from applying the convergence function to γ and θ.[4]

Condition 10 (precision enhancement) *Given any subset C of the N clocks with $|C| \geq N - F$, and clocks p and q in C, then for any readings γ and θ satisfying the conditions*

1. *for any l in C, $|\gamma(l) - \theta(l)| \leq x$*
2. *for any l, m in C, $|\gamma(l) - \gamma(m)| \leq y$*

[4]Note that the order of arguments to π are reversed from their order in Schneider's description [Sch87].

3. for any l, m in C, $|\theta(l) - \theta(m)| \leq y$

there is a bound $\pi(x,y)$, such that

$$|cfn(p,\gamma) - cfn(q,\theta)| \leq \pi(x,y) \tag{3.18}$$

The final condition of *accuracy preservation* bounds the distance between the value of $cfn(p,\theta)$ and the nonfaulty readings in θ.[5]

Condition 11 (accuracy preservation) *Given any subset C of the N clocks with $|C| \geq N - F$, and clock readings θ such that for any l and m in C, the bound $|\theta(l) - \theta(m)| \leq x$ holds, there is a bound $\alpha(x)$ such that for any q in C*

$$|cfn(p,\theta) - \theta(q)| \leq \alpha(x) \tag{3.19}$$

Schneider also proposes a condition called *monotonicity* that is actually not satisfied by several clock synchronization protocols though it is used heavily in Schneider's proofs. Fortunately, this condition turns out to be unnecessary in the derivation. The monotonicity condition asserts that if for each processor l, $\theta(l) \geq \gamma(l)$, then $cfn(p,\theta) \geq cfn(p,\gamma)$. The failure of the monotonicity condition for ICA is demonstrated in Section 5.

4 The Correctness Proof

The formal arguments are extremely delicate to carry out carefully and correctly due to the additional consideration of processor failure. The phenomenon of processor failure is usually dealt with casually in informal presentations, but adds significantly to the complexity of the formalization as well as the proof. A brief sketch of the proof is given below, and a few further details are provided in Appendix A.

To establish the main result, Theorem 3.1, we must show that the skew, or absolute difference, between the readings of any two nonfaulty clocks p and q at time t, given by $|VC_p(t) - VC_q(t)|$, is bounded by a quantity δ. The first step (Theorem 4.1) is to bound the skew at the instant when both p and q have resynchronized their clocks for the i'th time. This time, $t^i_{p,q}$, is $max(t^i_p, t^i_q)$. The skew at this instant can be bounded by a quantity δ_S that can be computed using the conditions of *translation invariance*, *precision enhancement*, and *reading error*, and the proof is by induction on i.

Theorem 4.1 *There is a bound δ_S such that for synchronization round i and any two nonfaulty processors p and q*

$$|IC^i_p(t^i_{p,q}) - IC^i_q(t^i_{p,q})| \leq \delta_S \tag{4.20}$$

[5] Footnote 7 in Schneider [Sch87] explains the choice of the terms *precision enhancement* and *accuracy preservation*. 'Precision' is defined as the closeness with which a measurement can be reproduced, whereas 'accuracy' is the proximity of the measurement to the actual value being measured. The virtual clocks represent various measurements of real time. *Precision enhancement* characterizes the closeness of these measurements to each other. *Accuracy preservation* can be seen as bounding the drift rate of the virtual clock with respect to real time.

We can now compute the quantity δ that bounds the skew in the interval $t^i_{p,q} \leq t < t^{i+1}_{p,q}$. The proof of Theorem 4.2 has two cases according to whether $t < min(t^{i+1}_p, t^{i+1}_q)$ or $t \geq min(t^{i+1}_p, t^{i+1}_q)$. The first case follows easily from *bounded drift* and *bounded interval*, and the second case requires *accuracy preservation* as well.

Theorem 4.2 *For any two nonfaulty clocks p, q, and* $t^i_{p,q} \leq t < t^{i+1}_{p,q}$,

$$|VC_p(t) - VC_q(t)| \leq \delta. \tag{4.21}$$

Note that for any $t \geq 0$, there is an i such that $t < t^i_{p,q}$. The main theorem then follows from the fact that Theorem 4.2 yields a skew bound δ for any t such that $0 \leq t < t^i_{p,q}$. We note of the various constraints on δ and δ_S that arise from the proofs in Appendix A:

1. $\pi(2\Lambda + 2\beta\rho, \delta_S + 2\rho(r_{max} + \beta) + 2\Lambda) \leq \delta_S$
2. $\delta_S + 2\rho r_{max} \leq \delta$
3. $\alpha(\delta_S + 2\rho(r_{max} + \beta) + 2\Lambda) + \Lambda + 2\rho\beta \leq \delta.$

5 ICA as an instance of Schneider's scheme

To gain some intuition into the SCS protocol, we demonstrate that the egocentric mean function from the Interactive Convergence Algorithm of Lamport and Melliar-Smith [LMS85] satisfies Schneider's conditions of translation invariance, precision enhancement, and accuracy preservation.

With the interactive convergence algorithm, the convergence function cfn_I takes the *egocentric mean* of p's estimate of the readings of the N clocks numbered from 0 to $N-1$, *i.e.*, any readings that are more than Δ away from p's own reading are replaced by p's own reading. This yields the definition

$$cfn_I(p, \theta) = \frac{\sum_{l=0}^{N-1} fix_p(\theta(l))}{N} \tag{5.22}$$

where

$$fix_p(x) = \begin{cases} x & \text{if } |x - \theta(p)| \leq \Delta \\ \theta(p) & \text{otherwise.} \end{cases}$$

Translation invariance follows from the observation that

$$fix_p((\lambda l\colon \theta(l) + t)(q)) = fix_p(\theta(q)) + t \tag{5.23}$$

and

$$\frac{\sum_{l=0}^{N-1}(\theta(l) + t)}{N} = \frac{\sum_{l=0}^{N-1}(\theta(l))}{N} + t \tag{5.24}$$

To demonstrate *precision enhancement*, we start with a set of processors C of cardinality $|C|$ greater than $N - F$. Let f be $N - |C|$. The hypotheses for precision enhancement are that for any l and m in C, $|\gamma(l) - \theta(l)|$ is bounded by x, and $|\gamma(l) - \gamma(m)|$ and $|\theta(l) - \theta(m)|$ are bounded by y. When $y \leq \Delta$, it can be shown that

$$|cfn_I(p, \gamma) - cfn_I(q, \theta)| \leq \frac{(N - f)x}{N} + \frac{2f\Delta + fx + fy}{N}. \tag{5.25}$$

The right-hand side of (5.25) is the required $\pi(x, y)$ in this case.

In the typical situation when the egocentric mean is computed, the quantity x representing the reading error is negligible, and y representing the clock skew is bounded by Δ. Since the skew following synchronization should be smaller than Δ, we can see that in Equation (5.25), the number of failed processors f should be below $N/3$. The derivation of $\pi(x, y)$ for the case when $y > \Delta$, is carried out in the mechanized proof.

To show that cfn_I satisfies accuracy preservation, it is sufficient to observe that if all the clocks in C (the C-clocks) are within x of each other, then the C-clocks can cause the egocentric mean to be at most $(N - f)x/N$ away from any nonfaulty clock. The clocks outside C can cause the egocentric mean to be up to $f \times (x + \Delta)/N$ away from a good clock. The total thus yields

$$\alpha(x) = x + \frac{f\Delta}{N}.$$

The final step is to demonstrate the failure of the monotonicity condition for ICA. The monotonicity condition mentioned at the end of Section 3.2 asserts that if for each processor l, $\theta(l) \geq \gamma(l)$, then $cfn(p, \theta) \geq cfn(p, \gamma)$. Let $\theta(p) = \gamma(p)$. Observe now that if there is some l such that $\theta(l) + \Delta < \theta(p)$, but with $\gamma(p) > \gamma(l) \geq \gamma(p) - \Delta$, then $fix_p(\theta(l)) > fix_p(\gamma(l))$ holds. So, it is possible to have $fix_p(\theta(l)) > fix_p(\gamma(l))$, even though we have $\theta(l) < \gamma(l)$.

6 Discussion

The formal verification of the proof of the SCS protocol required considerable effort. Our own experience suggests that the difficulty would have been comparable using other verification systems. The following discussion highlights some of the complexity of reasoning about fault-tolerant clock synchronization protocols, and examines some issues that are relevant to managing this complexity.

Formalization itself is one source of complexity. A number of issues are simply swept under the rug in informal descriptions. For example, informal presentations of fault-tolerant protocols treat failure in a somewhat casual way. Clocks that fail are regarded as always having been faulty. Note that our informal presentation above makes essentially the same simplifying assumption. The mechanized proof is a great deal more precise. Clocks are allowed to fail at any point in time. A great deal of care is taken to not build in any assumptions regarding the behavior of failed clocks save that their clock readings are unspecified functions of the standard global time. Under the assumption that clocks can fail at any time, the precise forms of the axioms are considerably more difficult to formulate. For example, the axioms `rts0` and `rts1` in Figure 2 are not at all the obvious formulations of the condition of ***bounded interval*** since p is constrained to be correct at time t. The obvious form of these axioms would assert p to be correct at t_p^{i+1} and would thus say little about time points other than t_p^i and t_p^{i+1}. We initially employed the "obvious" forms of these axioms and quickly found them inadequate during the course of the mechanized proof. These two axioms turned out to have some other inadequate variations as well. Other axioms formalizing clock behavior also posed similar challenges.

Clock drifts are another obvious and inherent source of difficulty in the proof. Virtually the entire proof consists of determining real time bounds on certain intervals and using these to derive bounds on the clock drifts over these intervals. The algebra involved in these calculations is very complex and has to be done with some care. It is conceivable that these calculations could be further automated so that these bounds can be computed rather than supplied by the user.

A key difference between our proof of the SCS protocol and the Rushby/von Henke proof of the ICA protocol is that they follow Lamport and Melliar-Smith in formalizing local clocks as functions from clock time to real time whereas local clocks are formalized here as functions from real time to clock time. As a consequence, our main theorem asserts that the skew between any two correct clocks at the same time instant is bounded, whereas their main theorem asserts a bound in the real time interval separating one correct clock's reading of time T from another correct clock's reading of time T. We were unable to conclusively establish either approach to be clearly superior. It seems likely that the best approach would be to simultaneously employ both notions of a local clock. The internal behavior of the each processor is governed by the local clock, and this can be mapped to its real time behavior. The synchronization between various processors is best done in terms of a real time frame of reference and subsequently mapped back to timing requirements on the behavior of the individual processors.

The proofs here use a dense notion of time; the time variables range over ordered fields. The proof has been carried out with very minor changes with respect to a discrete notion of time. One significant point ignored by the formalization here is that the *bounded drift* axiom can only be guaranteed to hold over a large enough interval of measurement. It would be easy to incorporate such a notion into the formalization and redo the proof. (The ability to easily redo proofs with minor perturbations to the assumptions is one of the significant benefits of mechanized formal verification.)

Though there are a number of variants of temporal logic [Pnu77] that capture real-time notions [AH89, EMSS89, Koy89], none of these seem adequately sophisticated for dealing with faults and local clocks with any special felicity. It appears that what is required of such a logic is the ability to easily translate from local time to global time assertions, to treat faultiness as a time-varying property, and to reason about the cardinalities of processors satisfying certain constraints. We are currently investigating the design of such a logic. It would also be interesting to examine whether the model-checking approaches [CG87] to the verification of distributed protocols can be fruitfully applied here.

7 Conclusions

Rigorously proving the correctness of distributed protocols is an extremely difficult task, with or without mechanical assistance. Fault-tolerant clock synchronization is an excellent example of a problem where the algorithms, though often simple, are not at all easily verified. In such cases, it is extremely important to have certain organizing principles which capture the common features of the various protocols with convincing generality. Schneider's schema for Byzantine clock synchronization provides such principles.

The formalization here revises Schneider's presentation in some small ways. Schneider's notion of a global signal to trigger resynchronization has been dropped because such a notion is difficult to instantiate for many protocols. Though the quantities r_{max} and r_{min}

have a different meaning from Schneider's, these differences ought not to matter in any of the bounds derived. The derivation we present is extremely tight, given the structure of the proof. Schneider's monotonicity condition is avoided the proofs here. This condition is used heavily by Schneider in his arguments, but it actually turns out to not hold for many protocols. The statement of accuracy preservation here is also slightly different here from that of Schneider.

The initial proof using EHDM and including the pencil-and-paper development took about a month. The conditions on clocks and convergence functions were fairly difficult to formalize due to the careful treatment of failure. Many of these statements were refined during the course of the proof. The proof itself has been considerably revised and improved since the first effort. Verifying that the egocentric mean function of ICA satisfied the conditions of *translation invariance, accuracy preservation*, and *precision enhancement*, took about two weeks. The complete proof involves about 182 theorems or lemmas.

The most useful feature of EHDM for this proof were the decision procedures for linear integer and rational inequalities and equalities. The proof is of course replete with long chains of inequality reasoning, and the decision procedures handled those steps in a fairly mechanical manner. The higher-order features of the language were also used to formalize the conditions of translation invariance, precision enhancement, and accuracy preservation, but such features were not essential to this proof.

Fault-tolerant distributed protocols are sufficiently delicate to warrant careful, formal, mechanized analysis. Such an analysis is possible with the existing technology for specification and verification. Schneider's presentation provides a valuable mathematical framework for the verification of synchronization protocols. The machine-checked proof of Schneider's protocol led to a precise formulation of the protocol and a closely reasoned proof. It is inconceivable that the same degree of logical rigor and preciseness could be achieved without computational assistance.

References

[AH89] R. Alur and T. A. Henzinger. A really temporal logic. In *30th IEEE Symp. on Foundations of Computer Science*, 1989.

[BY90] W. R. Bevier and W. D. Young. Machine checked proofs of the design and implementation of a fault-tolerant circuit. NASA Contractor Report 182099, Computational Logic, Inc., 1990.

[CG87] E. M. Clarke and O. Grumberg. Research on automatic verication of finite state concurrent systems. In *Annual Review of Computer Science*, pages 269–290. Annual Reviews, Inc., 1987.

[EMSS89] E. A. Emerson, A. K. Mok, A. P. Sistla, and J. Srinivasan. Quantitative temporal reasoning. In *Computer-Aided Verification*, 1989.

[Koy89] R. Koymans. *Specifiying message passing and time-critical systems with temporal logic.* PhD thesis, Eindhoven Univ. of Technology, 1989.

[LL84] J. Lundelius and N. A. Lynch. A new fault-tolerant algorithm for clock synchronization. In *Proc. of the Third ACM Symp. on Principles of Distributed Computing*, pages 75–88, 1984.

[LMS85] L. Lamport and P.M. Melliar-Smith. Synchronizing clocks in the presence of faults. *Journal of the ACM*, 32(1):52–78, January 1985.

[LSP82] Leslie Lamport, Robert Shostak, and Marshall Pease. The Byzantine generals problem. *ACM TOPLAS*, 4(3):382–401, July 1982.

[Pnu77] A. Pnueli. The temporal logic of programs. In *Proc. 18th Ann. IEEE Symp. on Foundations of Computer Science*, pages 46–57, 1977.

[Rus91] John Rushby. Formal specification and verification of a fault-masking and transient-recovery model for digital flight-control systems. Technical Report SRI-CSL-91-3, Computer Science Laboratory, SRI International, Menlo Park, CA, January 1991. Also available as NASA Contractor Report 4384, and to appear in Proceedings of the Symposium on Formal Techniques in Real Time and Fault Tolerant Systems, Nijmegen, Netherlands, January 1992.

[RvH91] John Rushby and Friedrich von Henke. Formal verification of the interactive convergence clock synchronization algorithm using EHDM. Technical Report SRI-CSL-89-3R, Computer Science Laboratory, SRI International, Menlo Park, CA, February 1989 (Revised August 1991). Also available as NASA Contractor Report 4239.

[RvHO91] John Rushby, Friedrich von Henke, and Sam Owre. An introduction to formal specification and verification using EHDM. Technical Report SRI-CSL-91-2, Computer Science Laboratory, SRI International, Menlo Park, CA, February 1991.

[Sch87] Fred B. Schneider. Understanding protocols for Byzantine clock synchronization. Technical Report 87-859, Department of Computer Science, Cornell University, Ithaca, NY, August 1987.

[Sha91] Natarajan Shankar. Mechanical verification of a schematic Byzantine clock synchronization algorithm. Nasa contractor report 4386, June 1991.

A Details of the Correctness Proof

The details of the proof of *bounded skew* are completed below by providing the proofs of Theorems 4.1 and 4.2.

Proof of Theorem 4.1. The proof is by induction on the round number i.

Base case: When $i = 0$, by (3.13) we have $t_p^0 = t_q^0 = 0$. The skew bound of δ_S follows from Definitions (3.3) and (3.1), and the *initial skew* condition.

Induction case: The induction hypothesis asserts that for every pair of nonfaulty processors, l and m

$$|IC_l^i(t_{l,m}^i) - IC_m^i(t_{l,m}^i)| \leq \delta_S \tag{A.26}$$

The goal is to establish for any pair of nonfaulty processors p and q, that

$$|IC_p^{i+1}(t_{p,q}^{i+1}) - IC_q^{i+1}(t_{p,q}^{i+1})| \leq \delta_S \tag{A.27}$$

Without loss of generality, assume that t_q^{i+1} precedes t_p^{i+1} so that $t_{p,q}^{i+1} = t_p^{i+1}$. Then Equation (3.6) and *translation invariance* yield

$$\begin{aligned} IC_q^{i+1}(t_p^{i+1}) &= cfn(q, \Theta_q^{i+1}) + PC_q(t_p^{i+1}) - PC_q(t_q^{i+1}) \\ &= cfn(q, (\lambda n\colon \Theta_q^{i+1}(n) + PC_q(t_p^{i+1}) - PC_q(t_q^{i+1}))) \end{aligned} \tag{A.28}$$

By Equation (3.5), we have

$$IC_p^{i+1}(t_p^{i+1}) = cfn(p, \Theta_p^{i+1}) \tag{A.29}$$

so that the required skew can be rewritten as

$$|cfn(q, (\lambda n\colon \Theta_q^{i+1}(n) + PC_q(t_p^{i+1}) - PC_q(t_q^{i+1}))) - cfn(p, \Theta_p^{i+1})|$$

This quantity can be bounded using *precision enhancement* with $(\lambda n\colon \Theta_q^{i+1}(n)+PC_q(t_p^{i+1})- PC_q(t_q^{i+1}))$ for γ and Θ_p^{i+1} for θ. The set C in *precision enhancement* is taken to be the subset of nonfaulty clocks as permitted by *bounded faults*. To satisfy Hypothesis 1 of *precision enhancement*, we need an x such that for any nonfaulty l,

$$|(\Theta_q^{i+1}(l) + PC_q(t_p^{i+1}) - PC_q(t_q^{i+1})) - \Theta_p^{i+1}(l)| \leq x.$$

The value $2\rho\beta + 2\Lambda$ can be substituted for x since by *bounded drift*,

$$|(PC_q(t_p^{i+1}) - PC_q(t_q^{i+1})) - (t_p^{i+1} - t_q^{i+1})| \leq \beta\rho \tag{A.30}$$

and by *reading error* and *bounded drift*, we get

$$\begin{aligned} &|(\Theta_p^{i+1}(l) - \Theta_q^{i+1}(l)) - (t_p^{i+1} - t_q^{i+1})| \\ \leq\ & 2\Lambda + |(IC_l^i(t_p^{i+1}) - IC_l^i(t_q^{i+1})) - (t_p^{i+1} - t_q^{i+1})| \\ \leq\ & 2\Lambda + \beta\rho \end{aligned} \tag{A.31}$$

The induction hypothesis is needed in order to satisfy Hypotheses 2 and 3 of the relevant instance of *precision enhancement*. For both hypotheses, we need a y such that for any nonfaulty processors k, l and m,

$$|\Theta_k^{i+1}(l) - \Theta_k^{i+1}(m)| \leq y \tag{A.32}$$

By *reading error*, the induction hypothesis (A.26), and Equations (3.15) and (3.10), we have

$$\begin{aligned}
&|\Theta_k^{i+1}(l) - \Theta_k^{i+1}(m)| \\
&\leq \; 2\Lambda + |IC_l^i(t_k^{i+1}) - IC_m^i(t_k^{i+1})| \\
&\leq \; 2\Lambda + |IC_l^i(t_{l,m}^i) - IC_m^i(t_{l,m}^i)| + 2\rho(t_k^{i+1} - t_{l,m}^i) \\
&\leq \; 2\Lambda + \delta_S + 2\rho(r_{max} + \beta)
\end{aligned} \tag{A.33}$$

so that the required y is $\delta_S + 2\rho(r_{max} + \beta) + 2\Lambda$. So by *precision enhancement*, if

$$\pi(2\Lambda + 2\beta\rho, \delta_S + 2\rho(r_{max} + \beta) + 2\Lambda) \leq \delta_S, \tag{A.34}$$

then

$$|IC_p^{i+1}(t_p^{i+1}) - IC_q^{i+1}(t_p^{i+1})| \leq \delta_S \tag{A.35}$$

thus completing the proof of Theorem 4.1. ∎

Proof of Theorem 4.2. Assume without loss of generality that $t_q^{i+1} \leq t_p^{i+1}$. The proof has two cases according to whether $t_{p,q}^i \leq t < t_q^{i+1}$ or $t_q^{i+1} \leq t < t_p^{i+1}$.

Case 1: Assuming $t_{p,q}^i \leq t < t_q^{i+1}$. From *bounded interval* we get $t - t_{p,q}^i \leq r_{max}$. By Equation (3.4), we get $VC_p(t) = IC_p^i(t)$ and $VC_q(t) = IC_q^i(t)$. Then by Equations (3.10), (3.3), and Theorem 4.1,

$$\begin{aligned}
&|VC_p(t) - VC_q(t)| \\
&\leq \; |VC_p(t_{p,q}^i) - VC_q(t_{p,q}^i)| + 2\rho r_{max} \\
&\leq \; \delta_S + 2\rho r_{max}
\end{aligned} \tag{A.36}$$

The bound δ should therefore be chosen so that

$$\delta_S + 2\rho r_{max} \leq \delta. \tag{A.37}$$

Case 2: Assuming $t_q^{i+1} < t < t_{p,q}^{i+1}$. In this interval, $VC_q(t) = IC_q^{i+1}(t)$, whereas $VC_p(t) = IC_p^i(t)$. The strategy here is to bound the skew at t_q^{i+1} and then compute the additional quantity by which the clocks can drift apart in the given interval. By Equations (3.5) and (3.4), we have

$$|VC_p(t_q^{i+1}) - VC_q(t_q^{i+1})| = |IC_p^i(t_q^{i+1}) - cfn(q, \Theta_q^{i+1})|. \tag{A.38}$$

By *reading error* and *accuracy preservation*, the quantity $|IC_p^i(t_q^{i+1}) - cfn(q, \Theta_q^{i+1})|$ can by bound by $\alpha(x) + \Lambda$, where for any pair of nonfaulty clocks l and m,

$$|\Theta_q^{i+1}(l) - \Theta_q^{i+1}(m)| \leq x. \tag{A.39}$$

As already seen in the derivation of Equation (A.32), that (A.39) holds with $\delta_S + 2\rho(r_{max} + \beta) + 2\Lambda$ for x. We then have

$$|VC_p(t_q^{i+1}) - VC_q(t_q^{i+1})| \leq \alpha(\delta_S + 2\rho(r_{max} + \beta) + 2\Lambda) + \Lambda. \tag{A.40}$$

We can now bound the skew over the interval $t_q^{i+1} \leq t < t_p^{i+1}$, by observing that $t_p^{i+1} - t_q^{i+1} \leq \beta$ by (3.12), and applying Equation (3.10) to derive the inequality,

$$|VC_p(t) - VC_q(t)| \leq \alpha(\delta_S + 2\rho(r_{max} + \beta) + 2\Lambda) + \Lambda + 2\rho\beta. \tag{A.41}$$

Therefore δ has to be chosen to satisfy

$$\alpha(\delta_S + 2\rho(r_{max} + \beta) + 2\Lambda) + \Lambda + 2\rho\beta \leq \delta. \tag{A.42}$$

Both cases of the proof of Theorem 4.2 have been completed. ■

This concludes the informal presentation of the proof.

B The EHDM Proof Highlights

This section contains the EHDM formalization of the conditions axiomatizing the behavior of clocks described in Section 3 and the statements of the key theorems. Figure 1 contains the type declarations for some of the variables and constants used in `clockassumptions`. The `clockassumptions` module makes use of the module `arith`, which contains the basic arithmetic facts, and `countmod`, which introduces a counting function. Nonfaultiness is expressed by the predicate `correct`.

The axioms constraining the physical behavior of the clock appear in Figure 2. Since we require μ to not exceed δ_S, the axiom `init` corresponds to *initial skew*. Axiom `correct_closed` asserts that a failed processor never recovers (see *bounded faults*). Axioms `rate_1` and `rate_2` together express the *bounded drift* condition. The axioms `rts0` and `rts1` capture the *bounded interval* condition. These axioms look strange because the variable t, needed to properly capture the correctness condition, appears in them but not in *bounded interval*. Most of the obvious ways of stating these axioms are either too restrictive or wrong. The axiom `rts2` captures *bounded delay*, and `synctime_0` is just *initial synchronization*. The condition of *nonoverlap* appears as an antecedent to the concluding theorem rather than as an axiom. In the LaTeX format below, multiplication is represented by $*$ as well as $\star$. These are synonymous, but the latter represents the uninterpreted form of multiplication whereas the former is interpreted by the linear arithmetic decision procedures of EHDM.

The definitions of the virtual clock and the interval clock in terms of the physical clock appear in Figure 3. These correspond to (3.1), (3.4), and (3.3), respectively.

The conditions on the convergence function appear in Figure 4. The axiom `Readerror` corresponds to the condition *reading error*. The axiom `correct_count` corresponds to *bounded faults*. The remaining correspondences should be self-evident.

The conclusion corresponding to Theorem 3.1 is the theorem `agreement` in Figure 5. The verified version of Theorem 4.1 is given in Figure 6, and that of Theorem 4.2 in Figure 7. The expression $t^i_{(p\Uparrow q)[i]}$ is an alternative notation for $t^i_{p,q}$ since $(p \Uparrow q)[i]$ represents p if $t^i_p \geq t^i_q$, and q otherwise.

clockassumptions: **Module**

Using arith, countmod

Exporting all with countmod, arith

Theory

process: **Type is** nat
event: **Type is** nat
time: **Type is** number
Clocktime: **Type is** number
$l, m, n, p, q, p_1, p_2, q_1, q_2, p_3, q_3$: **Var** process
i, j, k: **Var** event
x, y, z, r, s, t: **Var** time
X, Y, Z, R, S, T: **Var** Clocktime
γ, θ: **Var** function[process → Clocktime]
$\delta, \mu, \rho, r_{min}, r_{max}, \beta, \Lambda$: number
$PC_{\star 1}(\star 2), VC_{\star 1}(\star 2)$: function[process, time → Clocktime]
$t_{\star 1}^{\star 2}$: function[process, event → time]
$\Theta_{\star 1}^{\star 2}$: function[process, event → function[process → Clocktime]]
$IC_{\star 1}^{\star 2}(\star 3)$: function[process, event, time → Clocktime]
correct: function[process, time → bool]
cfn: function[process, function[process → Clocktime] → Clocktime]
π: function[Clocktime, Clocktime → Clocktime]
α: function[Clocktime → Clocktime]

Figure 1: Declarations from module clockassumptions

init: **Axiom** correct$(p, 0) \supset PC_p(0) \geq 0 \wedge PC_p(0) \leq \mu$

correct_closed: **Axiom** $s \geq t \wedge$ correct$(p, s) \supset$ correct(p, t)

rate_1: **Axiom** correct$(p, s) \wedge s \geq t \supset PC_p(s) - PC_p(t) \leq (s - t) \star (1 + \rho)$

rate_2: **Axiom** correct$(p, s) \wedge s \geq t \supset PC_p(s) - PC_p(t) \geq (s - t) \star (1 - \rho)$

rts0: **Axiom** correct$(p, t) \wedge t \leq t_p^{i+1} \supset t - t_p^i \leq r_{max}$

rts1: **Axiom** correct$(p, t) \wedge t \geq t_p^{i+1} \supset t - t_p^i \geq r_{min}$

rts_0: **Lemma** correct$(p, t_p^{i+1}) \supset t_p^{i+1} - t_p^i \leq r_{max}$

rts_1: **Lemma** correct$(p, t_p^{i+1}) \supset t_p^{i+1} - t_p^i \geq r_{min}$

rts2: **Axiom** correct$(p, t) \wedge t \geq t_q^i + \beta \wedge$ correct$(q, t) \supset t \geq t_p^i$

rts_2: **Axiom** correct$(p, t_p^i) \wedge$ correct$(q, t_q^i) \supset t_p^i - t_q^i \leq \beta$

synctime_0: **Axiom** $t_p^0 = 0$

Figure 2: Physical clock axioms

VClock_defn: **Axiom**
correct$(p, t) \wedge t \geq t_p^i \wedge t < t_p^{i+1} \supset VC_p(t) = IC_p^i(t)$

Adj: function[process, event $\rightarrow$ Clocktime] =
($\lambda\, p, i$: (**if** $i > 0$ **then** $cfn(p, \Theta_p^i) - PC_p(t_p^i)$ **else** 0 **end if**))

IClock_defn: **Axiom** correct$(p, t) \supset IC_p^i(t) = PC_p(t) + \mathrm{Adj}(p, i)$

Figure 3: Clock definitions

Readerror: **Axiom** $\text{correct}(p, t_p^{i+1}) \wedge \text{correct}(q, t_p^{i+1})$
$\supset |\Theta_p^{i+1}(q) - IC_q^i(t_p^{i+1})| \leq \Lambda$

translation_invariance: **Axiom**
$X \geq 0 \supset \mathit{cfn}(p, (\lambda\, p_1 \rightarrow \text{Clocktime}: \gamma(p_1) + X)) = \mathit{cfn}(p, \gamma) + X$

ppred: **Var** function[process → bool]
maxfaults: process
okay_Readpred: function[function[process → Clocktime], Clocktime,
function[process → bool] → bool] =
$(\lambda\, \gamma, Y, \text{ppred}: (\forall\, l, m: \text{ppred}(l) \wedge \text{ppred}(m) \supset |\gamma(l) - \gamma(m)| \leq Y))$
okay_pairs: function[function[process → Clocktime],
function[process → Clocktime], Clocktime,
function[process → bool] → bool] =
$(\lambda\, \gamma, \theta, X, \text{ppred}: (\forall\, p_3: \text{ppred}(p_3) \supset |\gamma(p_3) - \theta(p_3)| \leq X))$
N: process

N_0: **Axiom** $N > 0$

N_maxfaults: **Axiom** maxfaults $\leq N$

precision_enhancement_ax: **Axiom**
$\text{count}(\text{ppred}, N) \geq N - \text{maxfaults}$
$\wedge$ okay_Readpred$(\gamma, Y, \text{ppred})$
$\wedge$ okay_Readpred$(\theta, Y, \text{ppred})$
$\wedge$ okay_pairs$(\gamma, \theta, X, \text{ppred}) \wedge \text{ppred}(p) \wedge \text{ppred}(q)$
$\supset |\mathit{cfn}(p, \gamma) - \mathit{cfn}(q, \theta)| \leq \pi(X, Y)$

correct_count: **Axiom** $\text{count}((\lambda\, p: \text{correct}(p, t)), N) \geq N - \text{maxfaults}$

accuracy_preservation_ax: **Axiom**
okay_Readpred$(\gamma, X, \text{ppred})$
$\wedge\ \text{count}(\text{ppred}, N) \geq N - \text{maxfaults} \wedge \text{ppred}(p) \wedge \text{ppred}(q)$
$\supset |\mathit{cfn}(p, \gamma) - \gamma(q)| \leq \alpha(X)$

Figure 4: Conditions on Logical Clocks

agreement: **Lemma** $\beta \leq r_{min}$
$\wedge\ \mu \leq \delta_S \wedge \pi(2 * \Lambda + 2 * \beta \star \rho, \delta_S + 2 * ((r_{max} + \beta) \star \rho + \Lambda)) \leq \delta_S$
$\wedge\ \delta_S + 2 * r_{max} \star \rho \leq \delta$
$\wedge\ \alpha(\delta_S + 2 * (r_{max} + \beta) \star \rho + 2 * \Lambda) + \Lambda + 2 * \beta \star \rho \leq \delta$
$\wedge\ t \geq 0 \wedge \text{correct}(p, t) \wedge \text{correct}(q, t)$
$\supset |VC_p(t) - VC_q(t)| \leq \delta$

Figure 5: Main Theorem

okaymaxsync: function[nat, Clocktime → bool] =

$(\lambda i, X: (\forall p, q:$

$\quad \text{correct}(p, t^i_{p,q}) \wedge \text{correct}(q, t^i_{p,q})$

$\quad\quad \supset |IC^i_p(t^i_{p,q}) - IC^i_q(t^i_{p,q})| \leq X))$

lemma_2: **Lemma** $\beta \leq r_{min}$

$\quad \wedge\, \mu \leq X \wedge \pi(2 * \Lambda + 2 * \beta \star \rho, X + 2 * ((r_{max} + \beta) \star \rho + \Lambda)) \leq X$

$\supset$ okaymaxsync(i, X)

Figure 6: Skew immediately following resynchronization

okayClocks: function[process, process, nat → bool] =

$(\lambda p, q, i: (\forall t:$

$\quad t \geq 0 \wedge t < t^i_{(p \Uparrow q)[i]} \wedge \text{correct}(p, t) \wedge \text{correct}(q, t)$

$\quad\quad \supset |VC_p(t) - VC_q(t)| \leq \delta))$

lemma3_3: **Lemma** $\beta \leq r_{min}$

$\quad \wedge\, \mu \leq \delta_S \wedge \pi(2 * \Lambda + 2 * \beta \star \rho, \delta_S + 2 * ((r_{max} + \beta) \star \rho + \Lambda)) \leq \delta_S$

$\quad\quad \wedge\, \delta_S + 2 * r_{max} \star \rho \leq \delta$

$\quad\quad\quad \wedge\, \alpha(\delta_S + 2 * (r_{max} + \beta) \star \rho + 2 * \Lambda) + \Lambda + 2 * \beta \star \rho \leq \delta$

$\supset$ okayClocks(p, q, i)

Figure 7: Skew up to ith resynchronization

Formal Specification and Verification of a Fault-Masking and Transient-Recovery Model for Digital Flight-Control Systems*

John Rushby

Computer Science Laboratory
SRI International
Menlo Park CA 94025 USA

Abstract

We present a formal model for fault-masking and transient-recovery among the replicated computers of digital flight-control systems. We establish conditions under which majority voting causes the same commands to be sent to the actuators as those that would be sent by a single computer that suffers no failures. The model and its analysis have been subjected to formal specification and mechanically checked verification using the EHDM system.

Keywords: digital flight control systems, formal methods, formal specification and verification, proof checking, fault tolerance, transient faults, majority voting, modular redundancy

1 Introduction

Many modern airplanes and spacecraft are crucially dependent on digital flight control systems (DFCS), and so extreme reliabilities are required of those systems. The reliabilities required are beyond those that can be guaranteed for the individual digital devices that constitute the DFCS hardware and so it follows that some form of fault tolerance based on replication and redundancy is needed in order to achieve an underlying "hardware platform" of the required reliability. There are many configurations for redundant and replicated computer systems, and careful reliability analysis is required to evaluate the reliability provided by a given configuration and level of redundancy [3]. Such analyses show that suitably constructed N-modularly redundant systems (which we will call N-plexes for brevity) can achieve the desired reliability.

Within an N-plex, all calculations are performed by N identical computer systems and the results are submitted to some form of averaging or voting before being sent to the

*This research was supported by NASA Langley Research Center under contract NAS1 18969

actuators. Great care must be taken to eliminate single-point failures, so the separate computer systems (or "channels," as they are often called in fault-tolerant systems) will generally use different power supplies and be otherwise electrically and physically isolated to the greatest extent possible. In order that voting should not become a single point of failure, each channel usually has its own voter, and the voted values from each channel are then further voted or averaged at the actuators through some form of "force-summing." For example, different channels may energize separate coils of a single solenoid, or multiple hydraulic pistons may be linked to a single shaft [6, Figure 3.2–2].

Notice that although this approach provides protection against random hardware failures, there is no protection against design faults: any such faults in either the hardware or the software will be common to all members of the N-plex and all will fail together. In this paper, we do not address the issue of design faults in the hardware, nor in the application software that it runs. We are, however, very much concerned with the possibility of design faults in the redundancy-management software that harnesses the failure-prone individual components together as a fault-tolerant N-plex. Redundancy management presents a major challenge in the design of a fault-tolerant N-plex for DFCS. Instead of a single computer executing the DFCS software there will be several—which must coordinate and vote (or average) actuator commands, and tolerate faults among their own members. In addition to the replicated computers, sensors will be replicated also, and their values must be distributed in a Byzantine fault-tolerant manner [13]. The management of all this redundancy and replication adds considerable complexity to both the operating system (generally called an "executive" in process-control systems) and the application tasks. Indeed, there is evidence that redundancy management is sufficiently complex and difficult that it can become the *primary* source of unreliability in a DFCS [12, pp. 40–41].

Consequently, the overall goal of a research program led by NASA Langley Research Center, of which this work forms a part, is to develop a fault-tolerant architecture for DFCS using formal methods to provide a rigorous basis for documenting and analyzing design decisions. Ultimately, we hope to provide mechanically-checked formal specifications and verifications for the key components of a "Reliable Computing Platform" for DFCS, all the way from high-level requirements down to implementation details. Clearly, this is a major undertaking, so initially we are concentrating on some of the better-understood requirements and levels in the hierarchy.

The approach we are following, in common with most of those performing research in fault-tolerant systems for DFCS [7, 9, 11, 22], employs synchronized channels and exact-match voting; it also has much in common with the "state-machine" approach [19]. For exact-match voting, each channel must operate on the same data. Thus the computers cannot simply use their own private sensor readings, but must exchange sampled values with each other in a Byzantine fault-tolerant manner. Byzantine fault-tolerant algorithms are also required to synchronize the separate channels. In this way, every (working) computer begins each frame at approximately the same time as the others, and with the same set of sensor readings. Each computer will then run the same sensor selection and averaging algorithms, and the same control laws, and should therefore generate identical actuator commands. Exact-match majority voting of the actuator commands then suffices to mask faults among the redundant channels. Notice that this arrangement allows sensor failures to be distinguished from failures among the redundant computers: sensor failure is detected or masked by the diagnostic, averaging, and selection algorithms run by each computer, whereas failure of a computer is masked (and optionally detected) by the exact-match majority voting of their outputs.[1]

We and our colleagues at NASA and the other companies engaged in this program have been undertaking formal specification and verification of some of the key algorithms required to support the synchronized, exact-match voting approach to DFCS. These include Byzantine fault-tolerant clock synchronization algorithms [16, 20] and a Byzantine agreement algorithm [1] and circuits [2, 21]. The work described in this paper is a step towards the next higher layer in the modeling hierarchy: the layer that uses exact-match voting to provide fault-tolerance and transient-recovery. The model and results developed in the following sections have been formally specified and verified using our EHDM system [18].

2 Fault Masking and Transient Recovery in DFCS

Not all faults are equal: some are "hard" faults that permanently disable the afflicted channel; others are "soft" or "transient" faults from which recovery is possible. Examples of transient faults include "single event upsets" (SEUs), where a single bit of memory is flipped by a cosmic ray. These can be recovered by simply restoring the affected bit to its correct value. Experience indicates that transient faults are orders of magnitude more common than hard faults and it follows that overall reliability will be much greater—or, equivalently, much less redundancy will be required for a given level of reliability—if some attempt is made to recover channels that suffer transient faults.

There is no firm line between transient and hard faults considered in the abstract; what might be merely a transient fault to one system may be a hard fault to another that lacks the necessary recovery mechanisms. Fault-tolerant system architectures are designed and evaluated against explicitly stated fault models. For transient faults, we employ a fault model in which we distinguish two subclasses of faults.

State data faults are those in which the processor is working correctly (i.e., is synchronized and executing the right task), but its local state data are corrupted. If its state data were replaced with correct values, it would recover. In our formal model, the predicate $OK(i)(c)$ will indicate whether processor i has state data faults that can affect its computation of task c.

Control faults are those in which the processor is not working correctly (i.e., something other than, or additional to, a state data fault has occurred). In our formal model, the predicate $\mathcal{F}(i)(j)$ will indicate whether processor i suffers a control fault during the computation of the j'th task.

---

[1]A plausible and apparently simpler approach to redundancy management in an N-plex is the "asynchronous" design, in which the computers run fairly independently of each other: each computer samples sensors and evaluates the control laws independently. The triplex-redundant DFCS of the experimental AFTI-F16 was built this way, and its flight tests revealed some of the shortcomings of the approach [8, 12] (see [15, Chapter 1] for a summary). Asynchronous systems cannot distinguish accurately between the failure of a sensor and that of a computer, and may mistake the consequences of clock drift for either. Nonetheless, aircraft manufacturers continue to use this approach.

In our model, we think of control faults as happening spontaneously, and state data faults as the consequences of control faults. Faults such as SEUs, in which a single bit of state data is spontaneously corrupted, can be considered as instantaneous control faults: we imagine that the processor computes the wrong value but then immediately recovers, leaving a state data fault behind. Thus, reliability analysis must consider the arrival and repair rates of control faults (which are not considered here), and the recovery process for state data faults.

State data faults can be recovered by periodically replacing the state data maintained by each processor with a majority-voted version. It is not necessary to vote and replace all the state data, since many of them are refreshed by sampling sensors (i.e., some of the state data are "stored" in the airframe itself): only the data that are carried forward from one frame or cycle to the next (e.g., time-integrated data such as velocity and position) need to be voted. Even so, the quantity of state data maintained by a modern DFCS is considerable, and performance would be seriously degraded if all of it were voted at every opportunity. Accordingly, exposure is traded for performance and rather sparse voting patterns are preferred. Clearly, the less frequently a particular item of state data is voted, the longer will be the duration of the consequences of a fault that corrupts that item. Overall reliability will be determined by the fault arrival rate, the voting pattern, and the dataflow dependencies among control tasks and state data items.

In this paper we develop and formally specify a model that describes the operation of an N-plex with transient-recovery based on an arbitrary sparse voting pattern. We will formally verify a theorem concerning the conditions under which such a system masks faults successfully. A concrete instance of the theorem (for a specific data dependency graph and voting pattern) might be that the system is "safe" provided that at most two processors suffer control faults in any sequence of five successive frames. Markov or other methods of reliability analysis must be used to determine the overall reliability of the system, given assumptions about the arrival and repair rates of control faults [4].

3 The Fault-Masking Model

Our goal is to prove that, subject to certain conditions, an N-plex provides transient-recovery and fault masking for a certain class of faults. Our first requirement, therefore, is a benchmark model for correct, fault-free behavior, against which the efficacy of transient-recovery and fault masking in the N-plex may be judged. We take as our benchmark a model for the behavior of a fault-free process-control system. Our model for an N-plex will then compose N fault-prone versions of the basic model, together with some voting and recovery mechanisms, and our theorem will establish that the voted results of the N-plex equal those of the fault-free system (under suitable conditions).

We begin by describing our model for fault-free process control. This model is deliberately simple and abstract. It treats task executions as atomic actions and does not consider their duration in real time. Nor does it model the cyclic, structured pattern of task executions found in statically-scheduled frame-based systems. We exclude real time and assume atomicity (and later perfect synchronization) for simplicity: we prefer to contemplate one issue at a time and to add complexity in stages. We abstract away the concrete details of frame-based organization since they seem irrelevant to (indeed, limitations on) the results we seek. Our colleagues at NASA Langley have formulated a frame-based model [4,5]; the reconciliation between their model and ours is described in [15, Chapter 4].

3.1 A Model for Fault-Free Process Control

A process-control system manages some physical device by sending control signals to *actuators*. The values of the control signals are determined by calculations based on the values of *sensors* that monitor the device and on a record (maintained by the process-control system) of the *state* of the system. The process-control system is internally composed of computational *tasks* that are activated periodically in order to sample sensors, perform the necessary calculations, and send values to the actuators. Some tasks may also perform internal housekeeping functions. Because task activations may depend on the results of other task activations, there is a dataflow dependency among task activations that the execution schedule must take into account. The "slots" in the execution schedule are called *cells*; a process-control system requires a specification of which tasks are assigned to which cells, the dataflow relationships among cells, and the order in which cells are to be executed. These ideas are formalized in the following definitions.

We assume

- A set C of *cells*, and
- A relation $G \subseteq C \times (\mathbf{N} \times C)$ (where $\mathbf{N}$ denotes the natural numbers),

and we define

- $M \stackrel{\text{def}}{=} \{1, 2, \ldots, |C|\}$.

Cells correspond to the activations (or executions) of *tasks* (to be formally defined later) or the sampling of sensors; the relation G records the dataflow dependencies among task activations associated with cells: the interpretation of $(i,(n,j)) \in G$ is that the output of the task activation (or sensor sample) associated with cell i supplies the input for the n'th argument of the task activation associated with cell j. A simplified relation

- $\overline{G} \stackrel{\text{def}}{=} \{(i,j) | \exists n : (i,(n,j)) \in G\}$

captures just the basic dataflow dependencies among cells, without concern for which input of cell j it is that receives its data from i. We will ensure by conditions given later that $\overline{G}$ is a directed acyclic graph—so that there are no circularities in the dependencies among cells.

Note that the set C of cells comprises all the task activations performed during a single run of the system (which may extend for the entire lifetime of the system). It is therefore potentially unbounded (though finite) in size. For many (statically scheduled) process-control systems, the set C and its associated data dependency graph $\overline{G}$ will have a repetitive structure induced by the "unrolling" of a periodic, or cyclic, pattern of activity.

Cells with indegree zero in $\overline{G}$ are called *sensor* cells; those with outdegree zero are called *actuator* cells. The set of sensor cells is denoted C_S; that of actuators is denoted C_A. Nonsensor cells (including actuator cells) have a computational task associated with them and are called *active-task* cells. The set of active-task cells is denoted C_T and given by $C_T \stackrel{\text{def}}{=} C \setminus C_S$ (where \ denotes set difference).

Each task activation (or sensor sample) generates a value that is either communicated to an actuator or stored so that it will be available as input to later task activations. The system state records these stored output values. Formally, we define

- A set D of *domain* values, and
- A set of *states* $\mathcal{S} \subseteq C \rightarrow D$.

The data values computed, stored, and manipulated by the system are assumed to be drawn from the uninterpreted domain D. The system state is represented by a function from cells to this domain: if $\sigma \in \mathcal{S}$ is the instantaneous state of the system, and c is a cell, then $\sigma(c)$ denotes the output value stored for that cell. It may seem that a system satisfying this description must have a huge amount of storage in order to record the values of all task activations for all time. This is not so. Anticipating definitions that are given below, we observe that tasks are executed in a sequential order that respects the dependency ordering represented in the graph $\overline{G}$, and run to completion. There is no need to record a value for a cell that has not yet been executed, nor for one whose immediate successors in the relation $\overline{G}$ have already completed. (Although this result is intuitively obvious, its formal verification is an interesting exercise.)

Formalizing the notion of sequential execution, we introduce

- A bijection *sched*: $M \rightarrow C$, with
- Inverse *when*: $C \rightarrow M$.

The interpretation here is that the i'th task execution (or sensor sample) is the one associated with cell *sched*(i); conversely, the activity at cell c is the *when*(c)'th to be executed. We require that the order of execution respect the dataflow dependencies recorded in $\overline{G}$:

$$(i, j) \in \overline{G} \supset when(i) < when(j).$$

Notice that this requires that $\overline{G}$ is acyclic.

Active-task cells have some computational task associated with them, so we require

- A set $T \subseteq \mathcal{S} \rightarrow D$ of *task-functions*, and
- A function *task*: $C_T \rightarrow T$.

When an active-task cell c executes, the function *task*(c) is applied to the current state, say σ, yielding the result *task*(c)(σ). This is then stored in the system state as the value of cell c to yield a new state τ. That is,

$$\tau = \sigma \textbf{ with } [c := task(c)(\sigma)]$$

where **with** [...] denotes function modification (as in EHDM).[2] The only components of the system state that may influence the result are those of the immediate predecessors

[2]The notation f **with** $[x := a]$, where x is a value in the domain of f and a a value in the range, denotes a function with the same signature as f defined by

$$f \textbf{ with } [x := a](y) \stackrel{\text{def}}{=} \textbf{if } x = y \textbf{ then } a \textbf{ else } f(x).$$

of cell c in the dataflow dependency graph $\overline{G}$.[3] Formally, we state this as a requirement that the result be functionally dependent on just those values:

$$(\forall a : (a, c) \in \overline{G} \supset \sigma(a) = \tau(a)) \supset task(c)(\sigma) = task(c)(\tau).$$

Sensor cells store their results in the system state just like active-task cells. However, they take no input from the system state; instead, they sample properties of the external environment (including control inputs). These properties vary with time, so it might seem that sensors should be modeled as functions of real-time. In fact, this is unnecessary and inappropriate, since our model is not concerned with real-time properties such as absolute execution rates, but with those of sequencing and voting. We want to prove that if an N-plex gets the *same* sensor samples as an ideal fault-free system, then it will deliver the same actuator commands (despite the occurrence of faults). Thus, we need only model the sensor samples actually obtained, which can be done by modeling sensor samples as functions of position in the execution schedule (i.e., we use the number of cells executed as our notion of "time"). Thus we introduce

- A set $S \subseteq M \to D$ of *sensor-functions*, and
- A function $sensor: C_S \to S$.

When a sensor cell c executes, the sensor-function $s = sensor(c)$ samples the environment (at time $when(c)$) to yield the value $s(when(c))$. This is then stored in the system-state as the value of cell c.

Formally, the execution of cells is modeled by the function

- $step: \mathcal{S} \times C \to \mathcal{S}$

where

$$step(\sigma, c) \stackrel{\text{def}}{=} \sigma \textbf{ with } [c := \textbf{if } c \in C_S \textbf{ then } sensor(c)(when(c)) \textbf{ else } task(c)(\sigma)]$$

is the new state that results from executing the task of cell c in state σ at time $when(c)$.

We are interested in the state after the system has executed some number m of cells according to its schedule. This is modeled by the function

- $run: M \to \mathcal{S}$

where

$$\begin{aligned} run(0) &\in \mathcal{S}, \\ run(m+1) &\stackrel{\text{def}}{=} step(run(m), sched(m+1)). \end{aligned}$$

A variant is the function

- $runto: C \to \mathcal{S}$

where

$$runto(c) \stackrel{\text{def}}{=} run(when(c))$$

is the state of the system when execution of its schedule has reached cell c. Observe that $run(0)$, the initial state, is chosen arbitrarily.

[3]Operationally, the function $task(c)$ is applied to the tuple of values

$$(\sigma(c_1), \sigma(c_2), \ldots, \sigma(c_n))$$

where $(c_i, (i, c)) \in G$ and $n = indegree(c)$.

3.2 The N-plex Model

In this section, we admit the possibility that machines may fail and we introduce replication and voting to overcome that fallibility.

We assume a replicated system comprising $r \geq 3$ component systems of the type described in the previous section and we define

- $R \stackrel{\text{def}}{=} \{1, 2, \ldots, r\}$.

In the following, we will often refer to the component systems as "machines."

Component machines may fail and revive independently; at any time a machine is either "failed" or "working." This is specified by a function

- $\mathcal{F}: R \rightarrow (M \rightarrow \{T, F\})$

where $\mathcal{F}(i)(m)$ is T just in case component machine i is failed at time m.[4] Intuitively, a component machine i is failed at time m if it suffers a control fault at any point during execution of the task scheduled at time m. We know nothing at all about the behavior of failed component machines. Working (i.e., non-failed) machines correctly compute the function associated with the task scheduled at time m. However, the result computed may be incorrect if an earlier failure has caused the input data to be bad. A machine that is working correctly, but on bad data, has state data faults that will eventually be overcome through majority voting of state data.

States of the replicated machine are drawn from the set

- $\mathcal{R} \subseteq R \rightarrow \mathcal{S}$.

Thus, if $\rho \in \mathcal{R}$ is a replicated state, then $\rho(i)$ is the state of the i'th component machine, and $\rho(i)(c)$ is the value of cell c in that machine.

The components of a replicated machine behave much like a single machine, except that components may fail, and so they periodically *vote* their results. Thus we assume a set

- C_V of voted cells

and require

$$C_A \subseteq C_V \subseteq C_T$$

(that is, all actuator cells are voted, but no sensor cells are).[5]

Each execution step in the replicated machine takes place in two stages. In the first stage, each working component machine performs a single (ordinary) step. This is specified by the function

- $sstep: \mathcal{R} \times C \rightarrow \mathcal{R}$

[4]A function with range $\{T, F\}$ can be interpreted as the characteristic predicate of a set (this is how sets are defined in EHDM). Thus $\mathcal{F}(i)$ can be interpreted as the set of times when the i'th machine is failed during execution of the cell scheduled at that time.

[5]Sensor cells are not voted because we assume an underlying Byzantine fault-tolerant distribution mechanism which ensures that all working machines get the same sensor samples. This assumption is captured in the definition of the function *sstep*.

where

$$\neg\mathcal{F}(i)(when(c)) \supset sstep(\rho, c)(i) = step(\rho(i), c).$$

This definition states that a working component machine updates its own state in exactly the same way the unreplicated system model would, given the same state. Two important consequences of this definition may not be obvious:

- If cell c is a sensor cell, then the value of $step(\rho(i), c)$ is

  $$\rho(i) \textbf{ with } [c := sensor(c)(when(c))]$$

  (this comes from the definition of *step*). Note that the expression in the **with** clause is independent of the machine i; thus, as noted above, our model requires that all working machines get exactly the same sensor samples.

- If machine i is failed when execution of cell c should be performed, we know nothing whatsoever about the subsequent state of that machine, i.e., $sstep(\rho, c)(i)$. We do *not* assume merely that the value stored for cell c could be incorrect; we allow the whole state (of that machine) to be damaged or destroyed.

When a voted cell is executed, the working component machines each calculate the majority vote of the full set of all their individual results for that cell. This is specified by the function

- $vote: \mathcal{R} \times C \rightarrow \mathcal{R}$

where

$$\neg\mathcal{F}(i)(when(c)) \supset vote(\rho, c)(i) = \rho(i) \textbf{ with } [c := maj\{\!\!\{\rho(j)(c) | j \in R\}\!\!\}],$$

maj is the "majority" function, and $\{\!\!\{\rho(j)(c) | j \in R\}\!\!\}$ denotes the bag (multiset) of values recorded for cell c by all the component machines.[6]

As with the *sstep* function, we know absolutely nothing about the state of a failed component machine after a vote in which it should have participated. Another interesting element of this definition is that all working machines are specified to perform a majority vote on the *same* bag of values: this suggests they must not only read each other's values correctly, but they should agree on the values attributed to faulty components. These are precisely the requirements that "Byzantine agreement" (also known as "interactive consistency") algorithms are required to satisfy. It may seem, therefore, that any realization of this model should employ a Byzantine agreement algorithm to distribute the values to be voted among all of the component machines. This is unnecessary, however, since it is a majority vote that is being computed, and our results will establish that the good values comprise a majority. Thus, the values ascribed to failed processors are irrelevant, and the working processors do not, in fact, need to agree on those values. We do not prove this result here; we regard it as a proof obligation on the implementation.

[6]Note that *maj* is a *partial* function: it is undefined if an absolute majority of components do not agree on a value. Our results will always take care to establish conditions in which it is defined.

The overall behavior of the replicated machine is specified by the function

- $rstep: \mathcal{R} \times C \to \mathcal{R}$

which is simply the appropriate combination of the two steps above:

$$rstep(\rho, c) \stackrel{\text{def}}{=} \begin{cases} vote(sstep(\rho, c), c) & \text{if } c \in C_V \\ sstep(\rho, c) & \text{otherwise.} \end{cases}$$

Functions *rrun* and *rrunto* are defined analogously to the single machine case:[7]

- $rrun: M \to \mathcal{R}$,

is given by

$$\begin{aligned} rrun(0) &\stackrel{\text{def}}{=} (\lambda i: run(0)) \\ rrun(m+1) &\stackrel{\text{def}}{=} rstep(rrun(m), sched(m+1)) \end{aligned}$$

and

- $rrunto: C \to \mathcal{R}$

by

$$rrunto(c) \stackrel{\text{def}}{=} rrun(when(c)).$$

Notice that our model assumes that computation and voting are atomic, and that the components of the replicated machine are completely synchronous. These are idealizations of reality and we intend to explore more realistic assumptions in later work. They are adequate, however, for the purpose of the current investigation, where we are primarily concerned to develop the conditions under which majority voting successfully masks transient failures.

3.3 Fault Tolerance and Transient-Recovery

Our goal in this section is to show that, under certain conditions concerning the failure "pattern" $\mathcal{F}$, the replicated machine produces the same actuator behavior as the single machine, despite failures among the components of the replicated machine. Our requirements are that the majority-voted value for each actuator should be the correct value—that is, the value produced by a single fault-free system. In our model, actuator cells are voted, so that any nonfaulty component machine will set its own value for an actuator cell to that of the majority. Thus, the correctness statement can be rephrased as the requirement that the value computed for an actuator cell by any nonfaulty component machine should be the correct value.

We can state the condition that a component machine i have the correct value for cell c in terms of a predicate:

[7] Readers unfamiliar with higher-order logic may find the so-called "Curried" functions that we employ somewhat strange. Rather than the Curried application $rrun(m)(i)(c)$, they might prefer the application of a function with multiple arguments: $rrun(m, i, c)$. The advantage of our approach is that the separate components of the application have individual meaning and can be manipulated individually: $rrun(m)$ is the state of the replicated machine after m steps, $rrun(m)(i)$ is the state of the i'th component machine at that point, and $rrun(m)(i)(c)$ is the value stored for cell c in that state.

- *good-value*: $R \times C \rightarrow \{T, F\}$

where

$$good\text{-}value(i, c) \stackrel{\text{def}}{=} (rrunto(c)(i)(c) = runto(c)(c)).$$

We then seek a predicate

- *safe*: $C \rightarrow \{T, F\}$

such that

$$\forall c \in C_A, i \in R : (safe(c) \wedge \neg \mathcal{F}(i)(when(c)) \supset good\text{-}value(i, c)).$$

Intuitively, *safe*(c) will capture the conditions under which the replicated machine has enough working components, and those components have been working for long enough since their last failure, that good values form a majority and faults will be masked successfully.

If only actuator cells were voted, it would be trivial to derive the required result: *safe*(c) would be the condition that a majority of components have been working continuously since the very first cell through the computation and vote of cell c. That this condition is sufficient follows from the fact that working component machines given the same inputs produce the same results as each other; failed machines can produce anything (including nothing). Thus, the continuously working machines will agree among themselves at every voting stage and, since they are hypothesized to be in the majority, leave their states unchanged. Since actuator cells are voted, any machine that is working during the vote of an actuator cell will acquire the correct value from this continuously-correct majority.

To see that this condition is necessary, suppose that there has not been a majority of components working continuously since the beginning. Then a majority of machines have failed at some time or other prior to the execution of cell c. When they failed, they may have destroyed their system state. Since we are now assuming no votes other than at actuators (and actuators do not provide input to other cells), this corruption may persist even after a failed machine starts working again. Thus a failed machine cannot be guaranteed ever to recover fully. Since these machines are hypothesized to form a majority by the time cell c is executed, they could outvote the good machines at that point.

Without intermediate voting of state values, a component machine that suffers a transient failure may never fully recover, since there is no way for it to repair its state data. There are many possible strategies for intermediate voting: we can vote at every cell or only at certain cells, and we can vote the entire state, or just some portions of it, or just the value computed at that cell. Voting more data or voting more often than necessary can be very expensive, using up resources that could be put to better use. Early DFCS maintained very little state data and it was feasible to vote the entire state every frame. Modern systems maintain much more information and it is necessary to be more sparing in the frequency of voting, and in the quantity of data voted. Obviously there is a tradeoff here: voting less frequently, or less data at each vote, may increase the time taken to recover from transients, and thereby reduce the reliability of the system. Notice that particular voting strategies can be modeled within our framework by introducing additional voted tasks whose only function is to gather data for voting. For example "frame-based voting" (voting all the results computed in a given frame at the end of the frame) can be modeled by introducing an additional "voter" task scheduled at the end of each frame. All other tasks in the frame feed copies of their outputs into the voter task (which is the only voted task within the frame), and all subsequent tasks obtain their values from the voter task.

Clearly, overall reliability depends upon the relationship between the voting strategy, the fault arrival rate, and the dataflow dependencies in the system. We need to encode this relationship as the condition in the predicate *safe*. Intuitively, the condition must ensure that, for every cell, a majority of machines have been working for long enough since their last failure that they have acquired correct values (from sensor samples or votes) for data values that ultimately contribute to the value of cell c, and have computed all intermediate values correctly. Stating this condition formally requires some additional definitions.

We define

- $foundation: C \rightarrow \mathcal{P}(C)$, where $\mathcal{P}$ denotes *powerset*,

recursively as follows:

$$foundation(c) \stackrel{\text{def}}{=} \begin{cases} \{c\} & \text{if } c \in (C_S \cup C_V) \\ \{c\} \cup \bigcup\limits_{(b,c)\in\overline{G}} foundation(b) & \text{otherwise} \end{cases}$$

and

- $support: C \rightarrow \mathcal{P}(C)$

by

$$support(c) \stackrel{\text{def}}{=} \begin{cases} \{c\} \cup \bigcup\limits_{(b,c)\in\overline{G}} foundation(b) & \text{if } c \in C_V \\ foundation(c) & \text{otherwise.} \end{cases}$$

The *foundation* of a cell c consists of all those cells that directly or indirectly contribute input data to c by a path that does not pass through any (other) voted cells. Note that a voted or sensor cell is its own foundation.

Figure 1 gives a graphical representation of these concepts. In the figure, circles indicate cells, double circles indicate voted cells and the arrows indicate the flow of data (the arrow from cell A to cell D represents the arc $(A, D) \in \overline{G}$). The left to right position of cells on the page suggests the order in which they are executed. In this case, the foundation for cell J is just $\{J\}$ (since J is a voted cell), that for A is $\{A\}$ (since A is a sensor cell), and that for cell D is $\{A, C, D\}$.

The *support* for a nonvoted cell is simply the foundation for that cell; the support for a voted cell is the union of the foundations of all the cells that directly provide input to that cell. The intuition here is that if a machine computes correct values for all the cells in *support*(c), and if the machine keeps working, then the value eventually computed for cell c will be correct. In Figure 1, the supports for A and D equal their foundations, whereas the support for the voted cell J is $\{A, C, D, J\}$. A machine that is working throughout the support of cell J will compute the correct value for that cell: since it is working at

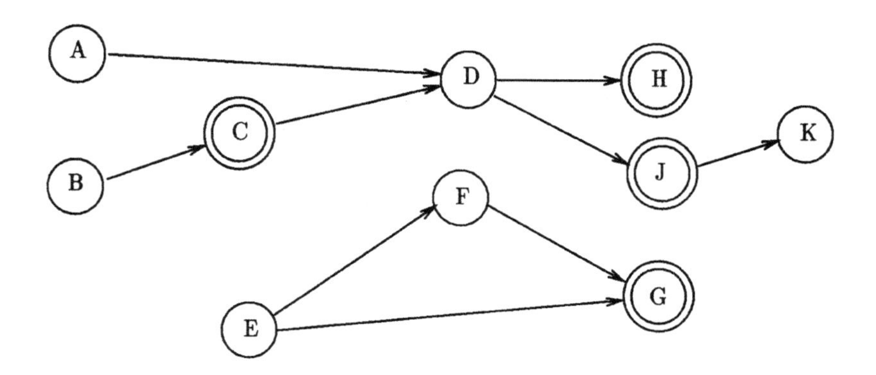

Figure 1: Example Dataflow Dependency Graph

sensor cell A, it will acquire the correct sample value from that sensor; since it is working at voted cell C, it will acquire the correct value for that cell during its majority vote, even if it had been failed earlier and had not computed the right value itself [8]; since it has the correct input values for cell D and is working at that cell, it will compute the correct output value; and since it has (from D) the correct input value for cell J, and is working at J, it will compute the correct value for J.

We need just a few more definitions. The function

- $\textit{committed-to}: C \rightarrow M$

is defined by

$$\textit{committed-to}(c) \stackrel{\text{def}}{=} min\{when(a) | a \in support(c)\}.$$

In the example of Figure 1, $\textit{committed-to}(J) = when(A)$. Once a machine reaches $\textit{committed-to}(c)$ in its schedule, it must keep working until $when(c)$ if it is to compute the correct value for cell c. Conversely, if it does keep working throughout this period, it will compute the correct value for cell c even if its own state data are corrupt at the beginning of the period. This is because all the data required to compute cell c are derived from either sensor samples or voted values that are acquired at or later than $\textit{committed-to}(c)$. Thus, provided enough other machines are working, this machine will acquire good values during the votes and sensor-samples and its own bad state data will not contribute to the result.

The function OK captures the condition under which a particular component machine has been working for "long enough" since its last fault that any bad state data values have been replaced by good values through votes and sensor samples—so that it is able to compute a good result for the current cell. Thus,

- $OK: R \rightarrow (C \rightarrow \{T, F\})$

[8] We are assuming here that enough machines were working correctly at c that correct values form the majority. We cannot give a characterization of the necessary condition yet, since we are in the process of developing the concepts that make its statement possible.

is defined by

$$OK(i)(c) \stackrel{\text{def}}{=} (\forall m : \textit{committed-to}(c) \leq m \leq \textit{when}(c) \supset \neg\mathcal{F}(i)(m)).$$

In other words, $OK(i)(c)$ is the condition which ensures that component machine i has no state data faults that can affect the value computed for cell c.

For the replicated machine to be *safe*, a majority of its components must be OK for every cell. We therefore introduce the function

- $MOK: C \rightarrow \{T, F\}$

(for Majority OK) defined as follows

$$MOK(c) \stackrel{\text{def}}{=} \exists \Theta \subseteq R : |\Theta| > r/2 \wedge (i \in \Theta \supset OK(i)(c)).$$

We then define the predicate *safe* as follows

$$\textit{safe}(c) \stackrel{\text{def}}{=} (\forall a : \textit{when}(a) \leq \textit{when}(c) \supset MOK(a)).$$

That is, the replicated machine is *safe* at cell c if, the condition MOK holds at c itself and at all cells evaluated earlier than c.

Now we can state and prove our main theorem. This "Consensus Theorem" is similar to lemmas of that name in [4].

Theorem 1 (Consensus Theorem) If $\textit{safe}(c)$, then

$$\forall j \in R : OK(j)(c) \supset \textit{good-value}(j, c).$$

Proof: The proof is by strong induction on $\textit{when}(c)$. The basis is the case $\textit{when}(c) = 1$, in which case c must be a sensor cell, and so

$$\textit{rrunto}(c)(j)(c) = \textit{sensor}(c)(1) = \textit{runto}(c)(c)$$

as required.

For the inductive step, suppose the theorem true for all cells a such that $\textit{when}(a) < \textit{when}(c)$ and let j be a component machine such that $OK(j)(c)$. If $c \in C_S$, the argument is the same as for the basis case, and so we consider $c \in C_T$ and consider a such that $(a, c) \in \overline{G}$. Since the result of c is a function of its inputs, the result will follow if we can demonstrate

$$\textit{good-value}(j, a).$$

There are two cases to consider.

Case 1: $a \in C_V$. It may not be that $OK(j)(a)$ and so we cannot appeal to the inductive hypothesis directly, but we do know that $MOK(a)$ and hence that a majority of machines exemplified by k (possibly not including j) satisfy $OK(k)(a)$. By the inductive hypothesis, $\textit{good-value}(k, a)$ for these machines. Now, we hypothesized $OK(j)(c)$ and hence $\neg\mathcal{F}(j)(a)$. It follows that during the voting stage of the execution of cell a, machine j will acquire the majority value for that cell, i.e., $\textit{good-value}(j, a)$, as required.

Case 2: $a \notin C_V$. A component machine i is OK for cell c if it is working throughout the period from *committed-to*(c) to $when(c)$. Observe that the support of a nonvoted cell a is a subset of any cell c to which it provides input. It follows that *committed-to*(a) can be no earlier than *committed-to*(c). We must also have $when(a) < when(c)$. Thus $OK(i)(c) \supset OK(i)(a)$ and the result then follows directly from the inductive hypothesis.

□

The result we seek follows from the Consensus Theorem:

Corollary 1 For $c \in C_A$, if $safe(c)$ then

$$\forall i \in R : \neg\mathcal{F}(i)(when(c)) \supset \mathit{good\text{-}value}(i, c).$$

Proof: The statement of the corollary implies $MOK(c)$, so there must exist $j \in R$ such that $OK(j)(c)$. The Consensus Theorem then supplies

$$\forall j \in R : OK(j)(c) \supset \mathit{good\text{-}value}(j, c)$$

which, on expanding the definition of *good-value*, gives

$$rrunto(c)(j)(c) = runto(c)(c).$$

Now $c \in C_A$, so c is a voted cell, and the definition of the voting function ensures,

$$\forall i, j \in R : (\neg\mathcal{F}(i)(when(c)) \wedge \neg\mathcal{F}(j)(when(c))) \supset rrunto(c)(i)(c) = rrunto(c)(j)(c),$$

since all working machines acquire the majority value as the result of voted cells. By definition, $OK(j)(c) \supset \neg\mathcal{F}(j)(when(c))$. Hence, for any $i \in R$ such that $\neg\mathcal{F}(i)(c)$,

$$rrunto(c)(i)(c) = rrunto(c)(j)(c) = runto(c)(c)$$

and we conclude *good-value*(i, c) as required. □

In words, the corollary states that each working component of the replicated machine computes the correct value for an actuator if a majority of machines is working throughout the period from *committed-to*(c) to $when(c)$ for each cell c in the schedule up to and including the actuator concerned.

4 Formal Specification and Verification

We have formally specified the model of the previous section in the language of EHDM [18], and formally verified the Consensus Theorem and its Corollary [15]. The specification language of EHDM is a simply-typed higher-order logic with fairly rich facilities for subtyping, dependent typing, modularity, and parameterization. The heart of the formal specification and verification is structured into 11 EHDM modules; in addition, 5 standard modules are used to specify sets, Noetherian induction, and other supporting concepts. The mechanically-checked proof of the Consensus Theorem and its Corollary is structured into 93 elementary lemmas, which can be checked in a total of 7 minutes on a Sun Sparcstation 2. The formal verification rests on a very simple axiomatic basis in addition to the theories (which include arithmetic) built in to the EHDM system itself: four axioms are required in the development of a theory of sets, one is required to state Noetherian induction, and 11 are used to develop the fault masking model. All other formulas are either definitions (these provide conservative extension in EHDM; 15 are used in the specification), or lemmas whose proofs have been checked mechanically.

The model of Section 3 was developed with specification in EHDM in mind; it is built from straightforward mathematical concepts and could therefore be transliterated more or less directly into EHDM. An example module is shown in the appendix (the typesetting was done automatically by EHDM); the formula called `the_result` corresponds to the Consensus Theorem in Section 3. Whereas the formal specification required only a few hours to construct, the formal verification required about three weeks of part-time work. Some of this time was required because the formal verification proves a number of subsidiary results that are glossed over in the ordinary mathematical presentation, and some of it was required because EHDM's proof support was not very efficient at the time the proof was performed (in particular, it lacked a rewriter). We regard the time taken for the verification to be excessive; three days would seem a more reasonable target, and we believe that this will be feasible in the next release of EHDM.

Even three days may seem excessive to some who observe that the proofs given in Section 3 require little more than a page to describe. It should be noted, however, that that proof itself took several days to develop and that it glosses over a number of points that are quite tricky when considered in detail. For example, it assumes that if the output of cell a is used as an input to cell c, then the value recorded for a immediately after it is computed will still be the same when it is accessed (possibly much later) in order to be used in the computation of c. In the case of the simple machine, this result is straightforward; it is less so in the case of the replicated machine (since failures must be accounted for). A rigorous demonstration of this property requires a proof by induction comparable to that required for the main theorem.

Finally, the mechanically verified theorem is stronger than that stated and proved in Section 3. The difference is that the formal specification of *safe(c)* requires only that the replicated machine be MOK for those cells a that transitively contribute input to c; the definition in Section 3, on the other hand, requires that the replicated machine be MOK for c and for all cells executed earlier than c. Clearly the cells that transitively contribute input to c must all be executed earlier than c, and so the second condition implies the first. The reason we used a stronger definition for *safe* (and hence obtained a weaker theorem) in the traditional mathematical presentation than we did in the formal specification and verification is that the stronger definition allows the Consensus Theorem to be proved by simple induction over the natural numbers, whereas the weaker definition requires a proof by Noetherian induction over the structure of the dataflow dependency graph. Noetherian induction is rather tricky to state and carry out in quasi-formal notation (and may not be familiar to all readers) and so we opted for the stronger notion of *safe*, and hence a weaker theorem, in the traditional development. In the truly formal notation of EHDM, it is no more difficult to perform Noetherian than simple induction, and so we used the definition for *safe* that gave the strongest theorem.

5 Conclusion

We have presented a formal model for fault-masking and transient-recovery among the redundant computers of a DFCS, and proved a condition on the dataflow dependencies in the task graph, the voting pattern, and the arrival of faults, which is sufficient to ensure that correct values are always in the majority among those sent to the actuators. The model and its theorem have been subjected to mechanically-checked specification and verification.

Unlike the verifications of clock synchronization algorithms [16,20], formal verification of our fault-masking model did not uncover flaws or errors in the conventional mathematical presentation. (We attribute this partly to the fact that analysis of the clock synchronization algorithms is very much more difficult, and partly to the fact that considerable experience in formal verification has given us a much refined appreciation whether an *informal* proof is valid.) Formal verification did provide a number of benefits, however: it forced us to attend to a number of details that were glossed over in Section 3, it precisely identified the axiomatic basis of the model, and it gave us a stronger theorem. Identification of the axiomatic basis is important, since it provides the requirements specification for the next lower level in the modeling hierarchy. It also allows us to identify, and we hope later to explore, some fundamental assumptions. For example, plurality voting works just as well as majority voting when a majority exists; we are now in a position to examine the precise properties required of the voting function.

The benefits of formal verification are bought at a price; we believe that price can be reduced substantially in the very near future with carefully-tuned theorem proving support (this is not the same as more powerful theorem proving support). Formal specification in a suitably rich specification language, however, exacts very little price at all, and exerts a beneficial discipline. We see no reason why it should not be routine. Indeed, the main point we wish to make is that modern verification systems can support a sufficiently rich fragment of mathematics that it is feasible to develop models of important elements of interesting computer systems directly in that mechanized fragment of mathematics, without compromising convenience or elegance of expression. We analyze the attributes that maximize convenience and utility of specification languages and verification systems elsewhere [14,17].

Our plans for the future include elimination of some of the simplifying assumptions of the present model. Accurate modeling of fault masking must recognize that the separate channels are not perfectly synchronized (the clock-synchronization algorithms keep the separate channels synchronized only within some small skew δ of each other), and that the communication and coordination of voting data takes a certain amount of time. A fault-tolerant system should take active measures to recover from transient control faults, in addition to the voting strategy for overcoming state data faults. The Mars system [10,11] is a good example of a system that provides such recovery. Our plans for the future also include formal analysis of selected mechanisms for recovering from transient control faults. In other current work, we are developing and formally verifying a hardware-assisted implementation of one of the clock-synchronization algorithms.

Summarizing the results of the AFTI-F16 flight tests, NASA engineer Mackall observed [12, pp. 40–41]

> "...qualification of such a complex system as this, to some given level of reliability, is difficult ... [because] the number of test conditions becomes so large that conventional testing methods would require a decade for completion. The fault-tolerant design can also affect overall system reliability by being made too complex and by adding characteristics which are random in nature, creating an untestable design.
>
> "...reducing complexity appears to be more of an art than a science and requires an experience base not yet available. If the complexity is required, a method to make system designs more understandable, more visible, is needed."

We hope that the work described here (and our larger program) contributes to the goal of developing testable designs, purged of "random characteristics," and which satisfy Mackall's plea for a method that will make designs "understandable, more visible."

Acknowledgements

We are grateful to Ricky Butler of NASA Langley Research Center for posing the challenge of applying formal methods to aspects of digital flight control systems, and for structuring the overall problem into manageable pieces. Our treatment of the problem tackled in this paper owes much to discussions with Ben Di Vito, and to his model for fault masking and transient recovery. Jim Caldwell provided valuable assistance and encouragement in the first stage of the formal verification reported here.

References

[1] W.R. Bevier and W.D. Young. Machine-checked proofs of a Byzantine agreement algorithm. Technical Report 55, Computational Logic Incorporated, Austin, TX, June 1990.

[2] W.R. Bevier and W.D. Young. The design and proof of correctness of a fault-tolerant circuit. In *2nd. International Working Conference on Dependable Computing for Critical Applications*, pages 107–114, Tucson, AZ, February 1991. IFIP WG. 10.4.

[3] Ricky W. Butler and Sally C. Johnson. The art of fault-tolerant system reliability modeling. NASA Technical Memorandum 102623, NASA Langley Research Center, Hampton, VA, March 1990.

[4] Ben L. Di Vito, Ricky W. Butler, and James L. Caldwell. Formal design and verification of a reliable computing platform for real-time control. NASA Technical Memorandum 102716, NASA Langley Research Center, Hampton, VA, October 1990.

[5] Ben L. Di Vito, Ricky W. Butler, and James L. Caldwell. High level design proof of a reliable computing platform. In *2nd. International Working Conference on Dependable Computing for Critical Applications*, pages 124–136, Tucson, AZ, February 1991. IFIP WG. 10.4.

[6] Carl S. Droste and James E. Walker. *The General Dynamics Case Study on the F16 Fly-by-Wire Flight Control System*. AIAA Professional Study Series. American Institute of Aeronautics and Astronautics. Undated.

[7] Richard E. Harper and Jaynarayan H. Lala. Fault-tolerant parallel processor. *AIAA Journal of Guidance, Control, and Dynamics*, 14(3):554–563, May-June 1991.

[8] Stephen D. Ishmael, Victoria A. Regenie, and Dale A. Mackall. Design implications from AFTI/F16 flight test. NASA Technical Memorandum 86026, NASA Ames Research Center, Dryden Flight Research Facility, Edwards, CA, 1984.

[9] R.M. Kieckhafer, C.J. Walter, A.M. Finn, and P.M. Thambidurai. The MAFT architecture for distributed fault tolerance. *IEEE Transactions on Computers*, 37(4):398–405, April 1988.

[10] H. Kopetz, H. Kantz, G. Grünsteidl, P. Puschner, and J. Reisinger. Tolerating transient faults in MARS. In *Digest of Papers, FTCS 20*, pages 466–473, Newcastle upon Tyne, UK, June 1990. IEEE Computer Society.

[11] Hermann Kopetz et al. Distributed fault-tolerant real-time systems: The Mars approach. *IEEE Micro*, 9(1):25–40, February 1989.

[12] Dale A. Mackall. Development and flight test experiences with a flight-crucial digital control system. NASA Technical Paper 2857, NASA Ames Research Center, Dryden Flight Research Facility, Edwards, CA, 1988.

[13] M. Pease, R. Shostak, and L. Lamport. Reaching agreement in the presence of faults. *Journal of the ACM*, 27(2):228–234, April 1980.

[14] John Rushby. Design choices in specification languages and verification systems. In Phillip Windley, editor, *Proceedings of the HOL Theorem Proving System and Applications Conference*, Davis, CA, August 1991. IEEE Computer Society.

[15] John Rushby. Formal specification and verification of a fault-masking and transient-recovery model for digital flight-control systems. Technical Report SRI-CSL-91-3, Computer Science Laboratory, SRI International, Menlo Park, CA, January 1991. Also available as NASA Contractor Report 4384.

[16] John Rushby and Friedrich von Henke. Formal verification of the interactive convergence clock synchronization algorithm using EHDM. Technical Report SRI-CSL-89-3R, Computer Science Laboratory, SRI International, Menlo Park, CA, February 1989 (Revised August 1991). Also available as NASA Contractor Report 4239.

[17] John Rushby and Friedrich von Henke. Formal verification of algorithms for critical systems. In *SIGSOFT '91: Software for Critical Systems*, New Orleans, LA, December 1991.

[18] John Rushby, Friedrich von Henke, and Sam Owre. An introduction to formal specification and verification using EHDM. Technical Report SRI-CSL-91-2, Computer Science Laboratory, SRI International, Menlo Park, CA, February 1991.

[19] Fred B. Schneider. Implementing fault-tolerant services using the state machine approach: A tutorial. *ACM Computing Surveys*, 22(4):299–319, December 1990.

[20] Natarajan Shankar. Mechanical verification of a schematic Byzantine fault-tolerant clock synchronization algorithm. Technical Report SRI-CSL-91-4, Computer Science Laboratory, SRI International, Menlo Park, CA, January 1991. Also available as NASA Contractor Report 4386.

[21] Mandayam Srivas and Mark Bickford. Verification of the FtCayuga fault-tolerant microprocessor system, volume 1: A case-study in theorem prover-based verification. NASA Contractor Report 4381, July 1991.

[22] John H. Wensley et al. SIFT: design and analysis of a fault-tolerant computer for aircraft control. *Proceedings of the IEEE*, 66(10):1240–1255, October 1978.

A Example Specification Module

correctness: **Module**

Using supports, sets[R], cardinality[R]

Exporting all with supports, sets[R]

Theory

i, j: **Var** R

a, c: **Var** C

m: **Var** M

OK: function[$R \rightarrow$ set[C]] =
$(\lambda\, i :$
$(\lambda\, c :$
$(\forall\, m : \text{committed\_to}(c) \leq m \wedge m \leq \text{when}(c) \supset \neg F(i)(m))))$

working: function[$C \rightarrow$ set[R]] == $(\lambda\, c : (\lambda\, i : \text{OK}(i)(c)))$

MOK: function[$C \rightarrow$ bool] = $(\lambda\, c : 2 * |\text{working}(c)| > |\text{fullset}[R]|)$

safe: **Recursive** function[$C \rightarrow$ bool] =
$(\lambda\, c : \text{MOK}(c) \wedge (\forall\, a : (a, c) \in \overline{G} \supset \text{safe}(a)))$ **by** when

correct: function[$C \rightarrow$ bool] =
$(\lambda\, c : (\forall\, j : \text{OK}(j)(c) \supset \text{rrunto}(c)(j)(c) = \text{runto}(c)(c)))$

the_result: **Theorem** safe(c) $\supset$ correct(c)

End correctness

On Fault–Tolerant Symbolic Computations

Bernard Delyon and **Oded Maler**

INRIA/IRISA, Rennes*

Abstract. In this paper we propose a model that captures the influence of noise on the correct behavior of a computing device. Within this model we analyze the relation between structural properties of automata and their immunity to noise. We prove upper– and lower–bounds on the effect of noise for various classes of finite automata. Our model, combining basic notions from algebraic automata theory and the theory of stochastic processes, can serve as a starting point for a rigorous theory of computational systems embedded in the real world.

1 Introduction

Traditional computer science models try to abstract away as many real–world features as possible. The symbols in the input tape of a Turing machine are immediately and precisely recognized, the durations of automaton transitions are much smaller than the arrival rate of input symbols, etc. This idealization is appropriate for investigating the ultimate limitations of discrete machines in calculating well–defined functions on sequences. Using these models it has been proved that certain computational problems are unsolvable while other are intractable.

When we consider "real" computers embedded in a physical environment, as in the case of robots, controllers or signal processors, the idealized model is not satisfactory anymore. Decades of research in computer vision and pattern recognition have shown us how hard it is to transduce sensory input signals into a's and b's, and from research in robotics and motor control we learn how complicated is the relation between a symbolic output such as "GET OBJECT" and its physical realization.

The goal of this research is to build a framework for comparing the "ideal" behavior of a discrete computational device with its behavior in "realistic" situations. The ideal behavior is the one usually studied in theoretical computer science models, that is, the behavior of a transition system (the language it accepts, its associated sequential function, etc.) when all inputs are correctly interpreted and all state transitions are performed correctly with a negligible duration. The bridge between the idealized and real world is

*INRIA/IRISA, Campus de Beaulieu, Rennes 35042, France. E-mail: maler@allegro.irisa.fr

built by introducing noise: with some probability the system takes a wrong transition. This noise can result from the physical properties of sensors, from limitations of classification algorithms, from unreliability of computational hardware or from insufficient speed of the computer with respect to the arrival rate of input symbols. Whatever the physical reason of the noise is, and no matter what its logical form is (omission, misclassification or duplication of symbols) in our model it is assumed to be reducible to a bound ϵ on the probability of making an error in taking a transition.

The noise transforms the original deterministic computational system into a probabilistic one over the same set of states. The deviation of the noisy system from the original "normative" behavior is defined as the probability that these two systems are in different states given the same input sequence of external events. The first class of systems we consider are finite–state automata and our main result is in establishing the relation between the properties of the original automaton and the distance between itself and its noisy version. It turns out that some classes of automata are less sensitive to noise than others. The significance of this work is in demonstrating a possible theoretical basis for the analysis of embedded systems and in linking together concepts and notions from automata theory, Markov processes and the theory of semigroups.

It should be noted that unlike other works on fault–tolerance, ours is not concerned with the design of computer architectures that minimize the effect of noise on arbitrary computations. In contrast, we try to classify computational tasks according to their inherent immunity to noise, and in particular according to whether they can be performed in a satisfactory manner in spite of temporary errors during execution.

The paper is organized as follows: in section 2 we define formally the noisy version of an automaton and the distance between the ideal and the noisy versions. In section 3, we calculate an upper bound on this distance for a class of automata whose associated transformation semigroup contains a reset. In section 4 we give a lower–bound on this distance for the complementary class of automata. In section 5 we conclude and mention briefly some relations with past and future work.

2 Ideal and Noisy Automata

2.1 The Effect of Noise

The finite–state automaton (see [HU79]) is the basic discrete computational device. It consists of an input alphabet, that is, a finite set of input stimuli it can get from the outside world, a set of internal states, and a transition function that defines the changes of state based on the current state and the input. The essential behavior of the automaton is the mapping of external sequences into sequences of internal states. When, for some reason, the automaton fails to take the correct transition (e.g., the current input arrived before the previous one is completely processed) we are at the risk that the relation between the input history and the internal state is no longer maintained. When we have a bound on the probability of such a fault, the situation can be viewed as if we work with a noisy version of our intended automaton which is just a probabilistic automaton in the sense of [Rab63, Paz70]. This notion is formalized below.

Definition 1 (Noisy Version) *Let $\mathcal{A} = (\Sigma, Q, \delta)$ be a deterministic automaton and let ϵ, $0 \leq \epsilon \leq 1$ be a probability. An ϵ–noisy version of $\mathcal{A}$ is any probabilistic automaton $\mathcal{A}' = (\Sigma, Q, \delta')$ where δ' is a time-invariant probabilistic transition function such that for every state q and input σ satisfying $\delta(q, \sigma) = q'$ we have $Pr\{\delta'(q, \sigma) = q'\} \geq 1 - \epsilon$* and consequently, $\sum_{q'' \neq q'} Pr\{\delta'(q, \sigma) = q''\} \leq \epsilon$.

An example of an automaton and its noisy version is depicted in figure 1. This particular pattern of noise is associated with a probability ϵ of omitting an input symbol and thus not performing a transition.

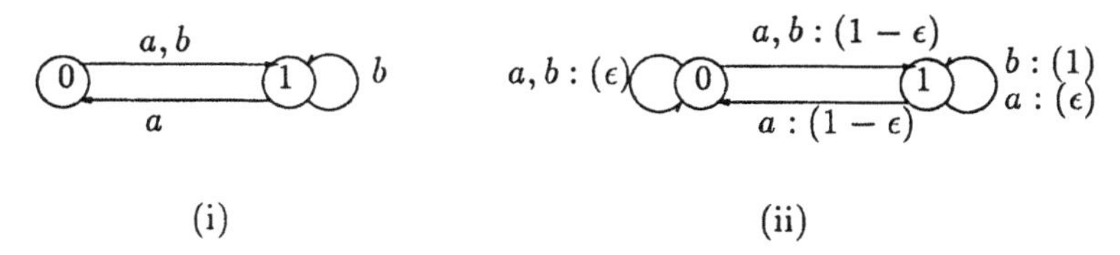

Figure 1: *(i) A deterministic automaton $\mathcal{A}$. (ii) A corresponding ϵ–noisy version $\mathcal{A}'$. The $(\sigma : p)$ labels on the arrows indicate that the corresponding transition is taken with probability p when the input is σ.*

Some insight concerning the nature of this "noisyfication" can be gained by employing the alternative description via state–vector and transition–matrix terminology. The current state can be represented by a probabilistic vector $\bar{v}$ where v_i is the probability that the current state is q_i, and every input letter σ in a stochastic automaton can be associated with a probabilistic transition matrix M^σ such that $M^\sigma_{ij} = Pr\{\delta(q_i, \sigma) = q_j\}$. In the deterministic case we are restricted to $0 - 1$ vectors and matrices. The matrices for $\mathcal{A}$ and $\mathcal{A}'$ appear in figure 2.

	M^a	M^b
$\mathcal{A}$	$\begin{pmatrix} 0 & 1 \\ 1 & 0 \end{pmatrix}$	$\begin{pmatrix} 0 & 1 \\ 0 & 1 \end{pmatrix}$
$\mathcal{A}'$	$\begin{pmatrix} \epsilon & 1-\epsilon \\ 1-\epsilon & \epsilon \end{pmatrix}$	$\begin{pmatrix} \epsilon & 1-\epsilon \\ 0 & 1 \end{pmatrix}$

Figure 2: *A matrix representation of $\mathcal{A}$ and $\mathcal{A}'$.*

2.2 The Distance Between Behaviors

We want to define some quantitative measures that express the "deviation" of the behavior of $\mathcal{A}'$ from the normative behavior of $\mathcal{A}$. For this purpose we must first define a distance function d on Q such that $d(q, q')$ expresses the significance of being incorrectly in q' instead of in q. Each individual sequence $w \in \Sigma^*$ induces a deterministic behavior on $\mathcal{A}$

and a stochastic behavior on $\mathcal{A}'$ both starting from some initial state q_0. The expected distance between the automata after reading w is defined as:

$$\rho_w(\mathcal{A}, \mathcal{A}') = \sum_{q' \neq \delta(q_0, w)} d(q, q') \cdot Pr\{\delta'(q_0, w) = q'\} \tag{1}$$

Using the discrete metrics on Q (that is, $d(q,q) = 0$ and $d(q,q') = 1$ iff $q \neq q'$) this definition reduces to:

$$\rho_w(\mathcal{A}, \mathcal{A}') = Pr\{\delta(q_0, w) \neq \delta'(q_0, w)\} \tag{2}$$

Next we consider, for every k, a probability distribution μ_k on all the input sequences of length k. This induces an expected distance measure ρ^k defined as:

$$\rho^k(\mathcal{A}, \mathcal{A}') = \sum_{w \in \Sigma^k} \mu_k(w) \cdot \rho_w(\mathcal{A}, \mathcal{A}') \tag{3}$$

Finally we consider $\{\mu_k\}_{k=1}^{\infty}$ as a sequence of probability distributions on $\{\Sigma^k\}_{k=1}^{\infty}$. The asymptotic expected distance between $\mathcal{A}$ and $\mathcal{A}'$ is

$$\rho(\mathcal{A}, \mathcal{A}') = \lim_{k \to \infty} \rho^k(\mathcal{A}, \mathcal{A}') \tag{4}$$

It is reasonable to assume additional restrictions on $\{\mu_k\}$, such as

$$\mu_k(w) = \sum_{\sigma \in \Sigma} \mu_{k+1}(w\sigma)$$

In the rest of this paper we will assume μ_k as induced by by a Bernoulli process: for every position in the sequence, the probability of a letter $\sigma_i \in \Sigma$ is a fixed probability p_i. We will also denote $\min\{p_i\}$ by $\hat{p}$.

2.3 Representation by Products

A useful conceptual tool for describing the joint behavior of two automata reacting to the same input is their product.

Definition 2 (Product of Probabilistic Automata) *Let $\mathcal{A}_1 = (\Sigma, Q_1, \delta_1)$ and $\mathcal{A}_2 = (\Sigma, Q_2, \delta_2)$ be two probabilistic automata. Their cartesian product $\mathcal{A}_1 \times \mathcal{A}_2$ is a probabilistic automaton $\mathcal{A} = (\Sigma, Q, \delta)$ where $Q = Q_1 \times Q_2$ and δ is a probabilistic transition function such that for every $(q_1, q_2), (p_1, p_2) \in Q$ and $\sigma \in \Sigma$*

$$Pr\{\delta((q_1, q_2), \sigma) = (p_1, p_2)\} = Pr\{\delta_1(q_1, \sigma) = p_1\} \cdot Pr\{\delta_2(q_2, \sigma) = p_2\}$$

In the special case of deterministic automata this definition reduces to the usual direct product. The product of $\mathcal{A}$ and $\mathcal{A}'$ from figure 1 appears in figure 3.

In terms of matrices this is equivalent to the following construction: for every $\sigma \in \Sigma$, let M_1^σ and M_2^σ be the corresponding matrices in $\mathcal{A}_1$ and $\mathcal{A}_2$ respectively. The matrix associated with σ in $\mathcal{A}_1 \times \mathcal{A}_2$ is defined as $M^\sigma = M_1^\sigma \otimes M_2^\sigma$ where $\otimes$ denotes the Kronecker product of the two matrices. The resulting matrices for $\mathcal{A}' \times \mathcal{A}$ in our example appear in figure 4.

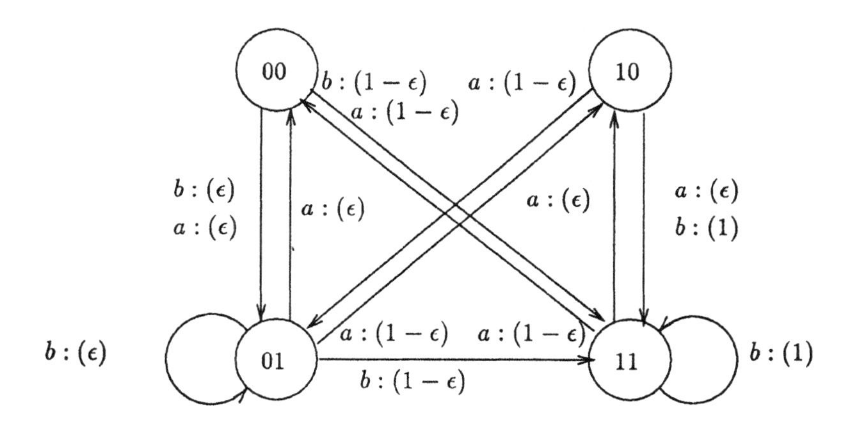

Figure 3: *The product $\mathcal{A}' \times \mathcal{A}$.*

$$M^a \qquad\qquad M^b$$

$$\begin{pmatrix} 0 & \epsilon & 0 & 1-\epsilon \\ \epsilon & 0 & 1-\epsilon & 0 \\ 0 & 1-\epsilon & 0 & \epsilon \\ 1-\epsilon & 0 & \epsilon & 0 \end{pmatrix} \quad \begin{pmatrix} 0 & \epsilon & 0 & 1-\epsilon \\ 0 & \epsilon & 0 & 1-\epsilon \\ 0 & 0 & 0 & 1 \\ 0 & 0 & 0 & 1 \end{pmatrix}$$

Figure 4: *A matrix representation of $\mathcal{A}' \times \mathcal{A}$. The indices of the rows and columns correspond to the pairs* $(0,0),(0,1),(1,0)$ *and* $(1,1)$.

Now we have a probabilistic automaton where all the trajectories ending in "diagonal" states, i.e., states in $\{(q,q) : q \in Q\}$, represent good behavior ($\mathcal{A}'$ agrees with $\mathcal{A}$) while the distance is the probability of going to the remaining states. Since we are assuming input sequences generated by a Bernoulli process we can replace the input-dependent matrices $M^{\sigma_1}, M^{\sigma_2}, \ldots, M^{\sigma_m}$ by a common averaged matrix

$$M = \sum_{i=1}^{m} p_i \cdot M^{\sigma_i}$$

representing the expected transition probabilities, namely an ordinary input-less Markov chain ([KS60]). In our example, by assuming probabilities p for a and $1-p$ for b, the probabilistic automaton of figure 4 becomes the chain of figures 5 and 6.

The distance between $\mathcal{A}$ and $\mathcal{A}'$ now becomes:

$$\rho(\mathcal{A}, \mathcal{A}') = \lim_{k \to \infty} \bar{v}_0 \cdot M^k \cdot \bar{u} \tag{5}$$

where $\bar{v}_0$ is a row vector indicating the initial state and $\bar{u}$ is a $0-1$ vector with 1's in the entries corresponding to non-diagonal states in $\mathcal{A}' \times \mathcal{A}$. In our example $\bar{v}_0 = (1,0,0,0)$ (if we consider 0 as the initial state of $\mathcal{A}$) and $\bar{u} = (0,1,1,0)^T$. The question we ask

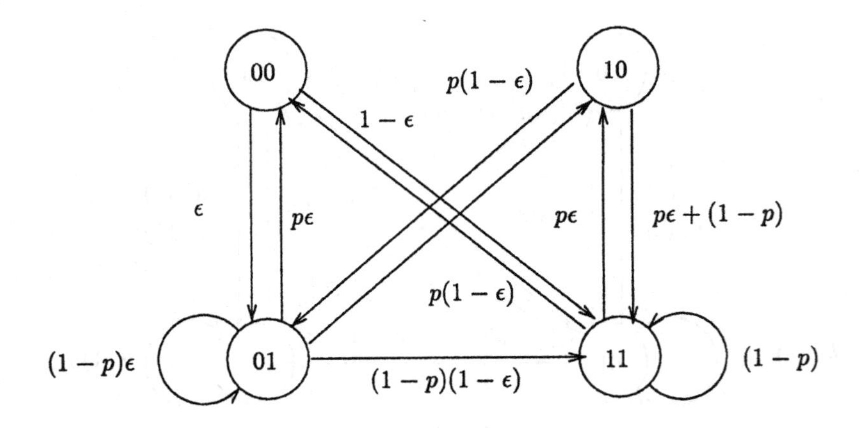

Figure 5: *The Markov chain associated with $\mathcal{A}' \times \mathcal{A}$.*

$$\begin{pmatrix} 0 & \epsilon & 0 & 1-\epsilon \\ p\epsilon & (1-p)\epsilon & p(1-\epsilon) & (1-p)(1-\epsilon) \\ 0 & p(1-\epsilon) & 0 & p\epsilon + (1-p) \\ p(1-\epsilon) & 0 & p\epsilon & 1-p \end{pmatrix}$$

Figure 6: *The expected transition probabilities of $\mathcal{A}' \times \mathcal{A}$.*

ourselves[1] is the following: *What is the relation between the structure of $\mathcal{A}$ and $\rho(\mathcal{A}, \mathcal{A}')$?*

3 Robustness of Synchronizing Automata

The essential observation underlying our results is that if for two states $q, q' \in Q$ there exists a sequence $w \in \Sigma^+$ that merges them (i.e., $\delta(q,w) = \delta(q',w) = q''$) then, whenever we are in an error configuration (q,q') or (q',q) in $\mathcal{A} \times \mathcal{A}'$, an application of w will bring us back to a correct diagonal configuration (q'',q''), and the effect of the past error will be cancelled. If we had such a transformation for every pair of states then we could recover every error with high probability.

Definition 3 (Synchronizing Automata) *An automaton is synchronizing if there exists a sequence $w \in \Sigma^*$ and a state q' such that for all $q \in Q$, $\delta(q,w) = q'$. We call w a reset[2] and let $\ell(\mathcal{A})$ denote the length of the minimal reset in $\mathcal{A}$ if there exists one or ∞ otherwise.* It can be shown that $\ell(\mathcal{A}) < \infty$ implies $\ell(\mathcal{A}) < |Q|^3$.

[1] Additional interesting measures can be introduced such as the average length of sequences of correct behavior.

[2] Resets are also known as synchronizing sequences, or synchronizers (see [BP85]). Note that if w is a reset then so is wu for every $u \in \Sigma^*$.

Claim 1 *An automaton is synchronizing if and only if every pair of states has a merging sequence.*

Sketch of proof: One direction is obvious by the definition of resets. The other can be proved inductively based on the following argument: Suppose w merges q_1 and q_2 but not necessarily q_3, that is, $\delta(q_1, w) = \delta(q_2, w) = q$ and $\delta(q_3, w) = q'$. But q and q' has a sequence w' that merges them, and $\delta(q_1, ww') = \delta(q_2, ww') = \delta(q_3, ww')$. Thus, if any pair of states has a merging sequence we can construct a global reset. ∎

Definition 4 (Reset Probability) *For every $k > 0$ we let $R(k)$ denote the probability that $w \in \Sigma^k$ is a reset.* Obviously if $\mathcal{A}$ is reset–free then $R(k) = 0$ for every k.

Claim 2 (Probability of Resets) *If $\mathcal{A}$ is synchronizing then*

1. *$R(\ell(\mathcal{A})) \geq \hat{p}^{\ell(\mathcal{A})}$.*
2. *Moreover, $\lim_{k\to\infty} R(k) = 1$.*

Sketch of Proof:

1. Trivial, follows from the existence of a reset induced by a sequence of length $\ell(\mathcal{A})$.
2. We take the transformation semigroup S associated with $\mathcal{A}$ and convert it into an automaton (Σ, S, γ) as is done in the proof of Cayley theorem. We replace the Σ–labeling of the edges by their corresponding probabilities and get a Markov chain over the space of transformations. The set of all resets, which is the minimal right ideal of S (see [Lal79]), is a terminal maximal strongly–connected component in this chain and its probability goes to one. ∎

Unfortunately we cannot make use of the asymptotic convergence of S to resets because as $|w|$ grows, the probability that $\delta(q, w) = \delta'(q, w)$ decreases. In fact we have a trade–off between an increasing probability of reset in $\mathcal{A}$ and a decreasing probability of an equivalent error–free behavior in $\mathcal{A}'$. Our main result is:

Theorem 3 (Robustness of Synchronizing Automata) *Let $\mathcal{A}$ be an synchronizing automaton with n states and let $\mathcal{A}'$ be its ϵ–noisy version. Then, for any $k \geq 0$,*

$$\rho(\mathcal{A}, \mathcal{A}') \leq \frac{1-\eta}{1-\eta+R\eta} \tag{6}$$

where $\eta = (1-\epsilon)^k$, $R = R(k)$. By letting $k = \ell(\mathcal{A})$ we obtain

$$\rho(\mathcal{A}, \mathcal{A}') \leq \frac{1-(1-\epsilon)^{\ell(\mathcal{A})}}{1-(1-\epsilon)^{\ell(\mathcal{A})}+\hat{p}^{\ell(\mathcal{A})}(1-\epsilon)^{\ell(\mathcal{A})}} \tag{7}$$

Sketch of Proof: We partition the state–space of $\mathcal{A} \times \mathcal{A}'$ into two sets, the "good" diagonal pairs G, and the "bad" error states B. We consider the transition probabilities between G and B after reading k symbols, fore some k, $k \geq \ell(\mathcal{A})$. The probability of staying in a diagonal state is at least the probability of having k non–noisy transitions while the probability of returning from B to G is the latter multiplied by the probability of a reset in $\mathcal{A}$. Thus for every t,

$$P_{t+k}(G) \geq (1-\epsilon)^k(P_t(G) + R(k)(1-P_t(G))) \tag{8}$$

The result follows from the well-known fact that if a positive sequence $\{x_k\}$ satisfies $x_{n+k} > \alpha x_k + \beta$, $0 < \alpha < 1$, then

$$\liminf_{n\to\infty} x_n > \frac{\beta}{1-\alpha}$$

∎

Corollary 4 *For every synchronizing automaton* $\mathcal{A}$

$$\lim_{\epsilon\to 0} \rho(\mathcal{A}, \mathcal{A}') = 0 \tag{9}$$

The significance of this result is in showing that the "global" accuracy of computations with resets can always be improved by decreasing the "local" noise. This means that computational tasks that fall into this category can be made nore and more reliable by improving the components realizing them, e.g, by using redundant sensors, faster processors, etc. In the next section we will see that in other cases the presence of a local noise, no matter how small, causes a large global deviation from the correct behavior.

4 Non–Robustness of Reset–Free Automata

After establishing an upper–bound on the distance for synchronizing automata, we would like to set a lower–bound for the complementary class of reset–free automata. In the special case of permutation automata, i.e., those in which all the input letters induce permutations, we have the following lower–bound:

Claim 5 (Non–Robustness of Permutation Automata) *Let* $\mathcal{A}$ *be any* n*–state permutation automaton* $(n > 1)$. *Then*

1. *For any noisy version* $\mathcal{A}'$ *such that for every* q, σ, $Pr(\delta'(q,\sigma) \neq \delta(q,\sigma)) \geq \epsilon$ *we have*

$$\rho(\mathcal{A}, \mathcal{A}') \geq \frac{1}{2} \tag{10}$$

2. *There exist an* ϵ*–noisy version* $\mathcal{A}'$ *such that*

$$\rho(\mathcal{A}, \mathcal{A}') \geq \frac{2n-1}{2n} \tag{11}$$

Sketch of Proof:

1. The proof is similar to the previous one. This time we note that the probability of a transition from G to B is at least ϵ while the probability of moving back from B to G is at most ϵ (because of the lack of any merging sequence). Thus we have a symmetric chain that converges to 1/2.

2. We use the same argument but consider a noise pattern such that every letter that induces a permutation in $\mathcal{A}$, induces, with probability ϵ, a permutation completely different from the original one in $\mathcal{A}'$. Thus, the probability of moving from G to B is the same but the probability of correcting an error decreases from ϵ to ϵ/n. ∎

If we look at n asymptotically we see that for large permutation automata there exists a noise pattern that can make them being wrong most of the time.

Our last result concerns the whole class of reset–free automata. The analysis here is a bit more complicated because the set B of non–diagonal states divides into two subsets: W containing all the pairs which are not mergable by any sequence, and U containing those that are correctable. The synchronizing case corresponds to $W = \emptyset$, while the permutation case corresponds to $U = \emptyset$. We will denote by $R'(k)$ the probability over Σ^k of those sequences leading from U to G.

Claim 6 (Non–Robustness of Reset–Free Automata) *For every reset–free automaton $\mathcal{A}$ with n states there exists an ϵ–noisy version $\mathcal{A}'$ such that*

$$\lim_{\epsilon \to 0} \rho(\mathcal{A}, \mathcal{A}') \geq \frac{1}{1 + \frac{1}{\hat{p}}} \tag{12}$$

Sketch of Proof: Our analysis is based on the following observations: 1) There exist at least two unmergable states q, q' and since the automaton is strongly–connected,[3] there exists at least one letter $\sigma \in \Sigma$ such that for some $q^* \in Q$ we have $\delta(q^*, \sigma) = q$. Then we define a noisy version in which $\delta'(q^*, \sigma) = q'$. This means that from (q^*, q^*) we can go to $(q, q') \in W$ with a probability not smaller than $\epsilon\hat{p}$. 2) The probability of leaving W in one step is smaller then ϵ (as in the permutation case). 3) The probability of going from G to W in k steps is at least the probability of getting from every $(q, q) \in G$ to (q^*, q^*) in $k-1$ steps multiplied by $\epsilon\hat{p}$. From all this we obtain:

$$P_{t+k}(W) \geq (1-\epsilon)^k P_t(W) + \epsilon\hat{p}P_t(G)\hat{p}^{k-1}(1-\epsilon)^{k-1} \tag{13}$$

and

$$P_{t+k}(G) \geq (1-\epsilon)^k [P_t(G) + R'(k)P_t(U)] \tag{14}$$

Summing up (13) and (14) and using the fact that $P_t(U) = 1 - P_t(G) - P_t(W)$, we obtain an equation that we treat like (8) in order to show that $P_t(U)$ is negligible when ϵ is small. Thus we can replace $P_t(G)$ by $1 - P_t(W)$ in (13), let $k = 1$ and obtain the result. ∎

5 Discussion

5.1 Past

In this paper we have built a model that captures an intuitive property of computations in the presence of noise: the longer is the past history upon which a computation depends, the larger is the probability of error. The essence of the model is in considering a class of

[3]When considering asymptotic probabilistic behavior we should only care about strongly–connected components.

probability distributions on Σ^* such that the notion of expected distance between behaviors becomes meaningful. This idea, inspired by an old paper on language identification ([Wha74]), is in contrast with traditional treatment of stochastic automata in computer science (see [Rab63, Paz71]) where such automata are used as acceptors of *individual* sequences whose probability of reaching a terminal state is above some threshold. This notion of expected correctness relative to some probability on the input also underlies the *PAC*-learnability model ([Val84]) and we believe that investigating its properties can contribute to the general shift from worst-case to average-case analysis of computational phenomena.

The notion of comparing the ideal and the noisy behavior appears already in Von Neumann's seminal paper ([VN56]). In that paper a similar question of obtaining global correctness in spite of local noise is discussed and the solution of redundancy is devised. It is interesting to note that Von Neumann considered local/global relations in space, i.e., some logical gates can be noisy but the output of the whole circuit is correct, while we consider the same relationship with respect to time. Another association which comes in mind is with Dijkstra's notion of self-stabilizing system ([Dij74]) where the system can go from any incorrect configuration into a correct one after finitely-many steps.

5.2 Future

We will mention briefly several research direction that can follow this work.

- Within this model one can give a finer characterization of the distance as a function of certain semigroup-theoretic properties of the automaton (e.g., the length of the longest chain in the group-free subsemigroups of S).

- The same kind of analysis can be applied to the behavior of machines having an infinite state-space such as push-down automata and Turing machines.

- It might be interesting to investigate specific classes of automata that arise in the modeling of realistic situations, for example, automata whose state-space is embedded in a metric space and the transitions have some arithmetical or geometrical interpretation. In such a case the distance between the behaviors will be more refined than the distance induced by the discrete metric.

- Another possibility is to investigate specific patterns of noise such as those that arise when the arrival rate of the inputs is high relative to the speed of the automaton. In this case, by using methods from queueing theory, some trade-offs between the speed of the automaton, the size of the input buffer and the expected noise can be established.

- In our model we have only considered the task of mapping classes of input histories into internal states. This model can be extended into a control model by specifying the dynamics of the environment, the structure of observations (the relation between the states of the environment and the input of the program), and the effect of the automaton's output on the environment. For a discrete environment, such an extension will enhance the recent models concerning the control of discrete-event dynamical systems (see [RW89]) by adding a robustness dimension. If on the other hand we consider automata interacting with a continuous environment we come into the realm of *hybrid* systems having much more intricate relationships between time, change, observation and noise. The modeling of such systems requires a broader synthesis of computational and control-theoretic models (see [MMP92] for a preliminary effort in this direction).

Acknowledgements

We would like to thank A. Benveniste, A. Juditsky, A. Pnueli and an anonymous referee for their attention and comments.

References

[BP85] J. Berstel and D. Perrin, Theory of Codes, Academic Press, New-York, 1985.

[Dij74] E.W. Dijkstra, Self-stabilizing Systems in Spite of Distributed Control, *Comm. of the ACM* 17, 643-644, 1974.

[HU79] J.E. Hopcroft and J.D. Ullman, *Introduction to Automata Theory, Languages and Computation*, Addison-Wesley, Reading, MA, 1979.

[KS60] J.G. Kemeny and J.L. Snell, *Finite Markov Chains*, Van Nostrand, New York, 1960.

[Lal79] G. Lallement, *Semigroups and Combinatorial Applications*, Wiley, New-York, 1979.

[MMP92] O. Maler, Z. Manna and A. Pnueli, From Timed to Hybrid Systems, to appear in: Proc. of the REX Workshop "Real-Time, Theory in Practice", Springer, Berlin, 1992.

[Paz70] A. Paz, *Introduction to Probabilistic Automata*, Academic Press, New-York, 1970.

[Rab63] M.O. Rabin, Probabilistic Automata, *Information and Control* 6, 230-245, 1963.

[RW89] P.J.G. Ramadge and W.M. Wonham, The Control of Discrete Event Systems, *Proc. of the IEEE* 77, 81-98, 1989.

[V84] L.G. Valiant, A Theory of the Learnable, *Comm. of the ACM* 27, 1134-1142, 1984.

[Wha74] R.M. Wharton, Approximate Language Identification, *Information and Control* 26, 236-255, 1974.

[VN56] J. Von Neumann, Probabilistic Logics and the Synthesis of Reliable Organisms from Unreliable Components, in *Automata Studies*, 205-228, Princeton University Press, Princeton, 1956.

Temporal Logic applied to Reliability Modelling of Fault-Tolerant Systems

Klaus D. Heidtmann
University of Hamburg, Dept. Computer Science
Bodenstedtstr. 16, D-2000 Hamburg 50, Germany

Abstract

Clearly, as more sophisticated fault-tolerant systems are developed, powerful formal techniques for modelling their reliability will be necessary. It is the intention of this paper to demonstrate the usefulness of temporal logic, an extension of the traditional Boolean logic, for formal specification and probabilistic analysis of fault-tolerant computer systems. This recognized and versatile formalism can be applied to describe dynamic behaviour in a simple and elegant fashion, which also supports reliability modelling and probabilistic reliability analysis. As will be shown, temporal logic provides a natural means for describing various forms of redundant resources and fault-tolerance mechanisms, which are frequently found in fault-tolerant computer systems. In this paper a formal technique which describes qualitative aspects of systems serves as a basis for a method that finally quantifies system behaviour.

Key Words

Temporal Logic, Formal Specification, Reliability, Modelling, Analysis, Fault Tolerance, Dynamic Redundancy, Deterministic Structure, Probabilistic Evaluation

1. Introduction

As systems become more and more complex, good methods for specifying and analysing the systems and their submodules become more and more important. Particularly reliability modelling including prediction, evaluation, and control is a vital problem for proper design, dependable operation, and effective maintenance of present and future computer systems with the indispensable attribute of fault tolerance. For these systems having various forms of redundancy, the reliability and availability is extremely sensitive to even small variations in certain parameters, and understanding and insight can only be gained by modelling and analysis using formal techniques.

In reliability modelling and analysis a system is considered to be operational if specified combinations of its components are operational. Two types of mathematical models are used to predict system reliability measures: combinational and Markov models. Combinatorial models are based on two-state (structure) functions to describe the causal relationship between component and system failures which is logical in nature and can be very intuitively illustrated by success or fault trees and reliability bock diagrams. Combinatorial methods can be applied to systems with a wide range of component types (e.g. various failure distributions) and a great number of components like computer communication networks but unfortunately they are unable to represent in detail the many dynamic aspects of fault-tolerant systems that can strongly affect system reliability like cold and warm redundancy as well as fault coverage. Markov models can capture these important features of fault tolerance but as each combination of up and down components has a corresponding state the size of the state space grows exponentially with the number of different components. This precludes the modelling of large systems. Other weakpoints of Markov models are the restriction to exponentially distributed holding times and information hiding .This means that structural properties determine the state definition but do not appear in the differential equations and transition diagrams. The presented approach is intended to reduce the gap between both types of models with an extension of the combinatorial model based on temporal logic.

Especially for modelling deterministic aspects of reliability and fault tolerance nearly no improved methods have been evaluated since the introduction of Boolean structure functions [4,7] perhaps with the exception of multivalued logical functions [6,12,13]. Reliability modelling of fault-tolerant computer systems considers probabilistic and deterministic properties. The structural relationship between a system and its components is deterministic in nature and reveals the causal connection between component and system states. The current description technique considers only a fixed moment of time and assumes that the actual state of a system depends only on the actual states of the components. This static relationship between component and system states is represented by the set of paths resp. cuts sets or by logical expressions. Obviously, Boolean logic can only express static properties and its expressiveness is insufficient to cover for instance changes during a time period. Without the ability to describe dynamic behaviour it is only of minor importance for todays fault-tolerant computer systems which mostly use dynamic redundancy and provide different mechanisms to recover from actual failure scenarios. For instance, as more dependable and intelligent systems are able to detect and locate faults, substitute faulty components and reconfigure, the structure of the system changes at times when faults occur and errors are detected. Temporal logic can integrate the static and dynamic characteristics of fault-tolerant systems to reflect the potential chronological evolution of the system. This ability is very valuable for the conceptualization, requirement specification, design and analysis of fault-tolerant computer systems which are often quite complex and prone to error.

Although originally developed for application in philosophy [15,22], temporal logic has been put forward as a useful formal tool for dealing with computer programs [8,17,23] and digital hardware [9,19]. In more detail, temporal logic is gaining recognition as an attractive and versatile formalism for rigorously specifying and reasoning about computer programs, digital circuits and message passing systems. These applications with a strong qualitative aspect consider time to be discrete. In contrast, dense time temporal logic has their relevance for modelling of continuous phenomena. Reliability and availability modelling refers to continuous random variables like lifetimes (life length), time to failure, time to repair etc. Hence, the presented approach uses the linear temporal logic of continuous time called modal system S4.3. Within this framework, one can express logical operators corresponding to time-dependent concepts such as "always" and "sometimes" and so it is possible to describe dynamic behaviour in a simple and elegant fashion. As will be shown, this temporal logic provides a natural means for describing dynamic notations as sequences of failures, switching mechanisms, reconfiguration after failure, transition from cold or dormant to active component state, substitution policies, recovery procedures, repair and maintenance strategies, and others.

First a brief introduction of temporal logic is presented insofar as it is needed for the purpose of reliability modelling including the definition of temporal structure function which characterizes the deterministic relationship between component and system failures. Then expressions of temporal logic for active and passive redundancy are explained followed by a discussion of switches and sequential systems. Subsequently, the concept of temporal structure function is illustrated by important classes of dynamic systems. Finally, it is shown how temporal specifications of structural reliability facilitate subsequent probabilistic analysis and make the reliability analysis of complex dynamic systems feasible.

2. Temporal Structure Function

Assume that a system consists of n components, which are numbered from 1 to n. So the set of components can be denoted by $N = \{1,2,...,n\}$. Usually, reliability models distinguish between only two states: a functioning and a failed state. This dichotomy confines the number of atomic propositions to one for each system component; i.e. component i is functioning, $i \in N$. This proposition, denoted by x_i, is true if the corresponding component is functioning, and it is false as long as component i remains in the failed state.

Similarly, the binary variable ϕ indicates the state of the system. It is true if the system is functioning and false if it failed. We assume that the system state is completely determined by the states of the components, so that we may write

$$\phi = \phi(x), \qquad \text{where} \qquad x = (x_1, \dots, x_n).$$

The function $\phi(x)$ is called the *(static) structure function* of the system [4].

Now time is introduced to this traditional static reliability model by the use of temporal logic. Given the proposition x_i, $i \in N$, the realization of x_i at τ, denoted by $x_i(\tau)$, is defined as the evaluation of x_i at time $\tau \in T = [0,t]$. It is true if component i operates at time τ and is false otherwise. The corresponding temporal statement X_i, $i \in N$, is true if component i operates at the present moment. Usually the present moment is equivalent to time 0 but in nested expressions this moment may also refer to a time $\tau \in T$ with $\tau > 0$. Expression

$$\Box X_i$$

is true if component i is functioning at the present time and henceforth (at least until time t), while

$$\Diamond X_i \quad \text{resp.} \quad \Diamond \neg X_i$$

is true if there exists a time between now and t in which component i operates resp. failed. For these temporal operators holds

$$\Box(X_1 \; X_2) \; = \; \Box X_1 \Box X_2$$

$$\Diamond(X_1 \vee X_2) \; = \; \Diamond X_1 \vee \Diamond X_2.$$

The involved temporal operator

$$\Diamond \Box X_i$$

is interpreted as follows: either x_i is true now and in the future or there exists a future moment from which on x_i will be true henceforth. Nested temporal operators can also represent sequencies of events, for instance

$$\Diamond(X_1 \; \Diamond(X_2 \; \Diamond X_3))$$

is true if there exists a present or future moment where x_1 is true and a subsequent moment where x_2 is true followed by a moment where x_3 holds. The truth of temporal expression

$$\Diamond X_1 \Diamond X_2$$

implies no sequence of events, i.e. it is not determined if x_1 or x_2 holds first.

Throughout this paper only the linear temporal logic of continuous time is used which covers the discussed dynamic problems. This logic called modal system S4.3 contains the following axiom of inversion [22].

$$\neg\Box X_i = \Diamond\neg X_i$$

$$\neg\Diamond X_i = \Box\neg X_i$$

The following operators are relevant when the present time is different from time 0, i.e. time 0 belongs to the past.

$$\blacksquare X_i$$

is true if component i is functioning from time 0 until now, while

$$\blacklozenge X_i \quad \text{resp.} \quad \blacklozenge\neg X_i$$

is true if there exists a time between 0 and now where component i was operational resp. failed. For the reason of symmetry the properties of these past operators correspond to those of the future operators already given above.

These operators can be interpreted as functions mapping T=[0,t], a subset of the real numbers, onto {true,false}, i.e. they apply a single logical value to the logical values of a component variable for all times $\tau \in T$. Now we assume that the state of the system is completely determined by the states of the components evaluated in the given time period T. Then we can define a function χ (χ is an abbreviation for χρονος, the ancient geek word for time) which determines the consecutive behaviour of the system depending on temporal statements for the component states. The function

$$\chi = \chi(X) \quad \text{where} \quad X = (X_1, \ldots, X_n)$$

mapping $\{true,false\}^n$ onto {true,false} is called the *temporal structure function* of the system [16]. It can be extended by additional variables, for instance

$$\chi = \chi(X,Y) \quad \text{where} \quad Y = (Y_1, \ldots, Y_n)$$

to express various operational modes for system components, e.g. when components are switched on but are not under load so that their failure rate differs from components which are switched off or under load.

The deterministic aspect of dependability as discussed until now is illustrated by examples. It is desirable to describe the causal relationship between system and component states exactly by logical expressions.

For the purpose of simplified probabilistic analysis the structure function can be transformed into disjoint products. Abrahams algorithm [1] complements single variables, for instance

$$x_1 x_2 \vee x_3 = x_1 x_2 \vee \neg x_1 x_3 \vee x_1 \neg x_2 x_3,$$

while the method proposed in [14] uses complemented subproducts,

$$x_1 x_2 \vee x_3 = x_1 x_2 \vee \neg(x_1 x_2) x_3.$$

Both methods and the corresponding computer programs [14] are applicable to temporal structure functions, i.e.

$$\begin{aligned} \Box X_1 \Box X_2 \vee \Box X_3 &= \Box X_1 \Box X_2 \vee \neg\Box X_1 \Box X_3 \vee \Box X_1 \neg\Box X_2 \Box X_3 \\ &= \Box X_1 \Box X_2 \vee \Diamond\neg X_1 \Box X_3 \vee \Box X_1 \Diamond\neg X_2 \Box X_3, \\ \Box X_1 \Box X_2 \vee \Box X_3 &= \Box(X_1 X_2) \vee \neg\Box(X_1 X_2) \Box X_3 \\ &= \Box X_1 \Box X_2 \vee \Diamond\neg(X_1 X_2) \Box X_3. \end{aligned}$$

3. Specification of Active and Passive Redundancy

Consider a system which requires only one component for operation. A second component is redundant and will be substituted when the original component fails. In case of active (hot) redundancy both components are switched on simultaneously and operate in parallel. If the first component fails system operation terminates if the second component has already failed or continues until the failure of the redundant component. Consequently, the system works as long as at least one component is always functioning which can be expressed by the following temporal structure function

$$\chi_1 = \Box X_1 \vee \Box X_2$$

This expression can be transformed to a disjunction of two disjoint terms:

$$\chi_1 = \Box X_1 \vee \Box X_2 \Diamond\neg X_1$$

In the general case of n-1 redundant components one obtains

$$\bigvee_{i \in N} \Box X_i$$

Passive (cold) redundancy means that the second component is switched off as long as the first one operates and that it is switched on for substitution not before the failure of the original component. If the spare is switched on but is not under load we speak of warm redundancy. In both cases the failure of the first component implies that the second component must work henceforth, i.e.

$$\begin{aligned} \chi_2 &= \Diamond\neg X_1 \Rightarrow \Box X_2. \\ &= \Box X_1 \vee \Diamond(\neg X_1 \Box X_2). \end{aligned}$$

A system with n-1 spares can be characterized as follows.

$$\Diamond\neg X_1 \Rightarrow \Diamond\neg X_2 \Rightarrow \dots \Rightarrow \Diamond\neg X_{n-1} \Rightarrow \Box X_n$$

The previous models for reliability using spares employ formulas which assume that the spares deteriorate as rapidly as the units being used or not at all. This could be too pessimistic resp. too optimistic, so other expressions that allow for different appropriate failure rates were developed. Sometimes the probability that dormant components also fail can not be neglected. We specify the corresponding deterministic aspect with atomic temporal predicates Y_i for each passive component i: dormant component i is functioning now. The failure of the first component implies that the second component must be functioning in the dormant state as long as X_1 operates $\Diamond(\neg X_1 \blacksquare Y_2)$ and that component 2 must work henceforth $\Box X_2$. So the above temporal structure function expands to

$$\chi_3 = \Diamond\neg X_1 \Rightarrow \blacksquare Y_2 \Box X_2$$

or to the disjoint form

$$\chi_3 = \Box X_1 \vee \Diamond(\neg X_1 \blacksquare Y_2 \Box X_2).$$

Now we specify the switching mechanism of the token-ring a well-known local computer network. Here a computer is attached to a loop of the network by means of an adapter card. One task of this hardware is to forward the received data to the next station. So any adapter failure would disrupt the network disabling all further communication. Hence, a switch shortcuts the loop and closes the ring bypassing the defect component and enabling further data transfer. This means that the network is operational as long as no failures occur or failures of attached stations are bypassed. The corresponding switch (component 3) called bypass mechanism must be intact until the switchover time, i.e. until the failure of the adapter (component 1).

$$\chi_4 = \Box X_1 \vee \Diamond(\neg X_1 \blacksquare X_3)$$

4. Sequential Systems

Until now switches of fault-tolerant systems are modelled by the coverage c. This is a probability and is not usable for deterministic modelling, which should be the basis for probabilistic evaluation. Classical deterministic reliability models like Boolean expressions are unable to cover sequential systems. An essential advantage of the presented temporal logic approach is that this common characteristic of fault-tolerant computers can be described in a simple and elegant fashion.

The system of figure 1 with cascading switches is used to demonstrate the essential property of sequential systems. Here the system state depends also on the sequence of component failures. The system consists of 3 processing elements (components 1,2,3) which are connected by switches (components 4,5). System operation begins with component 1. As soon as it fails component 4 switches on the next processing element (component 2). This is only possible if the switch (component 4) survives component 1. When the next processing element (component 2) fails the last processing element (component 3) is substituted by the second surviving switch (component 5).

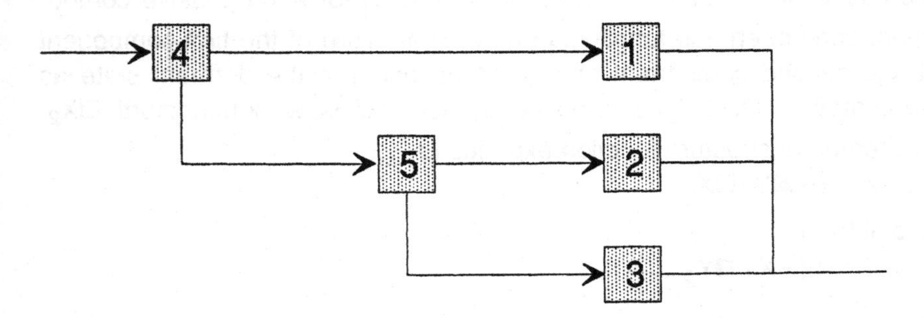

Fig. 1. Sequential system with cascading switches (components 4,5)

This system includes two sequences of component failures which imply system operation, i.e. the switch (component 4 resp. 5) fails after the failure of the corresponding processing element (component 1 resp. 2). The following temporal structure function of this sequential system illustrates this property within the brackets.

$$\chi_5 = \square X_1 \vee \lozenge(\neg X_1 \blacksquare X_4 \blacksquare Y_2 \square X_2) \vee \lozenge(\neg X_1 \blacksquare X_4 \blacksquare Y_2 \lozenge(\neg X_2 \blacksquare X_5 \blacksquare Y_3 \square X_3)).$$

The sequences of component failures which imply system failure can be derived from the good sequences above by negation. The bracket on the right side of the following equation represents a failure sequence which results in system failure, i.e. the switch (component 4) fails prior to the failure of the processing element (component 1) which he is intended to replace.

$$\neg\lozenge(\neg X_1 \blacksquare X_4) = \square X_1 \vee \lozenge(\neg X_4 \lozenge \neg X_1)$$

The term ultrahigh-reliability has been used to indicate the stringent reliability requirements demanded by these nonrepairable computer systems [11]. With such demands on the reliability measure, mathematical modelling is often the only means of assessing and comparing the reliability of two or more ultra-reliable systems. The ARIES model proposed in [20] is a unified model developed for evaluating fault-tolerant computer systems. Based on the Markov modelling technique it unifies most reliability models proposed earlier. We begin with a brief description of the ARIES model for closed fault-tolerant systems (see figure 2 and for a detailed description [3,20]). The following set of parameters (n,s,d,c_a,c_d) completely specifies the structure of the fault-tolerant system.

n = Initial number of modules in the active configuration
s = Number of spares modules
d = Number of degradations allowed in the active configuration
c_a = Recovery from active-module failure
c_d = Recovery from spare-module failure

Both active modules and spares can fail but we distinguish between them because the failure rate of the spares may be different from that of the active modules. It is possible for a module to fail without being detected. Spares are switched into the active configuration in a linear fashion. Also, there could be failed but undetected spares in the linear array. Such spares are called badly failed. The first badly failed spare in the linear array determines the number of working spares accessible to the fault-tolerant system. In the presence of a badly failed spare, the system can never be working with an active configuration less than n (n is the number of working modules in the active configuration) because ARIES conservatively assumes that the entire system fails if a badly failed spare is switched into the active configuration. This essentially means that the degradation capability is lost once the system has an undetected bad spare.

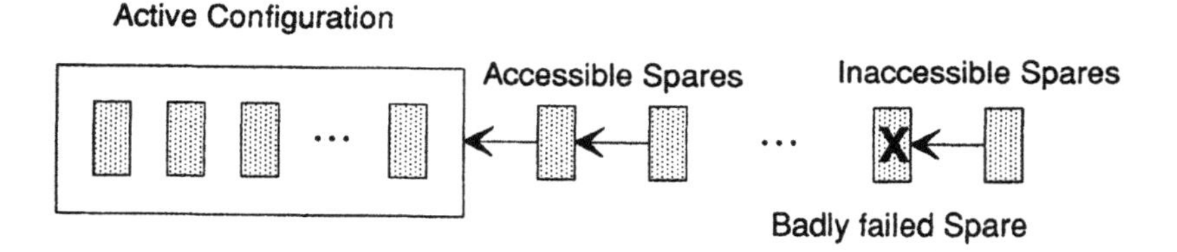

Fig. 2. Arrangement of spares in closed ARIES systems.

Now we discuss the simple example of one active module (called component 1) and two spares (component 2 and 3) with parameter set $(1,2,0,c_a,c_d)$. This fault-tolerant system is functioning if component 1 works all the time, $\Box X_1$. When component 1 fails, i.e. $\Diamond\neg X_1$, system operation continues if it recovers from active module failure C_{1a} and the first spare (component 2) is accessible $\blacksquare Y_2$ and works henceforth $\Box X_2$ or it fails $\Diamond\neg X_2 C_{2a}$ and component 3 is substituted working henceforth $\blacksquare Y_3 \Box X_3$ (see figure 3).

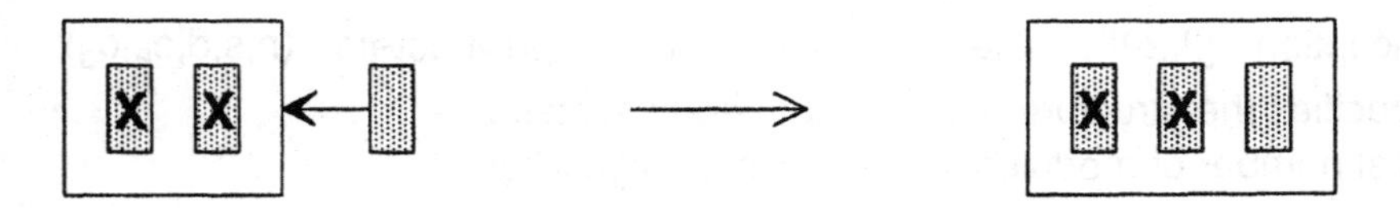

Fig. 3 System operation with two failed components (the third can be substituted)

The conjunction of the terms explained above yields the temporal structure function of the discussed fault-tolerant system.

$$\chi_6 = \Box X_1 \vee \Diamond(\neg X_1\, C_{1a} \blacksquare Y_2\, (\Box X_2 \vee \Diamond(\neg X_2\, C_{2a} \blacksquare Y_3\, \Box X_3))).$$

$$= \Box X_1 \vee \Diamond(\neg X_1\, C_{1a} \blacksquare Y_2 \Box X_2) \vee \Diamond(\neg X_1\, C_{1a} \blacksquare Y_2\, \Diamond(\neg X_2\, C_{2a} \blacksquare Y_3 \Box X_3))$$

The accessibility of component 2 is lost in situations where events occur in the following order. If component 2 fails badly, i.e. as a spare prior to its incorporation into the active configuration

$$\Diamond(\neg X_1\, \neg C_{1a} \blacklozenge \neg(Y_2 C_{2d})) \;=\; \Diamond(\neg(Y_2 C_{2d})\, \Diamond\neg(X_1\, C_{1a})),$$

then component 3 is inaccessible and the system fails (see figure 4).

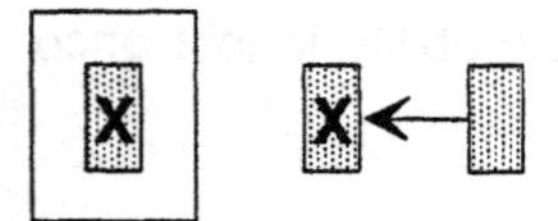

Fig. 4. Two failed components implying system failure (third cannot be substituted)

But when component 2 fails as an active component, i.e.

$$\Diamond(\neg X_1\, \neg C_{1a} \blacksquare Y_2\, \Diamond\neg(X_2 C_{2a})),$$

then the next spare component 3 can be substituted. If it is not operational at this moment, i.e. $\blacklozenge\neg(Y_3 C_{3d})$, or fails afterwards until time t, i.e. $\neg X_3$, then the system fails too. With these temporal terms deterministic specification of system failure is obtained as follows.

$$\neg\chi_6 = \Diamond(\neg(Y_2 C_{2d})\Diamond\neg(X_1 C_{1a})) \vee \Diamond(\neg X_1\, C_{1a} \blacksquare Y_2 \Diamond(\neg X_2\, \neg C_{2a} \blacklozenge \neg(Y_3 C_{3d})\, \Diamond\neg X_3))$$

5. Deterministic Modelling of Fault-Tolerant Systems

Now we add another active module to the example configuration to build a degradable system with parameters $(2,1,1,c_a,c_d)$. This means that the system works as long as one of the two active components 1 and 2 operates or as long as the spare (component 3) works after substitution for a failed active module.

$$\chi_7 = \Box X_1 \vee \Box X_2 \vee \Diamond((\neg X_1 \vee \neg X_2) C_{1a} \blacksquare Y_3 \Box X_3)$$

This can be transform to a sum of disjoint products as follows.

$$\chi_7 = \Box X_1 \vee \Box X_2 \Diamond(\neg X_1 C_{1a})$$
$$\vee \Diamond(\neg X_1 C_{1a} \blacksquare Y_3 \Box X_3 \Diamond(\neg X_2 C_{2a})) \vee \Diamond(\neg X_2 C_{2a} \blacksquare Y_3 \Box X_3 \Diamond(\neg X_1 C_{1a}))$$

The state in the Markov chain for ARIES closed system is an ordered pair where the first element denotes the number of active modules and the second element denotes the number of spares available to the system. It is assumed that there is exactly one absorbing state in the Markov chain, the failed state. ARIES assumes that the active configuration is a set of homogeneous modules and that all spares are identical. Note that in the Markov chain there are pairs of states with identical configuration (one state of each pair is underlined in the figure). These pairs are not merged together because underlined states correspond to a situation where the system has badly failed spares in the linear array. For systems which are not permitted to degrade, pairs with identical configurations are mergeable. Hence, the Markov chain of figure 3 for the simple from above has five states. The Markov chain of figure 5 for the degradable system encounters five states, one underlined.

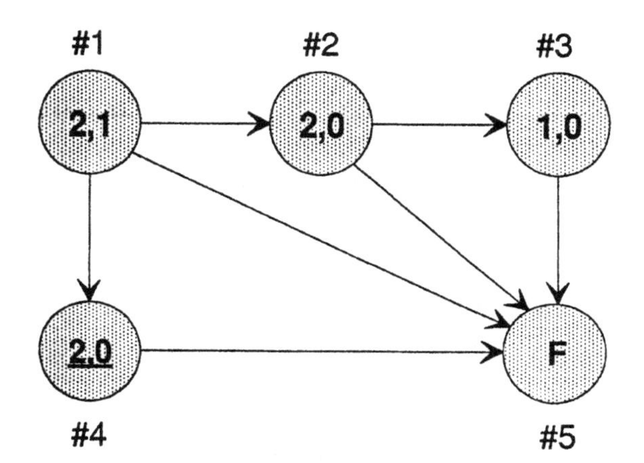

Fig. 5. ARIES Markov chain for system with parameter set $(2,1,1,c_a,c_d)$

The states are characterized by the following temporal terms.

#1 $\Box X_1 \Box X_2 \Box Y_3$

#2 $\Box X_1 \Box X_2 \Diamond(\neg Y_3 C_d) \vee \Box X_1 \Diamond(\neg X_2 C_{2a} \blacksquare Y_3 \Box X_3) \vee \Box X_2 \Diamond(\neg X_1 C_{1a} \blacksquare Y_3 \Box X_3)$

#3 $\Box X_1 \Diamond(\neg Y_3 C_d \Diamond(\neg X_2 C_{1a})) \vee \Box X_2 \Diamond(\neg Y_3 C_d \Diamond(\neg X_1 C_{1a}))$
$\vee \Box X_1 \Diamond(\neg X_2 C_{2a} \blacksquare Y_3 \Diamond(\neg X_3 C_{3a})) \vee \Box X_2 \Diamond(\neg X_1 C_{1a} \blacksquare Y_3 \Diamond(\neg X_3 C_{3a}))$

$$\vee \Diamond(\neg X_1 C_{1a} \blacksquare Y_3 \Box X_3 \Diamond(\neg X_2 C_{2a})) \vee \Diamond(\neg X_2 C_{2a} \blacksquare Y_3 \Box X_3 \Diamond(\neg X_2 C_{2a}))$$

#4 $\Box X_1 \Box X_2 \Diamond(\neg Y_3 \neg C_d)$, i.e. the spare failed badly.

The combination of the expressions for #1, #2, #3 results in χ_7. This shows that the easily derived temporal logic expression is much clearer than the representation based on the Markov chain. This fact will essentially facilitate the probabilistic evaluation in the final section.

TMR stands for triple modular redundant system, where a single faulty unit is masked by two faultfree components. This well-known structure can also be represented by a Boolean structure function which looks like the following formula without temporal operators.

$$\chi_8 = \Box X_1 \Box X_2 \vee \Box X_1 \Diamond\neg X_2 \Box X_3 \vee \Diamond\neg X_1 \Box X_2 \Box X_3$$

It is well-known that for long mission time even a single nonredundant component also called simplex system is more reliable than TMR systems. A variant of the TMR scheme, called TMR/simplex system, yields increased reliability by adopting the following strategy [10]. In a triplicated majority voted system, upon the first failure of a component, that unit is discarded: however, one of the two remaining good units is also discarded, the system from then on being operated in a simplex method. If for instance component 2 or 3 fails first then the system is identical with component 1. When component 1 fails, while both other units work, component 3 is also discarded and henceforth the system consists of unit 2. This consideration results in the following temporal structure function for TMR/simplex systems which cannot be expressed by a Boolean function without temporal operators.

$$\chi_9 = \Box X_1 \vee \Box X_2 \Diamond(\neg X_1 \blacksquare X_3)$$

A comparison to the TMR structure function with temporal operators

$$\chi_8 = \Box X_1 (\mathbf{\Box X_2} \vee \Diamond\neg \mathbf{X_2 \Box X_3}) \vee \Box X_2 \Diamond(\neg X_1 \blacksquare X_3 \mathbf{\Box X_3})$$

shows the additional conditions (bold) which imply reduced reliability. The temporal expression χ_D for the difference of both temporal structure function can easily be derived as follows.

$$\chi_D = \chi_9 \vee \neg\chi_8 = \Box X_1 \Diamond\neg X_2 \Diamond\neg X_3 \vee \Box X_2 \Diamond(\neg X_1 \Diamond\neg X_3).$$

The probabilistic evaluation of χ_D is presented in section 10 and quantifies the difference between the reliability of TMR and TMR/simplex system structure.

Another method to overcome poor reliability for long mission time is additional passive or cold redundancy. These systems combine both active and passive redundant components in a way called hybrid redundancy [5,18]. The first strategy is similar to the ARIES model except for the linear array of spares, i.e. after a failure in the active TMR configuration this structure is reconfigured as long as any operational spare is available. When the pool of spares is exhausted the system is functioning as long as at least two of the three active components remain in operation. Hence, this strategy replaces any failed active component immediately and masks faults as long as possible. Suppose for instance that a hybrid system has TMR structure and a single additional passive spare called component 4. With $N=\{1,2,3\}$ and $k \in N-\{i,j\}$ the temporal structure function of this system is given by

$$\chi_{10} = \chi_8 \vee \bigvee_{i \in N} \Box X_i \bigvee_{j \in N-\{i\}} \Diamond(\neg X_j \blacksquare Y_4 \Box X_4 \Diamond \neg X_k).$$

The second technique of hybrid redundancy reserves the spares until the degradation capacity of the active configuration is lost. In this irredundant active configuration spares are substituted for any failed component as long as possible. This means for the TMR system with a single spare (component 4) that the second failed component is replaced by this component.

$$\chi_{11} = \chi_8 \vee \bigvee_{i \in N} \Box X_i \bigvee_{j \in N-\{i\}} \Diamond(\neg X_j (\Diamond \neg X_k \blacksquare Y_4 \Box X_4)).$$

Now repair of components is regarded. Assume for instance that a system has two components. When the first component fails the second is switched on and substituted. Then the first one is repaired and can be substituted at the time where the second component fails if repair is finished. This situation is represented by the following temporal expression

$$\Diamond(\neg X_1 \Diamond(X_1 \Diamond \neg X_2))$$

So the temporal structure function of this system is given by

$$\chi_{12} = \Box X_1 \vee \Diamond(\neg X_1 X_2) \vee \Diamond(\neg X_1 \Diamond(\Box X_1 \Diamond \neg X_2))$$

Where cost prohibits sufficient fault tolerance to ensure continuous error-free operation, some amount of downtime for repair, replacement or maintenance is inevitable. In this case system failures can be tolerated if they occur infrequently and result in short system downtimes. A system is considered operational or available at a given time t if specified combinations of its components are operational at this time. The conditions that result in an available system are represented by the *temporal evolution function* which is true if the system is operational at time t. It is clear that whenever the temporal structure function χ is true, i.e. the system is continuously functioning from time 0 to t, the temporal evolution function is also valid, i.e. the system is functioning at time t. Actually additional mechanisms like repair can cause system operation at time t while operation was disrupted before. In this case χ is not valid but the temporal evolution function is true.

6. Probabilistic Evaluation of Temporal Expressions

In general, life length of components is random, and so we are led to a study of probabilistic aspects of fault-tolerant systems. Assume that L and L_i for $i \in N$ are the random variables of life length for the system and for component i. Functions

$$F(t) = Pr\{L \leq t\}, \quad F_i(t) = Pr\{L_i \leq t\}$$

are called life distributions of the system and of component i or failure distribution, where Pr represents a probability measure. f(t) and f_i (t) denote the corresponding densities. The *reliabilities* are defined as follows.

$$R(t) = 1 - F(t), \quad r_i(t) = 1 - F_i(t)$$

In the following the temporal structure function will be applied to derive the life distribution of the system from those of the components. Hence, our temporal model must be adapted to the life distributions used in the above probabilistic reliability model. For this reason we assume that the time period T lasts from time 0 to t, i.e. T=[0,t].
Now we can write straight forward

$$Pr\{\Box X_i = true\} = Pr\{L_i > t\} = r_i(t),$$
$$Pr\{\Diamond X_i = true\} = Pr\{L_i \leq t\} = F_i(t) \quad \text{for } i \in N \text{ and}$$
$$Pr\{\chi_j = true\} = Pr\{L > t\} = R_j(t).$$

If events are disjoint the probability of their union can be computed as the sum of the probabilities of the single events, e.g.

$$Pr\{\Box X_1 \vee \Diamond\neg X_1 \Box X_2 = true\} = Pr\{\Box X_1 = true\} + Pr\{\Diamond\neg X_1 \Box X_2 = true\}.$$

For statistically independent events the probability of their intersection is the product of the probabilities of the single events, e.g.

$$Pr\{\Diamond\neg X_1 \Box X_2 = true\} = Pr\{\Diamond\neg X_1 = true\} Pr\{\Box X_2 = true\} = F_1(t) r_2(t)$$

To transform other temporal terms may be some more complicated. Consider for instance the token-ring example χ_4 of section 3. Until the time when component 1 fails the switch named component 3 must be functioning. The failure of component 1 may occur at time τ between 0 and t. The failure probability of component 1 at time τ is represented by the density $f_1(\tau)$ and the probability that component 3 operates until this time is $r_3(\tau)$. So we obtain

$$\Pr\{\Diamond(\neg X_1 \blacksquare X_3) = \text{true}\} = \int_0^t f_1(\tau) r_3(\tau)\, d\tau$$

The probability of switch failure may depend on time. The approach with coverage can not cover time dependency.

A simple example of cold standby redundancy is given in the following. At the time say when component 1 fails, component 2 is switched on and substituted. This may occur at time τ between 0 and t. The failure probability of component 1 at time τ is represented by $f_1(\tau)$ and the probability that component 2 operates the rest of the time from time τ to t is given by $r_2(t-\tau)$. So we obtain

$$\Pr\{\Diamond(\neg X_1 \square X_2) = \text{true}\} = \int_0^t f_1(\tau) r_2(t-\tau)\, d\tau$$

Finally, the last two formulas are combined to consider that a passive component Y_2 may have a different life distribution r_{p2}. Assume that the distribution of the residual lifetime $L_2-\tau$ of component 2 starting at time τ with active state is denoted by

$$r_{\tau 2}(\theta) = \Pr\{L_2-\tau > \theta \mid L_2 > \tau\}$$

Now the probability of the second term in χ_3 can be derived as follows

$$\Pr\{\Diamond(\neg X_1 \blacksquare Y_2 \square X_2) = \text{true}\} = \int_0^t f_1(\tau) \Pr\{L_2 > \tau\} \Pr\{L_2-\tau > t-\tau \mid L_2 > \tau\}\, d\tau$$

$$= \int_0^t f_1(\tau) r_{p2}(\tau)\, r_{\tau 2}(t-\tau)\, d\tau$$

7. Probabilistic Analysis of Fault-Tolerant Systems

Applying these results to the system with passive (cold) redundancy specified in section 3 by the temporal structure function χ_2 yields with convolution *

$$R_2(t) = \Pr\{\square X_1 \vee \Diamond(\neg X_1 \square X_2) = \text{true}\}$$

$$= r_1(t) + \int_0^t f_1(\tau) r_2(t-\tau)\, d\tau = r_1(t) + f_1 * r_2(t).$$

For exponential life distributions with failure rate λ_i (i.e. $r_i(t) = e^{-\lambda_i t}$) this equals

$$R_2(t) = e^{-\lambda_1 t} + \lambda_1 e^{-\lambda_2 t} \frac{e^{(\lambda_2-\lambda_1)t}-1}{\lambda_2-\lambda_1}$$

and for identical components ($\lambda = \lambda_1 = \lambda_2$) this reduces to

$$R_2(t) = (1+\lambda t)e^{-\lambda t}.$$

The comparison with the system χ_1 using one active redundant component results in

$$R_1(t) = r_1(t) + F_1(t)\, r_2(t)$$
$$= e^{-\lambda_1 t} + (1-e^{-\lambda_1 t})e^{-\lambda_2 t} = (2-e^{-\lambda t})e^{-\lambda t}.$$

Considering a different failure rate δ for passive (dormant) components as in χ_3 and the "as good as new" property of the exponential distribution, i.e. $r_\tau(t)=e^{-\lambda t}$, together with the last formula of the previous section yields

$$R_3(t) = e^{-\lambda_1 t} + \lambda_1 e^{-\lambda_2 t} \frac{e^{(\lambda_2-\lambda_1-\delta_2)t}-1}{\lambda_2-\lambda_1-\delta_2}.$$

As a more complex example the probabilistic analysis of a closed fault-tolerant computer system according to the ARIES model is presented based on the temporal structure function χ_6. Coverage c is the conditional probability that, given the existence of a failure in the operational system, the system is able to recover and continue information processing with no permanent loss of essential information [2,10].

$$\begin{aligned} R_6(t) &= \Pr\{\Box X_1 = \text{true}\} + \Pr\{\Diamond(\neg X_1\, C_{1a} \blacksquare Y_2 \Box X_2) = \text{true}\} \\ &\quad + \Pr\{\Diamond(\neg X_1\, C_{1a} \blacksquare Y_2\, \Diamond(\neg X_2\, C_{2a} \blacksquare Y_3 \Box X_3)) = \text{true}\} \\ &= r_1(t) + \int_0^t c_{1a}\, f_1(\tau)\, r_{p2}(\tau)\, r_{\tau 2}(t-\tau)\, d\tau \\ &\quad + \int_0^t c_{1a}\, f_1(\tau)\, r_{p2}(\tau) \int_\tau^t c_{2a}\, f_2(\theta-\tau)\, r_{p3}(\theta)\, r_{\tau 3}(t-\theta)\, d\theta\, d\tau, \\ &= e^{-\lambda t}(1+c_a\lambda\delta^{-1}(1-e^{-\delta t}+c_a\lambda\delta^{-1}(1-e^{-\delta t}(\delta+(1-\delta)e^{-\delta\tau})))). \end{aligned}$$

Now examples using various techniques to combine active and passive redundancy are discussed to illustrate the usefulness of temporal structure functions for probabilistic analysis. First the reliability of the TMR/simplex system specified by χ_9 is evaluated.

$$R_9(t) = \Pr\{\Box X_1 = \text{true}\} + \Pr\{\Box X_2 \Diamond(\neg X_1 \blacksquare X_3) = \text{true}\}$$

$$= r_1(t) + r_2(t)\int_0^t f_1(\tau) r_3(\tau)\, d\tau$$

$$= e^{-\lambda_1 t} + \lambda_1 e^{-\lambda_2 t}\frac{1 - e^{-(\lambda_1+\lambda_3)t}}{\lambda_1+\lambda_3}$$

$$= e^{-\lambda t}\frac{3 + e^{-2\lambda t}}{2}$$

Previously, only this last expression with identical exponential life distributions was known [10]. As no formal specification method like the one introduced in this paper was available merely special expressions were discovered which could not serve for more general situations like the one above. Moreover, the comparison of TMR and TMR/simplex with the temporal expression χ_D is easy.

$$R_9(t) - R_{TMR}(t) = \Pr\{\chi_D = \text{true}\}$$

$$= r_1(t) F_2(t) F_3(t) + r_2(t)\int_0^t f_1(\tau)\,(F_3(t) - F_3(\tau))\, d\tau\,.$$

$$= e^{-\lambda t}\frac{3 - e^{-\lambda t}(6-5e^{-\lambda t})}{2}\,.$$

Hybrid redundancy is a widely used method to improve reliability and availability of computers by fault-tolerance. Temporal structure functions χ_m for m=10,11 imply

$$R_m(t) = R_{TMR}(t) + \sum_{i\in N} r_i(t) \sum_{j\in N-\{i\}} \int_0^t f_j(\tau)\, h_m(\tau)\, d\tau$$

with

$$h_{10}(\tau) = r_{p2}(\tau)\, r_{\tau 2}(t-\tau)\, (F_k(t)-F_k(\tau))\,,$$

$$h_{11}(\tau) = \int_\tau^t f_k(\theta)\, r_{p4}(\theta)\, r_{\tau 4}(t-\theta)\, d\theta$$

and with identical exponential life distributions

$$R_{10}(t) = R_{TMR}(t) + 6\lambda^2 e^{-2\lambda t}\frac{1-e^{-\lambda t}(\lambda+\delta-\lambda e^{-\delta t})\delta^{-1}}{\lambda+\delta}$$

$$R_{11}(t) = R_{TMR}(t) + 6\lambda^2 e^{-2\lambda t}\frac{e^{-\delta t}(\delta - \delta e^{-\lambda t} - \lambda) + \lambda}{\lambda+\delta}$$

The proposed method is also applicable to hybrid systems which achieve a higher degree of fault tolerance by extensive redundancy like NMR (N Modular Redundancy) or K-out-of-N (see e.g. [15]) with s additional passive spares for N>3 and s>1. A linear-time algorithm for the computation of NMR or K-out-of-N system reliability is given in [5]. It can be used to evaluate $R_{NMR}(t)$ instead of $R_{TMR}(t)$ in the previous expressions.

In all cases of exponential lifetime distributions for system components the solution for the reliability function R(t) is an analytical expression which is in the form of a weighted sum of pure negative exponentials. Hence, the solution is completely specified by a set of parameter pairs (parameters of the exponentials and the corresponding multipliers) which depend on the fault-tolerant system under evaluation.

The modelling of systems with high availability requirements such as telephone switching systems, general-purpose computer systems, transaction processing systems (e.g., airline reservation systems), and communication-network computers has received less attention. For systems like these, failed components can be repaired or replaced which may result in infrequently or short downtimes. So no mission-oriented measures like reliability can be applied and system availability is becoming an increasingly important factor in evaluating the behaviour of commercial computer systems. Commercial computer systems, which must be affordable as well as dependable, are normally designed for high availability.

Conclusion

In this paper an application of temporal logic for modelling and analysis of fault-tolerant computers was introduced. Obviously the presented approach results in a versatile and precise notation for formally specifying dynamic behaviour of fault-tolerant systems. The proposed formalism enables reliability and availability modelling in a simple and elegant fashion which was illustrated by numerous representative examples. Furthermore it was shown that the formal specification of fault tolerance using temporal logic supports reliability and availability analysis substantially. Thus the discussed technique shows how methods for quantification can profit from a formal foundation. For the purpose of practical use it seems promising to utilize the preceding investigation for the development of a CAE software tool assisting reliability and availability modelling and analysis of fault-tolerant computer systems. Moreover, it may be possible to combine the advantages of different types of reliability models including combinatorial and Markov models with the new temporal technique.

References

[1] Abraham J.A., An improved algorithm for network reliability, *IEEE Trans. Reliability*, vol. R-28, 1979 Apr, pp. 58-61

[2] Arnold T.F., The concept of coverage and its effect on the reliability model of a repairable system, *IEEE Trans. Computers*, vol. C-22, no. 3, 1973, pp. 251-254.

[3] Balakrishnan M., Raghavendra C.S., On reliability modeling of closed fault-tolerant computer systems, *IEEE Trans. Computers*, vol. C-39, no. 4, 1990, pp. 571-575.

[4] Barlow R.E., Proschan F., *Statistical Theory of Reliability and Life Testing*, Holt, Rinehart and Winston, New York, 1975

[5] Barlow R.E., Heidtmann K.D., Computing k-out-of-n structure reliability, *IEEE Trans. Reliability*, vol. R-33, 1984, pp. 322-323.

[6] Barlow R.E., Wu A.S., Coherent systems with multistate components, *Math. Operations Research*, vol. 3, 1978, pp. 275-281.

[7] Birnbaum Z.W., Esary J.D., Saunders S.C., Multi-component systems and structures and their reliability, *Technometrics*, vol. 3, no. 1, 1961, pp. 55-77

[8] Ben-Ari M., Manna Z., Pnueli A., The temporal logic of branching time, *Acta Inf.*, vol. 20, 1983, pp. 207-226

[9] Bochmann G.V., Hardware specification with temporal logic: An example, *IEEE Trans. Computers*, vol. C-31, no. 3, 1982, pp. 223-231

[10] Bouricius W.G. et al., Reliability modeling for fault-tolerant computers, *IEEE Trans. Computers*, vol. C-20, 1971, pp. 1306-1311

[11] Geist R.M., Trivedi K.S., Ultra-reliability prediction for fault-tolerant computers, *IEEE Trans. Computers*, vol. C-32, no. 12, 1983.

[12] Griffith W.S., Multistate reliability analysis, *J. Appl. Prob.*, vol. 17, 1980, pp 735-744.

[13] Heidtmann K.D., Reliability analysis of sequential two-state systems, *J. Inf. Processing & Cybernetics*, vol. 21, 10/11, 1985, pp. 547-555

[14] Heidtmann K.D., Smaller sums of disjoint products by subproduct inversion, *IEEE Trans. Reliability*, vol. R-38, 1989, pp 305-311.

[15] Heidtmann K.D., A class of noncoherent systems and their reliability analysis, *Dig. FTCS-11*, 1981, pp. 96-98

[16] Hughes G.E., Cresswell M.J., *An Introduction to Modal Logic*, Methuen, London, 1974

[17] Lamport L., Sometime is sometimes not never - On the temporal logic of programs, *J. ACM*, 1980, pp. 174-185

[18] Losq J., A highly efficient redundancy scheme: Self-purging redundancy, *IEEE Trans. Computers*, vol. C-25, no. 6, 1976, pp. 569-578

[19] Moszkowski B., A temporal logic for multilevel reasoning about hardware, *IEEE Computer*, vol. 18, no. 2, 1985, pp. 10-19

[20] Ng Y.W., Avizienis A.A., A unified model for fault-tolerant computers, *IEEE Trans. Computers*, vol. C-29, no. 11, 1980

[21] Prior A.N., *Time and Modality*, Oxford University Press, Oxford, 1957

[22] Rescher N., Urquhart A., *Temporal Logic*, Springer, New York, 1971

[23] Schwartz R.L., Melliar-Smith P.M., From state machines to temporal logic: Specification methods for protocol standards, *IEEE Trans. Communications*, vol. COM-30, no. 12, 1982, pp. 33-43

Specifying Asynchronous Transfer of Control

Padmanabhan Krishnan [1]
Department of Computer Science
University of Canterbury,
Christchurch 1, New Zealand
email:paddy@cosc.canterbury.ac.nz

Peter D. Mosses
Department of Computer Science
Ny Munkegade Building 540
Aarhus University
DK 8000, Aarhus C Denmark
email:pdm@daimi.aau.dk

1 Introduction

A principal requirement of a safety critical system is that it should be able to cope with errors and deficiencies in software and hardware. There are two main approaches in handling this viz., masking and recovery. Masking is usually achieved by replicating the hardware/software. One can either adopt strategies such as voting [Avi85] or treat part of the system as a shadow system and activate it when a fault occurs [HAH89]. Even if a subset of the components fail, the entire system can continue to function. The degree of replication depends on the criticality of the unit and the probability of failure. It is easy to see that such a technique cannot be adopted for large systems, as the cost would be prohibitively large. Recovery from hardware failures, usually results in reassigning the task on the failed unit to other unit(s) in the system. Recovery from software failures is achieved by transferring control to a recovery unit.

The general strategy for recovery can be described as follows. After a unit detects a malfunction, another unit is notified. The notified unit responds to the malfunction as soon as possible by taking appropriate action. The action it takes depends on the nature of the error and could affect other units in the system.

[Cri91] describes the various dimensions that are important in fault-tolerant computing. It does not appear to be possible to support all the issues directly in a single framework. However, one can provide a few primitives which can then be used to code the various detection/recovery techniques necessary. Asynchronous transfer of control is an important primitive and in this paper we concentrate on this aspect. As fault recovery is a high priority task, the communication between the detection unit and the handler is usually in the form of an interrupt. In this paper we describe a semantic framework for interrupts and show how different kinds of recovery actions can be specified. The model is an extension of the Action Notation [Mos90, Mos91], which supports various features including distributed computation (asynchronously communicating agents). However it does not support interrupts (or asynchronous transfer of control.)

[1] Author supported in part by grant 1787123 from The University of Canterbury, Christchurch and in part by the Danish Research Council

This paper is organized as follows. In the next section we present a brief overview of the Action Notation. In section 3 our model for interrupts is described. In section 4 the change to the operational semantics of Action Notation is described. In section 5, we present a few examples using the extended notation.

2 The Action Notation

The aim of Action Semantics, which has evolved from Abstract Semantic Algebras [Mos82], is to allow descriptions of realistic programming languages. It uses the Action Notation to specify elementary actions and techniques for combining them. Actions are objects which when performed process information and are used to represent semantics of programs. Actions can be combined using the action combinators to derive a compositional semantics.

Actions are classified into the following facets: 1) Control 2) Functional 3) Declarative 4) Imperative and 5) Communicative. We give a brief and informal introduction to the above facets.

The control actions include **complete**, **diverge**, **fail**, **escape**, **commit**. **complete** is an action that always terminates, while **diverge** never terminates. The **fail** action indicates abortive termination and is used to abandon the current alternative. The **commit** action corresponds to cutting away all alternatives, while **escape** corresponds to raising an exception.

The combinators include **or**, **and**, **and then** and **trap**. **or** represents non-deterministic choice. An alternative to the chosen action is performed when the chosen action fails (unless a **commit** has been performed.) **and** is an combinator which performs two actions with arbitrary interleaving. **and then** corresponds to sequential performance, while **trap** corresponds to handling the exception.

The functional actions process transient (as opposed to input/output) data and give/are given data. The actions include **give** D which yields the datum D, **regive** which gives any data given to it. **choose** D gives an element of the data of sort D. The principal combinator is **then**. A1 **then** A2 corresponds to functional composition, i.e., A2 is given the data produced by A1.

The declarative actions process scoped information. The actions include **bind** T **to** D, which produces a binding of token T to datum D and **rebind** which reproduces all the bindings it received. The combinators include **moreover**, **hence** and **before**. A1 **moreover** A2 corresponds to letting bindings produced by A2 override those produced by A1. A1 **hence** A2 restricts the bindings received by A2 to those produced by A1. A1 **before** A2 corresponds to letting bindings accumulate.

The imperative actions deal with storage (consisting of cells) which is stable information. The actions include **store** and **allocate**. The action **store** D1 **in** D2 stores the datum D1 in cell D2 while **allocate** D corresponds to the allocation of a cell of sort D.

The action notation also provides primitives to model parallelism. Agents form the basic unit of parallelism. The actions for this facet include **send** D whose effect is to send the message identified by D, **receive** D whose effect is to receive any message identified by D and **subordinate** D which corresponds to creating a agent of sort D which is then sent a message containing actions which are to be executed. As agents cannot share cells, it models virtual nodes or distributed memory systems.

The Action Notation may appear informal, but it has a formal signature and an operational semantics specified in [Mos90, Mos91]. A brief introduction to the notation and its formal semantics is presented in [Mos89]. See also [MW87, Wat87].

3 Interrupts

Interrupts can be considered as a command to a scheduler directing it to execute a certain subprogram, viz., the interrupt handler. They can be classified as either hardware interrupts or software interrupts. A hardware interrupt can be thought of as a command to the 'instruction scheduler' and changes the program counter asynchronously. Therefore, the handling of a hardware interrupt suspends all processes on the device. A software interrupt on the other hand 'suspends' only the process for which the interrupt is intended. Conceptually, there appears to be little difference between hardware and software interrupts. However, if a distinction between distribution and interleaving is made a distinction between hardware and software interrupts is necessary.

In the Action Notation, an agent represents a processing element and actions executed by different agents can overlap in time. Hence, agents can be used to model hardware components of the system. The notation does not directly support the notion of processes (as in operating systems.) But unnamed processes can be modeled. For example, the fork-join structure (the figure on the left) can be represented as the action on the right.

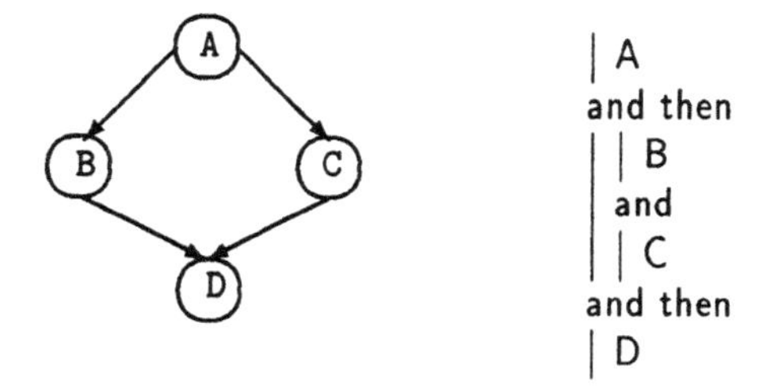

Interprocess communication between these unnamed processes can occur via shared variables. It can also occur via message passing if each process receives messages only of a particular sort and distinct processes operate on distinct sort of messages, i.e., the sort of message acts as process identifier. For example, B can execute **receive [For-B] message**, while C can execute **receive [For-C] message**. However, no direct naming scheme is supported by the notation.

Agents do not share memory and communicate solely via messages. That is, the Action Notation assumes a distributed memory model. Thus, hardware interrupts have to be modeled as messages. Modeling both hardware and software interrupts as messages gives a unified framework in which to study interrupts.

The interrupt handler can either be supplied by the unit raising the interrupt or can be fixed by the unit receiving the interrupt. As our aim is to describe fault-tolerant systems, we adopt the former option. The unit detecting the fault has a general idea of what went wrong and it pieces together a handler based on the information available. Thus, in our model interrupts are more like remote executions [SG90] than remote procedure call

[BN81]. However, this is not a recommendation for an implementation strategy; rather, it should be considered to be a technique for specifying interrupts.

3.1 Hardware Interrupts

A hardware interrupt is modeled as a special sort of message. The message contains the interrupt name and the procedure to be executed as the handler (an abstraction). The receiving agent proceeds as normal till it receives an interrupt message. It then executes the handler contained in the message. The interrupt handler should have the power to terminate the current computation. One technique is for the handler to escape or to fail and to specify the continuation between the handler and the rest of the computation as and then. This, however, results in an abnormal termination for the entire computation. If the handler terminated normally and the continuation was trap, the entire computation terminates normally and the original computation is aborted.

An interrupt handler can be activated 'asynchronously' with respect to the rest of the computation. For example, consider the following action: (A1 then A2) and (A3 then A4). While A1 and A3 are given the same transients, A2 and A4 receive their transients from A1 and A3 respectively. Consider the state where the action A1 has completed execution but not A3. In this state, the transients associated with A2 is not identical to the transients for A3 and A4. If an interrupt handler was invoked in this state and it has the power of altering the current transients, issues such as whether to discard the transients or overlay them need to be addressed. This unnecessarily complicates the semantics. To avoid these complications, the passing of transients and bindings from the handler to the rest of the computation is not supported. Therefore, if a handler is to affect the rest of the computation, it must alter the store or history (the stable state of the computation). If the continuation is trap, the original computation is resumed. Associated with the escape is a datum (identifying the cause), which is passed as a transient to the trap handler. As the original computation cannot receive any new transients, the data associated with the escape is lost. Thus escape in an interrupt handler is only a technique to restart the suspended process.

The semantics of such a computation is no different from the usual interleaving (and) semantics. From an implementation view point, this restriction is quite obvious. The transient data and binding represent register values and the execution of an interrupt handler requires the saving/restoring of registers and changes by the handler are to the memory.

As an interrupt handler indicates a high priority task, the handler must not be interleaved with the suspended computation. It must be finished before the rest of the task is resumed. Therefore, the continuation combinator for resumption is and then. The choice of the continuation combinator is made by the unit generating the interrupt as part of the interrupt message.

Note that if interleaving is to be permitted one can use the and combinator. However, the and combinator is not 'fair' due to which the intuition behind interrupts is lost. Such behavior can be simulated in the current version of the notation by sending a message which is removed and enacted by a polling loop. But the execution of this message (handler) is not guaranteed.

Usually, masking and unmasking accompany interrupts. Masking of a particular interrupt allows executing a piece of code without being affected by that interrupt, while unmasking makes the code interruptible. Masking is necessary for 1) atomicity and 2) predictability. Certain code fragments (such as data-base updates) may represent critical sections and should be executed 'atomically'. This can be achieved by masking all interrupts before executing the code and resetting them on completing the critical section. In real-time systems predictability is an important concern. Therefore, it is essential that an interrupt is not handled during a time critical computation.

As masking/unmasking is not supported in the Action Notation we define the following extensions. Define mask D (where D is a set of interrupt names) as setting a mask for all interrupts in D and unmask D as reseting the mask for interrupts D.

The operational semantics of interrupts should require that as soon as an unmasked interrupt message is detected, the agent's normal processing is suspended and the interrupt handler activated. To identify a message as a hardware interrupt we define a sort restriction interrupt _. For example, [interrupt][From-Disk] message identifies a sort of hardware interrupts called From-Disk. The operational semantics of the Action notation does not have a construct which forces the presence of an item in the buffer to execute an action. Hence, one has to change the operational semantics to force the execution of the handler. To avoid race conditions, the activation of an interrupt handler masks interrupts of the same name. This prevents the handler(s) from getting interrupted by the same interrupt before it can do any useful work. If the handler is to be interrupted by the same type of interrupt it explicitly unmasks the interrupt.

3.2 Software Interrupts

While hardware interrupt messages identify the agent to be interrupted, software interrupts only identify the sort of messages. There is no analog of names for processes. To identify processes which can be interrupted by software, we extend the action notation to include listening to D A, where D is a set of interrupt names and A the action. Intuitively, the execution of action A can be interrupted by any software interrupt in D. That is, the execution of listening to D A makes the execution of the action A sensitive to the interrupts named in D. As in the hardware case, the continuation can be and then or trap. Note that nesting of listening to is not the same as listening to of the union of the signal names. For example,

listening to D1
    listening to D2 A $\neq$ listening to (D1$\cup$D2) A .

This is because if a signal in D2 occurs the handler in the nested case can be preempted by a handler for a signal in D1. This is not the case in the union case.

A software interrupt should interrupt only the process for which it is destined. Therefore, the arrival of a software interrupt at an agent does not immediately force an asynchronous transfer of control to the handler. Only when the relevant process is executed is it interrupted. To identify a message as a software interrupt define a sort restriction signal _. For example, [signal][kill-9]message defines a sort of software interrupts called kill-9.

Software interrupts do not have the notion of masking and unmasking. Processes are susceptible to signals only if they indicate so. To avoid race conditions in software interrupts, the handler is impervious to signals unless it explicitly exposes itself. However, the original process does not lose its ability to be interrupted on resumption. This is because unlike hardware masking, software 'masking' does not change the stable state.

In the next section, we describe the changes to the operational semantics to support interrupts.

4 Operational Semantics

The main features of the operational semantics for the Action Notation are as follows. The global state of the distributed computation is captured by an entity of the form **processing** C S E, where C is the state of the communication medium (i.e., the messages that are sent but not yet delivered), S is called the **stating** component and represents the state transition(s) being performed and E is the set of agents that are active. The local state of an agent is denoted by **state A s h** where A is the action being executed along with the transients and bindings, s the storage and h the history. **step A s h c** which changes the state to **A s h**, and sends the message in **c** represents a local transition. **stepped** given a state gives the next **step**. An auxiliary function **propagated** is defined which handles the propagation of transients and bindings and termination details.

For example, the following rules help to define the semantics for **and**.

(1) **stepped state A1 s h :- step A1' s' h' c' ⇒**
 stepped state ⟦ A1 "and" A2 ⟧ :- propagated step ⟦ A1' "and" A2 ⟧ s' h' c'

(2) **stepped state A2 s h :- step A2' s' h' c' ⇒**
 stepped state ⟦ A1 "and" A2 ⟧ :- propagated step ⟦ A1 "and" A2' ⟧ s' h' c'

Recall that the **and** combinator defines the interleaved execution of two actions. The first rule states that if the state **A1 s h** can make a transition to the state **A1' s' h' c'**, ⟦**A1 "and" A2** ⟧ s h can make a transition to ⟦**A1' "and" A2** ⟧**s' h' c'**. The second rule specifies the progress of **A2**. Note that the ':' can be interpreted as '→' as in labeled transition systems.

4.1 Hardware Interrupts

The main transition rule for an agent is as follows.

(1) **stepped state A s h :- step A' s' h' c;**
communications of c' :- C' ;
stating state A' s' h's :- S'
 ⇒ step-stating processing C set (state A s h) E :-
  processing union(C, C') set(S') E .

The idea is that given a local state consisting of acting A with state s and history h, which can make a transition to state A' s' h' and communicate C', the global state is changed appropriately. To support interrupts the above transition rule is divided into two rules. The first transition rule is as above but with the additional check that the history has no pending interrupt message that can be handled, while the second rule activates an interrupt handler that is present in the buffer.

Before describing the transition rules, we need to address the issue related to masking. The masking vector is modeled as a cell (masking-vector), which can contain a set of interrupt names. The masking vector cannot be modeled as a transient or a binding as they are scoped information. The masking vector should be visible in all scopes and hence a part of the stable state. Masking an interrupt has the effect of adding the interrupt to the stored values, while umasking removes the interrupt from the stored set. The transition rules for masking/unmasking are

```
(1) evaluated D t b s h :- si : set;
    m' = union (s at masking-vector) si;
    s' = overlay(map masking-vector m',s)
        ⇒ stepped state ⟦ ⟦ mask D ⟧ t b ⟧ s h :-
              step ⟦ completed empty-map empty-map ⟧ s' h committed

(2) evaluated D t b s h :- si : set;
    m' = difference (s at masking-vector) si;
    s' = overlay(map masking-vector m',s)
        ⇒ stepped state ⟦ ⟦ unmask D ⟧ t b ⟧ s h :-
              step ⟦ completed empty-map empty-map ⟧ s' h committed
```

The main transition rule for an agent supporting interrupts is as follows.

```
(1) m = (s at masking-vector) ;
    nothing = [ not in m] [ interrupt] message & buffer of h;
    stepped state A s h :- step A' s' h' c;
    communications of c' :- C' ;
    stating state A' s' h's :- S'
        ⇒ step-stating processing C set (state A s h) E :-
              processing union(C, C') set(S') E

(2) m = (s at masking-vector);
    X: [not in m][interrupt]message & buffer of h ;
    H = Body (contents of X);
    Cont = Continuation (contents of X);
    I = Name (contents of X);
    h' = remove (X, h);
    s' = overlay(map masking-vector to union(m,I),s);
    stating state ⟦ ⟦ ⟦enact H ⟧ empty-map empty-map⟧ Cont A ⟧ s' h' :- S'
        ⇒ step-stating processing C set (state A s h) E :-
              processing C set(S') E
```

4.2 Software Interrupts

The transition rules local to a 'process' are of the from stepped state A s h :- step A' s' h' c'. As a software interrupt does not affect an 'unarmed' process, these rules need not be changed. We have to add rules to handle listening to D A which before executing A checks the current buffer for the presence of a relevant software interrupt.

(1) stepped state A1 s h :- step A' s' h' c';
nothing = [in D][signal]message & buffer of h;
⇒ stepped state ⟦ listening D A1 ⟧ s h :- step ⟦ listening D A' ⟧ s' h' c'

(2) X: [in D][signal]message & buffer of h;
H = Body (contents of X);
Cont = Continuation (contents of X);
h' = remove (X, h);
⇒ stepped state ⟦ listening D A1 ⟧ s h :-
step ⟦ ⟦ ⟦ H ⟧ empty-map empty-map⟧ Cont ⟦listening D A1⟧ ⟧ s h'

(3) stepped state ⟦ listening D ⟦ completed t b⟧ ⟧ s h :- propagated step ⟦ completed t b ⟧ s h

If there is no software interrupt, the 'process' continues to execute as usual. The presence of a relevant software interrupt activates the handler. The handler is given no datum or bindings (the empty-maps) to make the execution 'predictable'. The last rule specifies the termination behavior of the process.

In the next section we show how the extended Action Notation can be used. As the Action Notation has been primarily designed to define semantics of programming languages, we concentrate on language constructs which can be used in fault-tolerant systems.

5 Examples

Two examples are considered here. The first is the modeling of heart beats [KU87]; a simple technique in fault detection and recovery. The second is a semantics for the asynchronous 'and' suggested as an extension for Ada [RTA88].

Both these examples use time outs. This requires the specification of time in the notation. While the notation uses a definition of time for its operational semantics, it does not give access to the current time at the notation level. This is to obtain algebraic laws such as **complete and then** A is equal to A. If the access to time were allowed, **complete and then give current-time** will not be equal to **give the current-time**. This can be rectified by defining that the action **complete** takes 0 time but then one can do infinite actions in 0 time. Even if time were available, one cannot specify a timeout for an action A as A **and time-out** as the **and** is not fair. Therefore, the **time-out** action may never be executed. Thus we define our own definition of time and code time outs as necessary.

We model time as an agent, which broadcasts the 'current time' to the relevant agents. For this to map to the usual notion of time, the execution of broadcasting and the message transfer time must be 'regular'. The behavior of a a clock agent starting from an initial value of time and a fixed increment of time can be specified as follows. The clock agent first receives a message containing a list of agents which require a time service after which a message of sort Time containing the time is sent periodically.

```
Metronome Init Incr = | receive a message then
                      | | | bind %system-agents to contents of it
                      | | moreover
                      | | | bind %current-time to Init
                      hence
                      | unfolding
                      | | | Broadcast-time and then Step-Time
                      | | hence
                      | | | unfold .

Broadcast-time =
        give the datum bound to %system-agents then
        | unfolding
        | | check (it is the empty-list)
        | or
        | | check (it is not the empty-list) and then
        | | | give the datum bound to %current-time then
        | | | | send [to (head it)][Time][containing the datum] message and then
        | | | | | give (the tail of the list) then unfold .

Step-Time = give the datum bound to %current-time then
            | give the sum (Incr, the datum) then
            | | rebind moreover bind %current-time to it .
```

A local agent can obtain the current time by selecting the maximum of all the values of Time messages in the buffer and is specified below. We do not insist that the messages are deleted from the current buffer as various messages (from potentially different time agents) could be used to create a distributed time reference and specify clock synchronization [CAS86, ST87].

L is the empty-list ⇒ received-time L = 0 .

L is list(m:[Time]message) ; T: natural is (contents of m) ⇒ received-time L = T .

L is concatenation(l_1,l_2) ⇒ received-time L = maximum(received-time l_1, received-time l_2) .

current-time = received-time [Time] current-buffer .

5.1 Heart Beats

Heart beats [KU87] or watch dogs [KK88] is a common technique for fault detection. In this example we show how this technique can be modeled. We assume that there is a main process which needs service from another process which is replicated on a number of service agents (such as Proc1 and Proc2 etc). We assume that the computation starts by using Proc1. Furthermore, the standby agent to be used when the agent currently in use fails is determined from the current agent by a function Next. There is also a heart beat agent or a watch dog process, (HBC) which periodically sends a message to the service agent currently in use and delays for time Timeout. If an acknowledgement from the service agent is received, the heart beat agent continues as usual. However, if no acknowledgement is received, it assumes the service agent is no longer usable and thus interrupts the main process to reconfigure to use the standby process.

We define MPB as the main process, which initializes the system by storing the agent name Procl in the cell %service-agent and then executes the code MPC. MPC, in our example is an infinite loop consisting of performing an initial computation (indicated by Local-Task-1) followed by getting service and using the result obtained (indicated by Local-Task-2.) As we concentrate on the fault-tolerance aspect of the system and not on the computational aspects, we do not specify a behavior for Local-Task-1 and Local-Task-2. This is indicated by defining their behavior to be □. We specify a system where obtaining a service is atomic with respect to reconfiguration. (More elaborate schemes can be defined by generalizing the state information and the recovery mechanism.) Assume that MPB is executed on an agent called MP. The heart beat code (HBC) sends a Poll message to the service agent and then awaits a reply within time Timeout. If the timer expires, the waiting for acknowledgement is terminated (by the escape) and the MP agent is interrupted with the Reconfigure message. The message is an abstraction which when enacted alters the name of the service agent. The 'continuation' is and then as the original computation need not be abandoned.

```
Initialize = store Procl in %service-agent

MPB = Initialize before MPC

MPC = unfolding
        | Local-Task-1 and then
        | | Get-Service then
        | | | Local-Task-2 and then unfold

Local-Task-1 = □

Local-Task-2 = □

Get-Service = mask Reconfigure and then
              | | give the contents of %service-agent
              | then
              | | send [to the agent][Request]message and then
              | | receive[from the agent][Response]message and then
              | | unmask Reconfigure

HBC = unfolding
        | | Send-heart-beat and then
        | | Start-Timed-Check Timeout
        | trap
        | | | check (the datum is Okay) and then unfold
        | | or
        | | | check (the datum is Dead) and then
        | | | Change-agent and then unfold
```

```
Send-heart-beat = give the contents of %service-agent then
                  send [to the agent][Poll]message .

Start-Timed-Check D =
        give the sum (current-time, D) then
        | patiently
        | | | check (the current-time is less than it) and then
        | | | Is-Ack-Present
        | | or
        | | | check (the current-time is not less than it ) and then
        | | | give Dead then escape .

Is-Ack-Present = | give the contents of %service-agent then
                 | | choose [from the agent][Ack] message then remove it
                 and then
                 | give Okay then escape

Change-agent = | give the contents of %service-agent then
               | give Next it
               then
               | | send [to MP][interrupt][Reconfigure][Handler the agent ]message
               | and
               | | store the agent in %service-agent .

Next Ag = □

Handler Ag = Message(Body Ag,"and then")

Body Ag = abstraction
          | store Ag in %service-agent and then unmask Reconfigure

Service-agent = | unfolding
                | | receive [poll]message then
                | | | send[to sender of it][Ack]message and then unfold
                and
                | unfolding
                | | receive [Request]message then
                | | | send [to the sender of it][Response]message and then unfold .
```

5.2 Asynchronous And

The need for asynchronous transfer of control in Ada especially for mode changes has been discussed in [RTA88]. One of the proposals [Taf89] augments the select statement with an "and" clause. An example is

```
select
        delay D; Sd;
    or
        accept E1; S1;
```

```
        or
            accept E2; S2;
        and
            S3;
    end select
```

The informal meaning of this construct is as follows. On reaching the select alternative, if there is no pending entry call for the accepts or the delay is non zero, execution of statement S3 is started. However, if any of the other alternatives become open (i.e., the delay expires or an entry call is issued) before the execution of S3 is completed, the execution of the remainder of S3 is abandoned and the statement associated with the open alternative (delay/entry) is executed. In this section, we present a formal semantics for the above construct which also handles the situation where the calling task and the called task are distributed.

Since the semantics requires abandoning the current execution, when an entry call is detected, it is natural to translate an entry call as an interrupt. However, it should not affect the other tasks on the agent. Therefore, an entry call is a software interrupt. The entry call also sends the appropriate statement to be executed and other code to finish the execution of the select. The 'continuation' used is **trap** so that after the handler executes, the remainder of the code associated with the select is skipped. As the entries are interrupts, all statements except the select statement are impervious to interrupts. The select statement executes the "and" alternative such that it is sensitive to possible entry calls and timer interrupt. The issuer of the entry call or the timer interrupt sends a signal message to the agent executing the select statement.

The delay is modeled by a timer agent. It receives the duration of time to delay and the body to be executed when the delay expires. The timer agent is connected to the metronome in the system. It polls for the duration to exceed the specified delay and when the specified duration has elapsed it interrupts the agent that issued the delay. In keeping with the semantics of the asynchronous and, the continuation is **trap**. However, if an entry call is made before the delay expires, the timer should be reset. This is modeled by a signal **reset**.

Towards a formal description of the "and" construct we use the following abstract syntax fragment. It is not a complete semantics for the tasking model in Ada and should be considered only as an illustration. The semantic function **Establish** creates the necessary bindings in which the execution occurs. The bindings produced for the select statement contains the set of entries and a token representing the delay statement in it (which have to be unmasked **%possible-entries**) and a mapping of the entry names and the delay alternative to the statement to be executed as part of the interrupt handler. The semantic function **Execute** defines the dynamic behavior of the construct. The execution of the select statement proceeds as follows. The set of entry names is obtained (via **Establish**). The timer is set to the appropriate delay and the body of the and branch (S) is started in parallel with the delay. If an entry call (a software interrupt message) is detected, the body associated with it is executed. As the and/delay alternatives must be abandoned, the continuation is **trap** (we assume that the body does not have an abnormal termination).

comment: Partial Grammar

```
Statement          = Select | Entry-Call | Delay-Statement | □

Select             = ⟦ "select" Delay-Statement "or" Accepts "and" Statement ⟧

Entry-Call         = ⟦Identifier " " Identifier ⟧

Delay-Statement    = ⟦ "delay" Expression Statement⟧

Accepts            = ⟦ "accept" Entry ";" Statement ⟧ | ⟦ Accepts "or" Accepts ⟧

Establish ⟦ "select" Ds "or" As "and" S2 ⟧ =
        | Establish Ds before Establish As
        hence
        | rebind and bind %possible-entries to domain of current bindings

Establish ⟦ "delay" E S ⟧ = give closure abstraction Execute S then
                            bind %delay to it

Establish ⟦ "accept" E ";" S ⟧ = give closure abstraction
                                 | Reset-Timer and then Execute S
                                 then
                                 | bind (token of E) to it

Establish ⟦ A1 "or" A2 ⟧ = Establish A1 before Establish A2

Execute ⟦ "select" Ds "or" As "and" S2 ⟧ =
        | Establish ⟦ "select" Ds "or" As "and" S2 ⟧
        before
        | give the set bound to %possible-entries then
        | | listening to it
        | | | Execute Ds and then
        | | | Execute S2 and then Reset-Timer .

Execute ⟦ "delay" E S ⟧ = | | Evaluate E
                          | and
                          | | give the closure abstraction Execute S
                          then
                          | give Timer-Message(the datum #1,the datum #2) then
                          | send [to timer-agent][containing the datum] message .

Execute ⟦ T"." E ⟧ = send [to agent of T][signal][token of E][For token of E] message

Body E = closure abstraction
         enact the datum bound to (token of E)

For E = Message (Body E, "trap")
```

```
Timer-agent =
        receive [Timer-message] message then
        | give Delay(it) and
        | give Body(it) and
        | give sender(it)
        then
        | listening to set(reset)
        | | unfolding
        | | | | check (current-time is greater than the datum#1) and then
        | | | | send [to the datum#3 ][signal][%delay][containing For-delay] message
        | | | or
        | | | | check (current-time is not greater than the datum#1) and then
        | | | | unfold

For-delay = Message (the datum#3, "trap")

Reset-Timer = send [to timer-agent][signal][reset][containing Finish] message

Finish = Message(abstraction complete, "trap")
```

6 Conclusion

We have shown how the effect of asynchronous transfer of control can be specified. While the transfer of control occurs at one agent, a remote agent can cause it via message passing. We have developed our ideas within the Action Notation framework. While the Action notation has been used to describe semantics for realistic programming languages, it does not support interrupts. But with a few notational additions and a change to the operational semantics, we have been able to model interrupts.

Further research is necessary to develop a high level language in which fault-tolerance can be specified. Such a language could involve constructs such as "Normal-Processing *on-fault F* Recovery" (a generalization of the asynchronous and). The work described here provides a framework in which the semantics of such constructs can be defined. The semantics of the construct can be defined by translating F to an interrupt and defining a handler to transfer control from "Normal-Processing" to "Recovery".

In [Kri91], we have shown how the notation can be used to specify real-time systems. Thus the extended system can be used to describe fault-tolerant real-time systems. While we have shown two examples here, further experience is necessary to gauge the applicability of the ideas in describing the semantics of general fault-tolerant languages.

Acknowledgements

The authors thank Jens Palsberg for useful comments on the preliminary version of this paper. The authors also thank the anonymous referees for useful comments.

References

[Avi85] A. Avizienis. The N-version Approach to Fault-Tolerant Software. *IEEE Transactions on Software Engineering*, pages 1491–1501, Dec 1985.

[BN81] A.D. Birrell and B.J. Nelson. Implementing Remote Procedure Calls. *ACM Transactions on Computer Systems*, 2(4):39–59, February 1981.

[CAS86] F. Cristian, H. Aghili, and R. Strong. Clock synchronization in the presence of omission and performance faults and processor joins. In *Proceedings of the Sixteenth International Symposium on Fault-Tolerant Computing*, 1986.

[Cri91] F. Cristian. Understanding fault-tolerant distributed systems. *CACM*, 34(2), February 1991.

[HAH89] M. Hecht, J. Argon, and S. Hochhauser. A distributed fault-tolerant architecture for nuclear reactor control and safety functions. In *The 10th IEEE Real-Time Systems Symposium*, pages 214,221, 1989.

[KK88] R. Koymans and R. Kuiper. Paradigms for real-time systems. In M. Joseph, editor, *Proceedings of the symposium on Formal Techniques in Real-Time and Fault-Tolerant Systems: LNCS 331.* Springer Verlag, 1988.

[Kri91] P. Krishnan. Real-time Action. In *Euromicro Workshop on Real-Time Systems*, 1991.

[KU87] J. C. Knight and J. I. A. Urquhart. On the implementation and use of Ada an fault-tolerant distributed systems. *IEEE Trans. on Software Engineering*, 13(5):553–563, May 1987.

[Mos82] P. D. Mosses. Abstract semantics algebras. In D. Bjoerner, editor, *Proceeding of the IFIP TC2 Working Conference on Formal Description of Programming Concepts II*, pages 63–88. North Holland, 1982.

[Mos89] P. D. Mosses. Unified Algebras and Action Semantics. In *STACS 89, LNCS-349.* Springer Verlag, 1989.

[Mos90] P. D. Mosses. Action semantics. Technical report, DAIMI: Aarhus University, 1990.

[Mos91] P. D. Mosses. *Action Semantics.* Cambridge University Press (in the series Tracts in Theoretical Computer Science), to appear in 1991. Available as tech report from DAIMI:Aarhus University.

[MW87] P. D. Mosses and D. A. Watt. The use of Action Semantics. In *Proceeding of the IFIP TC2 Working Conference on Formal Description of Programming Concepts III, 1986.* North Holland, 1987.

[RTA88] *Proceedings of the 2nd International Workshop on Real-Time Ada issues*, volume 8(7). ACM, Ada Letters, 1988.

[SG90] J. Stamos and D. Gifford. Remote evaluation. *ACM Transactions on Programming Language and Systems*, 12(4):537–565, October 1990.

[ST87] T. K. Srikanth and S. Toueg. Optimal clock synchronization. *Journal of the Association of the Computing Machinery*, 34(3):626–645, July 1987.

[Taf89] T. Taft. Asynchronous event handling: Revision request. Technical report, Intermetrics Inc, March 1989.

[Wat87] D. A. Watt. An Action Semantics of standard ML. In *Mathematical Foundations of Programming Language Semantics: LNCS 298*, pages 572–598. Springer Verlag, 1987.

Protocol Design by Layered Decomposition

A compositional approach

Wil Janssen Job Zwiers

University of Twente,
Dep. of Computer Science
P.O. Box 217,
7500 AE Enschede, The Netherlands
E-mail:{janssenw,zwiers}@cs.utwente.nl

Abstract

A version of the two phase commit protocol is formally derived from its specification. The design starts with an *initial design phase* that properly reflects the *logical* structure of the protocol as a sequence of *layers*. Thereafter algebraic transformations are applied, resulting in an implementation that matches the physical structure of the network. Substantial use is made of an algebraic formulation of the communication closed layers design principle.

1 Introduction

Fault-tolerance for distributed systems requires the execution of complicated protocols. Examples are the various commit protocols for distributed database systems [BHG], the protocols for (re) building spanning trees for networks with nodes and interconnecting links that can fail [GHS][SR], and protocols for atomic broadcast [CASD]. The analysis of such protocols is often hampered by the fact that the *physical program structure*, in the form of a number of parallel executing sites each executing a sequential program, does not comply with the *logical structure* of the protocol. This protocol structure is often best understood as a sequential composition of a number of *phases*, where each phase consists of a number of actions distributed over the whole system. If we take for example protocols for *atomic broadcast* we have that every value broadcast arrives at every node in what might *logically* be seen as a single atomic action. This single action however consists of a number of actions executed at different nodes at different points in time. Therefore several broadcasts can execute simultaneously and overlap in time, although *conceptually* they are executed one after the other.

Thus, at first sight it appears that *compositional* methods, that exploit the program structure, are not of much help here. Critique along these lines on the compositional and process algebraic style of development can also be found in the work on *communication closed layers* by Elrad and Francez [EF], related work by Stomp and de Roever [SR], and in the work on Unity [CM].

We show in this paper that design principles such as communication closed layers can be formulated in a compositional style though. We illustrate this by deriving the well known Two Phase Commit protocol from its specification by combined use of assertional methods and algebraic transformation rules, one of which is actually a formulation of the communication closed layers principle as an algebraic law. The derivation starts with an initial design that closely follows the logical protocol structure that has the form of two phases, each of which can be decomposed into smaller subphases. The transformations applied thereafter yield a final program that matches the physical network structure, that has the form of distributed nodes connected via a spanning tree network.

The language and the underlying *partial order model* for distributed systems is a slightly improved version of the model introduced in [JPZ]. The model we use makes a sharp distinction between *logical precedence* and *temporal precedence* of events. This distinction is crucial to formulate the proper notion of *action refinement* and the so called *conflict composition operation*, both of which are used extensively in the development of the Two Phase Commit protocol.

The Two Phase Commit protocol is an example of *atomic commit protocols* that are used in distributed databases to guarantee *consistency* of the database. A distributed database consists of a number of sites connected by some network, where every site has a local database. Data are therefore distributed over a number of sites. In such a distributed database system *transactions* are executed which consist of a sequence of read and write actions. Reading and writing database items can be done by forwarding the action to the site where the item is stored. Terminating the transaction however involves *all* sites accessed in the transaction, as all sites must agree on the decision to be taken (to *commit* or to *abort*) in order to guarantee consistency. In the case of an abort all changes made by the transaction are discarded, in the case of a commit they are made permanent. A protocol that guarantees such consistency is called an atomic commit protocol.

Such an atomic commit protocol can be implemented by an algorithm consisting of two phases. A first phase in which all sites are requested to give their vote whether they can commit or not, and a second phase in which, after a decision has been taken, all sites are informed of the decision. This protocol is known as the *Two Phase Commit protocol.*

In this paper we derive an implementation of the Two Phase Commit protocol from its specification. The basic structure of the initial design is the composition of a number of actions. These actions are refined into *layers* consisting of a number of actions executed in parallel. Using the communication closed layers law and other algebraic laws we transform this system with a *layered structure* to a system with a *distributed structure* of the form

$$P_0 \parallel P_1 \parallel \cdots \parallel P_n$$

The transformation is correctness preserving and we can therefore reason in terms of the layered system, for which well known proof methods for sequential systems can be used!

The structure of this paper is as follows. After this introduction we informally introduce the language used throughout the paper, and the underlying model. We also give some properties of the language constructs introduced and provide algebraic laws used in the transformation and derivation process. After summarizing these laws we formulate our atomic commitment problem and derive an solution for it following the lines of the design trajectory sketched above.

Acknowledgement

We would like to thank Mannes Poel for fruitful discussions and his detailed comments on earlier verions of this paper.

2 The formalism

We describe a simple formalism that is used for the design of the two phase commit protocol. We discuss the language, its underlying model, and verification and transformation rules. The *language* that we use is intended to be appropriate for the *initial design stage* – during which we prefer to have no bias towards a certain network architecture – as well as for the description of the final program that *should* fit the network structure. The reason for having a single language rather than two separate languages, one for initial design and another for coding the final program, is that we aim at a *gradual* transformation from initial design towards final implementation, which requires a single language that can represent all stages, including intermediate ones. Since we introduce some rather unconventional language operators, which are difficult to appreciate without a basic knowledge of the underlying model, we start with a sketch of the latter.

The model

Basically, we describe the execution of distributed systems by *histories* h that consist of a partially ordered set of *events*. This model is related to the *pomset* model as introduced in [Pratt]. Typical examples of events that we actually use in this paper include send and receive actions and read and write operations to local or shared memory. The precise interpretation of an event e is determined by its *attributes* $a(e)$, some of which will be mentioned below. For each system many different histories are possible, due to different behaviour of the concurrent environment of the system and other causes of nondeterminism. Therefore a system *semantically* denotes a *set of possible histories*.

Events e and f that are *unordered* in some history h, are said to be independent. Potentially such events execute in parallel, i.e. at the same time or at overlapping time intervals. Within our design formalism there are two causes for *ordering* events which consequently do *not* execute in parallel:

- The first one is because e and f *conflict* in the sense that they both access a common resource that does not allow simultaneous access. The generic example (and the terminology) stems from conflicts between concurrent database actions[BHG] due to read and write operations to the same shared memory locations. When this happens e and f simply *cannot* execute (fully) in parallel and so must *logically* be ordered, which we denote as either as $e < f$ or as $f < e$, depending on which is the case. Only conflicting actions are ordered logically.
- The second cause is that actions are *temporally ordered* as the result of the use of language operators that explicitly require such ordering. Such operators are typically used in the last design stage where actions are actually *allocated* on specific processors, or to specific network nodes. Clearly, actions that should run on a single processors have to be ordered temporally. Temporal precedence of e over f is denoted by $e \ll f$.

Because of the sharp difference between logical and temporal precedence, conceptually from the point of view of a designer as well as from a more technical point of view, we use a formal semantic model where histories are structures of the form $(E, <, \ll)$, and where E is a set of events, with a dual ordering defined on it: $(E, <)$ is a directed acyclic graph (DAG), i.e., the transitive closure of $<$ is a *partial order* on E. $(E, \ll)$ is simply a partial order itself. The two ordering relations are weakly related. Temporal order obviously does not imply logical precedence. If two events e and f are *logically* ordered, say $e < f$, then they cannot be ordered *temporally* in the reversed direction, i.e.

$$e < f \Rightarrow f \not\ll e$$

Informally one can think of $e < f$ as e *influencing* f which cannot be the case if f completely precedes e in time. Any stronger relationship cannot be assumed; for instance from database serializability theory there are well known examples of atomic transactions e, f and g such that $e < f < g$, yet $g \ll e$!

Informal semantics and algebraic properties

The two main composition operators of the language, parallel composition and conflict composition, are defined purely in terms of logical precedence, i.e. no temporal order is enforced by these operators.

The histories for a *parallel composed* system S of the form $Q \parallel R$ can be described as follows. The events executed by S is some history h can be partitioned into subhistories h_Q and h_R that are possible histories for Q and R. Moreover, the logical precedence relation *between* h_R and h_Q is such that all conflicting events are logically ordered, where the direction of the precedences are nondeterministically chosen. This nondeterminism is constrained of course by the fact that logical precedence is an *order*, so cycles of the form $e_0 < e_1 < \cdots < e_0$ are not allowed.

Conflict composition can be considered an *asymmetric form of parallel composition*. For $Q \parallel R$ the logical precedence between conflicting actions of Q and R is nondeterministically determined. For conflict composition of Q and R, which is denoted by $Q \cdot R$, actions from Q take logical precedence over actions from R in case of conflicts. In the case of independent actions no order is enforced however, just as is the case for parallel composition.

Conflict composition should be compared with *sequential composition* of the form $Q\,;\,R$. This is somewhat like conflict composition except that we also enforce *temporal ordering* between Q and R actions: all Q actions temporally precede all R actions, regardless of conflicts. So whereas conflict composition admits parallel execution of certain actions, sequential composition does not. A sharp difference between the two forms of composition shows up when we consider Elrad and Francez' "communication closed layers" [EF]. The essence of comunication closed layers is that under certain conditions a parallel system $S \parallel T$ where S and T are sequential programs of the form $S_0\,;\,S_1$ and $T_0\,;\,T_1$, is "equivalent" to a sequential composition of "layers" $S_0 \parallel T_0$ and $S_1 \parallel T_1$, thus:

$$(S_0\,;\,S_1) \parallel (T_0\,;\,T_1) \equiv (S_0 \parallel T_0)\,;\,(S_1 \parallel T_1) \qquad (*)$$

The side condition is that there is no communication, or in our parlance *no conflict*, between actions from S_0 and T_1, nor should there be conflicts between action from S_1 and T_0. Generalized forms of this principle appear also in [SR]. The equivalence used in $(*)$ is sometimes called IO-equivalence, referring to the fact that although the histories of left hand and right hand sides of $(*)$ are not the same, the relation between initial and final states of the system *is* the same nevertheless. A problem with this equivalence is that it is not a congruence, so we cannot simply interchange left and right hand side of $(*)$ *within contexts!* Within our framework we can replace the sequential composition in $(*)$ by conflict composition however, resulting in the following algebraic law (with the same side conditions as for $(*)$).

$$(S_0 \cdot S_1) \parallel (T_0 \cdot T_1) = (S_0 \parallel T_0) \cdot (S_1 \parallel T_1) \qquad \text{(CCL)}$$

Note that we not only have a congruence, but even semantic *equality* here, which is to be understood as the fact that both sides of the equation admit exactly the same partial order based histories. (Semantically, a process S denotes a set of histories $[\![S]\!]$; consequently, an equation like $S = T$ is interpreted as $[\![S]\!] = [\![T]\!]$.) The derivation of the two phase commit protocol relies also on three simplified forms of the CCL law. They are obtained by substituting the **skip** process for S_0, S_1 or for S_1 and T_1. The **skip** process is the process that performs *no* action. It satisfies the following laws:

$$\mathbf{skip} \parallel P = P \parallel \mathbf{skip} = P$$

$$\mathbf{skip} \cdot P = P \cdot \mathbf{skip} = P$$

$$\mathbf{skip} \; ; P = P ; \; \mathbf{skip} = P$$

By simplification based on these laws we obtain the following derived laws: Let P, Q, and R be processes with no conflict between P and Q actions. Then:

$$\begin{array}{rcll} P \parallel (Q \cdot R) & = & Q \cdot (P \parallel R) & (i) \\ P \parallel (R \cdot Q) & = & (P \parallel R) \cdot Q & (ii) \\ P \parallel Q & = & P \cdot Q & (iii) \end{array}$$

We continue with a discussion of the *refinement operator* and the related *contraction operator*. Events as they occur in histories are to be regarded as "atomic" though that does not prevent us from *refining* such events into sub histories provided that the overall effect "appears" to be atomic. Let us make this precise. In interleaving models of concurrency, "atomicity" of the execution of S, as enforced for instance by the well known "atomic brackets" construct $\langle S \rangle$, is obtained by allowing no (potentially interfering) environment action in between any of the actions of S. For our partial order model this turns out to be too restrictive; environment actions executed *temporally* in between two S actions are harmless as long as the "in betweenness" does not hold on the level of the *logical* precedence relation. This means that if some environment event e conflicts with one or more $\langle S \rangle$ events, then it should either logically precede all of them or else logically follow all of them. In such cases it is possible to pretend that e either (logically) precedes or follows the whole of S, and so S appears to be atomic. These ideas form the basis for the refinement operation in the language, which is of the following form:

$$\mathbf{ref}\ a\ \mathbf{to}\ T\ \mathbf{in}\ S$$

Informally, occurrences of "atomic" action a in system S are refined to the process T. When T is executed, interference as described above is guaranteed to be absent, that is, conceptually one can regard this construct as substitution of $\langle T \rangle$ for a in S.

Our notion of refinement has the following attractive property. If we refine events a and b into T and U in a situation where a logically precedes b the result is not that all T events logically precede all U events; rather the T events logically precede those U events with which they *conflict*. We can paraphrase this by saying that inheritance of logical precedence is *conflict based*. The result of this is that refinement does distribute over conflict composition and over parallel composition, thus:

$$\mathbf{ref}\ a\ \mathbf{to}\ T\ \mathbf{in}\ (R \cdot S) = (\,\mathbf{ref}\ a\ \mathbf{to}\ T\ \mathbf{in}\ R) \cdot (\,\mathbf{ref}\ a\ \mathbf{to}\ T\ \mathbf{in}\ S)$$

$$\mathbf{ref}\ a\ \mathbf{to}\ T\ \mathbf{in}\ (R \parallel S) = (\,\mathbf{ref}\ a\ \mathbf{to}\ T\ \mathbf{in}\ R) \parallel (\,\mathbf{ref}\ a\ \mathbf{to}\ T\ \mathbf{in}\ S)$$

This is in contrast to the usual form of refinement in partial orders, e.g. as in [GG], where the order inheritance is full.

Inheritance of *temporal precedence* though is *full*, meaning that when actions a and b are refined as above, and a temporally precedes b, then *all* T events temporally precede *all* U events. As a result of this, refinement does also distribute over sequential composition:

$$\textbf{ref } a \textbf{ to } T \textbf{ in } (R\,;\,S) \;=\; (\textbf{ ref } a \textbf{ to } T \textbf{ in } R)\,;\,(\textbf{ ref } a \textbf{ to } T \textbf{ in } S)$$

Finally we discuss the notion of *contraction* for histories which is closely related to IO-equivalence already mentioned above. A history h in the shared memory model represents a number of local state transformations associated with the events of h. In the absence of interference from outside environment events we can determine the cumulative effect of all these local state transformations. The result is a history $contr(h)$, called the contraction of h, that contains a single atomic event that represents this cumulative, global state transformation. The semantics of the contraction operator $\mathbf{contr}(S)$ is to apply this contraction pointwise, i.e. to all histories of the system $\langle S \rangle$. (The atomic brackets are included here to guarantee absence of interference) The term $\mathbf{contr}(S)$ represents the IO-behaviour of S, in the absence of interference from outside. So IO-equivalence $S \equiv T$ can be formulated in our language as $\mathbf{contr}(S) = \mathbf{contr}(T)$.

IO-behaviour of a system can be *specified* by means of classical pre- and postconditions. We interpret a Hoare style formula of the form:

$$\{pre\}\; S\; \{post\} \qquad (**)$$

where *pre* and *post* are *state formulae* as usual, as follows. For each history h in $\mathbf{contr}(S)$ let $s_0(h)$ and $s(h)$ denote the initial and final state of the (unique) S event in h. Then (**) requires that if the initial state $s_0(h)$ satisfies precondition *pre* the corresponding final state $s(h)$ satisfies the postcondition *post*. Hoare style program verification for concurrent systems is more complicated than verification of sequential programs due to the possibility of interference. The classical proof system for shared variables by Owicki and Gries [OG] for instance includes extra interference freedom checks for assertions used in proof outlines. It has been shown by Apt and Olderog [AO] that for *restricted cases* it is possible to verify parallel programs relying on techniques for *sequential* programs however. This work relies on classical Hoare style verification in combination with program transformation based on IO-equivalence. We use similar techniques in the derivation of the two phase commit protocol, where we exploit the fact that conflict composition, although it does admit parallelism, behaves just like sequential composition when we apply the **contr** operation! This follows from the fact that the contraction of some history h can be determined without taking temporal ordering into account; logical precedence as such is sufficient to determine the cumulative state transformation associated with h. As a result we have the following algebraic law:

$$\mathbf{contr}(S \cdot T) \;=\; \mathbf{contr}(S\,;\,T)$$

This implies that to verify a pre-post specification for a program of the form $S \cdot T$ it suffices to verify the associated *sequential* program $S\,;\,T$. This is indeed the technique used in the verification of the protocol, where Hoare style proof outlines figure in the verification of systems built up with conflict composition, without any form of interference freedom test.

This technique can be clarified in a slightly different manner as follows. The conflict composition operators preserves "atomic" structure in the sense as expressed by the following two laws:

$$\langle S \cdot T \rangle \;=\; \langle \langle S \rangle \cdot \langle T \rangle \rangle$$

$$\mathbf{contr}(S \cdot T) = \mathbf{contr}(\ \mathbf{contr}(S) \cdot \mathbf{contr}(T))$$

Informally these laws say that to analyse the IO-behaviour of $S \cdot T$, we can pretend that the system reaches an intermediate state where all S actions have been executed and none of the T actions has yet been executed. Note that in terms of *temporal* precedence such an intermediate state need not occur at all; the picture is valid in terms of the *logical* precedence relation.

We conclude this section with a somewhat more detailed description of the shared memory model and the communication mechanism used in the description of the protocol.

Shared memory and communication

In our shared memory model the basic actions are *guarded assignments* of the form

$$b \& x_1, x_2, \ldots, x_m := exp_1, exp_2, \ldots, exp_m$$

Informally such an assignment is postponed until the guard b holds, whereafter the values of the expressions exp_i are assigned simultaneously to all x_i. So our guarded assignments are really limited forms of the well known **await** statement.

Conceptually each event corresponds to such an assignment, even when it actually corresponds to the cumulative effect of a whole history, and it is therefore basically a *state transformation*. The attributes of an event e thus include a *local* initial state $s_0(e)$ and a *local* final state $s(e)$. The domain of these states is called the *base* of the event, which consists of all variables either read or written.

In order to know what events are in conflict in our model we distinguish between variables read by an event and written by an event, i.e. we have two attributes $R(e)$ and $W(e)$ for every event, denoting the *read set* and *write set*. The base of e is the union of the read and write sets.We now define events to be conflicting iff one of them reads or writes some variable and the other event writes the same variable, i.e.

$$\mathit{conflict}(\ e, f\ ) \stackrel{\mathrm{def}}{=} (R(e) \cup W(e)) \cap W(f) \neq \emptyset \ \vee \ W(e) \cap (R(f) \cup W(f)) \neq \emptyset$$

Other models would also have been possible, for example also introducing read-read conflicts. The notion of conflict we use corresponds to the notion usually used in distributed database systems.

Although we work within a shared variables models it is still possible to define communication over channels. We can model *undirectional, asynchronous* channels by shared variables. A channel c is defined as a pair $(c.flag, c.val)$ where $c.flag$ is a boolean that is true iff a value is available on the channel, and $c.val$ the value to be read. Send and receive actions can now be modelled as guarded assignments:

$$\mathit{send}(c, v) \stackrel{\mathrm{def}}{=} (\neg c.flag) \&\ c.flag, c.val := true, v$$

$$\mathit{receive}(c, x) \stackrel{\mathrm{def}}{=} (c.flag) \&\ c.flag, x := false, c.val$$

We do not model deadlocks or blocking behaviour. For example a process consisting of two *receive* actions on the same channel would result in an empty history h.

3 Language and laws

We summarize the language introduced in the previous section and present the laws we use in the derivation and transformation of the Two Phase Commit protocol. First we present the syntax of the language:

$S \in \mathcal{P}rocess$

$$S \ ::= \ A \mid \textbf{skip} \mid S_0 \parallel S_1 \ \mid \ S_0 \cdot S_1 \ \mid \ S_0 \,;\, S_1 \ \mid \ \langle\, S \,\rangle \ \mid \ \textbf{contr}(S) \ \mid$$
$$\textbf{if } b \textbf{ then } S_0 \textbf{ else } S_1 \textbf{ fi} \ \mid \ \textbf{while } b \textbf{ do } S \textbf{ od} \ \mid \ \textbf{ref } a \textbf{ to } S_0 \textbf{ in } S_1$$

where

$$A \ ::= \ b \,\&\, x_1, \ldots, x_m := exp_1, \ldots, exp_m \ \mid \ send(c, v) \ \mid \ receive(c, x) \ \mid \ a$$

We proceed with a summary of algebraic transformation rules:

Commutativity and Associativity:

$$\begin{array}{rcll} P \parallel Q & = & Q \parallel P & \text{(COM)} \\ P \parallel (Q \parallel R) & = & (P \parallel Q) \parallel R & \text{(ASSOC1)} \\ P \cdot (Q \cdot R) & = & (P \cdot Q) \cdot R & \text{(ASSOC2)} \\ P \,;\, (Q \,;\, R) & = & (P \,;\, Q) \,;\, R & \text{(ASSOC3)} \end{array}$$

Skip:

$$\begin{array}{lclcll} \textbf{skip} \parallel P & = & P \parallel \textbf{skip} & = & P & \text{(SKIP0)} \\ \textbf{skip} \cdot P & = & P \cdot \textbf{skip} & = & P & \text{(SKIP1)} \\ \textbf{skip} \,;\, P & = & P \,;\, \textbf{skip} & = & P & \text{(SKIP2)} \end{array}$$

Communication Closed Layers:

Provided that there are no conflicts between S_0 and T_1, S_1 and T_0, and P and Q:

$$\begin{array}{rcll} (S_0 \cdot S_1) \parallel (T_0 \cdot T_1) & = & (S_0 \parallel T_0) \cdot (S_1 \parallel T_1) & \text{(CCL)} \\ P \parallel (Q \cdot R) & = & Q \cdot (P \parallel R) & \text{(CCL-L)} \\ P \parallel (R \cdot Q) & = & (P \parallel R) \cdot Q & \text{(CCL-R)} \\ P \parallel Q & = & P \cdot Q & \text{(Independence)} \end{array}$$

Refinement:

$$\begin{array}{rcll} \textbf{ref } a \textbf{ to } T \textbf{ in } a & = & \langle T \rangle & \text{(REF0)} \\ \textbf{ref } a \textbf{ to } T \textbf{ in } (R \parallel S) & = & (\textbf{ ref } a \textbf{ to } T \textbf{ in } R) \parallel (\textbf{ ref } a \textbf{ to } T \textbf{ in } S) & \text{(REF1)} \\ \textbf{ref } a \textbf{ to } T \textbf{ in } (R \cdot S) & = & (\textbf{ ref } a \textbf{ to } T \textbf{ in } R) \cdot (\textbf{ ref } a \textbf{ to } T \textbf{ in } S) & \text{(REF2)} \\ \textbf{ref } a \textbf{ to } T \textbf{ in } (R \,;\, S) & = & (\textbf{ ref } a \textbf{ to } T \textbf{ in } R) \,;\, (\textbf{ ref } a \textbf{ to } T \textbf{ in } S) & \text{(REF3)} \end{array}$$

Provided that a does not occur free in U, and b not in T:

$$\textbf{ref } a \textbf{ to } T \textbf{ in } \textbf{ ref } b \textbf{ to } U \textbf{ in } S \ = \ \textbf{ref } b \textbf{ to } U \textbf{ in } \textbf{ ref } a \textbf{ to } T \textbf{ in } S \quad \text{(REF4)}$$

Atomic brackets:

$$\begin{array}{rcll} \langle S \cdot T \rangle & = & \langle \langle S \rangle \cdot \langle T \rangle \rangle & \text{(ATOM1)} \\ \langle S \,;\, T \rangle & = & \langle \langle S \rangle \,;\, \langle T \rangle \rangle & \text{(ATOM2)} \end{array}$$

Contraction:

$$\begin{array}{rcll} \textbf{contr}(\langle S \rangle) & = & \textbf{contr}(S) & \text{(CONT1)} \\ \textbf{contr}(S \cdot T) & = & \textbf{contr}(S \,;\, T) & \text{(CONT2)} \\ \textbf{contr}(S \cdot T) & = & \textbf{contr}(\textbf{ contr}(S) \cdot \textbf{ contr}(T)) & \text{(CONT3)} \end{array}$$

4 The Two Phase Commit protocol

In this section we formally derive an implementation of an *Atomic Commit Protocol* using layer based reasoning and algebraic transformation techniques. The implementation derived is a version of the *Two Phase Commit Protocol.*

We start with an informal description of the requirements of an atomic commit protocol (ACP) and the context of ACP's in distributed databases ([BHG], [Raynal]). The general idea of the Two Phase Commit protocol (TPC) for a fully connected network is described after which we give a formal correctness proof of a layered implementation of TPC that is transformed to a (distributed) process oriented implementation by means of algebraic techniques. All the transformation steps preserve the functional correctness obliviating a new correctness proof for the second implementation.

This second implementation is then *refined* into an implementation that can handle more general network topologies, exploiting the resemblance of different processes in the protocol.

For a more extensive treatment of issues concerning distributed databases we refer to [BHG] which gives a clear exposition of the field.

4.1 Atomic Commitment

Distributed systems consist of a collection of *sites* that are connected via some network. The architecture of the network is not important for the moment, but we assume the network is connected. In a distributed (database) system an important concept is maintaining *consistency* of the different sites. Each site has a local database system, that maintains a part of the distributed database. In this paper we assume that data are stored at one site only, i.e. we do not deal with *replicated data.*

A *distributed transaction* consists of a sequence of reads and writes to database items. We assume every transaction T has a "home site"where it originated. Read and write requests are sent by the home site to the appropriate sites where they can be executed. Reading and writing of items is therefore – apart from routing problems, which we do not consider – not more difficult than in a centralized (non-distributed) database system.

The termination of a transaction T however is more involved as it concerns *all* sites that participated in T. Terminating a transaction is done by issuing a *Commit* or *Abort* operation to all sites where T accessed database items. Therefore an operation that one would logically like to view as a single, i.e. *atomic,* operation, involves a number of distributed sites that must agree upon the decision to be taken. Simply sending a message to a site stating that it should commit is not sufficient: it is possible that – for example due to failures of storage media or volatile memory – a site cannot store the changes to the database, and therefore cannot commit. The fact that a single site is not able to commit should result in the aborting the transaction, i.e. aborting at all sites involved, in order to keep the distributed database consistent.
A second problem is the possibility of *failures* in the network, which may result in sites that cannot communicate with the home site due to broken network links or failures of the site. This introduces the need for recovery protocols and timeouts in the protocols.

An algorithm that guarantees consistent termination of distributed transactions is called an *Atomic Commit Protocol.* Let us precisely state the requirements of an ACP. We assume the transaction involves a coordinator process C at the home site, and a set of participating processes P for all sites that were accessed. Every participating process has one vote: YES or NO, and every process can reach one of two decisions: COMMIT or ABORT. An ACP must observe the following requirements: ([BHG])

ACP1 *All processes that reach a decision reach the same one.*

ACP2 *A process cannot reverse its decision after it has reached one.*

ACP3 *The* COMMIT *decision can only be reached if all processes voted* YES.

ACP4 *If there are no failures and all processes voted* YES, *then the decision will be* COMMIT.

ACP5 *If there are no failures for sufficiently long, then all processes will reach a decision.*

In this paper we do not take into account the possibility of communication failures or site failures, i.e. we assume every message sent is eventually delivered and that sites are working correctly. This does not obliviate the necessity of an ACP: we still have to take into account system failures and media failures which can lead to the abortion of a transaction. The addition of communication failures and site failures is straightforward, but leads to the introduction of recovery algorithms in the protocol. For sake of brevity we treat the simpler case only.

A well-known example of an ACP is the *Two Phase Commit Protocol.* It is the simplest and most popular algorithm. In the absence of communication failures it informally goes as follows:

1. The protocol starts when the coordinator receives a VOTE_REQ message from the system.
2. The coordinator starts by sending VOTE_REQ to all participants.
3. When a participant receives a VOTE_REQ it decides what to vote and sends its vote (YES or NO) to the coordinator, according to whether it can commit or not.
4. The coordinator collects all votes, and decides COMMIT iff all votes including its own vote were YES. It sends the decision to every participant, after which it acts accordingly, that is, it commits or aborts.
5. When the participants receive the decision they act accordingly.

The protocol conceptually consists of two phases: a first phase in which all votes are gathered and a decision is made, and a second phase in which all processes are informed of the decision.

4.2 A derivation of a layered implementation

The usual way to present distributed algorithms is as a number of communicating processes running in parallel. Compared to sequential algorithms however, this complicates the correctness proof due to the fact that one has to take into account the possiblity of interference by processes running in parallel. Here we take the approach to model algorithms as the *composition of abstract actions.* These actions can be viewed as if they were executed atomically one after the other, although in fact they represent *layers* in which actions execute in parallel. This leads to a simple correctness proof, in which we can use well-known proof methods as for sequential processes ([OG], [Lamport]). Refining these abstract actions leads to an algorithm with a clear *layered* structure in agreement with the abstract view.

After we have constructed our initial design of the TPC protocol in the form of a *layered* system, we transform it so as to meet the *distributed* structure of the network. This is done by means of correctness preserving (algebraic) transformation steps. This allows for the combination of a simple correctness proof *plus* a distributed structure of the algorithm.

Specification of the protocol

We start of with the specification of the protocol. The set of sites involved in the transaction T is $\{S_i \mid i \in I\}$, where I={ 0,1,..., n} and S_0 is the home site, i.e. the site where the coordinator process is located.
Let dec_i be the variable containing the decision made by site S_i for $i \in I$, and let $vote_i$ be the variable containing the vote of process i. Let *dec* be the decision taken by the coordinator process. We assume the initial VOTE_REQ message is sent to the coordinator via the channel req_0. Furthermore we assume all other channels are initially empty. The latter assumption is left implicit. For channel $c = (c.flag, c.val)$ we define

$full(c) \stackrel{\text{def}}{=} c.flag$ and $val(c) \stackrel{\text{def}}{=} c.val$

This leads to the following specification of an ACP:

$$\{\ \forall_{0\le i\le n}\ (dec_i = \text{NONE} \wedge vote_i = \text{NONE})\ \wedge\ full(req_0) \wedge val(req_0) = \text{VOTE\_REQ}\ \}$$
$$ACP$$
$$\{\ (dec = \text{COMMIT} \vee dec = \text{ABORT}) \wedge (dec = \text{COMMIT} \Rightarrow \forall_{0\le i\le n}\ (vote_i = \text{YES})) \wedge \forall_{0\le i\le n}\ (dec_i = dec)\}$$

We now derive an implementation of the Two Phase Commit protocol that satisfies the ACP specification.

Decomposing the protocol

Our layered implementation consists of two phases which can be split up into a number of distinct actions. The specification of the phases is:

$$\{\ \forall_{0\le i\le n}\ (dec_i = \text{NONE} \wedge vote_i = \text{NONE})\ \wedge\ full(req_0) \wedge val(req_0) = \text{VOTE\_REQ}\ \}$$
$$Phase_1$$
$$\{\ (dec = \text{COMMIT} \Rightarrow \forall_{0\le i\le n}\ (vote_i = \text{YES}))\ \wedge\ \forall_{0\le i\le n}\ (vote_i = \text{YES} \vee vote_i = \text{NO})\}$$

and

$$\{\ (dec = \text{COMMIT} \Rightarrow \forall_{0\le i\le n}\ (vote_i = \text{YES}))\ \wedge\ \forall_{0\le i\le n}\ (vote_i = \text{YES} \vee vote_i = \text{NO})\}$$
$$Phase_2$$
$$\{\ (dec = \text{COMMIT} \vee dec = \text{ABORT}) \wedge (dec = \text{COMMIT} \Rightarrow \forall_{0\le i\le n}\ (vote_i = \text{YES})) \wedge \forall_{0\le i\le n}\ (dec_i = dec)\}$$

From this specification trivially follows that the (conflict) composition of the two phases satisfies the ACP specification.

The first phase consists of three subphases: a *request action* requesting participants for their votes, a *reply action* where the participants inform the coordinator of their votes, and a *decision action* where the coordinator decides whether the transaction commits or aborts. $Phase_1$ is the composition of these actions, i.e.

$$Phase_1 \triangleq Request \cdot Vote \cdot Decide$$

We use the variables msg_i to store the last message received by processes i, and variables $stable_i$ that states whether it is safe for process i to commit, for example because it has succeeded in pre-writing all items that were accessed in the transaction to stable storage ([BHG]). The implementation of these actions will be such, that the following proof outline is valid.

$$\{ \forall_{i \in I}(vote_i = \text{NONE} \wedge dec_i = \text{NONE}) \ \wedge \ full(req_0) \wedge val(req_0 = \text{VOTE\_REQ} \}$$
$$Request \cdot$$
$$\{ \forall_{i \in I}(vote_i = \text{NONE} \wedge dec_i = \text{NONE}) \ \wedge \ msg = \text{VOTE\_REQ} \}$$
$$Vote \cdot$$
$$\{ \forall_{i \in I}(vote_i \neq \text{NONE} \ \wedge \ (vote_i = \text{YES}) \Leftrightarrow stable_i \ \wedge \ (vote_i = \text{NO}) \Leftrightarrow \neg stable_i)\}$$
$$Decide$$
$$\{ \forall_{i \in I}(vote_i \neq \text{NONE} \ \wedge \ (vote_i = \text{YES}) \Leftrightarrow stable_i \ \wedge \ (vote_i = \text{NO}) \Leftrightarrow \neg stable_i) \ \wedge$$
$$((dec = \text{COMMIT} \ \wedge \ \forall_{i \in I}(vote_i = \text{YES})) \vee (dec = \text{ABORT} \ \wedge \ \exists_{i \in I}(vote_i = \text{NO}))) \}$$

The second phase consists of a single *effectuation* layer only, satisfying:

$$\{ ((dec = \text{COMMIT} \ \wedge \ \forall_{i \in I}(vote_i = \text{YES})) \vee (dec = \text{ABORT} \ \wedge \ \exists_{i \in I}(vote_i = \text{NO}))) \}$$
$$Effectuate$$
$$\{ ((dec = \text{COMMIT} \ \wedge \ \forall_{i \in I}(vote_i = \text{YES})) \vee (dec = \text{ABORT} \ \wedge \ \exists_{i \in I}(vote_i = \text{NO})))$$
$$\forall_{i \in I}(dec_i = dec)\}$$

Refining actions to layers

Most of the abstract actions are implemented as layers consisting of a number of send and receive actions in which communication between the coordinator and the participants takes place. The *Request* action however simple consists of a *receive* statement where the coordinator gets the request to gather votes. This results in: [1]

$$Request \triangleq \quad receive(req_0, msg) \qquad\qquad (\alpha_0)$$

The $Vote$ action consists of the composition of a number of layers. First $Vote$ requests in its turn the participants to send their votes; the participants receive the request, decide what to vote and reply accordingly; finally the coordinator gathers the votes and decides. Communication between the coordinator and the participants takes place via a set of channels req_i and rep_i. This results in the following layered implementation of the $Vote$ action (where $P = I - \{0\}$):

$$\begin{array}{lll} Vote \triangleq & (\|_{i \in P}\ send(req_i, msg)) \cdot & (\alpha_1) \\ & (\|_{i \in P}\ (receive(req_i, msg_i)\ ;\ \textbf{if}\ stable_i\ \textbf{then}\ v_i := \text{YES}\ \textbf{else}\ v_i := \text{NO}\ \textbf{fi}\ )) \cdot & (\alpha_2) \\ & (\|_{i \in P}\ send(rep_i, v_i)) \cdot & (\alpha_3) \\ & (\|_{i \in P}\ receive(rep_i, vote_i)) \cdot & (\alpha_4) \\ & \textbf{if}\ stable_0\ \textbf{then}\ vote_0 := \text{YES}\ \textbf{else}\ vote_0 := \text{NO}\ \textbf{fi} & (\alpha_5) \end{array}$$

Note that a parallel component in α_2 consists of the *sequential* composition of two actions: the participant checks whether it can commit only *after* it receives the request. This in

[1] every statement is numbered by indexed Greek characters, in order to be able to refer to it in the transformation.

order to prevent it from deciding what to vote and sending the vote unsollicited (as the two actions are not in conflict, a conflict composition would not suffice here).

The actions within a parallel composition can be executed truly in parallel, as they are non-conflicting. This implies that no interference can occur and so a proof outline proving the correctness of the implementation can be given in a straightforward way.

In the decision layer the variable *dec* is assigned its proper value, based upon the votes of all participants and the coordinator:

$$Decide \triangleq \ \mathbf{if}\ \forall_{i\in I}(vote_i = \text{YES})\ \mathbf{then}\ dec := \text{COMMIT}\ \mathbf{else}\ dec := \text{ABORT}\ \mathbf{fi} \qquad (\beta)$$

The final effectuation action can be split into two sublayers: the first informing all processes of the decision taken, and a second where the processes commit or abort (which in our example is modelled by assigning the correct value to a variable dec_i). This leads to:

$$\begin{aligned} Effectuate \triangleq\ & (\|_{i\in P}\ send(inform_i, dec)) \cdot && (\gamma_1)\\ & (\|_{i\in P}\ receive(inform_i, msg_i)) \cdot && (\gamma_2)\\ & (\|_{i\in P}\ \mathbf{if}\ msg_i = \text{ABORT}\ \mathbf{then}\ dec_i := \text{ABORT}\ \mathbf{else}\ dec_i := \text{COMMIT}\ \mathbf{fi}\ ) \cdot && (\gamma_3)\\ & dec_0 := dec && (\gamma_4) \end{aligned}$$

Again the correctness proof is straightforward.

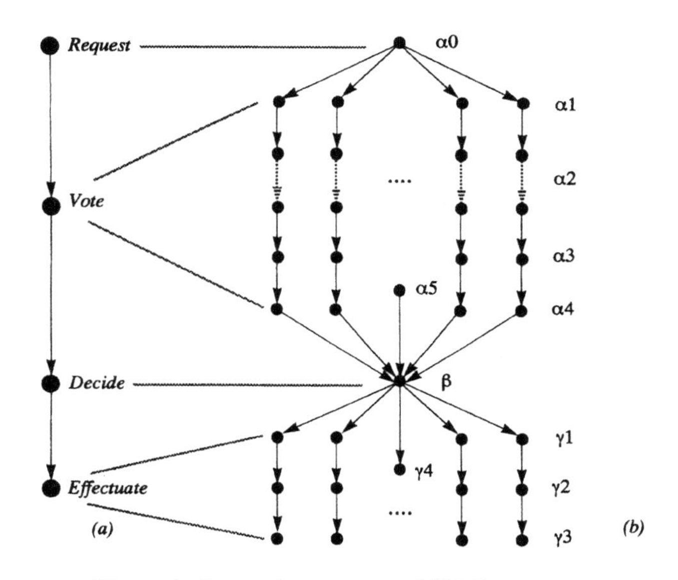

Figure 1: Semantic structure of TPC

The algorithm as a whole can now be described as:

$$\begin{aligned} TPC \triangleq \langle\ \mathbf{ref}\ \ & Request\ \mathbf{to}\ \alpha_0,\\ & Vote\ \mathbf{to}\ \alpha_1 \cdot \alpha_2 \cdot \alpha_3 \cdot \alpha_4 \cdot \alpha_5,\\ & Decide\ \mathbf{to}\ \beta,\\ & Effectuate\ \mathbf{to}\ \gamma_1 \cdot \gamma_2 \cdot \gamma_3 \cdot \gamma_4\\ \mathbf{in}\ \ & Request \cdot Vote \cdot Decide \cdot Effectuate\ \rangle \end{aligned}$$

The basic semantic partial order structure of the system before refinement and after refinement is sketched in fig. 1*(a)* and fig. 1*(b)* respectively. In these figures *logical precedence* is denoted by full arrows, whereas *temporal ordering* is denoted by dashed arrows.

4.3 Transforming layers to processes

From the refinement laws we described in section 3 it can be derived that the implementation we derived in the previous section is of the form:

$$TPC \triangleq \langle \alpha_0 \cdot \alpha_1 \cdot \alpha_2 \cdot \alpha_3 \cdot \alpha_4 \cdot \alpha_5 \cdot \beta \cdot \gamma_1 \cdot \gamma_2 \cdot \gamma_3 \cdot \gamma_4 \rangle$$

The derivation is as follows:

$$\begin{array}{ll} & TPC \\ = & \{ \text{ definition } TPC, \text{REF2 } \} \\ & \langle (\ \textbf{ref } Request \textbf{ to } \alpha_0 \textbf{ in } Request) \cdot \\ & (\ \textbf{ref } Vote \textbf{ to } \alpha_1 \cdot \alpha_2 \cdot \alpha_3 \cdot \alpha_4 \cdot \alpha_5 \textbf{ in } Vote) \cdot \\ & (\ \textbf{ref } Decide \textbf{ to } \beta \textbf{ in } Decide) \cdot \\ & (\ \textbf{ref } Effectuate \textbf{ to } \gamma_1 \cdot \gamma_2 \cdot \gamma_3 \cdot \gamma_4 \textbf{ in } Effectuate) \rangle \\ = & \{ \text{ REF0 } \} \\ & \langle\ \langle \alpha_0 \rangle \cdot \langle \alpha_1 \cdot \alpha_2 \cdot \alpha_3 \cdot \alpha_4 \cdot \alpha_5 \rangle \cdot \langle \beta \rangle \cdot \langle \gamma_1 \cdot \gamma_2 \cdot \gamma_3 \cdot \gamma_4 \rangle\ \rangle \\ = & \{ \text{ ATOM1 } \} \\ & \langle \alpha_0 \cdot \alpha_1 \cdot \alpha_2 \cdot \alpha_3 \cdot \alpha_4 \cdot \alpha_5 \cdot \beta \cdot \gamma_1 \cdot \gamma_2 \cdot \gamma_3 \cdot \gamma_4 \rangle \end{array}$$

□

We transform this – conceptually *sequentially phased* – algorithm to an algorithm of the form:

$$TPC' \triangleq \langle C \parallel (\parallel_{i \in P} P_i) \rangle$$

The final outcome of this transformation that we decribe in detail below, is as follows:

$$\begin{array}{lll} C \triangleq & receive(req_0, msg) \cdot & (\alpha_0) \\ & (\parallel_{i\in I}\ send(req_i, msg)) \cdot & (\alpha_1) \\ & (\parallel_{i\in I}\ receive(rep_i, vote_i)) \cdot & (\alpha_4) \\ & \textbf{if } stable_0 \textbf{ then } vote_0 := \text{YES} \textbf{ else } vote_0 := \text{NO} \textbf{ fi } \cdot & (\alpha_5) \\ & \textbf{if } \forall_{i\in I}(vote_i = \text{YES}) \textbf{ then } dec := \text{COMMIT} \textbf{ else } dec := \text{ABORT} \textbf{ fi } \cdot & (\beta) \\ & (\parallel_{i\in I}\ send(inform_i, dec)) \cdot & (\gamma_1) \\ & dec_0 := dec & (\gamma_4) \end{array}$$

and

$$\begin{array}{lll} P_i \triangleq & receive(req_i, msg_i)\ ;\ \textbf{if } stable_i \textbf{ then } v_i := \text{YES} \textbf{ else } v_i := \text{NO} \textbf{ fi } \cdot & (\alpha_2^i) \\ & send(rep_i, v_i) \cdot & (\alpha_3^i) \\ & receive(inform_i, msg_i) \cdot & (\gamma_2^i) \\ & \textbf{if } msg_i = \text{ABORT} \textbf{ then } dec_i := \text{ABORT} \textbf{ else } dec_i := \text{COMMIT} \textbf{ fi} & (\gamma_3^i) \end{array}$$

Next we show how the transformation can be carried out using the algebraic laws presented in the section 3 and a lemma concerning send and receive actions on channels:

Lemma 4.1
Let c be a channel used only once for a *send* and a *receive* action. Under the assumption that initially the channel c is empty (i.e. $\neg c.flag$ holds) we have that:

$$send(c, v) \cdot receive(c, x) \ = \ send(c, v) \parallel receive(c, x)$$

Proof:
follows directly from the assumptions and the definition of *send* and *receive* as guarded assignments. □

A straightforward corollary of this lemma is that it is also valid if we have the sequential composition of the receive action with a second (non-conflicting) action.

Lemma 4.2
Let c be a channel where only a single *send* and *receive* action are executed. Under the assumption that initially the channel c is empty and that P does not conflict with the *send* and *receive* actions we have that:

$$send(c,v) \cdot (receive(c,x)\ ;\ P) \ = \ send(c,v) \ \| \ (receive(c,x)\ ;\ P)$$

Proof: omitted □

Furthermore we use the following corollary combining lemma 4.2 and the Communication Closed Layers law.

Corollary 4.3
Let c_i be a set of channels which are only used once for send and receive actions. We then have that:

$$(\|_i\ send(c_i,v_i)) \cdot ((\|_i\ receive(c_i,x_i)\ ;\ P_i)) \ = \ (\|_i\ send(c_i,v_i)) \ \| \ (\|_i\ (receive(c_i,x_i)\ ;\ P_i))$$

Proof:

$$\begin{array}{ll} & (\|_i\ send(c_i,v_i)) \cdot (\|_i\ (receive(c_i,x_i)\ ;\ P_i)) \\ = & \quad \{\ \text{CCL}\ \} \\ & \|_i\ (send(c_i,v_i) \cdot (receive(c_i,v_i)\ ;\ P_i)) \\ = & \quad \{\ \text{lemma 4.2}\ \} \\ & \|_i\ (send(c_i,v_i) \ \| \ (receive(c_i,v_i)\ ;\ P_i)) \\ = & \quad \{\ \text{COM and ASSOC1}\ \} \\ & (\|_i\ send(c_i,v_i)) \ \| \ (\|_i\ (receive(c_i,x_i)\ ;\ P_i) \end{array}$$

This of course also holds with the P_i component removed, for we can always substitute **skip** for P_i. □

From now on we use the fact that parallel composition is commutative and associative without mentioning. By repeatedly applying these laws and lemmas we eventually get the algorithm TPC'.
In the following let α_i^j denote the action in α_i concerning participant P_j (and analogously for β and γ). We now derive:

$$\begin{array}{ll} & TPC \\ = & \quad \{\ \text{by definition}\ \} \\ & \langle \alpha_0 \cdot \alpha_1 \cdot \alpha_2 \cdot \alpha_3 \cdot \alpha_4 \cdot \alpha_5 \cdot \beta \cdot \gamma_1 \cdot \gamma_2 \cdot \gamma_3 \cdot \gamma_4 \rangle \\ = & \quad \{\ \text{corollary 4.3, for channels } req_i\ \} \\ & \langle \alpha_0 \cdot (\alpha_1 \ \| \ \alpha_2) \cdot \alpha_3 \cdot \ \ldots \ \cdot \gamma_4 \rangle \\ = & \quad \{\ \text{CCL-L}\ \} \\ & \langle ((\alpha_0 \cdot \alpha_1) \ \| \ \alpha_2) \cdot \alpha_3 \cdot \alpha_4 \cdot \alpha_5 \cdot \ \ldots \ \cdot \gamma_4 \rangle \\ = & \quad \{\ \text{corollary 4.3, for channel } rep_i\ \} \\ & \langle ((\alpha_0 \cdot \alpha_1) \ \| \ \alpha_2) \cdot (\alpha_3 \ \| \ \alpha_4) \cdot \alpha_5 \cdot \beta \cdot \ \ldots \ \cdot \gamma_4 \rangle \end{array}$$

$$
\begin{aligned}
&= \quad \{\text{ CCL }\} \\
&\quad \langle((\alpha_0 \cdot \alpha_1 \cdot \alpha_4) \parallel (\alpha_2 \cdot \alpha_3)) \cdot \alpha_5 \cdot \beta \cdot \gamma_1 \cdot \ldots \cdot \gamma_4\rangle \\
&= \quad \{\text{ CCL-R }\} \\
&\quad \langle((\alpha_0 \cdot \alpha_1 \cdot \alpha_4 \cdot \alpha_5 \cdot \beta) \parallel (\alpha_2 \cdot \alpha_3)) \cdot \gamma_1 \cdot \ldots \cdot \gamma_4\rangle \\
&= \quad \{\text{ corollary 4.3, for channels } \mathit{inform}_i \} \\
&\quad \langle((\alpha_0 \cdot \alpha_1 \cdot \alpha_4 \cdot \alpha_5 \cdot \beta) \parallel (\alpha_2 \cdot \alpha_3)) \cdot (\gamma_1 \parallel \gamma_2) \cdot \gamma_3 \cdot \gamma_4\rangle \\
&= \quad \{\text{ CCL }\} \\
&\quad \langle((\alpha_0 \cdot \alpha_1 \cdot \alpha_4 \cdot \alpha_5 \cdot \beta \cdot \gamma_1) \parallel (\alpha_2 \cdot \alpha_3 \cdot \gamma_2)) \cdot \gamma_3 \cdot \gamma_4\rangle \\
&= \quad \{\text{ Independence }\} \\
&\quad \langle(\alpha_0 \cdot \alpha_1 \cdot \alpha_4 \cdot \alpha_5 \cdot \beta \cdot \gamma_1) \parallel (\alpha_2 \cdot \alpha_3 \cdot \gamma_2) \cdot (\gamma_3 \parallel \gamma_4)\rangle \\
&= \quad \{\text{ CCL }\} \\
&\quad \langle(\alpha_0 \cdot \alpha_1 \cdot \alpha_4 \cdot \alpha_5 \cdot \beta \cdot \gamma_1 \cdot \gamma_4) \parallel (\alpha_2 \cdot \alpha_3 \cdot \gamma_2 \cdot \gamma_3)\rangle \\
&= \quad \{\text{ definition of } \alpha_i\text{, Communication Closed Layers }\} \\
&\quad \langle(\alpha_0 \cdot \alpha_1 \cdot \alpha_4 \cdot \alpha_5 \cdot \beta \cdot \gamma_1 \cdot \gamma_4) \parallel (\parallel_{i \in P}\, \alpha_2^i \cdot \alpha_3^i \cdot \gamma_2^i \cdot \gamma_3^i)\rangle \\
&= \quad \{\text{ by definition }\} \\
&\quad TPC'
\end{aligned}
$$

□

The final step now is to *allocate* the different parallel components in the algorithm TPC' to sites. This is done by replacing ***conflict composition*** by ***sequential composition*** which would result in a program consisting of a number of sequential processes executing in parallel (and a coordinator process with actions executing in parallel). Every (sequential) component can now be executed at a single site, where for the coordinator process we could even execute the actions in the parallel composition ***sequentially***. This transformation does not change the ***functional behaviour*** of the algorithm (due to law CONT2) but allows for a straightforward implementation in a conventional programming language.

We used different channels in the different layers of the algorithm as to make sure that the laws we want to apply hold on simple syntactic grounds. It is however possible to map the different channels to a single pair of (unidirectional) channels under certain circumstances, that can be characterized semantically. The exact formulation is not discussed in this paper however.

4.4 Refining the algorithm to other network topologies

If we look at the structure of the processes that are the result of the transformation, we can observe that there is a clear correspondance between the C process and the P_i processes. As we show below we can distinguish the same layers in both processes, except for the fact that the coordinator process does not reply to the system, but this is a simple extension. We therefore ***unify*** the different processes as a first step.

Unifying the processes

Basically all processes can been considered as the composition of five abstract actions, with analogous functions as in the layered decomposition. To specify the processes in this uniform way, we introduce some notation, that will also allow us to extend the protocol to other network structures.

Communication in our algorithm takes place only between the coordinator and the participants. In order to be able to describe the communication actions and their result for coordinator and participants in a uniform way we introduce some notation.
We define, for site i, the set of *successor sites* $S(i)$ as:

$S(0) = P$, and $S(i) = \emptyset$, if $0 < i \leq n$

Furthermore let $T(i)$ denote the set of successors of site i including i, that is:

$T(0) = I$, and $T(i) = \{i\}$, if $0 < i \leq n$

The process P_i at site i (for $0 \leq i \leq n$) can now be described as:

$P_i \triangleq Request_i \; ; Vote_i \cdot Reply_i \cdot Decide_i \cdot Effectuate_i$

such that

$\{ dec_i = \text{NONE} \wedge vote_i = \text{NONE} \wedge full(req_i) \wedge val(req_i) = \text{VOTE\_REQ} \}$
$Request_i$;
$\{ dec_i = \text{NONE} \wedge vote_i = \text{NONE} \wedge msg_i = \text{VOTE\_REQ} \}$
$Vote_i$ ·
$\{ (vote_i = \text{YES}) \Leftrightarrow (\forall_{j \in T(i)}(stable_j)) \wedge \forall_{j \in T(i)}(vote_j \neq \text{NONE}) \}$
$Reply_i$ ·
$\{ (vote_i = \text{YES}) \Leftrightarrow (\forall_{j \in T(i)}(stable_j)) \wedge \forall_{j \in I}(vote_j \neq \text{NONE}) \}$
$Decide_i$ ·
$\{ \forall_{j \in I}(vote_j \neq \text{NONE}) \wedge dec_i = dec_0 \wedge$
$(dec_0 = \text{COMMIT} \Leftrightarrow (\forall_{j \in I} vote_j = \text{YES})) \wedge (dec_0 = \text{ABORT} \Leftrightarrow (\exists_{j \in I} vote_j = \text{NO})) \}$
$Effectuate_i$
$\{ dec_0 \neq \text{NONE} \wedge \forall_{i \in I}(dec_i = dec_0) \wedge$
$(dec_0 = \text{COMMIT} \Leftrightarrow \forall_{i \in I}(vote_i = \text{YES})) \wedge (dec_0 = \text{ABORT} \Leftrightarrow (\exists_{j \in I} vote_j = \text{NO})) \}$

Here again the voting may only start *after* the request to do so was received. The process P_0 is the coordinator process. If we substitute $i = 0$ in the specification above, we clearly have that the pre- and postcondition satisfy the specification ACP. P_0 is therefore a correct implementation of an ACP.

The algorithm as a whole is now defined by:

$$TPC'' \triangleq \langle \textbf{ref } P_i \textbf{ to } \; Request_i \cdot Vote_i \cdot Reply_i \cdot Decide_i \cdot Effectuate_i \;\; \textbf{in} \; (\|_{i \in I} P_i) \rangle$$

The process above is given by means of abstract actions. In order to get to a implementable algorithm we have to *refine* the actions to processes in compliance with the outline above. This will in general give different refinements for different processes.

Refining the $Request_i$ is straightforward and equivalent in both cases:

$Request_i \triangleq receive(req_i, msg_i)$

The refinement of the $Vote_i$ action seems to be quite different for the two types of processes. For participant processes it is simple, as the specification boils down to:

$\{ dec_i = \text{NONE} \wedge vote_i = \text{NONE} \wedge msg_i = \text{VOTE\_REQ} \}$
$Vote_i$
$\{ vote_i \neq \text{NONE} \wedge (vote_i = \text{YES}) \Leftrightarrow stable_i \}$

This gives us:

$$Vote_i \triangleq \textbf{if } stable_i \textbf{ then } vote_i := \text{YES} \textbf{ else } vote_i := \text{NO} \textbf{ fi}$$

Voting is more intricate for the coordinator process, as could be expected because of the previous protocol. It can be shown that

$$\begin{aligned} Vote_0 \triangleq\ & (\|_{i \in S(0)}\, send(req_i, msg_i)) \cdot \\ & (\|_{i \in S(0)}\, receive(rep_i, vote_i)) \cdot \\ & \textbf{if } stable_0 \textbf{ then } vote_0 := \text{YES} \textbf{ else } vote_0 := \text{NO} \textbf{ fi} \end{aligned}$$

is a correct refinement, given the refinement of $Request_i$ and $Vote_i$ (for $i \neq 0$). Although this seems to be a statement completely different from the participant's case, it is in fact not! If we define

$$\|_{i \in \emptyset}\, S_i \;=\; \textbf{skip}$$

we have that both refinements are instances of the same pattern.

The *Reply* action can be implemented as **skip** in the coordinator process, as the postcondition of $Vote_0$ already implies the postcondition of $Reply_0$. In the case of the participants it can be implemented by a *send* action. The correctness of this refinement is not straightforward. We can prove the correctness by transforming the implementation we describe in this section to a layered structure, which corresponds to the reversed transformation we described above.

We therefore define:

$$Reply_0 \triangleq \textbf{skip}$$

$$Reply_i \triangleq send(rep_i, vote_i) \quad (i \neq 0)$$

Analogously we can derive:

$$Decide_0 \triangleq \textbf{if } \forall_{i \in I}\, (vote_i = \text{YES}) \textbf{ then } dec := \text{COMMIT} \textbf{ else } dec := \text{ABORT} \textbf{ fi}$$

$$Decide_i \triangleq receive(inform_i, msg_i) \quad (i \neq 0)$$

and finally:

$$\begin{aligned} Effectuate_i \triangleq\ & (\|_{j \in S(i)}\, send(inform_j, msg_i)) \cdot \\ & \textbf{if } msg_i = \text{COMMIT} \textbf{ then } dec_i := \text{COMMIT} \textbf{ else } dec_i := \text{ABORT} \textbf{ fi} \end{aligned}$$

The actual refinement step

The extension to spanning tree network architectures has now become – due to the formulation chosen above – very simple.

In the case of a (spanning) tree network not only the coordinator has successor sites, but every non-leaf node in the tree does. Furthermore we have to define $T(i)$ to be the ***subtree*** induced by site i, which is in agreement with the definition we gave above. Under this definition the implementation of the processes *and* the corresponding proof outline remain the same. This would informally result in a recursive "blow-up" of the statement that assign a value to $vote_i$ on the basis of $stable_i$. It can in fact be seen as a *refinement* of the algorithm derived above, where the $Vote_i$ action (for $i \neq 0$) is refined into a process similar to $Vote_0$!

The resulting algorithm can be viewed as an extension of the *Propagation of Information with Feedback* protocol (PIF) as originally described in [Segall] and used as an example of sequentially phased reasoning in [SR].

5 Conclusion and future work

In this paper we derived an implementation of an atomic commit protocol for different network topologies. One would like to consider other extensions of the algorithm, for example allowing a process to decide to ABORT individually after it voted NO and thereafter immediately sending an ABORT message to its successors. This would lead to a slightly more complicated structure of the algorithm and requires a generalized Communication Closed Layers law. The extension with communication failures can be done in an analogous way.

The underlying partial order model in principle allows for the use of real-time in our language. This would enable us to specify real-time properties of algorithms and analyze real-time protocols. At this moment we are studying various ways to indeed introduce real-time behaviour, which we plan to apply to correctness proofs of synchronous atomic broadcast protocols.

Applicability of the design principles given in this paper in other fields of computing science is also a subject of current investigations. Well known algorithms in parsing theory have already shown to be easily formalized using our framework [JPSZ].

References

[AO] K.R. Apt, E.-R. Olderog, *Verification of sequential and concurrent programs*, Springer, 1991.

[BHG] P.A. Bernstein, V. Hadzilacos and N. Goodman, *Concurrency Control and Recovery in Database Systems*, Addison-Wesley, 1987.

[CM] K.M. Chandy and J. Misra, *Parallel Program Design: A Foundation*, Addison-Wesley, 1988.

[CASD] F. Critian, H. Aghili, R. Strong, D. Dolev, Atomic Broadcast: From Simple Message Diffusion to Byzantine Agreement, *Proceedings 15th International Symposium on Fault-Tolerant Computing,* 1985.

[EF] T. Elrad and N. Francez, Decomposition of distributed programs into communication closed layers, *Science of Computer Programming 2*, 1982.

[GHS] R.T. Gallager, P.A. Humblet and P.M. Spira, A distributed algorithm for minimum-weight spanning trees, *ACM TOPLAS 5-1*, 1983.

[GG] R. J. van Glabbeek and U. Goltz, *Equivalence Notions for Concurrent Systems and Refinement of Actions*, Arbeitspapiere der GMD, Number 366, GMD, 1989.

[Hooman] J. Hooman, *Specification and Compositional Verification of Real-Time Systems*, Ph.D. Thesis, Eindhoven University of Technology, 1991.

[JPZ] W. Janssen, M. Poel and J. Zwiers, *Consistent alternatives of parallelism with conflicts*, Memorandum INF-91-15, University of Twente.

[JPZa] W. Janssen, M. Poel and J. Zwiers, Action Systems and Action Refinement in the Development of Parallel Systems, an Algebraic Approach, *proceedings CONCUR '91*, Springer LNCS 527, 1991.

[JPSZ] W. Janssen, M. Poel, K. Sikkel, and J. Zwiers, The Primordial Soup Algorithm, A Systematic Approach to the Specification and Design of Parallel Parsers, *Proceedings Computing Science in the Netherlands Conference,* 1991.

[Lamport] L. Lamport, The Hoare Logic of concurrent programs, *Acta Informatica 14*, 1980.

[OG] S. Owicki and D. Gries, An axiomatic proof technique for parallel programs, *Acta Informatica 6*, 1976.

[Pratt] V. Pratt, Modelling Concurrency with Partial orders, *International Journal of Parallel Programming 15*, 1986, pp. 33-71.

[Raynal] M. Raynal, *Distributed Algorithms and Protocols,* John Wiley & Sons, 1988.

[Segall] A. Segall, Distributed Network Protocols, *IEEE Transactions on Information Theory,* Vol. IT-29, 1983, pp. 23-35.

[SR] F.A. Stomp and W.P. de Roever, Designing distributed algorithms by means of formal sequentially phased reasoning, *Proc. of the 3rd International Workshop on Distributed Algorithms*, Nice, LNCS 392, Eds. J.-C. Bermond and M. Raynal, 1989, pp. 242-253.

[ZR] J. Zwiers and W.P. de Roever, Predicates are Predicate Transformers: a unified theory for concurrency, *Proc. of the conference on Principles of Distributed Computing*, 1989.

Scheduling in Real-Time Models

Reino Kurki-Suonio [1], *Kari Systä* [1] and *Jüri Vain* [2]

[1] Tampere University of Technology
Software Systems Laboratory
Box 553, SF-33101 Tampere, Finland
rks@cs.tut.fi, ks@cs.tut.fi

[2] Estonian Academy of Sciences
Institute of Cybernetics, Computer R&D Division
Akadeemia tee 21/1, 200108 Tallinn, Estonia
jyri@deseg.ioc.ew.su

Abstract

Interleaving semantics is shown to provide an appropriate basis also for the modeling of real-time properties. Real-time scheduling of interleaved actions is explored, and the crucial properties of such schedulings are analyzed. The motivation of the work is twofold: to make real-time modeling practical already at early stages of specification and design, and to increase the reliability and predictability of reactive real-time systems by improved insensitivity to changes in the underlying real-time assumptions.

Keywords: executable specifications, fairness, formal methods, interleaving model, joint actions, reactive systems

1. Introduction

When real-time aspects are ignored, there is wide agreement that interleaving models provide an appropriate basis for the modeling of reactive systems and for formal reasoning on their properties [18]. When attention is focused on real-time, this basis is often discarded as unrealistic, and models are adopted where state information is ignored [17]. It is obviously an unsatisfactory situation that incompatible approaches are advocated to deal with state properties and real-time aspects.

The purpose of this paper is to show that the common belief is unjustified that interleaving models would be inappropriate for real-time modeling. On the contrary, we show that modeling of state properties can be effectively used as a first approximation, on which real-time aspects can be added, and that the same system of reasoning can be used in all stages. Such a uniform approach has advantages especially when models of reactive systems are constructed stepwise, starting from a high level of abstraction, which should be the case in specification and design.

A key issue here is how actions are scheduled for real-time execution in the model. In particular, scheduling with maximal parallelism [20] turns out to be too restrictive for our purposes. Under a more nondeterministic policy of *undelayed scheduling*, any interleaved computation can be scheduled, which means that augmenting a model with real-time properties need not add new state properties. Two problems arise, however, which will be addressed in this paper: scheduling need not be able to enforce all liveness properties of the underlying interleaved model, and real-time requirements may be violated by such computations that could be excluded by a more practical scheduling policy.

The solution to the first problem is to transform the system into a form where the problem no longer arises [8, 9]. This means that such fairness assumptions that are too strong to be directly enforced are replaced by an explicit mechanism by which some weaker fairness assumptions are guaranteed to be sufficient. The same technique is proposed in this paper for also the second problem. Although this may require more work in specification and design, the reliability and predictability of the resulting programs are essentially improved by their better insensitivity to changes in the underlying real-time assumptions.

As the semantic basis we will use *joint action systems* as developed in [7-9]. The main motivation in developing the ideas has been to incorporate them in future versions of a specification language for reactive systems, DisCo (*Dis*tributed *Co*operation) [11-15, 21].

The structure of the rest of the paper is as follows. We start in Section 2 with an introductory example that illustrates the intuitive ideas of the paper. The semantic basis is introduced in Section 3. Section 4 gives a general definition of schedulings of interleaved actions, and discusses their crucial properties. Undelayed scheduling is introduced and contrasted to more practical scheduling policies in Section 5. Section 6 illustrates the ideas of the paper in terms of the Dining Philosophers example, and the paper ends with some concluding remarks in Section 7.

2. A Simple Example

Maximal parallelism [20, 10] is a realistic scheduling principle for real-time systems. According to it, all communication events occur at the earliest possible times, and no unnecessary waiting is modeled. Intuitively it is reasonable to assume that any implementation follows this principle. In this section we demonstrate, however, that this principle may be too restrictive to be used in proofs, and that a more general approach is therefore needed.

We start with a discussion taken from [17], where some opinions are expressed that we will challenge. (The reader might notice that [17] deals explicitly with analysis of distributed systems, while our focus is in model-based specification and design.)

As stated in [17], the effect of the CSP program

$$P{::}\ R!1;\ \|\ Q{::}\ R!2;\ \|\ R{::}\ [P?x \rightarrow Q?y\ []\ Q?x \rightarrow P?y]$$

can be described by the postcondition

$$(x = 1 \wedge y = 2) \vee (x = 2 \wedge y = 1)\ .$$

Adding a delay statement with $t > 0$,

$$P{::}\ \mathbf{delay}\ t;\ R!1;\ \|\ Q{::}\ R!2;\ \|\ R{::}\ [P?x \rightarrow Q?y\ []\ Q?x \rightarrow P?y]\ ,$$

and assuming maximal parallelism, the postcondition reduces to

$$x = 2 \wedge y = 1\ .$$

From this the authors continue: "Under the interleaving semantics, no matter what the value of t is, we cannot get the same postcondition since there can be an arbitrary delay between the execution of any two actions. In other words, the delay statement in the above program is meaningless if we use the interleaving semantics. This informal reasoning shows that the interleaving semantics is unrealistic for real-time concurrency."

On this conclusion and its premises we have a few remarks. Firstly, the authors take it for granted that we should be able to prove the stronger postcondition. With this we do not agree. Instead we find it better to let the semantics of the delay statement correspond to its intuitive meaning only: the next event for the current process will not happen before the delay has passed. Reading more into it has the drawback that slight changes in relative execution speeds may then affect the behaviors that can materialize. Although this is certainly true in practice, it seems harmful in a model for reasoning. Insensitivity to relative speeds is a significant advantage of interleaving semantics, and we see no reason to give it up completely.

A related point is the adoption of the maximal parallelism principle. The fact that it reflects the behavior of real implementations is not by itself sufficient to justify its use in reasoning. Some drawbacks become evident when we consider its application during the design process. Typically, a refinement that preserves safety and liveness properties decreases the nondeterminism of the system [1, 6]. Therefore, such a transformation may remove those computational paths that would have been selected under maximal parallelism. This means that properties that rely on maximal parallelism – whether they refer to real time or not – are much more difficult to preserve in stepwise refinement than properties that do not rely on this assumption.

Thirdly, the interleaving model is discarded in [17], as it allows "an arbitrary delay between the execution of any two actions". Since the interleaving model has no notion of real time, this is analogous to discarding arithmetics for counting apples, as it allows them to be of any color. The question that should be asked is whether the interleaving model is consistent with reasonable modeling of real-time properties; this is the question that we shall address in this paper.

As an informal introduction to our approach let us analyze the above CSP program in terms of interleaving semantics. For this purpose we first augment it with explicit labels for the program counters of the three processes:

P:: α: **delay** t; β: $R!1$; γ: || Q:: δ: $R!2$; ε: || R:: ζ: $[P?x \rightarrow \eta$: $Q?y$ [] $Q?x \rightarrow \theta$: $P?y]$; ι:

and rewrite it as a joint action system [7, 9] with the following atomic actions, where $P.pc$, $Q.pc$, and $R.pc$ denote the three program counters:

action $a1$ **by** P -- delay in P without explicit duration
when $P.pc = \alpha$ **do** $P.pc := \beta$ **end**
action $a2$ **by** P, R -- P sends 1 to R at ζ
when $P.pc = \beta \wedge R.pc = \zeta$ **do** $P.pc := \gamma$; $R.pc := \eta$; $x := 1$ **end**
action $a3$ **by** P, R -- P sends 1 to R at θ
when $P.pc = \beta \wedge R.pc = \theta$ **do** $P.pc := \gamma$; $R.pc := \iota$; $y := 1$ **end**
action $a4$ **by** Q, R -- Q sends 2 to R at η
when $Q.pc = \delta \wedge R.pc = \eta$ **do** $Q.pc := \varepsilon$; $R.pc := \iota$; $y := 2$ **end**
action $a5$ **by** Q, R -- Q sends 2 to R at ζ
when $Q.pc = \delta \wedge R.pc = \zeta$ **do** $Q.pc := \varepsilon$; $R.pc := \theta$; $x := 2$ **end**

Each action has an enabling guard that allows it to be executed when true. The effects of the actions are given as sequences of assignments. Starting with $P.pc = \alpha$, $Q.pc = \delta$, $R.pc = \zeta$, we get three possible sequences of actions, $\sigma1 = \langle a1, a2, a4\rangle$, $\sigma2 = \langle a1, a5, a3\rangle$, and $\sigma3 = \langle a5, a1, a3\rangle$.

Since the interleaving model has no notion of real time, it cannot be used to deduce any real-time properties. To impose them, let us consider how $\sigma1$, $\sigma2$, and $\sigma3$ could be executed in real

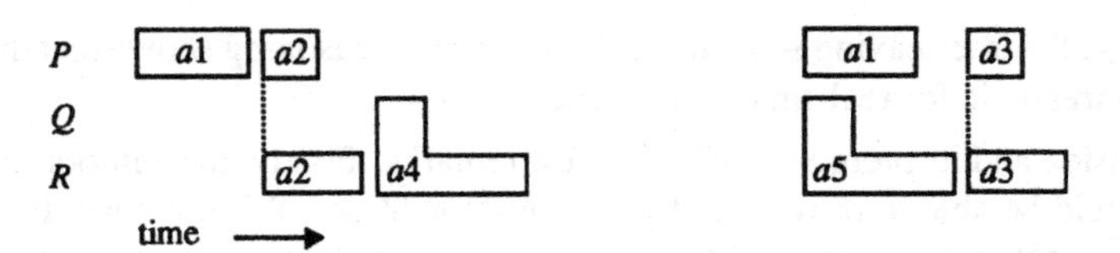

Figure 1. Possible timing diagrams for σ1 (left), and for σ2 and σ3 (right).

time. Assuming that there is a dedicated processor for each process, and that each action starts with a synchronized handshake by its participants, we could illustrate computations as in Figure 1, where arbitrary durations have been given for how long each participant is needed in each action. Instead of maximal parallelism we have applied a weaker principle of scheduling, where actions are scheduled in the order they appear in the interleaved computations, and each action is always scheduled to start as soon as is possible for all its participants.

Figure 1 imposes real-time properties on the underlying interleaved computations, but the original interleaving model is still also valid. Notice that several interleaved computations may lead to the same timing diagram, as is here the case for σ2 and σ3. This is exactly as it should be, since if Figure 1 represents what really happens in real time, σ2 and σ3 provide possible observations of the same distributed computation. The diagram on the left of Figure 1 corresponds to a unique interleaved computation, since common participants prevent observing *a*2 before *a*1, or *a*4 before *a*2.

3. Semantic Model

In this section we present the semantic model of *joint action systems*, on which the above informal action language was based.

The *global state* of a system is assumed to be partitioned into *objects* with local (program) variables. The combined contents of the local variables of one object are called its *local state.*[1] Systems are always considered as *closed systems* where the environment is also included. Therefore, the same kind of objects are used to model the state of the system, and the environment where the system is used.

Possible state changes in objects are expressed as *actions* $a \in A$. Each action *a* has a collection of *participant* objects *pa*, an *enabling guard ga*, which is a boolean expression that depends on the combined local states of *pa*, and a *body ba*, which expresses how the local states of *pa* are updated by the execution of *a*. Action *a* can be executed only in a state where *ga* is true, and it is assumed that *ba* then updates the local variables of *pa* in a deterministic manner, leaving other program variables intact.[2]

An action is said to be *enabled* when *ga* is true. A set of actions is said to be enabled when at least one of them is enabled. For simplicity we assume *A* always to contain a special *stuttering* action $i \in A$ for which *pi* is empty, *gi* is identically true, and *bi* has no affect on the state.

---

[1] Object-orientation in DisCo has led us to use the term "object" instead of "process". For the purposes of this paper this makes no difference, but, more generally, we should allow one process to be responsible for several objects.

[2] Termination of bodies is essential; determinism is assumed for simplicity.

We define:

A *joint action system* is a quadruple $S = (X, \Sigma_0, A, F)$ where X is a set of objects, Σ_0 is an initial condition for the global state, A is a set of actions, and F is a *fairness family*, which is a collection of *fairness sets* $f \subseteq A$.

A *computation* of a joint action system S is a non-empty, finite or infinite, alternating sequence of states and actions (ending in a state in the finite case), $c = \langle s_0, a_1, s_1, a_2, s_2, ...\rangle$, where s_i are states and a_i are actions such that

(i) the initial state s_0 satisfies Σ_0, and

(ii) for each occurrence a_i of an action, s_{i-1} satisfies ga_i, and states s_{i-1} and s_i differ only by the local states of pa_i being updated according to ba_i.

An infinite computation c is *fair* (with respect to all sets $f \in F$), if it satisfies the following condition:

(iii) if any of the fairness sets $f \in F$ is enabled in infinitely many states of c, then c contains infinitely many occurrences of actions in f.[1]

The set of all fair computations of S will be denoted by $C(S)$, or simply by C when S is known from context. Using *pcl* to denote prefix closure, $pcl(C)$ is the extension of C with all finite computations.

Since no real time is involved in S, its computations will also be called *logical* computations. When the identity of actions has no significance, logical computations can be given as sequences of states only, $c = \langle s_0, s_1, ...\rangle$. Similarly, when the initial state is known from context, it is sufficient to indicate only actions, $c = \langle a_1, a_2, ...\rangle$.

The significance in indicating the participants *pa* of an action is in guaranteeing syntactically that the local states of all other objects $x \notin pa$ remain unaffected by a. Two actions a and b are said to be *independent* if $pa \cap pb = \emptyset$. If a and b occur as consecutive actions in a computation, and a distributed implementation has allocated their participants to disjoint parts of the physical system, there may be no way for an external observer to determine the mutual order of a and b. Other parts of the computation being equal it is obvious that either both or none of the two orders give a correct computation. It is therefore reasonable to consider computations equivalent if they differ from each other in this way only. Two computations that can be transformed into each other by a finite number of such exchanges are finitely equivalent. This can be generalized as follows to situations where an infinite number of exchanges is needed.

Let two computations be called n-identical, if they are identical to each other as far as their first n actions (and the associated states) are concerned. Two computations are *partial-order equivalent* if, for any $n > 0$, each of them is finitely equivalent to a computation that is n-identical with the other. In particular, addition or deletion of stuttering keeps a computation within the same equivalence class.

Partial-order equivalence allows the view that interleaved computations are observations of "real" computations that are partial-order equivalence classes of such observations [19].

The most important properties of computations are *safety* and *liveness* properties of the state sequences involved. Safety properties are those that can be refuted by finite computations, where-

---

[1] This is also called *strong fairness*. We could have another family J for *justice* (which is also called *weak fairness*) [18]. With justice with respect to $f \in J$, f cannot be continually enabled without c containing infinitely many occurrences of actions in f. For simplicity we omit J in this paper.

as liveness properties are those that cannot. Each property (of a state sequence) is a combination of a safety property and a liveness property [3]. The actions of a joint action system determine safety properties only, while all liveness properties need to be given by fairness sets. Notice that even the fundamental liveness property, i.e., fairness with respect to the set of all non-stuttering actions, is not implicitly assumed.

An appropriate logic for reasoning on safety and liveness properties of reactive systems is *temporal logic of actions* [16]. The identity of actions and of participants is then abstracted away, which means that partial-order equivalence also remains outside this system for formal reasoning. Statistical properties are obviously also inexpressible, as they are not boolean properties possessed by all computations of a system.

When action participants are taken into consideration, the (partial-order) *equivalence closure* of a set C of computations will be denoted by $ecl(C)$. A property is called *equivalence robust* [4] if, for each equivalence class, it is possessed either by all or by none of the computations in the class. Since there is no objective way for an observer to distinguish between different computations in an equivalence class, all reasonable requirements for a system are equivalence robust. Therefore, if a specification in temporal logic of actions, for instance, determines a collection C of computations, then the larger set $ecl(C)$ can be equally well used instead.

4. Scheduled Systems

In order to discuss scheduling of actions, we have to distinguish between the system and its environment in a closed system.[1]

The objects of a system are partitioned into *internal* (system) and *external* (environment) *objects*, $X = X_S \cup X_E$, where the latter are used to model the state of the environment. This partitioning leads also to partitioning of non-stuttering actions into *system actions* and *environment actions*, $A = A_S \cup A_E \cup \{i\}$. System actions are those where all participants are internal objects; the others are environment actions. Intuitively, environment actions are those for which the environment is responsible. System actions model what the system does, and they cannot modify the environment directly. Communication between the system and its environment takes place through variables that are local to internal objects and are updated by both system and environment actions.

Furthermore, implementability requires that the fairness responsibilities of the system and the environment be separated. Therefore we assume that F is of the form $F = F_S \cup F_E$, where sets in F_S contain system actions only, and sets in F_E contain environment actions only. In addition, we assume fundamental liveness of the system part, i.e., $A_S \in F_S$, and that no environment action can disable any system action.

With these assumptions we can define what we mean by scheduling of actions. We use C to denote the set of all fair computations, and C^+ for those computations that are fair with respect to fairness sets in $\{A_S\} \cup F_E$. Obviously $C \subseteq C^+$, since fairness requirements are relaxed in C^+.

In accordance with [2], the generation and scheduling of computations can be interpreted as a two-person infinite game played by the system and the environment. The initial state s_0 is chosen by the environment, after which the environment and the system alternate moves, the environment taking the first move. In each environment move a finite (possibly empty) sequence

[1] Lamport [16] uses the term "closed system" to indicate that the system and its environment cannot be distinguished in a model. We use it in the weaker meaning that both are described together and in a uniform way.

of environment actions are executed, and a uniquely determined non-negative real number t_i (start time) is associated with each. In a system move, at most one system action is executed with a unique start time. In selecting their moves, both parties are restricted by the safety properties of S, which means that at each stage the game has produced some computation $c \in pcl(C^+)$ with an association of start times with its actions. For infinite games the players are restricted to produce computations in $ecl(C^+)$. (Stuttering actions can be assumed to be generated only in the end.) The system wins, if the produced infinite computation belongs to $ecl(C)$. The system loses (and the environment wins), if this is not the case.

This process leads to sequences of the form $\langle s_0, (a_1, t_1), s_1, (a_2, t_2), \ldots \rangle$ called *scheduled computations*. More precisely, we define:

> *Scheduling* is a partial mapping σ of C^+ to scheduled computations, with a non-empty domain $dom(\sigma)$, such that
>
> (i) whenever $\sigma(c)$ is defined, c and $\sigma(c)$ are identical with respect to occurrences of states s_i and actions a_i,
>
> (ii) if action a_i precedes a_j in c and $pa_i \cap pa_j \neq \emptyset$, then $t_j - t_i \geq 0$ in $\sigma(c)$,
>
> (iii) for two computations with a common prefix, cc', $cc'' \in dom(\sigma)$, the start times t_i associated with actions a_i in the common prefix c are identical in $\sigma(cc')$ and $\sigma(cc'')$,
>
> (iv) if $cc' \in dom(\sigma)$, and the prefix c ends in a state in which an environment action a is enabled, then there is some $cc'' \in dom(\sigma)$ such that c'' starts with a, and
>
> (v) if $cc' \in dom(\sigma)$, and the prefix c ends in a state in which some system action is enabled, then there is some $cc'' \in dom(\sigma)$ such that c'' starts with a system action.

By a *scheduled system* (S, σ) we now understand a joint action system S together with a scheduling σ. For each logical computation $c \in dom(\sigma)$, $\sigma(c)$ is a scheduled computation of (S, σ). Notice that all safety properties of S are still present in (S, σ), but liveness properties may be violated, since by definition $dom(\sigma)$ is not restricted to fair computations.

Some important properties of schedulings are defined as follows:

> Scheduling σ is *sound*, if for all partial-order equivalent pairs $c, c' \in C^+$ for which both $\sigma(c)$ and $\sigma(c')$ are defined, action occurrences that correspond to each other in $\sigma(c)$ and $\sigma(c')$ have identical start times.
>
> Scheduling σ is *complete*, if $C \subseteq dom(\sigma)$.
>
> Scheduling σ is *totally correct*, if $dom(\sigma) \subseteq ecl(C)$, i.e., σ allows only games where the system wins.
>
> Scheduling σ is *partially correct*, if it never leads (by a finite number of steps) to a situation where the environment has a winning strategy.[1] When not partially correct, σ is *incorrect.*

Sound scheduling expresses the intuitively natural view that partial-order equivalent computations are alternative observations of the same "real" computation, and, hence, there is no objective basis to schedule them differently. Complete scheduling introduces no additional logical

---

[1] Here we deviate slightly from standard terminology. We take partial correctness to mean that a program can be refined into a totally correct one by an arbitrary refinement, which may reduce both nondeterminism and the domain of nontermination [5]. Ordinarily only the latter kind of refinement would be allowed; here only the former kind is possible.

properties. Partial correctness guarantees that, no matter how the environment behaves, it is always possible for the system to win. Total correctness requires this to happen. We now have:

> **Proposition 1.** Any complete scheduling is partially correct.

This is true, since enforcing all fairness requirements on a complete scheduling makes the system always win.

> **Proposition 2.** If $ecl(C) = ecl(C^+)$, then any scheduling is totally correct.

This follows directly from the definitions.

> **Proposition 3.** If P is an equivalence robust property of S, and σ is totally correct, then P is also a property of (S, σ).

This is also obvious by the definition of total correctness.

5. Undelayed Scheduling

In practice, scheduling requires some execution agents. Here we assume each object to be such an agent. More generally, we could have an allocation mapping of objects to agents.

A scheduling augments computations with real-time properties. Proving that a model satisfies some real-time requirements must be based on assumptions on scheduling. A reasonable assumption is that scheduling is undelayed in the following sense:

> Scheduling is *undelayed*, if the start time of the next action is always determined to be as early as is possible for its participants.

Obviously, maximal parallelism is a special case of this.

To determine what the earliest possibilities for the participants are, we assume a duration $D(a, p) \geq 0$ for the participation of each participant $p \in pa$ in an action a. *Local time* can then be associated with each object x by an additional local variable T_x, which is initialized as 0. Whenever an action a is executed, its earliest start time is the maximum of the local times T_p of its participants $p \in pa$, and these variables T_p are updated by the execution of a to the values of this start time plus $D(a, p)$. This obviously leads to timing diagrams as the one given in Figure 1.

Since this unrestricted form of undelayed scheduling can be imposed on a joint action system S by adding local variables to indicate local times, the resulting scheduled system (S, σ) can be represented as an ordinary joint action system with these additional variables. Then there is no difference in reasoning about properties in S and (S, σ). In the following we shall use the terms *logical properties* and *real-time properties* to refer to state properties in S and in this representation of (S, σ), respectively. Obviously, a basic real-time safety property that is satisfied for all scheduled systems (S, σ) is that all local times are monotonically non-decreasing.

If we keep the original fairness family with this representation of (S, σ), all logical properties of S are included in the real-time properties of (S, σ), including all liveness properties. The resulting σ is therefore not only sound and complete, but also totally correct.

Given some durations $D(a, p)$, let σ_0 denote the unrestricted form of undelayed scheduling with the weakest possible fairness family $\{A_S\} \cup F_E$. Obviously, σ_0 is sound and complete, and at least partially correct. Furthermore, any undelayed scheduling σ can be obtained of it by reducing its domain, i.e., $\sigma(c) = \sigma_0(c)$ for all $c \in dom(\sigma) \subseteq dom(\sigma_0)$. If all logical computations in $dom(\sigma)$ are fair, then σ is totally correct. In particular, if σ_0 is totally correct, any undelayed scheduling can be used:

Proposition 4. If σ_0 is totally correct for S, then any undelayed scheduling σ is also totally correct for S.

This follows directly from the definition of total correctness.

Furthermore, all real-time properties of (S, σ_0) are satisfied by (S, σ) with any undelayed scheduling σ:

Proposition 5. For any real-time property P, if an assumption Q on durations implies that P holds in (S, σ_0), then for any undelayed scheduling σ, Q implies that P holds also in (S, σ).

This is a consequence of the fact that, whenever $\sigma(c)$ is defined, $\sigma(c) = \sigma_0(c)$.

Maximal parallelism is a frequently used form of undelayed scheduling. In our framework we can formulate it as follows:

Undelayed scheduling σ has the *maximal parallelism* property, if

(i) for each scheduled computation $\sigma(c) = \langle s, (a_1, t_1), s_1, (a_2, t_2), s_2, ...\rangle$, if a_{i+1} is a system action, and another system action a is enabled in state s_i with an earliest start time t, then either $pa_{i+1} \cap pa = \emptyset$, or $t_{i+1} \leq t$, and

(ii) σ is not defined for any logical computation c where, from some point on, a system action is continually enabled, but its participants are no longer involved in any action.

Since this principle reduces $dom(\sigma)$, σ is no longer complete, in general. As was argued in Section 2, relying on an incomplete scheduling policy may reduce the predictability of a program when its operational environment is modified. In addition, σ need no longer be even partially correct. This is easily shown by an example where maximal parallelism leads to continual "conspiration" against an action that should eventually be executed by fairness. An example of this will be discussed in Section 6.

This potential incorrectness of maximal parallelism can, of course, lead to two opposite conclusions. The one that is often made is to discard fairness as irrelevant for real-time concurrency. The other possibility, for which we argue, is to consider it as an approximation of enforcing real-time properties, and to keep it as the only means to force something to happen in the model. This leads, however, to questioning the significance of maximal parallelism as a basic assumption.

Another conventional mechanism for restricting scheduling, which can be used also in connection with maximal parallelism, is priorities. In practice, priorities are usually assigned to processes, which leads to the need for dynamic priorities. In joint action systems this drawback is avoided by associating priorities with actions in a way that forces immediate scheduling of an enabled action with highest priority:

Undelayed scheduling σ is *priority scheduling* if the following is true for each scheduled computation $\sigma(c) = \langle s, (a_1, t_1), s_1, (a_2, t_2), s_2, ...\rangle$: if a_{i+1} is a system action, and another system action a is enabled in state s_i with an earliest start time t, then either the priority of a_{i+1} is at least that of a, or $t_{i+1} \leq t$.

Priority scheduling is also incomplete and incorrect, in general, since selecting actions continually on the basis of priorities may lead to conflict with the given fairness requirements.

The general construction for undelayed scheduling proves the point that an interleaving model is not in conflict with real-time concurrency. It may be questioned, however, whether the liveness properties of an interleaving model are reasonable for practical scheduling. There are actually two different problems here. Firstly, fairness assumptions can easily be given that are unrealistic for practical implementation. Secondly, no fairness assumptions are sufficient, in general, to impose real-time bounds for the execution of actions.

The first problem may lead to a situation where the original fairness family must be replaced by something weaker, like the weakest possibility $\{A_S\} \cup F_E$, which may weaken total correctness of σ to partial correctness. This allows, however, further transformation of the system into a form where the set of logical computations is reduced so that simpler fairness assumptions are sufficient. In practice this means that some explicit policy is built into the system, by which the execution of some actions is prevented in certain situations. Such transformations, with sufficient conditions for correctness, have been explored in [8, 9].

The second problem can be solved by the same technique, since an action in a fairness set is forced to be executed at a given time instance, if no other action is then enabled, or if the enabled actions continually involve other objects only. Such a situation can again be achieved by suitable transformations that reduce the set of possible logical computations.

6. Example

As an example consider the dining philosophers problem, which we formulate in terms of joint actions as follows. There are n objects, $n > 1$, called philosophers, $P(0)$,..., $P(n\text{-}1)$, and n objects called forks, $F(0)$,..., $F(n\text{-}1)$. Fork $F(i)$ is said to be on the left of philosopher $P(i)$, and $F(i+1)$ is on his right, counting modulo n. The local state of each philosopher contains a boolean variable $P(i).h$ indicating whether he is hungry or not. For forks no local variables are introduced at this stage. The system is considered to be a closed system without an environment.

A high-level description of the system needs two kinds of actions for each philosopher, one for thinking, one for eating:

> **action** *think*(i) **by** $P(i)$
> **when** $\neg P(i).h$ **do** $P(i).h :=$ *true* **end**
> **action** *eat*(i) **by** $P(i)$, $F(i)$, $F(i+1)$
> **when** $P(i).h$ **do** $P(i).h :=$ *false* **end**

The implementation problems of synchronizing a philosopher with two forks are ignored in this description, resulting in a model where deadlocks do not arise. The simplest way to guarantee the expected liveness properties is to let the fairness family consist of all singleton sets of actions. With this assumption each non-hungry philosopher will eventually think and get hungry, and, since fairness is defined for purely sequential execution of actions, each hungry philosopher will eventually eat.

In this model a computation may contain an arbitrary number of actions *eat*(0), for instance, without any other eating actions intervening. Fairness forces, however, also other eating actions to appear eventually. If *eat*(2) is the first of those to appear, its scheduling does not depend on the number of preceding *eat*(0) actions. However, if *eat*(1) appears before it, fork $F(1)$ forces this *eat*(1) to be scheduled after all preceding *eat*(0) actions, and, because of fork $F(2)$, a subsequent *eat*(2) can only be scheduled to take place after that. This shows that the model allows arbitrarily long real-time intervals in which all the participants for an eating action are idle and ready for eating, but this is still not started. For thinking actions it is true, however, that no matter when *think*(i) appears, it will always be scheduled to take place immediately after the previous *eat*(i).

Obviously, similar idle periods would not arise with maximal parallelism. However, maximal parallelism could easily force a philosopher to starve by leading to continual "conspiration" by the two neighbors. This demonstrates a conflict with the given fairness requirements. Not only does the domain of maximal parallelism scheduling include unfair computations, but these cannot even be removed by any decrease in nondeterminism, which means that the scheduling is incorrect.

As discussed in [9], the above fairness assumption is unreasonably strong in the sense that it cannot be expected to be directly enforced in a distributed implementation. Superposition was therefore applied in [9] to refine this specification into an implementable form where a specific policy disables eating actions in certain situations. The same idea will be used here, but this time to achieve a specification with implementable real-time properties.

Let the minimal and maximal durations of thinking and eating be $Think_{min}$, $Think_{max}$, Eat_{min}, and Eat_{max}, respectively. For simplicity we restrict our discussion to the situation where an eating action always has the same duration for all its three participants, and where thinking never takes longer than eating, $Eat_{min} \geq Think_{max}$. Under these assumptions we want to enforce a policy under which we can prove a real-time bound for a hungry philosopher to start eating.

Let the variables for the local times of philosophers and forks be denoted by $P(i).t$ and $F(i).t$. For a hungry philosopher, $P(i).t$ shows when he last became hungry in real time. When an eating action is executed, the maximum of $P(i).t$, $F(i).t$, and $F(i+1).t$ indicates when eating starts in real time. Therefore, in addition to preserving starvation freedom, we want to be able to prove an invariant $P(i).h \Rightarrow \max(F(i).t, F(i+1).t) - P(i).t \leq b$ for all i and some bound b.

The particular policy that we shall investigate is one where no philosopher is allowed to eat again until both of his neighbors have also eaten. This can be enforced by superposition as follows. The local state of each fork is augmented with a boolean variable $F(i).m$ to indicate whether the fork is "marked" or not. This marking is then utilized in eating actions so that eating is possible only with a marked left fork and an unmarked right fork, and the forks are subsequently exchanged:

> **action** *eat*(i) **by** $P(i)$, $F(i)$, $F(i+1)$
> **when** $P(i).h \wedge F(i).m \wedge \neg F(i+1).m$ **do**
> $P(i).h := false$; $F(i).m := false$; $F(i+1).m := true$
> **end**

Of the initial state we assume that k of the forks are marked, $0 < k < n$, and that exactly those philosophers are hungry for whom the forks are ready for eating. The following invariant obviously holds:

> *Invariant 1.* It is always true that exactly k forks are marked.

As an instance of superposition, the transformation preserves all safety properties. As for liveness, it is easy to prove that fairness with respect to each individual action still guarantees that each philosopher eats infinitely often. In this case this fairness requirement is also feasible for distributed implementation [9], since conspiratorial behaviors are no longer possible.

The crucial safety properties for our purposes are the following:

> *Invariant 2.* (a) For a non-hungry philosopher $P(i)$ it is always true that if the left fork is unmarked, then the local times of the two objects are the same, $\neg P(i).h \wedge \neg F(i).m \Rightarrow F(i).t = P(i).t$.
>
> (b) Correspondingly, $\neg P(i).h \wedge F(i+1).m \Rightarrow F(i+1).t = P(i).t$.

Invariant 3. (a) For a non-hungry philosopher $P(i)$ it is always true that, if the left fork is marked, then its local time is at least that of the philosopher, $\neg P(i).h \wedge F(i).m \Rightarrow F(i).t \geq P(i).t$. Furthermore, if $F(i).t - P(i).t > j\, Eat_{max}$, then there are at least j+1 unmarked forks with local times at least $F(i).t$, $F(i).t - Eat_{max}, \ldots, F(i).t - j\, Eat_{max}$, respectively.

(b) Correspondingly, $\neg P(i).h \wedge \neg F(i+1).m \Rightarrow F(i+1).t \geq P(i).t$, and, if in this case $F(i+1).t - P(i).t > j\, Eat_{max}$, then there are at least j+1 marked forks with local times at least $F(i+1).t$, $F(i+1).t - Eat_{max}, \ldots, F(i+1).t - j\, Eat_{max}$, respectively.

Assume first that thinking takes no time, $Think_{min} = Think_{max} = 0$. Then the conditions in invariants 2 and 3 hold also for hungry philosophers, which can be proved directly by checking that they hold initially and are preserved by each execution of an action. Invariant 1 then implies that $\max(F(i).t, F(i+1).t) - P(i).t \leq \max(k, n-k)\, Eat_{max}$. With $k = \lfloor n/2 \rfloor$ or $\lceil n/2 \rceil$ this gives the value $\lceil n/2 \rceil\, Eat_{max}$ for the best upper bound b.

When non-zero durations are reintroduced to thinking, condition $Eat_{min} \geq Think_{max}$ ensures that the above invariants still hold, although the corresponding conditions for hungry philosophers have to be adjusted. The assumption on the initial state then guarantees that thinking time always reduces waiting, so that the final bound reduces to $\lceil n/2 \rceil\, Eat_{max} - Think_{min}$.

In conclusion, this example illustrates how strict time bounds can be proved with a technique that is well-known from non-timed behavioral models. An obvious advantage is that the proofs remain valid under any scheduling principles. On the other hand, it may seem that the policies by which real-time bounds can be guaranteed in this approach are too rugged to allow efficient scheduling in situations where that would be possible. Additional freedom for scheduling is, however, possible at the cost of a less distributed policy. It should be noticed that a similar cost is always involved in attempts to solve the problem by practical scheduling principles.

7. Concluding Remarks

We have shown how an interleaving model can be used as a basis for dealing also with real-time properties. This leads to a uniform approach where the same system of reasoning can be applied to both state properties and real-time properties, and real-time modeling is possible already at early stages of specification.

The unrestricted form of undelayed scheduling that is natural for this approach affects also the design method to be used. In particular, it suggests that for meeting any real-time requirements one should not rely on practical scheduling policies. This makes it possible to avoid implicit assumptions that are difficult to utilize in proofs and also cause sensitivity to innocent-looking modifications in a program.

The alternative that is suggested here is to build sufficient mechanisms into the system itself, so that such computations become logically impossible that would not meet the given real-time requirements. Instead of relying on implicit assumptions, all the required mechanisms are then made explicit in some stage of the design. This may mean more work, but we believe that this is more than compensated by the advantages: a uniform approach to stepwise design and the use of formal proofs, and improved insensitivity to changes in the assumptions of the real-time model.

Acknowledgments

The joint action approach was developed together with Ralph Back. Research on its use as a basis for a practical specification language was carried out in project DisCo, which was part of the FINSOFT programme of the Technology Development Centre of Finland, and was supported by four industrial partners.

References

[1] Abadi, M., Lamport, L., The existence of refinement mappings. Research Report 29, Digital Systems Research Center, 1990. To appear in *Theoretical Computer Science*.

[2] Abadi, M., Lamport, L., Composing specifications. Research Report 66, Digital Systems Research Center, 1990.

[3] Alpern, B., Schneider, F.B., Defining liveness. *Information Processing Letters 21*, 4 (Oct. 1985), 181-185.

[4] Apt, K.R., Francez, N., Katz, S., Appraising fairness in distributed languages. *Distributed Computing 2*, (Aug. 1988), 226-241.

[5] Back, R.J.R., A calculus of refinements for program derivations. *Acta Informatica 25*, (1988), 593-624.

[6] Back, R.J.R., Refinement calculus II: parallel and reactive programs. In *Stepwise Refinement of Distributed Systems: Models, Formalisms, Correctness*, LNCS 430, Springer-Verlag 1990, 67-93.

[7] Back, R.J.R., Kurki-Suonio, R., Decentralization of process nets with a centralized control. *Distributed Computing 3* (1989), 73-87. An earlier version in *Proc. 2nd ACM SIGACT-SIGOPS Symposium on Principles of Distributed Computing*, Montreal, Canada, Aug. 1983, 131-142.

[8] Back, R.J.R., Kurki-Suonio, R., Serializability in distributed systems with handshaking. In *Automata, Languages and Programming* (Ed. T. Lepistö and A. Salomaa), LNCS 317, Springer-Verlag 1988, 52-66.

[9] Back, R.J.R., Kurki-Suonio, R., Distributed cooperation with action systems. *ACM Trans. Programming Languages and Systems 10*, 4 (Oct. 1988), 513-554.

[10] Hooman, J., Widom, J., A temporal-logic based compositional proof system for real-time message passing. In *PARLE '89 Parallel Architectures and Languages Europe, Vol II* (Ed. E. Odijk, M. Rem and J.-C. Syre), LNCS 366, Springer-Verlag 1989, 424-441.

[11] Järvinen, H.-M., Kurki-Suonio, R., DisCo specification language: marriage of actions and objects. *Proc. 11th International Conference on Distributed Computing Systems*, Arlington, Texas, May 1991, IEEE Computer Society Press, 142-151.

[12] Järvinen, H.-M., Kurki-Suonio, R., Sakkinen, M., Systä, K., Object-oriented specification of reactive systems. *Proc. 12th International Conference on Software Engineering*, Nice, France, March 1990, IEEE Computer Society Press, 63-71.

[13] Kurki-Suonio, R., Operational specification with joint actions: serializable databases. To appear in *Distributed Computing*.

[14] Kurki-Suonio, R., Modular modeling of temporal behaviors. To appear in *Information Modelling and Knowledge Bases: Foundations, Theory, and Applications* (Ed. S. Ohsuga, H. Kangassalo, H. Jaakkola, K. Hori, N. Yonezaki). IOS Press, Amsterdam, 1991.

[15] Kurki-Suonio, R., Järvinen, H.-M., Action system approach to the specification and design of distributed systems. *Proc. Fifth International Workshop on Software Specification and Design. ACM Software Engineering Notes 14*, 3 (May 1989), 34-40.

[16] Lamport, L., A temporal logic of actions. Research Report 57, Digital Systems Research Center, 1990. A revised and extended version is in preparation.

[17] Liu, L.Y., Shyamasundar, R.K., Static analysis of real-time distributed systems. *IEEE Trans. on Software Engineering 16*, 4 (April 1990), 373-388.

[18] Pnueli, A., Applications of temporal logic to the specification and verification of reactive systems: a survey of current trends. In *Current Trends in Concurrency* (Ed. J.W. de Bakker, W.-P. de Roever and G. Rozenberg), LNCS 224, Springer-Verlag 1986, 510-584.

[19] Reisig, W., Partial order semantics versus interleaving semantics for CSP-like languages and its impact on fairness. In *Automata, Languages and Programming* (Ed. J. Paredaens), LNCS 172, Springer-Verlag 1984, 403-413.

[20] Salwicki, A., Müldner, T., *On the Algorithmic Properties of Concurrent Programs*. LNCS 125, Springer-Verlag 1981.

[21] Systä, K., A graphical tool for specification of reactive systems. *Proc. Euromicro '91 Workshop on Real-Time Systems*, Paris, France, June 1991, IEE Computer Society Press, 12-19.

A Temporal Approach to Requirements Specification of Real-Time Systems

Yogesh Naik
Department of Computer Science
University of Warwick
Coventry CV4 7AL, U.K.

Abstract

This paper describes a specification notation of temporal logic to describe the requirements of real-time systems. The notation is extended by a calculus of occurrences of predicates. Using the logic and the calculus we show that common real-time properties such as durations, number of occurrences, precedence and other properties can be described. It is then used to describe the IEEE 802 Token Bus specification.

1 Introduction

In recent years there has been considerable interest in specification of real-time systems. For a real-time system, one needs to specify requirements of responsiveness of events and satisfy timing constraints. The responsive requirement merely states a temporal relationship between events, say p and q. There is no need to specify when p and q must occur or what their durations must be. To specify timing constraints several alternative approaches have been suggested. These can roughly be grouped into two distinct approaches: *explicit clock* and *bounded operator*. In the *explicit clock* approach, no new operators are introduced. Instead to refer to time, a flexible variable T is used to denote the current value of a global clock. Various examples of this style of specification can be found in the literature, for instance RTTL [Ost89], GCTL [PH88], and XCTL [HLP90]. On the other hand, the *bounded operator* approach introduces new temporal operators to describe timing constraints. For each temporal operator (such as $\Diamond$, $\Box$) a new temporal operator is defined. For example, the temporal operator $\Diamond_{\leq\tau}\varphi$ is used to state that φ will eventually be true within τ time units from now. This approach is described in [Koy89]. A similar approach is taken in Timed CTL (TCTL) [ACD90], where for each temporal operator of Computation Tree Logic (CTL) [EC82] a new operator is defined with subscript to restrict its scope in time. There are various other approaches to real-time specification which donot fall into the above classification. A calculus of durations is introduced in [CHR91] to reason about timing constraints of time-critical systems. It uses a calculus of integrals of duration of states in an interval to state timing properties of a system. It is based on an interval temporal logic [MM83] in which integrals of durations of a state are considered as variables. It differs from other approaches because it doesnot make any explicit reference to time. An alternative approach where time is explicit is taken in RTL [JM86]. RTL uses an occurrence function which returns the time of occurrence of an event. Times of different events can be compared to state timing properties.

The main focus of the use of temporal logics and the need to extend its expressive power has been the specification and verification of programs [MP81, BKP84, HW89] but not in describing requirements of systems. Consider the following example.

Example : "The nodes on a token bus communicate by passing frames. The right of access to the network is regulated by using a token frame, which a node may hold for a maximum of 20 time slots. During the period a node holds the token, it can perform general maintenance by sending frames. For example, to grant an opportunity to other nodes to insert themselves in the ring, the token-holder sends a solicit-successor frame. If there is no response, it tries twice before taking any other action."

To describe the above example, one needs additional concepts and operators. For instance, the statement "a node may hold the token for a maximum of 20 time slots" requires a notion of duration of a predicate. Another example is the statement "If there is no response the node tries twice before taking any action" can be expressed more naturally if we can count the number of times the node sends the solicit-successor frame before it takes any action. Moreover, operators such as *during*, *precedes* etc. seem to be more appropriate for describing requirements.

The main contribution of this paper is to extend a temporal logic so that some additional features and operators such as *during*, duration, and number of occurrences can be easily expressed. Most specification approaches in computer science are based on the central notion that events are instantaneous. In specification of real-time systems, we abandon this notion in favour of events which may have extent in time. We can then specify common real-time properties such as duration, precedence, inclusion of events etc. We further introduce a concept of an occurrence of a predicate φ to denote a maximal interval by inclusion in which φ holds everywhere. By relating occurrences of predicates, we show that behaviours of systems can be described. This can be compared with temporal logics where an occurrence of a predicate is the time point at which it is true.

The rest of the paper is organised as follows. The syntax and the semantics of the logic is presented in the next section. The proof theory is given in Section 3. Section 4 gives the calculus of occurrences of predicates including behaviour rules for durations and number of occurrences. Section 5 takes a description of the IEEE 802 Token Bus and uses the logic to formally specify it.

2 The Temporal Framework

2.1 Syntax

The basic symbols of the language consists of constants, variables, propositions, functions and predicate symbols. We use the usual set of propositional connectives: negation ($\neg$), and implication ($\Rightarrow$) and the first-order universal quantifier ($\forall$) which is applied only to variables. The modal connectives we use are weak next ($\odot$), weak previous ($\oslash$), weak until (**U**) and weak since (**S**). Formally,

1. If φ and ψ are formulae, then so are $\neg\varphi$, and $\varphi \Rightarrow \psi$.
2. If x is a variable and φ is a formula, then $\forall x.\varphi$ is also a formula.
3. If φ and ψ are formulae, then so are $\odot\varphi$, $\varphi\mathbf{U}\psi$, $\oslash\varphi$, and $\varphi\mathbf{S}\psi$.

Other operators are defined syntactically using these connectives.

2.2 Semantics

We define a model $M = (I, \sigma, t)$, consisting of an interpretation function I, a mapping σ called behaviour and t is a time point.

- The interpretation function I specifies a nonempty set D and assigns constants, function and predicate symbols to elements in D. For a constant c, I assigns a fixed element in D, for a n-place function symbol I assigns a function $D_1 \times D_2 \ldots D_n$ into D, and for a n-place predicate symbol it assigns n-ary relation in D.
- Let T be an infinite domain of time values which is linearly ordered and closed under addition and $STATE$ be defined as a set of mappings from a set of variables VAR to a set of values VAL. i.e. it assigns values to variables.

  $$STATE = \{s | s : VAR \rightarrow VAL\}$$

  Then a behaviour σ is a mapping from T to $STATE$.

  Definition 2.1 **(Well-formedness)** In the rest of the thesis, we will consider only those behaviours which observe the following properties. These properties are essentially those which rule out the possibility of infinitely many state changes in a finite time i.e. finite variability (cf. [BKP86])

$$\forall t.((\exists t'.(t' > t \wedge \sigma(t') \neq \sigma(t))) \Rightarrow \exists t''.(t < t'' \wedge \sigma(t'') = \sigma(t') \\ \wedge \forall t'''.(t < t''' < t'' \Rightarrow \sigma(t''') = \sigma(t)))$$

The property states that if there is a future instant in which a change in state occurs then there is a first such time point.

$$\forall t.((\exists t'.(t' < t \wedge \sigma(t') \neq \sigma(t))) \Rightarrow \exists t''.(t'' < t \wedge \sigma(t'') = \sigma(t') \\ \wedge \forall t'''.(t'' < t''' < t \Rightarrow \sigma(t''') = \sigma(t)))$$

The second property is similar for the past instances.

- The time point $t \in T$ is the point at which the formula is being interpreted.

Given a well-formed formula A, its meaning is given inductively below. For readability, we have removed I from the definitions. The value of a term τ under M is denoted by $\tau|_t^\sigma$ with I implicit in the definition.

- For a constant

  $c|_t^\sigma = I(c)$

- For a variable

  $x|_t^\sigma = (\sigma(t))(x)$

- For an n-ary function

  $F(t_1, ..., t_n)|_t^\sigma = I(F)(t_1|_t^\sigma, ..., t_n|_t^\sigma)$

  i.e. the value given by the application of $I(F)$ to the value $(t_1, \ldots, t_n)$.

- For an n-ary predicate $\varphi(t_1, \ldots, t_n)$

  $(I, \sigma, t) \models \varphi(t_1, \ldots, t_n)$ iff $I(\varphi)(t_1|_t^\sigma, \ldots, t_n|_t^\sigma)$

  An n-ary predicate is true iff it holds at time t.

- For negation

  $(\sigma, t) \models \neg\varphi$ iff not $(\sigma, t) \models \varphi$

  $\neg\varphi$ holds iff φ does not hold.

- For implication

$(\sigma,t) \models \varphi \Rightarrow \psi$ iff $(\sigma,t) \models \varphi$ implies $(\sigma,t) \models \psi$

$\varphi \Rightarrow \psi$ holds iff φ implies ψ.

- For universal quantification

$(\sigma,t) \models \forall x.\varphi$ iff for every $d \in D$ $(\sigma,t) \models \varphi(d/x)$, where $\varphi(d/x)$ is obtained by substituting d for all free occurrences of x in φ

- For weak next

$$\begin{aligned}(\sigma,t) \models \odot\varphi \text{ iff } & \exists t'.(\, t < t' \wedge (\sigma,t') \models \varphi \wedge \sigma(t) \neq \sigma(t') \\ & \qquad \wedge(\, \forall t''.(t < t'' < t' \Rightarrow \sigma(t'') = \sigma(t)) \\ & \qquad\quad \vee \forall t''.(t < t'' < t' \Rightarrow \sigma(t'') = \sigma(t')))) \\ & \vee \forall t'.(t < t' \Rightarrow \sigma(t') = \sigma(t))\end{aligned}$$

$\odot\varphi$ holds now iff φ is true at some time t' in the future and the state at that instant is different from what it is now and does not change between now and t' or no such t' exists.

- For weak until

$$\begin{aligned}(\sigma,t) \models \varphi\mathbf{U}\psi \text{ iff } & \exists t'.(t \leq t' \wedge (\sigma,t') \models \psi \wedge \forall t''.(t \leq t'' < t' \Rightarrow (\sigma,t'') \models \varphi)) \\ & \vee \forall t'(t \leq t' \Rightarrow (\sigma,t') \models \varphi)\end{aligned}$$

$\varphi\mathbf{U}\psi$ is true iff there is a future instant in which ψ is true and φ holds holds continuously until that instant or φ holds in all future instances.

- For weak previous

$$\begin{aligned}(\sigma,t) \models \oslash\varphi \text{ iff } & \exists t'.(\, t' < t \wedge (\sigma,t') \models \varphi \wedge \sigma(t) \neq \sigma(t') \\ & \qquad \wedge(\, \forall t''.(t' < t'' < t \Rightarrow \sigma(t'') = \sigma(t)) \\ & \qquad\quad \vee \forall t''.(t' < t'' < t \Rightarrow \sigma(t'') = \sigma(t')))) \\ & \vee \forall t'.(t' < t \Rightarrow \sigma(t') = \sigma(t))\end{aligned}$$

$\oslash\varphi$ holds now iff φ is true at some time t' in the past and the state at that instant is different from what it is now and does not change in the interval between now and t' or no such t' exists.

- For weak since

$$\begin{aligned}(\sigma,t) \models \varphi\mathbf{S}\psi \text{ iff } & \exists t'.(\, t' \leq t \wedge (\sigma,t') \models \psi \wedge \forall t''.(t' < t'' \leq t \Rightarrow (\sigma,t'') \models \varphi)) \\ & \vee \forall t'(t' \leq t \Rightarrow (\sigma,t') \models \varphi)\end{aligned}$$

$\varphi\mathbf{S}\psi$ is true there is an instant in the past in which ψ is true and φ holds continuously since that instant or φ holds in all past instances.

A wff A is logically valid iff A is true for every model. This is denoted by

$$\models A$$

Following are syntactic abbreviations for temporal operators.

DF1. $\Box\varphi \triangleq \varphi \mathbf{U} \mathit{false}$	DP1. $\boxminus\varphi \triangleq \varphi \mathbf{S} \mathit{false}$
DF2. $\Diamond\varphi \triangleq \neg\Box\neg\varphi$	DP2. $\diamondminus\varphi \triangleq \neg\boxminus\neg\varphi$
DF3. $\varphi\mathcal{U}\psi \triangleq \varphi\mathbf{U}\psi \wedge \Diamond\psi$	DP3. $\varphi\mathcal{S}\psi \triangleq \varphi\mathbf{S}\psi \wedge \diamondminus\psi$
DF4. $\oplus\varphi \triangleq \neg\odot\neg\varphi$	DP4. $\ominus\varphi \triangleq \neg\oslash\neg\varphi$
DF5. $\varphi\mathcal{U}^{+}\psi \triangleq \varphi \wedge \varphi\mathcal{U}\psi$	DP5. $\varphi\mathcal{S}^{+}\psi \triangleq \varphi \wedge \varphi\mathcal{S}\psi$
DF6. $\varphi\mathbf{U}^{+}\psi \triangleq \varphi \wedge \varphi\mathbf{U}\psi$	DP6. $\varphi\mathbf{S}^{+}\psi \triangleq \varphi \wedge \varphi\mathbf{S}\psi$

3 Proof Theory

The proof theory consists of seven axioms each for propositional part of the future and past fragments.

Axioms

F1. $\vdash \Box(\varphi \Rightarrow \psi) \Rightarrow (\Box\varphi \Rightarrow \Box\psi)$	P1. $\vdash \boxminus(\varphi \Rightarrow \psi) \Rightarrow (\boxminus\varphi \Rightarrow \boxminus\psi)$
F2. $\vdash \odot(\varphi \Rightarrow \psi) \Rightarrow (\odot\varphi \Rightarrow \odot\psi)$	P2. $\vdash \oslash(\varphi \Rightarrow \psi) \Rightarrow (\oslash\varphi \Rightarrow \oslash\psi)$
F3. $\vdash \odot\neg\varphi \Rightarrow \neg\odot\varphi$	P3. $\vdash \oslash\neg\varphi \Rightarrow \neg\oslash\varphi$
F4. $\vdash \Box\varphi \Rightarrow \varphi \wedge \Box\odot\varphi$	P4. $\vdash \boxminus\varphi \Rightarrow \varphi \wedge \boxminus\oslash\varphi$
F5. $\vdash \Box(\varphi \Rightarrow \odot\varphi) \Rightarrow (\varphi \Rightarrow \Box\varphi)$	P5. $\vdash \boxminus(\varphi \Rightarrow \oslash\varphi) \Rightarrow (\varphi \Rightarrow \boxminus\varphi)$
F6. $\vdash \varphi\mathbf{U}\psi \Leftrightarrow (\psi \vee (\varphi \wedge \odot(\varphi\mathbf{U}\psi)))$	P6. $\vdash \varphi\mathbf{S}\psi \Leftrightarrow (\psi \vee (\varphi \wedge \oslash(\varphi\mathbf{S}\psi)))$
F7. $\vdash \Box\varphi \Rightarrow \varphi\mathbf{U}\psi$	P7. $\vdash \boxminus\varphi \Rightarrow \varphi\mathbf{S}\psi$

Axiom F1 states that if φ always implies ψ in the future then if φ holds in the future then so does ψ. Axiom F2 is similar to F1 for the "weak next" operator. Axiom F3 establishes that if there exists a next state in which $\neg\varphi$ holds then φ does not hold in the next state. Axiom F4 states that the present is part of the future. It also states that henceforth either φ is true in the next state or φ is always true. Axiom F5 states that if it is always the case that φ implies that in the next state φ holds then if φ holds now then φ always holds. Axiom F6 defines " weak until" recursively by stating that either ψ is true now or φ is true and in the next state $\varphi\mathbf{U}\psi$ holds. Axiom F7 states that "henceforth φ" implies $\varphi\mathbf{U}\psi$ holds. Axioms P1-P7 are symmetric to F1-F7 for the past fragment.

There are two inference rules : modus ponens, and $\Box\boxminus$ Introduction

Inference Rules

R1. Modus Ponens

$$\frac{\vdash \varphi, \vdash \varphi \Rightarrow \psi}{\vdash \psi}$$

R2. $\Box$ and $\boxminus$ Introduction

$$\frac{\vdash \varphi}{\vdash \Box\varphi \wedge \boxminus\varphi}$$

4 A Calculus of Occurrences

Let t_1 and t_2 be such that $t_1 \leq t_2$. Then, we say that intervals of the form $[t_1, t_2]$ are closed.

Definition 4.1 (Occurrence of a predicate) An occurrence of a predicate φ is defined as a maximal interval (by inclusion) in which φ holds everywhere in that interval.

Given a first order predicate φ, a time point t and a computation sequence σ, we define a function ρ which return a set of occurrences of a predicate φ which occur after the time point t.

$$\begin{aligned}\rho(\varphi, t, \sigma) = \{[t_1, t_2] \mid\, & t \leq t_1 \leq t_2 \wedge \forall t'.(t_1 \leq t' \leq t_2 \Rightarrow (\sigma, t') \models \varphi) \\ & \wedge \neg\exists t'.(t_2 < t' \wedge \forall t''.(t_2 < t'' < t'. \Rightarrow (\sigma, t'') \models \varphi)) \\ & \wedge \neg\exists t'.(t \leq t' < t_1 \wedge \forall t''.(t' < t'' < t_1 \Rightarrow (\sigma, t'') \models \varphi)\}\end{aligned}$$

Given a computation sequence σ, and a time point t, we define number of occurrences of a predicate as the cardinality of the set ρ.

$$\#(\varphi, t, \sigma) = \#\rho(\varphi, t, \sigma)$$

Let ρ' be a sequence obtained from ρ such all its members are in ρ' and they are ordered by precedence relation on intervals.

We define the duration of the ith occurrence of a predicate as the duration of ith member of ρ'.

$$D(\varphi, i, t, \sigma) = \begin{cases} t_2 - t_1 & \text{if } [t_1, t_2] \text{ is the } i^{th} \text{ occurrence of } \varphi \\ 0 & \text{otherwise} \end{cases}$$

An occurrence predicate is a predicate on number of occurrences of predicates $\varphi_1, \ldots, \varphi_m$. If $\mathcal{N}\varphi_1, \ldots, \mathcal{N}\varphi_m$ are free variables corresponding to the number of occurrences of predicates $\varphi_1, \ldots, \varphi_m$ and $OCC[\mathcal{N}\varphi_1, \ldots, \mathcal{N}\varphi_m]$ is a general predicate over the number of occurrences, then the semantics of an occurrence predicate is given by

$$(\sigma, t) \models OCC[\mathcal{N}\varphi_1, \ldots, \mathcal{N}\varphi_m] \text{ iff } OCC[\#(\varphi_1, t, \sigma), \ldots, \#(\varphi_m, t, \sigma)]$$

We can similarly define a predicate for the duration of an occurrence of a predicate and its semantics. If $\mathcal{D}(\varphi_1, i), \ldots, \mathcal{D}(\varphi_m, j)$ are free variables corresponding to predicates $\varphi_1, \ldots, \varphi_m$ and $DUR[\mathcal{D}(\varphi_1, i), \ldots, \mathcal{D}(\varphi_m, j)]$ is a general duration predicate, the semantics of a duration predicate is given by

$$(\sigma, t) \models DUR[\mathcal{D}(\varphi_1, i), \ldots, \mathcal{D}(\varphi_m, j)] \text{ iff } DUR[D(\varphi_1, i, t, \sigma), \ldots, D(\varphi_m, j, t, \sigma)]$$

Occurrence and duration predicates are true for a model $M = (\sigma, t)$ iff M satisfies them and they are false iff M does not satisfy them. They are valid iff they are true for all models.

4.1 Behaviour Rules

The following are valid rules for occurrences of predicates in the logic.

Rule S1 states that the number of occurrences of $false$ is zero.

S1. $\mathcal{N}(false) = 0$

Rule S2 states that the number of occurrences of any predicate is non-negative.

S2. $\mathcal{N}(\varphi) \geq 0$

S3 states that the difference in the number of occurrences of a predicate and its negation is either zero or one.

S3. $0 \leq |\mathcal{N}(\varphi) - \mathcal{N}(\neg\varphi)| \leq 1$

Rule S4 states that if henceforth φ is true then the number times it occurs is one.

S4. $\Box\varphi \Rightarrow (\mathcal{N}(\varphi) = 1)$

Rule S5 states that number of occurrences any predicate either remains the same or decreases in the next state.

S5. $(\mathcal{N}(\varphi) = m \wedge \odot(\mathcal{N}(\varphi) = n) \Rightarrow (m \geq n)$

Rule S6 is similar rule to S5 for the previous state.

S6. $(\mathcal{N}(\varphi) = m \wedge \oslash(\mathcal{N}(\varphi) = n) \Rightarrow (m \leq n)$

The total duration of a predicate is defined as

$$\mathcal{T}(\varphi) \triangleq \sum_{i=1}^{\mathcal{N}(\varphi)} \mathcal{D}(\varphi, i)$$

Rule S7 states the duration of false is zero.

S7. $\forall i.\mathcal{D}(false, i) = 0$

Rule S8 states that duration of any occurrence of a predicate is greater than or equal to zero.

S8. $\forall i.\mathcal{D}(\varphi, i) \geq 0$

Rule S9 states that total duration of any two individual predicates is the sum of durations of their disjunction and their conjunction.

S9. $\mathcal{T}(\varphi) + \mathcal{T}(\psi) = \mathcal{T}(\varphi \vee \psi) + \mathcal{T}(\varphi \wedge \psi)$

Some theorems about occurrences of predicates are given in Appendix A.2.

4.2 Relating Occurrences of Predicates

Given predicates φ and ψ, we use syntactic abbreviations to relate occurrences of predicates. Five operators are defined : *occurs before*, *during*, *occurs immediately before*, *precedes* and *includes*.

$\varphi \lhd \psi$ (*occurs before*) if true iff (i) the current occurrence of φ is followed by an occurrence of ψ with no occurrences of φ in between (ii) or the current occurrence of ψ is preceded by an occurrence of φ with no occurrences of ψ in between. It is described using two disjuncts ω_1 and ω_2. ω_1 states that φ is true now and will become false and remain false until ψ becomes true.

$$\omega_1 \triangleq \varphi \wedge (\varphi \mathcal{U}(\neg\varphi \mathcal{U}((\neg\varphi \wedge \neg\psi)\mathcal{U}^+\psi)) \vee (\varphi \mathcal{U}((\varphi \wedge \neg\psi)\mathcal{U}^+(\neg\varphi \wedge \psi))))$$

ω_2 states that ψ is true now and φ has been true sometime in the past and became and remained false until ψ became true.

$$\omega_2 \triangleq \psi \wedge (\psi \mathcal{S}(\neg\psi \mathcal{S}((\neg\psi \wedge \neg\varphi)\mathcal{S}^+\varphi)) \vee (\psi \mathcal{S}((\psi \wedge \neg\varphi)\mathcal{S}^+(\neg\psi \wedge \varphi))))$$

$\varphi \lhd \psi$ is then defined as

$$\varphi \lhd \psi \triangleq \omega_1 \vee \omega_2$$

$\varphi \bigtriangleup \psi$ (*during*) is true iff the current occurrence of φ occurs during an occurrence of ψ or the current occurrence of ψ has an occurrence of φ within it. It holds iff ψ is true now and either ω_1, ω_2, or ω_3 is true.

ω_1 is true iff φ is true now and became true after ψ became true and will become false before ψ becomes false.

$$\omega_1 \triangleq \varphi \wedge (\psi \mathbf{U} \neg\varphi) \wedge (\psi \mathbf{S} \neg\varphi)$$

ω_2 is true iff φ is false now and will become true and then false before ψ becomes false.

$$\omega_2 \triangleq \neg\varphi \wedge (\psi \mathcal{U}(((\varphi \wedge \psi)\mathbf{U}^+\neg\varphi)))$$

ω_3 is true iff φ is false now and became true and then false before ψ becomes false.

$$\omega_3 \triangleq \neg\varphi \wedge (\psi \mathcal{S}(((\varphi \wedge \psi)\mathbf{S}^+\neg\varphi)))$$

$\varphi \bigtriangleup \psi$ is then defined as

$$\varphi \bigtriangleup \psi \triangleq \psi \wedge (\omega_1 \vee \omega_2 \vee \omega_3)$$

$\varphi \unlhd \psi$ (*occurs immediately before*) is true iff (i) the current occurrence of φ occurs immediately before an occurrence of ψ (ii) or the current occurrence of ψ occurs immediately after an occurrence of φ. It is described using two disjuncts : ω_1 and ω_2.

ω_1 states that φ is true now and will become false when ψ becomes true.

$$\omega_1 \triangleq \varphi \wedge (\varphi \mathcal{U}((\varphi \wedge \neg\psi)\mathcal{U}^+(\neg\varphi \wedge \psi)))$$

ω_2 states that ψ is true now and became true when φ became false.

$$\omega_2 \triangleq \psi \wedge (\psi \mathcal{S}((\psi \wedge \neg\varphi)\mathcal{S}^+(\neg\psi \wedge \varphi)))$$

$\varphi \trianglelefteq \psi$ is then defined as

$$\varphi \trianglelefteq \psi \triangleq \omega_1 \vee \omega_2$$

Precedence ($\prec$) and inclusion ($\sqsubseteq$) are then defined using the following abbreviation.

$$\mathbf{always}\varphi \triangleq \Box\varphi \wedge \boxminus\varphi$$

Precedence ($\prec$) is defined as

$$\varphi \prec \psi \triangleq \mathbf{always}(\varphi \Rightarrow \varphi \lhd \psi) \wedge \mathbf{always}(\psi \Rightarrow \varphi \lhd \psi)$$

and *inclusion* ($\sqsubseteq$) as

$$\varphi \sqsubseteq \psi \triangleq \mathbf{always}(\varphi \Rightarrow \varphi \bigtriangleup \psi) \wedge \mathbf{always}(\psi \Rightarrow \varphi \bigtriangleup \psi)$$

Precedence is transitive while inclusion is transitive and reflexive.

Theorems related to the above operators are given in the Appendix A.1 .

5 Example

The example we are concerned with is described below. The informal specification is originally from [Pic86].

5.1 The Specification Statement

The token bus is a technique used for controlling access to a communication bus. Nodes on the bus are assigned logical positions in an ordered sequence so that they form a logical ring. Each node is assigned an address. All active nodes know the address of the active node preceding it and the one following it. The last member of the logical ring is followed by the

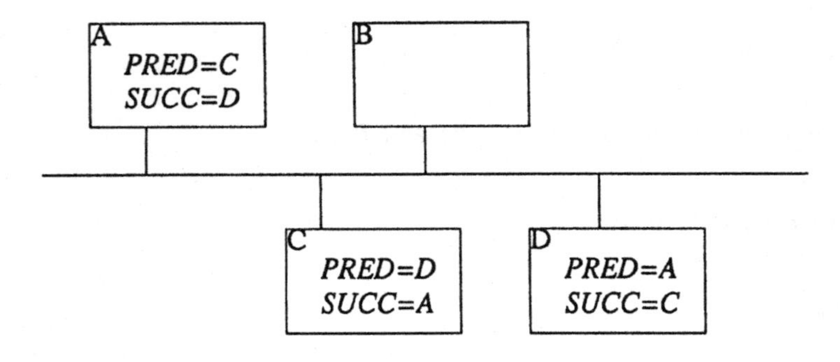

Figure 1: Token Bus

first. The physical position of a station on the bus is irrelevent and independent of the logical ordering. Figure 1 illustrates the bus network.

Communication between nodes is achieved by passing frames. There are six types of frames : token, solicit-successor, set-successor, resolve-contention, who-follows and claim-token. The token frame is used for regulating the right of access to the bus. It contains a destination address and the node receiving the token is granted access to the bus for a specified time after which it passes the token frame to the next node in the logical order. During the period a node holds the token, it can transmit frames, it may also drop itself from the ring, poll other nodes and send messages. An inactive node can only respond to polls or request for acknowledgement. As part of maintenance, the token bus provides following functions.

1. *Deletion of a node.* A node may voluntarily remove itself from the ring.
2. *Addition of a node.* Periodically, inactive nodes are granted an opportunity to insert themselves in the ring.
3. *Fault management.* A number of errors can occur. These include duplicate addresses (two nodes think it is their turn) and a broken ring (no node thinks that it is its turn).
4. *Ring initialisation.* When the ring is started up, or after the logical ring has broken down, it is reinitialised. A decentralised algorithm is used to sort out the logical order.

To save space only the deletion and addition of nodes is described in this paper ; the full example appears in [Nai91].

5.2 Formalisation of Requirements

There are six types of frames.

$$frame : \{token, sol\_succ, set\_succ, res\_cont, who\_follows, claim\}$$

Each node in the ring knows the address of its logical successor and predecessor. That is, given an address of a node, we define functions, *succ* and *pred* which return addresses of a node's successor and predecessor respectively.

$$succ : address \rightarrow address$$
$$pred : address \rightarrow address$$

We also denote the period during which a node A holds the token by a predicate $H(A)$. $H(A)$ is true iff a node with an address A holds the token and false otherwise. A node may send frames to other nodes. $S(F, A, X)$ is true iff the node with address A sends a frame F to a node with address X.

Predicates

$H(A)$: node with address A holds the token
$S(F, A, X)$: node with address A sends a frame F to a node with address X

We include the following general requirement about the duration each node may hold the token where c is some constant.

$$\forall x, i.\mathcal{D}(H(x), i) \leq c$$

5.2.1 Deletion of a node

If a node wishes to drop out, it waits until it receives the token, then sends a set-successor frame to its predecessor, instructing it to change its successor to be the token holder's successor.

Figure 2 illustrates the operation of allowing a node to drop out of the ring, using a timing diagram . It is formalised in stages below.

If a node wishes to drop out, it waits until it receives the token, then sends a set-successor frame to its predecessor.

$$S(set\_succ, A, pred(A)) \triangle H(A)$$

The predecessor splices to the token holder's successor after receiving the set-successor frame.

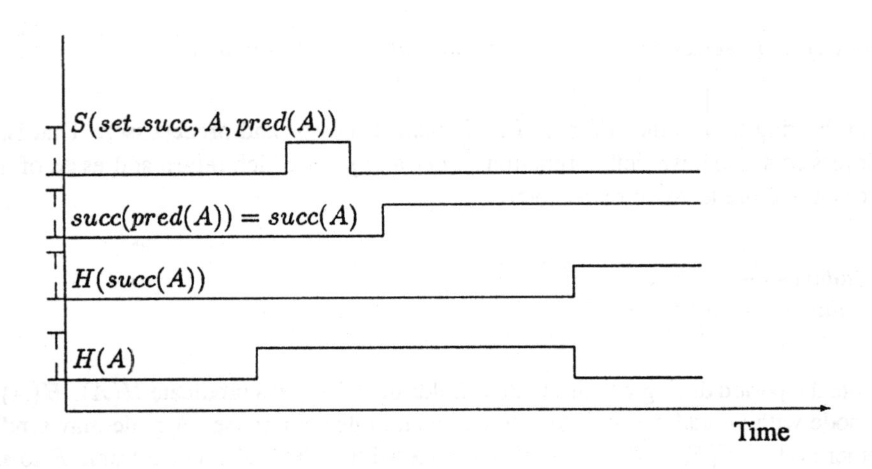

Figure 2: Timing diagram for the delete operation

$$S(set\_succ, A, pred(A)) \lhd (succ(pred(A)) = succ(A))$$

The token holder passes the token to its logical successor.

$$H(A) \unlhd H(succ(A))$$

The delete operation can now be described:

$$\begin{aligned} delete \triangleq \Box(S(set\_succ, A, pred(A)) \Rightarrow \\ &(S(set\_succ, A, pred(A)) \bigtriangleup H(A) \\ &\land S(set\_succ, A, pred(A)) \lhd (succ(pred(A)) = succ(A)) \\ &\land H(A) \unlhd H(succ(A)))) \end{aligned}$$

5.2.2 Addition of a node

Each node in the ring has the responsibility of periodically granting inactive nodes to enter the ring. While holding the token, the node issues a solicit-successor frame, inviting inactive nodes with an address between itself and the next node in the logical sequence to demand entrance. Let the set of nodes between the node A and the next node in the logical sequence be

$$I(A) \triangleq \{y | A < y < succ(A)\}$$

The token holder sends a solicit-successor frame to all the members of the above set.

$$send\_sol \triangleq \forall x \in I(A).S(sol\_succ, A, x)$$
$$solicit \triangleq send\_sol \Rightarrow (send\_sol \bigtriangleup H(A))$$

The transmitting node then waits for a response window. We denote a node A waiting for a response window by a predicate $W(A)$.

$$wait \triangleq send\_sol \Rightarrow (send\_sol \trianglelefteq W(A) \wedge W(A) \vartriangle H(A))$$

The duration of the wait period is 1 time slot.

$$\forall x, i. \mathcal{D}(W(x), i) = 1$$

Three events can occur.

1. *No response.* Nobody wants to enter the ring, in which case the token is transferred to the logical successor.

$$no\_resp \triangleq (\neg \exists x \in I(A).(S(set\_succ, x, A) \vartriangle W(A)) \wedge send\_sol \trianglelefteq W(A) \\ \wedge W(A)) \Rightarrow H(A) \trianglelefteq H(succ(A))$$

2. *One response.* There is exactly one response by issuing set-successor frame. The token holder sets its successor node to be the requesting node and transmits the token to it. The requestor sets its linkages accordingly and proceeds.

$$one\_resp \triangleq \exists! x \in I(A).(S(set\_succ, x, A) \vartriangle W(A) \wedge send\_sol \trianglelefteq W(A) \\ \wedge W(A)) \Rightarrow (H(A) \trianglelefteq H(x) \\ \wedge (S(set\_succ, x, A) \wedge succ(A) = y) \triangleleft \\ (succ(A) = x \wedge pred(x) = A \\ \wedge succ(x) = y)).$$

3. *Multiple Responses.* The token holder will detect a garbled response if more than one node demands entrance. The conflict is resolved as follows. The token holder transmits a resolve-contention frame and waits four demand windows.

$$send\_res \triangleq \forall x \in I(A). S(res\_cont, A, x)$$

We denote a node A waiting in a demand window by following state predicates.

$W_1(A)$: Node with address A waiting in demand window 1
$W_2(A)$: Node with address A waiting in demand window 2
$W_3(A)$: Node with address A waiting in demand window 3
$W_4(A)$: Node with address A waiting in demand window 4

The duration of each demand window is 1 time slot.

$$\forall x, i.\mathcal{D}(W_1(x), i) = \mathcal{D}(W_2(x), i) = \mathcal{D}(W_3(x), i) = \mathcal{D}(W_4(x), i) = 1$$

Each demanding node can respond in one of these windows. If a demanding node hears anything before its window comes up, it refrains from demanding. If the token holder receives a valid set-successor frame, it is in business. Otherwise it tries again, and only those nodes that responded the first time are allowed to respond this time. This process of resolving is defined as

$$\begin{aligned} \mathit{resolve} \triangleq\ & (W_1(A) \trianglelefteq W_2(A) \vee W_2(A) \trianglelefteq W_3(A) \vee W_3(A) \trianglelefteq W_4(A)) \\ & \wedge W_1(A) \vartriangle H(A) \wedge W_2(A) \vartriangle H(A) \wedge W_3(A) \vartriangle H(A) \\ & \wedge W_4(A) \vartriangle H(A) \\ & \wedge(\mathit{set\_w1} \Rightarrow \neg(\mathit{set\_w2} \wedge \mathit{set\_w3} \wedge \mathit{set\_w4})) \\ & \wedge(\mathit{set\_w2} \Rightarrow \neg(\mathit{set\_w3} \wedge \mathit{set\_w4})) \\ & \wedge(\mathit{set\_w3} \Rightarrow \neg \mathit{set\_w4}) \end{aligned}$$

set_w1, set_w2, set_w3, and set_w4 are defined as follows.

$$\begin{aligned} \mathit{set\_w1} &\triangleq \exists x \in I(A).(S(\mathit{set\_succ}, x, A) \vartriangle W_1(A)) \\ \mathit{set\_w2} &\triangleq \exists x \in I(A).(S(\mathit{set\_succ}, x, A) \vartriangle W_2(A)) \\ \mathit{set\_w3} &\triangleq \exists x \in I(A).(S(\mathit{set\_succ}, x, A) \vartriangle W_3(A)) \\ \mathit{set\_w4} &\triangleq \exists x \in I(A).(S(\mathit{set\_succ}, x, A) \vartriangle W_4(A)) \end{aligned}$$

The process is continued until one of the three events occur.

(a) A valid set-successor frame is received, and the token is passed.

$$\begin{aligned} \mathit{valid} \triangleq \exists! x \in I(A).(\ & S(\mathit{set\_succ}, x, A) \wedge S(\mathit{set\_succ}, x, A) \vartriangle W_1(A) \\ & \wedge H(A) \trianglelefteq H(x) \\ & \wedge (S(\mathit{set\_succ}, x, A) \wedge \mathit{succ}(A) = y) \triangleleft \\ & (\mathit{succ}(A) = x \wedge \mathit{pred}(x) = A \\ & \wedge \mathit{succ}(x) = y))) \end{aligned}$$

(b) No response is received and the token is passed to the logical successor.

$$\mathit{zero\_resp} \triangleq W_1(A) \wedge \neg \mathit{set\_w1} \wedge H(A) \trianglelefteq H(\mathit{succ}(A))$$

(c) The maximum retry count is reached and the token is passed to the logical successor. Let x be the number of retries when the resolve-contention frame is sent and max be the number of retries allowed.

$$\begin{aligned} \mathit{max\_resp} \triangleq\ & H(A) \trianglelefteq H(\mathit{succ}(A)) \wedge H(\mathit{succ}(A)) \\ & \wedge \mathcal{N}(W_1(A)) = x - \mathit{max} \end{aligned}$$

Let two or more nodes responding in the response window be defined as

$$many \triangleq \exists x, y.(S(set\_succ, x, A) \bigtriangleup W(A) \wedge S(set\_succ, y, A) \bigtriangleup W(A) \\ \wedge x \neq y)$$

$mult\_resp$ is then defined as

$$mult\_resp \triangleq (W(A) \wedge send\_sol \trianglelefteq W(A) \wedge many \wedge \mathcal{N}(W_1(A)) = x) \\ \Rightarrow (send\_res \bigtriangleup H(A) \wedge W(A) \trianglelefteq (send\_res \\ \wedge (resolve\mathcal{U}(valid \vee zero\_resp \vee max\_resp)))$$

The complete operation is as described below.

$$add \triangleq \Box(solicit \wedge wait \wedge (no\_resp \vee one\_resp \vee mult\_resp))$$

6 Conclusion

The logic described in this paper is a temporal logic without stuttering. It departs from the usual treatment by abandoning the central concept in most specification approaches that events are instantaneous. There are many examples where one needs to consider events which extend in time. For example, the delay action can not be described as instantaneous but has a duration. Properties described using first order logic also have duration. Occurrences of predicates were defined and used to describe behaviours of systems using operators such as *during*, *immediately before* etc. A calculus of occurrences of predicates was introduced in Section 4 to reason about the duration and number of occurrences.In Section 5, the IEEE 802 Token Bus was specified in the logic.

The use of intervals in requirements specifications in not new : for example, Zhou ChaoChen, Hoare and Ravn [CHR91] use an interval logic extended with a calculus of durations to specify timing properties, and RTL [JM86] to describe and analyse systems.

In duration calculus, timing constraints are described using integrated durations of states in an interval. A duration of a state predicate φ in a closed interval $[b, e]$ is defined by an integral $\int_e^b \varphi(t)dt$ and is denoted by a variable $\int\varphi$. It is easy to see that $\int true$ in any interval $[b, e]$ is $e - b$ and $\int false$ is zero. Given this definition, axioms are given which relate durations of different predicates and theorems proved from them.

A further notation is introduced which converts a state predicate to an interval one ($[\varphi]$). $[\varphi]$ is defined as

$$[\varphi] \triangleq \int\varphi = \int true \wedge \int true > 0$$

That is, $\lceil\varphi\rceil$ holds in an interval iff it holds throughout a nonempty interval. Point interval is defined as

$$\lceil\,\rceil \triangleq \int true = 0$$

The definition of $\lceil\varphi\rceil$ differs from an occurrence of φ in one respect. While $\lceil\varphi\rceil$ holds for any interval where φ holds throughout, an occurrence of φ denotes a maximal interval (by inclusion) where φ holds everywhere.

A further modal operator *chop* ($\varphi \frown \psi$) is introduced and defined in an interval which can be subdivided into two sub-intervals of which in the first φ holds and in the second ψ holds. Interval temporal operators *eventually* and *henceforth* are defined in terms of *chop*. The resulting logic is very expressive for stating timing requirements using durations. For example, the formula

$$\int true = 60 \Rightarrow \int busy(m) \leq 50$$

asserts that in any given hour, m is not busy for more than 50 minutes.

In RTL, the event-action model captures timing constraints which are transformed into an RTL formula. An RTL formula is formed using constants and an occurrence function. There are three kinds of constants: actions, events and integers. The occurrence function @ is introduced to capture the notion of time. Given an event e and a non negative integer i, the @ function returns the time of the ith occurrence of e. Using the occurrence function, one can specify requirements such as periodicity and duration of events. For example, to state that an action a occurs periodically every c time units, a formula of the form

$$\forall i \geq 1. @(\uparrow a, i+1) - @(\uparrow a, i) = c$$

is used. $\uparrow a$ and $\downarrow a$ are used to denote starting and finishing events marking start and finish of the action a respectively.

Duration of actions is specified similarly. The formula

$$\forall i \geq 1. @(\downarrow a, i) - @(\uparrow a, i) = c$$

asserts that the time of finishing of an action a is exactly c time units after it started.

Further notations are used to denote that a state predicate is true in an interval. For example, $\varphi[t_1, t_2]$ asserts that φ becomes true at t_1 and false at t_2, and $\varphi[t_1, t_2 >$ states that φ becomes true at t_1 and false after t_2. $\varphi[t_1, t_2]$ is the maximal interval φ is true everywhere and therefore (in our terminology) defines an occurrence of φ. However, there are two major differences between RTL and our approach. Firstly, RTL does not have any temporal operators and secondly, we donot refer explicitly to time.

This paper describes only preliminary ideas on using temporal logic to specify requirements. One of the aspects of future work is to investigate whether the primitives defined here can specify large class of real-time systems. Also an area of investigation is the relationship between the logic and programming languages such as a real-time extension of CSP.

Acknowledgements : I would like to thank my supervisor, Mathai Joseph for detail comments on earlier drafts and to Asis Goswami and Liu Zhiming for numerous technical discussions and ideas . Thanks are also due to the referees for detail technical comments and improvements.

A Theorems

A.1 Theorems about $\triangle$, $\trianglelefteq$, and $\triangleleft$

Following are some useful theorems stated without proofs.

I1. $\vdash \varphi \triangleleft \psi \Rightarrow (\varphi \vee \psi)$
I2. $\vdash \varphi \triangle \psi \Rightarrow \psi$
I3. $\vdash \varphi \trianglelefteq \psi \Rightarrow \varphi \triangleleft \psi$
I4. $\vdash \neg((\neg\varphi) \triangle \varphi)$
I5. $\vdash \varphi \triangle \psi \wedge \psi \triangle \omega \Rightarrow \varphi \triangle \omega$
I6. $\vdash \varphi \triangle (\psi \wedge \omega) \Rightarrow \varphi \triangle \psi \wedge \varphi \triangle \omega$
I7. $\vdash \varphi \triangle (\neg\psi) \Rightarrow \neg(\varphi \triangle \psi)$
I8. $\vdash (\varphi \vee \psi) \triangle \omega \Rightarrow \varphi \triangle \omega \vee \psi \triangle \omega$
I9. $\vdash \varphi \triangle \psi \vee \varphi \triangle \omega \Rightarrow \varphi \triangle (\psi \vee \omega)$
I10. $\vdash \neg(\varphi \trianglelefteq \varphi)$
I11. $\vdash \varphi \trianglelefteq (\psi \trianglelefteq \omega) \Leftrightarrow (\varphi \trianglelefteq \psi) \trianglelefteq \omega$
I12. $\vdash \varphi \trianglelefteq \psi \wedge \varphi \trianglelefteq \omega \Rightarrow \varphi \trianglelefteq (\psi \wedge \omega)$
I13. $\vdash (\varphi \vee \psi) \trianglelefteq \omega \Rightarrow \varphi \trianglelefteq \omega \vee \psi \trianglelefteq \omega$
I14. $\vdash \varphi \trianglelefteq (\psi \vee \omega) \Rightarrow \varphi \trianglelefteq \psi \vee \varphi \trianglelefteq \omega$
I15. $\vdash (\neg\varphi) \trianglelefteq \psi \Rightarrow \neg(\varphi \trianglelefteq \psi)$
I16. $\vdash \varphi \trianglelefteq (\neg\psi) \Rightarrow \neg(\varphi \trianglelefteq \psi)$
I17. $\vdash \neg(\varphi \triangleleft \varphi)$
I18. $\vdash (\varphi \vee \psi) \triangleleft \omega \Rightarrow \varphi \triangleleft \omega \vee \psi \triangleleft \omega$
I19. $\vdash \varphi \triangleleft (\psi \vee \omega) \Rightarrow \varphi \triangleleft \psi \vee \varphi \triangleleft \omega$

A.2 Theorems about Occurrence of Predicates

Following are some useful theorems stated without proofs.

M1. $\mathcal{T}(\varphi) + \mathcal{T}(\neg\varphi) = \mathcal{T}(true)$
M2. $\mathcal{T}(\varphi) \leq \mathcal{T}(true)$
M3. $\Box(\varphi \Rightarrow \psi) \Rightarrow \Box(\mathcal{T}(\varphi) \leq \mathcal{T}(\psi))$

References

[ACD90] R. Alur, C. Courcoubetis, and D. L. Dill. Model checking for real-time systems. In *Proceedings Symposium on Logic in Computer Science*, 1990.

[BKP84] H. Barringer, R. Kuiper, and A. Pnueli. Now you may compose temporal logic specifications. In *Proceedings of the 16th ACM Symposium on the Theory of Computing*, pages 51–63, Washington D.C., 1984.

[BKP86] H. Barringer, R. Kuiper, and A. Pnueli. A really abstract concurrent model and its temporal logic. In *Proceedings of the 13th ACM Symposium on Principles of Programming Languages*, pages 173–183, Florida, 1986.

[CHR91] Z. Chaochen, C.A.R. Hoare, and A.P. Ravn. A calculus of durations. Unpublished, 1991.

[EC82] E.A. Emerson and E.M. Clarke. Using branching time logic to synthesize synchronization skeletons. *Science of Computer Programming*, 2:241–266, 1982.

[HLP90] D. Harel, O. Lichtenstein, and A. Pnueli. Explicit clock temporal logic. In *Proceedings Symposium on Logic in Computer Science*, pages 402–413, 1990.

[HW89] J. Hooman and J. Widom. A temporal-logic based compositional proof system for real-time message passing. In *Lecture Notes in Computer Science 366*, pages 424–441. Springer-Verlag, Heidelberg, 1989.

[JM86] F. Jahanian and A. Mok. Safety analysis of timing properties in real-time systems. *IEEE Transactions on Software Engineering*, 12:890–904, 1986.

[Koy89] R. Koymans. *Specifying Message Passing and Time-Critical Systems with Temporal Logic*. PhD thesis, Department of Mathematics and Computing Science, Eindhoven University of Technology, Eindhoven, 1989.

[MM83] B. Moszkowski and Z. Manna. Reasoning in interval temporal logic. In *Lecture Notes in Computer Science 164*, pages 371–383. Springer-Verlag, Heidelberg, 1983.

[MP81] Z. Manna and A. Pnueli. Verification of concurrent programs: The temporal framework. In R.S. Boyer and J.S. Moore, editors, *The Correctness Problem in Computer Science*, pages 215–273. Academic Press, London, 1981.

[Nai91] Y. Naik. A temporal approach to requirements specification of real-time systems. To be submitted as a Technical Report, 1991.

[Ost89] J.S. Ostroff. *Temporal Logic for Real-time Systems*. Research Studies Press, 1989.

[PH88] A. Pnueli and E. Harel. Applications of temporal logic to the specification of real time systems (extended abstract). In *Lecture Notes in Computer Science 331*, pages 84–98. Springer-Verlag, Heidelberg, 1988.

[Pic86] R. L. Pickholtz, editor. *Local Area Networks and Multiple Access Networks*, chapter 1, pages 1–30. Computer Science Press, 1986.

RLucid, a general real-time dataflow language

John A. Plaice
Department of Computer Science, University of Ottawa
150 Louis Pasteur, Ottawa, Ontario, CANADA K1N 6N5
(Unceded Algonquin Land)
e-mail: plaice@csi.uottawa.ca

October 21, 1991

Abstract

We present the RLUCID programming language, which is LUCID with a new operator, called **before**. With **before**, merging two streams in a deterministic manner, according to the time of arrival, becomes a possibility. We present the semantics of RLUCID, and then show that it is a superset of LUSTRE, which was designed for the programming of real-time kernels. Finally, we show that real-time interfaces can be written in RLUCID, thereby making RLUCID, in some sense, a general real-time language.

1 Introduction

Reactive systems [10, 16] are systems which interact *continually* with their environments. They are used more and more, in fields as diverse as electronic gadgets, power plants, and aircraft and railroad control. Reactive systems include real-time systems, which are subject to hard real-time constraints, typically required response times or data sampling rates.

To program reactive systems, programming and specification languages must have the expressive power to describe not only what actions must be taken, but also when they must take place. Since the events arriving from the environment typically do so in a completely asynchronous manner, the most difficult part of programming such a system is often defining the synchronisations which must take place between events so that useful computations can be done.

When a program has been written, it would ideally be possible to determine in a static manner whether the program can be run with a bounded memory, as well as to determine whether the timing constraints that the program is subject to can be met.

Furthermore, time should be multi-form. There is no reason that the real-time clock of a computer should be the only way to define "time". Any input event, if it occurs repeatedly, can define its own version of time. It should be up to the program to figure out how the different versions of time should be combined.

Stream processing seems natural to reactive programming: a program takes as input streams of events, and generates as output new streams of events. For this reason, many attempts [1, 5, 6, 7, 12, 19, 20, 22] have been made to use dataflow languages to program or specify reactive or real-time systems.

LUCID [22] has been considered as a possible language for both programming and specifying real-time systems. Faustini and Lewis [7] describe a variant of LUCID with hiaton streams, where ordinary LUCID streams have 'hiatons' interspersed between the values when data is not available. However, the semantics is unclear: when does a hiaton need to be introduced into a stream?

Skillicorn and Glasgow [20] describe the use of LUCID for real-time specification. An existing LUCID program is examined to determine exactly how long it will take to execute the program. The result of this analysis is itself a LUCID program which must be run to get the results. But is is static analysis which is of interest, not dynamic analysis.

Barth, Guthery and Barstow [1] describe the STREAM MACHINE. This language for describing real-time systems consists of two parts: the first part is a "deterministic" dataflow language which can be used for the description of filters; the second part is a "nonderterministic" set of primitives for describing how to read signals from a set of input events, depending on the order of their arrival. However, It would be better if the two formalisms could be integrated.

Sheeran [19] describes RUBY, a language for describing perfectly synchronous systems. Only lock-step synchronous systems can be described.

Broy [5] defines a specification system with functions between streams containing values and ticks. Although more developed than the work done with LUCID and hiatons, the question remains, what generates the ticks?

Berry came up with a radically different approach to describing real-time systems: the synchronous approach [4]. The assumption is that if a program can always respond to its input before future input arrives, then the programming can be done by supposing that the (output) reaction occurs at *exactly* the same time as the (input) signal. The semantics of languages becomes greatly simplified. It behooves the implementer to ensure that the compiled code does actually meet the constraints, as it was supposed when the programmer first started to write.

The first synchronous language was ESTEREL [3]. It is an imperative language, where different components may run in parallel. Communication between these components is done in an instantaneous, broadcast manner. This communication is symbolically run by the compiler, and a finite state automaton is generated. The resulting code is very fast.

Two synchronous dataflow languages have also been devised: LUSTRE [6] and SIGNAL [12, 13]. LUSTRE programs are written functionally while SIGNAL programs are written relationally. There are several fast implementations of LUSTRE [15, 17, 18].

So that LUSTRE programs do not have to be perfectly synchronous, with all the input and output working in perfect synchrony, the concept of clock was added. With clocks, certain parts of a system are not necessarily always running. Certain parts may only run when certain variables (clocks) are true. To ensure that the clock system works, a static clock calculus, akin to a type checking algorithm, is used.

Synchronous languages are well adapted to describing the "kernel" of a reactive system, the innermost part which determines exactly what has to be done. But this kernel must receive its input from the external environment, often completely asynchronous. In fact, the typical manner for receiving input is through interrupts, an inherently asynchronous mechanism. So that a synchronous language can be used, an "interface" must be written in a host language between the environment and the kernel. But in many situations, the complex synchronisation of input events takes place within the interface, and so the power of the synchronous language is lost. What is needed is a language which is sufficiently rich to allow the programming of the complex part of interfaces.

This paper presents RLUCID, a programming language which manipulates timestamped streams. It is a variant of Ashcroft and Wadge's LUCID [22], with a timed semantics and a new operator, called `before`. It is designed so that it can be efficiently implemented, like LUSTRE and ESTEREL. It can handle all of the tasks that one would want of an interface.

2 RLucid

RLUCID programs are expressions defined using sets of mutually recursive equations. Free variables of an expression are its input.

2.1 Syntax

RLUCID is a member of the ISWIM [11] family of languages, as is shown by the syntax of the bare language:

E	$::=$	k	constant
	$\mid$	id	variable
	$\mid$	$(E, \ldots, E)$	tuple creation
	$\mid$	**select** (i, E)	component selection
	$\mid$	$id\ E$	function application
	$\mid$	E **where** Q **end**	

Q	$::=$	$id = E$	variable declaration
	$\mid$	$id\ id = E$	function declaration
	$\mid$	$Q\ Q$	simultaneous declaration

where k is a constant, id is a identifier, E is an expression, and Q is an equation.

There is an initial environment of functions, listed below. Although the abstract syntax is $id\ E$, the concrete syntax is as follows:

E **op** E
if E **then** E **else** E
next E
E **fby** E
E **before** E

op refers to any ordinary data operation. It is not necessarily binary. Among the data operations is the n-ary operator `||`, which creates tuples. The operators **next** and **fby** come from LUCID. **before** is the novel operator of RLUCID. `E1` **before** `E2` determines if the first value of `E1` arrived before the first value of `E2`. It is **before** that makes RLUCID into a real-time language.

2.2 Semantics

2.2.1 Semantic Domains

The set of possible timestamps $(\omega, \varnothing, \sqsubseteq)$ is a well-ordered set whose minimal element is $\varnothing$. τ will denote a type or the set of values associated with it. The *domain of simple streams* of type τ is the set

$$\mathbf{SS}_\tau \quad = \quad [\tau, \omega]^\infty \ni \phi.$$

The order on simple streams is the prefix order, which will be written $\leq$. Normally, timestamps will be monotone increasing.

Each element of a simple stream is a pair $[\upsilon, \chi]$, where υ is a value, and χ is a timestamp. Υ is used to select the value and X the timestamp:

$$\Upsilon[\upsilon, \chi] \quad = \quad \upsilon$$
$$\mathrm{X}[\upsilon, \chi] \quad = \quad \chi$$

A stream ϕ can be considered to be a function from the integers to value–timestamp pairs. The stream $([x_0, s_0], [x_1, s_1], \ldots, [x_n, s_n], \ldots)$ is understood to mean that value x_i occurs at time s_i.

A *tuple stream* is a tuple of streams: if $\phi_1, \phi_2, \ldots, \phi_n$ are streams, then $\phi_1 \bullet \phi_2 \bullet \ldots \phi_n$ is a tuple stream. The order of tuple streams is the order induced by the order on simple streams. The domain $\mathbf{S}$ of all streams is therefore:

$$\mathbf{S} \ = \ \sum_\tau \mathbf{SS}_\tau + \prod_{n \in \mathbf{N}} \mathbf{S}$$

The operator π_i is used to access the ith stream in the tuple stream ϕ. A tuple stream is not the same as a stream of tuples.

The domain of environments of identifiers is denoted by **Env** ($\zeta, \zeta' \in \mathbf{Env}$); an identifier can denote a stream or a function from streams to streams. The derived domains **Exp** and **Equ** are standard.

$$\begin{aligned} \mathbf{Env} &= id \rightarrow \mathbf{S} + id \rightarrow \mathbf{S} \rightarrow \mathbf{S} \\ \mathbf{Exp} &= \mathbf{Env} \rightarrow \mathbf{S} \\ \mathbf{Equ} &= \mathbf{Env} \rightarrow \mathbf{Env} \end{aligned}$$

2.2.2 Semantic functions

The semantics of an expression E is $[\![E]\!]\zeta_0$, where ζ_0 is the initial environment describing the basic functions (see next section). Constant streams are available at the beginning of time:

$$\begin{aligned} [\![k]\!]\zeta &= \lambda n \cdot [k, \emptyset] \\ [\![id]\!]\zeta &= \zeta(id) \\ [\![(E_1, \ldots, E_n)]\!]\zeta &= [\![E_1]\!]\zeta \bullet \ldots \bullet [\![E_n]\!]\zeta \\ [\![\texttt{select}\ (i, E)]\!]\zeta &= \pi_i([\![E]\!]\zeta) \\ [\![id\ E]\!]\zeta &= ([\![id]\!]\zeta)([\![E]\!]\zeta) \\ [\![E\ \texttt{where}\ Q]\!]\zeta &= [\![E]\!]\left(\mu\zeta' \cdot \zeta[[\![Q]\!]\zeta']\right) \end{aligned}$$

$$\begin{aligned} [\![id = E]\!]\zeta &= \zeta[[\![E]\!]\zeta \ / \ id] \\ [\![id_1\ id_2 = E]\!]\zeta &= \zeta[\lambda id_2 \cdot [\![E]\!]\zeta \ / \ id_1] \\ [\![Q_1\ Q_2]\!]\zeta &= [\![Q_1]\!]([\![Q_2]\!]\zeta) \end{aligned}$$

2.2.3 The initial environment

We give the semantics for an data operator of arity p. It is the pointwise application of the function. For all functions, the synchronous hypothesis means that output is available as soon as the inputs are available:

$$[\![\mathtt{op}_p]\!]\zeta_0 = \lambda\phi \cdot \lambda n \cdot \left[\mathbf{op}(\Upsilon(\pi_1\phi n), \ldots, \Upsilon(\pi_p\phi n)),\ \mathbf{max}(\mathrm{X}(\pi_1\phi n), \ldots, \mathrm{X}(\pi_p\phi n))\right]$$

$$[\![\mathtt{if}]\!]\zeta_0 = \lambda\phi \cdot \lambda n \cdot \begin{array}{ll} \text{if} & \Upsilon(\pi_1\phi n) \\ \text{then} & \left[\Upsilon(\pi_2\phi n),\ \mathbf{max}\left(\mathrm{X}(\pi_1\phi n), \mathrm{X}(\pi_2\phi n)\right)\right] \\ \text{else} & \left[\Upsilon(\pi_3\phi n),\ \mathbf{max}\left(\mathrm{X}(\pi_1\phi n), \mathrm{X}(\pi_3\phi n)\right)\right] \end{array}$$

$$[\![\mathtt{next}]\!]\zeta_0 = \lambda\phi \cdot \lambda n \cdot \left[\Upsilon(\phi(n+1)),\ \mathrm{X}(\phi(n+1))\right]$$

$$[\![\mathtt{fby}]\!]\zeta_0 = \lambda\phi \cdot \lambda n \cdot \begin{array}{ll} \text{if} & n = 0 \\ \text{then} & \left[\Upsilon(\pi_1\phi 0),\ \mathrm{X}(\pi_1\phi 0)\right] \\ \text{else} & \left[\Upsilon(\pi_2\phi(n-1)),\ \mathrm{X}(\pi_2\phi(n-1))\right] \end{array}$$

$$[\![\mathtt{before}]\!]\zeta_0 = \lambda\phi \cdot \lambda n \cdot \left[\mathrm{X}(\pi_1\phi 0) \sqsubseteq \mathrm{X}(\pi_2\phi 0),\ \mathbf{min}(\mathrm{X}(\pi_1\phi 0), \mathrm{X}(\pi_2\phi 0))\right]$$

2.3 Derived constructs

There are several RLUCID constructs which are heavily used which can be derived from the base language. Many of these constructs are defined as recursive functions.

- `first X` generates a constant stream, whose value is the first value of `X`:

```
first X = X fby first X
```

- `X wvr Y` generates the corresponding value of `X` every time that `Y` is true:

```
X wvr Y = if first Y then X fby (next X wvr next Y)
                      else (next X wvr next Y)
```

- `event X` generates a void (empty type) stream which is synchronous with `X`. It will be used to define other operations which do not care about the value of `X`:

```
event X = select (0, void||X)
```

- `A on C` generates a stream whose values are those of `A` but whose time-stamps are those of `C` if `A` is faster than `C`:

```
A on C = select(0, A||C)
```

  Suppose we wish to count the occurrences of an event `A`, exactly when they occur. This would be done as follows:

```
count on (void fby event A)
where
   count = 0 fby count + 1
end
```

  `count` counts the natural numbers infinitely fast. The `on` slows the counter to the rate of `A`.

- `choose X1:E1 X2:E2 ... Xn:En end` will execute the `Ei` corresponding to the first `Xi` which has arrived. It is defined as follows:

```
choose X1:E1 X2:E2 ... Xn:En end =
   if X1 before X2 and X1 before X3 and ... and X1 before Xn then E1
   else if X2 before X3 and ... and X2 before Xn then E2
   ...
   else En
```

- `last(A,C)` outputs the last value held by `A` every time that `C` occurs. It is defined as follows:

```
last(A,C) = current(A, C, first A)
where
   current(A, B, C) =
   choose
      A: current(next A, B, first A)
      B: C fby current(A, next B, C)
   end
end
```

Suppose that we wish to take the count of occurrences of **A** from above and use it whenever **B** turns up. The result would be:

```
last (occurs, event B)
where
   occurs = count on (void fby event A)
   count = 0 fby count + 1
end
```

- **MergeLeft** will generate a single stream from two input streams. Should the two send values at the same time, then the left one takes precedence:

```
MergeLeft(A,B) =
choose
   A: A fby MergeLeft(next A, B)
   B: B fby MergeLeft(A, next B)
end
```

- **SignalLeftFair** will generate a single stream from two input streams. However, the elements will be pairs of booleans, showing which stream generated the pair. Only one true value will be sent at a time:

```
SignalLeftFair(A,B) =
choose
   A||B: (true||false) fby (false||true) fby SignalLeftFair(next A, next B)
   A:    (true||false) fby SignalLeftFair(next A, B)
   B:    (false||true) fby SignalLeftFair(A, next B)
end
```

- **SignalBoth** will send two true values at a time, if necessary:

```
SignalBoth(A,B) =
choose
   A||B: (true||true) fby SignalBoth(next A, next B)
   A:    (true||false) fby SignalBoth(next A, B)
   B:    (false||true) fby SignalBoth(A, next B)
end
```

- **which** will return 1, 2 or 3, to signal which of its arguments has arrived:

```
which(A,B) =
choose
   A||B: 3 fby which(next A, next B)
   A:    1 fby which(next A, B)
   B:    2 fby which(A, next B)
end
```

2.4 Assertions

To improve specifications, as well as to give some aid to the compiler, the user can write assertions. The syntax is **assert** E, where E is a boolean expression. Here is a simple example:

```
assert synchro(A,B) and not synchro(C,D)
where synchro(X,Y) = which(A,B) = 3 end
```

It states that `A` and `B` will always appear together and that `C` and `D` will never appear together.

`not synchro` can be generalized to n inputs: `notsynchro(E1,...,En)` is defined as none of the `Ei` can occur simultaneously.

To be able to determine if it can run without bounded memory, the compiler must know the relative rates of arrival of different input. For example, if streams `A` and `B` are to be added, the assertion:

```
assert diff_A_B < 2 and diff_B_A < 2
where
   diff_A_B = last (occurs_A, event B) - occurs_B
   diff_B_A = last (occurs_B, event A) - occurs_A
   occurs_A = count on (void fby event A)
   occurs_B = count on (void fby event B)
   count = 0 fby count + 1
end
```

will tell the compiler that the two streams are never more than two values out of sync.

3 Examples

In this section, we present three RLucid examples.

- Consider the first order linear filter defined by:

$$\begin{aligned} y_0 &= init \\ y_{n+1} &= ay_n + bx_{n+1}, \quad n \geq 0 \end{aligned}$$

  In RLucid, it becomes:

```
filter(init, a, b, X)
where
   filter(init, a, b, X) = Y
   where
      Y = init fby (a*Y + b * next X)
   end
end
```

- Consider a stopwatch with three inputs:
  - `onOff` turns on the stopwatch if it is not running, and turns it off otherwise.
  - `reset` will reset the stopwatch if it is not running.
  - `quartz` is a signal to change the time if the stopwatch is running.

Here, R, Q, running and time are synchronous:

```
Stopwatch(onOff, reset, quartz) = time
where
   (R,Q) = SignalLeftFair(reset, quartz)
   alternate = false fby not alternate
   running = last((alternate on event onOff), event R))
   time = 0 fby if R and not running then 0
                else if Q and running then time + 1
                else time;
end
```

- Consider a mouse program with three inputs:
  - TICK is the TICK from a clock.
  - Delta is the number of TICKs in an interval.
  - CLICK signals that the mouse button has been pressed. Upon receipt of the CLICK, if after receipt of the next Delta clicks, another CLICK is received, the program outputs true, otherwise it outputs false.

One possible RLUCID solution is:

```
mouse(TICK, CLICK, Delta) = clicks(TICK, CLICK, 0, false, false)
where
   clicks(T, C, count, clicking, double) =
   choose
      C: clicks(T, next C, count, true, clicking)
      T: if clicking and (count = (Delta-1))
         then double fby clicks(next T, C, 0, false, false)
         else clicks(next T, C, (count+1) mod Delta, clicking, double)
   end
end
```

4 Lustre in RLucid

In this section, we present LUSTRE [6, 15, 14], and show that LUSTRE programs can be easily translated into RLUCID.

LUSTRE is a functional dataflow language, derived from LUCID, which was designed so that it could be implemented very efficiently. LUSTRE has been used to program automatic control examples [2], systolic algorithms [9] and hardware circuits [8]. The semantics and the prototype compiler were presented in the author's dissertation [15].

Aprogram which is going to be used in a real-time situation must run with a bounded memory. However, it is difficult with dataflow networks to prove that they can run with a bounded memory. LUSTRE's solution to this problem was to insist that the operands of data functions be perfectly synchronous.

However, LUCID's next is not compatible with this choice. One can not write

```
X = Y fby X + next Y;
```

To get around this problem, two new operators were introduced to replace fby and next:

$$\begin{aligned} \texttt{pre E} &= (\mathrm{nil}, e_0, e_1, \ldots, e_{n-1}, \ldots) \\ \texttt{E -> F} &= (e_0, f_1, f_2, \ldots, f_n, \ldots) \end{aligned}$$

LUCID's

```
X = 0 fby X+1
```

becomes

```
X = 0 -> pre X + 1
```

These two operators are sufficient to describe perfectly synchronous systems. But there are situations when it makes no sense to have a value in a stream. To allow for this, an operator to sample streams according to a boolean expression was introduced. `E when C` produces the stream consisting of values of `E` when the corresponding value of `C` is true.

Since streams are no longer necessarily synchronous, but the data operators are strictly synchronous, the notion of stream is changed to that of a *clocked stream*. A clocked stream consists of a sequence of values and a clock, which defines the sequence of instants at which those values appear. If the sequence of values is

$$(e_0, e_1, \ldots, e_n, \ldots)$$

the stream takes the value e_n at the nth instant of its clock.

A clock is either the base clock of the program (intuitively, the sequence of instants at which it is active), or a boolean stream. A boolean stream `B`, considered as a clock, defines the sequence of instants where its value is true.

So, if `E` and `C` are on the same clock, then `E when C` has clock `C`. Data operators are allowed to operate over streams of the same clock. With `when`, one can create cascades of ever coarser clocks, thereby forming a tree with the base clock at root.

In the denotational semantics given in the author's dissertation [15, 14], a clocked stream is a pair: a value stream and a clock. A clock, in turn, would be either the base clock or another stream.

The given translation only works for LUSTRE programs for which clocks have been inferred [14]:

$$\begin{array}{rcl}
\mathcal{L}(\texttt{base}) & \longrightarrow & \texttt{event on } arg \\
\mathcal{L}(k \texttt{ on } E) & \longrightarrow & k \texttt{ on } \mathcal{L}E \\
\mathcal{L}(E_1 \texttt{ op } E_2) & \longrightarrow & \mathcal{L}E_1 \texttt{ op } \mathcal{L}E_2 \\
\mathcal{L}(\texttt{if } E_1 \texttt{ then } E_2 \texttt{ else } E_3) & \longrightarrow & \texttt{if } \mathcal{L}E_1 \texttt{ then } \mathcal{L}E_2 \texttt{ else } \mathcal{L}E_3 \\
\mathcal{L}(\texttt{pre } E) & \longrightarrow & \texttt{nil fby } \mathcal{L}E \\
\mathcal{L}(E_1 \texttt{ -> } E_2) & \longrightarrow & \mathcal{L}E_1 \texttt{ fby next } \mathcal{L}E_2 \\
\mathcal{L}(E \texttt{ when } B) & \longrightarrow & \mathcal{L}E \texttt{ wvr } \mathcal{L}B \\
\mathcal{L}(\texttt{current } E) & \longrightarrow & \texttt{last}(\mathcal{L}E, \mathcal{L}C) \quad (\texttt{C} \text{ is statically determined})
\end{array}$$

where *arg* is an argument on the base clock. The translation for node declarations is straightforward, and follows directly from these rules.

5 Signal in RLucid

In this section, we present the basic operators of SIGNAL [13], and show that they can be translated into RLUCID. SIGNAL is a relational dataflow language, designed for signal processing.

The basic operators of SIGNAL are:

`Y := f(X1,...,Xn)`	(pointwise application of data functions)
`Y := X $ N`	(delay)
`Y := X when B`	(sampling)
`Y := U default V`	(merge with priority)
`P | Q`	(composition of processes)

`X $ N` delays the value of `X` `N` times. It can be defined in RLUCID by:

```
X $ N = if N=0 then X else 0 fby X $ (N-1)
```

`X when B` delivers the value of `X` whenever `X` and `B` are both present and the value of `B` is true. It can be defined in RLUCID by:

```
X when B =
choose
   X||B: if first B then X fby next X when next B
                    else next X when next B
   X:    next X when B
   B:    X when next B
end
```

`U default V` delivers the value of `U` if it is present; if it is not, and the value of `V` is available, it delivers the value of `V`. It can be defined in RLUCID by:

```
U default V =
choose
   U||V: U fby next U default next V
   U:    U fby next U default V
   V:    V fby U default next V
end
```

With these definitions, a SIGNAL program can be translated into an RLUCID program.

6 Conclusions

RLUCID is a synchronous dataflow language which can be used for programming reactive systems. The combination of recursive function definitions and the operator **before** suffice to define myriads of real-time functions. All LUSTRE and SIGNAL functions can be translated into RLUCID.

As we said in the introduction, synchronous programs must have an interface with the external environment, the real world, written in a host language. If the input are arriving through interrupts, they must somehow be converted into streams. A minimal interface would be for all interrupt handlers to simply put the values provided to them in queues, and that the synchronous language would take over from there.

RLUCID is sufficiently rich that interfaces can be minimal. Input does not need to be synchronized. To design a reactive program, a user can write a highly synchronous kernel, as is done in LUSTRE, and then use operators such as **SignalLeftFair** and **MergeLeft** to create an interface between the aforementioned queues and the kernel [21]. Or, a user can attempt to synchronize the input as little as possible. Or, as was done for the stopwatch, the two methods can be combined.

Future work entails developing an efficient implementation for RLUCID. To generate good code for all RLUCID programs would be difficult. In particular, the use of recursive functions which contain conditionals can easily create networks requiring unbounded memory. However, all of the functions which have been defined recursively above can be translated into finite state automata, responding to their input. The language can be restrained so that this is always the case, and the remaining question is how to prevent problems of unbounded memory within the network. A system akin to the clock calculus of LUSTRE, but more lenient, is being envisaged. If the compiler refused to compile a program, the user would have to use **last** to synchronize streams.

Once a good compiler is generated, then only one of the goals outlined in the introduction remains unsolved: actually proving that timing constraints are met. This would entail adding assertions in RLUCID which would refer to the *absolute* time between events. The compiler would then have to determine that the code it generated actually met those constraints. This should be done in any case to determine if the synchronous hypothesis is realistic.

References

[1] P. Barth, S. Guthery, and D. Barstow. The Stream Machine: A data flow architecture for real-time applications. In *Proc. 8th International Conference on Software Engineering*, London, U.K., 1985. IEEE. Also available as SYS-85-12 from Schlumberger-Doll Research, Ridgefield, CO.

[2] J.-L. Bergerand, P. Caspi, N. Halbwachs, and J. A. Plaice. Automatic control systems programming using a real-time declarative language. In *Proc. 4th IFAC/IFIP Symposium (SOCOCO '86)*, Graz, Austria, 1986.

[3] G. Berry and G. Gonthier. The ESTEREL synchronous programming language: design, semantics, implementation. *Science of Computer Programming*, 1988. To appear.

[4] G. Berry, S. Moisan, and J.-P. Rigault. ESTEREL: Towards a synchronous and semantically sound high level language for real time applications. In *IEEE Real-Time Systems Symposium*, pages 30–37, 1983.

[5] M. Broy. Functional specification of time sensitive communicating systems. In *Stepwise Refinement of Distributed Systems: Models, Formalisms, Correctness*, volume 430 of *Lecture Notes in Computer Science*, pages 153–179, Mook, the Netherlands, 1989. Springer-Verlag.

[6] P. Caspi, N. Halbwachs, D. Pilaud, and J. A. Plaice. LUSTRE: a declarative language for programming synchronous systems. In *Proc. 14th Annual ACM Symposium on Principles of Programming Languages*, pages 178–188, Munich, Germany, 1987.

[7] A. Faustini and E. Lewis. Toward a real-time dataflow language. In J. Stankovic and K. Ramamrithan, editors, *Hard Real Time Systems*, pages 139–145. IEEE, 1989.

[8] N. Halbwachs, A. Longchampt, and D. Pilaud. Describing and designing circuits by means of a synchronous declarative language. In *Proc. IFIP Working Conference from HDL Descriptions to Guaranteed Correct Circuit Designs*, Grenoble, France, 1986.

[9] N. Halbwachs and D. Pilaud. Use of a real-time declarative language for systolic array design and simulation. In *Proc. International Workshop on Systolic Arrays*, Oxford, U.K., 1986.

[10] D. Harel and A. Pnueli. On the development of reactive systems: logic and models of concurrent systems. In *Proc. NATO Advanced Study Institute on Logics and Models for Verification and SPecification of Concurrent Systems*, pages 477–498. Springer-Verlag, 1985.

[11] P. J. Landin. The next 700 programming languages. *Communications of the ACM*, 9(3):157–166, 1966.

[12] P. LeGuernic, A. Benveniste, P. Bournai, and T. Gautier. Signal: A data flow oriented language for signal processing. *IEEE-ASSP*, 34(2):362–374, 1986.

[13] P. LeGuernic, T. Gautier, M. LeBorgne, and C. LeMaire. Programming real time applications with SIGNAL. *Proceedings of the IEEE, Special Issue on Synchronous Programming.* To appear, 1991.

[14] J. Plaice. Nested clocks: The LUSTRE synchronous dataflow language. Technical Report TR-91-09, University of Ottawa Computer Science Dept, Ottawa, Canada, 1991. Submitted to *Science of Computer Programming*.

[15] J. A. Plaice. *Sémantique et compilation de LUSTRE, un langage déclaratif synchrone.* PhD thesis, Institut National Polytechnique de Grenoble, Grenoble, France, 1988.

[16] A. Pnueli. Specification and development of reactive systems. In *Proc. IFIP Congress 86*, pages 845–858. Elsevier North-Holland, 1986.

[17] P. Raymond. Compilation efficace d'un langage déclaratif synchrone : Le générateur de code LUSTRE-v3. PhD thesis, Institut National Polytechnique de Grenoble, defence expected in November 1991.

[18] F. Rocheteau and N. Halbwachs. POLLUX, a LUSTRE based hardware design environment. In P. Quinton and Y. Robert, editors, *Conference on Algorithms and Parallel VLSI Architectures II, Chateau de Bonas*, June 1991.

[19] M. Sheeran. muFP, a language for VLSI design. In *Proc. ACM Symposium on Lisp and Functional Programming*, pages 104–112, Austin, Texas, 1984.

[20] D. Skillicorn and J. Glasgow. Real Time specification using Lucid. *IEEE Transactions on Software Engineering*, 15(2):221–229, 1989.

[21] N. S. Vempati. Programming reactive systems using dataflow. Master's thesis, University of Victoria, Victoria, B.C., Canada, 1990.

[22] W. W. Wadge and E. A. Ashcroft. *Lucid, the Dataflow Programming Language*. Academic Press, 1985.

A Mechanized Theory for the Verification of Real-Time Program Code using Higher Order Logic

Rachel Cardell-Oliver
Computer Laboratory, University of Cambridge
DSTO, Australia

1 Introduction

This paper addresses the problem of verifying computer programs whose correct operation depends on interactions with an unco-operative real-time environment. We intend to verify properties such as

- **safety:** the behaviour of variable s and output port o in program p satisfy some predicate P of s and o at all times
- **timeliness:** the program p will always respond to input x within n time steps

The verification of program properties directly from a programming language semantics is demonstrated. Our semantics and verification proofs have been mechanized in the HOL theorem prover. In summary, the paper reports on the following:

1. The definition of a simple imperative programming language for real-time programs with asynchronous, single-buffered, non-blocking I-O.
2. The definition of a real-time semantics for this language in which the behaviour or meaning of a computer program is given by a set of histories, each of which represents the behaviour, over time, of a program variable or input or output port.
3. The mechanization of the semantics and a few proof tools in the HOL theorem prover.
4. The verification of safety and timeliness properties of a program which uses a handshaking protocol to interact with its environment. The verification proofs have been mechanically checked in the HOL system.

The paper is organised as follows. Section 2 states assumptions we have made and the relation of our work to others. Section 3 describes the syntax of our language and Section 4 its semantics. Section 5 discusses the mechanization of the semantics in the HOL system. Section 6 reports on the verification of two properties of a device which uses a handshaking protocol to interact with its environment, reading two values and outputting their sum. Future work and our conclusions are discussed in Section 7.

2 Assumptions

The programming language presented in this paper was designed in order to investigate techniques for the verification of real-time programs. To this end our language and semantics are succinct: a minimal set of commands and assumptions are used. In the terms of the British Ministry of Defence Interim Standard 055, (Part 1)/1 - 30.1.3 [MoD91]

> Safety critical software shall be designed so that it is easy to justify that it meets its specification in terms of both functionality and performance. This requirement may restrict the length and complexity of the software and inhibit the use of concurrency, interrupts, floating point arithmetic, recursion, partitioning and memory management.

We have assumed that programs will be executed on a simple processor for which the execution times of all commands and expressions is known *a priori*. That is, we assume knowledge of the instruction set of the processor and the compilation algorithm used. In order to achieve known execution times we also assume that there is no process scheduling, and no commands or expressions with unbounded execution time such as synchronized input-output, unbounded wait commands, etc. Instead our language supports asynchronous, single-buffered, non-blocking communication. Primitives such as synchronized communication with timeouts and multiple buffers could be built from our primitives if required. The issues of declarations and run-time errors are not considered in this paper.

Programs in our language are processes running on a dedicated processor and communicating with their environment asynchronously. Concurrent systems are specified by giving two or more programs to be executed on separate processors and by specifying, in higher order logic, the relationship between outputs from one program and the inputs of another.

Our semantics differs from other real-time programming language semantics in its treatment of time. By *real-time* we mean a time scale related to hardware time such as a processor clock, nano-seconds, microcode instruction times etc. Time is discrete and modelled by the natural numbers.

A different approach has been used by many authors for specifying real-time programs. For example, lower and upper bound estimates of computation time are proposed for programming languages in which primitives such as concurrency and synchronous communication are built into the the language [Haa81, Hoo87, Sha89, Ost89, Heh89]. The idea is to abstract from the complexities of execution environments so that, say, a program's verification proof might be valid for a range of different scheduling policies, communication rules etc. However, the cost of this abstraction is that much information about the global environment must be carried about in the program semantics in order to estimate, for example, the time one process must wait for another to communicate with it [Hoo87]. Our approach differs from this in that we make a program's execution model as simple as possible and make explicit all choices about the time taken to execute a program.

Our semantics has been mechanized in the HOL system: a theorem prover derived from the LCF system [GMW79] for a version of higher order logic based on Church's

EXPRESSIONS	
`Val` x	the value of the variable x
`NumConst` n	value of natural number n
`BoolConst` b	value of boolean b: `TRUE` or `FALSE`
`Plus` $e1$ $e2$	sum of $e1$ and $e2$
`Minus` $e1$ $e2$	natural number subtraction of $e2$ from $e1$
`Not` b	if b is `TRUE` then `FALSE` and vice versa
`And` $b1$ $b2$	logical conjunction of $b1$ and $b2$
`Or` $b1$ $b2$	logical disjunction of $b1$ and $b2$
`Eq` $e1$ $e2$	the value of $e1$ is the same as $e2$
`Lt` $e1$ $e2$	the value of $e1$ is less than $e2$
`TestInput` p	`TRUE` if input is available on channel p and `FALSE` otherwise

Figure 1: Syntax of Expressions

simple type theory [Chu40]. The HOL system checks intricate verification proofs. For example verification in HOL demonstrated (small) errors in early hand proof versions of the examples given in Section 6. More importantly HOL provides support for programming higher level proof tools on top of theories such as the semantics presented in this paper [Gor88, Gor91, CIN91]. Parts of the proofs presented in this paper were automated in HOL. Future plans include encoding the rules of a program logic for real-time programs as theorems in HOL, and using HOL tactics for user-guided synthesis of programs.

HOL has been used to define a number of program language semantics and program logics. Hale has defined a denotational semantics for a real-time programming language SAFE0 [Hal90]. The semantics have been used for the verification of a compiler for SAFE0. We have adapted the SAFE0 language but our style of semantics differs from Hale's. Gordon has proposed a logic for reasoning about a real-time programming language [Gor91]. Gordon's programming language has a simpler input-output model than our language but is in other respects similar. Its semantics is given in terms of the assembler language of a simple machine and then a programming logic is proposed to abstract from the resulting low level semantics. The HOL system has also been used for a mechanization of Hoare logic and a range of verification tools for that logic [Gor88].

3 Syntax

Figures 1 and 2 define the abstract syntax for the expressions and commands of our language. Precedence is indicated by parentheses, where necessary, in compound expressions and commands.

In order to simplify our semantics and verification strategies we shall consider only non-terminating real-time programs of the form

(`Loop` *cmd*)

in which the command *cmd* may not contain the `Loop` command. Although chosen for pragmatic reasons, this class of programs has proved sufficient for programming the ex-

COMMANDS	
`Skip`	do nothing
`Delay` *e*	evaluate *e* and then
	do nothing for time given by the value of *e*
`Assign` *x e*	copy the value of expression *e* to variable *x*
`Read` *p x*	copy input from port *p* to variable *x*
`Write` *p e*	output the value of *e* on port *p*
`If` *b c1 c2*	if the value of *b* is true then execute *c1* else *c2*
`Seq` *c1 c2*	execute *c1* and then *c2*
`Loop` *c*	execute command *c* forever

Figure 2: Syntax of Commands

amples we have considered so far. Definitions for a more general semantics are given in the Appendix.

In the examples of program code used in the paper we shall use the abbreviations `x:=e` for `Assign x e`, `c1;c2` for `Seq c1 c2`, `e1+e2` for `Plus e1 e2` and `e1=e2` for `Eq e1 e2`. We write `x` instead of `(Val x)`, `TRUE` instead of `(BoolConst TRUE)`, `3` instead of `(NumConst 3)` and so on.

4 Semantics

The meaning of all but the input-output commands in our language is similar to that of corresponding commands in other imperative languages such as Pascal. The treatment of input and output is based on that of SAFE0 [Hal90]. Writing a value to an output port means transmitting that value: there is no queuing or buffering of output. Reading from an input port returns the current value in the port buffer. If there is no input available then garbage may be copied from the port buffer. The expression `TestInput` distinguishes these cases: it is true only when valid data is available in the port buffer. Input values from the environment are copied into the buffer whenever they become available. Thus, a new value may overwrite an old before the old value has been read. Reading from an input port has the side effect of clearing the buffer. This prevents the same value from being read twice. It is the responsibility of programmers and verifiers to show, if required, that a program reacts sufficiently fast so as not to miss inputs and that the program uses `TestInput` to test whether valid input is available before performing a `Read` command.

Our semantics maps programs to their meanings where the meaning of any program is an interpretation in higher order logic of what it means to execute that program. A program's meaning will be described by histories: one history for each program identifier. For example, the semantics of the program

```
Loop (s:=s+1; Write (Port_1) s)
```

is given by the histories of the variable `s` and the output port `(Port_1)`.

A history for any program identifier gives the value of that identifier at any time during program execution. For example, in the above program each loop traversal takes 11 time units. Thus, if the variable `s` has value 0 at time 0 and the program starts execution at time 0, then

EXPRESSION E	$\mathcal{E}_\mathcal{V}$ E f t	$\mathcal{E}_\mathcal{T}$ E
`Val` x	f x t	1
`NumConst` n	n	1
`BoolConst` b	b	1
`Plus` $e1$ $e2$	$(\mathcal{E}_\mathcal{V}\ e1\ \mathsf{f}\ \mathsf{t}) + (\mathcal{E}_\mathcal{V}\ e2\ \mathsf{f}\ \mathsf{t})$	$(\mathcal{E}_\mathcal{T}\ e1) + (\mathcal{E}_\mathcal{T}\ e2) + 3$
`Minus` $e1$ $e2$	$(\mathcal{E}_\mathcal{V}\ e1\ \mathsf{f}\ \mathsf{t}) - (\mathcal{E}_\mathcal{V}\ e2\ \mathsf{f}\ \mathsf{t})$	$(\mathcal{E}_\mathcal{T}\ e1) + (\mathcal{E}_\mathcal{T}\ e2) + 3$
`Not` e	$\neg\ (\mathcal{E}_\mathcal{V}\ e\ \mathsf{f}\ \mathsf{t})$	$(\mathcal{E}_\mathcal{T}\ e) + 1$
`And` $e1$ $e2$	$(\mathcal{E}_\mathcal{V}\ e1\ \mathsf{f}\ \mathsf{t}) \wedge (\mathcal{E}_\mathcal{V}\ e2\ \mathsf{f}\ \mathsf{t})$	$(\mathcal{E}_\mathcal{T}\ e1) + (\mathcal{E}_\mathcal{T}\ e2) + 1$
`Or` $e1$ $e2$	$(\mathcal{E}_\mathcal{V}\ e1\ \mathsf{f}\ \mathsf{t}) \vee (\mathcal{E}_\mathcal{V}\ e2\ \mathsf{f}\ \mathsf{t})$	$(\mathcal{E}_\mathcal{T}\ e1) + (\mathcal{E}_\mathcal{T}\ e2) + 1$
`Eq` $e1$ $e2$	$(\mathcal{E}_\mathcal{V}\ e1\ \mathsf{f}\ \mathsf{t}) = (\mathcal{E}_\mathcal{V}\ e2\ \mathsf{f}\ \mathsf{t})$	$(\mathcal{E}_\mathcal{T}\ e1) + (\mathcal{E}_\mathcal{T}\ e2) + 1$
`Lt` $e1$ $e2$	$(\mathcal{E}_\mathcal{V}\ e1\ \mathsf{f}\ \mathsf{t}) < (\mathcal{E}_\mathcal{V}\ e2\ \mathsf{f}\ \mathsf{t})$	$(\mathcal{E}_\mathcal{T}\ e1) + (\mathcal{E}_\mathcal{T}\ e2) + 1$
`TestInput` p	$\exists$ e t1. t1 $\leq$ t $\wedge$ (e,t1) $\in$ (f p t)	1

Figure 3: Expression Values and Evaluation Times

- at time 8, **s** will have value 1 and
- between times 8 and 19 the value of **s** is 1 and
- at time 19, **s** will have value 2 and so on

The meaning of a program also depends on the context in which the program is executed. An execution context is defined by the time at which execution starts and by the values of program identifiers at that time.

Informally, our semantics takes programs and their contexts and returns a history for every identifier. These intuitive ideas will be formalised in Section 4.2.

4.1 Expression Semantics

Expressions are evaluated using the function $\mathcal{E}_\mathcal{V}$ which returns a natural number value. Although expressions may only return numerical values, natural numbers may be interpreted as booleans and vice versa where required: 0 is interpreted as false, and any other value (usually 1) as true. The definition of $\mathcal{E}_\mathcal{V}$ is given in Figure 3. The parameter f captures state information and will be explained in Section 4.2; t is the time at which the expression is evaluated. The semantics of `TestInput` relies on the semantics of reading and so we delay its discussion until Section 4.2.

Evaluating an expression has no side effects except the passing of time. The function $\mathcal{E}_\mathcal{T}$, defined in Figure 3, takes an expression and returns the time taken to evaluate it. Evaluation time depends on the time taken to copy partial results and to perform comparisons. In our language the time taken to evaluate an expression is independent of its evaluation context.

4.2 Command Semantics

The type name specifies all possible program identifiers. Identifiers may be variables, input ports or output ports. The type

$$\mathsf{behaviour} = \mathsf{name} \rightarrow \mathsf{history}$$

maps any program identifier to its history. A history maps times to values. Time is modelled by the natural numbers.

$$\text{history} = \text{time} \rightarrow \text{values}$$

The type values depends on the type of name whose history is being defined

- a program variable always has a natural number value.
- an output port has either a natural number value when the port is active or a null value when the port is not actively transmitting.
- an input port is specified by a set of natural number values and input times representing all inputs available on that port

To make the definitions given in the next sections more readable we shall not include type coercion functions for name and values where these can be deduced from the context. However, formal type coercions are used in the HOL mechanization of the semantics.

The semantics of commands is given by a function $\mathcal{M}$ which takes a command c of type cmd, a behaviour f and a starting time t and returns the command's meaning as a predicate on f and t. The predicate distinguishes the set of histories which could be generated by executing the command c at time t. The "state" in which execution occurs is given implicitly by the values of f v t for any identifier v of type name.

$$\mathcal{M} : \text{cmd} \rightarrow \text{behaviour} \rightarrow \text{time} \rightarrow \text{bool}$$

Given a command c, a behaviour function f, a starting time t, and any name v

1. If c is a terminating command then ($\mathcal{M}$ c f t) is
   - true if f v t, f v t+1, f v t+2, ..., f v t+n is the sequence of values of v when c is executed at time t and takes time n to execute, and
   - false otherwise
2. If c is a non-terminating command then ($\mathcal{M}$ c f t) is
   - true if f v t, f v t+1, f v t+2, ... is the infinite sequence of values of v when c is executed at time t, and
   - false otherwise

For the identifier s in the previous program example $\mathcal{M}$ returns a predicate[1] such as

$$\forall \text{t}. \ (\text{t MOD } 11 = 7) \Rightarrow (\text{f s (t+1)} = (\text{f s t})+1) \mid (\text{f s (t+1)} = (\text{f s t}))$$

which means that every 11 time units, starting at 8 (that is 7+1), the value of s is increased by 1. At all other times t the value of s at (t+1) is the same as the value of s at time t. The notation (b $\Rightarrow$ c1 | c2) means c1 if b is true and c2 if b is false.

We use the abbreviation **Stable** n v f t to mean that in the interval of time from t to t+n, for identifier v, the value of the history (f v) does not change. Stable intervals occur, for example, while an expression is being evaluated.

To return to the semantics of expressions, the functions in Figure 3 have types

[1] This is a slight simplification: $\mathcal{M}$ actually returns histories for all identifiers, not just s.

$\mathcal{E}_{\mathcal{V}}$: expr → behaviour → time → num
$\mathcal{E}_{\mathcal{T}}$: expr → time

Simple Commands

The Skip command gives no information about any history and takes no time to execute. That is, Skip is syntactic sugar and is not compiled to any instruction. T denotes truth. Delay evaluates an expression and then delays for the time given by the value of the expression. Assign evaluates the expression e and copies its value to the variable x.

$\mathcal{M}$ Skip f t = T

$\mathcal{M}$ (Delay e) f t = ∀ v:name. Stable $((\mathcal{E}_{\mathcal{T}}$ e) + $(\mathcal{E}_{\mathcal{V}}$ e f t)) v f t

$\mathcal{M}$ (Assign x e) f t =
  ∀ v:name.
    Stable $(\mathcal{E}_{\mathcal{T}}$ e) v f t ∧
    (v = x) ⇒ f v (t + $(\mathcal{E}_{\mathcal{T}}$ e) + 1) = $(\mathcal{E}_{\mathcal{V}}$ e f t)
        | Stable 1 v f (t + $(\mathcal{E}_{\mathcal{T}}$ e))

Communication

The Write command copies the value of an expression to an output port (c.f. Assign). Values copied to an output port are immediately transmitted; there is no buffering.

$\mathcal{M}$ (Write p e) f t =
  ∀ v:name.
    Stable $(\mathcal{E}_{\mathcal{T}}$ e) v f t ∧
    (v = p) ⇒ f v (t + $(\mathcal{E}_{\mathcal{T}}$ e) + 1) = $(\mathcal{E}_{\mathcal{V}}$ e f t)
        | Stable 1 v f (t + $(\mathcal{E}_{\mathcal{T}}$ e))

The history of input available on a particular port is modelled by a set of value, time pairs. If the pair (w,t) is an element of an input set for some port p this means that at time t the value w is placed in the buffer of port p. Initially the history of any input port is the set of all inputs which will be available on that port. This set is pruned by each execution of the Read command. Pruning represents the loss of values which, once read, can not be re-read. The semantics of Read in terms of this input set is defined by

1. search the current input set for the most recently available input
2. copy this value to the read variable (changes the history of f x)
3. prune the input set to prevent old values being re-read (changes the history of f p)

Step 1 is represented by the Inp function. For a Read performed at time t, if there exists a pair (w,t') in the input set for time t and t'≤t and there is no pair (w',u) such that u is between t' and t then the value w is returned by Inp. If there is no such w then Inp returns garbage: this represents reading an input buffer which does not contain valid data. The notation (ϵ x.P x) means choose an x which satisfies P x. If there is no such x then an arbitrary value of the same type as x is returned.

$$\text{Inp p f t} = \epsilon\,\text{w}.\ \exists \text{t1}.\ \text{t1} \leq \text{t} \land (\text{w,t1}) \in (\text{f p t}) \ \land \ \forall \text{tt}.\ \text{t1} \leq \text{tt} \land \text{tt} < \text{t} \Longrightarrow \neg(\exists\,\text{w1}.\ (\text{w1,tt}) \in (\text{f p t}))$$

For step 2 the value returned by Inp is copied to the input variable of the Read command. This is represented by updating the history of the variable x.

Step 3 is performed by updating the history of the port p to a new set of inputs in which all pairs in the old set with time fields before t are deleted. This prevents values from being re-read. The notation { x | P x } means the set of all x which satisfy P of x.

The Read command takes 1 time step to execute and thus for any identifiers except p and x the history generated is a stable interval of length 1.

$$\begin{aligned} &\mathcal{M}\ (\text{Read p x})\ \text{f t} = \\ &\ \forall\ \text{v:name}. \\ &\quad (\text{v} = \text{x}) \Rightarrow \text{f v (t + 1)} = \text{Inp p f t} \\ &\ \mid (\text{v} = \text{p}) \Rightarrow \text{f v (t + 1)} = \{(\text{w,t1}) \mid (\text{w,t1}) \in (\text{f p t}) \land \text{t} < \text{t1}\} \\ &\qquad\qquad \mid \text{Stable 1 v f t} \end{aligned}$$

The expression (TestInput p) is true if there is a value available for input on port p. The definition given in Figure 3 can be paraphrased: TestInput p is true if there exists a pair (w,t') in the set for port p where t' is less than or equal to the current time. Testing that TestInput is true before performing a Read command guarantees the Read command will not return garbage.

Command Execution Times

The function $\mathcal{C_T}$ of type behaviour → time → time returns the time taken to execute any terminating command given its execution context (a behaviour f and starting time t). The constants (+1) are a result of jumps and copying actions in the machine code which implements these commands. $\mathcal{C_T}$ is not defined for the Loop command since that does not terminate.

COMMAND c	$\mathcal{C_T}$ c f t
Skip	0
Delay e	$(\mathcal{E_T}\ \text{e}) + (\mathcal{E_V}\ \text{e f t})$
Assign x e	$(\mathcal{E_T}\ \text{e}) + 1$
Read p x	1
Write p e	$(\mathcal{E_T}\ \text{e}) + 1$
If b c1 c2	$((\mathcal{E_V}\ \text{b f t})$
	$\Rightarrow (\mathcal{E_T}\ \text{b}) + 1 + (\mathcal{C_T}\ \text{c1 f (t} + \mathcal{E_T}\ \text{b} + 1)) + 1$
	$\mid\ (\mathcal{E_T}\ \text{b}) + 1 + (\mathcal{C_T}\ \text{c2 f (t} + \mathcal{E_T}\ \text{b} + 1))$
Seq c1 c2	$(\mathcal{C_T}\ \text{c1 f t}) + (\mathcal{C_T}\ \text{c2 f (t} + (\mathcal{C_T}\ \text{c1 f t}))$

Compound Commands

The command Seq performs its two commands one after the other.[2] There is no time overhead for sequential composition of commands.

The If command evaluates its expression b, taking $\mathcal{E}_T$ b time steps. If ($\mathcal{E}_V$ b f t) is true then jump to command c1, execute it, and jump to the end of command. If ($\mathcal{E}_V$ b f t) is false then jump to c2 and execute that. Each jump takes 1 time step.

```
M (Seq c1 c2) f t = (M c1 f t) ∧ (M c2 f (t+(C_T c1 f t)))

M (If b c1 c2) f t =
let time-b-jump-done = (t + E_T b + 1) in
let time-c1-done = time-b-jump-done + (C_T c1 f time-b-jump-done) in
  ∀ v:name.
    (E_V b f t)
      ⇒ Stable (E_T b + 1) v f t ∧
         M c1 f time-b-jump-done ∧
         Stable 1 v f time-c1-done
      | Stable (E_T b + 1) v f t ∧
        M c2 f time-b-jump-done
```

The only non-terminating command in our language is the Loop command which executes its body forever. The command (Loop c) is defined as an infinite sequence of (If TRUE c Skip) commands, each executed in the context created by its predecessor. That is, Loop is a syntactic abbreviation for a non-terminating While loop. The semantics for a general (terminating or non-terminating) while loop is not difficult; however, it has not yet been included in our mechanized semantics. The Appendix shows how the semantics of While loops could be defined. The time at which each traversal of the loop begins is given by the primitive recursive function $\mathcal{L}_T$.

```
L_T c1 f t1 0 = t1 ∧
L_T c1 f t1 (SUC n) =
    (L_T c1 f t1 n) + 2 + (C_T com f ((L_T c1 f t1 n) + 2)) + 1

M (Loop c1) f t =
    ∀ N. M (If TRUE c1 Skip) f (L_T c1 f t N)
```

It should be noted that this semantic definition of Loop is simpler to evaluate than the traditional limit definition (Seq (If TRUE c Skip) (Loop c)).

5 Mechanization

The syntax and semantics described in Sections 3 and 4 can be defined (with trivial notational differences) as a theory in HOL. The types of expressions and commands

[2] As stated in Section 3 the semantics of Seq c1 c2 is undefined if c1 contains a Loop command. However, see also the Appendix which explains how these constraints may be removed.

are defined using HOL's primitive recursive types package and $\mathcal{M}$, $\mathcal{E}_{\mathcal{V}}$ etc are recursive functions on these types.

For verification we often require only the history of one identifier and not every possible history. We define $\mathcal{M}_{\mathsf{v}}$ to be the same as $\mathcal{M}$ except that it extracts just the history of v and not histories for all v.

$$\mathcal{M}_{\mathsf{v}}\ \mathsf{c\ f\ t} = \mathsf{P\ v\ f\ t} \quad \text{where} \quad \mathcal{M}\ \mathsf{c\ f\ t} = \forall\, \mathsf{v.\ P\ v\ f\ t}$$

Evaluating the semantics of a program gives a predicate which contains a conjunction of terms of the form (Stable n v f t) and (f v t+1 = X). In some cases a large amount of control history must be carried about in time parameters (see, for example, the definition of the `If` command) and this makes the predicates large, unreadable and thus difficult to use in verification proofs.

We have defined an interpretation function $\mathcal{I}$ which allows such predicates to be written in a concise form. An example is used to illustrate the definition of $\mathcal{I}$. The semantic definitions proposed in Sections 3 and 4 and the function x n = $\mathcal{L}_{\mathcal{T}}$(`s:=s+1`) f 0 n can be used to show

$$\begin{array}{l} \mathcal{M}_{\mathsf{s}}\ (\mathsf{Loop\ s{:}{=}s{+}1})\ \mathsf{f}\ 0 = \\ \forall\, \mathsf{n.\ Stable\ 7\ s\ f\ (x\ n)} \wedge \\ \quad \mathsf{f\ s\ (x\ n){+}8 = (f\ s\ (x\ n){+}2){+}1} \wedge \\ \quad \mathsf{Stable\ 1\ s\ f\ (x\ n){+}8} \end{array}$$

We abbreviate (Stable n v f t) to D n and (f v t+1 = X) to U X and use lists to indicate the order of such actions. The recursive function $\mathcal{I}$ interprets a list of Ds and Us as **Stable** and update actions as in the predicate above. That is, the predicate above is equivalent, by the definition of $\mathcal{I}$, to

$$\begin{array}{l} \mathcal{M}_{\mathsf{s}}\ (\mathsf{Loop\ s{:}{=}s{+}1})\ \mathsf{f}\ 0 = \\ \forall\, \mathsf{n.}\ \mathcal{I}\ [\mathsf{D\ 7;\ U\ (f\ s\ (x\ n){+}2){+}1;\ D\ 1}]\ \mathsf{s\ f\ (x\ n)} \end{array}$$

6 Verification Examples

In this section we demonstrate the use of the semantics of Section 4 to verify two properties of a real-time program. The example is adapted from a problem of Gordon [Gor91]: to implement a program which calculates the sum of two inputs using a handshaking protocol to interact with its environment. The environment can request the program whenever it is available and must then produce two inputs. The program waits for a request, reads two inputs then calculates the sum of the inputs, outputs them to the environment, and finally becomes available again. The environment's request and program's available signal are said to perform a *handshake* to co-ordinate input and output. Figure 4 describes the intended behaviour of the program and its environment.

The program is responsible for reading input after a request, for producing output and for setting the available signal correctly. We have assumed that the program and its environment are able to communicate using the variables `req` and `avail` although, strictly, all communication should occur through input-output ports. We claim (and in Sections 6.1 and 6.2 we prove) that the following program (`prog`) satisfies the program's obligations to the environment.

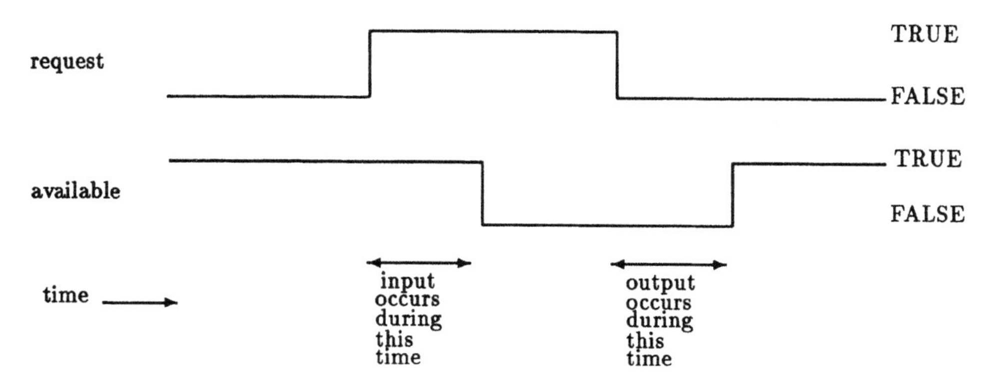

Figure 4: A Pictorial Specification of Handshaking

```
Loop (
 (If (And (req=TRUE) (avail=TRUE))
     (Read (Port_1) x; Read (Port_2) y; avail:=FALSE)
     Skip ) ;
 (If (And (req=FALSE) (avail=FALSE))
     (Write (Port_3) (x+y); avail:=TRUE)
     Skip) )
```

The environment is responsible for the correct setting of the request signal and for producing input at the correct time. The behaviour of the request signal is defined by the following constraints

- request is initially false
- if the program is available then request may become true otherwise it does not change its value
- if the program is not available then request may become false otherwise it does not change its value

We shall also assume that the program variable **avail** is initially true since otherwise the program could never become available. Formally, this environment is specified by ENV_OBLIG in terms of the histories of the program variables **avail** and **req**

$$\begin{aligned}
&\mathsf{ENV_OBLIG}\ \mathsf{f} \equiv \\
&\mathsf{f}\ \mathbf{avail}\ 0 = \mathsf{T}\ \wedge\ \mathsf{f}\ \mathbf{req}\ 0 = \mathsf{F}\ \wedge \\
&\forall\, \mathsf{t}.\ \neg(\mathsf{f}\ \mathbf{avail}\ \mathsf{t} = \mathsf{f}\ \mathbf{req}\ \mathsf{t}) \\
&\qquad \Rightarrow (\mathsf{f}\ \mathbf{req}\ \mathsf{t}{+}1 = \neg(\mathsf{f}\ \mathbf{req}\ \mathsf{t})) \vee (\mathsf{f}\ \mathbf{req}\ \mathsf{t}{+}1 = \mathsf{f}\ \mathbf{req}\ \mathsf{t}) \\
&\qquad \mid (\mathsf{f}\ \mathbf{req}\ \mathsf{t}{+}1 = \mathsf{f}\ \mathbf{req}\ \mathsf{t})
\end{aligned}$$

In order to prove safety and timeliness properties of the program and its environment we shall need to evaluate the history of the program variable **avail** observed from time 0. That is, we must evaluate $\mathcal{M}_{\mathbf{avail}}$ **prog** f 0. We use the following abbreviations (cc is the command body of the loop in **prog**)

$cc = $ `If And(req=TRUE)...; If And(req=FALSE)...`
$x\ n = \mathcal{L}_{\mathcal{T}}\ cc\ f\ 0\ n$
$T1\ f\ t = \mathcal{E}_{\mathcal{V}}$ `(And (req=TRUE)(avail=TRUE))` f t
$T2\ f\ t = \mathcal{E}_{\mathcal{V}}$ `(And (req=FALSE)(avail=FALSE))` f t

In HOL we have programmed an automatic proof procedure for generating theorems of the form ($\vdash \forall f.\ \mathcal{M}_v\ cmd\ f\ 0 = \ldots$). The following theorem is a tidied version of the output of this procedure

$$
\begin{array}{l}
\vdash \forall f.\ \mathcal{M}_{\texttt{avail}}\ \texttt{prog}\ f\ 0 = \\
\forall n. \\
\quad T1\ f\ (x\ n)+2 \wedge T2\ f\ (x\ n)+15 \\
\Rightarrow \mathcal{I}\ [D\ 13;\ U\ FALSE;\ D\ 16;\ U\ TRUE;\ D\ 2]\ \texttt{avail}\ f\ (x\ n) \\
\mid\ T1\ f\ (x\ n)+2 \wedge \neg(T2\ f\ (x\ n)+15) \\
\Rightarrow \mathcal{I}\ [D\ 13;\ U\ FALSE;\ D\ 10]\ \texttt{avail}\ f\ (x\ n) \\
\mid\ \neg(T1\ f\ (x\ n)+2) \wedge (T2\ f\ (x\ n)+10) \\
\Rightarrow \mathcal{I}\ [D\ 25;\ U\ TRUE;\ D\ 2]\ \texttt{avail}\ f\ (x\ n) \\
\mid\ \mathcal{I}\ [D\ 19]\ \texttt{avail}\ f\ (x\ n)
\end{array}
$$

A useful property of the loop time function x is that it is possible to express any time t in terms of a particular loop traversal: for any t there exist some n and m such that t is the m-th step within the n-th loop traversal. Formally,

$$\vdash \forall t.\ \exists m\ n.\ (\ (t = (x\ n)+m) \ \wedge\ (t < x(n+1))\ )$$

6.1 Safety: Preserving Correctness

A safety property is one that must hold at all times during program execution. There are a range of safety properties we might wish to prove about `prog`. We have chosen to verify the correctness of the handshaking mechanism used by our program. The correct behaviour of the program's available signal is specified by the predicate HANDSHAKE

$$
\begin{array}{l}
\mathsf{HANDSHAKE}\ f \equiv \\
\forall t.\ (f\ \texttt{avail}\ t = f\ \texttt{req}\ t) \\
\qquad \Rightarrow (f\ \texttt{avail}\ t+1 = \neg(f\ \texttt{avail}\ t)) \vee (f\ \texttt{avail}\ t+1 = f\ \texttt{avail}\ t) \\
\qquad \mid\ (f\ \texttt{avail}\ t+1 = f\ \texttt{avail}\ t)
\end{array}
$$

which states that if available and request are both true then the program may stop being available, and it may become available again when available and request are both false (c.f. Figure 4 and `ENV_OBLIG`).

We have proved the following theorem in HOL

REQUIRED TO PROVE:

$$
\begin{array}{l}
\forall f{:}behaviour. \\
\quad ENV\_OBLIG\ f \ \wedge\ \mathcal{M}_{\texttt{avail}}\ \texttt{prog}\ f\ 0 \ \Longrightarrow\ HANDSHAKE\ f
\end{array}
$$

PROOF:

In order to prove HANDSHAKE f assume that ENV_OBLIG f and $\mathcal{M}_{\mathbf{avail}}$ prog f 0 are true and deduce HANDSHAKE f. That is, we have

goal $\forall$ t. (f **avail** t = f **req** t)
$\Rightarrow$ (f **avail** t+1 = $\neg$(f **avail** t)) $\vee$ (f **avail** t+1 = f **avail** t)
| (f **avail** t+1 = f **avail** t)

to be proved given the assumptions **A1** and **A2** where

A1 ENV_OBLIG f

A2 $\mathcal{M}_{\mathbf{avail}}$ prog f 0

To prove that **goal** is satisfied at all times t we shall choose an arbitrary time such that t = x n + m.

There are four possible paths through the n-th loop of prog and we must examine each path in order to prove that all possible behaviours of prog satisfy HANDSHAKE. Since the proofs for all cases are similar we shall show just one

A Case T1 f (x n)+2 $\wedge$ $\neg$(T2 f (x n)+15)

The behaviour of program loop for this case is given by

$\mathcal{I}$ [D 13; U FALSE; D 10] avail f (x n)

Possible cases for the value of m are ($0 \leq$ m < 13),(m=13),($13 <$ m). In the first case, D 13 represents a stable interval and this means f **avail** t+1 = f **avail** t which satisfies the goal. In the second case, when m=13 then f **avail** t+1 is false. We know that f **avail** (x n)+2 = f req (x n)+2 since T1 is true. We know from ENV_OBLIG that the request signal cannot change its value whilst **avail** and **req** have the same value and so (f **req** t = f **avail** t). Thus when m=13 (f **req** t = f **avail** t) and (f **avail** t+1 = $\neg$(f **avail** t)) and so the goal is true as required. Finally when 13 < m then the stable interval denoted by D 10 implies f **avail** t+1 = f **avail** t. Thus, for all possible values of m the path [D13;U FALSE;D 10] satisfies our goal.

The proofs that other possible paths through the n-th loop satisfy HANDSHAKE f are similar. Much of this safety proof was automated in HOL by programming a proof procedure for the type of arguments given above.

END PROOF

6.2 Timeliness: Real-Time Response

We now consider the problem of proving that the program responds to an environment request in a timely manner, by proving there is a bound on the time taken to produce output. We prove that once the environment ceases to request the program, then the available signal will be reset (and output will be produced) in at most 34 time steps.

For the proof, let Q t stand for (f **avail** t) is true which is the condition we aim to establish. Let IOK f t0 t1 stand for the condition that the interval of time from t0 to t1 is greater that 34 time steps and during that interval the value of **req** is always false.

REQUIRED TO PROVE:

ENV_OBLIG f $\wedge$ $\mathcal{M}_{\texttt{avail}}$ prog f 0 $\Longrightarrow$ ($\forall$ t0 t1. IOK f t0 t1 $\Longrightarrow$ Q t1)

PROOF:

First we choose arbitrary m and n which satisfy (t1 = (x n)+m). Then we use the fact that the maximum length of one loop traversal is 33 time steps and IOK f t0 t1 to deduce that (t0 $\leq$ (x n)).

The proof is goal directed: we assume ENV_OBLIG f and $\mathcal{M}_{\texttt{avail}}$ prog f 0 and then look for conditions which will ensure Q t1. Then find conditions which ensure those and so on until we get to a condition, CC, which is always true. Then, since we can always start by assuming CC and then deduce Q t1, we have proved Q t1.

To prove Q t1 it is sufficient to show Q (x n) or 26$\leq$m since

- If Q (x n) then tests T1 and T2 will be false and interpreting [D 19] at (x n) ensures that (f avail t1 = f avail (x n)) and therefore Q t1
- If 26$\leq$m and Q (x n) is not true then T1 will be false and T2 will be true. Thus interpreting [D 25;U TRUE;D 2] ensures Q (x n)+m which is Q t1

We shall refer to the time at which T2 is evaluated in the n-th loop as T2time n where T2time n = (T1 f (x n)+2) $\Rightarrow$ (x n)+17 | (x n)+10

To prove Q (x n) it is sufficient to prove n=0 or 0 < n and either t0 occurs before T2time(n$-$1) or T2 is true in the (n$-$1)-th loop, since

- If n=0 then (x n = 0) and Q 0 by our initial assumptions.
- If 0<n and T2 is true in the (n$-$1)-th loop then interpreting [D 25;U TRUE;D 2] at (x n$-$1) or interpreting [D 15;U FALSE;D 16;U TRUE;D 2] at (x n$-$1) implies Q(x n).
- If 0<n and t0 occurs before T2time(n$-$1) then
  - if T2 is true in the (n$-$1)-th loop then Q (x n) as above
  - if T2 is false in the (n$-$1)-th loop then avail must be true at T2time(n$-$1) since T2 false means request is true or available is true by De Morgan's Laws and we know request is false from IOK f t0 t1. Thus, interpreting [D 19] at (x n$-$1) or interpreting [D 15;U FALSE;D 10] at (x n$-$1) when f avail (T2time(n$-$1)) is true implies Q (x n).

Finally, it remains to prove that the condition we have reached is always true. The condition, called CC, is

(0=n) $\vee$
(0<n $\wedge$ ((t0 < T2time(n$-$1)) $\vee$ (T2 true in the (n$-$1)-th loop))) $\vee$
(26$\leq$m)

and it is always true since

- For any n we know n=0 $\vee$ 0<n.

- If n=0 then CC is true whatever the value of the other disjuncts.
- If 0<n then we must show either (t0 < T2time(n−1)) ∨ (T2 is true in the (n−1)-th loop) or 26≤m.
- If (t0 < T2time(n−1)) ∨ (T2 is false in the (n−1)-th loop) then both disjuncts must be false so T2time(n−1) ≤ t0 and T2 is false in the(n−1)-th loop. IOK f t0 t1 ensures (t0+34<t1) and so we can deduce ((x n)+26 ≤ t1). Therefore, since t1=(x n)+m we know (26≤m) and thus CC is true.
- If (26≤m) is false then IOK ensures 0 < n and (t0 < T2time(n−1)) which means t0 < T2time(n−1) ∨ (T2 is true in the(n−1)-th loop) and thus CC is true.

Now since we have shown that CC is always true and that CC implies **Q t1** we have proved the goal Q t1 as required.

END PROOF

7 Discussion and Conclusions

In this paper we have presented a programming language for real-time programs and its semantics. We have mechanized the semantics in HOL and used it for program verification: safety and timeliness properties of a program with handshaking were performed in HOL. Our specification language and semantics were adequate for expressing and verifying these properties.

The proofs outlined in Section 6 have been formally proved in the HOL system. The HOL proofs are much more detailed than their outlines. These more detailed proofs involve a considerable amount of work but may identify errors in the proof outlines. For example, the proof outlined in Section 6.2 took about 4 days working in HOL, which included finding errors in assumptions in the original proof outline. The proof outlined in Section 6.1 took 1 day working in HOL. The safety proof was easier than the timeliness one partly because many steps could be automated in HOL by programming general proof procedures.

It would be very time consuming to verify programs larger than `prog` using the proof strategies of Section 6. We expected that it would be possible to use HOL to automate most of the steps in such proofs. However, although automation was feasible for safety proofs (Section 6.1) it proved very difficult for timeliness proofs (Section 6.2). The difference is that the safety proof involved simple tests on a small number of cases and both the test and stepping through the cases could be automated requiring only a small amount of user guidance for each case. The second proof could also have been performed automatically in this manner. However, we would have to test that every path through the program which satisfied IOK terminated with `avail` true. The number of cases involved is very large ($34 \times 4 \times 4 \ldots$) and thus even a small amount of user intervention for each case would be onerous. Timeliness properties can be more easily proved by other, non-automatic, means such as the theorem and proof of Section 6.2.

These observations about the complexity of verification proofs do not mean that such proofs can not be mechanized in the HOL system. Indeed, HOL is a good tool for experimenting with different types of proof strategy. Program verification rules can be proved as theorems in higher order logic using the formal semantics we have already defined in HOL. Examples of rules with which we are currently experimenting include

- Execution of command c in any context which satisfies P results in some condition Q (c.f. Hoare logic tuples).
- If condition P holds throughout any execution of command c1 and condition Q holds throughout command c2 and Q $\Longrightarrow$ P then P holds throughout c1;c2 or c2;c1.
- If any stable interval satisfies P at all times and command c satisfies P throughout its execution then the command (Loop c) executed at time 0 satisfies $\forall$ t.P t.

A further possibility for future work is to use the type of rules suggested above to perform user-directed program synthesis. For example, given the proof goal

$$\exists \text{ prog. } \forall \text{ f. } \mathcal{M} \text{ prog f 0} \Longrightarrow \forall \text{ t. P t}$$

we can use the loop rule above to reduce the goal to the subgoals of finding a command c which satisfies P during its execution and proving that any stable interval satisfies P. The process continues (e.g. using the sequence rule above) until all subgoals have been proved. Finally, prove $\exists$ prog. ... by choosing the program (Loop c) which has now been proved to satisfy ($\forall$ t. P t). The HOL system supports goal directed proof by keeping track of proved and unproved subgoals and by putting together the proofs that subgoal theorems imply their goal.

Acknowledgements

I would like to thank Mike Gordon, Juanito Camilleri and the referees for their helpful comments on drafts of this paper. Mike Gordon suggested the handshaking problem which is verified in this paper and discussed my semantics and his own program logic for real-time programs with me. The SAFE0 language and the proposed behaviour of its commands is due to Roger Hale whose first semantics for SAFE0 influenced mine. Jeff Joyce and participants in the UBC HOL Course held in Vancouver in May 1991 provided stimulating discussions about this work. I am grateful for financial support from the Australian Defence Science and Technology Organisation, the Australian Committee of the Cambridge Commonwealth Trust and an Overseas Research Studentship.

References

[Chu40] A. Church. A formulation of the simple theory of types. *Journal of Symbolic Logic*, 5, 1940.

[CIN91] A. Camilleri, P. Iverardi, and M. Nesi. Combining interaction and automation in process algebra verification. In *TAPSOFT '91*, number 494 in Lecture Notes in Computer Science, pages 283–296. Springer Verlag, April 1991.

[CO91] Rachel Cardell-Oliver. Using higher order logic for modelling real-time protocols. In *TAPSOFT '91*, number 494 in Lecture Notes in Computer Science, pages 259–282. Springer Verlag, April 1991.

[GMW79] M.J.C. Gordon, A.J.R.G. Milner, and C.P. Wadsworth. *Edinburgh LCF: a mechanized logic of Computation*, volume 78 of *Lecture Notes in Computer Science*. Springer Verlag, 1979.

[Gor88] Michael J.C. Gordon. Mechanizing programming logics in higher order logic. Technical Report 145, Computer Laboratory, University of Cambridge, September 1988. published in 1988 Banff Conf on Hardware Verification.

[Gor91] Mike Gordon. Hard real time. In *Fourth Refinement Workshop*, pages 00–00. BCS FACS and Logica Cambridge, January 1991.

[Haa81] Volkmar H. Haase. Real-time behaviour of programs. *IEEE Transactions on Software Engineering*, pages 494–501, September 1981.

[Hal90] Roger Hale. Safe/0. Draft Document, May 1990.

[Heh89] Eric C R Hehner. Real-time programming. *Information Processing Letters*, 30:51–56, 16 January 1989.

[Hoo87] Jozef Hooman. A compositional proof theory for real-time distributed message passing. In *PARLE: Parallel Architectures and Languages Europe. Volume II*, pages 315–332, June 1987. Springer Verlag Lecture Notes in Computer Science 259.

[MoD91] Ministry of Defence. Interim Defence Standard 00-55. The Procurement of Safety Critical Software in Defence Equipment. 5 April 1991

[Ost89] Johnathan S. Ostroff. *Temporal Logic for Real-Time Systems.* Research Studies Press, 1989.

[Sha89] Alan C. Shaw. Reasoning about time in higher-level language software. *IEEE Transactions on Software Engineering*, 15(7):875–889, Jul 1989.

APPENDIX: The Semantics of While Statements

The programming language and semantics presented in the paper allow only for non-terminating loops and disallow nested loops. This restriction was made simply to limit the number of cases which must be considered in verification proofs. However, it is not difficult to remove the restriction and include a general `While` statement in our language.

The syntax of a general `While` statement is

```
While b c
```

The behaviour of this command is: perform (`If b c Skip`) and then, if `b` was true, perform the `While` loop again, otherwise, if `b` was false then the command terminates. The semantics of this general `While` loop which may or may not terminate is similar to that of the non-terminating `Loop` command: the meaning of (`While b c`) is the meaning of the sequence of (`If b c Skip`) commands, each executed in the context of its predecessor, up to the first context (if there is one) in which b is false. We use the abbreviation

$$\mathsf{x\ n} = \mathcal{L}_{\mathcal{T}}\ \mathsf{c\ f\ t\ n}$$

in the formal definition of the semantics of the `While` statement

$$\begin{array}{l}
\mathsf{first_false\ b} = \epsilon\ \mathsf{N}.\ (\mathcal{E}_{\mathcal{V}}\ \mathsf{b\ f\ (x\ N)} = \mathsf{F}) \wedge (\forall\ \mathsf{N'}.\ \mathsf{N'}<\mathsf{N} \Longrightarrow (\mathcal{E}_{\mathcal{V}}\ \mathsf{b\ f\ (x\ N')} = \mathsf{T}))\\
\mathcal{M}\ (\mathtt{While\ b\ c})\ \mathsf{f\ t} =\\
\quad \exists\ \mathsf{N'}.\ (\mathcal{E}_{\mathcal{V}}\ \mathsf{b\ f\ (x\ N)} = \mathsf{F})\\
\qquad \Rightarrow (\forall\ \mathsf{N}.\ \mathsf{N} < (\mathsf{first_false\ b}) \Longrightarrow \mathcal{M}\ (\mathtt{If\ b\ c\ Skip})\ \mathsf{f\ (x\ N)})\\
\qquad \mid\ (\forall\ \mathsf{N}.\ \mathcal{M}\ (\mathtt{If\ b\ c\ Skip})\ \mathsf{f\ (x\ N)})
\end{array}$$

The possibility of non-terminating sub-commands in our generalised language means we need a new definition for the function $\mathcal{C}_{\mathcal{T}}$ which calculates the time taken to evaluate commands. Execution time is now one of two values, either a natural number, when then command terminates, or a special value, ∞ when the command does not terminate. The special addition operator, $\oplus$, calculates execution times:

$$n_1 \oplus n_2 \equiv n_1{+}n_2 \quad n_1 \oplus \infty \equiv \infty \quad \infty \oplus n_2 \equiv \infty \quad \infty \oplus \infty \equiv \infty$$

The old table for command execution times will be replaced with one in which all + are replaced by $\oplus$ and with the addition of the definitions

$$\begin{array}{l}
\mathcal{C}_{\mathcal{T}}\ (\mathtt{While\ b\ c})\ \mathsf{f\ t} =\\
\quad (\exists\ \mathsf{N'}.\ (\mathcal{E}_{\mathcal{V}}\ \mathsf{b\ f\ (x\ N')} = \mathsf{F})\ )\\
\qquad \Rightarrow (\mathsf{x\ (first_false\ b)}) - \mathsf{t}\\
\qquad \mid\ \infty\\
\mathcal{C}_{\mathcal{T}}\ (\mathsf{Loop\ c})\ \mathsf{f\ t1} = \infty
\end{array}$$

Finally, we modify the definition of $\mathcal{M}$ to take care of the meaning of the sequences of commands such as `c1;c2` where `c1` does not terminate in which case the meaning of `c1;c2` is simply the meaning of `c1`. We specify that the meaning of any command executed at time ∞ is the truth predicate (c.f. `Skip`). That is, the meaning of any command which follows a non-terminating command gives no extra information about the behaviour of a program since it will never be executed.

$$\mathcal{M}\ \mathsf{c\ f}\ \infty = \mathsf{T}$$

The change required in $\mathcal{M}$ to allow for terminating, non-terminating and nested `While` loops is an extra test in each of the definitions for $\mathcal{M}$ which were given in the paper. For example, the new semantics for the `Delay` command is

$$\begin{array}{l}
\mathcal{M}\ (\mathtt{Delay\ e})\ \mathsf{f\ t} =\\
\quad (\mathsf{t} = \infty) \Rightarrow \mathsf{T}\\
\qquad\qquad \mid\ (\forall\ \mathsf{v{:}name}.\ \mathsf{Stable}\ ((\mathcal{E}_{\mathcal{T}}\ \mathsf{e}){+}(\mathcal{E}_{\mathcal{V}}\ \mathsf{e\ f\ t}))\ \mathsf{v\ f\ t})
\end{array}$$

Specification and Verification of Real-Time Behaviour Using Z and RTL

C.J. Fidge
Key Centre for Software Technology
The University of Queensland
Queensland 4072 Australia

Abstract

Real-Time Logic is a formal notation for reasoning about temporal behaviour. Z is a general purpose specification language, but lacks explicit features for expressing real-time constraints. We show how these complementary methods can be formally unified. An approach to verification of real-time properties by deriving temporal information directly from the specification is then described.

1 Introduction

Timing requirements expressed in Real-Time Logic (RTL) complement functional specifications expressed in Z. Herein we show how these two techniques can be formally combined by defining the RTL notation in Z. An approach to verification of real-time properties based on these definitions is then described.

2 Background

2.1 Real-Time Logic

RTL is a first order predicate logic for reasoning about real-time behaviour. It was originally defined by Jahanian and Mok (1986), has been used as the formal basis for the specification language Modechart (Jahanian et al., 1988), and in a number of dialects. Decision procedures for certain classes of RTL formulae have been developed using graph-theoretic approaches (Jahanian and Mok, 1987; Jahanian and Stuart, 1988) and these have been exploited in a prototype verification tool (Stuart, 1990).

Systems modelled in RTL are assumed to consist of a number of time-consuming *actions*, which may modify the system state, and *external events* initiated by the environment. *State attributes* are predicates on the system state, used to describe properties of interest.

An *event* serves as a temporal marker. They may denote the time at which an action began, an action ended, an external event occurred, a state attribute became true, or a state attribute became false.

Since real-time specifications are frequently concerned with the period(s) of time during which some state attribute is true, RTL provides a shorthand notation, known as a *state predicate*, for expressing such intervals. For instance, $S[x, y]$ defines an interval such that attribute S became true at time x and subsequently false at time y. Other state predicates, e.g., for open-ended intervals, or times during which an attribute is false, are also available (Jahanian and Mok, 1986).

Central to RTL is the *occurrence function*, denoted @, which is a mapping from events and indices to times. The statement

$$@(e, i) = t$$

tells us that the ith occurrence of event e occurred at time t.

RTL formulae are constructed from standard arithmetic relations and algebraic expressions over time constants and the occurrence function, first order logical connectives, equality/inequality predicates and universal and existential quantifiers (over indices only, event variables are not allowed).

2.2 Z

Z is a set-theoretic specification language suitable for defining, and reasoning about, a wide range of systems. It was originally developed at Oxford University, and has been used in non-trivial "real world" projects. It is particularly suited to non-constructive requirements specifications. Semantics-preserving refinement techniques allow formal translation of Z specifications to executable code (e.g., King, 1990), and formal proof techniques are also well established (Diller, 1990).

The Z notation consists of two components: the schema language, a specification structuring technique, and a mathematical language. A Z *schema* is usually represented as a named box partitioned into two parts, variable declarations and optional predicates relating the variables:

```
┌─ Name ─────────────────────────────
│ Declarations
├──────────────
│ Predicates
└────────────────────────────────────
```

Unnamed *axiomatic definitions* use a similar notation:

```
│ Declarations
├──────────────
│ Predicates
```

A number of operators are available for composing schemata. The declarations and predicates in a schema may be "imported" into another by including the schema name in the declaration part. By convention such a name preceded by Δ denotes that the operation may change the values of variables in the imported schema.

When a Z specification is interpreted operationally, three uses of schema boxes are normally distinguished: declaration of (global) variables, definitions of the effect of performing certain operations, and definition of the initial system state(s).

Each *operation schema* defines the effect of an operation by describing the pre-condition (on the global state) required for the operation to occur, and the post-condition after its occurrence. By convention a primed variable name represents the value of the variable after the operation has been performed and an unprimed name represents its value before.

The set-theoretic *mathematical notation* used within schema boxes includes a vast selection of operators. The glossary briefly describes those used herein; see Diller (1990) for a thorough tutorial introduction. Relations, functions, sequences and bags are represented by sets of ordered pairs which can be manipulated via the set operators.

2.3 Motivation

Although Z can be used to model real-time requirements, there is little agreement on the best way to do this. In practice Z offers its users no guidance on how to tackle a given specification problem. A large proportion of the Z literature consists of specific case studies, but few generally-applicable guidelines exist.

A number of real-time systems have been specified in Z. Significantly, several different approaches have been used. Raymond et al. (1990) introduce a global auxiliary integer time variable and a special "tick" operation to effect the passage of time — real-time requirements are then expressed via temporal logic assertions on the auxiliary variable. King (1989) expressed timing requirements via a history function that maps times to the names of significant actions performed at that moment. Coombes (1990) defines a number of interval time operators, and uses these to define functions from time intervals to values for each variable, thus representing the values of variables as they change over time. Delisle and Garlan (1989), in their formal specification of an oscilloscope, model the input signal as a function from all times to voltage values, thus describing this variable in perpetuity and avoiding the need to model the passage of time. The same approach is adopted by Mahony and Hayes (1991).

No single approach has yet asserted itself as the best starting point for defining and reasoning about real-time behaviour in Z. However, by augmenting Z with the RTL notation, we intend to provide just such a discipline for operational Z specifications. Indeed, the way Z and RTL complement each other has been noted elsewhere in the literature (Burns and Wellings, 1990).

3 Representing RTL Formulae in Z

3.1 Discrepancies Between the RTL and Z Models

Some compromises are necessary due to fundamental differences between RTL and Z. RTL assumes actions may be composed of sub-actions, are time-consuming, and may occur in parallel; Z operations are atomic and only one can occur at a time. RTL has a notion of an external event, initiated by some outside agency; Z makes no formal distinction between internal and external operations — if the pre-condition of an operation is satisfied it is free to occur, but no particular entity is assumed to initiate it.

Therefore, in modelling RTL concepts in Z, we abandon the distinction between "start", "stop" and "external" events and use a single concept of an "operation" event. This is not to say that Z cannot model these concepts, but merely that the distinctions are not formally enforced in the definitions below. If desired, time-consuming actions can be represented by two related "start" and "stop" operations, and certain operations may be designated as being due to some entity in the environment, but these distinctions would be made only in the prose accompanying the formal specification.

3.2 Time Representation

In previous work absolute time values have been represented by natural numbers (e.g., Delisle and Garlan, 1989), reals (e.g., King, 1989), or time intervals (Coombes, 1990). Future work on the specification of distributed systems (which lack a central time-base) will also see a rôle for a partially ordered representation of time. However, for the purposes of the example in section 4 a discrete linear representation is sufficient, hence:

$$TIME \mathrel{\widehat{=}} \mathbb{N}$$

3.3 A History Model for Z

Since the occurrence function defines a temporal history of event occurrences, access to the history model underlying the operational interpretation of Z is required. We adopt the approach advocated by Duke and Smith (1989), and later Raymond et al. (1990), as a basis for their definitions of temporal logic in Z. In defining temporal behaviour it is necessary for Z to "talk about itself" so that we can reason about the schemata that form the functional specification.

Each operation schema defines a relation between its "before" and "after" states. Therefore let

$$OPS \subseteq State \leftrightarrow State$$

be the set of relations corresponding to the operations in the Z specification, uniquely identified by the name of the corresponding schema. (If the Z specification does not contain a single global "*State*" schema one can be constructed easily using schema composition operators.) We assume that the *Init* operation is included, adopting the approach of Diller (1990) who treats *Init* as a special operation with no "before" state. In the definitions below care is taken to avoid any reference to the before state of *Init*.

A RTL state attribute can be represented in Z as a schema that expresses some predicate on the global state, thus defining one or more states in which that attribute is deemed to be true. Let

$$ATTS \subseteq \mathbb{P}\, State$$

be the sets of states corresponding to each state attribute schema, again uniquely identified by the corresponding schema name.

We now define an execution history for a Z specification as a sequence of states, commencing in one of the initial states, where each state is transformed into its successor by one of the operations from the specification:

History

$$states : \mathrm{seq}_{\infty}\, State$$
$$actions : \mathrm{seq}_{\infty}\, OPS$$
$$times : \mathrm{seq}_{\infty}\, TIME$$

$$\mathrm{dom}\, states = \mathrm{dom}\, actions \wedge \mathrm{dom}\, actions = \mathrm{dom}\, times$$
$$actions(1) = Init \wedge states(1) \in \{\mathrm{post}\, Init\}$$
$$\forall i : \mathrm{dom}\, actions - \{1\} \bullet states(i-1)\ \underline{actions(i)}\ states(i)$$
$$\forall i : \mathrm{dom}\, times - \{1\} \bullet times(i-1) \leqslant times(i)$$

(For aesthetic reasons we use the RTL terminology "action" as a synonym for the longer Z "operation" from now on.) The first predicate above says that for any history the numbers of states reached, actions performed, and times recorded are equal. The second predicate says that the first operation performed is always *Init* and that the first state is a member of the set of "after" states defined by the *Init* schema. The third predicate says that all subsequent states are formed by application of the corresponding operation recorded in the *actions* sequence, i.e., the occurrence of operation i takes the system from state $i-1$ to state i (and is assumed to occur at time i). The fourth predicate asserts that time never goes backward.

3.4 The Occurrence Function

The first argument to the occurrence function is an event. In our Z-based model events may be the occurrence of an action (operation), a state attribute becoming true, or a state attribute becoming false. Since Z is strongly typed a disjoint union and type constructors are used to create the corresponding type:

$$\begin{aligned} EVENT ::= \ & Action\langle\langle OPS\rangle\rangle \\ | \ & BecomesTrue\langle\langle ATTS\rangle\rangle \\ | \ & BecomesFalse\langle\langle ATTS\rangle\rangle \end{aligned}$$

The RTL occurrence function is now defined as follows (for consistency with RTL we have taken the liberty of using "@" as an identifier):

OccurrenceFn

$$History$$
$$@ : EVENT \rightarrow (\mathrm{seq}_{\infty}\, TIME)$$

$$\forall a : OPS \bullet$$
$$\quad @\,(Action\ a) = (\mathrm{dom}(actions \rhd \{a\})) \upharpoonright times$$
$$\forall S : ATTS \bullet$$
$$\quad @\,(BecomesTrue\ S) = \{n : \mathrm{dom}\, states \mid states(n) \in S \wedge (n > 1 \Rightarrow states(n-1) \notin S)\} \upharpoonright times$$
$$\wedge$$
$$\quad @\,(BecomesFalse\ S) = \{n : \mathrm{dom}\, states \mid states(n) \notin S \wedge (n > 1 \Rightarrow states(n-1) \in S)\} \upharpoonright times$$

This concise definition was achieved by Currying the occurrence function. (Since a sequence is defined in Z as a function from a contiguous series of natural numbers starting at 1 to the items in the sequence the definition of @ above is equivalent to $EVENT \rightarrow (\mathbb{N} \nrightarrow TIME)$.) This makes partial application possible. Thus '@ e' is the sequence of all times at which event e occurred, whereas '@ e i' is the time of the ith occurrence of e (if one exists).

The two predicates above define the occurrence function in terms of the history. For each operation a in the Z specification the sequence of times at which it occurred is formed by finding all occurrences of a in *actions*, and using their indices to extract the corresponding *times*. (A single value denotes the time at which each operation occurred — this may be thought of as marking the time at which the state change took effect.) For each state attribute S the sequence of times at which it becomes true is formed by finding all *states* in which the attribute is true (i.e., in which the current state is an element of the set of states defined by the state attribute schema) and the attribute was false in the preceding state (or the current state is the first state), and extracting the corresponding *times* via the *states* indices. Similarly for *BecomesFalse*.

Since histories may be finite we can model terminating computations. (To simplify the formulae most published work on RTL assumes non-terminating computations only.) To allow for the possibility that the ith instance of some event e never occurs, the RTL formulae below make frequent use of the expression 'dom(@ e)' to access all indices in the history for event e. (A relational, rather than functional, notation offers an alternative approach to non-occurring events; Zedan (1990) notes that a general way of introducing timed events is via a predicate that accepts three arguments, e, i and t, and returns "true" iff the ith occurrence of event e took place at time t.)

3.5 State Predicates

The following relation defines the set of all intervals (ordered time pairs) at which some attribute S became true and subsequently false:

StatePredicates

$$OccurrenceFn$$
$$TrueFromTo : ATTS \leftrightarrow (TIME \times TIME)$$

$$\begin{aligned}
&\forall S : ATTS \bullet \\
&\quad TrueFromTo(\!| S |\!) = \\
&\quad \{x, y : TIME \mid x < y \land \\
&\qquad x \in \operatorname{ran}(@\ (BecomesTrue\ S)) \land \\
&\qquad y \in \operatorname{ran}(@\ (BecomesFalse\ S)) \land \\
&\qquad (\nexists z : TIME \mid z \in \operatorname{ran}(@\ (BecomesFalse\ S)) \bullet \\
&\qquad\qquad x < z \land z < y)\}
\end{aligned}$$

This definition closely mirrors that of Jahanian and Mok (1986). If other forms of state predicate are required their definitions follow similarly.

The above definition does not include any interval where S remains true at the end of an incomplete history, but as long as we confine our reasoning to all possible histories this is not a problem. (A slight redefinition is required if we expect S to be true at the end of a finite computation.) This schema assumes that each state attribute remains true for a non-zero period of time (c.f., Jahanian et al., 1988).

4 Specification Example

As a brief illustrative example consider the following representation of a simplex "stop-and-wait" protocol.

4.1 Functional Requirements

The global state of the system is modelled via two variables,

```
┌─ State ──────────────────────────────────
│ LinkCond : LinkState
│ ChanCond : ChannelState
└──────────────────────────────────────────
```

whose types are enumerated below,

$$\begin{aligned} \mathit{LinkState} &::= \mathit{Idle} \mid \mathit{Transmitting} \mid \mathit{Acking} \mid \mathit{Lost} \\ \mathit{ChannelState} &::= \mathit{Full} \mid \mathit{Empty} \end{aligned}$$

The link as a whole may be idle, transmitting a frame, acknowledging receipt of a frame, or may be recovering from the loss of a frame. The channel from sender to receiver may be either full or empty depending on whether a data frame is currently in transit or not.

Initially the link is idle and no message is in transit:

```
┌─ Init ───────────────────────────────────
│ State'
├───────────────
│ LinkCond' = Idle
│ ChanCond' = Empty
└──────────────────────────────────────────
```

Following the convention of Diller (1990), the primed schema and variable names denote values after the completion of the *Init* operation — no assumption is made about variable values before this initial operation.

There are five operation schemata defining the behaviour of the protocol. Firstly, to send a frame, the link must be idle. Afterwards the link will be "transmitting" and the channel occupied:

```
┌─ Send ───────────────────────────────────
│ ΔState
├───────────────
│ LinkCond = Idle ∧ LinkCond' = Transmitting
│ ChanCond' = Full
└──────────────────────────────────────────
```

Successful reception of a frame at the far end of the link is always followed by an acknowledgement:

Rec
$\Delta State$

$LinkCond = Transmitting \land LinkCond' = Acking$
$ChanCond' = Empty$

A successful acknowledgement readies the link to transmit another frame:

Ack
$\Delta State$

$LinkCond = Acking \land LinkCond' = Idle$
$ChanCond' = ChanCond$

While a frame is being transmitted an error on the channel may cause it to become irretrievably lost or corrupted:

Lose
$\Delta State$

$LinkCond = Transmitting \land LinkCond' = Lost$
$ChanCond' = ChanCond$

Rec and *Lose* both have the same pre-condition; the choice is nondeterministic.

Following a loss retransmission occurs:

Retrans
$\Delta State$

$LinkCond = Lost \land LinkCond' = Transmitting$
$ChanCond' = ChanCond$

Only those aspects important to the forthcoming discussion have been represented. For instance, since the values of messages transmitted does not alter the sequence of operations, they have not been modelled. Similarly, the mechanism whereby the loss of a frame is detected, or the possibility that an acknowledgement will be lost, have not been described.

For the purposes of the RTL formulae we may also need to define state attributes. These are represented via schemata that define some predicate on the *State* variables. For example, the attribute denoting the state of the system while a message is in transit is represented as:

InTransit
$State$

$ChanCond = Full$

4.2 Timing Requirements

Real-time constraints are expressed by augmenting the functional specification with RTL formulae. These formulae act as restrictions on the allowable histories.

For instance, the following schema expresses the safety property that no acknowledgement operation occurs while a message is currently in transit:

NoAckInTransit
OccurrenceFn

$$\begin{array}{l} \forall i : \mathrm{dom}(@\,(\mathit{BecomesFalse}\ \mathit{InTransit})) \bullet \\ \quad \nexists j : \mathrm{dom}(@\,(\mathit{Action}\ \mathit{Ack})) \bullet \\ \qquad @\,(\mathit{BecomesTrue}\ \mathit{InTransit})\ i < @\,(\mathit{Action}\ \mathit{Ack})\ j \wedge \\ \qquad @\,(\mathit{Action}\ \mathit{Ack})\ j < @\,(\mathit{BecomesFalse}\ \mathit{InTransit})\ i \end{array}$$

This can be read as "for all occurrences of state attribute *InTransit* becoming false there does not exist an *Ack* operation that occurred since *InTransit* last became true." In style this formula closely resembles those of Jahanian and Mok (1986), but is slightly more verbose since the quantified variables are given a finite (but undetermined) range.

Using a state predicate the same requirement can be expressed more concisely as:

NoAckInTransit2
OccurrenceFn

$$\begin{array}{l} \forall x, y : \mathit{TIME} \mid \mathit{InTransit}\ \underline{\mathit{TrueFromTo}}\ (x, y) \bullet \\ \quad \nexists t : \mathrm{ran}(@\,(\mathit{Action}\ \mathit{Ack})) \bullet x < t \wedge t < y \end{array}$$

In other words, "for all intervals during which attribute *InTransit* is true, there does not exist a time t during the interval at which an *Ack* operation occurs."

Another desirable safety requirement is that each *Rec* is followed by at least one *Ack*, before the next frame is sent, i.e.,

FramesAcknowledged
OccurrenceFn

$$\begin{array}{l} \forall i : \mathrm{dom}(@\,(\mathit{Action}\ \mathit{Send})) - \{1\} \bullet \\ \quad \exists j : \mathrm{dom}(@\,(\mathit{Action}\ \mathit{Ack})) \bullet \\ \qquad @\,(\mathit{Action}\ \mathit{Rec})\ (i-1) \leqslant @\,(\mathit{Action}\ \mathit{Ack})\ j \wedge \\ \qquad @\,(\mathit{Action}\ \mathit{Ack})\ j \leqslant @\,(\mathit{Action}\ \mathit{Send})\ i \end{array}$$

The following formula asserts that each frame is lost at most twice before being successfully received:

BoundedLosses
OccurrenceFn

$$\begin{array}{l} \forall i : \mathrm{dom}(@\,(\mathit{Action}\ \mathit{Lose})) - \{1, 2\} \bullet \\ \quad \exists j : \mathrm{dom}(@\,(\mathit{Action}\ \mathit{Rec})) \bullet \\ \qquad @\,(\mathit{Action}\ \mathit{Lose})\ (i-2) < @\,(\mathit{Action}\ \mathit{Rec})\ j \wedge \\ \qquad @\,(\mathit{Action}\ \mathit{Rec})\ j < @\,(\mathit{Action}\ \mathit{Lose})\ i \end{array}$$

Two types of timing requirement common in real-time work are periodic and sporadic constraints. A *periodic* constraint states that some action must be executed repeatedly at fixed intervals. A *sporadic* constraint states that an action must be executed following some triggering event, within a specified deadline. Both can be expressed in RTL (Jahanian and Mok, 1986). For instance, the following statement expresses the sporadic requirement that each frame is acknowledged within 17 units of absolute time from the time it was sent:

*MaxDelay*17
$OccurrenceFn$

$\forall i : \mathrm{dom}(@\,(Action\ Ack)) \bullet$
$\quad @\,(Action\ Ack)\ i \leqslant (@\,(Action\ Send)\ i) + 17$

(This formula assumes that the ith occurrence of *Ack* is always preceded by the ith occurrence of *Send*, a fact confirmed in section 5.1.)

5 Verification

This section presents an approach to verifying timing properties of a Z specification expressed in Real-Time Logic. Emphasis is placed on making maximal use of the temporal information implicit in the Z specification.

5.1 Deriving RTL Formulae from the Z Specification

The *OccurrenceFn* schema formally embeds RTL in Z, but gives us no insight into the temporal behaviour of a particular Z specification. To allow for explicit reasoning we now derive RTL formulae from the Z specification itself.

Firstly a number of useful relations can be defined. One of the most fundamental concepts is that of "potential causality" between operations:

$$Enables : OPS \leftrightarrow OPS$$

$$Enables = \{a, b : OPS \mid \mathrm{ran}\ a \cap \mathrm{dom}\ b \neq \{\,\}\}$$

For instance, in the example of section 4, *Retrans* <u>*Enables*</u> *Rec* since there exists a post-state of *Retrans* that is also a pre-state of *Rec*. The relation can be enumerated as follows:

$$\begin{aligned} Enables = \{&(Init, Send), (Send, Rec), (Send, Lose), (Lose, Retrans),\\ &(Retrans, Lose), (Retrans, Rec), (Rec, Ack), (Ack, Send)\} \end{aligned}$$

It is a non-transitive representation of which operation(s) *may* follow each operation. (However, even if x <u>*Enables*</u> y, there does not necessarily exist a history in which x is followed by y, since the possible histories considered *in toto* may preclude the necessary condition from ever arising.) Notice that *Lose* and *Retrans* enable one another — there is no limit on how many times each frame may be retransmitted.

The converse of *Enables* is *LeadsTo* in which the "nodes" are states:

$$\begin{array}{|l} \mathit{LeadsTo} : \mathit{State} \leftrightarrow \mathit{State} \\ \hline \mathit{LeadsTo} = \bigcup(\mathit{OPS} - \{\mathit{Init}\}) \end{array}$$

Since operations are already defined as *State* to *State* relations it is merely necessary to take their distributed union. (*Init* is excluded due its non-existent "before" state.)

Other simple relations include definitions of which operations *always* make particular state attributes true or false:

$$\begin{array}{|l} \mathit{MakesTrue} : \mathit{OPS} \leftrightarrow \mathit{ATTS} \\ \hline \mathit{MakesTrue} = \{a : \mathit{OPS};\ S : \mathit{ATTS} \mid \\ \qquad (\operatorname{ran} a \subseteq S) \wedge (a \neq \mathit{Init} \Rightarrow \operatorname{dom} a \cap S = \{\,\})\} \end{array}$$

$$\begin{array}{|l} \mathit{MakesFalse} : \mathit{OPS} \leftrightarrow \mathit{ATTS} \\ \hline \mathit{MakesFalse} = \{a : \mathit{OPS};\ S : \mathit{ATTS} \mid \\ \qquad (\operatorname{ran} a \cap S = \{\,\}) \wedge (a \neq \mathit{Init} \Rightarrow \operatorname{dom} a \subseteq S)\} \end{array}$$

Section 4 defines only one state attribute, *InTransit*, hence:

$$\begin{array}{l} \mathit{MakesTrue} = \{(\mathit{Send}, \mathit{InTransit})\} \\ \mathit{MakesFalse} = \{(\mathit{Init}, \mathit{InTransit}), (\mathit{Rec}, \mathit{InTransit})\} \end{array}$$

From these fundamental concepts more complex relations can be built. For instance, the following definition captures all pairs of operations a and b such that, if b occurs, it must have been preceded by exactly one occurrence of a (since b last occurred or the computation began):

$$\begin{array}{|l} \mathit{Caused} : \mathit{OPS} \leftrightarrow \mathit{OPS} \\ \hline \mathit{Caused} = \{a, b : \mathit{OPS} \mid \\ \qquad (a, b) \in (\{b\} \lhd \mathit{Enables})^{+} \wedge \\ \qquad (a, a) \notin (\{b\} \lhd \mathit{Enables})^{+} \wedge \\ \qquad (\forall c : \mathit{OPS} \mid (c, b) \in (\mathit{Enables} \rhd \{a\})^{+} \bullet \\ \qquad\qquad c = a \vee (a, c) \in (\mathit{Enables} \rhd \{a\})^{+})\} \end{array}$$

That is, a path must exist from a to b, a must not enable itself (without going via b), and all "predecessors" c of b must be successors of a. By removing elements from *Enables* before taking its transitive closure we avoid the problem of all operations indirectly enabling all others in an iterative computation such as this. For the stop-and-wait example this relation is:

$$\mathit{Caused} = \{(\mathit{Send}, \mathit{Rec}), (\mathit{Lose}, \mathit{Retrans}), (\mathit{Rec}, \mathit{Ack}), (\mathit{Send}, \mathit{Ack})\}$$

Another relation that will prove valuable in the forthcoming proofs is,

$$\begin{array}{|l} \mathit{Between} : (\mathit{OPS} \times \mathit{OPS}) \leftrightarrow \mathit{OPS} \\ \hline \mathit{Between} = \{a, b, c : \mathit{OPS} \mid \\ \qquad a \ \underline{\mathit{Caused}}\ b \wedge \\ \qquad (a, c) \in (\{b\} \lhd \mathit{Enables})^{+} \wedge \\ \qquad (c, b) \in (\mathit{Enables} \rhd \{a\})^{+} \bullet ((a, b), c)\} \end{array}$$

which defines all operations c that *may* occur in between occurrences of two operations a and b related by *Caused*. For the stop-and-wait example,

$$\begin{aligned} Between = \{&((Send, Rec), Lose), ((Send, Rec), Retrans),\\ &((Send, Ack), Lose), ((Send, Ack), Retrans),\\ &((Send, Ack), Rec)\} \end{aligned}$$

i.e., *Lose* and *Retrans* may appear one or more times between the ith occurrences of *Send* and *Rec*, etc.

Numerous other such relations are possible, of course. No attempt has been made to be exhaustive above. It is intended that such definitions be produced as required for particular proofs.

These relations have been chosen to highlight significant characteristics of the example. From them a number of axioms relating the Z specification to the occurrence function can be established.

$$\begin{aligned} &\forall S : ATTS \mid (\exists_1\, a : OPS \bullet a\ \underline{MakesTrue}\ S) \bullet \\ &\quad @\,(Action\ a) = @\,(BecomesTrue\ S) \end{aligned} \tag{A1}$$

By definition a makes S true whenever it occurs. Since no other operations possess this property all occurrences of S becoming true must correspond to an occurrence of operation a.

Similarly,

$$\begin{aligned} &\forall S : ATTS \mid (\exists_1\, a : OPS \bullet a\ \underline{MakesFalse}\ S) \bullet \\ &\quad @\,(Action\ a) = @\,(BecomesFalse\ S) \end{aligned} \tag{A2}$$

Given a $\underline{Caused}$ b all histories containing a and b have the form $\langle a, b, a, b, a \ldots \rangle$. Therefore, for any such history, the number of a's is always equal to, or one greater than, the number of b's, and a's and b's are strictly interleaved:

$$\begin{aligned} &\forall a, b : OPS \mid a\ \underline{Caused}\ b \bullet \\ &\quad \mathrm{dom}(@\,(Action\ b)) \subseteq \mathrm{dom}(@\,(Action\ a)) \\ &\quad \wedge \\ &\quad \mathrm{dom}(tail(@\,(Action\ a))) \subseteq \mathrm{dom}(@\,(Action\ b)) \\ &\quad \wedge \\ &\quad \forall i : \mathrm{dom}(@\,(Action\ b)) \bullet \\ &\qquad @\,(Action\ a)\ i \leqslant @\,(Action\ b)\ i \\ &\quad \wedge \\ &\quad \forall i : \mathrm{dom}(@\,(Action\ a)) - \{1\} \bullet \\ &\qquad @\,(Action\ b)\ (i-1) \leqslant @\,(Action\ a)\ i \end{aligned} \tag{A3}$$

The following axiom illustrates a property implied by the *Between* relation:

$$\begin{aligned} &\forall a, b, c : OPS \mid c \notin Between(\!|\, \{(a, b)\} \,|\!) \bullet \\ &\quad \forall i : \mathrm{dom}(@\,(Action\ b)) \bullet \\ &\qquad \nexists j : \mathrm{dom}(@\,(Action\ c)) \bullet \\ &\qquad\quad @\,(Action\ a)\ i < @\,(Action\ c)\ j \wedge \\ &\qquad\quad @\,(Action\ c)\ j < @\,(Action\ b)\ i \end{aligned} \tag{A4}$$

In other words no occurrence of operation c may appear between the ith occurrences of a and b.

Numerous other such axioms may be defined but those above suffice for the proofs given below.

5.2 Example Proofs

We can now offer rigorous proofs of the properties in section 4.2. *NoAckInTransit* can be established as follows.

Proof. By definition,

$$MakesTrue^{-1}(\!|\, \{InTransit\} \,|\!) = \{Send\} \tag{1}$$

$$MakesFalse^{-1}(\!|\, \{InTransit\} \,|\!) = \{Rec\} \tag{2}$$

From equations (1) and (2) and axioms $A1$ and $A2$ the *NoAckInTransit* predicate can be rewritten as

$$\begin{aligned} &\forall\, i : \text{dom}(@\,(Action\ Rec)) \bullet \\ &\quad \nexists j : \text{dom}(@\,(Action\ Ack)) \bullet \\ &\qquad @\,(Action\ Send)\ i < @\,(Action\ Ack)\ j \wedge \\ &\qquad @\,(Action\ Ack)\ j < @\,(Action\ Rec)\ i \end{aligned} \tag{3}$$

By definition,

$$Between(\!|\, \{(Send, Rec)\} \,|\!) = \{Lose, Retrans\} \tag{4}$$

Since *Ack* does not appear in (4), we can substitute *Send*, *Rec* and *Ack* for a, b and c in axiom $A4$, which immediately gives us equation (3). □

The proof itself is short, due to the effort already expended on the definitions in section 5.1, and convincing, since it makes use only of facts derived directly from the Z specification.

The proof of *FramesAcknowledged* is slightly more involved because the endpoints of interest, *Send* and *Ack*, are not related by *Caused* (because *Send* can be enabled by *Init*, as well as *Ack*).

Proof. By definition,

$$Send\ \underline{Caused}\ Ack \tag{1}$$

$$Rec\ \underline{Caused}\ Ack \tag{2}$$

From axiom $A3$ and (2),

$$\begin{aligned} &\forall\, i : \text{dom}(@\,(Action\ Ack)) \bullet \\ &\quad @\,(Action\ Rec)\ i \leqslant @\,(Action\ Ack)\ i \end{aligned} \tag{3}$$

From axiom $A3$ and (1),

$$\forall i : \mathrm{dom}(@\,(\mathit{Action\ Send})) - \{1\} \bullet \\ \quad @\,(\mathit{Action\ Ack})\,(i-1) \leqslant @\,(\mathit{Action\ Send})\,i \qquad (4)$$

We now need to relate the quantified variables in (3) and (4). From (1) and axiom *A3* we obtain the following invariant,

$$\mathrm{dom}(\mathit{tail}(@\,(\mathit{ActionSend}))) \subseteq \mathrm{dom}(@\,(\mathit{Action\ Ack})) \qquad (5)$$

This allows us to replace the expression 'dom(@ (*Action Ack*))' in equation (3) with 'dom(*tail*(@ (*Action Send*)))'. Adding 1 to each element in this set, and correspondingly subtracting 1 from all occurrences of i in the predicate, we obtain

$$\forall i : \mathrm{dom}(@\,(\mathit{Action\ Send})) - \{1\} \bullet \\ \quad @\,(\mathit{Action\ Rec})\,(i-1) \leqslant @\,(\mathit{Action\ Ack})\,(i-1) \qquad (6)$$

Together (4) and (6) satisfy the *FramesAcknowledged* predicate. □

A shorter, but perhaps less clear, proof of *FramesAcknowledged* is possible by defining a relation similar to *Between* that does not rely on the endpoints being related by *Caused*.

Finally, as an example involving absolute time, consider the following theorem,

$$\mathit{BoundedLosses} \Rightarrow \mathit{MaxDelay}17 \qquad (1)$$

To prove this we need more information about the absolute temporal behaviour of each of the operations. Assuming that the occurrence of each operation marks the endpoint of some time-consuming behaviour, let the following function define the separation of each operation from its predecessor(s):

$$\mathit{Duration} : \mathit{OPS} \rightarrow \mathit{TIME}$$

$$\mathit{Duration} = \{(\mathit{Init}, 0), (\mathit{Send}, 1), (\mathit{Lose}, 1), (\mathit{Retrans}, 5), (\mathit{Rec}, 1), (\mathit{Ack}, 2)\}$$

(The long delay associated with performing a *Retrans* operation includes the timeout period necessary to detect that an acknowledgement has not been returned to the sender.)

Further, given the knowledge that the stop-and-wait example is inherently sequential, we can define the following predicate on its histories:

$$\forall i : \mathrm{dom}\,\mathit{times} - \{1\} \bullet \\ \quad (\mathit{times}(i) - \mathit{times}(i-1)) = \mathit{Duration}(\mathit{actions}(i)) \qquad (2)$$

Proof. By definition,

$$\mathit{Between}(\!|\,\{(\mathit{Send}, \mathit{Ack})\}\,|\!) = \{\mathit{Lose}, \mathit{Retrans}, \mathit{Rec}\} \qquad (3)$$

We now need to know the maximum number of each of these operations that can appear between the ith occurrences of *Send* and *Ack*. By definition,

$$\mathit{Send}\ \underline{\mathit{Caused}}\ \mathit{Rec} \qquad (4)$$

which implies that at most one *Rec* can occur between successive *Send*'s (axiom *A3*). By definition,

$$\mathit{Lose}\ \underline{\mathit{Caused}}\ \mathit{Retrans} \tag{5}$$

which implies that the same number of *Lose* and *Retrans* operations can occur between successive *Send*'s. These two facts, and (2), provide the worst case separation of *Send*/*Ack* pairs:

$$\begin{array}{l} \forall\, i : \mathrm{dom}(@\ (\mathit{Action}\ \mathit{Ack})) \bullet \\ \quad (@\ (\mathit{Action}\ \mathit{Ack})\ i) - (@\ (\mathit{Action}\ \mathit{Send})\ i) = \\ \quad \mathit{Duration}(\mathit{Ack}) + \mathit{Duration}(\mathit{Rec}) + \\ \quad n * (\mathit{Duration}(\mathit{Lose}) + \mathit{Duration}(\mathit{Retrans})) \end{array} \tag{6}$$

where n is the maximum number of *Lose* operations that may occur between *Sends*. Substituting 2 for n, using assumption *BoundedLosses*, and the various *Durations*, yields,

$$\begin{array}{l} \forall\, i : \mathrm{dom}(@\ (\mathit{Action}\ \mathit{Ack})) \bullet \\ \quad (@\ (\mathit{Action}\ \mathit{Ack})\ i) - (@\ (\mathit{Action}\ \mathit{Send})\ i) \leqslant 15 \end{array} \tag{7}$$

thus satisfying $\mathit{MaxDelay}17$ and hence (1). □

6 Discussion

We were originally motivated by a need to unintrusively improve the usability of Z for real-time specifications. To this end we have allowed the timing model to "choose itself" by directly mapping between the RTL and Z formalisms wherever possible. The resultant model includes all of the operators normally found in RTL, except the '↑', '↓' and 'Ω' event constant constructors. Indeed the notation can be criticised for not being sufficiently restrictive; great care must be exercised in the choice of quantifiers in RTL formulae to properly exclude non-occurring events.

The relations of section 5.1, while satisfactory for the proofs considered, are equivalent only to conventional "reachability graphs". Jahanian and Stuart (1988) observe that such graphs are, in the general case, inadequate for real-time verification. They define "computation graphs", in which state nodes may be duplicated if their future histories differ, and then show how to use these graphs to verify two general classes of RTL formulae: inclusion/exclusion of intervals and maximum/minimum separation of points. Although the examples in section 5.2 are representative of both of these classes our methodology is not yet as powerful as that available for Modechart.

7 Conclusion

We have given a formal definition of Real-Time Logic concepts as they apply to specifications expressed in Z. This exercise has produced a concise notation for expressing temporal requirements in Z specifications. By Currying the occurrence function the definitions were made surprisingly compact. Finite histories have received consideration.

With these formal definitions in place an approach to verification of real-time properties was explored. Although still in its infancy this work leaves us cautiously optimistic about the potential for methods that exploit graph-theoretic manipulations of Z relations.

Acknowledgements. I wish to thank the anonymous referees for their many helpful comments on this work, and Andrew Lister for his guidance during the course of this project. This project was supported by an Australian Postdoctoral Research Fellowship and an Australian Telecommunications and Electronics Research Board Project Grant.

References

BURNS, A. and WELLINGS, A. (1990): *Real-Time Systems and their Programming Languages*, Addison-Wesley.

COOMBES, A. (1990): *An Interval Logic for Modelling Time in Z*, Technical report, University of York, Dept. Computer Science.

DELISLE, N. and GARLAN, D. (1989): Formally Specifying Electronic Instruments, *ACM SIGSOFT Eng. Notes*, 14(3), pp. 242–248.

DILLER, A. (1990): *Z: An Introduction to Formal Methods*, John Wiley and Sons.

DUKE, R. and SMITH, G. (1989): Temporal Logic and Z Specifications, *The Australian Computer Journal*, 21(2), pp. 62–66.

JAHANIAN, F., LEE, R., and MOK, A. (1988): Semantics of Modechart in Real Time Logic, In Shriver, B., editor, *Proc. 21st Annual Hawaii International Conference on System Sciences*, pp. 479–489.

JAHANIAN, F. and MOK, A. (1986): Safety Analysis of Timing Properties in Real-Time Systems, *IEEE Transactions on Software Engineering*, SE-12(9), pp. 890–904.

JAHANIAN, F. and MOK, A. (1987): A Graph-Theoretic Approach for Timing Analysis and its Implementation, *IEEE Transactions on Computers*, C-36(8), pp. 961–975.

JAHANIAN, F. and STUART, D. (1988): A Method for Verifying Properties of Modechart Specifications, *Proc. Real-Time Systems Symposium*, Alabama, pp. 12–21.

KING, P. (1989): *A Formal Specification of Signalling System Number 7 Link Layer*, Technical Report 101, University of Queensland, Key Centre for Software Technology.

KING, S. (1990): Z and the Refinement Calculus, In Bjorner, D., Hoare, C., and Longmaack, H., editors, *Proc. VDM'90*, v. 428 Lecture Notes in Computer Science, pp. 164–188. Springer-Verlag.

MAHONY, B. and HAYES, I. (1991): Using Continuous Real Functions to Model Timed Histories, *Proc. Sixth Australian Software Engineering Conference (ASWEC'91)*, Sydney.

RAYMOND, K., STOCKS, P., and CARRINGTON, D. (1990): *Using Z to Specify Distributed Systems*, Technical Report 181, The University of Queensland, Key Centre for Software Technology.

STUART, D. (1990): Implementing a Verifier for Real-Time Systems, *Proc. Real-Time Systems Symposium*, Florida, pp. 62–71.

ZEDAN, H. (1990): *Formal Modelling of Distributed Real-Time Systems*, Technical Report YCS 132, University of York, Dept. Computer Science.

Glossary of Z Mathematical Notation

$\mathbb{N}$	The set of natural numbers (non-negative integers).
$\mathbb{P}\, S$	Powerset: the set of all subsets of S.
$\mathrm{seq}_\infty\, X$	Set of finite or infinite sequences with elements drawn from X.
$m \,..\, n$	The set of integers between m and n inclusive.
$\langle a, \ldots, s \rangle$	Sequence of items from a to s.
$tail\, S$	Sequence resulting from removal of first element from sequence S.
$X \leftrightarrow Y$	The set of relations from X to Y.
$X \rightarrow Y$	The set of total functions from X to Y.
$X \mathrel{+\!\!\!\!\rightarrow} Y$	The set of partial functions from X to Y.
$X \times Y$	Cartesian product: the set of all ordered pairs from X and Y.
$f\ t$	The function f applied to t (left associative).
$x\ \underline{R}\ y$	Relation R as an infix operator: $(x, y) \in R$.
R^+	Non-reflexive transitive closure of R.
R^{-1}	Inverse of relation R.
$R(\!\mid S \mid\!)$	Relational image: all elements in R mapped to by elements of S.
$\mathrm{dom}\, R$	Domain of relation R (or indices of a sequence).
$\mathrm{ran}\, R$	Range of relation R (or items in a sequence).
$R \rhd T$	Range restriction of relation (or sequence) R to T.
$S \mathbin{\lhd\!\!\!-} R$	Domain subtraction: R less all pairs with first elements from S.
$R \mathbin{\rhd\!\!\!-} T$	Range subtraction: R less all pairs with second elements from T.
$S \upharpoonright A$	Sequence A restricted to items with indices from S.
$\{D \mid P \bullet t\}$	The set of t's such that P holds given declarations D.
$\{D \mid P\}$	The set of D's such that P holds.
$\forall D \mid P \bullet Q$	Universal quantification: for all D's such that P holds, Q holds.
$\forall D \bullet P$	Universal quantification: for all D's P holds.
$\exists D \bullet P$	Existential quantification: there exists D such that P holds.
$\nexists D \mid P \bullet Q$	There does not exist D such that P and Q hold.
$\nexists D \bullet P$	There does not exist D such that P holds.
$\exists_1 D \bullet P$	Unique existence: there is one D such that P holds.

TAM: A Formal Framework for the Development of Distributed Real-Time Systems*

D. J. Scholefield, H.S.M. Zedan
Formal Systems Research Group
Department Of Computer Science
University of York
Heslington, York, YO1 5DD
email: { djs,zedan }@ac.uk.york.minster

Abstract

The Temporal Agent Model (TAM) is a wide-spectrum development language for real-time systems. In TAM, limited resources are modelled by deriving release times and absolute deadlines from weakest pre-condition predicate transformers. In this paper the language syntax and semantics are described along with a number of examples.

1 Introduction

The formal development of a real-time system can be divided into three separate tasks; requirements specification, concrete design, and verification between the former two activities. In real-time systems, verification must take into account not only functional correctness but also the timeliness of results.

In many real-time formal theories, two independent languages are used for the specification and design tasks. For example, in Timed-CSP [Dav91] [Schn90] and Timed-CCS [TM89] first-order predicate logic is used as the requirements specification language, and a process algebra defines the concrete design language. Verification is undertaken by proving that the logic formulae satisfy the traces arising from the processes.

However, a problem inherent in such 'two tiered' theories is the lack of method by which suitable designs are arrived at. A combination of experience and guess-work must be used in order to develop processes, and then verification – a rather time consuming activity – is undertaken. If the verification fails, then the processes are re-formulated and verification is undertaken again. There are some methods which overcome this specific problem, Ostroff [Ost89] has devised a synthesis method which enables the user to derive designs from specifications, however the method is conservative (it does not always find a suitable design even when one may exist). Other methods suffer from similar drawbacks.

*Partially supported by UK SERC grant GR/F 35920 /4/1/1214

To overcome this problem we have devised the Temporal Agent Model (TAM), which is a theory centered upon a 'wide- spectrum' language in which both real-time requirements specifications and concrete designs may be formulated. A specification in TAM may be transformed step-by-step into a mixed program containing both specification statements and concrete design statements. Such transformations continue until a completely concrete program is produced which is guaranteed correct with respect to the original specification. The transformation process is known widely as refinement, and is already common in purely transformational systems [Mor90] [SETL85] [CIP85].

TAM uses conservative extensions to first-order predicate logic to express the time at which variables hold certain values. The logic is used both as a basis for a language statement, and to define the timed predicate transformers for the language's semantics. The concrete language statements are loosely based on Dijkstra's guarded command language [Dij76] with real-time extensions to cover deadlines, delays and concurrency.

The paper is structured as follows. In the next section the syntax of the TAM language is defined, and an informal semantics is given. In the following section, the semantics are defined and a number of properties of those semantics discussed. A few examples are presented and analysed. The conclusion discusses the definition of a refinement relation, and future work.

2 TAM Syntax

The modular abstraction mechanism in TAM is the *agent* which represents an individually schedulable task with an associated release time and deadline. TAM is a wide-spectrum language in which both specifications and concrete designs may be expressed. In TAM, both specifications and designs are referred to as agents.

2.1 Assumptions

A prime consideration in designing a real-time language is its' amenability to scheduling analysis, therefore in TAM the following assumptions have been made:

- Time is discrete, linear, and global. It is modelled by the natural numbers.
- TAM agents may only be reasoned about if they are guaranteed to terminate within finite time.
- Communication is achieved within TAM agents by reading and writing shared memory spaces called *shunts*. A shunt always contains a single value (resulting from the most recent write), and a time-stamp corresponding to the time of the most recent write. Shunt reading and writing is asynchronous. Shunts may only be written by a single agent, but may be read simultaneously by different agents.
- Computation, such as assignment, input, and output, may not occur instantaneously.

2.2 Agents

The syntax of a TAM agent is defined recursively:

$$\mathcal{A} ::= x := e \mid [t]\mathcal{A} \mid g \uparrow \mathcal{A} \mid \mathcal{A};\mathcal{A} \mid w : \Phi \mid \mu X.\mathcal{A}$$
$$\mid X \mid x \rightarrow s \mid x \leftarrow s \mid \mathcal{A}|\mathcal{A} \mid \mathcal{A}/s \mid (x)\mathcal{A} \mid \delta$$

The agent $x := e$ (where x is some variable name, and e is some expression on variable names) results in the value of e being bound to x. It is assumed that there exists a suitable expression language for e.

The agent $[t]\mathcal{A}$ is guaranteed to terminate within t time units of its release time. The value t must be a constant natural number.

The agent $g \uparrow \mathcal{A}$ (where g is some boolean variable or constant) evaluates g, and if 'true', releases the agent $\mathcal{A}$ immediately. Otherwise no computation takes place and the agent terminates immediately without changing any variables or shunts.

The agent $\mathcal{A};\mathcal{A}$ specifies a precendence relation between the agents. The second agent will not be released until the first has terminated.

The specification $w : \Phi$ defines a *frame* w as a list of variable and shunt names whose bindings may change during the execution of an agent whose specification is given as the first-order predicate logic formula Φ. This agent is discussed in more detail in the next section.

The agent $\mu X.\mathcal{A}$ is recursive; $\mathcal{A}$ is released, and upon encountering the agent variable X, $\mu X.A$ is released again.

The agent $x \rightarrow s$ outputs the value of the variable or constant x into the shunt s with the current time as a time-stamp. The output may occur at any time between the release and termination time of the agent.

The agent $x \leftarrow s$ inputs the time-stamp and value pair from the shunt s into the variable x. The input may occur at any time between the release and termination time of the agent.

The agent $\mathcal{A}|\mathcal{A}$ releases the two agents immediately, and terminates some arbitrary time after both agents terminate. The well-formed test on concurrent agent states that no agent may read a variable which may be changed in a concurrently executing agent. The test is defined in detail below

The agent $\mathcal{A}/s$ restricts the shunt s to the agent $\mathcal{A}$. No other agent may read or write to this shunt.

The agent $(x)\mathcal{A}$ restricts the variable x to the agent $\mathcal{A}$. On entry to $\mathcal{A}$ any existing binding to x is replaced by a non-deterministic binding. On termination, the original binding is restored.

The agent δ delays for at least one time unit before terminating, and does not change any variable or shunt binding. The shorthand δn is defined by:

$$\delta 1 =_{def} \delta \quad \delta n =_{def} \delta; \delta n - 1 \qquad \text{(where } n \geq 1\text{)}$$

2.3 Specifications

The first-order predicate logic is extended with terms of the form '$x@t$' which represent the value of the computation variable 'x' at time 't'. Such extensions are conservative (they may be defined by functions from natural numbers onto some value set). Shunts are defined similarly, with values sets constructed from pairs of natural numbers and some value set. Shunt values (pairs) may be accessed by the usual 'dot' accessors: $(s@t).1$ (time-stamp) and $(s@t).2$ (value).

The function '$vars$' is defined over first-order predicate logic formulae listing timed variables occurring in the formulae (and similarly, a function '$shunts$'). The functions '$vars^n$' (for natural numbers n) list variables which occur at time n, (similarly '$shunts^n$') (these functions are only well-formed for closed formulae).

In a specification statement $w : \Phi$, w lists those variables and shunts which may change during the execution of an agent which satisfies Φ. The set w is called a frame. The formula Φ may contain the free natural number variables t and t', representing the agent's release and termination time. Computation variables may only be referred to at times t and t', and shunts may only be referred to at times in $[t, t']$. The formula therefore represents a timed transition on computation variables and shunts.

The notation $w.vars$ refers to the computation variables in a frame, and $w.shunts$ refers to the shunts.

2.4 Well-Formed Conditions

Three functions are defined which list those variables or shunts which are changed or read by an agent.

The function $\mathcal{C}$ lists variables which may be changed by an agent:

$$\mathcal{C}(x := e) = \{x\} \quad \mathcal{C}(x \leftarrow s) = \{x\} \quad \mathcal{C}(\delta) = \emptyset \qquad \mathcal{C}(x \rightarrow s) = \emptyset$$

$$\mathcal{C}(w : \Phi) = w.vars \quad \mathcal{C}([t]\mathcal{A}) = \mathcal{C}(\mathcal{A}) \quad \mathcal{C}(\mathcal{A}|\mathcal{B}) = \mathcal{C}(\mathcal{A}) \cup \mathcal{C}(\mathcal{B})$$

$$\mathcal{C}(\mathcal{A}; \mathcal{B}) = \mathcal{C}(\mathcal{A}) \cup \mathcal{C}(\mathcal{B}) \quad \mathcal{C}(\mathcal{A}/s) = \mathcal{C}(\mathcal{A}) \quad \mathcal{C}((x)\mathcal{A}) = \mathcal{C}(\mathcal{A}) - x$$

$$\mathcal{C}(g \uparrow \mathcal{A}) = \mathcal{C}(\mathcal{A})$$

The function $\mathcal{R}$ lists those variables which may be read by an agent:

$$\mathcal{R}(x := e) = \mathcal{R}(e) \quad \mathcal{R}(x \leftarrow s) = \emptyset \quad \mathcal{R}(\delta) = \emptyset \qquad \mathcal{R}(x \rightarrow s) = \{x\}$$

$$\mathcal{R}(w:\Phi) = vars(\Phi) \quad \mathcal{R}([t]\mathcal{A}) = \mathcal{R}(\mathcal{A}) \quad \mathcal{R}(\mathcal{A}|\mathcal{B}) = \mathcal{R}(\mathcal{A}) \cup \mathcal{R}(\mathcal{B})$$

$$\mathcal{R}(\mathcal{A};\mathcal{B}) = \mathcal{R}(\mathcal{A}) \cup \mathcal{R}(\mathcal{B}) \quad \mathcal{R}(\mathcal{A}/s) = \mathcal{R}(\mathcal{A}) \quad \mathcal{R}((x)\mathcal{A}) = \mathcal{R}(\mathcal{A}) - x$$

$$\mathcal{R}(g \uparrow \mathcal{A}) = \mathcal{R}(g) \cup \mathcal{R}(\mathcal{A})$$

Note that the function $\mathcal{R}$ is overloaded on expressions and boolean guards.

The function $\mathcal{W}$ lists those shunts that an agent may write to:

$$\mathcal{W}(x := e) = \emptyset \quad \mathcal{W}(x \leftarrow s) = \emptyset \quad \mathcal{W}(\delta) = \emptyset \qquad \mathcal{W}(x \rightarrow s) = \{s\}$$

$$\mathcal{W}(w:\Phi) = w.shunts \quad \mathcal{W}([t]\mathcal{A}) = \mathcal{W}(\mathcal{A}) \quad \mathcal{W}(\mathcal{A}|\mathcal{B}) = \mathcal{W}(\mathcal{A}) \cup \mathcal{W}(\mathcal{B})$$

$$\mathcal{W}(\mathcal{A};\mathcal{B}) = \mathcal{W}(\mathcal{A}) \cup \mathcal{W}(\mathcal{B}) \quad \mathcal{W}(\mathcal{A}/s) = \mathcal{W}(\mathcal{A}) - s \quad \mathcal{W}((x)\mathcal{A}) = \mathcal{W}(\mathcal{A})$$

$$\mathcal{W}(g \uparrow \mathcal{A}) = \mathcal{W}(\mathcal{A})$$

Agents can not read or change variables which may be changed by concurrently executing agents, i.e. in $\mathcal{A}|\mathcal{B}$:

$$(\mathcal{C}(\mathcal{A}) \cup \mathcal{R}(\mathcal{A})) \cap \mathcal{C}(\mathcal{B}) = \emptyset$$

Agents can not write to shunts that may be written by concurrently executing agents, i.e. in $\mathcal{A}|\mathcal{B}$:

$$\mathcal{W}(\mathcal{A}) \cap \mathcal{W}(\mathcal{B}) = \emptyset$$

2.5 Implications and Examples

A system is described by a single agent which is assumed, without loss of generality, to be released at time 0. For all but the most simple of systems the agent will be concurrent, and all sub-agents will be released at time 0. Later release times are then specified by the use of the δ agent.

For example, consider the following agent:

$$[5](x \leftarrow s;\ \delta 3;\ x.2 \rightarrow s') \quad | \quad [2](0 \rightarrow s'')$$

Which denotes two concurrent agents that are released at time 0. The sub-agents of the left hand agent may be scheduled in any way which respects the precedences defined by the sequence operators and timing constraints (there is only one such schedule as the overall deadline provides five time units, and each of the five sub-agents needs a minimum of one time unit). The right hand agent may perform the write at either time=1 or time=2.

On a single processor, this system requires that the two left hand side communications take a maximum of one time unit each, but the right hand agent write may take up to two time units. A possible schedule (the most relaxed) is therefore[1]:

[1]The task of finding a schedule given deadline, release time, and communication information is in general NP-hard, and a sophisticated analysis tool is required. The TAM theory does not aim to provide

Agent	execution start time	deadline (absolute)
$x \leftarrow s$	0	1
$0 \rightarrow s''$	1	2
$x.2 \rightarrow s'$	4	5

The following agent is a specification of an operation that finds any index into a boolean array 'a' which corresponds to the position of a 'true' value:

$$\{i\} : \exists n \in [1, \#a](a@t[n]) \;\Rightarrow\; a@t[i@t']$$

(where $\#a$ denotes the length of the array). This agent may be specified in a more concrete manner by:

$$
\begin{aligned}
&(g)(n)(n := 1;\\
&\quad \mu X.(g := n > \#a;\\
&\qquad \neg g \uparrow (\{i, n\} :(a@t[n@t] \;\Rightarrow\; i@t' = n@t) \wedge\\
&\qquad\qquad (\neg a@t[n@t] \;\Rightarrow\; i@t' = i@t) \wedge\\
&\qquad\qquad (n@t' = (n@t) + 1); X))\\
&\quad)
\end{aligned}
$$

Or in a completely concrete agent:

$$
\begin{aligned}
&(g)(n)(n := 1;\\
&\quad \mu X.(g := n > \#a;\; \neg g \uparrow (g := a[n];\\
&\qquad\qquad g \uparrow i := n;\\
&\qquad\qquad n := n + 1; X)) \quad)
\end{aligned}
$$

TAM makes no distinction between agents based upon their level of abstraction.

More interesting, are agents of the form:

$$(b \uparrow \delta);\; (\neg b \uparrow (\delta; \delta))$$

where timing information is determined at run time as a result of computation. This is a very flexible and powerful part of the TAM language, but care must be taken to ensure that schedulability is not compromised. In the case of variable delays the problem is one of efficiency (the worst-case must be assumed), but with agents of the form:

$$(b \uparrow [n]\mathcal{A});\; (\neg b \uparrow [m]\mathcal{B}) \qquad \text{(where n,m are constants)}$$

the problem is much more complex. In general, dynamically changing deadlines prove difficult to guarantee, and the use of the TAM language may have to be restricted in order to ensure that programs are amenable to schedulability analysis.

---

such schedules, only the information needed to perform the analysis

Consider the following agent:

$\mathcal{A}/s \mid \mathcal{B}/s$

The act of restricting a shunt is the same as renaming within the agent being restricted (along with some semantic transformations which we will not consider here). Therefore, this agent is exactly equivalent to:

$\mathcal{A}[s'/s]/s' \mid \mathcal{B}[s''/s]/s''$

The question of when the shunt or variable is restricted is therefore not relevant, only for which agent it is restricted for.

Consider the agent:

$\prod_{n \in [0,t]} [n]\delta n$

where '$\prod$' represents indexed concurrency and t is some very large natural number (i.e. distant time). This agent also uses no resources, but executes for large amount of time (until the time t) and eventually terminates. This agent is perfectly well-behaved, and may be reasoned about.

Consider the agent:

$\text{false} \uparrow x := x \;\mid\; x := y + 1$

which performs assignments to the variable x in parallel. This agent is ill-formed as it refers to the variable x in both parts of the concurrent agent. Even though one of the agents will never actually use x, it remains ill-formed (the well-formed test is pessimistic).

The following agent is not ill-formed:

$y := x \;\mid\; z := x$

(both agents refer to x but they are both reads.)

3 TAM Semantics

The semantics of TAM are described by predicate transformers over agents, execution intervals, post-conditions (which describe the functional and timing requirements of an agent), and communication side-effects (which describe the outputs caused by an agent).

3.1 Post-Conditions, Pre-Conditions and Environments

A post-condition is a first-order predicate logic formula over the value of computation variables at the termination time of an agent, and shunt values at any times. The post-condition therefore represents the user's requirements, and the time at which those requirements must be met.

For example, the post-condition:

$$x@5 \geq 0$$

requires that upon termination of an agent (which must be before or at time=5) the value of the computation variable x must be greater than, or equal to, zero. The post-condition:

$$s@8.1 \geq 0$$

asserts that the shunt s must be written to with some value before time=8 (either by the agent itself, or by some agent which is executing concurrently).

Similarly, a pre-condition is a first-order predicate logic formula over the value of computation variables at the release time of an agent, and shunt values at any time, along with a communication side-effect. The pre-condition represents the required initial state, and communications that are needed in order to guarantee some post-condition holds at the termination of an agent.

A communication side-effect is a first-order predicate logic formula which describes the outputs caused by an agent. It is used to discharge input requirements of concurrently executing agents.

Every agent has a corresponding environment which is a set of shunt names listing those shunts which may be changed by the agent.

3.2 Predicate Transformers

A predicate transformer describes the way in which a post-condition and a communication side-effect are syntactically transformed into the weakest pre-condition that needs to hold in order to guarantee the post-condition.

A transformer is a function over pairs of first-order predicate logic formulae, a release time, a termination time, an agent, and an environment, i.e:

$$wp_{\delta}^{u}(\mathcal{A},\ \phi,\psi) : w$$

where 'δ' is the relase time, 'u' is the termination time ($\delta \leq u$), '$\mathcal{A}$' is some agent, 'ϕ' is the post-condition, 'ψ' is the side-effect, and 'w' is an environment. The result is a pre-condition formula and a side-effect formula.

The functional part of the pair of formulae is referred to by $(\phi,\psi).f$, and the communication side-effect as $(\phi,\psi).c$, i.e:

$$(\phi,\psi) \ \equiv \ ((\phi,\psi).f,\ (\phi,\psi).c)$$

3.3 Supporting Concurrency

The following predicates appear commonly in the transformer rules.

The predicate $stable(s,\delta,u)$ asserts that the shunt s does not change value within the interval $[\delta,u]$:

$$stable(s,\delta,u) \ =_{def} \ \bigwedge_{\sigma\in[\delta,u]} s@\sigma = s@(\sigma-1)$$

(where $0 - 1 = 0$). The predicate *stable* is extended over sets:

$$all\_stable(S, \delta, u) \ =_{def} \ \bigwedge_{s \in S} stable(s, \delta, u)$$

The predicate $write(v, s, \delta, u)$ asserts that the value v is written to the shunt s sometime in the interval $[\delta + 1, u]$. The shunt remains stable at other times:

$$write(v, s, \delta, u) \ =_{def}$$

$$\bigvee_{\sigma \in [\delta+1, u]} (s@\sigma = (\sigma, v) \ \wedge \ stable(s, \delta, \sigma - 1) \ \wedge \ stable(s, \sigma + 1, u))$$

3.4 Transformer Definitions

In the following predicate transformers, the notation $[\![\phi]\!]^u_\delta$ represents the simultaneous replacement of all computation variables at time u in ϕ with the same variables at time δ (but *not* shunts), e.g:

$$[\![x@u = (y@\delta) + 1 \ \wedge \ s@u = (u, y@u) \ ]\!]^u_\delta \ \equiv \ x@\delta = (y@\delta) + 1 \ \wedge \ s@u = (u, y@\delta)$$

(where s is a shunt)

3.4.1 Assignment

$$wp^u_\delta(x := e, \phi, \psi) : w \ =$$

$$(\delta < u) \ \wedge \ [\![\phi[e@\delta/x@u]]\!]^u_\delta, \ (\delta < u) \ \wedge \ [\![\psi[e@\delta/x@u]]\!]^u_\delta \ \wedge \ all\_stable(w, \delta, u)$$

The expression e is evaluated with the bindings found at time δ. The result is bound to x at or before time u. The evaluation will take at least one time unit, and no shunts in the agent's environment are changed.

3.4.2 Deadline

$$wp^u_\delta([t]\mathcal{A}, \phi, \psi) : w \ = \ wp^n_\delta(\mathcal{A}, [\![\phi]\!]^u_n, [\![\psi]\!]^u_n) : w \qquad (\text{where } n = \text{minimum}(\delta + t, u))$$

The deadline declaration shortens the execution window only if the time given (t) is less than that already available in $[\delta, u]$.

3.4.3 Guard

$$wp^u_\delta(g \uparrow \mathcal{A}, \phi, \psi) : w \ =$$

$$(g@\delta \Rightarrow wp^u_\delta(\mathcal{A}, \phi, \psi) : w.f) \ \wedge \ (\neg g@\delta \Rightarrow [\![\phi]\!]^u_\delta),$$

$$(g@\delta \Rightarrow wp^u_\delta(\mathcal{A}, \phi, \psi) : w.c) \ \wedge \ (\neg g@\delta \Rightarrow [\![\psi]\!]^u_\delta \ \wedge \ all\_stable(w, \delta, u))$$

The guard is a boolean variable and therefore takes no time to evaluate. If it is true, then the execution of the guarded agent takes place immediately. If not, then the guarded agent terminates immediately, and no variables or shunt values are changed.

3.4.4 Sequential

$$wp_\delta^u(\mathcal{A};\mathcal{B},\phi,\psi):w \;=$$

$$\bigvee_{\sigma\in[\delta,u]} wp_\delta^\sigma(\mathcal{A}, wp_\sigma^u(\mathcal{B},\phi,\psi):w):w.f, \quad \bigvee_{\sigma\in[\delta,u]} wp_\delta^\sigma(\mathcal{A}, wp_\sigma^u(\mathcal{B},\phi,\psi):w):w.c$$

The sequential agent requires that there is at least one intermediate time in the execution window $[\delta, u]$ before which $\mathcal{A}$ will terminate and after which $\mathcal{B}$ is released successfully.

3.4.5 Specification

$$wp_\delta^u(w:\Phi,\phi,\psi):w' \;=$$

$$\#\Phi+\delta\le u \;\wedge\; [\![\forall w.vars@u(\exists w.shunts(\Phi[\delta/t][u/t'])\Rightarrow\phi)]\!]_\delta^u,$$

$$\exists(w.vars - vars^u(\psi))@u(\Phi[\delta/t][u/t']) \;\wedge\; \psi[y@\delta/y@u]_{(\text{all } y \text{ not in } w)}$$

$$\wedge\; stable(w'-w,\delta,u)$$

The quantifiers $\forall$ and $\exists$ are extended over computation variables and shunts by the definition:

$$\forall x@t(\phi) \;=_{def}\; \forall x(\phi[x/x@t])$$

$$\exists x@t(\phi) \;=_{def}\; \exists x(\phi[x/x@t])$$

These definitions are extended over sets of variables and shunts in the usual way. Variables and shunts which are quantified without specific times are assumed to be quantified for all times at which they occur in the formula.

The notation $\#\Phi$ denotes the *degree* of the specification Φ, which is the least amount of time which is needed to guarantee that the specification is feasible (logically consistent) for any execution interval. In order to define the degree of a specification we first need to define timing-feasibility.

A formula Φ, with natural number variables t and t' occurring free (but which is otherwise closed) is timing-feasible under n if n is the smallest natural number such that for every interval $[\delta, u]$ there exists an interpretation I such that:

$$I \models (\delta+n\le u)\Rightarrow\Phi[\delta/t][u/t']$$

The degree of a specification is therefore the specification's timing-feasibility number.

3.4.6 Recursion

$$wp_\delta^u(\mu X.\mathcal{A},\phi,\psi):w \;=$$

$$\bigvee_{n\in\mathcal{N}} wp_\delta^u(\mathcal{A}^n(\text{abort}),\phi,\psi):w.f, \quad \bigvee_{n\in\mathcal{N}} wp_\delta^u(\mathcal{A}^n(\text{abort}),\phi,\psi):w.c$$

Where $\mathcal{A}^0(\mathcal{B}) \;=_{def}\; \mathcal{B} \qquad \mathcal{A}^{n+1}(\mathcal{B}) \;=_{def}\; \mathcal{A}[\mathcal{A}^n(\mathcal{B})/X]$

(if X is the recursion variable).

Where abort $=_{def}$ $[0]\delta$

3.4.7 Output

$$wp^u_\delta(x \rightarrow s, \phi, \psi) : w \;=$$

$$\delta < u \;\wedge\; [\![\phi]\!]^u_\delta,\; write(x@\delta, s, \delta, u) \;\wedge\; all\_stable(w - s, \delta, u)$$

The output agent requires at least one time unit in which to execute (this condition is implicitly asserted in the communication side-effect because an empty disjunction in the 'write' predicate gives rise to the predicate 'false').

3.4.8 Input

$$wp^u_\delta(x \leftarrow s, \phi, \psi) : w \;=$$

$$\delta < u \;\wedge\; [\![\textstyle\bigwedge_{\sigma \in [\delta+1,u]} \phi[s@\sigma / x@u]]\!]^u_\delta, \qquad \delta < u \;\wedge\; all\_stable(w, \delta, u)$$

The input requires at least one time unit in which to execute, and asserts that all of the values held in the shunt s in the appropriate interval must fulfill the pre-condition. The shunts remain unchanged during an input.

3.4.9 Concurrency

$$wp^u_\delta(\mathcal{A}|\mathcal{B}, \phi, \psi) : w \;=$$

$$\exists X^*(wp^u_\delta(\mathcal{A}, \phi_\mathcal{A}, \psi_\mathcal{A}) : w - \mathcal{W}(\mathcal{B}).f \;\wedge\; wp^u_\delta(\mathcal{B}, \phi_\mathcal{B}, \psi_\mathcal{B}) : w - \mathcal{W}(\mathcal{A}).f),$$

$$\exists Y^*(wp^u_\delta(\mathcal{A}, \phi_\mathcal{A}, \psi_\mathcal{A}) : w - \mathcal{W}(\mathcal{B}).c \;\wedge\; wp^u_\delta(\mathcal{B}, \phi_\mathcal{B}, \psi_\mathcal{B}) : w - \mathcal{W}(\mathcal{A}).c)$$

The division of the functional predicate and communication predicate is dictated by the variables occurring in each agent. In both cases, the conjunction of the two predicates, along with the quantified variables found in the two sets X^* and Y^* must be equivalent to the original. In addition, the predicates may only refer to variables not changed by concurrently executing agents.

3.4.10 Restriction

$$wp^u_\delta(\mathcal{A}/s, \phi, \psi) : w \;=$$

$$\text{discharge}(wp^u_\delta(\mathcal{A}, \phi, \psi) : w) \;\wedge\; \neg(s \in shunts(\text{discharge}(wp^u_\delta(\mathcal{A}, \phi, \psi) : w))),$$

$$\exists s(wp^u_\delta(\mathcal{A}, \phi, \psi) : w.c)$$

The restriction of a shunt removes any shunt writes from the environment, and discharges any obligations arising in the functional pre-condition on that shunt. If the shunt occurs within the post-condition or side-effect, then it is renamed to a unique shunt name within the agent.

The discharge function is defined over predicates thus:

$$\text{discharge}(\phi, \phi') \;=\; \text{the weakest formula such that:}$$

$$\text{discharge}(\phi, \phi') \wedge \phi' \Rrightarrow \phi.$$

3.4.11 Local Variable

$$wp_\delta^u((x)\mathcal{A}, \phi, \psi) : w =$$

$$\exists x @ \delta(wp_\delta^u(\mathcal{A}, \phi, \psi) : w.f), \quad \exists x @ \delta(wp_\delta^u(\mathcal{A}, \phi, \psi) : w.c)$$

The restriction of a local variable removes any changes to that variable from the functional and communication predicates. If the variable occurs in the post-condition or side-effect, then it is renamed to a unique variable within the agent.

3.4.12 Delay

$$wp_\delta^u(\delta, \phi, \psi) : w =$$

$$\delta < u \wedge [\![\phi]\!]_\delta^u, \ \delta < u \wedge [\![\psi]\!]_\delta^u$$

The delay agent requires at least one time unit in which to execute, and does not change variables or shunts.

3.5 Examples

Consider the simple agent $x := x + 1$ executing in a window $[3, 8]$, with an empty environment, then:

$$wp_3^8(x := x + 1, x@8 \geq 1, \text{true}) : \emptyset \quad = \quad x@3 \geq 0, \text{true}$$

i.e. if the value of x at time=3 is at least zero, then incrementing x in the interval $[3, 8]$ will guarantee that the value of x at time=8 is at least one. The transformation is as follows:

$$wp_3^8(x := x + 1, x@8 \geq 1, \text{true}) : \emptyset$$

$$\equiv \ 3 < 8 \wedge [x@8 \geq 1[(x@3) + 1/x@8]]_3^8, \ 3 < 8 \wedge \mathit{all\_stable}(\emptyset, 3, 8)$$

$$\equiv \ [\![(x@3) + 1 \geq 1]\!]_3^8, \ \ \mathit{all\_stable}(\emptyset, 3, 8)$$

$$\equiv \ [\![x@3 \geq 0]\!]_3^8, \ \ \text{true}$$

$$\equiv \ x@3 \geq 0, \ \ \text{true}$$

Consider the agent:

$$\{x\} : x@t' = f(y@t)$$

which assigns the value of $f(y)$ at the release time, to the variable x at the termination time. This agent is not equivalent to the agent:

$$x := f(y)$$

The specification agent may take no time in which to execute (it has a degree of zero). In particular, in the case when $f(y@t) = x@t$, then no action need be taken. In the assignment agent the evaluation will always take at least one time unit.

Consider the agent:

$$[1](x \to s|y \leftarrow s)/s$$

The input and output to shunt s is constrained by the deadline to occur at release-time+1. This guarantees the assignment of the value in x to $y.2$ and the time of the transfer to $y.1$. The following transformation demonstrates the assignment within an execution window of $[0,5]$:

$$wp_0^5([1](x \to s|y \leftarrow s)/s, y@5.2 = x@0, \text{true}) : \emptyset$$

$$\equiv \quad wp_0^1((x \to s|y \leftarrow s)/s, y@1.2 = x@0, \text{true}) : \emptyset$$

The pre-discharge formulae is calculated first:

$$\equiv \quad \exists v(wp_0^1(x \to s, v = x@0, \text{true}) : \emptyset.f \ \wedge \ wp_0^1(y \leftarrow s, y@1.2 = v, \text{true}) : \emptyset.f)$$

$$\wedge\ \exists z(wp_0^1(x \to s, z = x@0, \text{true}) : \emptyset.c \ \wedge \ wp_0^1(y \leftarrow s, z = y@1.2, \text{true}) : \emptyset.c)$$

Individually:

$$wp_0^1(x \to s, v = x@0, \text{true}) : \emptyset.f \ \equiv \ v = x@0$$

$$wp_0^1(x \to s, v = x@0, \text{true}) : \emptyset.c \ \equiv \ s@1 = (1, x@0)$$

$$wp_0^1(y \leftarrow s, v = y@1.2, \text{true}) : \emptyset.f \ \equiv \ s@1.2 = v$$

$$wp_0^1(y \leftarrow s, v = y@1.2, \text{true}) : \emptyset.c \ \equiv \ \text{true}$$

Back in conjunction gives:

$$\exists v(v = x@0 \ \wedge \ s@1.2 = v), \ \exists z(s@1 = (1, x@0))$$

$$\equiv \quad s@1.2 = x@0, s@1 = (1, x@0)$$

The discharge on the shunt 's' results in:

$$\text{discharge}(s@1.2 = x@0, s@1 = (1, x@0)) \ \equiv \ \mathit{true}$$

$$(\text{because } s@1 = (1, x@0) \Rightarrow s@1.2 = x@0)$$

Giving the overall pre-condition:

$$\equiv \quad \text{true}, \exists s(s@1 = (1, x@0))$$

$$\equiv \quad \text{true}, \text{true}$$

3.6 Properties of The Predicate Transformers

We are primarily interested in two properties: distribution of conjunction and distribution of entailment. Both properties are used in the proof of monotonicity of a refinement relation defined later. The proofs of the following two theorems are available from the authors.

3.6.1 Distribution of Conjunction

Theorem 1: $wp_\delta^u(\mathcal{A},\ \phi \wedge \phi', \psi) : w \ \equiv\ wp_\delta^u(\mathcal{A},\ \phi, \psi) : w \ \wedge\ wp_\delta^u(\mathcal{A},\ \phi', \psi) : w$

Where $(\phi, \psi)\ \wedge\ (\phi', \psi')\ =_{def}\ (\phi \wedge \phi', \psi \wedge \psi')$

3.6.2 Implication Distribution (Monotonicity)

Theorem 2: $\quad \phi \Rrightarrow \phi' \ \Rightarrow\ wp_\delta^u(\mathcal{A},\ \phi, \psi) : w \Rrightarrow wp_\delta^u(\mathcal{A},\ \phi', \psi) : w$

Where $(\phi, \psi) \Rrightarrow (\phi', \psi')\ =_{def}\ (\phi \Rrightarrow \phi')\ \wedge\ (\psi \Rrightarrow \psi')$

4 Practical Real-Time Issues in TAM

In designing a formalism for real-time systems, a major objective is to provide a practical framework in which common real-time issues may be easily expressed. Such issues include the ability to read a global clock, specify periodic tasks, and provide time-outs on communication events. In the following examples the notation:

$$\odot_n(\mathcal{A}) \quad (\text{where } n \geq 1)$$

represents the iteration operator defined by:

$$\odot_1(\mathcal{A})\ =_{def}\ \mathcal{A} \qquad \odot_{n+1}(\mathcal{A})\ =_{def}\ \mathcal{A}; \odot_n(\mathcal{A})$$

4.1 Reading The Global Clock

TAM allows access to the global real-time clock in the time-stamps of the shunts. An agent called *Clock* which updates the time-stamp of a boolean shunt called 'time' on each clock-tick is easily derived. All TAM agents must execute for a finite time, and so an arbitrary duration for the clock agent of 'n' time units is used:

$$Clock\ =_{def}\ [n](\odot_n(\text{true} \rightarrow \text{time}))$$

Any agent requiring the current time may read the time-stamp of the shunt 'time'. An alternative solution is for the agent which requires the time to write and read a private shunt:

$$(true \rightarrow time;\ x \leftarrow time)/time$$

However, the latter solution is more time consuming (it takes a minimum of two time units in which to execute, rather that the one time unit it would take to read the Clock shunt). The Clock agent requires resources continuously, but it may be possible to map this simple agent onto a hardware device.

4.2 Periodic Tasks

A periodic task should be released once in every given period. In the following example, an agent form is defined which is parameterised over an agent (which represents the task), a period 'p', a deadline 'd', and a maximum number of invocations 'n':

$$Period(\mathcal{A}, p, d, n) \;=\; [n \times p](\odot_n([d]\mathcal{A}|\delta p))$$

The deadline $[n \times p]$ forces each release of $\mathcal{A}$ to occur every p time units. In this example it must also be the case that deadline $\leq$ period.

4.3 Time-Outs

A time-out defines the maximum length of time for which an agent is prepared to wait for an event before it decides on an alternative course of action. An 'event' is defined as the writing of the value 'true' into a boolean shunt. A particularly simple timeout agent form for shunt s, positive agent $\mathcal{A}$ (the agent which is executed if the event occurs on time), negative agent $\mathcal{B}$ (which is executed if the event does not occur on time), and time t is defined as:

$$Timeout(s, t, \mathcal{A}, \mathcal{B}) \;=_{def}\; [t](\delta(t-1); x \leftarrow s);\; (x.2 \uparrow \mathcal{A}); (\neg x.2 \uparrow \mathcal{B})$$

This timeout agent is inefficient; if the shunt is written to in the first time-unit after the timeout's release then the positive agent will still not be released until some time after the timeout period. More efficient timeouts can certainly be imagined.

5 Conclusion

In this paper the Temporal Agent Model (TAM) was introduced, a formal framework within which a distributed, reliable, real-time system may be specified and reasoned about in a compositional fashion.

TAM may be seen as a wide-spectrum language which is capable of expressing a real-time system at a range of levels of abstraction within a unified syntax. The abstract primitives enable the user to specify the logical and temporal behaviour of a program without recourse to developing a particular algorithm. The concrete primitives provide a typical imperative programming language based on assignment and flow control structures. The abstract and concrete primitives may be arbitrarily intermixed to provide an extremely flexible syntax in which programs can be expressed in a manner which suits the user's chosen development method.

Wide-spectrum languages are particularly amenable to refinement methods. Program refinement is a process of transformation which increases a particular metricated property, but which respects functionality. Usually the metricated property is 'abstraction', and refinement is used as a generator of implementations from specifications.

The transformation process often involves two separate considerations, operational refinement, which transforms the algorithmic parts of the program, and data refinement which transforms the representation of data structures.

In process algebra, refinement is sometimes seen as an exercise in process simulation [Mil89] [Tof90]. Data refinement is not carried out because most process algebras do not have data structures (they are purely abstract languages). However, both data and operational refinement is considered in Morgan's refinement calculus, which combines first-order predicate logic specifications over state transitions within Dijkstra's guarded command language [MRG88] [Mor90]. Refinement laws are derived which preserve functionality and reduce abstraction. Furthermore, Back and Wright [BW90] have automated proof obligations in a refinement calculus by implementing a guarded command language's semantics using the Higher Order Logic interpreter (HOL).

Nevertheless, there is little reference to refinement for real-time systems in the literature. It is clear that refinement in real-time systems should also preserve temporal properties along with the functional ones, and in the remainder of this section we investigate how this might be achieved.

Delays provide information to the scheduler about an agent's resource requirements (i.e. that no resources are needed for a specific amount of time). It would initially seem intuitive that a refinement of a TAM agent which delays for some time would be the same agent which delays for a longer period:

$$(\delta t; \mathcal{A} \sqsubseteq \delta t + t'; \mathcal{A})$$

where '$\sqsubseteq$' represents refinement (read as 'is refined by').

However, there are two problems with this. Firstly, the agent A might require the use of a resource at the specific time provided by t (an input occurs before a shunt is overwritten for example), and in the refined agent this resource would not be available at that time. Secondly, it would mean that the refinement relation was not monotonic, a result which would preclude compositional refinement.

It may also seem intuitive to shorten deadlines. Again, this gives us problems with resource management. If an agent is given a deadline, then it is possible for that agent to execute until the end of that deadline and still meet the specification. If an agent is relying on an input which is only possible at the instant before a deadline, then the agent needs all available execution time in order to meet the specification – shortening that deadline will mean that the specification is inconsistent.

Therefore, refinement respects timing information *absolutely* and only functional requirements may be weakened. A refinement relation is defined as follows:

$$\mathcal{A} \sqsubseteq \mathcal{B} \text{ iff } \quad \forall \delta, u, \ \forall \phi, \psi \ \forall w$$
$$wp^u_\delta(\mathcal{A}, \ \phi, \psi) : w.f \ \equiv\rangle \ wp^u_\delta(\mathcal{B}, \ \phi, \psi) : w.f$$
$$\wedge \ wp^u_\delta(\mathcal{A}, \ \phi, \psi) : w.c \ \equiv \ wp^u_\delta(\mathcal{B}, \ \phi, \psi) : w.c$$

At present a refinement calculus is being developed which will enable the user to generate concrete code (without specifications) by using a set of syntactic substitution rules. These rules will form an equational theory which is sound with respect to the wp^u_δ semantics.

6 Acknowledgement

The authors wish to acknowledge fruitful discussion with colleagues in the 'Real-Time Systems: The Next Generation' project team at the University of York, and especially Ken Tindell and Clive Adams.

7 References

[BW90] R. Back, J. Wright. "Refinement Concepts Formalised in Higher Order Logic". BCS-FACS Vol 2. No. 3. 1990.

[Dij76] E. Dijkstra. "A Discipline Of Programming". Prentice-Hall. 1976.

[MRG88] C. Morgan, K. Robinson, P. Gardiner. "On the refinement calculus". Technical Report PRG-70. Programming Research Group. 1988

[Mil89] R. Milner. "Communication and Concurrency". Prentice-Hall. 1989.

[Ost89] J. Ostroff. "Temporal Logic for Real-Time Systems". Research Studies Press. 1989.

[Mor90] C. Morgan. "Programming from Specifications". Prentice-Hall. 1990.

[Tof90] C. Tofts. "Timed Concurrent Processes". BCS-Leicester Workshop in Semantics for Concurrency. Springer-Verlag 1990.

[CIP85] The CIP Language Group, "The Munich Project CIP. Vol1", LNCS-183. 1985.

[Dav91] J. Davies, "Specification and Proof in Real-Time Systems". Oxford University, Ph.D Thesis PRG-93. 1991.

[Har88] D. Harrison, "RUTH: A Real-Time Applicative Language", University of Stirling Ph.D. 1988.

[Hoo91] J. Hooman, "Specification and Compositional Verification of Real-Time Systems". Ph.D. Thesis, Eindhoven University. 1991.

[Schn90] S. Schneider, "Correctness and Communication in Real-Time Systems". Oxford University, Ph.D. Thesis PRG-84. 1990.

[SETL85] E. Schonberg, D. Shields, "From Prototype To Efficient Implementation: a Case Study Using SETL and C". Courant institute of mathematical sciences, Dept of computer science, New York University. 1985.

[TM89] C. Tofts, F. Moller. "A Temporal Calculus Of Communicating Systems". Edinburgh University, Technical Report LFCS-89-104.

An Attempt to Confront Asynchronous Reality to Synchronous Modelization in the ESTEREL Language[1]

Martin Richard
Olivier Roux

Laboratoire D'Automatique de Nantes
(Unité de Recherche Associée CNRS 823)
Ecole Centrale de Nantes
1, rue de la Noë
44072 Nantes cedex 03
France

{richard, roux}@lan01.ensm-nantes.fr

October 1991

1. INTRODUCTION

The term "reactive systems" [9] has been introduced to qualify systems that react to inputs coming repetitively from controlled environment and produce outputs to that environment. In reactive systems we found real-time controlling process, automatisms of various control (distribution of money, man-machine interface, video games), etc. An important feature of reactive systems is their intrinsic determinism: they always react identically to the same sequence of inputs.

We assist now at the development of several languages or formalisms specifically adapted to reactive systems. We can mention ESTEREL [1], LUSTRE [4], SIGNAL [3], the Statecharts [8] and ELECTRE [6] [10] which is asynchronous. Except the last one, these languages forget the asynchronous concept that is at the base of all classical languages to replace it by a synchronous assumption. This assumption has two fundamental features. First, we suppose that the outputs are absolutely synchronous with the inputs, so their computations "take a null time". Second, we suppose that the time is discrete and we observe the system at some precise instants (of null duration). The occurrences of some events coming before an observation instant are only effective at the next instant. Therefore, they will be considered as simultaneous.

But the real world is sometimes absolutely asynchronous. It is impossible for two physical events to be produced at the same time because they occurred at an instant (null time interval). At least, they are not simultaneously caught by the system.

[1] Research supported by the Natural Sciences and Engineering Research Council of Canada and by the French Coordinated Research Project C^3 and Automatique-C2A.

In this paper, we are mostly interested in confronting an asynchronous context with a synchronous language, namely ESTEREL. Our aim is to bring out a taxonomy of the events and to suggest how to express in ESTEREL each class of event behaviours we pointed out.

After a quick review of the fundamental features of the language in the section 2, we attempt to find solutions to problems encountered at the time of the programming of an asynchronous real-time application. To do that, we propose in the third section a classification of the events to find a systematic solution to implement each of them in ESTEREL. The goal of this approach is to control the size of the automaton produced by the ESTEREL compiler, i.e. to avoid the combinatory explosion of the number of states which comes quickly out of the machine capacity. These problems and these solutions are presented in the fourth section of this paper.

Eventually we give some results about an ESTEREL experiment on a large real-time application in order to give an idea of the sizes of the source code and the automaton. This work is the first step of a more ambitious one the aim of which is to translate asynchronous expressions of real-time applications into synchronous programs written in the ESTEREL language.

2. RECALL ABOUT ESTEREL

2.1 Synchronous assumption of ESTEREL

The ESTEREL language is a parallel imperative language which possesses a mathematical semantics and a complete implementation based on the specific synchronism assumption. Our first task is to describe the ESTEREL vision of reactive systems. First of all we see them as "interrupting" systems where the environment controls the program.

With this synchronous assumption, each reaction is assumed to be instantaneous and atomic in the whole meaning of the world. We also assume that the underlaying machine "takes no time" to execute basic operations (assignment, addition, etc.) on data or operations present in a sequence of instructions, process support and process communication. First, a reaction takes no time from the ESTEREL point of view to the external environment which stays unchanged. Second, each process "takes no time" with respect to any other process. In the synchronous languages, the process communication is broadcast; at each time, all the processes are supposed to have the same view upon the environment as well as upon the other processes. A statement takes time if and only if it tells explicitly that it requires time. For example the statement:

```
delay 3 SECOND; delay 2 SECOND
```

is equivalent to the sequence:

```
delay 5 SECOND
```

It would not be the same in a classical language because we do not know the duration of the operator ";".

An ESTEREL program has several inputs which are not synchronized with each other. An ESTEREL program reacts instantaneously to the reception of input signals by emitting itself output signals towards its environment. In the same time the occurrence of an input signal is the only cause of the ESTEREL program reactions. The program stays idle as long as it does not receive input signals; it has no internal clock, unlike the asynchronous systems.

Obviously, one may wonder about the realism of such an assumption from the implementation aspect. It appears that synchronous parallel programs may be compiled into efficient and sequential finite automata which, on the other hand, are very efficient on line and, on the other hand, leads to a foreseeable deterministic behaviour.

2.2 The signals

Since they are the the only mean to make programs evolve, signals may be considered as clock ticks. A signal `S` may also have a persistent value of some type that can be accessible at any time in an ESTEREL program by the expression "`?S`". The signal value can only be modified by the emission of a new value to the environment.

The signals are emitted to the environment by the statement "**`emit`**". The syntax is:

```
emit S(expr)
```

where `expr` is an expression the value of which is assigned to the signal `S` as soon as the statement is executed. This expression is optional.

The signals are fleeting. It means that they are not memorized and either they are taken into account as soon as they are emitted or their occurrence is forgotten. Only the persistent value, if it has one, is modified and can be consulted.

It is possible to declare particular relations between signals. Declaration of relations allows one to restrict input events in a module. These relations may be used for two reasons. First, specifications may require that two signals never appear at the same time. Second, they greatly reduce the input combination for the compiler, usually from an exponential complexity to a more practical linear one.

It is also possible to declare that signals are local to a part of code. Their scope are limited inside a module or inside two or several specific modules. For example, a signal synchronizing two modules will be declared local to these two modules, thus limiting its scope. Recall that in ESTEREL local signals never produce code, i.e. they do not appear in the resulting automata since they are serialized at the compilation.

2.3 A few instructions and constructors

Some language features among the most important ones are presented here because they will be used in the following.

• The await instruction: this instruction is used to wait for an occurrence. Nothing is done until occ drops in, then it simply terminates.

```
await occ
```

• The watching construct: the watching construct is used to limit the duration of an instruction inst.

```
do
    inst
watching occ
```

The construct starts by executing inst, and if the occurrence of occ takes place before inst terminates, inst is immediately aborted and the construct terminates; otherwise the construct terminates when inst does.

• The timeout clause: often it is not sufficient to simply interrupt the instruction in a watching construct, and one wishes to take extra steps if a time overrun occurs; a timeout clause may be added to the watching for this purpose. Here is the syntax of such a timeout clause:

```
do
     inst1
watching  occ
timeout
     inst2
end
```

Here, $inst_1$ is executed normally, but it will be interrupted and $inst_2$ will be executed if occ happens before $inst_1$ completes; thus this instruction terminates when $inst_1$ does, or when $inst_2$ does if a timeout has occurred.

3.Nature of events

Our programming experiment with the ESTEREL language, leads us to a first classification of events in order to delimit their natures and to ease their expressions in this language. We are not interested in the signal aspects, we are only interested in their functions. This is the reason why we classify the events and not the signals. We pointed out seven event categories which maybe, are neither exhaustive nor independent but are fundamental for their asynchronous or synchronous meaning.

- interrupt events;
- priority events;
- starting events;
- supervision events;
- state events;
- choice events
- synchronization events.

3.1 Interrupt events

In an application, some events may come up as they were foreseen but not really waited for. These events often divert the normal execution for the benefit of a momentary treatment as when an alarm comes. This is the typical case of an event specifying that a tank overflows.

Let us consider, for example, a treatment unit where a mixing tank must be cleaned in while each product changes. The event that produces the start of the cleaning will stops the product inflow in the mixing tank during the whole cleaning time but the rest of the system normally works without knowing what is happening at the mixing tank. This short-cut is produced when the process managing the arrival valve of the mixing tank receives an interrupt event. This event was noi typically waited for but was the object of a kind of attention. Since the occurrences of these events are not accurately waited for, we cannot wait for the corresponding signal. In ESTEREL, we will enclose the interrupting code within a watchdog that has the following frame:

```
module Watchdog:
input alarm;

do

      .
      .
      .

watching alarm
{timeout
      . . . }
end.
```

the `timeout` is optional.

3.2 Priority events

Sometimes an interrupt event occurs while a treatment which was started previously by another event. Then, if we consider this new event has a greater priority, we must interrupt the work and launch a temporary treatment in order to deal with this new emergency. We will call priority events the interrupt events which have the capability of preempting procedures that have been activated by other interrupt events.

Let us come back to the example presented in the previous section. It is easy to imagine that, because of a valve failure, the tank may overflow during the cleaning time. This overflow will produce an interrupt event of greater priority than that of the cleaning. This event must be treated with a greater priority than the first one. We should suspend the first treatment for the benefit of the last one. It is easy to imagine some events with a greater priority than the last one; for example the complete shut down following an electrical failure. We must set this priority during the program writing so as to avoid problems when this situation takes place.

In ESTEREL, such events the occurrences of which cannot be predicted are watched at by the **watching** constructor. In order to establish a priority, just embed these **watching** one inside the other since the most exterior **watching** has the greatest priority in ESTEREL:

```
module Some_kind_of_treatment:
input cleaning, overflow, shut_down;
output . . .

do
    do
        do
            .
            .
            .
        watching cleaning
        timeout
            . . .
        end
    watching overflow
    timeout
        . . .
    end
watching shut_down
timeout
    . . .
end.
```

3.3 Starting events

A process is not necessary always ready to execute as soon as the system starts its execution. It may happen that it needs some previous conditions are satisfied before it is able to start its execution. Then, at first, it will only wait for a signal telling that it can start its work. We call start events all events coming from to a signal that must wait for a process before starting its execution.

For example, a process which has to mix some products must wait until it reaches a minimal level in its tank before starting because a too high speed could break the engine. This signal will be considered as a starting event. These events have the particularity to come from signal stemming from the environment or from another process. We can imagine an electronic probe detecting the minimal level of a tank or another process calculating continuously the quantity of liquid in the tank, which emits the signal.

In ESTEREL, these events may simply be pure signals (not valuated) emitted by a process at the appropriate moment when another process waits for this signal on an `await` statement, before it can execute.

3.4 Supervision events

A system is composed of some sensors specifying the state of the system. These detectors maintain a watchfulness on the whole by emitting some signals when necessary. These events are not always systematically treated by the software layer but tell about some change in the whole system.

All events produced by a signal giving information about the system state are considered as supervision events. For example a signal telling that a wagon is right over its target, or an electronic call sending a signal each time an object goes in front of it, is a supervision event.

Since these events are foreseeable and sometimes waited for, it is sufficient to put a simple `await` in ESTEREL for detecting their comings.

These events come from signals stemming generally from the environment unless they have been converted and retransmitted by a process watching it. Recall that in ESTEREL signals are always fleeting (i.e. they are present at an instant but they do not persist). This limitation raises problems when the supervision event is not waited for. This will generate a new class called the state event class.

3.5 State events

During the operation, it may be necessary to know about the state of a part of the system at some instant. In order to do that, the concerned components must provide informations about their states to a process asking for them. We call state events all the exchange of informations needed to know about the state of one system component.

For example, we can imagine a process controlling a tank which must fill containers and another controlling a robot to load and unload the containers under the tank. The robot may also be busy with some other tasks. When the robot is ready again to work with the tank, it must know the state of the container under the tank. It will ask the question "Are the containers filled?". When it gets the answer, it can decide whether to unload the container or to do something else and later return. We present a proposal for the implementation in ESTEREL of a mechanism which can take the state event in the section §4.2.4 .

3.6 Choice events

It is common in a system to have some processes able to choose with whom they will work or on which they will execute. In order to do this choice, the concerned components have to indicate when they are ready. If only one component indicates that it is ready, then the process work will go to this one; the choice is implicit. On the other hand, if more than one component indicate they are ready, the process has to choose explicitly on what component the treatment will be done. In all cases we name choice events the events generated by components for choosing it.

For example, a press waiting until one out of five tins to press, is full, will receive a choice event when the first tin makes its availability known. The waiting will be broken and the press will do an implicit choice if only one tin has indicated that it is available. The choice will be explicit and static if, between two observation instants, more than one tin have indicated their availabilities.

The implementation of the choice event in ESTEREL will depend on the kind of explicit choice that we want to do. If we want to go to the nearest tin ready or if we want to establish some priority at the programming like:

```
module Choice:
input Ready_tin1, Ready_tin2, Ready_tin3,
      Ready_tin4, Ready_tin5;
output . . .

await
    case Ready_tin1 do . . .
    case Ready_tin2 do . . .
    case Ready_tin3 do . . .
    case Ready_tin4 do . . .
    case Ready_tin5 do . . .
end.
```

where the script apparition order establishes the signal chosen when more than one signal occur between two observation instants. What would happen to a signal sent by a tin to tell that it is ready if the press is busy with another tin? The signal will be lost. This problem is detailed and solutions are proposed in the section §4 of this paper.

3.7 Synchronization events

First we consider as synchronization events all the events necessary for two independent processes to synchronize like in the rendez-vous. Second, all the events managing the access to a critical section are also synchronization events. This category collect the events exchanged between two processes which want to meet knowing that one is waiting for the other and the events exchanged between two or more processes that want to get in a critical section.

For example, consider two wagons wishing to go at a unique loading position. A wagon wishing to take possession of the critical resource will ask first the permission to the second wagon. If it obtains the permission, it will indicates that it takes possession of it and it must also indicate that it does not allow the second wagon any more to take it at its turn. All the communications between these two wagons are considered to be synchronization events. In ESTEREL, these synchronization events are expressed by what we call state processes and temporal conditionals that we will detail in the sections §4.2.3 and §4.2.5 .

These events have the particularity to be produced, in ESTEREL, by local signals, i.e. these signals do not come from the environment but are declared local to a part of the application code. They will not appear in the resulting automaton because they are serialized at the time of the compilation.

3.8 Comments

One can notice that the first four event classes are synchronous and very straight forward. This is the reason why we easily find an implementation solution in ESTEREL for each of them. But for the three last classes of events which of an asynchronous nature, it is necessary to create a specific mechanism expressible in the synchronous language ESTEREL. These implementation solutions and the details in relation to these classes of events are presented in the next section.

4. PROGRAMMING IN ESTEREL

4.1 ENCOUNTERED PROBLEMS

ESTEREL is not a usual programming language accompanied by signals handling primitives to synchronize the processes. One must not be trapped by using such constructors as the repetitives, the conditionals as they are used in sequential languages. From our experience in programming with ESTEREL, we want to highlight the encountered problems. The ESTEREL language will not be exhaustively presented here, the reader might refer to [1], [2] and [7] for a more complete description.

4.1.1 State memorization of the components

As it was already said, the signals are not memorized in ESTEREL, i.e. if nobody is waiting for the emitted signal, it will pass in the drop and will be definitely lost. Only its persistent value, if it has one, will be upgraded and available for the execution continuation. The signals have a synchronous character, a synchronization and logic clock function.

The problem

The loose of a signal may become a problem when we have two parallel processes working asynchronously, one related to the other but that are dependent at some level. Let-us have a look to an example to show how to synchronize two processes.

Imagine on one side a tank filled with liquid which must be unloaded in a container. On another side, we have a robot which is in charge of placing a container under the tank valve and then wait it until it is full to remove it and to put a new one (figure 1). For the safe of high efficiency or simply because the application requires it, we want the tasks to run concurrently but in this case, how the tank may know if it is ready to be filled? The first idea of a programmer will be to do such that the robot send a signal each time it puts a container. But what does happen when the tank is busy in adjusting its liquid level? What does happen if the robot signal is emitted when the tank does not wait for it? The signal is simply lost and no one knows about it. The tank that was busy will wait for a signal which will never come. The problem is the same on the other side since the tank must indicate to the robot that the container is filled and it can remove it. If the robot is not at its waiting position and the tank indicates to it that it has a container ready to remove, the signal will also be irreparably lost. The robot will wait indefinitely a signal telling that it can remove the container and the tank will also wait indefinitely a new container. This is a deadlock state which draws us to find new methods to synchronize the two processes.

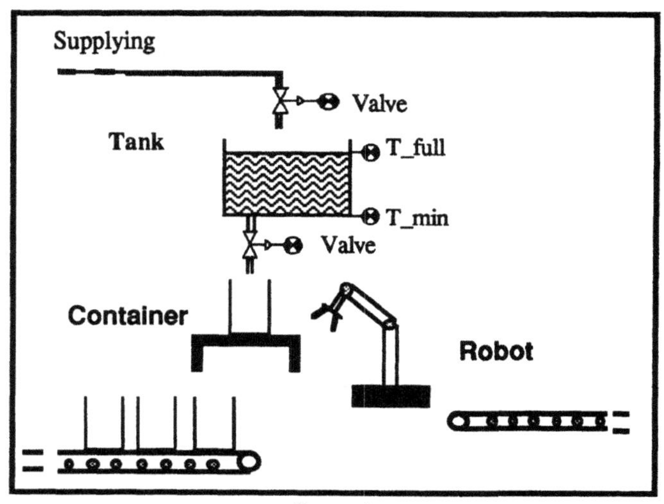

figure 1

This synchronization problem of two parallel processes in a asynchronous environment is the one that caused us the worst problem.

4.1.2 Combinatory explosion of the number of states

A second problem appeared when programming in ESTEREL. It deals with the combinatory explosion of the number of states of the automaton produced by the compiler. Recall that the ESTEREL compiler translates the parallel source code ESTEREL in a sequential state finite automaton. Unfortunately, one who writes the program of a large application is early constrained by the computer capacity. The generated automaton has indeed very quickly more than one thousand states! When translated in C code, it often reaches more then 5 Mb so that it may be impossible to compile and to get the executable code.

The problem is even more striking when one writes modules independently and combines them after. For example, two modules of respectively 5 and 17 states compiled together leads to 49 states. This is not too large but it is nevertheless a first warning. The problem increases when the two previous modules are compiled with a third one composed of 68 states. It produces the frightening result of 3217 states which are a subset of the cartesian product of the initial sets of states. This subset may change according to the used programming methods, the connection degree of the module (local signals exchange) and the number of signals exchanged with the environment by each modules.

It is quite impossible, at the beginning, to have a rough idea of the size of the automaton resulting of the code. Thus, it may be very amazing. For example, a first compilation may produce 200 states when only the emission of a local signal launching a neighbouring module appears to have been forgotten. When corrected the program may then lead to a 15 000 states automaton.

4.1.3 Instantaneous loop

Recall that the time assumption imposes that an instruction is executed within a null duration except if it specifies that it takes some time. This means that only the wait signal statement takes some actual time in ESTEREL. Therefore the loop constructor `loop ... end` is executed in a null time. When the body of a loop also takes no time, it may happen that an infinite loop has to run instantaneously, which is inconceivable. This is the case of the following example:

```
module Instantaneous_loop
output P, R:

loop
    emit P;
    emit R;
end.
```

or much more difficult to point out:

```
module Instantaneous_loop:
input S, J, Q, I;
output P, R;
loop
    present S then
        await J;
        emit  P
    else
        present Q then
            await I;
            emit R
        end
    end
end.
```

where if S and Q are not present there is not any time statement to be executed. This is the reason for the presence of an infinite loop in a null time or instantaneous loop. The compiler checks every loop and reports all anomalies found. Fortunately these errors are not too serious and for most of them, they are easy to fix. In the previous example, it would be enough to add:

```
else await case S case Q end
```

at the second `present` for if the signals S an Q are absent, they are waited for.

4.1.4 Causality errors problems

These problems are the most underhand because they have no equivalent in the usual programming languages. The compiler detects a causality error when the program has several meanings (non-determinism) or not any meaning. For example:

```
present S else emit S end
```

requires the signal S to be present and absent at the same "tick" of logic clock, that has no meaning and must be rejected.

The correct condition is as follows: a signal should not be read if it can still be written. In **n** branches of a parallel statement, if one of the awaited signals also appears as possibly emitted, the program is rejected. For example:

```
        present S then emit T end
    ||
        present T then emit S end
```

must be rejected because it has possibly two meanings, what infringes the rule of determinism which is is fundamental in the language.

These rules warrant the integrity of the ESTEREL semantics but the programmer has little chance to detect these causality errors before the compilation and avoid them. Often, the correction of these errors is impossible and requires to rewrite the entire program.

4.1.5 Complex algorithms implementation

This point does not constitute a problem but rather a limitation issued from the combinatory explosion of the number of states. In some applications, it would be desirable to refine the algorithms of synchronization to obtain a maximal use of the resources. But practically, we are often limited by the size of the automaton produced which becomes impossible to implement.

Look at an example where it has supplying tube of some product which moves on a rail above five tins which must form and warm up the product. The warm up times are different for each tin.

We can think of implementing an algorithm which allows the tube to go to the first ready tin or to the nearest tin having signalled its readiness. But practically some algorithms would be sufficiently complex to make the resulting automaton burst. First because a lot of signals are exchanged between various modules and next because the automaton serialization must take care of all the combinatory possible arrival of signals. We have whether a huge automaton (more than 15 000 states!) or it would be too big to be generated. We will be forced to use an algorithm easier to implement in ESTEREL, loosing a lot of efficiency. One possibility is to implement this algorithm in the host language, i.e. in the language produced by the ESTEREL compiler (C or ADA at the moment), but the synchronization problems would be difficult to solve. The point is still open.

4.2 TOWARDS A SOLUTION

The first thing to do in view of the tackled problem is to do the functional decomposition. This method is natural and well known but should not lead to an organic decomposition. As it is shown in [5], the organic decomposition where all physic processes are modelized by an ESTEREL module is obviously less efficient than a functional decomposition. As it was already mentioned, an important problem to solve is to find a way to memorize the state of a component to be accessible at any time by a process asking for it. All the work of this section consists in finding a solution to this major problem in ESTEREL. We will show that the solution is not unique.

4.2.1 Use of shared variables for memorization

A straightforward idea for a programmer confronted to the problem of the state memorization of components is to use state variables. Unfortunately, this method will not be employed simply since it is impossible, in ESTEREL, to share variables in several branches of a parallel statement. The limitation is obvious since no mechanism of control is available to warrant the integrity of the information.

4.2.2 Anarchical use of conditionals with valued signals

Since variables cannot be shared, it would be possible to use valuated signals to know about a component state. Therefore, signals would be used as global variables, i.e. reached by all the processes. The idea seems good but it requires introduction of a lot of conditionals to verify the state of these pseudo-variables. Let us come back to the first example where we had a tank filling containers and a robot loading and unloading these containers. Watch at the filling module operation.

```
module filling:
input container_load (boolean),
      container_full ;
output open_valve, close_valve;
loop
    if (? container_load) then
        emit open_valve;
        await container_full;
        emit close_valve
    else
        await container_load do
            emit open_valve;
            await container_full;
            emit close_valve
        end
    end
end.
```

In this example, the signal `container_load` is used as a global boolean variable. Each time a container has to be filled, it must be checked whether or not the container is actually present. If so, it is filled and this processing is restarted. Otherwise, it is waited for the next occurrence of the signal which will obviously be true (there is no need to check it). The mechanism is simple and natural but it has many and sometimes disastrous drawbacks.

Drawbacks

The biggest drawback of using conditionals is that they multiply by two or more the code of the resulting automaton since they always produce, on the contrary of **present**, a dynamic test in the automaton. Moreover, they require the programmer to copy out the code to execute. If in the part of code, signals are emitted to launch other processes, the number of states of the automaton will be multiplied by a ratio even greater since the compiler writes out the code of the module to launch at each place where the sending out of the launching signal may happen. If these conditionals are often used in the code, the automaton grows up very fast.

We have to know that the compiler does not evaluate the boolean expression in the conditional. It means that the conditional ends when one of these branches does. We must therefore pay attention to loop containing conditionals to avoid a branch with an **IF** terminating instantaneously. Otherwise we would have an infinite loop in a null time. Moreover, because the conditionals are not evaluated at the time of the compilation, it can produce the rejection of programs semantically correct for the programmer. Observe the following:

```
module Reject:
input OK (boolean), S, P;
output K, O;
loop
    if (? OK ) then
        emit K;
        await S;
        emit O
    end;
    if (? OK) else
        await P
    end
 end.
```

Since the bodies of the IF alternatives do not include statement spending time, this program will be rejected by the compiler because it detects an instantaneous loop. Indeed these few lines of code are a bit trivial but one can see some cases much more subtle where the cause of rejection is not obvious. The programmer believes and sees that its program might execute but the compiler refuses it! For all these reasons, the warning from the ESTEREL designers about the use of conditional statements is quite justified. Its use must be restricted and such a memorization of the component states must be avoided.

4.2.3 Blocking state process

In our new attempt to solve the problem of memorization of the state of a component we now introduce a method already given by Eve Coste-Manière [5], namely the memory. We generalized it and we call this the state process..

The aim of the state process is to receive the signal indicating that a resource is available and to signal it at a process asking for it. If the state process first receives a request for information, it will wait until it receives the signal indicating that a resource is available. Then it replies. The requesting process is suspended during this time. On a schema we have:

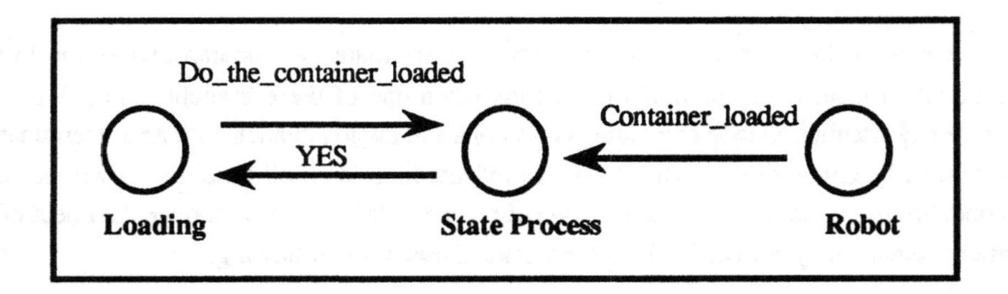

The state process can reply to the loading module from the moment where the robot has indicated to it that a container was loaded. Thus, we avoid the main problem of taking a signal when a process is not waiting for it. The state processes is in charge of watching, i.e. receiving signal, even if the process is not ready to receive it. Now, no signal is lost and the synchronization is ensured between the two processes. The mechanism may be compared with the "rendez-vous", except that the module signalling the availability of the resource does not wait, it sends signal and then does something else. The two modules may appear like the following:

```
module state_process:
input Do_the_container_loaded, Container_loaded;
output YES;

loop
    [
        await Do_the_container_loaded
    ||
        await Container_loaded
    ];
    emit YES
end.
```

```
module Loading:
input YES;
output Do_the_container_loaded;

loop
     .
     .

     emit Do_the_container_loaded;
     await immediate YES;

     .
     .

end.
```

where the **immediate** allows one to watch for the occurrence of the signal YES at the same ESTEREL clock tick where it was emitted.

This method introduces a lot of waiting statement in parallel for each of the state processes. Even if a signal does not logically occur, its presence is checked at each ESTEREL instant. Because of the asynchronous character of the application, nobody knows when the signal will occur. It is obvious that the verification is not done without cost. Moreover the exchange of signals between the modules requesting the resource, must be added.

Fortunately, this method is undoubtedly less costly (from the number of states point of view) that the use of classical conditionals but it creates a lot of useless states and transitions never used in the automaton. This solution remains the best, at our knowledge in some complex cases but there are some simple cases where the method introduced at the section §4.2.5 is clearly less costly. In order to differentiate the simple method using state processes, for the generalization, we will name blocking state process the one present in this section.

4.2.4 Generalization of the state process

One may meet some cases much more complex where one must record the state of a component. These complex cases may require a mechanism which allows one to know the state of components without waiting for a specific state. One no longer wants that the state process be suspended.

Let us come back to the example where we deal with two wagons wishing to go to a unique loading station. The station is a shared resource to be actually controlled. We will create a process to manage this resource. This process will give the permission to a wagon to go to the loading station. In a round robin way, the managing process has to ask each wagon if it is ready to move. The answer must not be delayed. The state process of a wagon must answer `YES_I_am_ready` or `I_am_not_ready`. In the case where the wagon is not ready, the managing process may ask the same question to the second wagon. Thus, the resource utilization is always maximal. The most efficient way to memorize the state of a wagon is to use an informative state process for each wagon. On a schema we have:

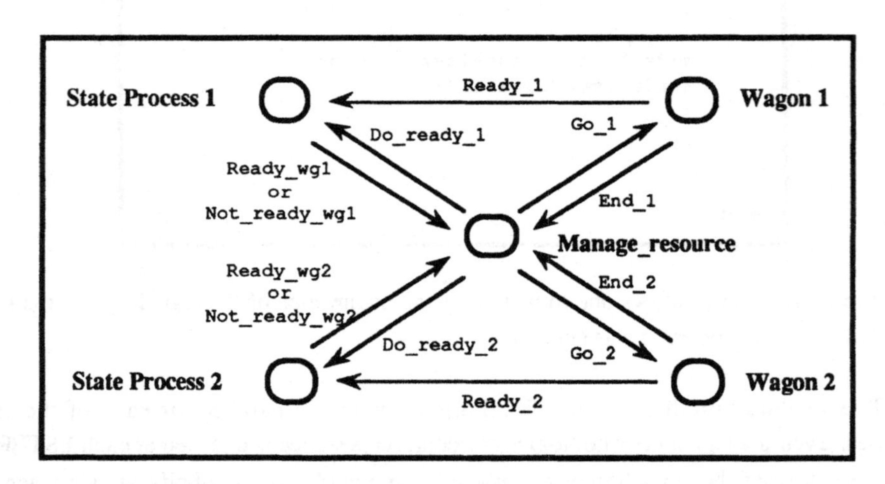

Unfortunately this mechanism is very costly, from the number of states concern, but in some cases it remains the best and often the only solution. The implementation of this generalization of state process may take several forms.

4.2.5 Temporal conditional

This method is merging the two previous ones. In fact, it uses conditionals like the state process. It has the form of a conditional but behaves like the state process without having additional process. Coming back to the first example with the loading module and the robot, we see what the code now looks like.

```
module Loading:
input Container_loaded (boolean);
output . . .

loop
    .
    .
    if (? Container_loaded) else await immediate Container_loaded end;
    .
    .
end.
```

The loading module checks the availability of the resource and it will wait until. The robot will only preserve the consistency of the signal value `container_load` so that the whole system keeps its consistency.

The main advantage of this method compared to the blocking state process, and the only functional difference, is that it eliminates the repeated wait of signals. We verify at only one moment that the resource is available and only if it is not, we wait until it occurs. There is no more useless waits and we eliminate all the useless states of the automaton which were introduced before by the state process. We keep the functionalities of the state process without its drawbacks and the form of the classical conditional, also without its associated drawbacks. In introducing the conditional, we paid attention not to put code inside. Not to put code inside conditionals avoids both the problem of the copied code and the problem of the automaton size. The conditional no more creates useless states, it adds only a dynamic test which is unavoidable.

The method works very well but unfortunately it cannot be generalized as the state process because it requires to add code inside the conditional and this would generate a too important automaton.

4.2.6 A well-balanced mixture of the introduced methods

The two methods we must hold are: the state process (the generalization) and the temporal conditional.

We will use the temporal conditional when we memorize the state of a component and wait for a precise state. Thus, the blocking state process is forgotten since it is less efficient. Nevertheless there are exceptions not well defined where the blocking state process produces less states in the resulting automata than the temporal conditional. This last point requires some explanations.

In the case where we only have to inquire about the state of a component, the generalization of the state process is the only mechanism at our disposal at the moment.

4.3 A few results of an application[2]

The results presented in this paper come from the development in ESTEREL of a large application having enough varied problems to be interesting. The process specification is due to Yvon Trinquet and Anne-Marie Leclech-Deplanche [12]. The aim of this section is not to describe the solution of the application but to give the reader an idea about the ESTEREL code and the produced automaton when programming a large real-time application.

The manufacturing process is constituted of three phases. The first one manufactures soluble bars using five tins over which a tube and a press circulate. The second one does the measuring and the mix of two liquids to which are added some soluble bars coming from the first unit. The last phase packages the product made in the second unit by putting it in a container used for the transport. The details of the process are available in [12]. All the examples in this paper are stemming from this application. We found the following problems:

- mutual exclusion;
- synchronization in asynchronous context;
- shared resources;
- protection of critical section

where each kind of events shown in §3 are present around elements as different as:

- robots;
- tanks;
- moving wagons;
- mixer;
- measure tank;
- conveyor belt;
- tins;
- oven;
- etc. . . .

[2] Recent improvements in the ESTEREL compiler:
A new ESTEREL compiler was very recently available, which brings some improvements to help in programming with the language. There is a new predefined input signal *tick*, which is by definition present in all reactions. Some persistency may now be given to a signal but the occurrences may still be memorized only once.

```
loop                          loop
      emit S;          -              emit S;          -        sustain S;
      await tick;             each tick;
end
```

The new **sustain** statement makes it possible to emit continuously a signal. These new primitives of the language were too recently introduced so that we were not able to use them in our real-time application. Nevertheless, it seems very useful to avoid some state processes or temporal conditionals.

By using the methods introduced in this paper, we succeed in making a simulator for each of the three phases building up the application. Nevertheless, it must be noticed that we abandoned the implementation of complex algorithms which would improve the execution for the reasons mentioned in §4.1.5 and because we included only two tins, out of five, in the first phase. As soon as we try to add the third tin, the ESTEREL compiler produces an automaton with more than 15 000 states! This is the table showing in a comparative way the resulting automata for each of the phases and the corresponding ESTEREL code:

	Information \ Phases	Bar	Product	Conditioning
Automata	states	3 001	145	2 776
	transitions	58 444	3 714	72 626
	actions	1 984	767	4 224
	simulator size	1.5 Mb	172 Kb	1.2 Mb
ESTEREL Code	modules	6	3	13
	input signals	12	11	20
	output signals	19	17	26
	code lines	186	132	361

5. CONCLUSION

Programming in ESTEREL a large real-time application with various specific problems issued from the domain of applications (automatic manufacturing process) shows the problem of an asynchronous reality opposed to a synchronous expression. Since the application has an asynchronous nature and the language was synchronous, we therefore searched to develop systematic methods to adapt synchronous programming to asynchronous context. In order to help us and to define accurately the problems, we classified the events by their nature so as to find a systematic implementation for each class.

We found systematic solutions for most of the problems but some persist and look like a shadow over the programmer who wants to use ESTEREL. If memorizing the state of a component is no more a problem, controlling the number of states produced by the compiler remains one. The programmer will always be haunt by the risk to make the automaton explose but, maybe, he is better driven than before.

This study comes within the framework of the development of the ELECTRE language we designed. ELECTRE is a reactive language which is intended to the expression of asynchronous applications. That is to say:

- actions have a no-null execution duration and they may be either resumed or restarted and they may be assigned a non-interruptible property
- events are complex entities which may be assigned properties so that either they are fleeting or they have single or multiple occurrences memorization.

Moreover, its operational semantics makes it possible to compile the ELECTRE programs into Finite State Automata. We are currently working on the translation from ELECTRE to ESTEREL.

6. BIBLIOGRAPHY

[1] G. Berry, G. Gonthier: *The Esterel Synchronous Programming Language: Design Semantics, Implementation;* Technical Report, ENSMP/INRIA, 1989.

[2] G. Berry, P. Couronné, G. Gonthier: *Programmation synchrone des systèmes réactifs: Le langage Esterel;* Technical Report, ENSMP/INRIA, 1987.

[3] A. Benveniste, P. Bournai, T. Gautier, P. Le Guernic: *Signal: A Data-Flow Oriented Language for Signal Processing;* IEEE Trans. on ASSP, ASSP-34, N°2, 1986, pp. 362-374.

[4] P. Caspi, N. Halbwachs, D. Pilaud, J. Plaice: *Lustre: A Declarative Language for Programming Synchronous Systems;* 14th ACM Symp. on Principles of Programming Languages, Munich, january 1987.

[5] E. Coste-Manière: *Synchronisme and asynchronisme dans la programmation des systèmes robotiques: apport du langage ESTEREL and de concepts objets* ; Doctor Thesis, Ecole des Mines de Paris, juillet 1991.

[6] D. Creusot, J. Perraud, O. Roux: *Le Système Electre;* Intern Report, LAN/ENSM n° 89.01, january, 1989.

[7] *ESTEREL V3: Language Reference Manual (V3.1r3);* CISI Ingénierie, december 1989.

[8] D. Harel: *Statecharts: A Visual Approach to Complex Systems;* Weizmann Institute of Science, Rehovot, Israël, 1984.

[9] D. Harel, A. Pnueli: *On the development of reactive Systems* ; NATO ASI Series vol. F13, Logics and Models of Concurrent Systems, Springer-Verlag Berlin Heidelberg, 1985.

[10] J. Perraud, O.Roux, M. Huou: *Operational Semantics of a kernel of the language ELECTRE* ; Theoretical Computer Science, N°100, november 1992, to appear.

[11] M. Richard: *Etudes des Langages Réactifs Synchrones: Application dans le langage ESTEREL* ; Master of Science, Computer Science Department of the Université Laval (Québec), april 1991.

[12] Y. Trinquet, A-M. Leclech-Deplanche: *Description d'un procédé de fabrication;* IUT of Nantes, G.E.I.I., march 1990.

The Real-Time Behaviour of Asynchronously Communicating Processes

Frank de Boer and Jozef Hooman
Eindhoven University of Technology
P.O. Box 513, 5600 Eindhoven
The Netherlands

October 23, 1991

Abstract

To describe the real-time behaviour of asynchronously communicating processes, a denotational semantics is formulated for an asynchronous version of CSP. In this real-time programming language channels are modeled by ("first in/first out") buffers to which messages are send and from which messages are read. The real-time reactive input/output behaviour of programs is formalized by a natural compositional semantics which is based on a simple and intuitive notion of timed traces of observable events. First we formulate a denotational semantics under the assumption that read actions of a program are observable. Next we abstract from these read events and give a modified semantics which is shown to be fully abstract with respect to this new notion of observable behaviour.

1 Introduction

In this paper we address the problem of a compositional characterization of the real-time behaviour of asynchronously communicating processes.

Much research has been focused on the theoretical foundations of concurrent systems of processes which interact by means of synchronous communication, as exemplified by CCS ([Mil80]). In general, a compositional semantics of such systems requires encoding of some branching information by trees, failure sets, in order to describe global non-determinism and the deadlock behaviour: simple traces of observable events are not sufficient ([BHR84,BKO88]). These models have been extended to characterize real-time versions of languages based on synchronous communication. Reed and Roscoe ([RR86]) give a hierarchy of timed models, based on a complete metric space structure. A fully abstract timed failure semantics for an extended CSP language has been developed in [GB87]. A fully abstract semantics for a real-time version of CSP based on linear histories appeared in [HGR87].

Recently, concurrent systems of asynchronously communicating processes gained interest. We mention the semantic research on Concurrent Logic Languages ([dBP90,dBP91, GCLS88,GMS89,Sar89,SR90]), and the work on ACSP ([JHJ90,JJH90]), an asynchronous version of CSP. For concurrent systems based on asynchronous communication as described by Dataflow, Concurrent Logic Languages and ACSP, it has been shown that for a compositional semantics some notion of traces of observable events suffices, no additional branching information is necessary ([dBP90,dBP91,JJH90,JK89,Jon85,Kok87]). In [BKPR91], it is shown that indeed this observation applies to any asynchronous concurrent system. The main objective of this paper is to develop a natural, fully abstract, compositional semantics of the real-time behaviour of asynchronously communicating processes, which is based on a intuitive notion of traces of observable events.

We study an asynchronous version of CSP, in which channels are modeled by ("first in/first out") buffers to which messages are send and from which messages are read. The operations of sending and receiving a message are independent, they can take place at different times. Real-time is incorporated in the programming language by delay-statements which suspend the execution for a certain period of time. Such a delay-statement is also allowed in the guard of a guarded command (similar to a delay-statement in the select-construct of Ada, [Ada83]). This enables us to program a time-out, that is, to restrict the waiting period for a communication and to execute an alternative statement if no communication is possible within a certain number of time units. The real-time behaviour is specified with respect to the assumption of maximal parallelism, which models the situation that every process has its own processor. Furthermore, we assume that both the operations of sending and receiving along a channel take some fixed amount of time. The time a send operation takes can be interpreted as the time it takes for a message to arrive at the buffer. Only when a message is available in the buffer it can be read by another process.

In the semantics a program is viewed as a black box connected with the outside world by means of input and output channels. Information is supplied by the environment via input channels, and the results are communicated to the environment via output channels. What we observe of the activity of a system are its reactions in time to inputs provided by the environment. First we consider these reactions to consist of both reading from an input channel and sending along an output channel. Then we show how to abstract from the read events, thus specifying a system solely in terms of its timed reactive input/output behaviour.

One of the main aims of the paper is to point out a basic difference between synchronous and asynchronous real-time, namely, that for a compositional semantics of asynchronous real-time we do not need to record additionally on what channels a process wants to read. In asynchronous real-time a description of a process in terms of its interactions with the environment provides all the necessary information both for compositionality and maximal progress. This basic difference we expect to be of relevance for the development of an axiomatic characterization of asynchronous real-time in, for example, temporal logic.

Our paper is organized as follows. The real-time programming language is introduced in section 2. In section 3 we give a compositional characterization of the timed reactive behaviour of a system described by the language, which includes events indicating when a message is read from an input channel. Next we define in section 4 a new notion of observable behaviour by abstracting from these read events and we discuss full abstaction. In the section 5 we formulate a fully abstract semantics for the timed reactive input/output behaviour of a system. The paper ends with a conclusion and a discussion of future work.

2 Syntax

Let *TIME* be some countable time domain of positive values and ∞ a special symbol, $\infty \notin TIME$. Let *CHAN* be a (nonempty) set of channel names. Let $I\!N$ be the set of natural numbers (including 0). The syntax of our programming language is given in Table 1.

Table 1: Syntax Programming Language

Statement	$S ::=$	**skip** $\mid$ **delay** d $\mid$ $c!!$ $\mid$ $c??$ $\mid$ $S_1; S_2$ $\mid$ G $\mid$ $\star G$ $\mid$ $S_1 \| S_2$	$d \in TIME$, $c \in CHAN$
Guarded-Command	$G ::=$	$[[]_{i=1}^{n} c_i?? \rightarrow S_i \; [] \; \mathbf{delay}\; d \rightarrow S]$	$d \in TIME \cup \{\infty\}$, $n \in I\!N$, $c_i \in CHAN$

We often use $[[]_{i=1}^{n} c_i?? \rightarrow S_i]$ as an abbreviation for $[[]_{i=1}^{n} c_i?? \rightarrow S_i \; [] \; \mathbf{delay}\; \infty \rightarrow S]$, if $n > 0$. If $n = 0$ in $[[]_{i=1}^{n} c_i?? \rightarrow S_i \; [] \; \mathbf{delay}\; d \rightarrow S]$ then we write $[\mathbf{delay}\; d \rightarrow S]$. Informally, the statements of this language have the following meaning:

Atomic statements

- **skip** terminates immediately.
- **delay** d suspends the execution for d time units.
- $c!!$ is used to send a signal along incident channel c. (In this paper we do not consider the value transmitted but only the fact of the transmission.) We assume that communication is asynchronous, that is, a send statement is never blocked (it need not wait for a communication partner). Furthermore, we assume that there is a (possibly infinite) buffer between sender and receiver.

- $c??$ is used to receive a signal along incident channel c. If there are no messages in the buffer (the number of send statements on c equals the number of receive statements on c), then $c??$ is suspended until a sending process executes a corresponding $c!!$.

Compound statements

- $S_1; S_2$ indicates sequential composition of statements S_1 and S_2.
- Guarded command $[[]_{i=1}^{n} c_i?? \rightarrow S_i \; [] \; \mathbf{delay} \; d \rightarrow S]$ is executed as follows: wait at most d time units until some input guard $c_i??$ becomes enabled, that is, until communication can actually occur along one of the c_i because its buffer is no longer empty. As soon as at least one of the c_i-communications is possible (before d time units have elapsed), one of these communications (non-deterministically chosen) is performed and thereafter the corresponding S_i is executed. If no guard becomes enabled within d time units after the start of the execution of the command, then S is executed.
- $\star G$ indicates repeated execution of guarded command G. Since we do not consider boolean guards in this paper, execution of $\star G$ never terminates.
- $S_1 \| S_2$ indicates parallel execution of S_1 and S_2 according to the maximal parallelism model.

Let $DCHAN$ be the set of channels extended with directional elements: $DCHAN = CHAN \cup \{c!! \mid c \in CHAN\} \cup \{c?? \mid c \in CHAN\}$.

Definition 2.1 (Channels Occurring in Statement) The set of (directional) channels occurring in a statement S, notation $dch(S)$, is defined as follows:

- $dch(\mathbf{skip}) = dch(\mathbf{delay}\; d) = \emptyset$,
- $dch(c!!) = \{c, c!!\}$, $dch(c??) = \{c, c??\}$,
- $dch(S_1; S_2) = dch(S_1) \cup dch(S_2)$,
- $dch([[]_{i=1}^{n} c_i?? \rightarrow S_i \; [] \; \mathbf{delay}\; d \rightarrow S]) = \bigcup_{i=1}^{n} dch(c_i??) \cup dch(S_i) \cup dch(S)$,
- $dch(\star G) = dch(G)$,
- $dch(S_1 \| S_2) = (dch(S_1) \cup dch(S_2)) \setminus (dch(S_1) \cap dch(S_2))$.

2.1 Syntactic Restrictions

The following syntactic constraints are imposed upon statements to guarantee that a channel connects exactly two processes:

- For $S_1; S_2$ we require that, for all $c \in CHAN$, $c! \in dch(S_1)$ implies $c? \notin dch(S_2)$, and $c? \in dch(S_1)$ implies $c! \notin dch(S_2)$.
- For a guarded command $G \equiv [[]_{i=1}^{n} c_i? \rightarrow S_i \; [] \; \mathbf{delay}\; d \rightarrow S_0]$ we require

- for all $i \in \{1, \ldots, n\}$ that $c_i! \notin dch(G)$, and
- for all $i, j \in \{0, 1, \ldots, n\}$, $i \neq j$, and $c \in CHAN$ that $c! \in dch(S_i)$ implies $c? \notin dch(S_j)$, and $c? \in dch(S_i)$ implies $c! \notin dch(S_j)$.

• For $S_1 \| S_2$ we require $dch(S_1) \cap dch(S_2) \subseteq CHAN$.

For instance, the programs $c!!; c??$, $[c_1?? \rightarrow d?? \,[]\, c_2?? \rightarrow c_1!!]$, $[c_1?? \rightarrow d?? \,[]\, c_2?? \rightarrow d!!]$, and $c!! \| c!!$ do not satisfy all of these requirements.

2.2 Basic Timing Assumptions

To determine the real-time behaviour of programs we have to make assumptions about:

- The execution time of atomic statements. In general, we will have bounds on the execution time. In this paper we assume that there is a fixed constant which gives the execution time, but the framework can be easily adapted to the more general case. Thus, for simplicity, we assume that a **delay** d statement takes exactly d time units. Furthermore we assume given a constant $T_{comm} > 0$ such that both the operations of sending and receiving along the channel c, when viewed as communications with the buffer associated with c, take T_{comm} time units. So when a process sends a message along a channel c it can be read only after T_{comm} time units. Note that the execution of an input statement may include a waiting period when no mesage is available. Since this waiting period depends on the environment of the communication statement, no assumptions can be made about its length.

- The overhead time required for compound programming constructs. Here we assume that there is no overhead for compound statements.

- How long a process is allowed to wait with the execution of enabled statements. That is, we have to make assumptions about the progress of actions. In this paper we consider the *maximal parallelism model* to represent the situation that each parallel process has its own processor. This implies maximal progress, that is, any enabled action will be executed as soon as possible. Hence, a process never waits with the execution of a local, non-communication, command or with an output statement. An input command can cause the process to wait, but only when its buffer is empty; as soon as the buffer is no longer empty the communication must take place. Thus the maximal parallelism model implies a minimal waiting period.

3 Compositional Semantics

3.1 Computational Model

Our model describes the responses in time of a system to communications on its input channels. More precisely, our formal model of real-time communication behaviour will record at each point of time along which input channels data sent by the environment has become available, along which output channels the process has sent, and finally, along which input channels the process has read a message. For simplicity, the time domain used in the semantics equals the domain *TIME* which is used in the syntax. In this paper we take a dense time domain; we assume that $TIME = \{\tau \in \mathbb{Q} \mid \tau > 0\}$, i.e., the positive rationals. Note that we do not observe the communications along internal channels.

For notational convenience, we use a special value ∞ with the usual properties such as $\infty \notin TIME$, for all $\tau \in TIME \cup \{\infty\}$, $\tau \leq \infty$, $\tau + \infty = \infty + \tau = \infty$, $\tau - \infty = \infty - \tau = \infty$, and $MAX(\infty, \tau) = MAX(\tau, \infty) = \infty$.
Let $\wp(A)$ denote the power set of a set A, that is, the set consisting of all subsets of A.

A model of a real-time computation is defined as follows:

Definition 3.1 (Model) Let $\tau_0 \in TIME \cup \{\infty\}$, and $(0, \tau_0] = \{\tau \in TIME \mid 0 < \tau \leq \tau_0\}$.
A *model* (of a real-time computation) is a mapping

$\sigma : (0, \tau_0] \rightarrow EVENT$,

where $EVENT = \{(I, O, R) \mid I, O, R \subseteq CHAN \text{ and } I \cap O = \emptyset\}$.
The set of all models is denoted by MOD.

Definition 3.2 (Duration of a Model) For a model σ with domain $(0, \tau_0]$, the *duration* of σ, denoted $|\sigma|$, is defined as $|\sigma| = \tau_0$.

Informally, a model σ represents the communication behaviour during the execution of a program at each point of time before and at $|\sigma|$, the termination time. If $|\sigma| < \infty$ then σ models a computation which terminates at time $|\sigma|$. Furthermore, $|\sigma| = \infty$ iff σ models a non-terminating computation. For any $\tau \leq |\sigma|$ we have $\sigma(\tau) = (I, O, R)$. Henceforth we use $\sigma(\tau).I$, $\sigma(\tau).O$, and $\sigma(\tau).R$ to denote, respectively, the first, second, and third component of $\sigma(\tau)$. If σ represents a computation of a program S then these fields have the following intuitive meaning.

- $c \in \sigma(\tau).I$ if at time τ a new message sent by the environment of S along channel c has become available. In general, there will be a difference in time between the moment at which the environment starts sending the message and the moment at which it becomes available for S.

- $c \in \sigma(\tau).O$ if S has sent a message along c and this message can be read by the enviroment starting at time τ.

- $c \in \sigma(\tau).R$ if S has just read a message from channel c at time τ. Thus τ represents the termination time of the read action.

Note that 0 is not included in the domain of a model. This has been done to avoid any overlap at sequential composition.

Henceforth, we use the following definitions.

Definition 3.3 (Model Concatenation) Define *concatenation* of two models σ_1 and σ_2, denoted $\sigma_1 \cdot \sigma_2$, as follows:

$$|\sigma_1 \cdot \sigma_2| = |\sigma_1| + |\sigma_2| \text{ and } \sigma_1 \cdot \sigma_2(\tau) = \begin{cases} \sigma_1(\tau) & \text{for all } \tau \leq |\sigma_1| \\ \sigma_2(\tau - |\sigma_1|) & \text{for all } \tau,\ |\sigma_1| < \tau \leq |\sigma_1| + |\sigma_2| \end{cases}$$

For two sets of models, Σ_1 and Σ_2, we define $\Sigma_1 \cdot \Sigma_2$ as the set $\{\sigma_1 \cdot \sigma_2 \mid \sigma_1 \in \Sigma_1 \wedge \sigma_2 \in \Sigma_2\}$.

Note that, for all models σ_1, σ_2, and σ_3,

- if $|\sigma_1| = \infty$ then $\sigma_1 \cdot \sigma_2 = \sigma_1$, and
- concatenation of models is associative, i.e., $(\sigma_1 \cdot \sigma_2) \cdot \sigma_3 = \sigma_1 \cdot (\sigma_2 \cdot \sigma_3)$; thus we can omit the brackets and write $\sigma_1 \cdot \sigma_2 \cdot \sigma_3$.

Definition 3.4 (Prefix) Define σ' to be a prefix of σ, notation, $\sigma' \preceq \sigma$, whenever $|\sigma'| \leq |\sigma|$ and, for all $\tau \leq |\sigma'|$, $\sigma'(\tau) = \sigma(\tau)$. Define $\sigma' \prec \sigma$ as an abbreviation of $\sigma' \preceq \sigma \wedge \sigma' \neq \sigma$.

Definition 3.5 (Buffer) A buffer is represented by a mapping which assigns to each channel an integer, indicating for an input channel the number of unread messages in the buffer. For infinite computations we use the special value ∞. Let Z be the set of integers. Define

$$BUF = CHAN \rightarrow Z \cup \{\infty\},$$

with typical element $b, b_1, \ldots$. For $b_1, b_2 \in BUF$, we define $b_1 + b_2$ such that $(b_1 + b_2)(c) = b_1(c) + b_2(c)$. Let b_∞ denote the buffer that assigns ∞ to every channel: $b_\infty(c) = \infty$, for each channel c. Note that $b + b_\infty = b_\infty$, for every buffer b.

Definition 3.6 (Buffer of Model) For every model σ we define a buffer $Buf(\sigma) \in BUF$ as follows: if $|\sigma| = \infty$ then $Buf(\sigma) = b_\infty$, and otherwise

$$Buf(\sigma)(c) = \#\{\tau \mid c \in \sigma(\tau).I\} - \#\{\tau \mid c \in \sigma(\tau).R\}.$$

Thus $Buf(\sigma)$ records for each input channel c the difference between the number of input events of the environment and the number of read events. Hence it represents the influence of σ on the buffers for the input channels.

Definition 3.7 (Variant of a Buffer) The *variant* of a buffer b with respect to a set of channels $cset \subseteq CHAN$ and an integer $z \in Z$, denoted by $(b|cset \mapsto z)$, is defined by

$$(b|cset \mapsto z)(c) = \begin{cases} z & \text{if } c \in cset \\ b(c) & \text{if } c \notin cset \end{cases}$$

Our semantic function will map statements into the set $BUF \rightarrow \wp(MOD)$. The intended meaning of $\sigma \in F(b)$, with $F \in BUF \rightarrow \wp(MOD)$, is that σ represents a computation starting from a state in which the number of messages present in the buffer of an input channel c is given by $b(c)$.

3.2 Semantic Function

In this section we give a compositional semantics for our programming language. The meaning of a program S, denoted by $\mathcal{M}(S)$, associates with each element $b \in BUF$, a set of models representing the possible computations of S starting at time 0 from a state in which for each input channel c the contents of its buffer is given by $b(c)$. Define $\mathcal{M}(S)(b) = \emptyset$ for any program S which is syntactically incorrect, i.e., does not satisfy the syntactic restrictions from section 2.1. For a syntactically correct program S and a buffer $b \in BUF$ we define $\mathcal{M}(S)(b)$ by induction on the structure of S according to the grammar in Table 1.

Skip

A skip-statement terminates immediately, which is expressed by a model that has termination time 0.

$$\boxed{\mathcal{M}(\textbf{skip})(b) = \{\sigma \mid |\sigma| = 0\}}$$

Delay

To define the meaning of a delay-statement we use the following definition of an "idling" computation:

Definition 3.8 Define

$IDLE(\sigma)$ iff for all $\tau \leq |\sigma|$: $\sigma(\tau).O = \sigma(\tau).R = \emptyset$.

Note that this predicate does not restrict $\sigma(\tau).I$, since that would restrict the environment, the output actions of which cannot be controlled by the process. A statement **delay** d terminates after exactly d time units.

$$\boxed{\mathcal{M}(\textbf{delay } d)(b) = \{\sigma \mid IDLE(\sigma) \text{ and } |\sigma| = d\}}$$

Thus during a delay period only communication events on input channels may occur, the process itself idles.

Send

To describe the semantics of a send statement $c!!$ we introduce the following definition:

Definition 3.9 For every channel c define

$$SEND(c) = \{\sigma \mid \text{for all } \tau < |\sigma|\text{: } \sigma(\tau).O = \sigma(\tau).R = \emptyset,\ \sigma(|\sigma|).O = \{c\},\ \sigma(|\sigma|).R = \emptyset, \text{ and } |\sigma| = T_{comm}\}$$

The semantics of $c!!$ is then simply defined as

$$\boxed{\mathcal{M}(c!!)(b) = SEND(c)}$$

Receive

To define the meaning of an input statement $c??$ we consider the following two possibilities. If the execution of $c??$ starts with a non-empty buffer of channel c, i.e., $b(c) > 0$, then the execution of $c??$ consists of reading a message from the buffer. This is modeled by the following definition:

Definition 3.10 For every channel c we define

$$READ(c) = \{\sigma \mid \text{for all } \tau < |\sigma|\text{: } \sigma(\tau).O = \sigma(\tau).R = \emptyset,\ \sigma(|\sigma|).O = \emptyset,\ \sigma(|\sigma|).R = \{c\}, \text{ and } |\sigma| = T_{comm}\}$$

On the other hand, if the buffer of channel c is initially empty, i.e., $b(c) = 0$, then the execution of $c??$ consists of a waiting period followed by, in case a message has arrived, a period during which the message is read. The notion of a waiting period is defined as follows:

Definition 3.11 For every buffer b and channel c we define

$$WAIT_b(c) = \{\sigma \mid IDLE(\sigma) \text{ and } ARR_{b,c}(\sigma)\}$$

where $ARR_{b,c}(\sigma)$ iff

for all $\sigma' \prec \sigma$: $(b + Buf(\sigma'))(c) = 0$ and $|\sigma| < \infty$ implies $(b + Buf(\sigma))(c) > 0$.

So during a waiting period the process idles and waits for a message to arrive. The waiting period is finite if a message has arrived, which is indicated by $(b + Buf(\sigma))(c) > 0$. Note that the waiting period is infinite in case no message arrives.
The semantics of an input statement $c??$ is now defined as follows:

$$\boxed{\mathcal{M}(c??)(b) = WAIT_b(c) \cdot READ(c)}$$

Observe that if $b(c) > 0$ then $(b + Buf(\sigma'))(c) > 0$, and thus $WAIT_b(c) = \{\sigma \mid |\sigma| = 0\}$. Hence $b(c) > 0$ implies $WAIT_b(c) \cdot READ(c) = READ(c)$.

Sequential Composition

To describe the semantics of the sequential composition of two statements we introduce the following semantic operator:

Definition 3.12 For $F_1, F_2 \in BUF \rightarrow \wp(MOD)$ we define $SEQ(F_1, F_2) \in BUF \rightarrow \wp(MOD)$ by:

$$SEQ(F_1, F_2)(b) = \{\sigma_1 \cdot \sigma_2 \mid \sigma_1 \in F_1(b) \text{ and } \sigma_2 \in F_2(b + Buf(\sigma_1))\}$$

Thus a model $\sigma \in SEQ(F_1, F_2)(b)$ can be decomposed as the concatenation of two models σ_1 and σ_2, with $\sigma_1 \in F_1(b)$ and σ_2 being a computation of S_2 which starts from the buffer b updated by the input and read events of σ. Note that if S_1 does not terminate then $|\sigma_1| = \infty$, and thus $\sigma_1 \cdot \sigma_2 = \sigma_1$. Using the SEQ operator, sequential composition is straightforward:

$$\boxed{\mathcal{M}(S_1; S_2) = SEQ(\mathcal{M}(S_1), \mathcal{M}(S_2))}$$

Guarded Command

In order to define the semantics of a guarded statement $G \equiv [\,[]_{i=1}^{n} c_i?? \rightarrow S_i\, []\, \mathbf{delay}\, d \rightarrow S]$ we first introduce the following abbreviations. Let $\bar{c} = \{c_1, \ldots, c_n\}$. $WAIT(\bar{c}, c_i, d)$ gives for every buffer b a set of models representing a waiting period for input, which takes less than d time units, and the availability of a message on channel c_i before d:

$WAIT(\bar{c}, c_i, d)(b) = \{\sigma \mid IDLE(\sigma),\ AVAIL(b, \bar{c}, c_i, \sigma), \text{ and } |\sigma| < d\}$, where

$AVAIL(b, \bar{c}, c_i, \sigma)$ iff

for all $c \in \bar{c}$, for all $\sigma' \prec \sigma$: $(b + Buf(\sigma'))(c) = 0$ and $(b + Buf(\sigma))(c_i) > 0$.

$TIMEOUT(\bar{c}, d)(b) = \{\sigma \mid IDLE(\sigma),\ |\sigma| = d, \text{ and for all } c \in \bar{c},\ (b + Buf(\sigma))(c) = 0\}$

Then define

$$\boxed{\begin{array}{rcl} \mathcal{M}(G) & = & \bigcup_{i=1}^{n} SEQ(WAIT(\bar{c}, c_i, d), \mathcal{M}(c_i??; S_i)) \\ & & \cup\ SEQ(TIMEOUT(\bar{c}, d), \mathcal{M}(S)) \end{array}}$$

Iteration

Every computation of $\star G$ either consists of a finite number of computations from G where the last one is non-terminating, or it consists of an infinite number of computations from G. Hence, the semantics of the iteration construct is defined recursively as follows:

$$\boxed{\mathcal{M}(*G) = SEQ(\mathcal{M}(G), \mathcal{M}(*G))}$$

This recursive definition is justified by defining $\mathcal{M}(*G)$ as the greatest fixed point of the monotonic operator Γ on the complete lattice $(BUF \rightarrow \wp(MOD)) \rightarrow (BUF \rightarrow \wp(MOD))$ which transforms F into $SEQ(\mathcal{M}(G), F)$. The ordering of the complete lattice is given by the pointwise extension to functions of set inclusion.

Parallel Composition

In order to define the semantic operator corresponding to the parallel composition $S_1 \| S_2$ we first introduce the following definitions.

Definition 3.13 We define a predicate $NoOutputLost(\sigma_1, \sigma_2)$ to guarantee that no messages sent are lost. This predicate expresses that at each moment the input events of σ_1 are required to contain the output events of σ_2 and vica versa. Formally,

$NoOutputLost(\sigma_1, \sigma_2)$ iff for all $\tau \leq MIN(|\sigma_1|, |\sigma_2|)$,

$$\sigma_1(\tau).O \subseteq \sigma_2(\tau).I \text{ and } \sigma_2(\tau).O \subseteq \sigma_1(\tau).I.$$

The predicate $SameEnv(\sigma_1, \sigma_2)$ expresses that the models σ_1 and σ_2 make the same assumptions about input events generated by the environment.

$SameEnv(\sigma_1, \sigma_2)$ iff for all $\tau \leq MIN(|\sigma_1|, |\sigma_2|)$,

$$\sigma_1(\tau).I \setminus \sigma_2(\tau).O = \sigma_2(\tau).I \setminus \sigma_1(\tau).O.$$

Observe that if $NoOutputLost(\sigma_1, \sigma_2)$ and $SameEnv(\sigma_1, \sigma_2)$ then for $\tau \leq MIN(|\sigma_1|, |\sigma_2|)$, $(\sigma_1(\tau).I \cup \sigma_2(\tau).I) \setminus (\sigma_1(\tau).O \cup \sigma_2(\tau).O) = \sigma_1(\tau).I \setminus \sigma_2(\tau).O = \sigma_2(\tau).I \setminus \sigma_1(\tau).O$, since $\sigma_i(\tau).I \cap \sigma_i(\tau).O = \emptyset$.

Definition 3.14 Define the extension σ^+ of a computation σ as follows: $|\sigma^+| = \infty$, and

$$\sigma^+(\tau) = \begin{cases} \sigma(\tau) & \text{if } \tau \leq |\sigma| \\ (\emptyset, \emptyset, \emptyset) & \text{otherwise} \end{cases}$$

Next we define the parallel operator on models.

Definition 3.15 We define $\sigma_1 \| \sigma_2$ such that $|\sigma_1 \| \sigma_2| = MAX(|\sigma_1|, |\sigma_2|)$ and, for all $\tau \leq |\sigma_1 \| \sigma_2|$,

$$\begin{aligned} (\sigma_1 \| \sigma_2)(\tau).I &= (\sigma_1^+(\tau).I \cup \sigma_2^+(\tau).I) \setminus (\sigma_1^+(\tau).O \cup \sigma_2^+(\tau).O) \\ (\sigma_1 \| \sigma_2)(\tau).O &= \sigma_1^+(\tau).O \cup \sigma_2^+(\tau).O \\ (\sigma_1 \| \sigma_2)(\tau).R &= \sigma_1^+(\tau).R \cup \sigma_2^+(\tau).R \end{aligned}$$

This leads to the following definition of parallel composition on sets of models.

Definition 3.16 For $\Sigma_1, \Sigma_2 \subseteq MOD$, we define
$\Sigma_1 \| \Sigma_2 = \{\sigma_1 \| \sigma_2 \mid \sigma_1 \in \Sigma_1 \wedge \sigma_2 \in \Sigma_2 \wedge \mathit{NoOutputLost}(\sigma_1, \sigma_2) \wedge \mathit{SameEnv}(\sigma_1, \sigma_2)\}$.

Finally we define an operator which models the closure and hiding of a set of (internal) channels *cset*, that is, no external inputs can influence the actions on these channels and the read and output events of a program on channels in *cset* are hidden.

Definition 3.17 Let Σ be a set of models and $cset \subseteq CHAN$, then $Hide(cset, \Sigma)$ deletes from those computations of Σ in which there occur no input events on channels of *cset*, the output events and read events on *cset*. The set *cset* is intended to represent the set of internal channels. Formally,

$$\begin{aligned} Hide(cset, \Sigma) = \{\sigma \mid & \text{ there exists a } \sigma_0 \in \Sigma \text{ such that } |\sigma_0| = |\sigma|, \text{ for all } \tau \leq |\sigma|: \\ & \sigma(\tau).I \setminus cset = \sigma_0(\tau).I,\ \sigma(\tau).O = \sigma_0(\tau).O \setminus cset, \text{ and} \\ & \sigma(\tau).R = \sigma_0(\tau).R \setminus cset\} \end{aligned}$$

Note that $\sigma(\tau).I \setminus cset = \sigma_0(\tau).I$ implies $\sigma_0(\tau).I \cap cset = \emptyset$, i.e., there are no external inputs on channels from *cset* in σ_0.
Define $jchan = dch(S_1) \cap dch(S_2)$, i.e., the set of joint channels connecting S_1 and S_2. The semantics of parallel composition can now be formulate now as:

$$\boxed{\mathcal{M}(S_1 \| S_2)(b) = Hide(jchan, \mathcal{M}(S_1)(b|jchan \mapsto 0) \parallel \mathcal{M}(S_2)(b|jchan \mapsto 0))}$$

4 Abstraction from Read Events

In the previous section we have defined a compositional characterization of some notion of observable behaviour, which includes the observation of when a system reads from a channel. However, one might argue that reading from a channel should be considered as an internal action. In this section we discuss the problem of formulating a fully abstract compositional characterization of the observable behaviour which involves only the input and output events. First we define a new notion of observable behaviour, using the following definition.

Definition 4.1 For a model $\sigma \in MOD$, we define its restriction to the I- and O-fields, denoted by $\sigma \ominus R$, as $|\sigma \ominus R| = |\sigma|$, and for all $\tau \leq |\sigma|$, $(\sigma \ominus R)(\tau).I = \sigma(\tau).I$,
$(\sigma \ominus R)(\tau).O = \sigma(\tau).O$.

We associate with each statement S and buffer b a new observable behaviour by restricting each computation of $\mathcal{M}(S)(b)$ to input and output events. Formally,

Definition 4.2 (Observable Behaviour) For any statement S and buffer b, we define the observable behaviour of S with respect to b, denoted by $\mathcal{O}(S)(b)$, as

$$\mathcal{O}(S)(b) = \{\sigma \ominus R \mid \sigma \in \mathcal{M}(S)(b)\}$$

To define full abstraction, the notion of a *context* is used, as defined by Table 2. Observe

Table 2: Syntax Context

$C ::=$	$[\,] \mid$ **skip** $\mid$ **delay** $d \mid c!! \mid c?? \mid$	$d \in TIME,\ c \in CHAN$
	$C_1; C_2 \mid G \mid \star G \mid C_1 \| C_2$	
$G ::=$	$[[\!]_{i=1}^{n}\ c_i?? \rightarrow C_i\ [\!]$ **delay** $d \rightarrow C]$	$d \in TIME \cup \{\infty\}$,
		$n \in I\!N,\ c_i \in CHAN$

that the only new construct is $[\ ]$ which serves as an "open place" for which we can substitute statements to obtain a program. We often denote a context by $C[\,]$ to emphasize that there is an open place, and use $C[S]$ to denote the context $C[\,]$ where every occurrence of $[\,]$ is replaced by S.

Definition 4.3 (Full Abstraction) A semantic function $\mathcal{M}$ is *fully abstract* with respect to a notion of observable behaviour $\mathcal{O}$ if for every two statements S_1 and S_2,
$\mathcal{M}(S_1) = \mathcal{M}(S_2)$ iff for every context $C[\,]$; $\mathcal{O}(C[S_1]) = \mathcal{O}(C[S_2])$.

We show that the semantics $\mathcal{M}$ is *not* fully abstract with respect to the observable behaviour $\mathcal{O}$ from defintion 4.2. Consider, for some $d \in TIME$, the statements

$$S_1 \equiv c_1??; *[\textbf{delay}\ d \rightarrow \textbf{skip}]$$

and

$$S_2 \equiv c_2??; *[\textbf{delay}\ d \rightarrow \textbf{skip}]$$

Clearly, $\mathcal{M}(S_1) \neq \mathcal{M}(S_2)$. However, for every context $C[\,]$, $\mathcal{O}(C[S_1]) = \mathcal{O}(C[S_2])$. Hence the semantics $\mathcal{M}$ is too distinguishing with respect to abstraction from read events. On the other hand, the semantics $\mathcal{O}$ identifies too many programs and hence is not compositional. Consider, for instance, the statements

$$S_1 \equiv \textbf{delay}\ T_{comm}; [c?? \rightarrow \textbf{skip}\,[\!]\,a?? \rightarrow \textbf{skip}] \| a!!$$

and

$$S_2 \equiv \textbf{delay}\ 2 \times T_{comm}$$

Observe that $\mathcal{O}(S_1) = \mathcal{O}(S_2)$, since these two programs have the same termination time and do not perform any output action. But $\mathcal{O}(S_1; c??) \neq \mathcal{O}(S_2; c??)$, because for a buffer b with $b(c) = 1$ we have that $\mathcal{O}(S_1; c??)(b)$ contains a $\hat{\sigma}$ with $|\hat{\sigma}| = \infty$ (by performing the $c??$ action in S_1), whereas for any $\sigma \in \mathcal{O}(S_2; c??)(b)$ we have $|\sigma| < \infty$. Thus we can find a context, namely $C[\ ] \equiv [\ ]; c??$, such that $\mathcal{O}(S_1) = \mathcal{O}(S_2)$, but $\mathcal{O}(C[S_1]) \neq \mathcal{O}(C[S_2])$. Hence $\mathcal{O}$ is not fully abstract with respect to $\mathcal{O}$! The problem is that $\mathcal{O}$ is not compositional, i.e., the meaning of a compound programming construct cannot be defined in terms of the meaning of the components. In the next section we formulate a fully abstract semantics by giving a compositional description of $\mathcal{O}$.

5 A Fully Abstract Semantics

In this section we define a semantic function $\mathcal{M}'$ which is fully abstract with respect to $\mathcal{O}$. To formulate this semantics, which abstracts from read events, we introduce the following new notion of a model:

Definition 5.1 Let $EVENT_{I/O}$ denote the set $\{(I,O) \mid I,O \subseteq CHAN\}$. The set of computational models, $MOD_{I/O}$, is then defined by
$MOD_{I/O} = \{\hat{\sigma} \mid \hat{\sigma} : (0,\tau_0] \to EVENT_{I/O}, \text{ with } \tau_0 \in TIME \cup \{\infty\}\ \}$.

We use the same notation for both the elements of MOD and $MOD_{I/O}$ (although for clarity we often use $\hat{\sigma}$, $\hat{\sigma}_1$, ... for elements of $MOD_{I/O}$). Actually, we can view $MOD_{I/O}$ as a subset of MOD, by identifying an element of $MOD_{I/O}$ with an element of MOD for which at each moment the set of read events is empty. Thus we can use earlier definitions for elements of MOD also for models from $MOD_{I/O}$. Note that if $\sigma \in MOD$ then $\sigma \ominus R \in MOD_{I/O}$.

Definition 5.2 Define $EnvOut(\hat{\sigma}) \in BUF$ by
$EnvOut(\hat{\sigma})(c) = \#\{\tau \mid c \in \hat{\sigma}(\tau).I\}$, for all c.

To formulate a compositional description of $\mathcal{O}$ we introduce the following extension of the notion of a model.

Definition 5.3 Define MOD_B as follows:
$MOD_B = \{(\hat{\sigma}, b) \mid \hat{\sigma} \in MOD_{I/O}, b \in BUF, \text{ and if } |\hat{\sigma}| = \infty \text{ then } b = b_\infty\}$

The intended meaning of a pair $(\hat{\sigma}, b)$ is that b specifies for each channel the contents of its buffer after the computation $\hat{\sigma}$. In case $\hat{\sigma}$ is infinite this additional information about the contents of the buffers is irrelevant and set to ∞.

We now describe how to modify the semantics $\mathcal{M}$ to a compositional and fully abstract (with respect to $\mathcal{O}$) semantics $\mathcal{M}'$ which associates with a statement and a buffer a set of computations of MOD_B. First we give a non-compositional definition of $\mathcal{M}'$.

Definition 5.4 Consider a program S and a buffer b. Observe that after a terminating computation $\sigma \in \mathcal{M}(S)(b)$ the resulting buffer is given by $b + Buf(\sigma)$. This leads to the following definition.

$$\mathcal{M}'(S)(b) = \{(\sigma \ominus R, b + Buf(\sigma)) \mid \sigma \in \mathcal{M}(S)(b)\}$$

Note that for $(\hat{\sigma}, b_1) \in \mathcal{M}'(S)(b)$ we have that $b_1 = b_\infty$ iff $b = b_\infty$ or $|\hat{\sigma}| = \infty$.

Then we show that $\mathcal{M}'$ can be defined compositionally by providing a semantic operator for each syntactic operator from the syntax of the programming language. The proof that this new semantic operator indeed corresponds to $\mathcal{M}'$ is only given for sequential composition. The proofs for other constructs are similar.

Sequential Composition

Define the following semantic operator for sequential composition:

$$SEQ_B(F_1, F_2)(b) = \{(\hat{\sigma}_1 \cdot \hat{\sigma}_2, b_2) \mid \text{there exists a } b_1 \text{ such that } (\hat{\sigma}_1, b_1) \in F_1(b) \text{ and } (\hat{\sigma}_2, b_2) \in F_2(b_1)\}$$

The next lemma expresses that this operator corresponds to sequential composition.

Lemma 5.5 For two statements S_1 and S_2 we have

$$\boxed{\mathcal{M}'(S_1; S_2) = SEQ_B(\mathcal{M}'(S_1), \mathcal{M}'(S_2))}$$

Proof: We use the following properties which can be proved easily:
$Buf(\sigma_1) + Buf(\sigma_2) = Buf(\sigma_1 \cdot \sigma_2)$ and $(\sigma_1 \ominus R) \cdot (\sigma_2 \ominus R) = (\sigma_1 \cdot \sigma_2) \ominus R)$.

$\mathcal{M}'(S_1; S_2)(b) =$

$\{(\sigma \ominus R, b + Buf(\sigma)) \mid \sigma \in \mathcal{M}(S_1; S_2)(b)\} =$

$\{(\sigma \ominus R, b + Buf(\sigma)) \mid \sigma \in SEQ(\mathcal{M}(S_1), \mathcal{M}(S_2))(b)\} =$

$\{(\sigma \ominus R, b + Buf(\sigma)) \mid \sigma = \sigma_1 \cdot \sigma_2 \text{ with } \sigma_1 \in \mathcal{M}(S_1)(b) \text{ and } \sigma_2 \in \mathcal{M}(S_2)(b + Buf(\sigma_1))\} =$

$\{((\sigma_1 \cdot \sigma_2) \ominus R, b + Buf(\sigma_1 \cdot \sigma_2)) \mid \sigma_1 \in \mathcal{M}(S_1)(b) \text{ and } \sigma_2 \in \mathcal{M}(S_2)(b + Buf(\sigma_1))\} =$

$\{((\sigma_1 \cdot \sigma_2) \ominus R), b_2) \mid \sigma_1 \in \mathcal{M}(S_1)(b), \sigma_2 \in \mathcal{M}(S_2)(b + Buf(\sigma_1)), \text{ and}$
$b_2 = b + Buf(\sigma_1) + Buf(\sigma_2)\} =$

$\{((\sigma_1 \ominus R) \cdot (\sigma_2 \ominus R), b_2) \mid \sigma_1 \in \mathcal{M}(S_1)(b), \sigma_2 \in \mathcal{M}(S_2)(b + Buf(\sigma_1)), \text{ and}$
$b_2 = b + Buf(\sigma_1) + Buf(\sigma_2)\} =$

$\{(\hat{\sigma}_3 \cdot \hat{\sigma}_4, b_2) \mid \text{there exist } b_1, \sigma_1, \text{ and } \sigma_2 \text{ such that } \hat{\sigma}_3 = \sigma_1 \ominus R, \sigma_1 \in \mathcal{M}(S_1)(b),$
$b_1 = b + Buf(\sigma_1), \hat{\sigma}_4 = \sigma_2 \ominus R, \sigma_2 \in \mathcal{M}(S_2)(b_1), b_2 = b_1 + Buf(\sigma_2)\} =$

$\{(\hat{\sigma}_3 \cdot \hat{\sigma}_4, b_2) \mid \text{there exists a } b_1 \text{ such that } (\hat{\sigma}_3, b_1) \in \mathcal{M}'(S_1)(b) \text{ and } (\hat{\sigma}_4, b_2) \in \mathcal{M}'(S_2)(b_1)\} =$

$SEQ_B(\mathcal{M}'(S_1), \mathcal{M}'(S_2))(b)$ □

Guarded Command

To give a compositional formulation of the semantics of a guarded statement we use the following definitions. Let $G \equiv [\|_{i=1}^{n} c_i?? \rightarrow S_i \| \textbf{delay } d \rightarrow S]$.
$WAIT_B(\bar{c}, c_i, d)(b) = \{(\hat{\sigma}, b + EnvOut(\hat{\sigma})) \mid IDLE_B(\hat{\sigma}),\ AVAIL_B(b, \bar{c}, c_i, \hat{\sigma}), \text{ and } |\hat{\sigma}| < d\}$,
where

$IDLE_B(\hat{\sigma})$ iff for all $\tau \leq |\hat{\sigma}|$, $\hat{\sigma}(\tau).O = \emptyset$,

$AVAIL_B(b, \bar{c}, c_i, \hat{\sigma})$ iff

for all $c \in \bar{c}$, for all $\hat{\sigma}' \prec \hat{\sigma}$: $(b + EnvOut(\hat{\sigma}'))(c) = 0$ and $(b + EnvOut(\hat{\sigma}))(c_i) > 0$.

$TIMEOUT_B(\bar{c}, d)(b) = \{(\hat{\sigma}, b) \mid IDLE_B(\hat{\sigma}),\ |\hat{\sigma}| = d$, and
for all $c \in \bar{c}$, $(b + EnvOut(\hat{\sigma}))(c) = 0\}$.

Lemma 5.6 For a guarded command G we have

$$\boxed{\begin{array}{rcl} \mathcal{M}'(G) & = & \bigcup_{i=1}^{n} SEQ_B(WAIT_B(\bar{c}, c_i, d), \mathcal{M}'(c_i??; S_i)) \\ & \cup & SEQ_B(TIMEOUT_B(\bar{c}, d), \mathcal{M}'(S)) \end{array}}$$

Iteration

The semantics of the iteration construct can be defined as follows.

Lemma 5.7 For the iteration $\star G$ we have

$$\boxed{\mathcal{M}'(*G) = SEQ_B(\mathcal{M}'(G), \mathcal{M}'(*G))}$$

Parallel Composition

For $\hat{\sigma}_1, \hat{\sigma}_2 \in MOD_{I/O}$, define $NoOutputLost_B(\hat{\sigma}_1, \hat{\sigma}_2)$ and $SameEnv_B(\hat{\sigma}_1, \hat{\sigma}_2)$ similar to definition 3.13. We define $\hat{\sigma}_1 \parallel \hat{\sigma}_2$ as in definition 3.15, omitting the clause for the R-field. For two buffers b_1 and b_2 we define $b_1 \parallel b_2$ as b_∞ if $b_1 = b_\infty$ or $b_2 = b_\infty$. Otherwise we have $(b_1 \parallel b_2)(c) = MIN(b_1(c), b_2(c))$, for all c. (Notice that each input channel of a parallel system can be accessed by at most one component. Thus either $b_1(c)$ or $b_2(c)$ is less than the number of input events on c.) This leads to the following definition of parallel composition on sets of models.

Definition 5.8 For Σ_1, $\Sigma_2 \subseteq MOD_B$, define

$$\Sigma_1 \|_B \Sigma_2 = \{(\hat{\sigma}_1 \| \hat{\sigma}_2, b_1 \parallel b_2) \mid (\hat{\sigma}_1, b_1) \in \Sigma_1 \wedge (\hat{\sigma}_2, b_2) \in \Sigma_2 \wedge NoOutputLost_B(\hat{\sigma}_1, \hat{\sigma}_2) \wedge SameEnv_B(\hat{\sigma}_1, \hat{\sigma}_2)\}.$$

The hiding operator is modified as follows:

Definition 5.9 Let $\Sigma \subseteq MOD_B$ and $cset \subseteq CHAN$, then $Hide_B(b, cset, \Sigma)$ hides from those computations of Σ in which there occur no input events on channels of *cset*, the output events on *cset*. Furthermore, with respect to the buffer associated with a computation the channels of *cset* are hidden by setting them to the buffer before execution b plus the new inputs by the environment. The set *cset* is intended to represent the set of internal channels. Formally,

$$
\begin{aligned}
Hide_B(b, cset, \Sigma) = \{(\hat{\sigma}_1, b_1) \mid\ & \text{there exists a } (\hat{\sigma}_0, b_0) \in \Sigma \text{ such that } |\hat{\sigma}_1| = |\hat{\sigma}_0|, \\
& \text{for all } \tau \leq |\hat{\sigma}_1|\text{: } \hat{\sigma}_1(\tau).I \setminus cset = \hat{\sigma}_0(\tau).I \\
& \qquad \hat{\sigma}_1(\tau).O = \hat{\sigma}_0(\tau).O \setminus cset \text{ and} \\
& b_1(c) = \begin{cases} b_0(c) & \text{if } c \notin cset \text{ or } b_0(c) = \infty \\ b(c) + EnvOut(\hat{\sigma}_1)(c) & \text{if } c \in cset \text{ and } b_0(c) \neq \infty \end{cases} \\
\}
\end{aligned}
$$

Define $jchan = dch(S_1) \cap dch(S_2)$. The semantics of parallel composition can now be formulated as follows.

Lemma 5.10 For any two statements S_1 and S_2,

$$\boxed{\mathcal{M}'(S_1 \| S_2)(b) = Hide_B(b, jchan, \mathcal{M}'(S_1)(b|jchan \mapsto 0) \|_B \mathcal{M}'(S_2)(b|jchan \mapsto 0))}$$

5.1 The Proof of Full Abstraction

We prove the following theorem stating that the equivalence induced by $\mathcal{M}'$ equals the congruence generated by $\mathcal{O}$, i.e., $\mathcal{M}'$ is fully abstract with respect to $\mathcal{O}$.

Theorem 5.11 For every two statements S_1 and S_2 we have

$$\mathcal{M}'(S_1) = \mathcal{M}'(S_2) \text{ iff for every context } C[\,];\ \mathcal{O}(C[S_1]) = \mathcal{O}(C[S_2]).$$

Proof:

$\Rightarrow$:

Define an abstraction function α, which gives the observable events of a model from MOD_B by removing the buffer, as follows: $\alpha \in (BUF \rightarrow MOD_B) \rightarrow (BUF \rightarrow MOD_{I/O})$ with

$$\alpha(F)(b) = \{\hat{\sigma} \in MOD_{I/O} \mid \text{there exists a } b' \text{ such that } (\hat{\sigma}, b') \in F(b)\}.$$

First observe that, for every statement S, we have $\alpha(\mathcal{M}'(S)) = \mathcal{O}(S)$, since
$\alpha(\mathcal{M}'(S))(b) = \{\hat{\sigma} \in MOD_{I/O} \mid \text{there exists a } b' \text{ such that } (\hat{\sigma}, b') \in \mathcal{M}'(S)(b)\} =$
$\{\hat{\sigma} \in MOD_{I/O} \mid \text{there exists a } \sigma \text{ such that } \hat{\sigma} = \sigma \ominus R, \text{ and } \sigma \in \mathcal{M}(S)(b)\} =$
$\{\sigma \ominus R \in MOD_{I/O} \mid \sigma \in \mathcal{M}(S)(b)\} = \mathcal{O}(S)(b)$

Then we have (note that the first implication follows by the compositionality of $\mathcal{M}'$):

$$\begin{array}{ll} \mathcal{M}'(S_1) = \mathcal{M}'(S_2) & \Rightarrow \\ \mathcal{M}'(C[S_1]) = \mathcal{M}'(C[S_2]) & \Rightarrow \\ \alpha(\mathcal{M}'(C[S_1])) = \alpha(\mathcal{M}'(C[S_2])) & \Rightarrow \\ \mathcal{O}(C[S_1]) = \mathcal{O}(C[S_2]) & \end{array}$$

$\Leftarrow$:

This direction is proved by contradiction. Suppose $\mathcal{M}'(S_1) \neq \mathcal{M}'(S_2)$.
We prove that there exists a context $C[\ ]$ such that $\mathcal{O}(C[S_1]) \neq \mathcal{O}(C[S_2])$. By the assumption, there exists a buffer b such that $\mathcal{M}'(S_1)(b) \neq \mathcal{M}'(S_2)(b)$. Without loss of generality we may assume that there exists a model $(\hat{\sigma}_1, b_1) \in \mathcal{M}'(S_1)(b) \setminus \mathcal{M}'(S_2)(b)$.

Observe that then S_1 is syntactically correct, since $\mathcal{M}'(S_1)(b) \neq \emptyset$ and thus $\mathcal{M}(S_1)(b) \neq \emptyset$. Further note that if S_2 is syntactically not correct, then $\mathcal{M}(S_2)(b) = \emptyset$, hence $\mathcal{O}(S_2)(b) = \emptyset$, for any b, and thus $\mathcal{O}(S_1) \neq \mathcal{O}(S_2)$. Hence, in the sequel we can assume that both S_1 and S_2 are syntactically correct.

Define, for a statement S, the set of input channels of S as

$IN(S) = \{c \mid c?? \in dch(S)\}$.

Note that if $c \notin IN(S)$ then for any $\sigma \in \mathcal{M}(S)(b)$ we have $\#\{\tau \mid c \in \sigma(\tau).R\} = 0$ and thus $Buf(\sigma)(c) = \#\{\tau \mid c \in \sigma(\tau).I\}$.

Suppose $IN(S_1) \neq IN(S_2)$. Without loss of generality we can assume $c \in IN(S_1) \setminus IN(S_2)$. Then $S_1; c!!$ is syntactically not correct, whereas $S_2; c!!$ is syntactically correct. Hence $\mathcal{M}(S_1; c!!)(b) = \emptyset$ and $\mathcal{M}(S_2; c!!)(b) \neq \emptyset$, for any buffer b. Thus, using $C[\,] \equiv [\,]; c!!$, we have $\mathcal{O}(C[S_1]) \neq \mathcal{O}(C[S_2])$.

Now assume $IN(S_1) = IN(S_2)$. Since $(\hat{\sigma}_1, b_1) \in \mathcal{M}'(S_1)(b) \setminus \mathcal{M}'(S_2)(b)$, we have that

$$\text{there exists a model } \sigma_1 \in \mathcal{M}(S_1)(b) \text{ with } \hat{\sigma}_1 = \sigma_1 \ominus R \text{ and } b_1 = b + Buf(\sigma_1) \quad (1)$$

$$\text{for all } \sigma_2 \in \mathcal{M}(S_2)(b),\ \hat{\sigma}_1 \neq \sigma_2 \ominus R \text{ or } b_1 \neq b + Buf(\sigma_2) \quad (2)$$

Let $IN(S_1) = IN(S_2) = \{c_1, \ldots, c_n\}$. Consider

$S_3 \equiv c_1??; \ldots; c_1??; c_2??; \ldots\ldots; c_{n-1}??; c_n??; \ldots; c_n??$,

where the number of statements $c_i??$ is equal to $b(c_i)$ for $i = 1, \ldots, n$. Let

$S_4 \equiv [\,[\,]_{i=1}^{n} c_i?? \rightarrow \textbf{skip} [\,] \textbf{delay}\ d \rightarrow a!!]$,

for some $d \in TIME$ and a channel a such that $a \notin dch(S_1) \cup dch(S_2)$.
Let $C[\,]$ be the context $[\,]; S_3; S_4$. Define a model $\sigma_0 \in MOD$ as follows:
$\sigma_0 = \sigma_1 \cdot \sigma_3 \cdot \sigma_4$ with σ_1 as in (1), and σ_3, σ_4 defined by

- $\sigma_3 \in \mathcal{M}(S_3)(b_1)$ such that, for all $\tau \leq |\sigma_3|$, $\sigma_3(\tau).I = \emptyset$;
- $|\sigma_4| = d + T_{comm}$, for all $\tau \leq |\sigma_4|$, $\sigma_4(\tau).I = \sigma_4(\tau).R = \sigma_4(\tau).O = \emptyset$, $\sigma_4(|\sigma_4|).I = \sigma_4(|\sigma_4|).R = \emptyset$, and $\sigma_4(|\sigma_4|).O = \{a\}$.

Observe that, for $i = 1, \ldots, n$, $(b_1 + Buf(\sigma_3))(c_i) = 0$. Furthermore, $\sigma_4 \in \mathcal{M}(S_4)(b_4)$ iff $b_4(c_i) = 0$, for all $i = 1, \ldots, n$. Hence $\sigma_4 \in \mathcal{M}(S_4)(b_1 + Buf(\sigma_3))$. Since $\sigma_1 \in \mathcal{M}(S_1)(b)$, $b_1 = b + Buf(\sigma_1)$, and $\sigma_3 \in \mathcal{M}(S_3)(b_1)$, we have $\sigma_1 \cdot \sigma_3 \cdot \sigma_4 \in \mathcal{M}(S_1; S_3; S_4)(b)$. Thus $\sigma_0 \in \mathcal{M}(C[S_1])(b)$, and hence $\sigma_0 \ominus R \in \mathcal{O}(C[S_1])(b)$.

We prove that $\sigma_0 \ominus R \notin \mathcal{O}(C[S_2])(b)$ by contradiction. Suppose $\sigma_0 \ominus R \in \mathcal{O}(C[S_2])(b)$. Then there exists a $\bar{\sigma}_0 \in \mathcal{M}(C[S_2])(b)$ such that $\sigma_0 \ominus R = \bar{\sigma}_0 \ominus R$. Thus $|\sigma_0| = |\bar{\sigma}_0|$ and $\bar{\sigma}_0 = \sigma_2 \cdot \bar{\sigma}_3 \cdot \bar{\sigma}_4$ with $\sigma_2 \in \mathcal{M}(S_2)(b)$, $\bar{\sigma}_3 \in \mathcal{M}(S_3)(b + Buf(\sigma_2))$, and $\bar{\sigma}_4 \in \mathcal{M}(S_4)(b + Buf(\sigma_2) + Buf(\bar{\sigma}_3))$. Since $\bar{\sigma}_4(|\bar{\sigma}_4|).O = \sigma_4(|\sigma_4|).O = \{a\}$, we have $\bar{\sigma}_4 \in \mathcal{M}(\textbf{delay } d \rightarrow a!!)(b + Buf(\sigma_2) + Buf(\bar{\sigma}_3))$. Hence $\bar{\sigma}_4 = \sigma_4$ and $(b + Buf(\sigma_2) + Buf(\bar{\sigma}_3))(c_i) = 0$, for $i = 1, \ldots, n$. Thus $Buf(\bar{\sigma}_3) \neq b_\infty$. By the definition of S_3 this implies $Buf(\bar{\sigma}_3))(c_i) = -b_1(c_i)$, for $i = 1, \ldots, n$. Together with $b + Buf(\sigma_2) + Buf(\bar{\sigma}_3))(c_i) = 0$, this leads to $(b + Buf(\sigma_2))(c_i) = b_1(c_i)$, for $i = 1, \ldots, n$. Hence $|\hat{\sigma}_3| = (b_1(c_1) + \ldots + b_n(c_n)) \times T_{comm}$, and thus $|\hat{\sigma}_3| = |\sigma_3|$. Since $|\sigma_1 \cdot \sigma_3 \cdot \sigma_4| = |\sigma_2 \cdot \bar{\sigma}_3 \cdot \bar{\sigma}_4|$ and $\sigma_4 = \bar{\sigma}_4$, this implies $|\sigma_1| = |\sigma_2|$. Using $(\sigma_1 \cdot \sigma_3 \cdot \sigma_4) \ominus R = (\sigma_2 \cdot \bar{\sigma}_3 \cdot \bar{\sigma}_4) \ominus R$ we obtain $\sigma_1 \ominus R = \sigma_2 \ominus R$. Thus, for $c \notin IN(S_1) = IN(S_2)$ we have $(b + Buf(\sigma_2))(c) = (b + Buf(\sigma_1))(c) = b_1(c)$. Together with $(b + Buf(\sigma_2))(c_i) = b_1(c_i)$, for $i = 1, \ldots, n$, this implies $b + Buf(\sigma_2) = b_1$. Since $\sigma_1 \ominus R = \sigma_2 \ominus R$, (1) leads to $\hat{\sigma}_1 = \sigma_2 \ominus R$ and we have a contradiction with (2). □

6 Conclusions

We developed a compositional semantics of the real-time reactive input/output behaviour of systems of asynchronously communicating processes. The model is based on a simple notion of traces of observables. This is to be contrasted with systems based on synchronous communication which require more elaborate structures to encode some necessary branching information.

Future research will be directed towards the development of a temporal proof system along the lines of [HW89]. Another interesting line of research is the investigation of possible real-time extensions of other languages based on asynchronous communication, like dataflow and concurrent logic languages.

References

[Ada83] The programming language Ada. *Reference manual*, 1983.

[BHR84] S.D. Brookes, C.A.R. Hoare, and W. Roscoe. A theory of communicating sequential processes. *J. Assoc. Comput. Mach.*, 31:560–599, 1984.

[BKO88] J.A. Bergstra, J.W. Klop, and E.-R. Olderog. Readies and failures in the algebra of communicating systems. *SIAM J. Comp.*, 17(6):1134–1177, 1988.

[BKPR91] F.S. de Boer, J.N. Kok, C. Palamidessi, and J.J.M.M. Rutten. The Failure of failures in a paradigm for asynchronous communication. In *Proceedings of Concur '91*, Amsterdam 1991.

[dBP90] F.S. de Boer and C. Palamidessi. On the asynchronous nature of communication in concurrent logic languages: A fully abstract model based on sequences. In J.C.M. Baeten and J.W. Klop, editors, *Proc. of Concur 90*, volume 458 of *Lecture Notes in Computer Science*, pages 99–114, The Netherlands, 1990. Springer-Verlag. Full version available as report at the Technische Universiteit Eindhoven.

[dBP91] F.S. de Boer and C. Palamidessi. A fully abstract model for concurrent constraint languages. *Proc. of TAPSOFT 91*, 1991. Also available as technical report, Centre for Mathematics and Computer Science (CWI), Amsterdam.

[GCLS88] R. Gerth, M. Codish, Y. Lichtenstein, and E. Shapiro. Fully abstract denotational semantics for Concurrent Prolog. In *Proc. of the Third IEEE Symposium on Logic In Computer Science*, pages 320–335. IEEE Computer Society Press, New York, 1988.

[GMS89] H. Gaifman, M. J. Maher, and E. Shapiro. Reactive Behaviour semantics for Concurrent Constraint Logic Programs. In E. Lusk and R. Overbeck, editors, *North American Conference on Logic Programming*, 1989.

[GB87] A. Boucher and R. Gerth. A timed failures model for extending communicating processes. In *Proceedings in the 14th International Colloquium on Automata, Languages and Programming*, pp. 95-114. LNCS 267, Springer-Verlag, 1987.

[Hoa78] C.A.R. Hoare. Communicating sequential processes. *Communications of the ACM*, 21(8):666–677, 1978.

[HW89] J. Hooman and J. Widom. A temporal-logic based compositional proof system for real-time message passing. In *Parallel Architectures and Languages Europe*, Vol. 2., pp. 424-441. LNCS 366, Springer-Verlag, 1989.

[HGR87] C. Huizing, R. Gerth, and W.P. de Roever. Full abstraction of a real-time denotational semantics for an OCCAM-like language. In *Proceedings of the 14th ACM Symposium on Principles of Programming Languages*, pp. 223-237, 1987.

[JHJ90] M.B. Josephs, C.A.R. Hoare, and He Jifeng. A theory of asynchronous processes. Technical report, Oxford University Computing Laboratories, 1990.

[JJH90] He Jifeng, M.B. Josephs, and C.A.R. Hoare. A theory of synchrony and asynchrony. In *Proc. of IFIP Working Conference on Programming Concepts and Methods*, pages 459–478, 1990.

[JK89] B. Jonsson and J.N. Kok. Comparing two fully abstract dataflow models. In *Proc. Parallel Architectures and Languages Europe (PARLE)*, volume 379 in Lecture Notes in Computer Science, pages 217–235, 1989.

[Jon85] B. Jonsson. A model and a proof system for asynchronous processes. In *Proc. of the 4th ACM Symp. on Principles of Distributed Computing*, pages 49–58, 1985.

[Kah74] G. Kahn. The semantics of a simple language for parallel programming. In *Information Processing 74: Proc. of IFIP Congress*, pages 471–475, New York, 1974. North-Holland.

[Kok87] J.N. Kok. A fully abstract semantics for data flow nets. In J.W. de Bakker, A.J. Nijman, and P.C. Treleaven, editors, *Proc. Parallel Architectures and Languages Europe (PARLE)*, volume 259 of *Lecture Notes in Computer Science*, pages 351–368. Springer Verlag, 1987.

[Kok89] J.N. Kok. Traces, histories and streams in the semantics of nondeterministic dataflow. In *Proceedings Massive Parallellism: Hardware, Programming and Applications*, 1989. Also available as report 91, Abo Akademi, Finland, 1989.

[Mil80] R. Milner. *A Calculus of Communicating Systems*, volume 92 of *Lecture Notes in Computer Science*. Springer-Verlag, New York, 1980.

[Ree89] G. Reed. A hierarchy of domains for real-time distributed computing. In *Mathematical Foundations of Programming Semantics*, pp. 80-128. LNCS 442, Springer-Verlag.

[RR86] G. Reed and A. Roscoe. A timed model for Communicating Sequential Processes. In *Proceedings of ICALP '86*, pp. 314-323. LNCS 226, Springer-Verlag, 1986.

[Sar89] V.A. Saraswat. *Concurrent Constraint Programming Languages.* PhD thesis, January 1989. To be published by the MIT Press.

[SR90] V.A. Saraswat and M. Rinard. Concurrent constraint programming. In *Proc. of the seventeenth ACM Symposium on Principles of Programming Languages*, pages 232–245. ACM, New York, 1990.

Asynchronous Communication in Real Space Process Algebra

J.C.M. Baeten*
Department of Software Technology, CWI,
P.O.Box 4079, 1009 AB Amsterdam, The Netherlands
and
Programming Research Group, University of Amsterdam,
P.O.Box 41882, 1009 DB Amsterdam, The Netherlands

J.A. Bergstra
Programming Research Group, University of Amsterdam,
P.O.Box 41882, 1009 DB Amsterdam, The Netherlands
and
Department of Philosophy, Utrecht University,
Heidelberglaan 2, 3584 CS Utrecht, The Netherlands

A version of classical real space process algebra is given in which messages travel with constant speed through a three-dimensional medium. It follows that communication is asynchronous and has a broadcasting character. A state operator is used to describe asynchronous message transfer and a priority mechanism allows to express the broadcasting mechanism. As an application, a protocol is specified in which the receiver moves with respect to the sender.

Key words & Phrases: process algebra, real time process algebra, real space process algebra, asynchronous communication, broadcasting, state operator, priorities.
Note: Partial support received by ESPRIT basic research action 3006, CONCUR, and by RACE contract 1046, SPECS. This document does not necessarily reflect the views of the SPECS consortium.

1. INTRODUCTION.

Our aim is to extend real time process algebra (see [BAB91a]) to classical (i.e. non-relativistic) real space process algebra (see [BAB91b]) in such a way that motion of processes can be taken into account. Although a rigorous proof is absent, it seems impossible to express communication between processes moving in three-dimensional space with the primitives of [BAB91a] and [BAB91b]. The difficulty arises because both papers use synchronous communication whereas motion of processes seems to call for asynchronous communication: if process P at place/time (x,t) sends d to Q, then Q may receive d at place/time (y,r) provided $|x - y| = v \cdot (r - t)$. Here $|x - y|$ is the distance between x and y and v is the message transmission velocity in the medium connecting P and Q. If P and Q are at rest we find $r = |x - y|/v + t$ and this formula can be used in process expressions in the style of [BAB91a] and [BAB91b].

However, if y depends on time, the equation $|x(t) - y(r)| = v \cdot (r - t)$ has to be solved in order to determine the time r at which message d will be received. In general, no closed form can be found for r. Even worse, the motion of Q (and so y) may depend on actions of the system after t (the time at which d is sent).

---

* Author's current affiliation: Department of Computer Science, Eindhoven University of Technology, P.O. Box 513, 5600 MB Eindhoven, The Netherlands.

In order to study communication between moving processes we assume that messages travel with speed v through space in all directions. I.e. after d is sent at (x,t) it travels through space as an expanding spherical wave. For simplicity we assume that the message can be detected at any distance. It is probably not difficult to describe a decrease in loudness of the signal together with a threshold mechanism for receivers.

Now according to [BEKT85] asynchronous communication in process algebra can be modeled using auxiliary operators. Subsequently, these operators have been studied in detail in [BAB88] where they are called state operators. In [BAB91c], it is indicated how any state operator definition of untimed process algebra can be extended to real time process algebra over finitely many locations, provided a partial ordering on locations is given. Here, we will consider special state operators (as in [BEKT85]) for which the action and effect functions depend on place and time, but that satisfy a commutativity requirement for actions that happen at the same instant of time.

The state operators λ_V^c (state operator λ of asynchronous channel c, e.g. a radio frequency, in state V) are parametrised by finite collections V of triples $\langle d, x, t\rangle$, d a message in D, x a point in space ($x \in \mathbb{R}^3$), t a time (a non-negative real, $t \in \mathbb{R}^{\geq 0}$). If $\langle d, x, t\rangle \in V$, this indicates that $\lambda_V^c(X)$ provides an environment for X in which along channel c a datum d was sent at place x and time t. The asynchronous send actions $c{\uparrow}d(x,t)$ will have the effect that $\langle d, x, t\rangle$ is added to V. In the scope of λ_V^c a process X can perform the action $c{\downarrow}d(y,r)$ (asynchronous read along channel c of datum d at place y and time r) provided $|y - x| = v{\cdot}(r - t)$.

Because the message d, after being sent by $c{\uparrow}d(x,t)$, travels in a spherical wave, it can be received more than once and hence the communication mechanism is of a broadcasting nature. As pointed out in [BE85] (see also [BAW90]), a broadcasting mechanism in process algebra calls for the use of the priority operator of [BABK86]. In the real time and space case this priority operator will express maximal progress with respect to some actions as well.

We can summarise this discussion as follows: P and Q, traveling through space, can communicate by performing actions $c{\uparrow}d(x,t)$ and $c{\downarrow}d(y,r)$. This works in a context $\lambda_V^c(P \| Q)$. Unsuccessful asynchronous reads are blocked by the action function of the state operator. In order to ensure successful reception of the messages a priority operator $\theta_{c{\Downarrow}D}$ is needed (here $c{\Downarrow}D$ contains all effectuated $c{\downarrow}d$ actions). This operator gives priority to all asynchronous communications at port c. We will allow synchronous communications as well. Unsuccessful synchronous reads and sends at channels in H are blocked by the encapsulation operator ∂_{srHD} ($srHD$ contains all synchronous send and receive actions at ports in H). Thus we are led to process expressions of the form:

$$\partial_{srHD} \circ \theta_{c{\Downarrow}D} \circ \lambda_V^c(P \| Q).$$

For instance, a concurrent alternating bit protocol with moving sender and receiver will take the following form:

$$\partial_{srHD}\big(\theta_{1{\Downarrow}D} \circ \lambda_V^1(P_1 \| Q_1) \ \| \ \theta_{2{\Downarrow}D} \circ \lambda_V^2(P_2 \| Q_2)\big).$$

After studying the asynchronous communication primitives we provide a brief discussion on how to model asynchronous message transfer on the basis of the synchronous communication primitives of ACP$\rho\sigma$. We find that process creation [BE90] and mode transfer [BE89] are helpful in the description of certain asynchronous media. Real time and space versions of these process constructors are provided and various examples are given.

So, besides giving an account of asynchronous communication in a real space setting, this paper also gives the real time equations for several additional features of ACP: the priority operator, state operators with uncountable state space, process creation and both mode transfer operators. Except for the mode transfer operators, these features are all covered in chapter 6 of [BAW90]. The mode transfer operators were not included there, because their equations are not fully satisfactory in the untimed case.

The real time setting provides a clearer picture and the equations given below seem perfectly adequate to us.

By now, there is much work on process algebras that encorporate notions of time (see e.g. [RR88], [MT90], [J91a]). However, there is not much work that also involves real space or the use of locations. Besides papers already mentioned, we only know of JEFFREY [J91b], MURPHY [MU91].

Finally, we remark that we only consider *concrete* process algebra here: there is no concept of a silent or empty step.

ACKNOWLEDGEMENT. We thank Willem Jan Fokkink (CWI Amsterdam) for pointing out that an earlier version of equation TH3 in 4.1 was incorrect. We thank Alan Jeffrey (Chalmers University) for stimulating discussions on maximal progress and priority locks.

2. REAL SPACE PROCESS ALGEBRA.

We give a brief review of classical real space process algebra as introduced in [BAB91b], but we use the timed deadlock of [BAB91a,BAB91c] instead of the untimed deadlock of [BAB91b]. The difference with [BAB91b] is that there, we had a finite set of locations, and here, every element of three-dimensional space is a location. We give an operational semantics in the style of KLUSENER [K91].

2.1 ATOMIC ACTIONS.

We start from a set A of (symbolic) atomic actions. The set of atomic actions with space and time, AST is now generated by

$$\{a(x,t) \mid a \in A, x \in \mathbb{R}^3, t \in \mathbb{R}^{\geq 0}\} \cup \{\delta(t) \mid t \in \mathbb{R}^{\geq 0}\}.$$

It will be useful to consider also the set of atomic actions with only space parameter, i.e. the set AS generated by

$$\{a(x) \mid a \in A, x \in \mathbb{R}^3\} \cup \{\delta\}.$$

We use $a(x)(t)$ as an alternative notation for $a(x,t)$.

2.2 MULTI-ACTIONS.

Multi-actions are process terms generated by actions with space parameter and the *synchronisation function* &. Multi-actions contain actions that occur synchronously at different locations. For α,β,γ elements of AS, we have the following conditions on the synchronisation function (table 1). Further, $x \in \mathbb{R}^3$, $a,b \in A$.

$\alpha \,\&\, \beta = \beta \,\&\, \alpha$	LO1
$\alpha \,\&\, (\beta \,\&\, \gamma) = (\alpha \,\&\, \beta) \,\&\, \gamma$	LO2
$\delta \,\&\, \alpha = \delta$	LO3
$a(x) \,\&\, b(x) = \delta$	LO4

Table 1. Synchronisation function on AS.

Using the axioms of table 1, each multi-action can be reduced to one of the following two forms:

- δ,
- $a_1(x_1) \,\&\, \ldots \,\&\, a_n(x_n)$, with all points x_i different, all $a_i \in A$.

Next, we have the *communication function* $\mid$. In table 2, $a,b,c \in A$, $\alpha,\beta,\gamma \in AS$. In order to state axiom CL7, we need an auxiliary function locs, that determines the set of points (locations) of a multi-action.

$a \mid b = b \mid a$	C1
$a \mid (b \mid c) = (a \mid b) \mid c$	C2
$\delta \mid a = \delta$	C3
$\alpha \mid \beta = \beta \mid \alpha$	CL1
$\alpha \mid (\beta \mid \gamma) = (\alpha \mid \beta) \mid \gamma$	CL2
$\delta \mid \alpha = \delta$	CL3
$a(x) \mid b(x) = (a \mid b)(x)$	CL4
$a(x) \mid (b(x) \,\&\, \beta) = (a \mid b)(x) \,\&\, \beta$	CL5
$(a(x) \,\&\, \alpha) \mid (b(x) \,\&\, \beta) = (a \mid b)(x) \,\&\, (\alpha \mid \beta)$	CL6
$locs(\alpha) \cap locs(\beta) = \varnothing \Rightarrow \alpha \mid \beta = \alpha \,\&\, \beta$	CL7
$locs(\delta) = \varnothing$	LOC1
$locs(a(x)) = \{x\}$	LOC2
$x \notin locs(\alpha), locs(\alpha) \neq \varnothing \Rightarrow$ $locs(a(x) \,\&\, \alpha) = locs(\alpha) \cup \{x\}$	LOC3

Table 2. Communication function on AS.

2.3 TIMED MULTI-ACTIONS.

It is now straightforward to extend the definition of the synchronisation and communication functions to timed multi-actions. In table 3, $\alpha,\beta \in AS$.

$t \neq s \Rightarrow \alpha(t) \mid \beta(s) = \delta(\min(t,s))$	CL8
$\alpha(t) \mid \beta(t) = (\alpha \mid \beta)(t)$	CL9

Table 3. Communication function on AST.

2.4 BASIC PROCESS ALGEBRA.

Process algebra (see [BEK84, BAW90]) starts from a given *action alphabet*, here AST. Elements of AST are constants of the sort P of *processes*. The theory Timed Basic Process Algebra with Deadlock (BPA$\rho\delta$) has two binary operators $+,\cdot$: $P \times P \rightarrow P$; + stands for alternative composition and $\cdot$ for sequential composition. Moreover, there is the additional operator $\gg$: $\mathbb{R}^{\geq 0} \times P \rightarrow P$, the *(absolute) time shift*. $t \gg X$ denotes the process X starting at time t. This means that all actions that have to be performed at or before time t are turned into deadlocks because their execution has been delayed too long.

$X + Y = Y + X$	A1
$(X + Y) + Z = X + (Y + Z)$	A2
$X + X = X$	A3
$(X + Y) \cdot Z = X \cdot Z + Y \cdot Z$	A4
$(X \cdot Y) \cdot Z = X \cdot (Y \cdot Z)$	A5

Table 4. BPA.

$\alpha(0) = \delta(0) = \delta$	ATA1
$\delta(t) \cdot X = \delta(t)$	ATA2
$t < r \Rightarrow \delta(t) + \delta(r) = \delta(r)$	ATA3
$\alpha(t) + \delta(t) = \alpha(t)$	ATA4
$\alpha(t) \cdot X = \alpha(t) \cdot (t \gg X)$	ATA5
$t < r \Rightarrow t \gg \alpha(r) = \alpha(r)$	ATB1
$t \geq r \Rightarrow t \gg \alpha(r) = \delta(t)$	ATB2
$t \gg (X + Y) = (t \gg X) + (t \gg Y)$	ATB3
$t \gg (X \cdot Y) = (t \gg X) \cdot Y$	ATB4

Table 5. Additional axioms of BPA$\rho\delta$.

BPA$\rho\delta$ has the axioms from table 4 and 5 ($\alpha \in AS$). The letter A in the names of the axioms in table 5 refers to *absolute time* (versions with relative time were also considered in [BAB91a], but are not treated here).

2.5 ALGEBRA OF COMMUNICATING PROCESSES.

An axiomatization of parallel composition with communication uses the left merge operator $\mathbin{⌊\!⌊}$, the communication merge operator $\mid$, and the encapsulation operator ∂_H of [BK84]. Moreover, two extra auxiliary operators introduced in [BAB91a] are needed: the ultimate delay operator and the bounded initialization operator.

The ultimate delay operator U takes a process expression X, and returns an element of $\mathbb{R}^{\geq 0}$. The intended meaning is that X can idle before U(X), but X can never reach time U(X) or a later time by just idling.

$X \parallel Y = X ⌊\!⌊ Y + Y ⌊\!⌊ X + X \mid Y$	CM1	$U(\alpha(t)) = t$	ATU1
$\alpha(t) ⌊\!⌊ X = (\alpha(t) \gg U(X)) \cdot X$	ATCM2	$U(\delta(t)) = t$	ATU2
$(\alpha(t) \cdot X) ⌊\!⌊ Y = (\alpha(t) \gg U(Y)) \cdot (X \parallel Y)$	ATCM3	$U(X + Y) = \max\{U(X), U(Y)\}$	ATU3
$(X + Y) ⌊\!⌊ Z = X ⌊\!⌊ Z + Y ⌊\!⌊ Z$	CM4	$U(X \cdot Y) = U(X)$	ATU4
$(\alpha(t) \cdot X) \mid \beta(r) = (\alpha(t) \mid \beta(r)) \cdot X$	CM5'		
$\alpha(t) \mid (\beta(r) \cdot X) = (\alpha(t) \mid \beta(r)) \cdot X$	CM6'	$r \geq t \Rightarrow \alpha(r) \gg t = \delta(t)$	ATB5
$(\alpha(t) \cdot X) \mid (\beta(r) \cdot Y) = (\alpha(t) \mid \beta(r)) \cdot (X \parallel Y)$	CM7'	$r < t \Rightarrow \alpha(r) \gg t = \alpha(r)$	ATB6
$(X + Y) \mid Z = X \mid Z + Y \mid Z$	CM8	$(X + Y) \gg t = (X \gg t) + (Y \gg t)$	ATB7
$X \mid (Y + Z) = X \mid Y + X \mid Z$	CM9	$(X \cdot Y) \gg t = (X \gg t) \cdot Y$	ATB8
$\partial_H(a) = a$ if $a \notin H$	D1		
$\partial_H(a) = \delta$ if $a \in H$	D2		
$\partial_H(\alpha \,\&\, \beta) = \partial_H(\alpha) \,\&\, \partial_H(\beta)$	ASD		
$\partial_H(\alpha(t)) = (\partial_H(\alpha))(t)$	ATD		
$\partial_H(X + Y) = \partial_H(X) + \partial_H(Y)$	D3		
$\partial_H(X \cdot Y) = \partial_H(X) \cdot \partial_H(Y)$	D4		

Table 6. Remaining axioms of ACPρ.

The bounded initialization operator is also denoted by $\gg$, and is the counterpart of the operator with the same name that we saw in the axiomatization of BPA$\rho\delta$. With $X \gg t$ we denote the process X with its behaviour restricted to the extent that its first action must be performed at a time before $t \in \mathbb{R}^{\geq 0}$.

The axioms of ACPρ are in tables 1 through 6. In table 6, $H \subseteq A$, $\alpha, \beta \in AS$, $a \in A_\delta$.

2.6 OPERATIONAL SEMANTICS.

We describe an operational semantics for ACPρ following KLUSENER [K91] and [BAB91c], a reformulisation of the original semantics in [BAB91a]. We have a binary relation $\xrightarrow{\mu}$ and a unary relation $\xrightarrow{\mu}\surd$ on closed process expressions for each $\mu \in AST$. In case $\mu \neq \delta(t)$, the extension of the relations is found as the least fixed point of a simultaneous inductive definition.

$$r > 0 \Rightarrow \alpha(r) \xrightarrow{\alpha(r)} \surd$$

$$\frac{x \xrightarrow{\alpha(r)} x'}{x+y \xrightarrow{\alpha(r)} x',\ y+x \xrightarrow{\alpha(r)} x'} \qquad \frac{x \xrightarrow{\alpha(r)} \surd}{x+y \xrightarrow{\alpha(r)} \surd,\ y+x \xrightarrow{\alpha(r)} \surd}$$

$$\frac{x \xrightarrow{\alpha(r)} x'}{x \cdot y \xrightarrow{\alpha(r)} x' \cdot y} \qquad \frac{x \xrightarrow{\alpha(r)} \surd}{x \cdot y \xrightarrow{\alpha(r)} r \gg y}$$

$$\frac{x \xrightarrow{\alpha(r)} x',\ r > s}{s \gg x \xrightarrow{\alpha(r)} x'} \qquad \frac{x \xrightarrow{\alpha(r)} \surd,\ r > s}{s \gg x \xrightarrow{\alpha(r)} \surd}$$

$$\frac{x \xrightarrow{\alpha(r)} x',\ r < U(y)}{x \| y \xrightarrow{\alpha(r)} x' \| (r \gg y),\ x \mathbin{\|\!\_} y \xrightarrow{\alpha(r)} x' \| (r \gg y),\ y \| x \xrightarrow{\alpha(r)} (r \gg y) \| x'}$$

$$\frac{x \xrightarrow{\alpha(r)} \surd,\ r < U(y)}{x \| y \xrightarrow{\alpha(r)} r \gg y,\ x \mathbin{\|\!\_} y \xrightarrow{\alpha(r)} r \gg y,\ y \| x \xrightarrow{\alpha(r)} r \gg y}$$

$$\frac{x \xrightarrow{\alpha(r)} x',\ y \xrightarrow{\beta(r)} y',\ \alpha \mid \beta = \gamma \neq \delta}{x \| y \xrightarrow{\gamma(r)} x' \| y',\ x \mid y \xrightarrow{\gamma(r)} x' \| y'} \qquad \frac{x \xrightarrow{\alpha(r)} \surd,\ y \xrightarrow{\beta(r)} \surd,\ \alpha \mid \beta = \gamma \neq \delta}{x \| y \xrightarrow{\gamma(r)} \surd,\ x \mid y \xrightarrow{\gamma(r)} \surd}$$

$$\frac{x \xrightarrow{\alpha(r)} x',\ y \xrightarrow{\beta(r)} \surd,\ \alpha \mid \beta = \gamma \neq \delta}{x \| y \xrightarrow{\gamma(r)} x',\ y \| x \xrightarrow{\gamma(r)} x',\ x \mid y \xrightarrow{\gamma(r)} x',\ y \mid x \xrightarrow{\gamma(r)} x'}$$

$$\frac{x \xrightarrow{\alpha(r)} x',\ r < t}{x \gg t \xrightarrow{\alpha(r)} x'} \qquad \frac{x \xrightarrow{\alpha(r)} \surd,\ r < t}{x \gg t \xrightarrow{\alpha(r)} \surd}$$

$$\frac{x \xrightarrow{\alpha(r)} x',\ ats(\alpha) \cap H = \varnothing}{\partial_H(x) \xrightarrow{\alpha(r)} \partial_H(x')} \qquad \frac{x \xrightarrow{\alpha(r)} \surd,\ ats(\alpha) \cap H = \varnothing}{\partial_H(x) \xrightarrow{\alpha(r)} \surd}$$

Table 7. Action rules for non-δ actions for ACPρ.

The inductive rules for the operational semantics are similar to those used in structured operational semantics. We list the rules for non-δ multi-actions. In table 7, we have $\alpha,\beta,\gamma \in AS - \{\delta\}$, $r > 0$ (we never allow timestamp 0!), $s,t \geq 0$, x,x',y,y' are closed process expressions. $ats(\alpha)$ is the set of atomic actions occurring in the multi-action α.

2.7 DELTA-TRANSITIONS.

Now we construct a transition system for a term as follows: first generate all transitions involving non-δ actions using the inductive rules of table 7. Then, for every node (term) p in this transition system we do the following: first determine its ultimate delay $U(p) = u$ by means of table 5 and the additional axioms below. Then, if the ultimate delay is larger than the supremum of the time stamps of all outgoing transitions, we add a transition

$$p \xrightarrow{\delta(u)} \surd.$$

Otherwise, we do nothing (add no transitions).

We see that the action rules for parallel composition, and the determination of δ-transitions, make use of the ultimate delay operator. We add the axioms in table 8 so that the ultimate delay can be syntactically determined for every closed term.

$U(t \gg X) = \max\{t, U(X)\}$	ATU5
$U(X \parallel Y) = \min\{U(X), U(Y)\}$	ATU6
$U(X \mathbin{\lfloor\!\lfloor} Y) = \min\{U(X), U(Y)\}$	ATU7
$U(X \mid Y) = \min\{U(X), U(Y)\}$	ATU8
$U(X \gg t) = \min\{t, U(X)\}$	ATU9
$U(\partial_H(X)) = U(X)$	ATU10

Table 8. Ultimate delay axioms.

2.8 BISIMULATIONS.

A *bisimulation* on is a binary relation R on closed process expressions such that ($\mu \in AST$):

i. for each p and q with $R(p, q)$: if there is a step μ possible from p to p', then there is a closed process expression q' such that $R(p', q')$ and there is a step μ possible from q to q'.

ii. for each p and q with $R(p, q)$: if there is a step μ possible from q to q', then there is a CPE p' such that $R(p', q')$ and there is a step μ possible from p to p'.

iii. for each p and q with $R(p, q)$: a termination step μ to $\surd$ is possible from p iff it is possible from q.

We say expressions p and q are *bisimilar*, denoted $p \leftrightarrow q$, if there exists a bisimulation on closed process expressions with $R(p,q)$. In [K91] it is shown that bisimulation is a congruence relation on closed process expressions, and that closed process expression modulo bisimulation determine a model for ACPρ. Indeed, this model is isomorphic to the initial algebra. The advantage of this operational semantics is, that it allows extensions to models containing recursively defined processes.

It is also possible to give an explicit graph model for ACPρ, see [BAB91c].

2.9 INTEGRATION.

An extension of ACPρ (called ACPρI) that is very useful in applications is the extension with the integral operator, denoting a choice over a continuum of alternatives. I.e., if V is a subset of $\mathbb{R}^{\geq 0}$, and v is a variable over $\mathbb{R}^{\geq 0}$, then $\int_{v \in V} P$ denotes the alternative composition of alternatives P[t/v] for t ∈ V (expression P with nonnegative real t substituted for variable v). For more information, we refer the reader to [BAB91a] and [K91]. The operational semantics is straightforward (table 9, μ ∈ AST).

$\dfrac{x(t) \xrightarrow{\mu} x',\ t \in V}{\int_{v \in V} x(v) \xrightarrow{\mu} x'}$	$\dfrac{x(t) \xrightarrow{\mu} \surd,\ t \in V}{\int_{v \in V} x(v) \xrightarrow{\mu} \surd}$

Table 9. Action relations for integration.

We will not provide axioms for the integral operator here (and refer the reader to [BAB91a] and [K91]), except for the axiom for the ultimate delay operator:

$$U(\int_{v \in V} P) = \sup\{U(P[t/v]) : t \in V\} \qquad \text{ATU11.}$$

3. ASYNCHRONOUS COMMUNICATION.

3.1 DEFINITION.

Let C be a finite collection of port names, and let D be a finite set of data. As special symbolic atomic actions we introduce, following [BEKT85], for c ∈ C, d ∈ D:

- $c{\uparrow}d$ potential send of message d along asynchronous channel c
- $c{\Uparrow}d$ effectuated send of message d along asynchronous channel c
- $c{\downarrow}d$ potential receive of message d along asynchronous channel c
- $c{\Downarrow}d$ effectuated receive of message d along asynchronous channel c.

We define $c{\Downarrow}D = \{c{\Downarrow}d \mid d \in D\}$, and likewise for the other actions.

Besides these asynchronous communication actions, we have the standard synchronous communication actions, for k ∈ H, d ∈ D:

- sk(d) send message d at synchronous port k
- rk(d) receive message d at synchronous port k
- ck(d) communicate message d at synchronous port k.

We define srHD = {sk(d), rk(d) | k ∈ H, d ∈ D}, cHD = {ck(d) | k ∈ H, d ∈ D}.

3.2. STATE OPERATOR.

The state operator was introduced in [BAB88]. It keeps track of the global state of a system, and is used to describe actions that have a side effect on a state space. In [BAB91c], it was shown how to define a state operator on real time process algebra with finitely many locations. There, an ordering on locations was needed in order to get a right definition for multi-actions. Here, we define a specific state operator, on a domain with infinitely many locations, that does not require this ordering.

The state operator comes equipped with two functions: given a certain state and an action to be executed from that state, the function action gives the resulting action and the function effect the resulting state. Different from In [BAB91c], here it will be needed that the action and effect function also depend on the place and time of an action. Hence, these functions are not defined on A, but on AST.

For each asynchronous channel $c \in C$, we will have a state operator λ^c. The set of states $\mathbb{V}$ is the collection of all finite subsets of $D \times \mathbb{R}^3 \times \mathbb{R}^{\geq 0}$. Moreover, v is a given constant.

The state operator λ^c_V ($V \in \mathbb{V}$) has functions

$\text{action}_c: AST \times \mathbb{V} \rightarrow AST \qquad \text{effect}_c: AST \times \mathbb{V} \rightarrow \mathbb{V}$

given by:

$\text{action}_c(c\!\uparrow\! d(x,t), V) = c\Uparrow d(x,t)$

$\text{effect}_c(c\!\uparrow\! d(x,t), V) = V \cup \{\langle d, x, t\rangle\}$

$\text{action}_c(c\!\downarrow\! d(y, r), V) = c\Downarrow d(y, r)$ if there is $\langle d, x, t\rangle \in V$ such that $|y - x| = v\cdot(r - t)$ and $r \neq t$

$\text{action}_c(c\!\downarrow\! d(y, r), V) = \delta(r)$ otherwise

$\text{effect}_c(c\!\downarrow\! d(y, r), V) = V$

$\text{action}_c(a(x,t), V) = a(x,t)$ for all a not of the form $c\!\uparrow\! d$ or $c\!\downarrow\! d$

$\text{effect}_c(a(x,t), V) = V$ for all a not of the form $c\!\uparrow\! d$ or $c\!\downarrow\! d$.

Now these functions are extended to multi-actions in the obvious way:

$$\text{action}_c((a(x)\ \&\ \alpha)(t), V) = \text{action}_c(a(x,t), V)\ \&\ \text{action}_c(\alpha(t), V)$$

$$\text{effect}_c((a(x)\ \&\ \alpha)(t), V) = \text{effect}_c(a(x,t), \text{effect}_c(\alpha(t), V)).$$

3.3. LEMMA. Let $c \in C$, $V \in \mathbb{V}$, $\mu,\nu \in AST$. Then $\text{effect}_c(\mu\ \&\ \nu, V) = \text{effect}_c(\nu\ \&\ \mu, V)$.

3.4 DEFINITION.

The defining equations for the state operator are now straightforward (cf. [BAB88]). In table 10, $V \in \mathbb{V}$, $c \in C$, $\mu \in AST$, x,y processes.

$\lambda^c_V(\mu) = \text{action}(\mu, V)$	SO1
$\lambda^c_V(\mu\cdot x) = \text{action}(\mu, V)\cdot\lambda^c_{\text{effect}(\mu,V)}(x)$	SO2
$\lambda^c_V(x + y) = \lambda^c_V(x) + \lambda^c_V(y)$	SO3
$\lambda^c_V(\int_{t\in T} P) = \int_{t\in T} \lambda^c_V(P)$	SO4

Table 10. State operator.

It is equally straightforward to give action rules for the operational semantics ($\mu,\nu \in AST$, $\mu,\nu \neq \delta(r)$).

$\dfrac{x \xrightarrow{\mu} x',\ action_c(\mu,V)=\nu\neq\delta(r)}{\lambda^c_V(x) \xrightarrow{\nu} \lambda^c_{effect_c(\mu,V)}(x')}$	$\dfrac{x \xrightarrow{\mu} \surd,\ action_c(\mu,V)=\nu\neq\delta(r)}{\lambda^c_V(x) \xrightarrow{\nu} \surd}$

Table 11. Action relations for the state operator.

In order to deal with δ-transitions, we add the axiom

$$U(\lambda^c_V(x)) = U(x) \qquad\qquad \text{ATU12.}$$

Then, we determine the existence of δ-transitions as before.

3.5 EXAMPLE.

Let us consider two processes S and R at fixed locations, communicating through an asynchronous channel 2 (see fig. 1). 1 and 3 are synchronous ports. w_0 and w_1 are system delay constants. Suppose the distance between S and R is equal to the distance a message travels in 1 time unit.

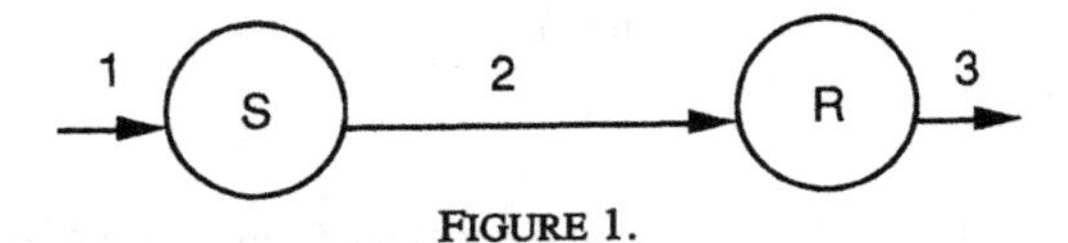

FIGURE 1.

We have the following specifications:

$$S(r) = \int_{t>r} \sum_{d\in D} r1(d)(x,t)\cdot 2{\uparrow}d(x,t+w_0)\cdot S(t+2w_0) \qquad (r \geq 0)$$

$$R(r) = \int_{t>r} \sum_{d\in D} 2{\downarrow}d(y,t)\cdot s3(d)(y,t+w_1)\cdot R(t+2w_1) \qquad (r \geq 0).$$

The system is now described by the expression $\lambda^2_\varnothing(S(0) \parallel R(0))$.

After a certain $2{\uparrow}d(x,t_0)$-action at time t_0, the triple $\langle d, x, t_0\rangle$ is added to the state. Then, action $2{\downarrow}d(y, t_0+1)$ is transformed into $2{\Downarrow}d(y, t_0+1)$, and all other $2{\downarrow}d(y,t)$ for $t \neq t_0+1$ are renamed to $\delta(t)$. However, the presence (in a sum context) of $\delta(t)$ with $t > t_0+1$ allows R to bypass the option $2{\Downarrow}d(y, t_0+1)$, wait too long before trying to receive, and then it is too late, and no further action will be possible. Thus, we have to enforce that the action $2{\Downarrow}d$ occurs as soon as it is possible.

This is a kind of *maximal progress* or *maximal liveness* assumption (cf. [RR88]). We will ensure that communication takes place as soon as possible by use of the priority operator of [BABK86]. A similar operator was used in JEFFREY [J91c].

4. PRIORITIES.

Let us start by recalling the priority axioms from [BABK86]. There, an operator θ is defined, that cancels out all actions in a sum context, that do not have maximal priority. Priorities are given by a partial ordering on actions. Here, we have a very simple ordering: we have a certain set $H \subseteq A$, and all actions from H have priority over actions from outside H. In example 3.5, H is the set $2{\Downarrow}D$.

4.1 DEFINITION.
Let a set $H \subseteq A$ be given. The *priority operator* θ_H has the axioms in table 12 ($\alpha \in AS$).

$\theta_H(\alpha(t)) = \alpha(t)$	TH1
$\theta_H(x + y) = \theta_H(x) \lhd_H y \; + \; \theta_H(y) \lhd_H x$	TH2
$\theta_H(\alpha(t)\cdot x) = \alpha(t)\cdot\theta_H(t \gg x)$	TH3
$\theta_H(\int_{v\in V} P) = \int_{v\in V} (\theta_H(P) \lhd_H \int_{t\in V-\{v\}} P[t/v])$	TH4

Table 12. Priority operator.

The priority operator is axiomatised by means of the *unless operator* $\lhd_H$. In $x \lhd_H y$, a starting action from H in y will cancel all starting actions in x with a later timestamp, and all starting actions not from H with the same timestamp. The unless operator can be axiomatised directly, but we will not do so here. Instead, we will define two other operators (one of which is a generalisation of the bounded initialisation operator) that are useful in their own right, and allow to define the unless operator easily.

4.2 MINIMAL DELAY WITH PRIORITY.
The operator D_H gives the minimal delay with priority, i.e. $D_H(x)$ determines the infimum of all times at which x can perform an initial action in H. If x can perform no initial action in H, we will set $D_H(x) = \infty$. In table 13, $H \subseteq A$, $a \in A$, $\alpha,\beta \in AS$, $x,y \in P$, X a process with time variable t.

$D_H(\delta(t)) = \infty$		MDP1
$D_H(a(t)) = \infty$	if $a \notin H$	MDP2
$D_H(a(t)) = t$	if $a \in H$	MDP3
$D_H((\alpha \,\&\, \beta)(t)) = \min\{D_H(\alpha(t)), D_H(\beta(t))$		MDP4
$D_H(x + y) = \min\{D_H(x), D_H(y)\}$		MDP5
$D_H(x\cdot y) = D_H(x)$		MDP6
$D_H(\int_{t\in T} X) = \inf\{D_H(X) \mid t \in T\}$		MDP7

Table 13. Minimal delay with priority.

4.3 BOUNDED INITIALISATION WITH PRIORITY.
Next, we define the operator $\gg_H$. This operator is just like the bounded initialisation operator of 2.5, but if the timestamp left coincides with the time right, actions from H will survive, but actions outside H will not. In table 14, $H \subseteq A$, $r \geq 0$, $a \in A$, $\alpha,\beta \in AS$, $x,y \in P$, X a process with time variable t. We see that the bounded initialisation operator $\gg$ is just $\gg_\varnothing$.

$x \gg_H \infty = x$		BIP1
$a(t) \gg_H r = a(t)$	if $t < r$ or $t = r$ and $a \in H$	BIP2
$a(t) \gg_H r = \delta(r)$	if $t > r$ or $t = r$ and $a \notin H$	BIP3
$(\alpha \,\&\, \beta)(t) \gg_H r = (\alpha(t) \gg_H r) \,\&\, (\beta(t) \gg_H r)$		BIP4
$(x + y) \gg_H r = (x \gg_H r) + (y \gg_H r)$		BIP5
$(x\cdot y) \gg_H r = (x \gg_H r)\cdot y$		BIP6
$(\int_{t\in T} X) \gg_H r = \int_{t\in T} (X \gg_H r)$		BIP7

Table 14. Bounded initialisation with priority.

4.4 UNLESS OPERATOR.

Now with the use of these minimal delay and bounded initialisation operators with priority the unless operator can be defined easily:

$$x \lhd_H y = x \gg_H D_H(y).$$

The typical axioms for the unless operator (see [BABK86]) can now be derived.

Returning to example 3.5, we see that the correct expression for the system is as follows:

$$\theta_2 \Downarrow_D \circ \lambda^2_\emptyset(S(0) \parallel R(0)).$$

4.5 EXAMPLE.

A consequence of axiom TH4 is that we have the following identity: $\theta_H(\int_{t\in(1,2)} h(t)) = \delta(1)$, if $h \in H$. This is because each h action is canceled by one with lesser timestamp. We can call this phenomenon a *priority deadlock*. The equation can be generalized to the following: $\theta_H(x + \int_{t\in(1,2)} h(t)) = \theta_H(x) \gg_H 1$ for all x.

4.6 EXAMPLE.

We now discuss a larger example, a protocol transmitting data via a mobile intermediate station.

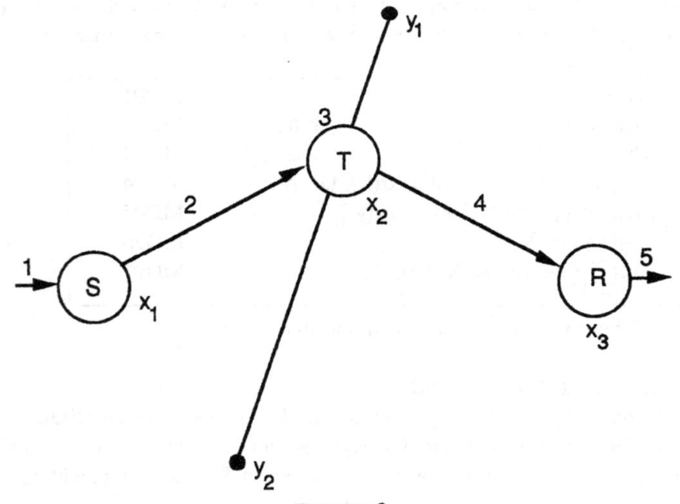

FIGURE 2.

In figure 2, we have a sender S at location x_1, a receiver R at location x_3, and a transmitter T that is moving on a line between locations y_1 and y_2, starting at y_1 at time t=0. The location of the transmitter is given by the following formula:

$$x_2(t) = y_1 + (\tfrac{1}{2} - \tfrac{1}{2}\cos(\omega t))\cdot(y_2 - y_1).$$

The transmitter consists of two parts, a receiving part and a sending part. These parts are interconnected by a synchronous communication port 3. 1 and 5 are also synchronous ports, and 2 and 4 are asynchronous channels. We have system delay constants w_1, w_2, w_3, w_4.

The components are now given by the following specification:

$$S(r) = \int_{t \geq r} \sum_{d \in D} r1(d)(x_1, t) \cdot 2{\uparrow}d(x_1, t+w_1) \quad \cdot S(t + 2w_1) \qquad (r \geq 0)$$

$$T_R(r) = \int_{t \geq r} \sum_{d \in D} 2{\downarrow}d(x_2(t), t) \cdot s3(d)(x_2(t+w_2), t+w_2) \quad \cdot T_R(t + 2w_2) \qquad (r \geq 0)$$

$$T_S(r) = \int_{t \geq r} \sum_{d \in D} r3(d)(x_2(t), t) \cdot 4{\uparrow}d(x_2(t+w_3), t+w_3) \quad \cdot T_S(t + 2w_3) \qquad (r \geq 0)$$

$$R(r) = \int_{t \geq r} \sum_{d \in D} 4{\downarrow}d(x_3, t) \cdot s5(d)(x_3, t+w_4) \quad \cdot S(t + 2w_4) \qquad (r \geq 0).$$

Now the system is given by the identity:

$$SYS = \partial_{sr3D}\big(\theta_{2{\Downarrow}D} \circ \lambda^2_{\emptyset}(S(0) \parallel T_R(0)) \parallel \theta_{4{\Downarrow}D} \circ \lambda^4_{\emptyset}(T_S(0) \parallel R(0))\big).$$

The protocol works well if the data enter at a sufficiently low frequency.

4.7 THE REPLACEMENT OPERATOR: STRUCTURED NOTATION FOR A MOBILE PROCESS.

Example 4.6 demonstrates a process T that is mobile in the sense that its actions take place at locations that vary in time. In this section, we will have a closer look at the description of such mobile processes.

Let $f: \mathbb{R}^3 \times \mathbb{R}^{\geq 0} \rightarrow \mathbb{R}^3$ be a continuous function. The operator ρ_f modifies each action $a(x, t)$ of a process P into $a(f(x,t), t)$. Thus, at time t, all spatial coordinates are translated according to the mapping $f_t: x \mapsto f(x, t)$. The equations for ρ_f are very simple (table 15).

$\rho_f(\delta(t)) = \delta(t)$	RRN1
$\rho_f(a(x, t)) = a(f(x,t), t)$	RRN2
$\rho_f((\alpha \,\&\, \beta)(t)) = \rho_f(\alpha(t)) \,\&\, \rho_f(\beta(t))$	RRN3
$\rho_f(x + y) = \rho_f(x) + \rho_f(y)$	RRN4
$\rho_f(x \cdot y) = \rho_f(x) \cdot \rho_f(y)$	RRN5
$\rho_f(\int_{t \in T} X) = \int_{t \in T} \rho_f(X)$	RRN6

Table 15. Replacement operator.

ρ_f is a renaming but in fact it affects spatial coordinates only, and not action labels. We prefer to call ρ_f a *replacement operator*.

In example 4.6, we can rephrase the definition of T as follows: let x be a point in space. Then we can define

$$f(x, t) = x + (\tfrac{1}{2} - \tfrac{1}{2}\cos(\omega t)) \cdot (y_2 - y_1).$$

Now put

$$T^*_R(x, r) = \int_{t \geq r} \sum_{d \in D} 2{\downarrow}d(x, t) \cdot s3(d)(x, t+w_2) \cdot T_R(t + 2w_2),$$

$$T^*_S(x, r) = \int_{t \geq r} \sum_{d \in D} r3(d)(x, t) \cdot 4{\uparrow}d(x, t+w_3) \cdot T_S(t + 2w_3).$$

Now the processes in 4.6 are given by: $T_R(t) = \rho_f(T^*_R(y_1, t))$, $T_S(t) = \rho_f(T^*_S(y_1, t))$.

Using this notation we obtain a better modular structure. The system specification becomes:

$$SYS = \partial_{sr3D}\big(\theta_{2{\Downarrow}D} \circ \lambda^2_{\emptyset}(S(0) \parallel \rho_f(T^*_R(y_1, 0))) \parallel \theta_{4{\Downarrow}D} \circ \lambda^4_{\emptyset}(\rho_f(T^*_S(y_1, 0)) \parallel R(0))\big).$$

Now suppose that the intermediate process T oscillates in a direction perpendicular to the segment (y_1, y_2) as well: let two points z_1, z_2 be given such that $(y_1, y_2) \perp (z_1, z_2)$. Define

$$g(x, t) = x + (\tfrac{1}{2} - \tfrac{1}{2}\cos(\omega^* t))\cdot(z_2 - z_1),$$

then the combined motion of T is given by the expressions $\rho_g \circ \rho_f(T^*_R(y_1, 0))$, $\rho_g \circ \rho_f(T^*_S(y_1, 0))$.

Thus, superposition of motions corresponds to composition of the corresponding replacement operators.

4.8 SIGNAL INTENSITY.

We proceed by discussing some variations in the communication mechanism. First of all, suppose we want to describe asynchronous communication with a decrease in signal intensity.

Assume that data are sent with intensity i_0 and that intensity decreases proportionally to the square of the distance traveled. So if i_0 is the signal intensity at distance 1 from the source, the signal intensity at distance is $a \cdot i_0 \cdot {}^{-2}$ for some constant $a \geq 0$. Let the receiving actions be equipped with an additional real parameter j that indicates at which intensity a signal can be received. Then the action function will allow $c\downarrow(d, j)(y, r)$ provided there exist x,t such that d was sent at (x, t) and

i. $|y - x| = v \cdot (r - t)$
ii. $a \cdot i_0 \cdot |y - x|^{-2} = j$.

A receiver that allows to receive messages between intensities lo, hi and between times fi, la will show integration over the intensity interval:

$$\text{Receiver} = \int_{j \in [lo, hi]} \int_{t \in [fi, la]} \sum_{d \in D} c\downarrow(d, j)(y(t), t) \cdot P(d, t).$$

4.9 IMPENETRABLE OBJECTS.

Next, we look at how to model a signal transmission space with an impenetrable object. Let B(2, z) be the closed ball with radius 2 and center z. In this case, the action function works as follows:

$action_c(c\uparrow d(x,t), V) = \delta(t)$ if $x \in B(2, z)$ (i.e. if $|x - z| \leq 2$)
$action_c(c\uparrow d(x,t), V) = c\Uparrow d(x,t)$ if $x \notin B(2, z)$
$effect_c(c\uparrow d(x,t), V) = V$ if $x \in B(2, z)$
$effect_c(c\uparrow d(x,t), V) = V \cup \{\langle d,x,t\rangle\}$ if $x \notin B(2, z)$

$action_c(c\downarrow(d,j)(y,r), V) = c\Downarrow d(y,r)$ if there is $\langle d, x, t\rangle \in V$ such that $|y - x| = v\cdot(r - t)$ and $a \cdot i_0 \cdot |y - x|^{-2} = j$ and the line segment from y to x does not intersect B(2, z)
$action_c(c\downarrow(d,j)(y,r), V) = \delta(r)$ otherwise.

As usual, $c\downarrow d$ actions have no effect on the state, so the effect function does not change the state.

4.10 REFLECTION.

A more complex example is obtained if one allows reflection of the signal. Carrying on with the previous example, we allow the signal to be reflected at the surface of the ball B(2, z). We assume perfect reflection without loss of signal intensity. Like before, we allow that a signal (or various signals) is simultaneously delivered at (y,r) through different paths. This can happen even with one source. It is possible that a signal d is sent by executing $c\uparrow d(x,t)$ and it reaches (y,r) along two different paths, both containing a reflection. In the case of the ball B(2, z) this is not possible of course, but with two balls this could happen indeed. The action function remains the same in the case of $c\uparrow d(x,t)$ but for $c\downarrow(d,j)(y,r)$ we get:

$action_c(c{\downarrow}(d,j)(y,r), V) = c{\Downarrow}d(y,r)$

if (1) there is $\langle d, x, t\rangle \in V$ such that $|y - x| = v\cdot(r - t)$ and $a\cdot i_0\cdot |y - x|^{-2} = j$ and the line segment from y to x does not intersect $B(2, z)$;

or (2) there is $\langle d, x, t\rangle \in V$ and w on the boundary of $B(2, z)$ such that $|y - w| + |w - x| = v\cdot(r - t)$, $a\cdot i_0\cdot(|y - w| + |w - x|)^{-2} = j$, x,y,z,w are in one plane, and (z,w) bisects the angle (x,w,y);

$action_c(c{\downarrow}(d,j)(y,r), V) = \delta(r)$ otherwise.

A state operator based on this action function describes an environment in which asynchronous communication through a non-trivial physical medium is supported. We remark that if the sending process also receives reflected signals, we are dealing with remote sensing, rather than asynchronous communication.

Interesting complications arise if we intend to model an asynchronous communication where, firstly, there are several independently moving objects that are impenetrable for signals but that do allow reflections with varying reflection coefficients, and, secondly, these objects have their motion guided by actions (in a discontinuous way) and by laws of classical mechanics between actions. This is in fact what happens if a sensor is used to assist a robot in navigating through a number of independently moving objects, each having its own surface characteristics.

It is our impression that the language of classical real space process algebra with priority and state operators is not sufficient to model the sensors needed for robot navigation in a dynamic environment.

5. Asynchronous message passing expressed in terms of synchronous communication.

In many cases, asynchronous message passing can be expressed using the available synchronous communication mechanism of ACPρ. In this approach, which appears in many forms throughout the literature, a process is introduced that represents the medium through which the data are being transported.

In this section we will provide various examples of such processes, representing an asynchronous transport medium. It turns out that a real time version of the process creation mechanism of [BE90] is useful to define a variety of such processes in a uniform way. For that purpose we describe process creation first, a mechanism which is of independent interest also.

It should be noted that the use of the state operator in the previous sections has been motivated by our inability to describe the particular form of asynchronous message transfer of example 4.6 in the more traditional fashion of this section. This in no way excludes that an elegant description of example 4.6 based on synchronous communication is possible, but our search has been without success.

5.1 Process creation.

For the description of process creation, we assume that we have special disjoint subsets $cr(D) = \{cr(d) \mid d \in D\}$ and $\overline{cr}(D) = \{\overline{cr}(d) \mid d \in D\}$ within the set of symbolic atomic actions A, with $cr(d) \mid a = \overline{cr}(d) \mid a = \delta$ for all $d \in D$ and $a \in A_\delta$.

Further, we assume the existence of a function $\phi: D \times \mathbb{R}^3 \times \mathbb{R}^{\geq 0} \to P$, and we require ϕ to be defined by means of guarded recursion equations. Like in [BE90], the function ϕ determines a process to be created from action $cr(d)$. In the real time case, $\phi(d, x, t)$ represents the process created from the timed action $cr(d)(x, t)$.

Next we use, as in the symbolic case, an operator E_ϕ which enforces process creation from $cr(d)$ actions occurring in its scope. The equations of E_ϕ are in table 16. In order to state these axioms, we need some extra notation. We need to be able to determine the set $CR(\alpha) = \{\langle d, x\rangle \in D \times \mathbb{R}^3 \mid cr(d)$ occurs in $\alpha\}$ for a multi-action α. This set can be defined recursively as follows:

$$CR(a(x)\ \&\ \alpha) = CR(\alpha) \qquad \text{if } a \notin cr(D)$$
$$CR(cr(d)(x)\ \&\ \alpha) = \{\langle d, x\rangle\} \cup CR(\alpha).$$

Also, we need the notation $\overline{\alpha}$ for the multi-action α, where all $cr(d)$ actions are changed into $\overline{cr}(d)$ actions, i.e.

$$\overline{a(x)\ \&\ \alpha} = a(x)\ \&\ \overline{\alpha} \qquad \text{if } a \notin cr(D)$$
$$\overline{cr(d)(x)\ \&\ \alpha} = \overline{cr}(d)(x)\ \&\ \overline{\alpha}.$$

$E_\phi(\alpha(t)) = \alpha(t)$	if $CR(\alpha) = \varnothing$	PCT1
$E_\phi(\alpha(t)) = \overline{\alpha}(t)\cdot E_\phi(\underset{\langle d,x\rangle\in CR(\alpha)}{\Vert} \phi(d,x,t))$	otherwise	PCT2
$E_\phi(\alpha(t)\cdot X) = \alpha(t)\cdot E\phi(X)$	if $CR(\alpha) = \varnothing$	PCT3
$E_\phi(\alpha(t)\cdot X) = \overline{\alpha}(t)\cdot E_\phi(X \Vert \underset{\langle d,x\rangle\in CR(\alpha)}{\Vert} \phi(d,x,t))$	otherwise	PCT4
$E_\phi(x + y) = E_\phi(x) + E_\phi(y)\}$		PCT5
$E_\phi(\int_{t\in T} X) = \int_{t\in T} E_\phi(X)$		PCT6

Table 16. Process creation.

The operational semantics of the process creation operator now follows easily on the basis of the axioms. Note that the process creation mechanism can be used in real time (without space) just as well. All examples of [BE90] can be adapted to a real time setting. We concentrate here on examples concerning message transport media.

5.2 MESSAGE HANDLER.

The following recursive specification defines a process that, upon receiving an input, creates a message handler to take care of the input.

$$M(r) = \int_{t>r+t_0} \sum_{d\in D} r(d)(x(t), t) \cdot cr(d)(x(t+t_1), t+t_1) \cdot M(t+t_1) \qquad (\text{for } r>0).$$

Here t_0 is a parameter that determines a minimum delay between the creation of the message handler and reception of a new input, whereas t_1 determines the delay between reception of a datum and creation of the corresponding handler. $x(t)$ is some function from $\mathbb{R}^{\geq 0}$ to $\mathbb{R}^3$ that determines the input location of the medium. We assume that the handler is created at that very same logical location but with a delay of t_1 (i.e. at $x(t + t_1)$).

Next, we provide possible functions ϕ_n: $D \times \mathbb{R}^3 \times \mathbb{R}^{\geq 0} \to P$ that determine the handler created. This leads to an asynchronous message transport medium $M_n = E_{\phi_n}(M)$ for each case. In all cases, $y(x, r)$ is the output location of the medium at time $t + r$ if at time t the input location is x. Thus, input at (x, t) leads to output at $(y(x, \Delta), t+\Delta)$ for some $\Delta > 0$.

CASE 1: $\phi_1(d, x, t) = s(d)(y(x, \Delta_1), t+\Delta_1)$.

In this case data need a constant time Δ_1 to travel through the medium. Moreover, the medium introduces no errors or omissions.

CASE 2: $\phi_2(d, x, t) = s(d)(y(x, \Delta_2(t)), t+\Delta_2(t))$.

In this case, the transmission time depends on t. This happens e.g. if the output location is moving relative to the input location. Notice that the difference with example 4.6 is that there, due to the broadcasting nature of the mechanism, the output location is not known in the same way. In particular, the function $\Delta_2(t)$ is not easy to define in that example, even if one assumes that broadcasting is irrelevant and each message is received exactly once. This is caused by the fact that in example 4.6 the output location (i.e. location (x, t) at which an action $c\Downarrow d(x, t)$ happens) must be derived from the combined behavior of *two* processes.

CASE 3: $\phi_3(d, x, t) = \int_{r\in(\Delta_1-\varepsilon_1,\Delta_1+\varepsilon_1)} s(d)(y(x, r-t), r).$

M_3 is like M_1, be it that there is a tolerance ε_1 in the arrival time. If $\varepsilon_1 > \frac{1}{2}(t_0 + t_1)$, complications can arise because M_3 may deliver different messages at the same time and place. Assuming $\varepsilon_1 < \frac{1}{2}(t_0 + t_1)$, this unwanted interference is not possible.

CASE 4: $\phi_4(d, x, t) = \int_{r\in(\Delta_2(t)-\varepsilon_1,\Delta_2(t)+\varepsilon_1)} s(d)(y(x, r-t), r).$

This example adds an arrival time tolerance to example 2.

Next, we assume that the orbit of a message d traveling from (x,t) to $(y(x,\Delta), t+\Delta)$ is given by $(z(r-t), r)$, so in particular, we have $(x,t) = (z(0), t+0)$ and $(y(x,\Delta), t+\Delta) = (z(\Delta), t+\Delta)$.

CASE 5: $\phi_5(d, x, t) = s(d)(y(x, \Delta_1), t+\Delta_1) + \int_{r\in(0,\Delta_1)} lost(z(r), t+r) \cdot s(\perp)(y(x,\Delta_1), t+\Delta_1).$

This example allows the datum to be changed into $\perp$ anywhere during transmission. In particular, M_5 keeps track where a datum gets lost. In the next example, M_6 will not deliver lost data.

CASE 6: $\phi_6(d, x, t) = s(d)(y(x, \Delta_1), t+\Delta_1) + \int_{r\in(0,\Delta_1)} lost(z(r), t+r).$

CASE 7: $\phi_7(d, x, t) = s(d)(y(x, \Delta_1), t+\Delta_1) + \int_{r\in(0,\Delta_1)} lost(z(r), t+r) \cdot cr(d)(x(t+r+\varepsilon_2), t+r+\varepsilon_2).$

M_7 will spontaneously retransmit lost data (very unlikely in practice). Note that this protocol can lead to collisions of different cr actions.

5.3 A FAULTY QUEUE FOR ASYNCHRONOUS MESSAGE TRANSFER.

In this section we will start out from the following description of a queue, that transports data in D from x to y in time Δ.

$$Q = \int_{t>0} \sum_{d\in D} r(d)(x,t) \cdot (s(d)(y, t+\Delta) \parallel Q).$$

This queue is the classical real space absolute time version of example 10.1 of [BAB91a], with port names 1,2 replaced by locations x,y.

We intend to describe a related queue that shows faults. In order to do this, the mode transfer operators $\rightarrow$ and $\hookrightarrow$ of [BE89] are casted in a real time and space setting. Informally, $X \rightarrow Y$ is a process that behaves like X but can at any time start behaving like Y (provided it decides to do so before X has terminated). $X \hookrightarrow Y$ is like $X \rightarrow Y$ with the additional constraint that its first action must be taken from X.

Using the mode transfer operator (to be discussed in detail below), a faulty version of the queue can be given as follows: let $r \in \mathbb{R}$, $n \in \mathbb{N}$, define

$Y(r, n) = s(\perp)(y, n\cdot r) \cdot Y(r, n+1).$

$Y(r, n)$ will produce erroneous signals at regular intervals. Now define Q' by

$$Q' = Q \to \sum_{n \in \omega} Y(r, n).$$

This process allows the queue to transmit data until it switches to mode Y.

Finally, we describe a queue that can be put back in the original mode by means of an action restart(z, t). Here, z is a location from which the queue is controlled.

$$Q'' = Q \to \Big(\sum_{n \in \omega} Y(r, n) \hookrightarrow \int_{t>0} \text{restart}(z, t) \cdot Q''\Big).$$

5.4 MODE TRANSFER.

The equations for the mode transfer operators are as follows ($\alpha \in AS$).

$\alpha(t) \to X = \alpha(t) + (X \gg t)$		MTT1
$\alpha(t) \cdot X \to Y = \alpha(t) \cdot (X \to Y) + (Y \gg t)$		MTT2
$(X + Y) \to Z = (X \to Z) + (Y \to Z)$		MTT3
$(\int_{t \in T} X) \to Y = \int_{t \in T} (X \to Y)$	(t not free in Y)	MTT4
$\alpha(t) \hookrightarrow X = \alpha(t)$		DMTT1
$\alpha(t) \cdot X \hookrightarrow Y = \alpha(t) \cdot (X \to Y)$		DMTT2
$(X + Y) \hookrightarrow Z = (X \hookrightarrow Z) + (Y \hookrightarrow Z)$		DMTT3
$(\int_{t \in T} X) \hookrightarrow Y = \int_{t \in T} (X \hookrightarrow Y)$	(t not free in Y)	DMTT4

Table 17. Mode transfer.

5.5 REMARK.

The equations for mode transfer in [BE89] are as follows:

$$a \to X = a + X$$
$$\delta \to X = \delta$$
$$a \cdot X \to Y = a \cdot (X \to Y) + Y$$
$$(X + Y) \to Z = (X \to Z) + (Y \to Z).$$

In particular the equation $\delta \to X = \delta$ is not obvious. However, if a system has deadlocked, it seems impossible to recover from that deadlock in a context $X \to Y$. In the real time case, we see a more differentiated picture:

$$\delta(2) \to a(3) \cdot b(4) = \delta(2)$$

while $\delta(2) \to a(1) \cdot b(4) = \delta(2) + a(1) \cdot b(4)$.

Likewise $c(2) \to a(3) \cdot b(4) = c(2)$ and $c(2) \to a(1) \cdot b(4) = c(2) + a(1) \cdot b(4)$.

5.6 REMARK.

We remark that the signal/observation mechanism of [BAB91d] can also be casted in a real time (and space) setting. In that case, this mechanism will provide yet another way to express asynchronous communication.

6. CONCLUSION.

We conclude that we have introduced a setting in which asynchronous communication can be adequately described in classical real space process algebra. This allows to describe communication between processes moving in space (e.g. communication with a satellite).

We claim that we can also describe such communications in the relativistic real space process algebra of [BAB91b]. In that case, because of the presence of the 0 process, the priority operator is not necessary in the description (all timed deadlocks are equated to 0).

REFERENCES.

[BAB88] J.C.M. BAETEN & J.A. BERGSTRA, *Global renaming operators in concrete process algebra,* Information & Computation 78 (3), 1988, pp. 205-245.

[BAB91a] J.C.M. BAETEN & J.A. BERGSTRA, *Real time process algebra,* Formal Aspects of Computing 3 (2), 1991, pp. 142-188. (Report version appeared as report P8916, Programming Research Group, University of Amsterdam 1989.)

[BAB91b] J.C.M. BAETEN & J.A. BERGSTRA, *Real space process algebra,* in Proc. CONCUR'91, Amsterdam (J.C.M. Baeten & J.F. Groote, eds.), Springer LNCS 527, 1991, pp. 96-110.

[BAB91c] J.C.M. BAETEN & J.A. BERGSTRA, *The state operator in real time process algebra,* report P9104, Programming Research Group, University of Amsterdam 1991. To appear in Proc. REX Workshop Real-Time: Theory in Practice, Mook 1991.

[BAB91d] J.C.M. BAETEN & J.A. BERGSTRA, *Process algebra with signals and conditions,* report CS-R9103, CWI Amsterdam 1991. To appear in Proc. NATO Summer School, Marktoberdorf 1990 (M. Broy et al., eds.), Springer Verlag.

[BABK86] J.C.M. BAETEN, J.A. BERGSTRA & J.W. KLOP, *Syntax and defining equations for an interrupt mechanism in process algebra,* Fund. Inf. IX (2), 1986, pp. 127-168.

[BAW90] J.C.M. BAETEN & W.P. WEIJLAND, *Process algebra,* Cambridge Tracts in Theor. Comp. Sci. 18, Cambridge University Press 1990.

[BE85] J.A. BERGSTRA, *Put and get, primitives for synchronous unreliable message passing,* report LGPS 3, Dept. of Philosophy, Utrecht University 1985.

[BE89] J.A. BERGSTRA, *A mode transfer operator in process algebra,* report P8808b, Programming Research Group, University of Amsterdam 1989.

[BE90] J.A. BERGSTRA, *A process creation mechanism in process algebra,* in: Applications of Process Algebra (J.C.M. Baeten, ed.), Cambridge Tracts in TCS 17, Cambridge University Press 1990, pp. 81-88.

[BEK84] J.A. BERGSTRA & J.W. KLOP, *Process algebra for synchronous communication,* Inf. & Control 60, 1984, pp. 109-137.

[BEKT85] J.A. BERGSTRA, J.W. KLOP & J.V. TUCKER, *Process algebra with asynchronous communication mechanisms,* in: Proc. Seminar on Concurrency (S.D. Brookes, A.W. Roscoe & G. Winskel, eds.), Springer LNCS 197, 1985, pp. 76-95.

[J91a] A. JEFFREY, *A linear time process algebra,* report 61, Programming Methodology Group, Chalmers University 1991, in Proc. CAV 91, Aalborg 1991.

[J91b] A. JEFFREY, *Observation spaces and timed processes,* in Proc. CONCUR'91, Amsterdam (J.C.M. Baeten & J.F. Groote, eds.), Springer LNCS 527, 1991, pp. 332-345.

[J91c] A. JEFFREY, *Translating timed process algebra into prioritized process algebra,* Programming Methodology Group, Chalmers University 1991.

[K91] A.S. KLUSENER, *Completeness in real time process algebra,* report CS-R9106, CWI Amsterdam 1991. Extended abstract in Proc. CONCUR'91, Amsterdam (J.C.M. Baeten & J.F. Groote, eds.), Springer LNCS 527, 1991, pp. 376-392.

[MT90] F. MOLLER & C. TOFTS, *A temporal calculus of communicating systems,* in: Proc. CONCUR'90, Amsterdam (J.C.M. Baeten & J.W. Klop, eds.), Springer LNCS 458, 1990, pp. 401-415.

[MU91] D.V.J. MURPHY, *Testing, betting and timed true concurrency,* in Proc. CONCUR'91, Amsterdam (J.C.M. Baeten & J.F. Groote, eds.), Springer LNCS 527, 1991, pp. 439-454.

[RR88] G.M. REED & A.W. ROSCOE, *A timed model for communicating sequential processes,* TCS 58, 1988, pp. 249-261.

Translating timed process algebra into prioritized process algebra

ALAN JEFFREY

ABSTRACT. A process algebra with priority and time is presented, based on linear timed CCS. It is given a set of axioms and a transition system semantics. The axioms are then shown to be equivalent to strong bisimulation over the transitions. It is shown how to translate any timed process into the untimed algebra, by time-stamping each action. Thus, maximal progress is shown to be a specific example of maximal priority.

1 Introduction

In [Wan90, Wan91], Wang introduced a timed variant of Milner's Calculus for Communicating Systems (CCS) [Mil80, Mil89]. In [Jef91] the notion of time was generalized to that of a totally ordered monoid, and a complete axiomatization was provided. In this paper, we investigate a variant of timed CCS with the added notion of *prioritized events*. It turns out that the notion of time is not actually required, as we can translate the timed calculus into a calculus with priority but no notion of time.

In constructing a process algebra with priorities, there are at least three different approaches that can be taken:

- Associate priority with *choice*, as in [Cam90]. This is similar to the approach taken to stochastic choice in many of the probabilistic algebras [LS89, Chr90, HJ90, Tof90], and indeed priority can be seen as an extreme example of stochastic choice. If $P +_p Q$ offers the environment choice between P and Q, but in the event of ambiguity chooses Q with probability p, then prioritized choice can be seen as $P +_0 Q$ [SS90].
- Associate priority with *events*, as in [CH88, BBK86]. This uses a priority function π from events to priorities. If a has higher priority than b, then $a.P + b.Q$ can perform an a, and can perform a b if not offered an a. To represent this, we replace the CCS silent action $\tau.P$ with many silent actions $\tau p.P$, one for each priority p. Then if we hide a from $a.P + b.Q$, we get $\tau\pi_a.P + b.Q$, which is equivalent to $\tau\pi_a.P$. This is the approach taken here.
- Associate priority with events *locally*. Wang (in conversation) has suggested that rather than having a fixed priority function, that the priorities for actions only need be given by the hiding operator. For example, if we write $P \setminus_p a$ to mean 'P, hiding a and giving it priority p', then $(a.P + b.Q) \setminus_p a \setminus_q b$ is equivalent to $\tau p.(P \setminus_p a \setminus_q b) + \tau q.(Q \setminus_p a \setminus_q b)$. Then if p is a higher priority

Presented at the Nijmegen Symposium on Real-Time and Fault-Tolerant Systems, 1992.
Author's address: Department of Computer Sciences, S-412 96 Göteborg, Sweden.
E-mail: jeffrey@cs.chalmers.se.

than q, this is equivalent to $\tau p.(P \setminus_p a \setminus_q b)$. It seems that providing a local priority function in this way allows for a simpler semantics (our rather clumsy definition of $P \downarrow p$ below can be given much more smoothly), but unfortunately we cannot translate a timed language into it, as there is no 'maximal priority' equivalent to the 'maximal progress' law for time.

In constructing this calculus, a number of design decisions were made, similar to those in [Jef91]:

- We are interested in modelling maximal progress. This is probably the most important decision in this calculus, and is the reason for investigating priority in the first place. Maximal progress states that if a process can perform some internal computation, then it does so immediately, rather than idling. For example, we will write $\varepsilon t.P$ for P delayed by time t, so maximal progress can be expressed as the law $\tau p.P + \varepsilon t.Q = \tau p.P$ when $t \neq 0$. Later, we shall see that maximal progress is an example of maximal priority.
- We allow instantaneous prefix, so the process $a.P$ can perform an a at time 0, and then become P immediately. This has the advantage of allowing a complete axiomatization using the expansion laws: for example, $a.0 \mid\mid\mid b.0$ is equivalent to $a.b.0 + b.a.0$. Unfortunately, this means we have specified that we have an interleaving semantics, and opens up the long-running argument (see for example [P+90]) of 'true' versus 'false' concurrency.
- We allow time-stop processes, that is, processes which can perform no actions and will not allow time to pass. Time-stop processes occur in many timed algebras [MT90, NS90, BB91, HR91, Jef91, Wan91], usually as a result of unguarded recursion. In this paper, we do not allow unguarded recursion, but we can still produce time-stop processes with infinite summation. If, for example, our time domain is the real numbers, then

$$\sum\{\varepsilon t.\tau p.P \mid t \neq 0\}$$

cannot perform any actions at time 0, but also cannot allow any time to pass, since if the process were to delay by time $u \neq 0$, this would ignore the internal action available at time $u/2$, which contradicts maximal progress. So this process cannot perform any actions, and cannot allow time to pass, so it is a time-stop. Similarly, the process

$$\sum\{\tau p.P \mid p \in \phi\}$$

is a priority-stop process if ϕ is a non-empty set of priorities with no maximal element. This is the prioritized equivalent of a time-stop process, and is to be expected if we can to translate every timed process into a prioritized one.

- We are using a rather non-standard syntax, made up in equal parts of CCS and Hoare's Communicating Sequential Processes (CSP) [BHR84, Hoa85]. This is because we are looking at bisimulation (for which the CCS prefix and summation operators are suited) but also need the CSP synchronization operation $P \mid\mid Q$ in order to translate the timed algebra into the prioritized algebra.

This paper also opens up the interesting question 'If time is an example of priority, then why investigate time specifically, rather than priority in general?' Some possible answers to this include:

- A model designed specifically for time will probably be much simpler.
- Proof methods exist for reasoning about timed processes, such as temporal logic and the proof methodologies for timed CSP [Sch90, Dav91].
- The extra structure that time gives a model allows topological methods for finding fixed points for recursions [RR87].

2 Assumptions

We are going to define a timed process algebra similar to CCS and CSP, so we need the same notion of an atomic, instantaneous action.

ASSUMPTION. *$\mathbf{A}$ is a nonemtpy set, ranged over by a, b and c.*

We are interested in providing a notion of prioritized actions, so we also need a totally ordered set of priorities.

ASSUMPTION. *$(\mathbf{P}, \preccurlyeq)$ is a total order, ranged over by p and q.*

For example, two simple priority domains are $\mathbf{1}$ (where every action has the same priority) and $\mathbf{2}$ (where actions are either of high priority or low priority).

DEFINITION. $(\mathbf{1}, \preccurlyeq) = (\{0\}, \{(0,0)\})$

DEFINITION. $(\mathbf{2}, \preccurlyeq) = (\{0,1\}, \{(0,0),(0,1),(1,1)\})$

Since we require silent events (such as $\tau.P$ in CCS) to have priorities, we require a different silent action τp for each priority p. The process $\tau p.P$ means 'perform a silent action with priority p and become P immediately.'

DEFINITION. $\mathbf{E} = \mathbf{A} \cup \{\tau p \mid p \in \mathbf{P}\}$ *is ranged over by α, β and γ.*

We require every event α to have a priority π_α. For example, if $\pi_\alpha \prec p$, then α has lower priority than p, so $\tau p.P + \alpha.Q$ should be equivalent to $\tau p.P$.

ASSUMPTION. $\pi : \mathbf{E} \to \mathbf{P}$ *is such that* $\pi_{\tau p} = p$.

We are also interested in modelling time, so we require a set of times. Since we are dealing with relative time, rather than absolute time [BB91], we need to be able to add delays together. For example, $\varepsilon t.P$ means 'wait for t units of time, then become P', so we would expect $\varepsilon t.\varepsilon u.P$ to be equivalent to $\varepsilon(t+u).P$. This is the same notion of time as used in [Jef91], except we do not require the time domain to be continuous. Examples of suitable time domains include $(\mathbb{N}, +, 0)$, $([0,\infty), +, 0)$, $([0,\infty) \cap \mathbb{Q}, +, 0)$, and $(\mathbf{1}, +, 0)$.

ASSUMPTION. *$(\mathbf{T}, +, 0)$ is a monoid, ranged over by t, u, and v, such that $t + u = t + v \Rightarrow u = v$, and where the relation $t \leqslant v$ iff $\exists u \,.\, t + u = v$ is a total order.*

When we come to give the operational semantics, a transition will either be labelled by an action a, a silent event τp, or a non-zero delay εt.

DEFINITION. $\mathbf{L} = \mathbf{E} \cup \{\varepsilon t \mid t \neq 0\}$ *is ranged over by ρ and σ.*

In this calculus, we allow infinite summation $\sum \mathcal{P}$, for a set of processes $\mathcal{P}$. However, if we were to allow any set of processes, we would discover that the class of all processes did not form a set, so we need to put a limit λ on the size of sets $\mathcal{P}$ we are allowed to sum over. In addition, we would like the law $\sum\{\sum\mathcal{P}_i \mid i \in I\} = \sum\bigcup\{\mathcal{P}_i \mid i \in I\}$ to hold, so the set $\bigcup\{\mathcal{P}_i \mid i \in I\}$ should be smaller than λ, given that I and each $\mathcal{P}_i$ is smaller than λ—this is the definition of a regular cardinal.

ASSUMPTION. *λ is an uncountable regular cardinal strictly larger than $2^{|\mathbf{L}|}$. (A cardinal λ is regular iff $|\mathcal{I}| < \lambda$ and $\forall I \in \mathcal{I} . |I| < \lambda$ implies $|\bigcup\mathcal{I}| < \lambda$.)*

We also need a set of variables with which to define recursion. We are going to allow infinite mutual recursions, so we have to be a bit careful to make sure we never 'run out' of variables. We can do this by insisting that the set of variables has regular cardinality, and is larger than $|\mathbf{L}|^\lambda$.

ASSUMPTION. *$\mathbf{V}$ is a set, ranged over by x, y and z, such that $|\mathbf{V}|$ is a regular cardinal, strictly greater than $|\mathbf{L}|^\lambda$.*

3 Vectors

In this paper, we will be defining recursive terms using infinite vectors of terms rather than the usual single fixed point $\mu x . E$. This has useful theoretical implications—we can give our set of laws purely in terms of equational reasoning rather than requiring inequational approximations [Hen88]. It also allows us to define infinite-state processes like a counter, which could not otherwise be defined directly.

$$\begin{aligned} COUNT_0 &\mathrel{\widehat{=}} up.COUNT_1 \\ COUNT_{n+1} &\mathrel{\widehat{=}} up.COUNT_{n+2} + down.COUNT_n \end{aligned}$$

We can give such infinite recursions in terms of a vector of terms, that is a partial mapping from variables to terms. For example, the counter process could be given as the vector $COUNT$:

$$\begin{aligned} x_0 &\mapsto up.x_1 \\ x_{n+1} &\mapsto up.x_{n+2} + down.x_n \end{aligned}$$

In general, to make sure we do not 'run out' of free variables, we insist that every recursion is smaller than $|\mathbf{V}|$.

DEFINITION. *A vector of X is a partial function $V : \mathbf{V} \to X$, where $|V| < |\mathbf{V}|$.*

We can then define substitution in the usual way, except that we can perform infinitely many substitutions at once, by substituting with a vector. For example, $E[F/x]$ is $E \lhd \{x \mapsto F\}$.

DEFINITION. *$E \lhd V$ is E, with any free variable $x \in \operatorname{dom} V$ replaced by Vx (with the usual renaming to avoid binding variables).*

We can also substitute one vector with another, in such a fashion that $E \lhd (V \lhd W)$ is $(E \lhd V) \lhd W$, up to renaming bound variables.

DEFINITION. $(V \lhd W)x \equiv \begin{cases} Vx \lhd W & \textit{if } x \in \operatorname{dom} V \\ Wx & \textit{otherwise} \end{cases}$

For example, $COUNT \lhd COUNT$ is the vector:

$$\begin{aligned} x_0 &\mapsto up.(up.x_2 + down.x_0) \\ x_1 &\mapsto up.(up.x_3 + down.x_1) + down.up.x_1 \\ x_{n+2} &\mapsto up.(up.x_{n+4} + down.x_{n+2}) + down.(up.x_{n+2} + down.x_n) \end{aligned}$$

If we have a partial function $f : \mathbf{V} \rightarrow \mathbf{V}$ we can treat it as a simple vector, and use it to rename the variables of a vector. For example, if f maps x_n to y_n, then $(COUNT \lhd f) \circ f^{-1}$ is:

$$\begin{aligned} y_0 &\mapsto up.y_1 \\ y_{n+1} &\mapsto up.y_{n+2} + down.y_n \end{aligned}$$

We can use this later on as a way of renaming the variables of a vector.

4 Syntax

We can now give the syntax for our language, $SYN_{(\mathbf{A},\mathbf{P},\mathbf{T})}$. For example,

- $SYN_{(\mathbf{A},\mathbf{1},\mathbf{1})}$ is an untimed, unprioritized calculus, similar to CCS or CSP.
- $SYN_{(\mathbf{A},\mathbf{1},\mathbb{N})}$ is unprioritized and has natural time, similar to [HR90, MT90].
- $SYN_{(\mathbf{A},\mathbf{1},[0,\infty))}$ is unprioritized and has real time, similar to [Sch91, Wan91].
- $SYN_{(\mathbf{A},\mathbf{2},\mathbf{1})}$ is untimed, and has two priorities, similar to [CH88].
- $SYN_{(\mathbf{T}\times\mathbf{A},\mathbf{T}\times\mathbf{P},\mathbf{1})}$ is untimed, but all of its events are time-stamped. We shall see later on that $SYN_{(\mathbf{A},\mathbf{P},\mathbf{T})}$ can be modelled by $SYN_{(\mathbf{T}\times\mathbf{A},\mathbf{T}\times\mathbf{P},\mathbf{1})}$.

This language is a mixture of timed CCS (prefix and summation) and timed CSP (parallelism and interleaving).

DEFINITION. *$SYN_{(\mathbf{A},\mathbf{P},\mathbf{T})}$ is the language given by:*

$$E \ ::= \ a.E \mid \tau p.E \mid \varepsilon t.E \mid \sum \mathcal{E} \mid E \parallel E \mid E \interleave E \mid f[E] \mid x \mid \mu V x$$

where

- *E, and F are terms,*
- *$\mathcal{E}$ and $\mathcal{F}$ are sets of terms of cardinality strictly less then λ,*
- *$f : \mathbf{E} \rightarrow \mathbf{E}$ has $f(\tau p) = \tau\pi_{f(\tau p)}$, and $\pi_\alpha \prec \pi_\beta \Rightarrow \pi_{f\alpha} \prec \pi_{f\beta}$.*
- *V and W are vectors of guarded terms (a term is guarded iff every free variable is inside a prefix $a.E$ or $\tau p.E$).*

Let P, Q and R range over processes (that is terms with no free variables) and let $\mathcal{P}$ and $\mathcal{Q}$ range over sets of processes strictly smaller than λ.

The main difference between this syntax and Wang's timed CCS is that $a.P$ represents the process that can perform an a only at time 0. If $a.P$ idles, it will deadlock—it will still allow time to pass (unlike the prefix in [MT90]) but it will perform no more actions. However, we can define Wang's $a@t.P$ (which can perform an a at any time t and become P_t) using infinite sum.

DEFINITION. $a@t.P \equiv \sum\{\varepsilon t.a.P_t \mid t \in \mathbf{T}\}$

We can also define Hoare's RUN process, which can perform any action at any time, and is a unit of synchronization.

DEFINITION. $RUN \equiv \mu x \,.\, \sum\{\varepsilon t.a.x \mid t \in \mathbf{T} \wedge a \in \mathbf{A}\}$

This uses the single fixed point operator $\mu x \,.\, E$ which we can define in terms of our infinite operator $\mu V x$.

DEFINITION. $\mu x \,.\, E \equiv \mu\{x \mapsto E\}x$

At the moment we can give a fixed point of a vector of expressions, for example if $V \equiv \{x \mapsto a.y, b \mapsto b.x\}$, then $\mu V x$ is equivalent to $a.b.a.b \ldots$ This can be generalized to a fixed point vector, for example μV is $\{x \mapsto \mu V x, y \mapsto \mu V y\}$, which is equivalent to $\{x \mapsto a.b.a.b \ldots, y \mapsto b.a.b.a \ldots\}$.

DEFINITION. *μV is the vector such that $(\mu V)x \equiv \mu V x$.*

We can also define some other standard operators in terms of our calculus, such as finite summation and deadlock.

DEFINITION. $P + Q \equiv \sum\{P, Q\}$

DEFINITION. $\mathbf{0} \equiv \sum\emptyset$

We can also define the CSP hiding operator $P \setminus A$, since we are allowed to use alphabet renaming $f[P]$ to rename visible actions to invisible events.

DEFINITION. $P \setminus A \equiv hide_A[P]$ *where* $hide_A\,\alpha = \begin{cases} \tau\pi_\alpha & \text{if } \alpha \in A \\ \alpha & \text{otherwise} \end{cases}$

5 Laws

We can provide a set of laws, which we shall show are equivalent to strong bisimulation. The summation operator $\sum$ behaves in the normal fashion of CCS—$\sum\mathcal{E}$ offers the environment the choice between the initial events of the terms in $\mathcal{E}$, and allows time to pass if all the terms in $\mathcal{E}$ allow time to pass.

LAW (SUM UNIT). $\vdash \sum\{E\} = E$

LAW (SUM HOMOMORPHISM). $\vdash \sum\{\sum\mathcal{E}_i \mid i \in I\} = \sum\bigcup\{\mathcal{E}_i \mid i \in I\}$

We inherit the delay laws from [Wan91, Jef91], including maximal progress, which says that internal events should happen as soon as possible.

LAW (ZERO DELAY). $\vdash \varepsilon 0.E = E$

LAW (TIME CONTINUITY). $\vdash \varepsilon t.\varepsilon u.E = \varepsilon(t+u).E$

LAW (TIME DETERMINACY). $\vdash \varepsilon t.\sum\mathcal{E} = \sum\{\varepsilon t.E \mid E \in \mathcal{E}\}$

LAW (MAXIMAL PROGRESS). *If* $t \neq 0$ *then* $\vdash \tau p.E = \tau p.E + \varepsilon t.F$

To match the maximal progress law for time, there is a law saying that high priority internal events always happen in preference to low priority events.

LAW (MAXIMAL PRIORITY). *If* $\pi_\alpha \prec p$ *then* $\vdash \tau p.E = \tau p.E + \alpha.F$

Since we have two parallelism operators, $E \,||\, F$ and $E \,|||\, F$, there are two expansion laws, similar to those in [Wan91, Jef91]. The first says that in $E \,||\, F$, E and F have to synchronize their actions.

Law (synchronization expansion). *If*

$$\begin{aligned} E_t &\equiv \sum\{\varepsilon(t_i - t).a_i.E_i \mid i \in I\} + \sum\{\varepsilon(t_i - t).\tau p_i.E_i \mid i \in I'\} \\ F_u &\equiv \sum\{\varepsilon(u_j - u).b_j.F_j \mid j \in J\} + \sum\{\varepsilon(u_j - u).\tau q_j.F_j \mid j \in J'\} \end{aligned}$$

then

$$\begin{aligned} \vdash E_0 \,||\, F_0 \; = \; & \sum\{\varepsilon t_i.a_i.(E_i \,||\, F_j) \mid i \in I \wedge j \in J \wedge t_i = u_j \wedge a_i = b_j\} \\ & + \sum\{\varepsilon t_i.\tau p_i.(E_i \,||\, F_{t_i}) \mid i \in I'\} \\ & + \sum\{\varepsilon u_j.\tau q_j.(E_{u_j} \,||\, F_j) \mid j \in J'\} \end{aligned}$$

where $\varepsilon(t - u).\alpha.E \equiv \begin{cases} \varepsilon v.\alpha.E & \textit{if } u + v = t \\ \mathbf{0} & \textit{if } t < u \end{cases}$

The second says that $E \,|||\, F$ allows E and F to perform events independently.

Law (interleaving expansion). *If*

$$\begin{aligned} E_t &\equiv \sum\{\varepsilon(t_i - t).\alpha_i.E_i \mid i \in I\} \\ F_u &\equiv \sum\{\varepsilon(u_j - u).\beta_j.F_j \mid j \in J\} \end{aligned}$$

then

$$\vdash E_0 \,|||\, F_0 \; = \; \sum\{\varepsilon t_i.\alpha_i.(E_i \,|||\, F_{t_i}) \mid i \in I\} + \sum\{\varepsilon u_j.\beta_j.(E_{u_j} \,|||\, F_j) \mid j \in J\}$$

We inherit the laws of alphabet relabelling from untimed CSP—$f[P]$ relabels all the events of a process. The only differences are:

- The timing of a process is not affected by relabelling.
- Visible actions can be relabelled to internal events.
- Internal events can be relabelled to different internal events.

Note that since f has to be $\prec$-monotonic, it cannot affect the relative priorities of the events that are relabelled. For example, if $\pi_\alpha \prec p$, then

$$\vdash f[\tau p.E] = f[\tau p.E + \alpha.F] = f[\tau p.E] + f[\alpha.F] = f(\tau p).f[E] + f\alpha.f[F]$$

This is consistent, since f is $\prec$-monotonic, so $f\alpha \prec \pi_{f(\tau p)}$, so:

$$\vdash f(\tau p).f[E] + f\alpha.f[F] = f(\tau p).f[E]$$

Law (renaming prefix). $\vdash f[\alpha.E] = (f\alpha).f[E]$

Law (renaming delay). $\vdash f[\varepsilon t.E] = \varepsilon t.f[E]$

Law (renaming plus). $\vdash f[\sum\mathcal{E}] = \sum\{f[E] \mid E \in \mathcal{E}\}$

Recursion can be unfolded, for example $\vdash \mu x \,.\, a.x = a.(\mu x \,.\, a.x)$.

Law (unfolding). $\vdash \mu V = V \lhd \mu V$

We can also rename recursion, with a partial function $f : \mathbf{V} \to \mathbf{V}$. For example, if $f = \{y \mapsto x\}$, then $\{x \mapsto a.x\} \circ f \equiv \{y \mapsto a.x\} \equiv \{y \mapsto a.y\} \lhd f$ so, using recursion renaming, $\vdash \mu x \,.\, a.x = \mu y \,.\, a.y$.

Law (renaming). *If* $\vdash V \circ f = W \lhd f$ *then* $\vdash \mu V \circ f = \mu W$.

Finally, each vector has a unique fixed point, so if V is a fixed point of W, then it is equal to μV. For example, $\vdash \mu x \,.\, a.x = a.a.(\mu x \,.\, a.x)$ so, using recursion induction, $\vdash \mu x \,.\, a.x = \mu x \,.\, a.a.x$.

LAW (INDUCTION). *If* $\vdash V = W \lhd V$ *then* $\vdash V = \mu W$

We will see below that these laws are equivalent to strong bisimulation on closed terms.

6 Semantics

We shall give $SYN_{(\mathbf{A},\mathbf{P},\mathbf{T})}$ a transition system semantics in the style of Milner [Mil80, Mil89] and Plotkin [Plo81]. This semantics is similar to that of timed CCS [Wan91], linear timed CCS [Jef91] and timed CSP [Sch91]. It consists of three types of transitions:

- $P \xrightarrow{a} P'$ means P can perform a a action and become P'.
- $P \xrightarrow{\tau p} P'$ means P can perform an internal event τp and become P'.
- $P \xrightarrow{\varepsilon t} P'$ means P can idle for time t and become P'.

However, as Wang observed, we cannot give this transition relation directly, as (for example) the transition rule for when $f[P]$ can delay has to be:

$$\frac{P \xrightarrow{\varepsilon t} P' \quad f[P] \downarrow t}{f[P] \xrightarrow{\varepsilon t} f[P']}$$

Here, $Q \downarrow t$ means that Q allows time t to pass—in this case, it is possible that $P \downarrow t$ but not $f[P] \downarrow t$. For example, $a.P \downarrow t$, but $(a.P \backslash \{a\}) \not\downarrow t$ because $a.P \backslash \{a\}$ is capable of performing an internal event, and so cannot idle. So we need to find a way of defining $P \downarrow t$, and its equivalent for priorities, $P \downarrow p$.

DEFINITION. *The initial actions of a guarded term (ignoring maximal progress and maximal priority) are:*

$$\begin{aligned}
(\alpha.E)^0 &= \{(0,\alpha)\} \\
(\varepsilon t.E)^0 &= \{(t+u,\alpha) \mid (u,\alpha) \in E^0\} \\
(\textstyle\sum \mathcal{E})^0 &= \bigcup \{E^0 \mid E \in \mathcal{E}\} \\
(E \,||\, F)^0 &= \{(t,p) \mid (t,p) \in E^0 \cup F^0\} \cup (E^0 \cap F^0) \\
(E \,|||\, F)^0 &= E^0 \cup F^0 \\
(f[E])^0 &= \hat{f}[E^0] \\
(\mu V x)^0 &= (V x)^0
\end{aligned}$$

where $\hat{f}(t,\alpha) = (t, f\alpha)$.

So P^0 gives the initial actions P would be capable of if maximal progress and maximal priority did not intervene. From this, we can define $P \downarrow p$ and $P \downarrow t$. Informally, $P \downarrow p$ iff P will allow an event with priority p to happen. For example, $(\tau p.0)^0 = \{(0,\tau p)\}$, so $\tau p.0 \downarrow q$ iff $p \preccurlyeq q$. Similarly, $P \downarrow t$ iff P will allow time t to pass. For example, $\varepsilon u.\tau p.0 \downarrow t$ iff $t \leqslant u$.

DEFINITION. $P \downarrow p$ *iff* $\forall q \succ p \,.\, (0, \tau q) \notin P^0$.

$$\frac{}{\alpha.P \xrightarrow{\alpha} P} \qquad \frac{P \xrightarrow{\alpha} P'}{\varepsilon 0.P \xrightarrow{\alpha} P'} \qquad \frac{P \xrightarrow{\alpha} P' \quad \sum\mathcal{P} \downarrow \pi_\alpha}{\sum\mathcal{P} \xrightarrow{\alpha} P'} [P \in \mathcal{P}]$$

$$\frac{P \xrightarrow{a} P' \quad Q \xrightarrow{a} Q'}{P \,||\, Q \xrightarrow{a} P' \,||\, Q'} \qquad \frac{P \xrightarrow{\tau p} P' \quad Q \downarrow p}{P \,||\, Q \xrightarrow{\tau p} P' \,||\, Q} \qquad \frac{Q \xrightarrow{\tau p} Q' \quad P \downarrow p}{P \,||\, Q \xrightarrow{\tau p} P \,||\, Q'}$$

$$\frac{P \xrightarrow{\alpha} P' \quad Q \downarrow \pi_\alpha}{P \,|||\, Q \xrightarrow{\alpha} P' \,|||\, Q} \qquad \frac{Q \xrightarrow{\alpha} Q' \quad P \downarrow \pi_\alpha}{P \,|||\, Q \xrightarrow{\alpha} P \,|||\, Q'}$$

$$\frac{P \xrightarrow{\alpha} P' \quad f[P] \downarrow \pi_{f\alpha}}{f[P] \xrightarrow{f\alpha} f[P']} \qquad \frac{Vx \lhd \mu V \xrightarrow{\alpha} P'}{\mu V x \xrightarrow{\alpha} P'}$$

TABLE 1. Event transitions for $SYN_{(\mathbf{A},\mathbf{P},\mathbf{T})}$.

DEFINITION. *$P \downarrow t$ iff $\forall u < t \,.\, (u, \tau p) \notin P^0$.*

We can use $P \downarrow p$ and $P \downarrow t$ in defining the transitions of $SYN_{(\mathbf{A},\mathbf{P},\mathbf{T})}$.

DEFINITION. *The transition system $(SYN_{(\mathbf{A},\mathbf{P},\mathbf{T})}, \mathbf{L}, \longrightarrow)$ is given in Tables 1 and 2, where $\mathcal{P} \xrightarrow{t} \mathcal{Q}$ iff $\forall P \in \mathcal{P} \,.\, \exists Q \in \mathcal{Q} \,.\, P \xrightarrow{\varepsilon t} Q$ and $\forall Q \in \mathcal{Q} \,.\, \exists P \in \mathcal{P} \,.\, P \xrightarrow{\varepsilon t} Q$.*

For example, the transitions of $SYN_{(\mathbf{A},\mathbf{1},\mathbf{1})}$ are those of CCS and CSP [BRW86], and those of $SYN_{(\mathbf{A},\mathbf{1},[0,\infty))}$ are those of timed CCS [Wan91, Jef91] and timed CSP [Sch91]. We can then define a bisimulation on this transition system.

DEFINITION. *$\mathcal{R}$ is a bisimulation iff $P \,\mathcal{R}\, Q$ implies*

- *if $P \xrightarrow{\sigma} P'$ then there exists Q' such that $Q \xrightarrow{\sigma} Q'$ and $P' \,\mathcal{R}\, Q'$,*
- *if $Q \xrightarrow{\sigma} Q'$ then there exists P' such that $P \xrightarrow{\sigma} P'$ and $P' \,\mathcal{R}\, Q'$, and*
- *$P \downarrow p$ iff $Q \downarrow p$.*

$P \sim Q$ iff there exists a bisimulation $\mathcal{R}$ such that $P \,\mathcal{R}\, Q$.

DEFINITION. *$E \sim F$ iff for all closed $E \lhd V$ and $F \lhd V$, $E \lhd V \sim F \lhd V$.*

The only slight difference between this and the standard definition of bisimulation [Mil89] is the condition that $P \downarrow p$ iff $Q \downarrow p$. This is required, because although time-stop processes can be told apart by their transitions ($P \downarrow t$ iff $P \xrightarrow{\varepsilon t}$) priority-stop processes cannot. For example, if $P_p \equiv \sum\{\tau q.0 \mid q \prec p\}$ then in $SYN_{(\mathbf{A},[0,\infty),\mathbf{T})}$, P_4 has the same transitions as P_2 (that is, none at all). But $P_4 + \tau 3.0$ cannot perform an $\tau 3$ event, whereas $P_2 + \tau 3.0$ can. We can only tell apart P_2 and P_4 by observing that $P_4 \not\downarrow 3$, but $P_2 \downarrow 3$.

LEMMA 1. *$\sim$ is a bisimulation.*

PROOF. A variant on the proof of [Mil89]. □

LEMMA 2. *$\sim$ is a congruence.*

PROOF. Construct a bisimulation for each operator. □

$$\frac{}{a.P \xrightarrow{\varepsilon t} 0} \qquad \frac{}{\varepsilon(t+u).P \xrightarrow{\varepsilon t} u.P} \qquad \frac{P \xrightarrow{\varepsilon u} P'}{\varepsilon t.P \xrightarrow{\varepsilon(t+u)} P'}$$

$$\frac{\mathcal{P} \xrightarrow{\varepsilon t} \mathcal{Q}}{\sum\mathcal{P} \xrightarrow{\varepsilon t} \sum\mathcal{Q}} \qquad \frac{P \xrightarrow{\varepsilon t} P' \quad Q \xrightarrow{\varepsilon t} Q'}{P \,||\, Q \xrightarrow{\varepsilon t} P' \,||\, Q'} \quad \frac{P \xrightarrow{\varepsilon t} P' \quad Q \xrightarrow{\varepsilon t} Q'}{P \,|||\, Q \xrightarrow{\varepsilon t} P' \,|||\, Q'}$$

$$\frac{P \xrightarrow{\varepsilon t} P' \quad f[P] \downarrow t}{f[P] \xrightarrow{\varepsilon t} f[P']} \qquad \frac{Vx \lhd \mu V \xrightarrow{\varepsilon t} P'}{\mu V x \xrightarrow{\varepsilon t} P'}$$

TABLE 2. Time transitions for $SYN_{(\mathbf{A},\mathbf{P},\mathbf{T})}$.

7 Consistency and completeness

We now have a set of laws for $SYN_{(\mathbf{A},\mathbf{P},\mathbf{T})}$, and a transition system semantics. We would like to know that they are equivalent. To begin with, we can show that the laws are consistent, by tedious construction.

THEOREM 3 (CONSISTENCY). *If* $\vdash P = Q$ *then* $P \sim Q$.

PROOF. Construct a bisimulation for each law. □

The proof of completeness is slightly trickier, and follows along standard lines [Bro83, Mil84, Jef91], altered slightly to fit with our laws for infinite recursion. To begin with, we can define a very large infinite recursion $\mu\Omega$ which will contain every process. In order to construct this, we need variables x_P such that $x_P = x_Q$ iff $P \sim Q$—we will then construct Ω so that $P \sim \mu\Omega x_P$.

DEFINITION (ASSUMING THE AXIOM OF CHOICE). $X \subset \mathbf{V}$ *is such that for any* $x_P, x_Q \in X$, $x_P = x_Q$ *iff* $P \sim Q$.

We then need to make sure that $|X| < |\mathbf{V}|$, otherwise we would not be able to construct the vector Ω.

LEMMA 4. $|X| \leqslant |\mathbf{L}|^\lambda$

PROOF. Let $\mathbb{T}$ be the infinite tree with λ arcs from each node. This tree has $\lambda + \lambda^2 + \lambda^3 + \cdots = \lambda$ arcs. Thus, there are $|Z|^\lambda$ different labellings of $\mathbb{T}$ with any set Z, so there are $(|\mathbf{L}|+1)^\lambda$ labellings of $\mathbb{T}$ with $\mathbf{L}_\perp$. Given $\mathbf{L}_\perp$-labellings T and U, we can define the transition system $T \xrightarrow{\sigma} U$ iff there is a σ-arc from the root of T to the root of U. Every process P is bisimilar to a $\mathbf{L}_\perp$-labelling of T, so there are at most $(|\mathbf{L}|+1)^\lambda$ processes up to bisimulation, so $|X| \leqslant (|\mathbf{L}|+1)^\lambda$. Since $|\mathbf{L}| > 1$, $(|\mathbf{L}|+1)^\lambda \leqslant (|\mathbf{L}|^2)^\lambda = |\mathbf{L}|^{2\lambda} = |\mathbf{L}|^\lambda$, so $|X| \leqslant |\mathbf{L}|^\lambda$. □

The vector Ω is then defined so that $\mu\Omega x_P \sim P$.

DEFINITION. Ω *is defined:*

$$\begin{aligned} \Omega x_Q \equiv\ & \sum\{\varepsilon t.\alpha.0 \mid Q \not\downarrow t\} \\ & + \sum\{\varepsilon t.\alpha.0 \mid Q \xrightarrow{\varepsilon t} R \not\downarrow \pi_\alpha\} \\ & + \sum\{\varepsilon t.\alpha.x_R \mid Q \xrightarrow{\varepsilon t}\xrightarrow{\alpha} R\} \end{aligned}$$

All we have to do now is show that $\vdash P = \mu\Omega x_P$. To begin with, we can define a summand to be a process of the form given in Ω.

DEFINITION. *E is a summand iff it is of the form*

$$\begin{aligned}&\textstyle\sum\{\varepsilon t.\alpha.0 \mid E \not\xrightarrow{t}\}\\ &\quad+\textstyle\sum\{\varepsilon t.\alpha.0 \mid E \xrightarrow{\varepsilon t} F \not\xrightarrow{\pi_\alpha}\}\\ &\quad+\textstyle\sum\{\varepsilon t.\alpha.F \mid E \xrightarrow{\varepsilon t}\xrightarrow{\alpha} F\}\end{aligned}$$

Then we can show that every guarded term can be converted into a summand.

LEMMA 5. *For any guarded E there is a summand F such that $\vdash E = F$.*

PROOF. A structural induction on E. □

We can use this to define a large mutual recursion for any process P, containing a variable y_Q for any Q which P can reach. To define this, we need the set $\vec{P}$ containing all the processes P can reach.

DEFINITION. *$\vec{P}$ is the smallest set such that $P \in \vec{P}$ and if $Q \in \vec{P}$ and $Q \xrightarrow{\varepsilon t}\xrightarrow{\alpha} R$ then $R \in \vec{P}$.*

We can then show that we can define a recursion of size $\vec{P}$, since $|\vec{P}| < \lambda$.

LEMMA 6. $|\vec{P}| < \lambda$

PROOF. An induction on the definition of $\vec{P}$. □

We can now show that every P can be proved equal to $\mu\Omega x_P$.

LEMMA 7. $\vdash P = \mu\Omega x_P$

PROOF (USING THE AXIOM OF CHOICE). Let Y be a set of variables $\{y_Q \mid Q \in \vec{P}\}$ such that $y_P = y_Q$ iff $P \equiv Q$. Define:

$$\begin{aligned} f y_Q &= x_Q\\ V y_Q &\equiv Q\\ W y_Q &\equiv \textstyle\sum\{\varepsilon t.\alpha.0 \mid Q \not\xrightarrow{t}\}\\ &\quad+\textstyle\sum\{\varepsilon t.\alpha.0 \mid Q \xrightarrow{\varepsilon t} R \not\xrightarrow{\pi_\alpha}\}\\ &\quad+\textstyle\sum\{\varepsilon t.\alpha.y_R \mid Q \xrightarrow{\varepsilon t}\xrightarrow{\alpha} R\}\end{aligned}$$

Then by Lemma 5, we can show $\vdash V y_Q = Q = W y_W \lhd V$, so $\vdash V = W \lhd V$, so $\vdash V = \mu W$. By definition, $W y_Q \lhd f \equiv \Omega(f y_Q)$, so $W \lhd f \equiv \Omega \circ f$, so $\vdash \mu W = \mu\Omega \circ f$. In particular, $\vdash P = V y_P = \mu W y_P = (\mu\Omega \circ f) y_P = \mu\Omega x_P$. □

Thus, we have converted every process into a normal form, and so we can show completeness of the laws.

THEOREM 8 (COMPLETENESS). *If $P \sim Q$ then $\vdash P = Q$.*

PROOF. Since $P \sim Q$, $x_P = x_Q$, so $\vdash P = \mu\Omega x_P = \mu\Omega x_Q = Q$. □

And so we have shown our laws equivalent to bisimulation.

8 Translating time into priority

The reason for investigating prioritized actions is that we can model $SYN_{(\mathbf{A},\mathbf{P},\mathbf{T})}$ by the untimed calculus $SYN_{(\mathbf{T}\times\mathbf{A},\mathbf{T}\times\mathbf{P},\mathbf{1})}$. We do this by time-stamping every action, so we model the process that can perform an a at time t by one which performs the composite action (t, a). Similarly, we time-stamp all the priorities, and define a new priority ordering for $\mathbf{T} \times \mathbf{P}$.

DEFINITION. *$(t,p) \preccurlyeq (u,q)$ iff $t > u$ or $t = u \wedge p \preccurlyeq q$.*

So (t,p) has higher priority than (u,q) iff t happens before u, or they happen at the same time, and p has higher priority than q. This mirrors the fact that maximal progress comes before maximal priority—for example, it is still the case that $\varepsilon t.\tau p.P = \varepsilon t.\tau p.P + \varepsilon u.\tau q.Q$ if $t < u$ or if $t = u$ and $p \succ q$. We can also define the new priority function π for $SYN_{(\mathbf{T}\times\mathbf{A},\mathbf{T}\times\mathbf{P},\mathbf{1})}$.

DEFINITION. *$\pi_{(t,\alpha)} = (t, \pi_\alpha)$*

Using these, we can translate any term from $SYN_{(\mathbf{A},\mathbf{P},\mathbf{T})}$ into $SYN_{(\mathbf{T}\times\mathbf{A},\mathbf{T}\times\mathbf{P},\mathbf{1})}$.

DEFINITION. *$[\![\cdot]\!] : SYN_{(\mathbf{A},\mathbf{P},\mathbf{T})} \to SYN_{(\mathbf{T}\times\mathbf{A},\mathbf{T}\times\mathbf{P},\mathbf{1})}$ is defined:*

$$
\begin{aligned}
[\![a.E]\!] &\equiv (0,a).[\![E]\!] \\
[\![\tau p.E]\!] &\equiv \tau(0,p).[\![E]\!] \\
[\![\varepsilon t.E]\!] &\equiv add_t[[\![E]\!]] \\
[\![\textstyle\sum\mathcal{E}]\!] &\equiv \textstyle\sum\{[\![E]\!] \mid E \in \mathcal{E}\} \\
[\![E \parallel F]\!] &\equiv [\![E]\!] \parallel [\![F]\!] \\
[\![E \mid\mid\mid F]\!] &\equiv ([\![E]\!] \mid\mid\mid [\![F]\!]) \parallel [\![RUN]\!] \\
[\![f[E]]\!] &\equiv \hat{f}[[\![E]\!]] \\
[\![x]\!] &\equiv x \\
[\![\mu V x]\!] &\equiv \mu[\![V]\!]x
\end{aligned}
$$

where $add_t(u,\alpha) = (t+u,\alpha)$.

For example,

- $a.\varepsilon 1.b.\varepsilon 1.c.0$ translates to $(0,a).\,add_1[(0,b).\,add_1[(0,c).0]]$, which is bisimilar to $(0,a).(1,b).(2,c).0$, and can perform an a at time 0, then a b at time 1, then a c at time 2, as expected.
- $\tau p.0 + \varepsilon 1.b.0$ translates to $\tau(0,p).0 + (1,b).0$, and since $\pi_{(1,b)} \prec (0,p)$, this is bisimilar to $\tau(0,p).0$. Since this is the translation of $\tau p.0$, we have an example of how the translation copes with maximal progress.
- $\varepsilon 1.a.0 \mid\mid\mid \varepsilon 2.b.0$ translates to $((1,a).0 \mid\mid\mid (2,b)) \parallel [\![RUN]\!]$, which is bisimilar to $(1,a).(2,b).0 + (2,b).0$. Since this is the translation of $\varepsilon 1.a.\varepsilon 1.b + \varepsilon 2.b$, this is an example of how the translation copes with interleaving. Note that this requires us to synchronize $[\![P]\!] \mid\mid\mid [\![Q]\!]$ with $[\![RUN]\!]$, otherwise $[\![P \mid\mid\mid Q]\!]$ would be capable of the nonsensical behaviour 'perform b at time 2 then a at time 1'.

We can now show that $P \sim Q$ iff $[\![P]\!] \sim [\![Q]\!]$, so we have indeed modelled $SYN_{(\mathbf{A},\mathbf{P},\mathbf{T})}$ in $SYN_{(\mathbf{T}\times\mathbf{A},\mathbf{T}\times\mathbf{P},\mathbf{1})}$. In fact, it is easier to prove this result about the laws for $\vdash P = Q$.

To begin with, we require the following propositions:

PROPOSITION 9. $\vdash id[P] = P$

PROPOSITION 10. $\vdash f[g[P]] = (f \circ g)[P]$

PROPOSITION 11. *If* $\forall a \,.\, fa \in \mathbf{A}$, *then* $\vdash f[P \parallel Q] = f[P] \parallel f[Q]$.

PROPOSITION 12. $\vdash P \parallel (Q \parallel R) = (P \parallel Q) \parallel R$

From these, consistency can be shown mechanically.

THEOREM 13 (CONSISTENCY). *If* $\vdash P = Q$ *then* $\vdash [\![P]\!] = [\![Q]\!]$.

PROOF. Equational reasoning. □

The result for completeness is a bit trickier, and requires the observation that $[\![\mu\Omega x_P]\!]$ is still a normal form.

LEMMA 14. *If* $[\![\mu\Omega x_P]\!] \sim [\![\mu\Omega x_Q]\!]$ *then* $x_P = x_Q$.

PROOF. Construct the relation $\mathcal{R} = \{(P, Q) \mid \mu\Omega x_P \sim \mu\Omega y_P\}$, then show that it is a bisimulation. □

Then we can prove completeness by converting processes into normal form.

THEOREM 15 (COMPLETENESS). *If* $\vdash [\![P]\!] = [\![Q]\!]$ *then* $\vdash P = Q$.

PROOF. $\vdash [\![\mu\Omega x_P]\!] = [\![P]\!] = [\![Q]\!] = [\![\mu\Omega x_Q]\!]$, so $x_P = x_Q$, so $\vdash P = \mu\Omega x_P = Q$. □

And so we have shown $SYN_{(\mathbf{T}\times\mathbf{A},\mathbf{T}\times\mathbf{P},\mathbf{1})}$ is a model for $SYN_{(\mathbf{A},\mathbf{P},\mathbf{T})}$.

References

[BB91] J. C. M. Baeten and J. A. Bergstra. Real time process algebra. *Formal Aspects Comp. Sci.*, 3:142–188, 1991.

[BBK86] J. C. M. Baeten, J. A. Bergstra and J. W. Klop. Syntax and defining equations for an interrupt operator in process algebra. *Fund. Inform.*, 9(2):127–168, 1986.

[BHR84] S. D. Brookes, C. A. R. Hoare, and A. W. Roscoe. A theory of communicating sequential processes. *J. Assoc. Comput. Mach.*, 31(3):560–599, 1984.

[Bro83] S. D. Brookes. *A Model for Communicating Sequential Processes.* D.Phil. thesis, Oxford University, 1983.

[BRW86] S. D. Brookes, A. W. Roscoe, and D. J. Walker. An operational semantics for CSP. Submitted for publication, 1986.

[Cam90] Juanito Camilleri. *Priority in Process Calculi.* PhD thesis, Cambridge University, 1990.

[Cam91] Juanito Camilleri. A conditional operator for CCS. In J. C. M. Baeten and J. F. Groote, editors, *Proc. Concur 91*, pages 142–156. Springer-Verlag, 1991.

[CH88] R. Cleveland and M. Hennessy. Priorities in process algebra. In *Proc. LICS 88.* The Computer Society, 1988.

[Chr90] Ivan Christoff. Testing equivalences and fully abstract models for probabilistic processes. In *Proc. Concur 90*, pages 126–141. Springer Verlag, 1990. LNCS 458.

[Dav91] Jim Davies. *Specification and Proof in Real-time Systems.* D.Phil. thesis, Oxford University, 1991.

[Hen88] M. Hennessy. *Algebraic Theory of Processes.* MIT Press, 1988.

[HJ90] Hans Hansson and Bengt Jonsson. A calculus for communicating systems with time and probabilities. Swedish Institute for Computer Science, 1990.

[Hoa85] C. A. R. Hoare. *Communicating Sequential Processes.* Prentice-Hall, 1985.

[HR90] M. Hennessy and T. Regan. A temporal process algebra. Technical Report 2/90, CSAI, University of Sussex, 1990.

[HR91] M. Hennessy and T. Regan. A process algebra for timed systems. Submitted for publication, 1991.

[Jef91] Alan Jeffrey. A linear time process algebra. In Kim G. Larsen and Arne Skou, editors, *Proc. CAV 91.* Springer-Verlag, 1991. To appear in LNCS.

[LS89] Kim G. Larsen and Arne Skou. Bisimulation through probabilistic testing. In *Proc. 16th ACM Symp. on Principles of Programming Languages*, 1989.

[Mil80] Robin Milner. *A Calculus of Communicating Systems.* Springer-Verlag, 1980. LNCS 92.

[Mil84] Robin Milner. A complete inference system for a class of regular behaviours. *J. Comput. System Sci.*, 28:439–466, 1984.

[Mil89] Robin Milner. *Communication and Concurrency.* Prentice-Hall, 1989.

[MT90] F. Moller and C. Tofts. A temporal calculus of communicating systems. In *Proc. Concur 90*, pages 401–415. Springer-Verlag, 1990. LNCS 458.

[NS90] X. Nicollin and J. Sifakis. The algebra of timed processes ATP: Theory and application. Technical Report RT-C26, Laboratoire de Génie Informatique de Grenoble, 1990.

[P+90] Vaughan Pratt et al. Modelling concurrency with partial orders. Concurrency mailing list, *concurrency@theory.lcs.mit.edu*, October 1990.

[Plo81] Gordon Plotkin. A structural approach to operational semantics. Technical Report DAIMI-FN-19, Computer Science Dept., Århus University, 1981.

[RR87] G. M. Reed and A. W. Roscoe. Metric spaces as models for real-time concurrency. In *Proc. 3rd Workshop on the Mathematical Foundations of Programming*, pages 331–343. Springer-Verlag, 1987. LNCS 298.

[Sch90] S. A. Schneider. *Communication and Correctness in Real-time Distributed Computing.* D.Phil. thesis, Oxford University, 1990.

[Sch91] Steve Schneider. An operational semantics for timed CSP. In *Proc. Chalmers Workshop on Concurrency.* Dept. Computer Sciences, Chalmers University, 1991.

[SS90] Scott A. Smolka and Bernhard Steffen. Priority as extremal probability. In *Proc. Concur 90*, pages 456–466. Springer Verlag, 1990. LNCS 458.

[Tof90] Chris Tofts. Relative frequency in a synchronous calculus. Technical report ECS-LFCS-90-108, LFCS, University of Edinburgh, 1990.

[Wan90] Wang Yi. Real-time behaviour of asynchronous agents. In *Proc. Concur 90*, pages 502–520. Springer-Verlag, 1990. LNCS 458.

[Wan91] Wang Yi. CCS + time = an interleaving model for real time systems. In J. Leach Albert, B. Monien, and M. Rodríguez, editors, *Proc. ICALP 91*, pages 217–228. Springer-Verlag, 1991. LNCS 510.

Operational Semantics for Timed Observations

Yolanda Ortega-Mallén
Sección de Informática y Automática
Facultad de Matemáticas
Universidad Complutense, 28040 Madrid, Spain*

Abstract

Timed Observations is a semantical framework suitable to study the behaviour of timed concurrent processes. For each process we observe its external behaviour consisting of *timed traces* (the visible actions performed by the process and related to the instant time when they are executed), *refusals* (actions refused by a process after executing a trace), and the possibly *divergent* timed traces (traces that can be extended by an unbounded sequence of internal actions). Thus the model has a timed failures divergences semantics, and was already presented in [OdF90b] as a denotational semantics, which was applied to TCSP. In the present paper we concentrate on giving an equivalent operational characterization.

1 Introduction

Nowadays it is unnecessary to talk about the importance of timed models in modern applications and programming projects. It is sufficient to point out the big proliferation of timed models over the last few years. Some of them are denotational-oriented [KSdR*85, RR87], some others are operational-oriented [QAF89,MT89,Yi90] or algebraic-oriented [NRSV90,BB90,HR90]. As a small contribution to this "hot" topic, *Timed Observations* was presented in [OdF90]. Important features of that model are:

- the *discrete* nature of time,
- the *non-instantaneousness* of actions, which are associated some non-zero duration time (in time units),
- the *explicit concurrency* expressed by actions multisets (called here bags), in contraposition to interleaving semantics,
- the assumption that actions can *wait indefinitely* before starting their execution, in order to synchronize with other actions.

*Part of the work was developed during a two-month sojourn of the author as collaborator in the RWTH-Aachen (Lehrstuhl für Informatik II) (G).

Considering that a concurrent process is not an isolated item, but part of a whole system of interacting machines and users, what is interesting about a process is its external behaviour, i.e. what an external observer (a user or another process) can notice. In our model (Timed Observations), the observer not only notices the visible actions performed by the process (traces), but he also relates each action to the instant when it is executed (timed traces). Moreover, he is an active observer and tries to guess the future behaviour of the process by asking it to perform some actions from a set (refusals as in [BR85]). In order to differentiate *divergence* (unbounded sequence of internal actions) from *deadlock* (doing nothing), we ideally suppose that divergence is observable. Thus the model has a timed failures divergences semantics.

In [OdF90] the model was presented as a denotational semantics which was applied to TCSP (Theory of Communicating Sequential Processes [HBR81]). Now in the present paper, an operational approach is developed, providing both a different characterisation of the model, and a relationship between these two kinds of semantics.

In order to make things easier, we first restrict our operational approach to finite processes. In this way, we can concentrate on the timed traces and forget about the divergences (as these can only appear for recursive processes). Afterwards we study the problems introduced by the recursive calls in the definition of processes, concluding that we must restrict our model to *guarded processes*. However, our definition of guarded processes is more natural, as guards are not lost when applying the hiding operator (thanks to the non-zero duration of actions).

The paper is structured as follows: section 2 is a brief introduction to the basic notions and definitions for Timed Observations. In section 3 the corresponding denotational semantics for TCSP-processes is given. In section 4 the operational version is presented. The paper concludes with a brief survey on related work, commenting on some timed models that consider an operational approach too.

2 Timed Observations

Only the main definitions and concepts of the model are given here. For more motivation, intuitive explanations and details (and proofs of the results presented in this section and the following one), the reader is referred to [OdF90,Ort90].

2.1 Action bags

We shall fix the set of *actions* that processes may perform: the *finite alphabet* $\mathcal{A}$. Actions are indivisible but non-instantaneous, for this purpose we assume a *duration* function $d : \mathcal{A} \longrightarrow \mathbb{N}^+$ which associates to each action the time it needs to complete its execution (in temporary units). $\mathbb{N}^+$ represents the non-zero integers.

As actions are no longer instantaneous, hidden actions do not disappear completely from the trace, because its duration time will remain reflected.

Some way of expressing the possibility of having several actions executing simultaneously is needed. Therefore we will consider finite *action bags* (or multisets).

Def. 2.1 *The set of* **bags over an alphabet** $\mathcal{A}$ *is:* $\mathcal{B}(\mathcal{A}) = \{B : \mathcal{A} \longrightarrow \mathbb{N}\}$.

We define a the **empty bag** *by* $B_\emptyset \stackrel{\text{def}}{=} \forall a \in \mathcal{A}, B_\emptyset(a) = 0$; *and the set of* **non-empty bags** *as* $\mathcal{B}^+(\mathcal{A}) = \mathcal{B}(\mathcal{A}) - \{B_\emptyset\}$.

When the alphabet is understood, it may be omitted obtaining $\mathcal{B}, \mathcal{B}^+$. □

We will use action bags to generalize the notion of failure [BR85], which embodies traces and refusals.

2.2 Timed traces

We will relate each action ocurrence to the instant when it is produced (more exactly to the instant when it starts, as actions are not instantaneous), obtaining *timed traces.*

Def. 2.2 *The set of* **timed traces** *is* $\mathcal{TT} = \{t : \mathbb{N}^+ \longrightarrow \mathcal{B}\}$.

An special and useful trace is the **empty trace** $t_\emptyset \stackrel{\text{def}}{=} \forall i \in \mathbb{N}^+, t_\emptyset(i) = B_\emptyset$. □

Def. 2.3 *For each* $t \in \mathcal{TT}$ *we define*

- *its first non-empty time instant* $\inf(t)$ *by:* $\inf(t_\emptyset) = 0$, $\inf(t) = \min\{n \mid t(n) \neq B_\emptyset\}, t \neq t_\emptyset$.
- *its last non-empty time instant* $\sup(t)$ *by:* $\sup(t_\emptyset) = 0$, $\sup(t) = \sup\{n \mid t(n) \neq B_\emptyset\}, t \neq t_\emptyset$.

□

Def. 2.4 *The set of* **finite timed traces** *is* $\mathcal{TTF} = \{(t,n) \mid t \in \mathcal{TT} \wedge n \geq \sup(t)\}$.

For each $tf = (t,n) \in \mathcal{TTF}$ *we define:* $trace(tf) = t$ *and* $end(tf) = n$. □

Notice that $end(tf) = 0$ implies $trace(tf) = t_\emptyset$. The trace $(t_\emptyset, 0)$ corresponds to the initial state of the process, when the observer has not yet started to observe.

2.3 Timed failures

A refusal is a *finite* set of non-empty action bags.

Def. 2.5 *The set of* **refusals** *is* $\mathcal{R} = \{r \in \mathcal{PF}(\mathcal{B}^+)\}$.[1] □

Def. 2.6 *The set of* **timed failures** *is* $\mathcal{FT} = \{\langle tf, r\rangle \mid tf \in \mathcal{TTF} \wedge r \in \mathcal{R}\}$.

For each $f = \langle tf, r\rangle \in \mathcal{FT}$ *we define:* $T(f) = tf$ *and* $R(f) = r$. □

2.4 Divergence

A process is said to *diverge* when it is involved in an unbounded sequence of internal actions. Thus a divergent trace is one after which infinitely many actions may occur, even if it takes infinitely long to do it.

The divergence in our model is not catastrophic; i.e. the feasibility of divergence is not necessarily permanent, it may *disappear* as the process evolves by performing further actions and choosing a non-diverging path. Therefore, we do not identify a possibly diverging process with *chaos* (the process which has every possible behaviour), as many denotational models do ([Bro83,OH86,TV89] among others).

[1] $\mathcal{PF}$ is the finite power set.

2.5 Timed observations

Putting all the above concepts together, the observable behaviour of a process consists of a timed failure set plus a divergence set (timed traces which can be extended with an infinite sequence of internal actions).

Def. 2.7 *The set of* **timed observations** *is:*

$$\mathcal{O} = \{\langle F, D\rangle | F \subseteq \mathcal{FT} \wedge D \subseteq \mathcal{TTF} \wedge D \subseteq T(F)\}$$

For each $ob = \langle F, D\rangle \in \mathcal{O}$ *we define:* $F(ob) = F$, $D(ob) = D$ *and* $T(ob) = T(F(ob))$.
□

2.6 Notation and operations

We shall now give more notation and operations related to action bags and timed traces.

bag generated by $a \in \mathcal{A}$: $B_a \stackrel{\text{def}}{=} \forall b \in \mathcal{A}, B_a(b) = \begin{cases} 0 & \text{if } b \neq a \\ 1 & \text{if } b = a \end{cases}$

bags containing $a \in \mathcal{A}$: $\mathcal{B}_a(\mathcal{A}) = \{B \in \mathcal{B}(\mathcal{A}) | B(a) > 0\}$,

bags different from B_a: $\mathcal{B}_{-a}(\mathcal{A}) = \mathcal{B}(\mathcal{A}) - \{B_a\}$.

2.6.1 Operations on bags

Let $a \in \mathcal{A}$; $A \subseteq \mathcal{A}$; $B, B_1, B_2 \in \mathcal{B}(\mathcal{A})$; $n \in \mathbb{N}$.

Alphabet: $\mathcal{A}(B) = \{a \in \mathcal{A} | B(a) > 0\}$.

Partial order: $B_1 \leq_{\mathrm{B}} B_2 \stackrel{\text{def}}{=} \forall a \in \mathcal{A}, B_1(a) \leq B_2(a)$.

Addition: $B_1 +_{\mathrm{B}} B_2 \stackrel{\text{def}}{=} \forall a \in \mathcal{A}, B_1 +_{\mathrm{B}} B_2(a) \stackrel{\text{def}}{=} B_1(a) + B_2(a)$.

Hiding: $B \backslash a \stackrel{\text{def}}{=} \forall b \in \mathcal{A}, B \backslash a(b) = \begin{cases} B(b) & \text{if } b \neq a \\ 0 & \text{if } b = a \end{cases}$

Restriction: $B \lceil A \stackrel{\text{def}}{=} \forall b \in \mathcal{A}, B \lceil A(b) = \begin{cases} B(b) & \text{if } b \in A \\ 0 & \text{if } b \notin A \end{cases}$

Synchronization: It is only defined when $B_1 \lceil A = B_2 \lceil A$,

$$B_1 \oplus_A B_2 \stackrel{\text{def}}{=} \forall a \in \mathcal{A}, B_1 \oplus_A B_2(a) = \begin{cases} B_1(a) & \text{if } a \in A \\ B_1(a) + B_2(a) & \text{if } a \notin A \end{cases}$$

Two synchronized actions are reduced to only one action.

2.6.2 Operations on traces

Let $A \subseteq \mathcal{A}; n \in \mathbb{N}^+; B \in \mathcal{B}; t \in \mathcal{TT}; tf, t_1, t_2 \in \mathcal{TTF}$.

Alphabet: $\mathcal{A}(tf) = \bigcup_{1 \leq i \leq end(tf)} \mathcal{A}(trace(tf)(i))$.

Adding traces: $t_1 +_T t_2 \stackrel{\text{def}}{=} \begin{cases} \forall i \in \mathbb{N}^+, trace(t_1 +_T t_2)(i) = trace(t_1)(i) +_B trace(t_2)(i) \\ end(t_1 +_T t_2) = \max\{end(t_1), end(t_2)\} \end{cases}$

Adding a bag to a trace:

$$tf +_{TB} (B, n) \stackrel{\text{def}}{=} \begin{cases} \forall i \in \mathbb{N}^+, trace(tf +_{TB} (B, n))(i) = \begin{cases} trace(tf)(i) & \text{if } i \neq n \\ trace(tf)(i) +_B B & \text{if } i = n \end{cases} \\ end(tf +_{TB} (B, n)) = \max\{end(tf), n\} \end{cases}$$

We can define now the trace including only one action: $t_{n,a} \stackrel{\text{def}}{=} trace((t_\emptyset, 0) +_{TB} (B_a, n))$.

Moving along time:

$$\text{mov}(tf, n) \stackrel{\text{def}}{=} \begin{cases} \forall i \in \mathbb{N}^+, trace(\text{mov}(tf, n))(i) = \begin{cases} trace(tf)(i - n) & \text{if } i > n \\ B_\emptyset & \text{if } i \leq n \end{cases} \\ end(\text{mov}(tf, n)) = end(tf) + n \end{cases}$$

Concatenation: $t_1 \cdot t_2 \stackrel{\text{def}}{=} t_1 +_T \text{mov}(t_2, end(t_1))$.

Stretch: $Stretch(tf)^2$,

$Stretch(t_\emptyset, n) = \{(t_\emptyset, m) | m \geq n\}$,

$Stretch((t, n) +_{TB} (B, n + 1)) =$

$\{tf +_{TB} \sum_{k=1}^{m} (B_k, n_k) | \sum_{k=1}^{m} B_k = B \wedge n_k \geq end(tf) + 1 \wedge tf \in Stretch(t, n)\}$.

Actions in a bag may be "stretched" and executed in different time instants. The only restriction for it, is to keep the relative order between bags in the original trace.

Hiding: $tf \backslash a \stackrel{\text{def}}{=} \begin{cases} \forall i \in \mathbb{N}^+, trace(tf \backslash a)(i) = trace(tf)(i) \backslash a \\ end(tf \backslash a) = end(tf) \end{cases}$

Restriction to an action set: $tf \lceil A \stackrel{\text{def}}{=} \begin{cases} \forall i \in \mathbb{N}^+, trace(tf \lceil A)(i) = trace(tf)(i) \lceil A \\ end(tf \lceil A) = end(tf) \end{cases}$

Initial interval: $tf \lceil n \stackrel{\text{def}}{=} \begin{cases} \forall i \in \mathbb{N}^+, trace(tf \lceil n)(i) = \begin{cases} trace(tf)(i) & \text{if } i \leq n \\ B_\emptyset & \text{if } i > n \end{cases} \\ end(tf \lceil n) = n \end{cases}$

Synchronization: it is only defined for traces verifying $t_1 \lceil A = t_2 \lceil A$.

$$t_1 \oplus_A t_2 \stackrel{\text{def}}{=} \begin{cases} \forall i \in \mathbb{N}^+, trace(t_1 \oplus_A t_2)(i) = trace(t_1)(i) \oplus_A trace(t_2)(i) \\ end(t_1 \oplus_A t_2) = end(t_1) \ (= end(t_2)) \end{cases}$$

Operators on finite timed traces can also be applied to non-finite timed traces.

[2] This operation is borrowed from [TV89].

2.7 Specification space

In order to develop a denotational semantics we need a suitable domain, which we shall call the **specification space** $\langle \mathcal{S}, \leq_{\mathrm{S}} \rangle$, where $\mathcal{S} = \{ob \in \mathcal{O} | ob$ verifies [P1] – [P8]$\}$ is the **set of specifications**, and $ob_1 \leq_{\mathrm{S}} ob_2 \stackrel{\mathrm{def}}{=} F(ob_2) \subseteq F(ob_1) \wedge D(ob_2) \subseteq D(ob_1)$, for $ob_1, ob_2 \in \mathcal{O}$, is the **partial order over timed observations.**

- [P1] $(t_\emptyset, 0) \in T(ob)$
- [P2] $(t_1, n_1) \cdot (t_2, n_2) \in T(ob) \Rightarrow (t_1, n_1) \in T(ob)$
- [P3] $(t, n) \in T(ob) \wedge r \in \mathcal{R} \Rightarrow$
  $(\exists B \in r : (t, n) +_{\mathrm{TB}} (B, n+1) \in T(ob)) \vee \langle (t, n), r \rangle \in F(ob)$
- [P4] $\langle tf, r \rangle \in F(ob) \wedge r' \subset r \Rightarrow \langle tf, r' \rangle \in F(ob)$
- [P5] $tf \in D(ob) \Rightarrow \forall m \geq \sup(tf) : tf \lceil m \in D(ob)$
- [P6] $\langle tf, r \rangle \in F(ob) \wedge tf' \in \mathit{Stretch}(tf) \Rightarrow \langle tf', r \rangle \in F(ob)$
- [P7] $tf \in D(ob) \wedge tf' \in \mathit{Stretch}(tf) \Rightarrow tf' \in D(ob)$
- [P8] $B \leq_{\mathrm{B}} B' \wedge \langle tf, r \cup \{B\} \rangle \in F(ob) \Rightarrow \langle tf, r \cup \{B, B'\} \rangle \in F(ob)$

Props. 2.1 $\langle \mathcal{S}, \leq_{\mathrm{S}} \rangle$ *is a* Complete Partial Order *(CPO).* □

3 Denotational semantics for TCSP-processes

In the previous section we have only given the semantic domain, but we still must provide, for each syntactic operator in the programming language, a continuous operator over this domain.

We consider the following set of syntactic operators for TCSP-processes: [3]

$$\Sigma_1 = \{stop\} \cup \{a \rightarrow | a \in \mathcal{A}\} \cup \{\sqcap, \Box\} \cup \{\|_A | A \subseteq \mathcal{A}\} \cup \{\backslash a | a \in \mathcal{A}\}$$

Def. 3.1 *A* **TCSP-process** *is a recursive term over* Σ_1 $(P \in REC(\Sigma_1))$*:*

$$P ::= stop \mid a \rightarrow P \mid P \sqcap P \mid P \Box P \mid P \|_A P \mid P \backslash a \mid \xi \mid \mu\xi.P$$

where ξ *belongs to some process identifier set Id.* □

Only recursive closed processes in $CREC(\Sigma_1)$ (without free identifiers) will be considered.

A semantic operator is associated to each syntactic operator. The resulting specification has two components, the first one expressing the timed failures and the second, the divergences. Let $Sp, Sp_1, Sp_2 \in \mathcal{S}$,

---

[3]They represent inaction, action prefix, internal and external choice, synchronization on an action set, and hiding respectively.

1. $\mathcal{S}_{stop}[\![\cdot]\!] = \begin{cases} \{\langle(t_\emptyset, m), r\rangle | m \in \mathbb{N} \wedge r \in \mathcal{R}\} \\ \emptyset \end{cases}$

2. $\mathcal{S}_{a\rightarrow}[\![Sp]\!] = \begin{cases} \{\langle(t_\emptyset, m), r\rangle | m \in \mathbb{N} \wedge r \in \mathcal{PF}(\mathcal{B}^+_{-a})\} \\ \cup\{\langle(t_{n,a}, m), r\rangle | n \in \mathbb{N}^+ \wedge n \leq m < nd \wedge r \in \mathcal{R}\} \\ \cup\{\langle(t_{n,a}, nd)\cdot tf, r\rangle | n \in \mathbb{N}^+ \wedge \langle tf, r\rangle \in F(Sp)\} \\ \\ \{(t_{n,a}, m) | n \in \mathbb{N}^+ \wedge n \leq m < nd \wedge (t_\emptyset, 0) \in D(Sp)\} \\ \cup\{(t_{n,a}, nd)\cdot tf \,| n \in \mathbb{N}^+ \wedge tf \in D(Sp)\} \end{cases}$

   where $nd = n + d(a) - 1$.

3. $\mathcal{S}_{\sqcap}[\![Sp_1, Sp_2]\!] = \begin{cases} F(Sp_1) \cup F(Sp_2) \\ D(Sp_1) \cup D(Sp_2) \end{cases}$

4. $\mathcal{S}_{\square}[\![Sp_1, Sp_2]\!] = \begin{cases} \{\langle(t_\emptyset, m), r\rangle \in F(Sp_1) \cap F(Sp_2)\} \\ \cup\{\langle tf, r\rangle \in F(Sp_1) \cup F(Sp_2) | trace(tf) \neq t_\emptyset\} \\ \\ D(Sp_1) \cup D(Sp_2) \end{cases}$

5. $\mathcal{S}_{\|_A}[\![Sp_1, Sp_2]\!] = \begin{cases} \{\langle tf_1 \oplus_A tf_2, r\rangle | \langle tf_i, r_i\rangle \in F(Sp_i), i = 1,2 \wedge r \in r_1 \oplus_A r_2\} \\ \{tf_1 \oplus_A tf_2 | \; (tf_1 \in D(Sp_1) \wedge tf_2 \in T(Sp_2)) \\ \qquad \vee (tf_1 \in T(Sp_1) \wedge tf_2 \in D(Sp_2))\} \end{cases}$

   Where $r_1 \oplus_A r_2 \stackrel{\text{def}}{=} \mathcal{PF}(\{B \in \mathcal{B}^+ | \forall B_1, B_2 : B = B_1 \oplus_A B_2 \Rightarrow B_1 \in r_1 \vee B_2 \in r_2\})$.

   The global process can refuse a bag only if every possible combination to perform this bag, by both components in cooperation, is refused by one component or the other.

6. $\mathcal{S}_{\backslash a}[\![Sp]\!] = \begin{cases} \{\langle tf\backslash a, r \cup r'\rangle | \langle tf, r\rangle \in F(Sp) \wedge r' \in \mathcal{PF}(\mathcal{B}_a)\} \\ \{tf\backslash a | \; \exists tf' : \mathcal{A}(tf') \subseteq \{a\} \wedge tf\cdot tf' \in D(Sp) \\ \qquad \vee \exists \{t_i\}_{i \in \mathbb{N}^+} : \forall i(\mathcal{A}(t_i) = \{a\}) \wedge tf\cdot t_1 \cdot \ldots \cdot t_i \in T(Sp)\} \end{cases}$

All the above given operators are well-defined, and they preserve properties $[P1]-[P8]$. All of them, except for the hiding operator, are continuous on the specification space. Nevertheless the hiding operator is monotonic and, when restricted to *guarded processes* (i.e. every recursive call must be preceded by the execution of at least one action), he have no problems.

Def. 3.2 Well-defined expression E:

$$E ::= stop \mid a \rightarrow E \mid E \sqcap E \mid E \square E \mid E \|_A E \mid E\backslash a \mid \xi \mid \mu\xi.PG$$

Guarded process $PG \in RECG(\Sigma_1)$:

$$PG ::= stop \mid a \rightarrow E \mid PG \sqcap PG \mid PG \square PG \mid PG \|_A PG \mid PG\backslash a \mid \mu\xi.PG$$

□

In distinction to untimed models, the hiding operator does not affect the guards in processes, as the guard is the duration of the action and not its name.

For guarded processes, the semantics of the recursive processes is defined, as usual, by means of the fixed-point theory.

4 Operational semantics

In operational semantics, the "global" meaning or behaviour of processes is not obtained in a direct way, but some rules are provided to describe each possible *evolution* (or operation) of a process, so that at each step we will know a little more about the process under observation. Collecting together every evolution describing each possible behaviour for a process, for a bounded but arbitrarily large time interval, we will obtain a global and complete description of that process.

4.1 Labeled transition systems

The usual way to represent the evolution of a process is by means of a *labeled transition system.*

Def. 4.1 *A* **labeled transition system** *is a triple* $\langle S, A, \{\xrightarrow{a}, a \in A\}\rangle$, *where*

- S *is the set of system states,*
- A *is the set of transition labels, and*
- $\xrightarrow{a} \subseteq S \times S$ *is the transition relation between states.*

□

4.2 Finite processes

We shall start with an operational semantics for processes without recursive calls, so that we can forget the problem of divergence for the moment.

4.2.1 Extended transition systems

In order to express the internal choice operator, we will need something a bit more powerful than simple labeled transition systems: *extended transition systems* (see [Hen88]), with two kinds of transitions:

- $\xrightarrow{a}$ where a is a label, corresponds to the usual transitions, and
- $\rightarrowtail$ does not have any associated label.

In our system states will correspond to processes, and labels to action bags. Therefore, the transition $\xrightarrow{B}$ represents the evolution of a process after executing the action multiset B. In order to define the operational semantics more easily, we will assume that the execution of any action lasts only one instant, so that we will simulate longer actions by means of a *delay* operator. In this way, each labeled transition corresponds to the passing of one time instant.

On the other hand, $\succ\!\!\longrightarrow$ represents an internal transition, which in our timed model is not equivalent to an internal action, as the latter has a time duration, whereas the internal transitions are instantaneous.

As commented above, we shall extend the set of syntactic operators for TCSP with a family of delay operators, which will represent the intermediary states obtained when executing actions longer than one time instant: $\Sigma_2 = \Sigma_1 \cup \{delay(\ell) | \ell \in \mathbb{N}\}$.

Def. 4.2 *The* **operational semantics** *of finite TCSP processes is given by means of the extended transition system* $\mathcal{FLTS} = \langle FIN(\Sigma_2), \mathcal{B}(\mathcal{A}), \xrightarrow{B} : B \in \mathcal{B}(\mathcal{A}), \succ\!\!\longrightarrow \rangle$*, where* $FIN(\Sigma_2)$ *is the set of finite processes defined over the operator set* Σ_2*, and the transition rules are given in table 4.2.1.*

Notation :

- $\mathcal{B}^*$ is the set of finite strings of bags and ε is the empty string,
- $\succ\!\!\longrightarrow^*$ represents a finite (possibly empty) sequence of internal transitions,
- $P\xrightarrow{B}$ means that there exists some $P' : P\xrightarrow{B}P'$,
- $P\overset{B}{\not\longrightarrow}$ means that there exists no $P' : P\xrightarrow{B}P'$,

and analogously for $P\succ\!\!\longrightarrow$ and $P\succ\!\!\not\longrightarrow$.

Internal transitions induce the concept of process *stability*.

Def. 4.3 *A process P is* **stable** *when it can perform no internal transition:*

$$Stable(P) \text{ iff } P\succ\!\!\not\longrightarrow$$

□

When dealing with finite processes, the stability is no problem at all, because every process can "become" stable, as the following proposition states.

Props. 4.1 *For* $P \in FIN(\Sigma_2)$ *there exists no unbounded sequence* $P = P_0\succ\!\!\longrightarrow P_1 \succ\!\!\longrightarrow \ldots$
Proof : It is easily proven by structural induction. □

Notice that if we replaced (D1) and (D2) by $\overline{delay(0, P)\succ\!\!\longrightarrow P}$, then the previous proposition would not be true.

(St) $$\frac{}{stop \xrightarrow{B_\emptyset} stop}$$

(Pr1) $$\frac{}{a \rightarrow P \xrightarrow{B_\emptyset} a \rightarrow P}$$

(Pr2) $$\frac{}{a \rightarrow P \xrightarrow{B_a} delay(d(a)-1, P)}$$

(D1) $$\frac{P \xrightarrow{B} P'}{delay(0,P) \xrightarrow{B} P'}$$

(D2) $$\frac{P \rightarrowtail P'}{delay(0,P) \rightarrowtail P'}$$

(D3) $$\frac{}{delay(\ell+1, P) \xrightarrow{B_\emptyset} delay(\ell, P)}$$

(In1) $$\frac{}{P \sqcap Q \rightarrowtail P}$$

(In2) $$\frac{}{P \sqcap Q \rightarrowtail Q}$$

(Ex1) $$\frac{P \rightarrowtail P'}{P \square Q \rightarrowtail P' \square Q}$$

(Ex2) $$\frac{Q \rightarrowtail Q'}{P \square Q \rightarrowtail P \square Q'}$$

(Ex3) $$\frac{P \xrightarrow{B} P'}{P \square Q \xrightarrow{B} P'}, B \in \mathcal{B}^+$$

(Ex4) $$\frac{Q \xrightarrow{B} Q'}{P \square Q \xrightarrow{B} Q'}, B \in \mathcal{B}^+$$

(Ex5) $$\frac{P \xrightarrow{B_\emptyset} P' \wedge Q \xrightarrow{B_\emptyset} Q'}{P \square Q \xrightarrow{B_\emptyset} P' \square Q'}$$

(Sy1) $$\frac{P \rightarrowtail P'}{P \|_A Q \rightarrowtail P' \|_A Q}$$

(Sy2) $$\frac{Q \rightarrowtail Q'}{P \|_A Q \rightarrowtail P \|_A Q'}$$

(Sy3) $$\frac{P \xrightarrow{B} P' \wedge Q \xrightarrow{C} Q'}{P \|_A Q \xrightarrow{B \oplus_A C} P' \|_A Q'}, B \lceil A = C \lceil A$$

(H1) $$\frac{P \rightarrowtail P'}{P \backslash a \rightarrowtail P' \backslash a}$$

(H2) $$\frac{P \xrightarrow{B} P'}{P \backslash a \xrightarrow{B \backslash a} P' \backslash a}$$

Table 1: Transition rules

4.2.2 Timed observations

Now we will see the kind of observations, which can be extracted from the given system. First of all we will abstract from internal transitions, and we will concentrate only on the labels:

Def. 4.4 *Let* $P \in FIN(\Sigma_2), B \in \mathcal{B}, s \in \mathcal{B}^*$, $P \overset{\varepsilon}{\Longrightarrow} P'$ iff $P \succ\!\!\longrightarrow^* P'$,
$P \overset{B\cdot s}{\Longrightarrow} P'$ iff $\exists Q, Q' : P \overset{\varepsilon}{\Longrightarrow} Q \overset{B}{\Longrightarrow} Q' \overset{s}{\Longrightarrow} P'$.
□

Notice that we always have $P \overset{\varepsilon}{\Longrightarrow} P$. Now we shall associate a timed trace to each label string.

Def. 4.5 $TF : \mathcal{B}^* \longrightarrow \mathcal{TTF}$ *is defined by*

$$TF(B_1 \ldots B_n) \overset{\text{def}}{=} (t, n) \in \mathcal{TTF} \text{ with } \forall i > n, t(i) = B_\emptyset \wedge \forall i : 1 \leq i \leq n, t(i) = B_i$$

□

Obviously we have $TF(\varepsilon) = (t_\emptyset, 0)$ and $TF(s_1 \cdot s_2) = TF(s_1) \cdot TF(s_2)$.

Next we define the transitions in terms of traces instead of bag sequences.

Def. 4.6 *Let* $P \in FIN(\Sigma_2), tf \in \mathcal{TTF}$*:* $P \overset{tf}{\Longrightarrow} P'$ iff $\exists s \in \mathcal{B}^* : TF(s) = tf \wedge P \overset{s}{\Longrightarrow} P'$.
□

As an immediate consequence we have the following lemma:

Lemma 4.2 *Take* $P, Q, R \in FIN(\Sigma_2); tf_1, tf_2 \in \mathcal{TTF}$,

$$P \overset{(t_\emptyset, 0)}{\Longrightarrow} P,$$
$$(\exists Q : P \overset{tf_1}{\Longrightarrow} Q \overset{tf_2}{\Longrightarrow} R) \Leftrightarrow P \overset{tf_1 \cdot tf_2}{\Longrightarrow} R.$$

Proof : It is straightforward. □

A more interesting lemma refers to the possibility of every stable process to perform transitions labeled by the empty bag.

Lemma 4.3 $\forall P \in FIN(\Sigma_2) : Stable(P) \Rightarrow P \overset{B_\emptyset}{\longrightarrow}$.
Proof : By structural induction,

- $stop, a \rightarrow P$ and $delay(\ell, P), \ell > 0$ are stable and verify the lemma clearly.
- $P \sqcap Q$ is unstable.
- $delay(0, P)$ and $P \backslash a$ are stable iff P is stable. In that case, by the i.h. we have $P \overset{B_\emptyset}{\longrightarrow}$, and this transition is possible for the original processes too, because $B_\emptyset \backslash a = B_\emptyset$.
- $P \Box Q$ and $P \|_A Q$ are stable iff P and Q are stable. The reasoning is analogous to the previous case.

□

The following corollary states that every process can let the time pass performing no visible actions.

Corollary 4.4 $\forall P \in FIN(\Sigma_2) : P \stackrel{B_\emptyset}{\Longrightarrow}$.
Proof : By proposition 4.1 and lemma 4.3. □

The observations generated by this transition system will include only timed failures (as there are no infinite processes, divergence cannot appear).

Def. 4.7 *OBS* : $FIN(\Sigma_2) \longrightarrow \mathcal{P}(\mathcal{FT})$ *is defined by*

$$OBS(P) = \{\langle tf, r\rangle | \exists P' : Stable(P') \wedge P \stackrel{tf}{\Longrightarrow} P' \wedge \forall B \in r, P' \stackrel{B}{\not\longrightarrow}\}$$

Notice that these r cannot include the empty bag thanks to the lemma 4.3. □

The concept of refusal would be lost if the achieved states were unstable. The idea behind it is that we cannot ask for the reaction of a process that has not yet decided how to behave.

The observations obtained through this operational semantics ($OBS(P)$) for finite TCSP processes coincide with the timed failures of the denotational semantics given before ($F(\mathcal{S}[\![P]\!])$). In order to prove this we will need some previous results.

Lemma 4.5 *If* $P \stackrel{tf}{\Longrightarrow} P'$ *and* $\langle tf', r\rangle \in OBS(P')$, *then* $\langle tf \cdot tf', r\rangle \in OBS(P)$.
Proof : By lemma 4.2. □

As a corollary we infer that internal transitions add no information to the observations.

Corollary 4.6 *If* $P \succ\!\!\longrightarrow^* P'$ *then* $OBS(P') \subseteq OBS(P)$.

Another interesting fact is that the process $delay(0, P)$ is equivalent, with respect to observations, to the process P.[4]

Lemma 4.7 $OBS(delay(0, P)) = OBS(P)$.
Proof : Obvious, because $delay(0, P)$ has exactly the same transitions as P. □

Finally, we introduce a technical lemma on the transitions related to the external choice, which reflects the fact that the selection is not made until the beginning of the first visible action.

Lemma 4.8 $P_1 \Box P_2 \stackrel{(t_\emptyset, n)}{\Longrightarrow} R$ *iff* $\exists P'_i : P_i \stackrel{(t_\emptyset, n)}{\Longrightarrow} P'_i, (i = 1, 2) \wedge R = P'_1 \Box P'_2$.
Proof : By induction on n:

[4]Moreover, we could have defined the family of delay operators as $delay(\ell)$ with $\ell > 0$, but then the transitions for the prefix operator would be more complex.

1. $P_1 \Box P_2 \stackrel{(t_\emptyset,0)}{\Longrightarrow} R \Leftrightarrow P_1 \Box P_2 \stackrel{\epsilon}{\Longrightarrow} R \Leftrightarrow P_1 \Box P_2 \succ\!\!\longrightarrow^* R \Leftrightarrow \exists P_i' : P_i \succ\!\!\longrightarrow^* P_i', (i = 1,2) \wedge R = P_1' \Box P_2'$.

2. $P_1 \Box P_2 \stackrel{(t_\emptyset,n+1)}{\Longrightarrow} R \Leftrightarrow \exists Q : P_1 \Box P_2 \stackrel{(t_\emptyset,n)}{\Longrightarrow} Q \stackrel{(t_\emptyset,1)}{\Longrightarrow} R$. By the i.h. $\exists P_i' : P_i \stackrel{(t_\emptyset,n)}{\Longrightarrow} P_i', (i = 1,2) \wedge Q = P_1' \Box P_2'$.

   On the other hand, $\exists P_i' : P_i \stackrel{(t_\emptyset,n+1)}{\Longrightarrow} P_i' \Leftrightarrow \exists Q_i : P_i \stackrel{(t_\emptyset,n)}{\Longrightarrow} Q_i \stackrel{(t_\emptyset,1)}{\Longrightarrow} P_i'$; and by the i.h. $P_1 \Box P_2 \stackrel{(t_\emptyset,n)}{\Longrightarrow} Q_1 \Box Q_2$.

□

Props. 4.9 *If $P \in FIN(\Sigma_1)$, then $OBS(P) = F(\mathcal{S}[\![P]\!])$.*
Proof : By induction on the structure of the term P.

1. For $P = stop$ it is obvious.

2. Consider $a \rightarrow P$,

   - suppose $\langle tf, r\rangle \in F(\mathcal{S}[\![a \rightarrow P]\!])$, we have three possible cases:

     (a) $tf = (t_\emptyset, m), r \in \mathcal{PF}(\mathcal{B}^+_{-a})$, then we have

     $$a \rightarrow P \stackrel{B_\emptyset^{(m}}{\Longrightarrow} a \rightarrow P \stackrel{B}{\not\longrightarrow}, \forall B \in r \wedge Stable(a \rightarrow P)$$

     (b) $tf = (t_{n,a}, m)$, with $0 \le n \le m < n + d(a) - 1 = nd$ and $r \in \mathcal{R}$,

     $$a \rightarrow P \stackrel{B_\emptyset^{(n-1}}{\Longrightarrow} a \rightarrow P \stackrel{B_a}{\Longrightarrow} delay(d(a)-1, P) \stackrel{B_\emptyset^{(m-n}}{\Longrightarrow} delay(nd - m, P).$$

     As $nd - m > 0$, $delay(nd - m, P)$ is stable and it has no transitions except for the empty bag.

     (c) $tf = (t_{n,a}, nd) \cdot tf$ with $\langle tf, r\rangle \in F(\mathcal{S}[\![P]\!]) = OBS(P)$ by the i.h.. We then have $a \rightarrow P \stackrel{B_\emptyset^{(n-1}}{\Longrightarrow} \stackrel{B_a}{\Longrightarrow} \stackrel{B_\emptyset^{(d(a)-1}}{\Longrightarrow} delay(0, P)$. We use then lemmas 4.7 and 4.5.

   - Suppose now $\langle tf, r\rangle \in OBS(a \rightarrow P)$, then by definition:

     $$\exists P' : a \rightarrow P \stackrel{tf}{\Longrightarrow} P' \stackrel{B}{\not\longrightarrow}, \forall B \in r \wedge Stable(P').$$

     Let us see the stable states reachable by $a \rightarrow P$, and denoted by P':

     (a) $P' = a \rightarrow P \Rightarrow tf = (t_\emptyset, m)$ and $r \in \mathcal{PF}(\mathcal{B}^+_{-a})$,

     (b) $P' = delay(\ell, P), \ell > 0 \Rightarrow tf = (t_{n,a}, m)$ with $0 \le n \le m < nd$ and $r \in \mathcal{R}$,

     (c) $a \rightarrow P \stackrel{(t_{n,a,nd})}{\Longrightarrow} delay(0, P) \stackrel{tf'}{\Longrightarrow} P'$, we use again lemmas 4.5 and 4.7 obtaining $\langle tf', r\rangle \in OBS(P) = F(\mathcal{S}[\![P]\!])$ (by the i.h.).

3. Consider $P \sqcap Q$,

   - $F(\mathcal{S}[\![P \sqcap Q]\!]) = F(\mathcal{S}[\![P]\!]) \cup F(\mathcal{S}[\![Q]\!]) = OBS(P) \cup OBS(Q) \subseteq OBS(P \sqcap Q)$ by corollary 4.6.

- $P \sqcap Q$ is unstable and its only transitions are internal, leading to P or Q. Therefore $OBS(P \sqcap Q) \subseteq OBS(P) \cup OBS(Q)$.

4. For $P \Box Q$ there are two facts we must take into account:

$$Stable(P \Box Q) \Leftrightarrow Stable(P) \wedge Stable(Q)$$

$$\forall B \in \mathcal{B}^+, P \Box Q \overset{B}{\not\longrightarrow} \Leftrightarrow P \overset{B}{\not\longrightarrow} \wedge Q \overset{B}{\not\longrightarrow}$$

When the trace is empty, we only need to use the lemma 4.8. When the trace is not empty we have $(*)$ $tf = (t_\emptyset, m) \cdot tf'$, with $m = \inf(tf) - 1$ and $trace(tf')(1) \neq B_\emptyset$,

- $\langle tf, r\rangle \in F(\mathcal{S}[\![P]\!]) \cup F(\mathcal{S}[\![Q]\!]) = OBS(P) \cup OBS(Q)$. Assuming that the failure belongs to $F(\mathcal{S}[\![P]\!])$, then for some P' and P'' we have $P \overset{(t_\emptyset,m)}{\Longrightarrow} P' \overset{tf'}{\Longrightarrow} P''$. By lemma 4.5 we have $\langle tf', r\rangle \in OBS(P')$. On the other hand, [P1] and [P6] imply $(t_\emptyset, m) \in T(\mathcal{S}[\![Q]\!]) = T(OBS(Q))$, and therefore $Q \overset{(t_\emptyset,m)}{\Longrightarrow} Q'$. By lemma 4.8 and the fact $(*)$ we obtain $P \Box Q \overset{(t_\emptyset,m)}{\Longrightarrow} P' \Box Q' \overset{tf'}{\Longrightarrow} P''$.
- $\langle tf, r\rangle \in OBS(P \Box Q)$; the fact $(*)$ and lemma 4.8 imply that $P \Box Q \overset{(t_\emptyset,m)}{\Longrightarrow} P' \Box Q'$ and $P \overset{(t_\emptyset,m)}{\Longrightarrow} P'$ for some P' and Q'. And then $\langle tf', r\rangle \in OBS(P') = F(\mathcal{S}[\![P']\!])$ (or the same for Q, Q').

5. For $P_1 \|_A P_2$ the proof is cumbersome to write, but conceptually easy if we have in mind the following facts:

$$Stable(P_1 \|_A P_2) \Leftrightarrow Stable(P_1) \wedge Stable(P_2)$$

$$P_1 \|_A P_2 \overset{B}{\not\longrightarrow} \Leftrightarrow \forall B_1, B_2 : (B = B_1 +_{\mathrm{B}} B_2, B_1 \lceil A = B_2 \lceil A), \P_1 \overset{B_1}{\not\longrightarrow} \vee P_2 \overset{B_2}{\not\longrightarrow}$$

and that every state reachable by $P_1 \|_A P_2$ is of the form $P_1' \|_A P_2'$.

6. For $P \backslash a$ we must consider that $Stable(P) \Leftrightarrow Stable(P \backslash a)$, and $B \backslash a \in (\mathcal{B} - \mathcal{B}_a)$.

□

4.3 Recursive processes

When dealing with recursive processes, there appear two problems, both related to the internal behaviour (any kind of non-externally visible activity of a process). For one thing we must consider the possibility of *divergence* (unbounded sequence of internal actions), for another thing the fact that a process may pass through *an unbounded amount of unstable states* (unbounded sequence of internal transitions).

Def. 4.8 *We say that P* **may not get stable** *($P \uparrow$) if there exists some unbounded sequence*

$$P = P_0 \rightarrowtail P_1 \rightarrowtail \ldots$$

□

In the system given for finite processes, internal actions do not originate special transitions (like the transition $\xrightarrow{\tau}$ in [Mil80] and others). The names of hidden actions disappear from the corresponding bags, and if in some bag there remain only internal actions, we obtain the empty bag. Therefore, if we want to detect the possibility of divergence, we should find some way to detect internal actions, i.e., we must leave some *flag* when the last action in a bag is hidden. Thereby, instead of obtaining an empty bag, we will get a special bag which will behave like the empty bag, but with respect to divergence.

Unstability introduces problems when registering observations, or more exactly, when computing refusals. If we carefully analyse the transition rules in the finite system, we observe that internal transitions are only generated by the internal choice operator. The other operators only propagate these internal transitions through its arguments. In the new transition system recursive calls will generate internal transitions too. Therefore, we can have unbounded sequences of internal transitions caused by an unbounded sequence of internal choices and/or recursive calls. Once again, the restriction to guarded processes will avoid all these problems (see props.4.11).

Def. 4.9 *New label sets:*

Timed labels $\mathcal{L} = \mathcal{B}(\mathcal{A}) \cup \{B_\tau\}$, *where* B_τ *is the special bag we had already mentioned representing the beginning of hidden actions.*

Invisible labels $Null = \{B_\emptyset, B_\tau\}$.

Notation : $\mathcal{L}^*$ *and* $Null^*$ *represent the corresponding sets of label strings.* □

We shall extend the operations defined over $\mathcal{B}(\mathcal{A})$ (in the appendix) to $\mathcal{L}$:

Restriction : $B_\tau\lceil A = B_\emptyset$.

Synchronization : $B_\tau \oplus_A B_\emptyset = B_\emptyset \oplus_A B_\tau = B_\tau$, and for each $B \in \mathcal{B}^+$: $B_\tau \oplus_A B = B \oplus_A B_\tau = B$.

Hiding : $B_\tau \backslash a = B_\tau$. Moreover, if $B \in \mathcal{B}_a$ then $B\backslash a = B_\tau$ (instead of $B_\emptyset$).

From now on we write $P \overset{B_\emptyset}{\not\longrightarrow}$ when with our previous notation we had $P \overset{B_\emptyset}{\not\longrightarrow} \wedge P \overset{B_\tau}{\not\longrightarrow}$.

4.3.1 New transition system

The new transition system for the semantics of eventually recursive processes is:

$$\mathcal{LTS} = \langle CREC(\Sigma_2), \mathcal{L}, \xrightarrow{\alpha} : \alpha \in \mathcal{L}, \rightarrowtail \rangle$$

Its transition rules are mostly the same as for $\mathcal{FLTS}$, but naturally extended to consider the new label. Only rule (Ex5) needs a little change:

$$\text{(Ex5)} \quad \frac{P \xrightarrow{\eta} P' \wedge Q \xrightarrow{\nu} Q'}{P \Box Q \xrightarrow{\eta \oplus \nu} P' \Box Q'} ; \eta, \nu \in Null$$

And we add a rule for recursion :

$$\text{(Rec)}\ \frac{}{\mu\xi.P \rightarrowtail P[\mu\xi.P/\xi]}$$

Next we extend the function to transform label sequences into traces, in order to transform the label B_τ into the empty bag.

Def. 4.10 $TF : \mathcal{L}^* \longrightarrow \mathcal{TTF}$,

$$TF(B_1 \ldots B_n) \stackrel{\text{def}}{=} (t, n) \in \mathcal{TTF} \text{ with } \forall i > n, t(i) = B_\emptyset \wedge \forall i : 1 \leq i \leq n, t(i) = \begin{cases} B_i & \text{if } B_i \in B \\ B_\emptyset & \text{if } B_i = L \end{cases}$$

□

The definitions for $\stackrel{\omega}{\Longrightarrow}$, $\stackrel{tf}{\Longrightarrow}$ and $OBS(P)$ are analogous to the ones given in the previous section for finite processes. In the same way, lemmas 4.3, 4.2, 4.5, 4.8 and 4.7, and the corollary 4.6 are preserved and extended for $\mathcal{L}$ and $CREC(\Sigma_2)$. The corollary 4.4 is no longer true, because the proposition 4.1 is false. However, we have the following corollary:

Corollary 4.10 $\forall P \in CREC(\Sigma_2) : P \stackrel{B_\emptyset}{\Longrightarrow} \vee P \uparrow$. □

Therefore the corresponding refusals are still in $\mathcal{R}$. Moreover we can state the following proposition, which guarantees that guarded processes never reach unstable states.

Props. 4.11 *If $P \in CRECG(\Sigma_1)$ and P' is a state reachable by P (there exists some transition sequence from P to P'), then $\neg(P' \uparrow)$.*

Proof : It is obvious that in order to reach a recursive call we must pass through guards; and each internal transition due to some $\sqcap$ leads to a process syntactically more simple.
□

4.3.2 Divergence

We shall define now the divergence observations for the system. These are traces with the possibility of being extended with some unbounded sequence of internal actions.

Def. 4.11 *For each $\alpha \in \mathcal{L}, \omega \in \mathcal{L}^*$, we define $|\omega|_\alpha$ as the number of labels α in the string ω:*

$$\begin{aligned} |\varepsilon|_\alpha &= 0 \\ |\beta \cdot \omega|_\alpha &= \begin{cases} |\omega|_\alpha + 1 & \text{if } \alpha = \beta \\ |\omega|_\alpha & \text{if } \alpha \neq \beta \end{cases} \end{aligned}$$

□

Def. 4.12 *P is said* **to diverge** *if it has an unbounded sequence of null transitions with an unbounded number of labels B_τ: Diverg(P)* iff $\exists \{s_i\}_{i \in \mathbb{N}^+} : s_i \in Null^* \wedge |s_i|_{B_\tau} = i \wedge P \stackrel{s_i}{\Longrightarrow}$.
□

Def. 4.13 *$DIV : CREC(\Sigma_2) \longrightarrow \mathcal{P}(TT\mathcal{F})$ is defined by*

$$DIV(P) = \{tf \mid \exists P' : Diverg(P') \wedge P \stackrel{tf}{\Longrightarrow} P'\}$$

□

We will give some lemmas to prove the relation between these new observations and the divergent traces in the denotational semantics:

Lemma 4.12 *If $P \stackrel{(t_\emptyset,m)}{\Longrightarrow} P'$ and $Diverg(P')$, then $Diverg(P)$.*

Proof : $P \stackrel{(t_\emptyset,m)}{\Longrightarrow} P' \Rightarrow \exists s \in Null^*, P \stackrel{s}{\Longrightarrow} P'$. By the definition of $Diverg(P')$ we have $Diverg(P)$. □

As a corollary we have that internal transitions do not affect the divergence observations.

Corollary 4.13 *If $P \rightarrowtail^* P'$ then $DIV(P') \subseteq DIV(P)$.*

Process $delay(0, P)$ is still equivalent to process P with respect to divergences.

Lemma 4.14 $DIV(delay(0, P)) = DIV(P)$.

We shall present now the main result for this section:

Props. 4.15 *If $P \in CRECG(\Sigma_1)$, then $\mathcal{S}[P] = \langle OBS(P), DIV(P)\rangle$.*

Proof : It is proven by structural induction on the term P. Following the same steps as in the proof for proposition 4.9, we obtain the result concerning the failures part (except for the recursive calls, which should be more carefully studied). We see now what happens with the divergences:

1. For $P = stop$ it is evident.

2. For the prefix operator $(a \rightarrow P)$, we have two kinds of traces, those in which action a has started but not finished, and those in which the process P has started. In the first case, we apply the lemma 4.12; in the second case we apply the i.h..

3. For the choice operators the proof is immediate. Although in the case of the external choice we must take into account the fact that processes are supposed to be guarded and then the corollary 4.10 can be applied together with the props.4.11. If one of the processes did not come to get stable, the whole process would be unable to make null transitions, and divergences would be lost.

4. Something analogous to the previous case happens when considering synchronization, as it is required that both components make transitions simultaneously. On the other hand, it is sufficient that one of the components should have the possibility of divergence.

5. Concerning the hiding operator, it is obvious that the same divergences are obtained by one method or the other (for consider the definitions and corollary 4.10).

In relation to recursive calls, the reasoning is the usual in these cases. As we are dealing with guarded processes, with each iteration, more information about the recursive process is obtained. When the denotational semantics was defined, we remarked that the semantics of a recursive term is its least fixed point. With the rule (Rec) we approximate this least fixed point. □

5 Related Work

In the introduction of the present paper, some timed models developed by other colleagues were mentioned. Some of them were already largely commented and compared with *Timed Observations* in [OdF90,Ort90]. Therefore, this time we concentrate only on the aspects concerning the operational approach. Under this perspective, although the model presented in [RR87] is the closest one to ours, we have no notice of any operational study of it[5].

The algebraic timed models ATP [NRSV90] and ACP_ρ [BB90] are both based on ACP ([BK85]), and both present algebraic and operational rules. In ATP time is treated like a special action, which needs the synchronization of every process in the system. Thus the progress of time is represented by the action χ, in a similar way to our $B_\emptyset$ in the operational system. On the other hand, in ACP_ρ time is seen as a new component in the system. The state of a process is represented by a pair formed by a process expression and an instant time; and a time stamp representing the instant in which each action is executed is added to atomic actions, which are instantaneous (like in ATP). The transition system includes three relations between states: the *step*-relation, labeled with an action with its time stamp ($a(t)$), representing the execution of a at instant t; the *terminate*-relation, labeled with a timed action too, represents again the execution of the given action at the given time, but now the state reached is a deadlocked one (there are no more actions and no more waiting); and the *idle*-relation representing the progress of time (from any instant to any later one).

Compared with our transition system, the first significant difference is the time component in the states. We must point out the *absolute* nature of time in ACP_ρ, well reflected in the time stamps added to the atomic actions. In contraposition, in our model, the origin of time is relative to the beginning of each process (when considering a system composed by several processes, the system is seen as a process and the origin is unique for all the components), and the state time is deduced by the actions executed (including hidden and empty actions). The authors develop in the same paper a variant of the model with what they call *relative time*, where time stamps associated with actions are relative to its enabling ocurrence [6], but this is only seen as an abstraction, very convenient to write process specifications. Operational semantics for this version still refers to absolute times in states.

The idle-relation is a consequence of the instantaneous nature of actions. The need for the termination-relation is not obvious, because the difference between the deadlock process and the termination state is quite subtle. In any case, our deadlock process *stop*

[5]It seems there exists some work on this precise topic by Schneider (PRG, Oxford/UK), but still not published.

[6]Like our *time-outs* in [OdF90].

may idle indefinitely. Our internal transition relation was mainly introduced for the sake of the internal choice operator, but neither of the algebraic models has such an operator.

In a similar way to the relative time version of ACP_ρ, the Timed Calculus presented in [QAF89] - a timed interpretation of LOTOS [ISO87] - considers that each action is given a time stamp representing the precise instant in which the action is executed, always relative to the previous action. In the transition system these stamped actions are incorporated again, but preserving their relative nature.

TCCS [MT89], the Temporal Process Algebra presented in [HR90] and CCS+Time in [Yi90] are timed versions of CCS ([Mil80]). TPA introduces a new action σ representing the time unit (like χ in ATP or $B_\emptyset$ in our model), and actions can arbitrarily delay their execution like in our model.

In contraposition, in TCCS actions are executed immediately, but new operators are incorporated to the language, like $(t)P$ (a delay of t time units) and $\delta.P$ (representing the indefinitely delay of P). The states in the transition system are just process expressions, and two relations are considered: the usual one labeled with actions, and representing the execution of actions; and a relation labeled with time units, representing the progress of time, which could be seen as a generalization of $\xrightarrow{B_\emptyset}$.

CCS+Time is similar to the other two timed CCS models commented above, but considers continuous time instead of discrete time (like the others and our model). It incorporates a general delay operator (like the one in TCCS) and a modified prefix operator $a@t.P$, similar to our interpretation of the prefix operator $a \rightarrow P$, with the originality of introducing the time variable t recording the exact time delay before executing the action a, so that the behaviour of P can be dependent on this time variable t. The arbitrary delay is only applicable to non-internal actions. In the corresponding transition system, there is no distinction between actions and idling, because time events $\epsilon(d)$ (representing an empty action which lasts for d time units) are added to the label set.

All the above mentioned timed models (except for [RR87]) consider instantaneous actions and are based on interleaving semantics, in contraposition to our model.

In ACP_ρ, the Timed Calculus for LOTOS, TCCS and CCS+Time, equivalence relations based on the *bisimulation* notion [Par81] are studied. We have not done the same, because our main interest when developing the operational semantics was just to obtain another characterisation of the original model, with a more "operational" (i.e. intuitive) flavour. In this way, we somehow complete the theoretical study of the model *Timed Observations*, which now can be considered from three different points of view: denotational, operational and algebraic [7].

Of course, an interesting line of future work would be the study of congruence relations (like bisimulation or testing) on this operational setting.

6 Acknowledgements

I want to thank Prof. K. Indermark (Lehrstuhl für Informatik II / RWTH-Aachen) and the whole department for their support and kind hospitality during those highly productive

[7] In [OdF91], a correct and complete proof system for the denotational approach is provided.

months. Thanks also to D. de Frutos for his help and suggestions. Thanks to Paz too.

References

[BB90] J.C.M. Baeten and J.A. Bergstra. *Real Time Process Algebra.* Technical Report, CWI, Amsterdam, 1990.

[BK85] J.A. Bergstra and J.W. Klop. Algebra of communicating processes with abstraction. *TCS*, (37):77–121, 1985.

[BR85] S.D. Brookes and A.W. Roscoe. An improved failures model for communicating processes. In *Pittsburgh Seminar on Concurrency*, Springer Verlag, 1985. LNCS 197.

[Bro83] S.D. Brookes. *A Model for Communicating Sequential Processes.* PhD thesis, Oxford Univ., 1983.

[HBR81] C.A.R. Hoare, S.D. Brookes, and A.W. Roscoe. *A Theory of Communicating Sequential Processes.* Technical Report, Programming Research Group / Oxford Univ. (UK), 1981. Tech. Monograph PRG-16.

[Hen88] M. Hennessy. *Algebraic Theory of Processes.* MIT Press, 1988.

[HR90] M. Hennessy and T. Regan. *A Temporal Process Algebra.* Technical Report, University of Sussex (UK), 1990.

[ISO87] ISO. *Information Processing Systems - Open Systems Interconnection - Lotos: a Formal Description Technique based on the Temporal Ordering of Observational Behaviour.* 1987. DIS8807.

[KSdR*85] R. Koymans, R.K. Shyamasundar, W.P. de Roever, R. Gerth, and S. Arun-Kumar. Compositional semantics for real-time distributed computing. In *Conference on Logics of Programs*, Springer Verlag, 1985. LNCS 193.

[Mil80] R. Milner. *A Calculus of Communicating Systems. LNCS 92*, Springer Verlag, 1980.

[MT89] F. Moller and C. Tofts. *A Temporal Calculus of Communicating Systems.* Technical Report, Dept.of Computer Science / Univ.Edinburgh (UK), 1989. ECS-LFCS 89-104.

[NRSV90] X. Nicollin, J.L. Richier, J. Sifakis, and J. Voiron. ATP-an algebra for timed processes. In M. Broy and C.B. Jones, editors, *TC2-Working Conference on Programming Concepts and Methods*, North-Holland, 1990.

[OdF90] Y. Ortega-Mallén and D. de Frutos-Escrig. Timed observations: a semantic model for real-time concurrency. In M. Broy and C.B. Jones, editors, *Programming Concepts and Methods*, North-Holland, 1990.

[OdF91] Y. Ortega-Mallén and D. de Frutos-Escrig. A complete proof system for timed observations. In Abramsky and Maibaum, editors, *TAPSOFT'91*, Springer-Verlag, 1991. LNCS 493, vol.1.

[OH86] E.R. Olderog and C.A.R. Hoare. Specification-oriented semantics for communicating processes. *Acta Informatica*, (23):9–66, 1986.

[Ort90] Y. Ortega-Mallén. *En Busca del Tiempo Perdido.* PhD thesis, Fac. CC. Matemáticas, Univ.Complutense de Madrid, 1990.

[Par81] D.M.R. Park. Concurrency and automata on infinite sequences. In *5 th. G.I. Conference*, Springer Verlag, 1981. LNCS 104.

[QAF89] J. Quemada, A. Azcorra, and D. Frutos. *A Timed Calculus for LOTOS.* Technical Report, DIT, E.T.S.I. Telecomunicación / Univ. Politécnica de Madrid (Spain), 1989.

[RR87] G.M. Reed and A.W. Roscoe. Metric spaces as models for real-time concurrency. In *Mathematical Foundations of Programming Language Semantics*, Springer Verlag, 1987. LNCS 298.

[TV89] D. Taubner and W. Vogler. The step failure semantics and a complete proof system. *Acta Informatica*, (27):125–156, 1989.

[Yi90] Wang Yi. CCS + time = an interleaving semantic model for real time systems. November 1990. Chalmers University of Technology, S-41296 Goteborg, Sweden.

REAL–TIMED CONCURRENT REFINEABLE BEHAVIOURS

David Murphy
GMD F2G2, Schloss Birlinghoven, Postfach 1240,
5205 Sankt Augustin, Germany.
E–mail: david@gmdzi.gmd.de

David Pitt
Department of Computational and Mathematical Sciences,
University of Surrey, Guildford, GU2 5XH. England.
E–mail: D.Pitt@mcs.surrey.ac.uk

ABSTRACT. The purpose of this paper is to present a real–timed concurrency theory incorporating true concurrency and event refinement. The theory is based on the occurrences of actions; each occurrence or *event* has a start and a finish. Causality is modelled by assigning a strict partial order to these starts and finishes, while timing is modelled by giving them reals.

The theory is presented in some detail. All of the traditional notions found in concurrency theories (such as conflict, confusion, liveness, fairness and so on) are seen to be expressible. Four notions of causality arise naturally from the model, leading to notions of *securing*. Three of the notions give rise to underlying event structures, demonstrating that our model generalises Winskel's.

Various causality–preserving bisimulations are introduced as notions of equivalence of timed causal structures. The nonatomicity of occurrences means that events can be refined into structures and, dually, substructures can be abstracted into single events. The equivalences presented are shown to be congruences of event refinement.

Time ... is that aspect of motion which makes possible the enumeration of successive states.

Aristotle

1. ON CONCURRENCY THEORIES

Concurrency theory is like detective work — a science of observation and deduction. One of its central concerns is describing the behaviour of distributed systems. Hence it can be seen as a part of natural philosophy, being interested in happenings in the world. This paper presents a concurrency theory from that point of view, giving a model with a physical bias. Our aim will be to describe structures of consisting of occurrences of nonatomic actions in a distributed system. We will be concerned with when those occurrences happen (timing) and where they happen (distribution). (We shall *not* assume that these systems have anything to do with computation; computation can be seen as an *interpretation* we place on happenings, not an inherent property of them.)

The central concern, then, will be the accurate description of behaviour. It is important to have concurrency theories with sufficient descriptive power to be able to describe the very rich and subtle behaviours of real systems; if a theory cannot describe a certain behaviour, such as metastability, or synchronisation failure, then a designer can never be sure that a system designed using that theory is free from that kind of behaviour.

Models related to the one presented here can be found in Rensink's [27] and Schettini & Winkowski's [29], while an associated process algebra is discussed in [21]. Comprehensive discussion of the semantics of the model presented here and example specifications are given in [18] and [20].

The rest of the paper is organised thus; in the remainder of this section some more introductory and motivational material is given. The next section introduces the basic causal and temporal formalism of the model, while the following one presents more complex notions of causality, and uses them to show that our model is a generalisation of Winskel's. Several causality–preserving bisimulations are then introduced, giving some notions of equivalence. Event refinement is treated in the last section, and our equivalences are shown to be congruences of it.

1.1. The Place of Descriptive Models. Olderog [22] and Pnueli [23] both argue that highly descriptive theories are at the bottom of the design hierarchy; one might begin a specification of a distributed system with a descriptively-poor but terse theory, such as linear time temporal logic [23], and move on to a process algebra, such as CSP [12], for detailed design. The model advocated here is one of the most expressive concurrency theories: it lies at the bottom of the hierarchy along with other 'true concurrency' models; Pratt's pomsets [25], Petri nets [26] and Winksel's event structures [37].

We should not expect these descriptive models to be abstract specification tools; their rôle is to analyse behaviours. Our area of application, then, is real-time control systems, asynchronous hardware and systems with similarly-involved behaviours.

1.2. Timing. Our model will be *timed*, because timing constraints are sometimes a fundamental part of a concurrent system. (See the discussion in Davies' thesis [7] or Joseph's [15]. A survey of timed models can be found in Fidge & Pilling's [9] or Jeffrey's [14]. We also recommend the rather more mathematical approach of Kasangian and Labella [16].)

One of the advantages of treating time is that it gives another perspective on the theory; our constructions should make sense both temporally and causally. This justifies our creation of a new model rather than imposing time on an extant one; we want to understand the natural constructions, rather than how to time the constructions we already have. Moreover, this approach has the great advantage of giving a simple notion of event refinement, furthering the work of Aceto [2], Darondeau & Degano [6], Goltz & van Glabbeek [32] and Vogler [34].

2. Basic Ideas

In this section we shall outline the basics of a partial-order model of real–timed concurrency. Causality will be modelled as a strict partial order over some set of transitions, and timing will be introduced.

We will consider as given a set of $\mathbf{T}$ of transitions, endowed with a strict partial order $<_c$ (i.e. an irreflexive and transitive relation). The relationship $x <_c y$ is interpreted as the transition x somehow causing the transition y.

Partial order models, of which Winskel's event structure model [37] and Gischer's pomset model [10] are both good examples, have a notion of *concurrency*. If neither $x <_c y$ nor $y <_c x$ nor $x = y$ then x and y are *unrelated*. They could be transitions in different places, or transitions simultaneous with each other, or neither; it doesn't matter. All that is certain is that they do not causally affect each other; they are on different *paths*. A model has (so called 'true') concurrency if there is a concept of two happenings being causally unrelated. Causal information can be seen as specifying abstract location information, since the absence of a causal relationship between two happenings means that they could be distributed: we expand on this point later.

2.1. Adding Time. Timing is becoming an issue that is much considered in the mainstream of concurrency theory, witness the recent flurry of activity [1], [8], [11], [16], [28]. In a timed model *when* something happens is as important as *what* happens, so timing should be introduced with the same status as causality. (This perspective is discussed at length in [18].)

We begin by assuming a notional, omniscient observer who records the time that each event happens.* Naïvely, it is clear that events in nature, like having a twenty-fifth birthday, or eating this apple, have durations, so we will associate an event with an interval of the reals, and so consider both the time an event starts, and the time that it finishes.

2.2. Occurrences and their Causality. Suppose that some finite set of actions $\mathbf{E}$ is of interest. These actions will be things that can happen more than once, like having a birthday, or eating an apple. Each time an action occurs, it will acquire a label taken from a (countable) set $\mathbf{L}$ so that the pair $(l, e), l \in \mathbf{L}, e \in \mathbf{E}$ is unique. This gives us a universe of occurrences. Obviously not all pairs (l, e) will be relevant; some occurrences may never happen.

An occurrence of an action, – a pair $a = (l, e)$, – will be referred to as an 'event,' following the usual CCS–community usage. Each event has a start and a finish; subscripts will be used for these. Following our previous loose terminology, we will refer to the starts and finishes of events as *transitions*, and use a_t for one of a_s or a_f.

Definition 1 (Transitions). A set of transitions $\mathbf{T}$ is a set endowed with a strict partial order $<_c$ (representing causality). Additionally we require:

- The beginnings of events cause their ends; $\forall\, a_s, a_f \in \mathbf{T} \,.\, a_s <_c a_f$.
- There is a distinguished event $*$, known as the *silent event*, the start of which causes everything, and whose finish everything causes.

The silent event $*$ can be thought of as an 'on light.' Once we see $*_s$, we know that the structure is active, while $*_f$ tells us that everything is over.

It may be important to have events beginning at the same time as the on light goes on, so we need to introduce a new relation, *causal equality*, written $=_c$. Two events are related by $=_c$ if they are caused by the same events and cause the same events.

---

* This is *not* a global clocks assumption, as the observer's clock cannot influence transitions, only record their times. Even in highly relativistic situations this is not an unreasonable assumption; other observers may disagree with our observer about the times things happen, but their observations can be deduced from those of our observer given information about their relative motions; see [19].

Definition 2 (Causal Equality). The relation $=_c$ is the largest (coarsest) congruence over $<_c$ defined by

$$(a_t =_c a'_t) \iff \forall b_t \in \mathbf{T}\,.\,(a_t <_c b_t \iff a'_t <_c b_t) \;\wedge\; (b_t <_c a_t \iff b_t <_c a'_t)$$

This clearly gives us a congruence that is, moreover, distinct from $<_c$;

$$(a_t =_c b_t) \implies (a_t \not<_c b_t) \wedge (b_t \not<_c a_t)$$

In order to be like an on light, the silent event must encompass all transitions, so we require additionally

$$\forall a_t \in \mathbf{T}\,.\,(*_s <_c a_t \;\vee\; *_s =_c a_t)$$
$$\forall a_t \in \mathbf{T}\,.\,(a_t <_c *_f \;\vee\; a_t =_c *_f)$$

This allows us to confine our attention to *connected* structures.

The set of possible occurrences of actions or events will be denoted by **LE**;

$$\mathbf{LE} \subseteq \mathbf{L} \times \mathbf{E} \cup \{*\}$$

Thus $(l, e) \in \mathbf{LE}$ means there may be an lth occurrence of action e. Our structures will always contain at least $*$, as we require $* \in \mathbf{LE}$. It may be helpful to think of $\mathbf{L} \times \mathbf{E}$ as a universe of events, and **LE** as a subset of those that are of interest at once.

The relationship between the sets of transitions, labels and actions is

$$\mathbf{T} = \{a_s \mid a \in \mathbf{LE}\} \cup \{a_f \mid a \in \mathbf{LE}\}$$

Notice that $\leq_c$ ($= <_c \cup =_c$) is a partial order on $\mathbf{T}$.

2.3. Paths in Time. Causality has been modelled by endowing transitions, members of $\mathbf{T}$, with a strict partial order $<_c$. Timing will be dealt with by assigning them a real number. This leaves us with a branching time model incorporating both timing and causality.

Causal information is abstract *spatial* information, since where something is influences what it can cause and what can affect it.† We want to record information about when things happen and how they are related; the when is time, and the how is space. Thus knowing temporal and causal information about a transition gives us a point in *spacetime.*

Definition 3 (Spacetime). A point in spacetime is a pair (r, a_t) where r is a real number and a_t is a transition. For a given set of transitions, $\mathbf{T}$, the set of all possible points in space time, **ST** say, will thus be $\mathbb{R} \times \mathbf{T}$. Points in spacetime will be symbolised by t with a subscript, so $t_s = (r_s, a_s)$ is the point the event a started, and t_t is either the point in **ST** that a started, or the point it finished.

By thinking of transitions as the starts and finishes of events, we can bring all our concepts together, defining a *task* to be a triple $(a, \underline{\text{begin}}(a), \underline{\text{end}}(a))$ consisting of an event, $a \in \mathbf{LE}$, and the points in spacetime it starts and finishes. The set of all tasks will be

† One motivation here is relativistic systems, or any distributed system where signals may take a significant time to propagate, as in [24]. In such situations a_t can only affect b_t if it is close enough to it in spacetime that a signal can travel between the two at less than the speed of light; see [19].

TK $\subset$ **LE** $\times$ **ST** $\times$ **ST**. (Here 'task' is used in the spirit of Gischer: his [10] introduces the term.) We will write $a \equiv [t_s, t_f]$ for $a \in \mathbf{LE}, t_s, t_f \in \mathbf{ST}$ to indicate that the task (a, t_s, t_f) associates the event a with the 'interval' in spacetime $[t_s, t_f]$, i.e. $\underline{\text{begin}}(a) = t_s$ and $\underline{\text{end}}(a) = t_f$.

Summary; given a set of occurrences of actions **LE**, we get a set of transitions **T**; their starts and finishes. Abstract spacetime is constructed from the reals and these transitions $\mathbf{ST} = \mathbb{R} \times \mathbf{T}$. The timing functions $\underline{\text{begin}}, \underline{\text{end}}$ are thus given; they tell us where and when transitions happen. From this information we construct tasks; events together with their starting and finishing spacetimes. Obviously, given a set of tasks **TK**, we can recover the underlying set of occurrences **LE**. We will abuse notation in future, and write **LE** for the set of events underlying a given **TK**, **LE**$'$ for that underlying **TK**$'$ and so on.

Definition 4 (Relations on Spacetime). Suppose we have events a, b with timings $a \equiv [(r_s, a_s), (r_f, a_f)]$, $b \equiv [(s_s, b_s), (s_f, b_f)]$. Then we can define spacetime relations

$$(r_t, a_t) <_s (s_t, b_t) \iff r_t < s_t \ \wedge\ a_t <_c b_t$$
$$(r_t, a_t) =_s (s_t, b_t) \iff r_t = s_t \ \wedge\ a_t =_c b_t$$

The relations $=_s$ and $<_s$ on spacetime represent both causality and timing, so that $t_t <_s u_t$ means that the instant in spacetime t_t happens before the instant u_t and causes it. We write $t_t \leq_s u_t$ iff $t_t <_s u_t$ or $t_t =_s u_t$. (The sensitive reader may be irritated by our use of t for s or f; they may be mollified by the realisation that the first equation above, written out in full, is the tedious long combination $(r_s, a_s) <_s (s_s, b_s) \iff r_s < s_s \ \wedge\ a_s <_c b_s$ and $(r_s, a_s) <_s (s_f, b_f) \iff r_s < s_f \ \wedge\ a_s <_c b_f$ and $(r_f, a_f) <_s (s_s, b_s) \iff r_f < s_s \ \wedge\ a_f <_c b_s$ and $(r_f, a_f) <_s (s_f, b_f) \iff r_f < s_f \ \wedge\ a_f <_c b_f$.)

There are some things that we require of our relations:

- Things can't happen before their causes; $(a_t <_c b_t) \implies (r_t < s_t)$. Hence, causal ordering implies temporal ordering.
- Transitions that are causally-equal must be temporally-equal; $a_t =_c b_t \implies r_t = s_t$.

2.4. Incomparability. A rather primitive notion of concurrency is supported by the model as presented thus far. It seems reasonable to say that two tasks are concurrent if there is some point in spacetime belonging to one that is not related to a point of the other. We can imagine an instance where a and b are concurrent, but their starts are necessarily simultaneous, as in $P \overset{\text{def}}{=} a \parallel b$, where we might expect $a_s =_s b_s$. This consideration motivates

Definition 5 (Incomparability). Suppose t_t, $u_t \in \mathbf{ST}$. Then t_t and u_t are said to be *incomparable*, written $t_t \ \underline{\text{inc}}\ u_t$, iff

$$t_t \not<_s u_t \ \wedge\ u_t \not<_s t_t \ \wedge\ t_t \neq_s u_t$$

while two tasks, $a \equiv [t_s, t_f]$ and $b \equiv [u_s, u_f]$ are said to be incomparable if a pair of their times is incomparable, that is, iff

$$t_s \ \underline{\text{inc}}\ u_s \ \vee\ t_f \ \underline{\text{inc}}\ u_s \ \vee\ t_s \ \underline{\text{inc}}\ u_f \ \vee\ t_f \ \underline{\text{inc}}\ u_f$$

If two tasks a and b are incomparable then we write $a \ \underline{c}\ b$, and abuse this notation for their associated events.

The model presented here will be called the *interval event structure* (I.E.S.) model, because it is based on Winskel's event structures, and because events are associated with intervals of time.

Definition 6 (Elementary Interval Event Structures). An elementary I.E.S. is a triple $(\mathbf{TK}, <_s, =_s)$ consisting of a set of tasks, $\mathbf{TK}$, a spacetime order $<_s$, and a spacetime equality $=_s$.

Note that it may be desirable to have undefined times in the model (to model imperfect knowledge, for instance). In this case we could take spacetime to be $(\mathbb{R} \cup \{\top\}) \times \mathbf{T}$ instead of $\mathbb{R} \times \mathbf{T}$. We could then extend the order by making undefined times in the future, so that $\forall r \in \mathbb{R} . r < \top$ and $\top < \top$; see [20]. This extension would also be useful if we wanted to express recursive timed processes as interval event structures; in that case the only transition with spacetime $\top$ would be $*_f$.

3. The Model Refined

In this section the model will be further developed, and its descriptive power examined by way of several examples.

It is useful to be able to express the causal relations between tasks somewhat more abstractly than is possible using $<_s$ and $=_s$. For this reason we introduce, in the first subsection, four *causal orders* on tasks. Then, nondeterminism is introduced via the notion of *conflict*. Several examples demonstrate some of the behavioural features that are expressible in the model. Finally, the causal orders are used to relate I.E.S.s to event structures and to Lamport's model of distributed systems.

3.1. Orders on Tasks.

Definition 7. Suppose that $a \equiv [t_s, t_f]$, $b \equiv [u_s, u_f]$; then there are four possible causal relationships:

If $t_f \leq_s u_s$, then a is *interior-causal* of b, written $a \sqsubseteq_i b$.

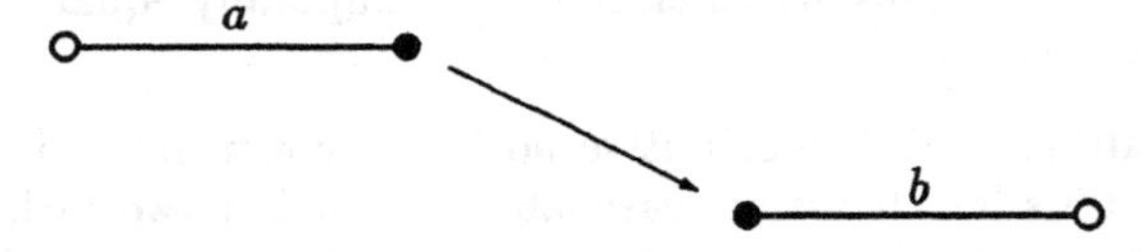

If $t_s \leq_s u_s$, then a is *head-causal* of b, written $a \sqsubseteq_h b$.

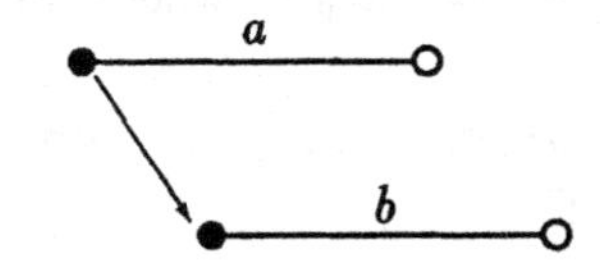

If $t_f \leq_s u_f$, then a is *tail-causal* of b, written $a \sqsubseteq_t b$.

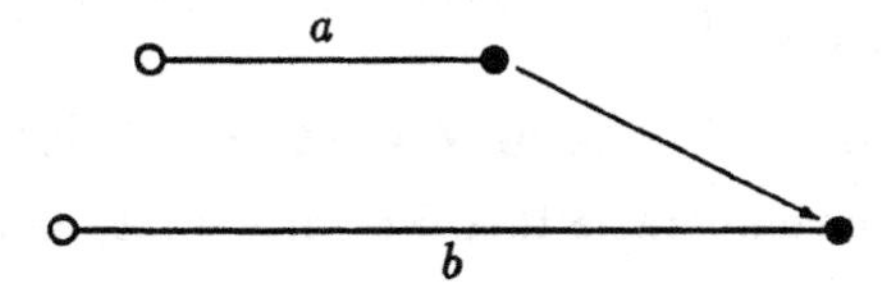

If $t_s \leq_s t_f$, then a is *exterior-causal* of b, written $a \sqsubseteq_e b$.

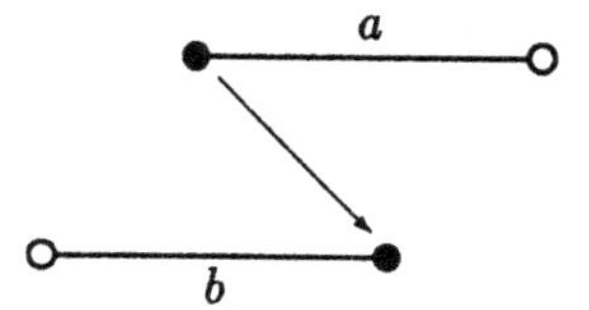

We will write $\sqsubseteq_j$ for one of $\sqsubseteq_i, \sqsubseteq_h, \sqsubseteq_t, \sqsubseteq_e$, and call the relations $\sqsubseteq_j$ *j-morphisms.*

The j-morphism $\sqsubseteq_i$ is a strict partial order. (Starts causing finishes, section 2.2, gives us

$$a \equiv [(r_s, a_s), (r_f, a_f)] \implies r_s < r_f$$

so $a \not\sqsubseteq_i a$ (and irreflexivity is vacuously true). Transitivity is obvious.) The j–morphisms $\sqsubseteq_h$ and $\sqsubseteq_t$ are preorders, but $\sqsubseteq_e$ is not in general transitive.

There is a simple relationship between the j-morphisms; they form a poset ordered by inclusion, as in Figure 1. Notice that this poset is not a lattice under union, since $a \sqsubseteq_h b$ and $a \sqsubseteq_t b$ do not together necessarily guarantee $a \sqsubseteq_i b$. However, if $a \sqsubseteq_j b$ and $\sqsubseteq_j \subset \sqsubseteq_k$ then $a \sqsubseteq_k b$.

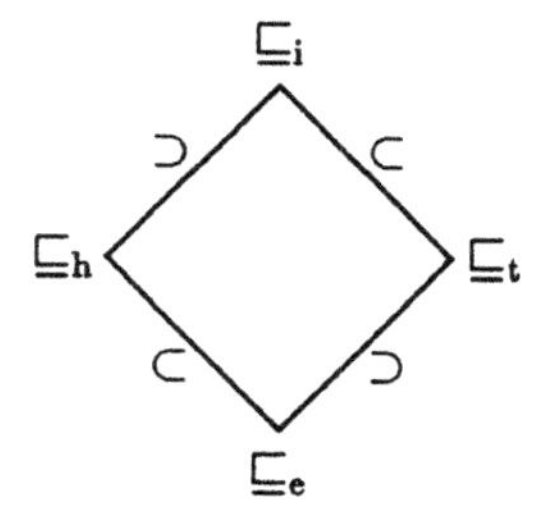

FIGURE 1. The relationship between the j–morphisms

Definition 8 (Securing). A sequence of tasks $\langle a_1, a_2, \ldots, a_n \rangle$ *j-secures* a task a just when $a_1 \sqsubseteq_j a_2 \sqsubseteq_j \ldots a_n \sqsubseteq_j a$.

3.2. Conflict. Conflict will be modelled by the common partial-order model technique of introducing a set of *consistent sets* of events, $\underline{\mathrm{Con}} \subseteq \wp(\mathbf{LE})$. An element of $\underline{\mathrm{Con}}$ is a set of events, all of which can happen in the same history, so if $\{a, b\} \in \underline{\mathrm{Con}}$, then there is some execution in which the events a and b both occur. The $\underline{\mathrm{Con}}$-set tells us where choice happens — where the branches of branching time branch.

If a set of events A is consistent, then any subset of it should also be consistent, so we have the restriction on possible $\underline{\mathrm{Con}}$-sets

$$A \in \underline{\mathrm{Con}} \ \& \ B \subseteq A \implies B \in \underline{\mathrm{Con}}$$

Two events are in conflict, $a \# b$, just when there is no consistent set containing them, so we recover binary conflict $\#$ from $\underline{\mathrm{Con}}$ by defining

$$a \# b \iff \{a, b\} \notin \underline{\mathrm{Con}}$$

Furthermore, it seems sensible to allow two events to be in conflict only if they are incomparable; if they are completely causally related, then it is hard to know how they could be in conflict, so we also require of Con that

$$a \# b \implies a \underline{c} b$$

This implies that if two events are completely causally related, then they are in essentially the same Con-sets. Say $a_t \sim_s b_t \iff (a_t \leq_c b_t \vee b_t \leq_c a_t)$.

$$\forall c \in \underline{\text{Con}} . (a \in c \wedge a_s \sim_s b_s \wedge a_f \sim_s b_f) \implies c \cup \{b\} \in \underline{\text{Con}}$$

We will require that Con covers **LE**, $\bigcup \underline{\text{Con}} = \textbf{LE}$ and that structures have finite choice. To impose this latter restriction, we need the concept of a *left–closed* Con–set.

Definition 9. A Con–set c is left–closed if it contains all consistent events starting before some point in time, i.e. if $\forall c' \in \underline{\text{Con}}, b \in \textbf{LE}$.

$$c' = c \cup \{b\} \implies (\forall a \in c . a \sqsubseteq_h b)$$

The set of all left–closed Con–sets of a given structure will be written LCon.

Our requirement of finite nondeterminism then just amounts to the set of left–closed Con-sets which extend a given one being finite, so $\forall c \in \text{L}\underline{\text{Con}}$.

$$\{ c' \mid c' \in \text{L}\underline{\text{Con}}, \exists a \in \textbf{LE} . c' = c \cup \{a\} \} \quad \text{is finite}$$

The introduction of conflict permits a more sophisticated definition of concurrency. Two tasks, (and by abuse their underlying events) a and b, will be *concurrent*, written $a \underline{co} b$, iff they are incomparable and not in conflict;

$$a \underline{co} b \iff a \underline{c} b \wedge \{a, b\} \in \underline{\text{Con}}$$

Definition 10 (Interval Event Structures). An interval event structure $\Im$ is a quadruple $(\textbf{TK}, \underline{\text{Con}}, <_s, =_s)$ consisting of a set of tasks, **TK**, a set of consistent sets, Con, a spacetime order $<_s$ and a spacetime equality $=_s$. The class of all interval event structures will be written **IES**.

Sometimes we will be interested in ignoring the causal and conflict structure, and considering just the real times of events. For this purpose, a *linear* interval event structure, or L.I.E.S. will suffice;

Definition 11 (Linear Interval Event Structures). A linear I.E.S $\Im$ is a triple (**TK**, $<, =$) consisting of a set of tasks, **TK**, and the usual $<$ and $=$ on $\mathbb{R}$. Obviously here if $\underline{\text{begin}}(a) = (r, a_t)$ only r will be meaningful.

A linear structure can be thought of as a model of an observation of a single execution of an I.E.S. A complete history of one such execution should contain no branching information, having just all the events from a *maximal* Con-set, together with the order on their times;

Definition 12 (Maximal Con-Sets). A Con-set c is said to be maximal with respect to an I.E.S. $\Im = (\textbf{TK}, \underline{\text{Con}}, <_s, =_s)$ if

$$(c' \in \underline{\text{Con}} \wedge c \subseteq c') \implies c = c'$$

The set of all maximal Con-sets of an I.E.S. $\Im$ is written $\mathcal{M}(\Im)$. Notice that the 'on light' $*$ goes on and off in every complete history; $\forall c \in \mathcal{M}(\Im) . * \in c$.

3.3. Examples. The expressive power of the interval event structure model will now be demonstrated. In what follows assume that a and b are tasks, that $a \equiv [(r_s, a_s), (r_f, a_f)]$ and that $b \equiv [(s_s, b_s), (s_f, b_f)]$. We can then express the following behavioural features;[‡]

- Sequentiality; the two tasks are sequential if $a \sqsubseteq_i b$. That is, a starts wholly before b and is on the same path.
- Temporal overlap; if $\{a, b\} \in \underline{\text{Con}}$ and $[r_s, r_f] \cap [s_s, s_f] \neq \emptyset$ then a and b overlap in time.
- Concurrency; a and b are concurrent if $a \; \underline{\text{co}} \; b$. Notice that temporal overlap does not imply concurrency, nor does concurrency imply temporal overlap.
- Local simultaneity; a good example of temporal overlap without concurrency is the situation $a \sqsubseteq_h b$ and $b \sqsubseteq_t a$ but not $a \; \underline{\text{co}} \; b$ or $a \# b$. Here a and b are coexisting on the same path.
- The N-poset. This is an important object, since the absence of it embedded in a given poset guarantees that the poset can be decomposed; see [10] for details. Given tasks a, b, c, and d, this poset can be described by $a \sqsubseteq_i b$, $c \sqsubseteq_i d$ and $a \sqsubseteq_i d$. This structure displays sequentiality (between a and b for instance), and incomparability (between a and c).
- Asymmetric confusion; an I.E.S. displaying asymmetric confusion can be formed from the three tasks a, b and c with $a \sqsubseteq_i b$, $a \; \underline{\text{co}} \; c$ and $b \# c$. (The correct Con-set to describe this situation is $\{\{a\}, \{b\}, \{a, b\}, \{c\}, \{a, c\}\}$.)
- An I.E.S. without conflict inheritance. Winskel demands that conflict is inherited, that is, if $a \# b$ and $a \sqsubseteq_i c$ then $c \# b$. We will not require conflict inheritance, so that we can describe situations like this; $a \sqsubseteq_i b$, $c \sqsubseteq_i d$, $a \sqsubseteq_i d$, $c \sqsubseteq_i b$, $a \# c$ but $b \; \underline{\text{co}} \; d$.
- An unstable I.E.S. $a \# b$, $a \sqsubseteq_i c$ and $b \sqsubseteq_i c$. Here the Con-set is $\{\{a\}, \{c\}, \{a, c\}, \{b\}, \{b, c\}\}$. Confused situations are not necessarily unstable, but unstable situations are always confused.

3.4. Realisable Structures. If one is only interested in modelling systems that we can actually build, then it makes sense to restrict the class of I.E.S.s considered. If an I.E.S. $\Im = (\mathbf{TK}, \underline{\text{Con}}, <_s, =_s)$ satisfies the axioms (1) and (2) below, then it is said to be *realisable*:

$$\forall c \in \underline{\text{Con}}, \forall r_1, r_2 \in \mathbb{R} \, . \quad \{ \, a \mid a \in c, \, a \equiv [(r_s, a_s), (r_f, a_f)], \, [r_s, r_f] \cap [r_1, r_2] \neq \emptyset \, \} \text{ is finite} \quad (1)$$

(The structure is only doing a finite amount at once.)

$$\forall a_t \in \mathbf{T} \, . \, \{ \, b_t \mid b_t <_c a_t, a_t \in \mathbf{T}, \{a, b\} \in \underline{\text{Con}} \, \} \text{ is finite} \quad (2)$$

(Everything is finitely caused.) Failure to observe these rules may lead to the construction of *Zeno machines*; machines which can exhibit an infinite number of distinguishable events in a finite time. Joseph's [15] discusses these kinds of structures more comprehensively; see also [30].

---

‡ The terminology of this section is that of the Petri net community. See [26] for details.

3.5. Relating I.E.S.s to Other Models. We begin by showing that I.E.S.s are generalisations of the event structures of [37], to which the reader is referred for relevant definitions.

Proposition 13. Define a new order on tasks $a \preccurlyeq b \iff a \sqsubseteq_{\mathrm{i}} b \vee a = b$. This is a partial order since $\sqsubseteq_{\mathrm{i}}$ is a strict partial order. The structure $(\mathbf{ST}, \#, \preccurlyeq)$ is a prime event structure.

If we ban simultaneous transitions, i.e. for $t_t, u_t \in \mathbf{ST}$

$$t_t = (r, a_t) \wedge u_t = (r, b_t) \implies t_t = u_t$$

then both $(\mathbf{ST}, \#, \sqsubseteq_{\mathrm{h}})$ and $(\mathbf{ST}, \#, \sqsubseteq_{\mathrm{t}})$ are prime event structures. (The ban is needed to make the relations $\sqsubseteq_{\mathrm{h}}$ and $\sqsubseteq_{\mathrm{t}}$ partial orders.)

The reader whose taste is for event structures of [36] instead of those of [37] could define for $c \in \underline{\mathrm{Con}}$

$$c \vdash_{\mathrm{h}} a \iff \langle a_1, a_2, \ldots, a_n \rangle \ h\text{–secures } a \wedge c = \{a_1, a_2, \ldots, a_n\}$$

and similarly for $\vdash_{\mathrm{t}}$. Then, without simultaneous transitions, both the structures $(\mathbf{ST}, \underline{\mathrm{Con}}, \vdash_{\mathrm{h}})$ and $(\mathbf{ST}, \underline{\mathrm{Con}}, \vdash_{\mathrm{t}})$ are event structures.

Proof. By expanding the defn.s and checking the conditions in [36], [37]. □

We now move on to discuss the relationship of I.E.S.s to Lamport's model of distributed systems [17]:

Recall from Abraham's discussion [1] that a Lamport structure is a triple $(\mathbf{TK}, \longrightarrow, \dashrightarrow)$ consisting of a set **TK** endowed with a 'wholly proceeds' relation $\longrightarrow$ and a 'can affect' relation $\dashrightarrow$. In our framework, we read 'wholly proceeds' $\longrightarrow$ as $\sqsubseteq_{\mathrm{i}}$ and express 'can affect' via

$$a \dashrightarrow b \iff r_s \leq s_f$$

Thus Lamport structures can be expressed as interval event structures.

A central feature of [17] is the global time axiom;

$$a \longrightarrow b \iff \neg(b \dashrightarrow a)$$

Reinterpreting this we have

Proposition 14. The Global Time axiom forbids any incomparable events. This follows since it is equivalent to

$$a \sqsubseteq_{\mathrm{i}} b \iff r_f < s_s$$

so the axiom allows simultaneity and overlap in time but forbids both concurrency.

4. Equivalence of Interval Event Structures

It is time to answer the question 'when are two interval event structures the same ?' We will do this by giving a notion of when one structure is as good as another for a given purpose. Gives us a notion of *equivalence* of structures; each will 'do' for the other. (An alternative equivalence for I.E.S.s, based on testing, is given in [20].)

The notions of equivalence presented here are similar to (but were developed independently from) the 'history–preserving bisimulation' of [31] and the 'ST–bisimulation' of [33]. Also related, and of particular interest, given our interpretation of transitions as abstract locations (section 2), are the 'distributed bisimulations' of [4] and [5].

The notion of a *bisimulation* has been very productive for CCS; it seems reasonable to investigate what it might mean in the world of I.E.S.s. Two CCS processes bisimulate each other when, loosely, whatever state one gets into, the other can always get into a state in which the same actions are possible next and the resulting states are similarly related.

For I.E.S.s, there is no fundamental notion of state, and we have four notions of causality (the $\sqsubseteq_j$s), so a different definition is needed. A notion, similar to bisimulation, but definable over I.E.S.s, is *j-homomorphism*. Given two I.E.S.s $\Im = (\mathbf{TK}, \underline{\mathrm{Con}}, <_s, =_s)$ with silent task $*$, and $\Im' = (\mathbf{TK}', \underline{\mathrm{Con}}', <_s', =_s')$ with $*'$, there is an *j-homomorphism* f between them iff f is a function from $\mathbf{TK}$ to $\mathbf{TK}'$ which preserves structure, i.e. if:

- Silent tasks are preserved; $f(*) = *'$.
- The j-order is preserved; $a, b \in \mathbf{TK}$ with $a \sqsubseteq_j b$ implies that $f(a) \sqsubseteq_j' f(b)$. (Here $\sqsubseteq_j$ comes from $\leq_s$ and $\sqsubseteq_j'$ from $\leq_s'$.)
- Previously consistent sets must remain consistent.
  If $c \in \underline{\mathrm{Con}}$ then $\{ f(a) \mid a \in c \} \in \underline{\mathrm{Con}}'$.

Hence, if there is a j-homomorphism from $\Im$ to $\Im'$, there will be a matching j-ordering in $\Im'$ for any in $\Im$, (although this may be the trivial one $a \sqsubseteq_j b$) so that $\Im'$ can be thought of as *simulating* $\Im$. The homomorphisms then, will indicate (causal bi–) simulation.

If there are j–homomorphisms $f : \Im \to \Im'$ and $g : \Im' \to \Im$ then we say $\Im$ and $\Im'$ are $\sqsubseteq_j$*–equivalent* write $\Im \sim_j \Im'$.

Proposition 15. The equivalence of I.E.S.s $\cong_j$, defined for any j by

$$\cong_j \;=\; \bigcup \{ \sim_j \mid \sim_j \text{ a } j\text{-homomorphism} \}$$

is the largest j-homomorphism and is an equivalence relation on **IES**.

Proof. As usual; see, for example [2]. □

Definition 16. Given two I.E.S.s, $\Im$ and $\Im'$, as usual, a *temporal homomorphism* is a function, f from $\mathbf{TK}$ to $\mathbf{TK}'$ such that f is an i-homomorphism and $\forall\, a \in \mathbf{LE}$ if $a \equiv [t_s, t_f]$ and $f(a) \equiv [t_s', t_f']$ then $[t_s', t_f'] \supseteq [t_s, t_f]$.

If there are temporal homomorphisms $f : \Im \to \Im'$ and $g : \Im' \to \Im$ then we say $\Im$ and $\Im'$ are *temporally equivalent* and write $\Im \sim_\tau \Im'$.

Proposition 17. The equivalence of I.E.S.s $\cong_\tau$, defined by

$$\cong_\tau \;=\; \bigcup \{ \sim_\tau \mid \sim_\tau \text{ a temporal homomorphism} \}$$

is the largest temporal homomorphism and is an equivalence relation on **IES**.

Proof. As usual; see, for example, [2]. □

Notice that our notions of equivalence range from the coarse (such as e–equivalence, which identifies processes which are not trace equivalent) to the fine (such as temporal equivalence, which is stronger than history–preserving bisimulation).

5. Equivalence and Event Refinement

In the introduction we mentioned that timed models where events are associated with intervals of time (i.e. where they are non-atomic) have a clean notion of event refinement. This section is devoted to justifying this claim.

Firstly we will indicate when a single task can be refined by an I.E.S. Then we investigate when a substructure can be replaced by a single task. It is shown that these two operations are inverses of each other, giving a duality between hiding and event refinement. Finally we show that our notions of equivalence are congruences of event refinement.

5.1. Event Refinement. Suppose $a \equiv [(r_s, a_s), (r_f, a_f)]$ is an event in an I.E.S. $\Im$, and $\Im^a = (\mathbf{TK}^a, \underline{\mathrm{Con}}^a, <_s{}^a, =_s{}^a)$ is another I.E.S. There are three conditions which ensure that $\Im^a$ can refine a:

(i) The 'duration' of $\Im^a$ should be that of a. One can think of a task as defining an interval of time within which any refinement of it must lie.

(ii) $\Im^a$ should not interfere *sub rosa* with $\Im$.

(iii) $\Im^a$ should have 'end points' able to inherit the causality of a. Thus everything related to a will now be related to a suitable end point. See the figure below; the arrows represent causality.

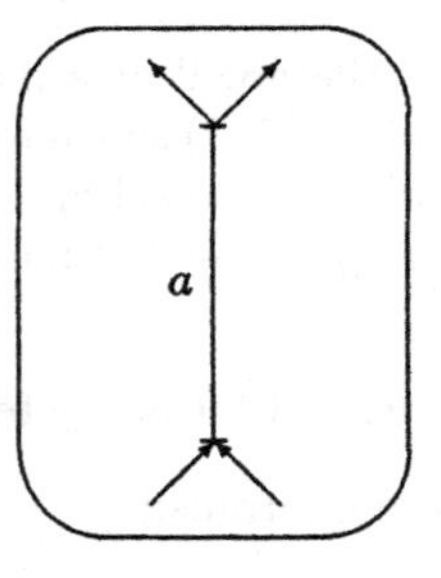

An event a in $\Im$

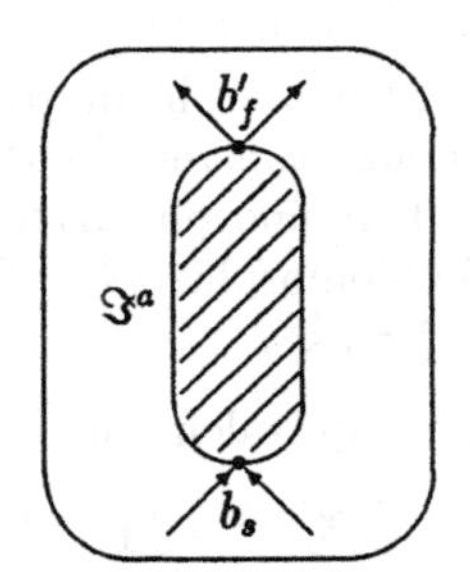

$\Im$ with $\Im^a$ refining a

Figure 2. A task refinement

Definition 18 (Task Refinement). An I.E.S. $\Im^a = (\mathbf{TK}^a, \underline{\mathrm{Con}}^a, <_s{}^a, =_s{}^a)$ with silent task $*^a$ can refine $a \equiv [(r_s, a_s), (r_f, a_f)]$ iff

(i) $\underline{\mathrm{begin}}(*^a) = (r, *^a_s)$ and $r = r_s$. Furthermore $\underline{\mathrm{end}}(*^a) = (r', *^a_f)$ and $r' = r_f$.

(ii) $\mathbf{LE} \cap \mathbf{LE}^a = \emptyset$.

(iii) $\exists b, b' \in (\mathbf{LE}^a - \{*^a\})$ such that $\underline{\mathrm{begin}}(b) =_s \underline{\mathrm{begin}}(*^a)$, $\underline{\mathrm{end}}(b') =_s \underline{\mathrm{end}}(*^a)$. Notice this gives $\forall c^a \in \mathcal{M}(\Im^a) \,.\, \{b, b'\} \subseteq c^a$.

The I.E.S. with $\Im^a$ refining a, which is defined only if $\Im^a$ can refine a, is written $\Im\left[\dfrac{\Im^a}{a}\right]$.

The refined structure $\Im' = (\mathbf{TK}', \underline{\mathrm{Con}}', <_s', =_s')$ is defined thus:

(i) $\mathbf{LE}' \stackrel{\mathrm{def}}{=} (\mathbf{LE} - \{a\}) \cup (\mathbf{LE}^a - \{*^a\})$, with timings inherited from **TK** and $\mathbf{TK}^a$.

(ii) Define the $*^a$-less $\underline{\mathrm{Con}}$-sets of $\Im^a$ by $\underline{\mathrm{Con}}^{-a} = \{c^a - \{*^a\} \mid c^a \in \underline{\mathrm{Con}}^a\}$.
Then $\underline{\mathrm{Con}}'$ will be the $\underline{\mathrm{Con}}$-sets of $\Im$ that didn't have a in them, together with those that did without a and with any $\underline{\mathrm{Con}}$-set of $\Im^a$:

$$\underline{\mathrm{Con}}' \stackrel{\mathrm{def}}{=} \{c \mid c \in \underline{\mathrm{Con}}, a \notin c\} \cup \{c - \{a\} \cup c^a \mid c \in \underline{\mathrm{Con}}, a \in c, c^a \in \underline{\mathrm{Con}}^{-a}\}$$

(iii) The definition of $<_s'$ (and the analogous case for $=_s'$) is rather messy; every occurrence of a_s must be rewritten by one of b_s in $<_s$, and similarly for a_f and b'_f. Furthermore, every occurrence of $*^a_s$ or $*^a_f$ in $<_s{}^a$ must be similarly rewritten.
Define $g\, a_t = a_t \left[\frac{b_s}{a_s}\right]\left[\frac{b'_f}{a_f}\right]$ for $a_t \in \mathbf{T}$ and $h\, a^a_t = a^a_t \left[\frac{b_s}{*^a_s}\right]\left[\frac{b'_f}{*^a_f}\right]$ for $a^a_t \in \mathbf{T}^a$. (Here $[]$ on transitions denotes syntactic substitution as usual; $a[\frac{b}{c}]$ is b if $a = c$ and a otherwise). Then the required order $<_s'$ is given by taking the transitive closure of the relation $\preccurlyeq$ defined by

$$\begin{aligned} \preccurlyeq \; = \; & \{((r_t, g\, a_t), (s_t, g\, a'_t)) \mid ((r_t, a_t), (s_t, a'_t)) \in <_s\} \cup \\ & \{((r_t, h\, a_t), (s_t, h\, a'_t)) \mid ((r_t, a_t), (s_t, a'_t)) \in <^a_s\} \end{aligned}$$

Clause (iii) has the advantage that no causality is lost; anything that was related to a will now be similarly related to the whole refining structure. (This would not have been the case if we had allowed refining structure to have durations less than that of what they refine.)

5.2. Substructure Abstraction. The dual of task refinement is *substructure abstraction*. If a substructure of an I.E.S. is only connected to the rest of the structure it sits in by two 'end point' transitions, then we can replace it. Thus consider

Definition 19 (Abstractable Substructures). Consider two I.E.S.s, $\Im = (\mathbf{TK}, \underline{\mathrm{Con}}, <_s, =_s)$ and $\Im^b = (\mathbf{TK}^b, \underline{\mathrm{Con}}^b, <_s{}^b, =_s{}^b)$. $\Im^b$ is a *abstractable substructure* of $\Im$ iff

(i) The non-silent events of $\Im^b$ are a subset of those of $\Im$.

$$\mathbf{LE}^b - \{*^b\} \subseteq \mathbf{LE}$$

(ii) The $\underline{\mathrm{Con}}$-sets of $\Im^b$ are just those of $\Im$ that refer to things in $\mathbf{LE}^b$:

$$\underline{\mathrm{Con}}^b = \{(c \cap \mathbf{LE}^b) \cup \{*^b\} \mid c \in \underline{\mathrm{Con}}\}$$

(iii) $\Im^b$ must only be connected to the rest of $\Im$ by 'end points.'
$\exists b, b' \in \mathbf{LE} \,.\, \forall a^b \in \mathbf{LE}^b, a \in (\mathbf{LE} \cap \mathbf{LE}^b)\,.$

$$\begin{aligned} (a_t <_s a^b_t) &\implies (a_t <_s b_s \leq_b a^b_t) && \wedge \\ (a^b_t <_s a_t) &\implies (a^b_t \leq_b b'_f <_s a_t) && \wedge \\ (a_t =_s a^b_t) &\implies (a_t =_s b_s =_s a^b_t) \vee (a_t =_s b'_f =_s a^b_t) \end{aligned}$$

(iv) The 'end points' really are first and last events; $\underline{\mathrm{begin}}(b) =_s \underline{\mathrm{begin}}(*^b)$, $\underline{\mathrm{end}}(b') =_s \underline{\mathrm{end}}(*^b)$.

Definition 20 (Substructure Abstraction). Consider an I.E.S. $\Im = (\mathbf{TK}, \underline{\mathrm{Con}}, <_s, =_s)$ and a new event $a \notin \mathbf{LE}$. Suppose that $\Im$ has an abstractable substructure $\Im^b$ with end points b, b'.

We can define an I.E.S. with $\Im^b$ *abstracted* by a, written $\Im\left[\frac{a}{\Im^b}\right]$, to be the I.E.S. $\Im' = (\mathbf{TK}', \underline{\text{Con}}', <_s', =_s')$:

(i) $\mathbf{LE}' \stackrel{\text{def}}{=} \mathbf{LE} - (\mathbf{LE}^b - \{*^b\}) \cup \{a\}$ with $a \equiv [(r_s, b_s), (s_f, b_f')]$ and all the other timings acquired from $\mathbf{TK}$.

(ii) For $\underline{\text{Con}}'$ we need to replace any $a' \in \mathbf{LE}^b$ by the abstracting event, so define

$$k(a') = \begin{cases} a & \text{if } a' \in \mathbf{LE}^b, \\ a' & \text{otherwise.} \end{cases}$$

Then

$$\underline{\text{Con}}' \stackrel{\text{def}}{=} \{ k(c) \mid c \in \underline{\text{Con}} \}$$

(iii) Next we need to rewrite spacetimes from $\Im^b$ with a suitable spacetime from an end point. For a transition t_t of an event $a' \in \mathbf{LE}$ define

$$i(t_t) = \begin{cases} t_t & \text{if } a' \notin \mathbf{LE}^b, \\ (r_s, b_s) & \text{otherwise.} \end{cases}$$

$$j(t_t) = \begin{cases} t_t & \text{if } a' \notin \mathbf{LE}^b, \\ (s_f, b_f') & \text{otherwise.} \end{cases}$$

Then for $<_s'$ we take the reflexive closure of

$$\preccurlyeq \stackrel{\text{def}}{=} \{ (i(t_t), j(s_t)) \mid (t_t, s_t) \in <_s \}$$

And similarly for $=_s'$.

Proposition 21. Suppose $\Im$ is an I.E.S. with $a \in \mathbf{LE}, b \notin \mathbf{LE}$, and $\Im^b$ is an abstractable substructure of $\Im$. Then (as proved in [18] by expanding the definitions)

$$\left(\Im\left[\frac{b}{\Im^b}\right]\right)\left[\frac{\Im^b}{b}\right] = \Im \qquad \text{and} \qquad \left(\Im\left[\frac{\Im^a}{a}\right]\right)\left[\frac{a}{\Im^a}\right] = \Im$$

Notice that in the I.E.S. framework we can separate the internal structure of events quite cleanly from the structures that contain them. This means that we can deal with event refinement, – delving into this internal structure, – separately from functional refinement, – structuring the descriptions of causality and interaction. It is this latter notion that has dominated consideration of refinement in concurrency thus far, with event refinement being seen as unclean. Our results seem to suggest that event refinement, at least seen as 'looking more closely' rather than as 'program development', can be natural in some models.

5.3. The equivalences are congruences of refinement. The recent interest in event refinement ([6], [32], [34]) centers on notions of equivalence that are congruences of refinement. Ours are:

Proposition 22. All of the notions of equivalence presented are congruences of event refinement.

Proof. Consider $\Im \cong \Im'$ where $\cong$ is one of the equivalence relations. Suppose that an task a is refined by $\Im^b$ in both $\Im$ and $\Im'$. (Trivially if a is in either then it is in the other.)

Observe first that all we have to show for the case of $\cong_j$ is that the 'end points' b, b' of the refining structure behave properly, that is, for $d \in \mathbf{LE}$ and $<''_c$ the order of the refined structure:

$$\begin{array}{rcl} a_s <_c d_t & \Longleftrightarrow & b_s <''_c d_t \\ d_t <_c a_s & \Longleftrightarrow & d_t <''_c b_s \\ a_f <_c d_t & \Longleftrightarrow & b'_f <''_c d_t \\ d_t <_c a_f & \Longleftrightarrow & d_t <''_c b'_f \end{array}$$

since no causality 'sticks out' of $\Im''$ into the refined structures except through the end points. The result then follows by construction.

For $\cong_\tau$ note that a_s of $\Im$ and b_s of $\Im\left[\frac{\Im^a}{a}\right]$ are contemporaneous, and similarly a_f, b'_f.

Hence if there is a temporal equivalence from $\Im$ to $\Im'$ there must be one between $\Im\left[\frac{\Im^b}{a}\right]$ and $\Im'\left[\frac{\Im^b}{a}\right]$. □

6. Concluding Remarks

We have presented a generalisation of Winskel's event structures that seems rich enough to describe most of the phenomena of interest in timed concurrency theory: it thus has a claim to usefulness as a descriptive formalism for detailed behaviour. Furthermore, it supports a natural notion of event refinement, and has a notion of equivalence that is a congruence of it.

Further Work. The interval event structures presented here are only one possible start towards a theory of real–timed true concurrency; much more is needed. Other characterisations of our causal bisimulations are desirable, for instance, as is a better understanding of the *categories* of interval event structures briefly treated in [18]. (We have in mind a treatment similar to [3] or [35].)

Parallel to our work is that of Schettini and Winkowski, who concentrate on more computational aspect of timed event structures; another topic of interest is the relationship of their model to ours. This would give us a better insight into the infinite behaviour of our structures.

There has recently been some interesting process on algebraic approaches to event refinement by Zwiers et al. [13]; under certain circumstances they allow much more respecification of causality after refinement than we do. It would be interesting to weaken our notion of event refinement to match their suggestion: the consequence would be a loss of the duality between abstraction and refinement, but this might be matched by a gain in usefulness.

Acknowledgments. Our thanks are due to Alan Jeffrey, Stefano Kasangian, Axel Poigné, Tim Regan, Arend Rensink and Mike Shields for helpful conversations. Part of this work was funded by the Science and Engineering Research Council of Great Britain; the first author acknowledges their support and that of a Royal Society European Fellowship.

Bibliography

1. U. Abraham, S. Ben-David, and M. Magidor, *On global time and inter-process communication*, in Semantics for Concurrency (M. Kwiatkowska, M. Shields, and R. Thomas, Eds.), (1990), Springer-Verlag Workshops in Computing, Leicester.
2. L. Aceto, *Action-refinement in process algebras*, Ph.D. thesis, Department of Computer Science, University of Sussex, 1991, available as Technical Report Number 3/91.
3. M. Bednarczyk, *Categories of asynchronous systems*, Ph.D. thesis, Department of Computer Science, University of Sussex, 1987.
4. G. Boudol, I. Castellani, M. Hennessy, and A. Kiehn, *Observing localities*, Technical Report 4/91, Department of Computer Science, University of Sussex, 1991.
5. I. Castellani and M. Hennessy, *Distributed bisimulations*, Journal of the ACM, Volume 10 (1989), Pp. 887–911.
6. P. Darondeau and P. Degano, *Event structures, causal trees, and refinements*, in Mathematical Foundations of Computer Science (B. Rovan, Ed.), (1990), Springer-Verlag LNCS 452.
7. J. Davies, *Specification and proof in real-time systems*, D.Phil. thesis, Oxford University Computing Laboratory, available as technical report Number 93, 1991.
8. J. Davies and S. Schneider, *An introduction to timed CSP*, Technical Report Number 75, Oxford University Computing Laboratory, 1989.
9. C. Fidge and M. Pilling, *Specification languages for real-time software*, Technical Report 156, Key Centre for Software Technology, Department of Computer Science, University of Queensland, 1990.
10. J. Gischer, *The equational theory of pomsets*, Theoretical Computer Science, Volume 61 (1989), Pp. 199–224.
11. M. Hennessy and T. Regan, *A temporal process algebra*, Technical Report 2/90, Department of Computer Science, University of Sussex, 1990.
12. C. Hoare, *Communicating sequential processes*, International series on computer science, Prentice-Hall, 1985.
13. W. Janssen, M. Poel, and J. Zwiers, *Action systems and action refinement in the development of parallel systems*, in the Proceedings of Concur, (1991), Springer-Verlag LNCS 527.
14. A. Jeffrey, *Timed process algebra* $\neq$ *time* $\times$ *process algebra*, Technical Report 79, Programming Methodology Group, Chalmers University, 1991.
15. M. Joseph and A. Goswami, *Relating computation and time*, Technical Report RR 138, Department of Computer Science, University of Warwick, 1985.
16. S. Kasangian and A. Labella, *On continuous real time agents*, in Mathematical Foundations of Programming Semantics, (1991), Springer-Verlag LNCS 469.
17. L. Lamport, *On interprocess communication. part I: Basic formalism*, Distributed Computing, Volume 1 (1986), Pp. 77–85.
18. D. Murphy, *Time, causality, and concurrency*, Ph.D. thesis, Department of Mathematics, University of Surrey, 1989, available as Technical Report CSC 90/R32, Department of Computing Science, University of Glasgow.
19. ———, *The physics of observation; a perspective for concurrency theorists*, Bulletin of the EATCS, Volume 44 (1991), Pp. 192–200.
20. ———, *Testing, betting and timed true concurrency*, in the Proceedings of Concur, (1991), Springer-Verlag LNCS 527.
21. ———, *Timed process algebra, Petri nets, and event refinement*, in Refinement (J. Morris, Ed.), (1991), Springer-Verlag Workshops in Computing.
22. E-R. Olderog, *Nets, terms and formulas*, Cambridge Tracts in Theoretical Computer Science, Volume 23, 1991.
23. A. Pnueli, *Linear and branching structures in the semantics and logics of reactive systems*, in Automata, Languages and Programming (W. Brauer, Ed.), (1986), Springer-Verlag LNCS 194.

24. V. Prasad, *A calculus of broadcasting systems*, in the Proceedings of TAPSOFT (CAAP) (S. Abramsky, Ed.), (1991), Springer-Verlag LNCS 494.

25. V. Pratt, *The pomset model of parallel processes: Unifying the temporal and the spatial*, in Proceedings of the CMU/SERC Workshop on Analysis of Concurrency, (1984), Springer-Verlag LNCS 197.

26. W. Reisig, *Petri nets: An introduction*, EATCS Monographs on theoretical computer science, Springer-Verlag, 1985.

27. A. Rensink, *Elementary structures of non-atomic events*, Technical Report 89-65, Faculteit der Informatica, Universiteit Twente, 1990.

28. M. Roncken and R. Gerth, *A denotational semantics for synchronous and asynchronous behaviour with multiform time*, in Semantics for Concurrency (M. Kwiatkowska, M. Shields, and R. Thomas, Eds.), (1990), Springer-Verlag Workshops in Computing, Leicester, 1990.

29. A. Schettini and J. Winkowsi, *Towards an algebra for timed behaviours*, Technical report, Institute of Computer Science, Polish Academy of Sciences, 1990, to appear in Theoretical Computer Science.

30. J. van Benthem, *The logic of time*, D. Reidel, 1983.

31. R. van Glabbeek and U. Goltz, *Equivalence notions for concurrent systems and refinement of actions*, in Mathematical Foundations of Computer Science, (1989), Springer-Verlag LNCS 379.

32. ———, *Refinement of actions in causality based models*, in Stepwise Refinement of Distributed Systems (J. de Bakker, Ed.), (1990), Springer-Verlag LNCS 430, Proceedings of REX 1989.

33. R. van Glabbeek and F. Vaandrager, *Petri net models for algebraic theories of concurrency*, in the Proceedings of the PARLE conference (J. de Bakker, Nijman, and Treleaven, Eds.), (1987), Springer-Verlag LNCS 259.

34. W. Vogler, *Bisimulation and action refinement*, Technical Report 342/10/90A, Institut für Informatik, Technische Universität München, 1990.

35. G. Winskel, *Synchronization trees*, Theoretical Computer Science, Volume 34 (1985), Pp. 34–84.

36. ———, *Event structures*, in Petri Nets: Central Models and Their Properties (W. Brauer, W. Reisig, and G. Rozenberg, Eds.), (1986), Springer-Verlag LNCS 254, Proceedings of Advances in Petri Nets.

37. ———, *An introduction to event structures*, in Linear Time, Branching Time and Partial Order in Logics and Models for Concurrency (J. de Bakker, W. de Roever, and G. Rozenberg, Eds.), (1989), Springer-Verlag LNCS 354, Proceedings of REX 1988.

Stepwise development of model-oriented real-time specifications from action/event models

Hans Toetenel Jan van Katwijk

Delft University of Technology
Faculty of Technical Mathematics and Informatics
P.O. Box 356, 2600 AJ Delft The Netherlands
email: toet@dutiaa.tudelft.nl

October 22, 1991

Abstract

Two notations for specifying real-time systems are presented: an analysis language A/EL and a design language MOSCA. An abstract syntax for an *action/event language* (A/EL) is presented that offers constructions on a high abstract level for describing behavioural and timing requirements for real-time systems. Next the *model-oriented specification language* MOSCA is shortly introduced. This language is based on VDM-SL [12] and CCS [20] with extensions for describing structure and timing issues. It is shown how specifications in A/EL can be transformed into MOSCA specifications guided by some simple design principles.

1 Introduction

MOSCA[25] [1] is an experimental language extending the Vienna Development Method specification language VDM-SL [2] [12] in order to increase the applicability of the language in the area of distributed, parallel and real-time systems. Plain VDM-SL is not adequate for these application area's since it lacks facilities to specify multiple threads of control while furthermore, it does not support the notion of time within specifications.

In this paper we describe an experiment within the scope of a series of experiments [27] aimed at deriving requirements for languages such as MOSCA to be useful in both the *description* and *development* of real-time systems. Our research is a logical extension to other previous research, concerning the place of VDM in various approaches for software development [22], [21], [28].

[1] Mosca is an acronym for *M*odel-*O*riented *S*pecification of *C*ommunicating *A*gents.

[2] The specification language for VDM for which a ISO standard is currently being developed (ISO SC22/WG19) in cooperation with the British Standardisation Institute.

1.1 The experiment

Real-time systems have in general a complex set of characteristics, and therefore need to be described from different viewpoints. In [9] Deutsch proposes three different viewpoints that are related to the major parties involved in system development: the customer, the user and the implementor. The models representing these points of view are, respectively, the requirements model, the operations-concept model and the implementation model. According to Deutsch

- the software requirements model describes the essence of the software part of the system by grouping external and internal I/O, internal flow/transformation, state transitions;
- the operations concept model defines processing scenarios through the system from external input to eventual system output, using stimulus/response paradigms;
- the implementation model details the physical design.

An action/event description expresses a high level of abstraction and is independent of any process structure or communication protocol. It describes the joint effects that the processes are to accomplish together including constraints upon the reaction on events. In this paper we concentrate on a possible mapping from A/EL specifications to MOSCA (executable) specifications. We believe that such an executable MOSCA specification of the software requirements provides a suitable starting point for the implementation of the software embedded within a real-time system. On one hand simulation and prototyping might provide insight in further characteristics of the system on the other hand, software development can be done by formal refinements in two different directions, one aimed at proces-refinement and one at data-refinement. The description of the transformation from A/EL to MOSCA centers around (i) the process creation from action chains and (ii) the distribution of state information to the process control. The activities result in a MOSCA specification presenting a more concrete view of the system specified in A/EL. From the MOSCA specification a set of RTL ([11]) axioms may be derived that contain its timing behaviour. The safety assertions concerning the action/event system can then be checked against the MOSCA system.

Throughout the paper the VDM notation is used as recording device (e.g. for the abstract syntaxes of the action/event language, and MOSCA fragments).

In this paper we start with the abstract syntax of our action/event language In section 3 we introduces MOSCA by giving an outline of the language, presenting the constructs relevant for this particular research. In section 4 we briefly introduce guidelines and principles for the transformation of an A/E specification into a correponding MOSCA specification. The method is illustrated with a short example presented in section 5 and we conclude this paper with a summary of our results6.

1.2 Related Work

In [11], Jahanian and Mok introduce an action/event language together with Real-Time Logic (RTL) which is used to record and assert safety assertions of the system expressed. Assertions of the system, expressed as RTL formulas are valid when they can be derived from the RTL axioms that describe the system. Furthermore, they present an outline of a transformation scheme to transform an action/event specification into a set of RTL axioms. They developed a decision procedure based on the work of Bledsoe and Hines ([5]) to resolve this problem.

Dasarthy[8] provides an approach oriented towards specifying time and time intervals in software requirements specifications. This approach is furthermore used by Diaz-Gonzales and Urban in [10].. Their action/event language is sufficiently rich to act as basis for a symbolic execution. ITL, the Interval Temporal Logic, a logic which is based on the temporal logic (TL) designed by Manna and Pnueli [18] is used as a basis for analysis. In [24] a survey of applications of TL is the area of reactive systems is given.

Kurki-Suonio & Järvinnen report on *DisCo* [14], a language from which a direct translation into the Ada language [29] is done.

As action/event systems are really state-based systems, they can be visualised through state-transition diagrams and modelled by state-machines. A notation, directly oriented towards state-machines is RT-CDL, in the work of Liu [16], [17]. Their language is an executional action/event based specification language based on ESTEREL [2].

What is noteworthy is that the notations and approaches mentioned above, are primarily used to express the *operation concepts view* of the system under construction. Although simulation of the the behaviour of the systems by executing the representation (either directly or by simulation) has got attention, none of the reported approaches addresses the problem of extracting the software requirements and implementing the software part of the real-time system.

2 The Action/Event Language

In this section we briefly discuss the abstract syntax of our action/event language under investigation. The specification is given as a series of VDM domain definitions in which we assume a basic domain for time T, the values of which being units of time. This time domain may be dense or discrete, depending on the choice of applications.

The central notions are *primitive actions*, *composite actions*, *state predicates*, *events*, and *timing constraints*. A specification of a system consists of a 5-tuple

$$\langle AT, ET, CTT, SPT, BEH \rangle$$

where AT is the collection of named actions, CTT a computing time specification for each primitive action, SPT the state predicate table, ET the events and BEH a set of executing actions together defining the behavioural and timing issues of the system. Furthermore, *Id* is a collection of identifiers.

$$\begin{aligned} \textit{System} :: \quad AT &: \textit{Id} \xrightarrow{m} \textit{Action} \\ ET &: \textit{Id} \xrightarrow{m} \textit{Event} \\ CTT &: \textit{Id} \xrightarrow{m} \mathsf{T} \\ SPT &: \textit{Predicate} \\ BEH &: \textit{ExecuteStatement}\textsf{-set} \end{aligned}$$

The *state* of a specified system consists of a set of state predicates, each predicate associating a boolean value to an assertion about the system. Examples of such predicates might be "IO device active" or "Power Down in Progress" or "Message Sended". Assertions describing the state of physical components are e.g. "Close Door Button Pressed",

"Liftdoor Open" etc. These assertions are modelled by a set of predicate names – value pairs, modelled as a map *Predicate*

$$Predicate = Id \xrightarrow{m} StateComponentValue$$

$$StateComponentValue = \{\text{DISABLED}, \text{TRUE}, \text{FALSE}, \text{DC}\}$$

The values range over the set of possible state component values {DISABLED, TRUE, FALSE, DC} where TRUE and FALSE have their usual meaning, DC denotes a don't care situation, which actually may either be TRUE or FALSE, and DISABLED denotes a state which is disjoint of the other three values. Its purpose is in modeling initial states of the system.

The *SPT* (state predicate table) component of a system specification records the actual values of all state predicates. Its initial value must be supplied by the specifier. *SPT* defines a state space of $4^{\text{card rng } SPT}$ elements, which is constrained by the causality relations induced by the post conditions of the primitive actions and the events.

A primitive action models a specific unit of work in the system. The effect of a primitive action might be a change in the system's state, leading to a change in state predicates. The effect of the action on the state of the system, modelled by the *SPT* component of the system, is specified by the *post* component of the action.

$$Action = PrimitiveAction \mid CompositeAction$$

$$PrimitiveAction :: \; pre : [Predicate] \quad post : [Predicate]$$

It is assumed that each primitive action has a bounded execution time which is statically known, which is recorded in the compution time table *CTT*. For any combined action in the system, the duration in time units can be computed this way.

Primitive actions can be guarded by a state predicate. If the value of the state predicate is TRUE the action is enabled, otherwise the action is void.

$$CompositeAction :: \; pre : [Predicate] \quad action : Component^{+}$$

$$Component = Id \mid ParallelComposition$$

$$ParallelComposition = Id\text{-set}$$

$$\text{inv-}ParallelComposition(ps) \;\triangleq\; \text{card } ps \geq 2$$

Primitive and composite actions are syntactically represented by their unique name. They are collected in the Action Table, which relates the name of the action with its behaviour with respect to *the state* of the system. In the sequel we let $a, b, \ldots$ range over the set of actions in *AT*.

Primitive actions can be grouped into composite actions. A composite action is modelled as a *sequence* of components. A composite action expresses a partial ordering over the involved components. The ordering is fixed by the relative position of the components in the sequence. E.g. the composite action $[a, b, c]$ denotes an action consisting of the sequential execution of its components a, b and c. Each component is either an action or a set of actions from which the execution of its elements proceeds in parallel. Different levels of abstraction can be establised through the usage of composite actions as components. E.g. the composite action

$$[a, \{[b, c], d\}, e]$$

denotes an action starting with the execution of action a, followed by the parallel execution of the components $[b, c]$ and d, followed by the execution of action e. Composite actions can be guarded.

$$\begin{aligned} \mathit{Event} :: \mathit{atrigger} &: [\mathit{Id}] \\ \mathit{post} &: [\mathit{Predicate}] \\ \mathit{timing} &: [\mathit{TimingEvent}] \end{aligned}$$

$$\begin{aligned} \mathit{TimingEvent} :: \quad \mathit{delay} &: \mathbf{T} \\ \mathit{separation} &: \mathit{SeparationSet} \end{aligned}$$

$$\begin{aligned} \mathit{SeparationSet} :: \quad \mathit{target} &: \mathit{Id}\text{-set} \\ \mathit{interval} &: \mathbf{T} \end{aligned}$$

An events acts as a marker in time, this marking is significant in describing the behaviour of the system. Examples of events are

- the arrival of an external event,
- the start of an action,
- the end of an action, or
- the change of a state predicate value.

An event not being a start event, a stop event, or or a state-changing event, is considered to be an external event. External events arrive from outside the system as viewed by the operations concept model. Examples of such events are "stop button pushed", or "fuel pump started", or "hardware timer time-out" or "posted io request", the behaviour of human controllers etc. In the event table ET, only external events are recorded.

An action can be triggered by an event, in which case the event is called *sporadic* and is supposed to occur within a specified delay starting from the occurence of the triggering event. Occurrence of an event may mask other occurrences of a similar event for a fixed period of time, called *separation*.

Furthermore, actions can might occur periodically in which case a *condition* specifies the values of the state predicates under which the execute statement should be active. If the state values in the range of the condition are DISABLED the execute statement is said to be disabled also.

$$\begin{aligned} \mathit{ExecuteStatement} :: \mathit{action} &: \mathit{Id} \\ \mathit{timing} &: \mathit{TimingConstraint} \end{aligned}$$

$$\mathit{TimingConstraint} = \mathit{PeriodicConstraint} \mid \mathit{SporadicConstraint}$$

$$\begin{aligned} \mathit{PeriodicConstraint} :: \mathit{condition} &: \mathit{Predicate} \\ \mathit{period} &: \mathbf{T} \\ \mathit{deadline} &: \mathbf{T} \end{aligned}$$

$$\begin{aligned} \mathit{SporadicConstraint} :: \quad \mathit{etrigger} &: \{\mathit{Id}\} \mid \mathit{Predicate} \\ \mathit{delay} &: \mathbf{T} \\ \mathit{separation} &: \mathbf{T} \end{aligned}$$

The behaviour of the system is captured by a set of execute statements. An execute statement specifies an action and a specification of the particular type of the behaviour, which can be periodic or sporadic. A *sporadic behaviour* is specified by a *sporadic constraint*, consisting of *the events* that might trigger the behaviour, a delay, and a minimum separation time. We use the term *sex* to indicate a set of execute statements, describing some sporadic behaviour. Such a sex can be triggered by all possible events like and/or combinations of external events and start/stop events of atomic actions or state-changing events.

A *periodic behaviour* is specified by a periodic constraint, consisting of a condition ranging over the *Predicate* domain, a period, denoting the minimum timespan of the period and a deadline, specifying a timedelay measured from the beginning of the period, with which the execution of the action may be delayed before starting. An execute statement with a periodic behaviour is further denoted by the term *pex*.

The behaviour of the system is captured by (i) the current state values, (ii) the set of periodic execute statements guarded by the current state values, (iii) the set of sporadic execute statements, activated by events resulting from executing actions.

Without going into further detail, the following assumptions are made:

- primitive actions are all atomic, and their execution can not be interrupted;
- the chain of actions specified in a pex is executed completely before the value of the condition is checked to decide whether the chain is started again, regardless any state change that might have taken place during execution of the pex.

3 The MOSCA language

The MOSCA specification language builds on VDM-SL and the process algebra approach, in particular CCS [19], [20]. MOSCA embeds the model-oriented specification language of VDM by using it as the process algebra's value manipulation language. The resulting language is further extended with capabilities to describe *structure* and *time*.

A MOSCA specification can describe four aspects of complete systems of communicating processes: their data-containment, their functional behaviour, their process-structure and their time characteristics.

For a survey of the concrete and abstract syntax of the full MOSCA languge the reader is referred to [26]. This description also contains a specification of the semantics, based on the dynamic semantic model for VDM-SL [15] and Plotkin's labelled transition systems [23] extended to handle time in a manner inspired by Wang Yi's work [31], [30]. In this presentation we concentrate on the MOSCA constructs supporting process-definition and the handling of time.

The basic element in the MOSCA model of a system is an *agent*, the MOSCA equivalent of a *process*.

AgDef ::	*heading*	:	*AgHeading*
	ports	:	*PortDef*
	behaviour	:	*AgBehaviour**
	localdefs	:	*Definitions*

The definition of an agent may contain local definitions for types, values, states, functions and operations. Its behaviour is specified by a sequence of *behaviour specifications.*

A system consists of a composition of agents and its behaviour is specified as a composition of behaviour expressions.

$$\mathit{AgHeading} :: \begin{array}{rl} \mathit{vpart} & : [\mathit{VType}] \\ \mathit{spart} & : [\mathit{StateDef}] \end{array}$$

$$\mathit{PortDef} :: \mathit{ports} : \mathit{Id} \xrightarrow{m} (\mathit{Cap} \times [\mathit{VType}])$$

$$\mathit{Cap} = \{\text{SYN}, \text{IN}, \text{OUT}\}$$

$$\mathit{AgBehaviour} :: \begin{array}{rl} \mathit{valpattern} & : \mathit{Pattern} \\ \mathit{body} & : \mathit{BExpr} \end{array}$$

The *AgHeading* part contains both the data, acting as the state of the agent, and protocols for accessing this state.

Ports can be used for synchronization without any value passing, or for value passing. The *ports* part defines the communication ports associated with the agent. For each port the capability, the direction of the dataflow, and the type of the values that might pass the ports is specified.

The behaviour of an agent is fixed through a series of *AgBehaviour* constructions, each consisting of an value pattern and a behaviour expression. The value pattern - referring to patterns of values appearing at the ports - enables case analysis over the behaviour constructions by matching the pattern against the value bound to the valuepart.

The concrete representation of the notation, used throughout this paper, is inspired by Milner's notation [20] and the mathematical concrete syntax of VDM-SL. Notational differences between CCS and MOSCA are e.g.found in CCS' prefix operator '.', MOSCA '$\odot$', and summation, '+' versus '$\oplus$'. The prefix construction in MOSCA is an extension to the CCS notation, which (i) enables the specification of a more differentiated communication model by having synchronization connections added and (ii) is capable of handling timing issues.

$$\mathit{Prefix} :: \begin{array}{rl} \mathit{timing} & : \mathit{TimeHandling} \\ \mathit{act} & : \mathit{Action} \\ \mathit{res} & : \mathit{BExpr} \end{array}$$

$$\mathit{TimeHandling} = \mathit{TimeVarActivation}^{+}$$

$$\mathit{TimeVarActivation} :: \mathit{timevar} : \mathit{Id}$$

$$\mathit{Action} = \mathit{SynAct} \mid \mathit{InAct} \mid \mathit{OutAct} \mid \mathit{IdleAct}$$

$$\mathit{SynAct} :: \mathit{label} : \mathit{Id}$$

$$\mathit{InAct} :: \begin{array}{rl} \mathit{label} & : \mathit{Id} \\ \mathit{pattern} & : \mathit{Pattern} \end{array}$$

$$\mathit{OutAct} :: \begin{array}{rl} \mathit{label} & : \mathit{Id} \\ \mathit{result} & : \mathit{VExpr} \end{array}$$

$$\mathit{IdleAct} :: \mathit{timespan} : \mathit{TimeExpr}$$

An example of an agent definition is given in specification 3.1.

Example 3.1 In specification 3.1 we use sequence patterns to give a model for a simple unbounded buffer in which natural numbers can be buffered. The specification of the agent *Buffer* contains a single value part meant to hold sequences of natural numbers, $\mathbb{N}^*$. The agent offers two ports, one input port in that accepts a number, and an output

MOSCA 3.1 Buffer 1

$Buffer\ \langle \mathbf{N}^* \rangle$
ports in $\mathtt{in}$: $\mathbf{N}$
out $\overline{\mathtt{out}}$: $\mathbf{N}$

$$Buffer\ \langle [\,] \rangle \triangleq \mathtt{in}(x) \odot Buffer\ \langle [x] \rangle$$

$$Buffer\ \langle [x] ^\frown s \rangle \triangleq \overline{\mathtt{out}}(x) \odot Buffer\ \langle s \rangle \oplus \mathtt{in}(y) \odot Buffer\ \langle [x] ^\frown s ^\frown [y] \rangle$$

port $\overline{\mathtt{out}}$ that delivers a value from the buffer. The sequence patterns '$[\,]$', and '$[x]^\frown s$' are used to mark the two cases of behaviour for the buffer. An empty buffer can only accept values at the input port, whereas a non-empty buffer may accept a value for buffering and may deliver a value from the buffer to the surrounding system. The two sets of values matching these two patterns are clearly disjoint. □

The behaviour expression given in the first agent behaviour construction is a typical example of a prefix expression. It consists of the action $\mathtt{in}(x)$, and the agent service expression $Buffer\ \langle [x] \rangle$. In the second behaviour construction a choice expression is used, indicating the willingness to accept values both through ports in and out.

In the next example we will apply a product type, rather than the sequence type applied in the previous example. In specification 3.2 the *Buffer* agent is given a counting device that contains the maximum number of elements in the buffer during its lifetime.

Example 3.2 The agent *CBuffer* offers three ports: an input port and an output port similar to the ports in the agent of the previous example *Buffer*, and one additional output port that yields the number indicating the maximum length of the buffer during its lifetime so far.

MOSCA 3.2 Counting Buffer

$CBuffer\ \langle \mathbf{N}^* \times \mathbf{N} \rangle$
ports in $\mathtt{in}$: $\mathbf{N}$
out $\overline{\mathtt{out}}$: $\mathbf{N}$
out $\overline{\mathtt{cnt}}$: $\mathbf{N}$

$$CBuffer\ \langle ([\,], max) \rangle \triangleq \mathtt{in}(x) \odot CBuffer\ \langle ([x], \mathsf{if}\ max > 0\ \mathsf{then}\ max\ \mathsf{else}\ 1) \rangle$$
$$\oplus\ \overline{\mathtt{cnt}}(max) \odot CBuffer\ \langle ([\,], max) \rangle$$

$$CBuffer\ \langle ([x]^\frown s, max) \rangle \triangleq \overline{\mathtt{out}}(x) \odot CBuffer\ \langle (s, max) \rangle$$
$$\oplus\ \mathtt{in}(y) \odot \mathsf{let}\ newbuf = [x]^\frown s^\frown [y]\ \mathsf{in}$$
$$\mathsf{if}\ \mathsf{len}\ newbuf > max$$
$$\mathsf{then}\ CBuffer\ \langle (newbuf, \mathsf{len}\ newbuf) \rangle$$
$$\mathsf{else}\ CBuffer\ \langle (newbuf, max) \rangle$$
$$\oplus\ \overline{\mathtt{cnt}}(max) \odot CBuffer\ \langle ([x]^\frown s, max) \rangle$$

The agent behaviour constructions for *CBuffer* use product patterns. The first behaviour construction shows that an empty *CBuffer* may accept an element for buffering,

$$\mathtt{in}(x) \odot CBuffer\ \langle ([x], \mathsf{if}\ max > 0\ \mathsf{then}\ max\ \mathsf{else}\ 1) \rangle$$

and may offer the length so far through

$$\overline{\texttt{cnt}}(max) \odot CBuffer\ \langle([\,], max)\rangle$$

In the second behaviour expression the value of the counter is set using an agent *if* construction. □

The body of the second behaviour contains an agent *let* construction introducing the name '*newbuf*' as a shorthand for the result of the prefix 'in(y)'. An agent *if* expression is used to compute the correct buffer length, which is either the length of the previous buffer *max*, or the length of the new buffer contents *newbuf*.

The last examples illustrate the capabilities of timehandling in MOSCA. Time handling is based on ideas in Timed CCS, [7] and and TCCS [30]. The time model shows the following characteristics:

- there is no external clock, no ticking device to register time.
- passing of time is measured related to *actions*: from start of an action to end of an action. We let

  $$a, \star t \odot P$$

  denote an agent waiting for the environment to synchronize on a and then become $P[d/t]$ in doing so. d represents the time delay from the activation of the agent until synchronization on a takes place, and $P[d/t]$ indicates the expression denoted by P, in which all ocurrences of the free variable t are replaced by the value d.
- time delay results from *idle* actions. The transition

  $$idle(x) \odot P \xrightarrow{\epsilon(y)} idle(x\text{-}y) \odot P$$

  is valid for each $y \leq x$ where x and y are time expressions, and $\epsilon(y)$ is the action label expressing the amount of time spent idling.
- time is only spent by idling and by waiting for synchronization on the ports.

We illustrate the time model by two examples.

Example 3.3 Stopwatch A particular model of a stopwatch models a stopwatch with two buttons for control, one for starting and one for stoping the stopwatch, and one device for reading the time value. Reading can be done only when the stopwatch is stopped. In specification 3.3 we present a MOSCA model. After starting the stopwatch, the agent SW is ready to accept `stop`. The time elapsed between start and stop is registered in time variable t. □

Time delay is measured in time units from a time-domain **T**, which can be dense or discrete, but has a least element 0. In the sequel we assume the time domain to be the domain $\mathbf{R}^+$ of the positive reals.

MOSCA 3.3 Stopwatch

SW

ports syn **start**

syn **stop**

out $\overline{\texttt{display}}$: T

$$SW \triangleq \texttt{start} \odot \texttt{stop}, \star t \odot \overline{\texttt{display}}(t) \odot SW$$

MOSCA 3.4 Realistic Stopwatch

values

tick : T = 0.01

end

$RSW \langle T \rangle$

ports syn **ss**

syn **reset**

out $\overline{\texttt{display}}$: T

$$RSW \langle t \rangle \triangleq \texttt{ss} \odot RSWR \langle t \rangle$$

$$RSWR \langle t \rangle \triangleq$$
$$(idle(tick) \odot RSWR \langle tick + t \rangle) \oplus$$
$$\texttt{ss}, \star d \odot idle(tick - d) \odot RSWS \langle t + d \rangle \oplus$$
$$\overline{\texttt{display}}(t), \star d \odot idle(tick - d) \odot RSWR \langle tick + t \rangle$$

$$RSWS \langle t \rangle \triangleq \texttt{ss} \odot RSWR \langle t \rangle \oplus \texttt{reset} \odot RSW \langle 0 \rangle$$

Example 3.4 Another Stopwatch A more realistic stopwatch, the specification of which is given in specification 3.4, can be started and stopped with one button, reset with another and constantly read. Its ticking value is e.g. preset to 1 centisecond. The state of the stopwatch is one of three values, ready for a fresh start — RSW, ticking — $RSWR$, or ready to be restarted after a previous ticking status, retaining the current reading — $RSWS$. In the expression

$$(idle(tick) \odot RSWR \langle tick + t \rangle) \oplus \texttt{ss}, \star d \odot idle(tick - d) \odot RSWS \langle t + d \rangle$$

either the idle action runs to completion, whereafter the first prefix will be active, or within the timespan of a tick the start/stop button is pressed, after which the second prefix will be activated. □

4 The Transformation

A major design activity in the development of real-time systems is the transformation of action/event specifications specifying input/output relations and their constraints, into operational specifications, specifying a software system's model. In this section we discuss the transformation of a operations concept model of a real-time system into a requirements model, encoded in MOSCA that forms the basis for implementation. Only action/event models that behave in a decent way will have any chance to be transformed into meaningful MOSCA models. As with any notation, the specifier remains responsable for the creation of meaningful models. The A/EL notations as discussed earlier provide means to validate the safety assertions.

The transformation involves (i) adding explicit dataflow information, including datatype definitions and binding of names to datatypes, (ii) adding explicit controlflow by grouping actions into processes, establish process interfaces and process interconnection structures. One of the main design paradigms is the transformation of state-based control to control by data communication.

Static analysis of the action/event model AES will reveal

- which pex's have an associated *condition* which will always take the value TRUE,
- which pex's do have an associated *condition* being FALSE forever, or
- which pex's are active/suspended intermittingly, through manipulation of the *condition* by other events or actions.
- which sex's are never activated, by absense of their activating event, or
- which sex's are activated by presence of their activating event.

The transformation constructs a MOSCA model MS from the action/event model AES. Since nested composite actions require at least some form of itertion or recursion over the proposed steps, it is assumed, for the sake of simplicity, that the structure of composite actions is flattened into a one–level structure, and nested sets of parallel compositions unraveled into one set of parallel actions.

This procedure transforms the system item AT into a set of primitive actions and linear composite actions. The mapping involves the following steps.

STEP 1 — Refine primitive actions by (i) fixing the attributes for each entity associated with the actions, and (ii) adding datatype information for these attributes. Assign to each system property described by a state-predicate a *data type*, thus transforming state predicates into state values containing data. The predicates are transformed into control constructions embedded within MOSCA agents.
Classify all primitive actions on whether their data manipulation involves only internal data elements or results in external data manipulation. Actions manipulating only internal data elements are mapped onto MOSCA functions or operations, actions with external effects are mapped onto MOSCA prefix expressions.

STEP 2 — Create action/event relations.

1. Setup a MOSCA action/reaction-pair relation ARR for all primitive actions with external effect. Associate with each action *act* a MOSCA action *mact* and a MOSCA reaction *mract*.
2. Next setup a pair of relations $T : (Action \times Event)$ and $IAB : (Event \times Action)$. T defines "triggers" and IAB the relation "is activated by". Through inspection of both relations we find (i) the external events that are not causally related to any action, i.e. the set AE of autonomous events in the system and (ii) the set of isolated actions IA, actions that are not in the domain of T.

The relations *ARR*, T and *IAB* form the basis for the next step.

STEP 3 — Setup MOSCA agents either as *actors* or *reactors*.

1. Anchor chains of action/reaction activities to actors and reactors by mapping all pex's to actors and sex's to reactors.
2. Anchor external events that control state values that act as guard to pex's onto actor/reactor compositions by translation into MOSCA prefix constructions in both actor and reactor.
3. Anchor external events that trigger sex's by translation into MOSCA prefix constructions in actors and reactors.
4. Chain the events in AE, modeled as MOSCA actions, to appropiate actors/reactors already in the specification. Create new reactors for the remaining actions. Both relations T and IAB induce a partial ordering $\preceq$ on the set of events.

These transformations eliminate all state predicate values that are used exclusively for pex control.

STEP 4 — Add timing control of pex's and sex's to all associated MOSCA agents. Here the most complex part is the realisation of masked external events. If appropiate this may be accomplished by bundling the control threads of all involved reactors.

STEP 5 — Transform guarded actions in AES to conditional behaviour expressions in agents. This step involves a case analysis over all remaining state predicate values. For each value it must be decided whether it can be mapped onto a control by communicating condition information, or whether it should remain a global accesable state value.

Several aspects of the different steps can be formalised completely. A partial mapping from the abstract syntax of the action/event language to the abstract syntax of MOSCA is presented in [27].

Step 5 is the most elusive part of the transformation. It involves distribution of the global state of the AES to separate agents in MS and the best we can offer here is a set of guidelines to the designer.

In practice the steps may proceed in parallel, including iterations, until all constructions are transformed. In the next section we give an example that illustrates various aspects of the steps in detail.

5 The Example

The case study is adapted from [1] where the problem described in figure 5 was set for the real-time systems domain. Although this description is to be considered far from complete, as pointed out by Wood in [32], it illustrates the specific real-time behaviour of reactive systems.

The action/event specification is presented in the appendix. It is documented in a informal tabular style, mapping it onto the abstract syntax is straight forward. The

The controller of an oil hot water home heating system regulates in-flow of heat, by turning the furnace on and off, and monitors the status of combustion and fuel flow of the furnace system, provided the master switch is set to HEAT position. The controller activates the furnace whenever the home temperature t falls below $t_r - 2$ degrees, where t_r is the desired temperature set by the user. The activation procedure is as follows:

1. the controller signals the motor to be activated.
2. the controller monitors the motor speed and once the speed is adequate it signals the ignition and oil valve to be activated.
3. the controller monitors the water temperature and once the temperature is reached a pre-defined value it signals the circulation valve to be opened. The heater water then starts to circulate through the house.
4. a fuel flow indicator and an optional combustion sensor signal the controller if abnormalities occur. In this case the controller signals the system to be shut off.
5. once the home temperature reaches $t_r + 2$ degrees, the controller deactivates the furnace by first closing the oil valve and then, after 5 seconds, stopping the motor.

Further the system is subject to the following constraints:

1. minimum time for furnace restart after prior operation is 5 minutes.
2. furnace turn-off shall be indicated within 5 seconds of master switch shut off or fuel flow shut off.

Figure 1: Informal requirements of Heating System

example stresses the usage of state predicates. Timing characteristics of primitive actions are assumed to be given. In creating the action/event specification assumptions have been made concerning the duration of periods of pex's. Also a mask is put on the master switch events to prohibit malfunction of the system due to very fast on/off switching of the system. The action/event specification is then stepwise transformed into a MOSCA specification. VDM offers constructions on a broad range of abstraction, from highly abstract implicit specifications to concrete explicit specifications. The data structures in the MOSCA specification are kept extremely simple. The control associated with the internal activities of the agents is explicitly stated. All auxiliary functions and operations are assumed. The emphasis is put on the transformation of the actions and pex's and sex's to MOSCA agents, and not specially directed to show off VDM. Together with Z, VDM is currently the most widespread known and used model-oriented specification method, from which most important characteristics are very well documented; see for example [4], [3], [13].

In composite actions the parallel composition of actions is denoted through the $||$ operator. Action sequencing is denoted by the ';' notation. The transformation from the action/event specification into MOSCA involves many detailed inspections of the action/event specification. **STEP 1** opens the transformation by adding model-oriented datatype information to the actions in the system. This results in an *Primitive action Attribute Table.*

Primitive Action Attibute Table			
Id	Attributes	Int	Ext
IMV	*velocity* : $\mathbb{N}$, *normal_value* : $\mathbb{N}$	x	
IWT	*water_temp* : $\mathbb{N}$, *water_working_temp* : $\mathbb{N}$	x	
IHT	*home_temp* : $\mathbb{N}$, *desired_home_temp* : $\mathbb{N}$	x	
SMO, *STM*	*motor_status* : {RUNNING, STOPPED}		x
AFI			x
AOV, *COV*	*valve_position* : {OPEN, CLOSED}		x
OCV, *CCV*	*valve_position* : {OPEN, CLOSED}		x
AMM			x
AWM			x
*SHT*5			x

The table enumerates for each action the involved attributes, and states whether the effect of the action involves internal data or external data. The attributes are sometimes shared between actions. The following table state the type of the remaining state attributes.

State Table	
HHSAC	*hhsac* : $\mathbb{B}$

Only one state predicate remains. It will be used in the main control of the system, the thermometer. All other predicates can be expressed using the attributes associated with the actions. These results finalise **STEP 1**.

Next we setup the *ARR*, *T*, and *IAB* relations.

Action Reaction Relation Table		
A/E action	MOSCA action	MOSCA reaction
SMO	start_motor	start_motor
AFI	ignite_furnace	ignite_furnace
AOV	open_oil_valve	open_oil_valve
OCV	open_water_valve	open_water_valve
AMM	motor_mon_start	motor_mon_start
AWM	water_mon_start	water_mon_start
COV	close_oil_valve	close_oil_valve
*SHT*5	start_timer	start_timer
STM	stop_motor	stop_motor
CCV	close_water_valve	close_water_valve

The names of the MOSCA actions can be equal to the reactions, as they will appear in different agents. They will introduce synchronization points between actors and reactors.

$$\begin{aligned} T \;=\; & \{(IMV, MVRNV{:=}\,\text{TRUE}), \\ & (IWT, WTRVW{:=}\,\text{TRUE}), \\ & (IHT, HTWDS{:=}\,\text{TRUE}), \\ & (IHT, HTWDS{:=}\,\text{FALSE}), \\ & (AMM, MVRNV{:=}\,\text{FALSE}), \\ & (AWM, WTRWV{:=}\,\text{FALSE}), \\ & (SHT5, HT5I)\} \end{aligned}$$

$$\begin{aligned} IAB \; = \; & \{(MVRNV{:}=\text{FALSE}, PHASE2), \\ & (HTWDS{:}=\text{FALSE}, PHASE1), \\ & (WTRVW{:}=\text{TRUE}, PHASE3), \\ & (HTWDS{:}=\text{TRUE}, PHASE4), \\ & (HT5I, PHASE5), \\ & (MSTO, PHASE4),\} \end{aligned}$$

Let $\Delta_i(R)$ denote the projection function delivering the set of all i^{th} elements of all relation tuples in a relation R. The set of autonomous actions AE becomes

$$AE = \mathsf{dom}\, ET - \Delta_2(T) = \{FFE, FCE, MSTH, MSTO\}$$

and the set of isolated action IA becomes

$$IA = \mathsf{dom}\, AT - \Delta_1(T) = \{SMO, AFI, AOV, OCV, COV, STM, CCV\}$$

This finalises **STEP 2** of the transformation.

The following design analysis captures the three final steps in the transformation. Time control (**step 4** is added in parallel with the main design.

Let's start **STEP 3** of the transformation by concentrating on the motor in the heating system. The properties *velocity* and *normal_value* are associated with the motor, where *velocity* is obtainded through the primitive action IMV, and *normal_value* is a constant. Both attributes are modelled by natural numbers.

state *motor_mon* **of**
  velocity : $\mathbb{N}$
  normal_value : $\mathbb{N}$
end

Actions involved with manipulating the motor are $\{IMV, SMO, AMM, STM\}$, partially ordered by T and IAB to $SMO \preceq AMM \preceq IMV \preceq STM$. The motor is started by SMO, an action with external effect (it occurs in ARR). As such we need a control thread issuing the action and a control thread reacting to it. Assume an actor $PHASE1$ that issues the MOSCA equivalent of SMO. The reactor will become $MOTOR\_CONTROL$. The actual starting of the motor is modelled through the state value *status*.

state *motor_control* **of**
  status : {RUNNING, STOPPED}
end

$MOTOR\_CONTROL \triangleq$ `start_motor` $\odot$ *motor_control.status*: = RUNNING ...

$PHASE1 \triangleq$ ... `start_motor` $\odot$ ...

SMO, the first action in *PHASE*1, is followed by *AMM* which initiates pex nr. 1 concerning *IMV*. The state predicate *MVRNV* controls the pex, when FALSE the pex is activated, on becoming true (an event) it initiates sex nr. 2. *AMM* initiates pex nr. 1 as it sets the predicate to FALSE. This event is modelled by the MOSCA action motor_mon_start, in both actor and reactor.

$MOTOR\_MON \triangleq$ `motor_mon_start` $\odot \ldots$

$PHASE1 \triangleq \ldots$ `start_motor` $\odot$ `motor_mon_start` $\odot \ldots$

The action *IMV* has internal effects only and is mapped onto the MOSCA operation *get_motor_velocity*. The realisation of the pex becomes

$MON\_VELOCITY \triangleq$
get_motor_velocity();
if *motor_mon.velocity* < *motor_mon.normal_value*
then *idle*(1) $\odot$ *MON_VELOCITY*
else ... $\odot$ *MOTOR_MON*

where it is assumed that the execution of *get_motor_velocity* and the conditional expression take a neglectable amount of time. The *idle*(1) action controls the period of the pex. Monitoring stops when the motor reaches its normal value, signalled by the action ... in the else part. Here again we are modelling a A/E state predicate change event. The actor is *MON_VELOCITY*, the reactor is the sex associated with the event, the action leading to the event is e.g. `motor_ready`.

$MON\_VELOCITY \triangleq$
get_motor_velocity();
if *motor_mon.velocity* < *motor_mon.normal_value*
then *idle*(1) $\odot$ *MON_VELOCITY*
else `motor_ready` $\odot$ *MOTOR_MON*

$PHASE2 \triangleq$ `motor_ready` $\odot \ldots$

After `motor_ready` has been signalled, the agent becomes ready to restart. This completes the specification of *MON_VELOCITY*. The agent *MOTOR_MON* directly activates *MON_VELOCITY*, and *MOTOR_MON* itself is activated by synchronization. Adding this information results in

$MOTOR\_MON \triangleq$ `motor_mon_start` $\odot$ *MON_VELOCITY*

Finally the A/E action *STM* may occur. It is activated by composite action *PHASE*5, and reacted upon by *MOTOR_CONTROL*:

$PHASE5 \triangleq$ `stop_motor` $\odot \ldots$

$$MOTOR\_CONTROL \triangleq$$
$$(\texttt{start\_motor} \odot motor\_control.status{:=} \text{RUNNING}; MOTOR\_CONTROL)$$
$$\oplus (\texttt{stop\_motor} \odot motor\_control.status{:=} \text{STOPPED}; MOTOR\_CONTROL)$$

This completes the design of the motor. It results in two MOSCA agents for which the full specification, including their interfaces are presented in specification 5.1.

MOSCA 5.1 Furnace Motor

MOTOR_CONTROL
state *status* : {RUNNING, STOPPED}

ports syn **start_motor**
syn **stop_motor**

$$MOTOR\_CONTROL \triangleq$$
$$(\texttt{start\_motor} \odot motor\_control.status{:=} \text{RUNNING}; MOTOR\_CONTROL)$$
$$\oplus (\texttt{stop\_motor} \odot motor\_control.status{:=} \text{STOPPED}; MOTOR\_CONTROL)$$

MOTOR_MON
state *velocity* : $\mathbb{N}$
normal_value : $\mathbb{N}$

ports syn **motor_mon_start**
syn **motor_ready**

shares *MOTOR_MON*, *MON_VELOCITY*

$$MOTOR\_MON \triangleq \texttt{motor\_mon\_start} \odot MON\_VELOCITY$$

$MON\_VELOCITY \triangleq$
get_motor_velocity();
if $motor\_mon.velocity() < motor\_mon.normal\_value$
then $idle(1) \odot MON\_VELOCITY$
else **motor_ready** $\odot$ *MOTOR_MON*

The next design is dedicated to the water in the system. It involves the second pex, concerning *IWT* and the sex that initiates *PHASE*3. Again we need an actor to issue the **water_mon_start** action and a reactor for it.

$$PHASE2 \triangleq \texttt{motor\_ready} \odot \ldots \odot \texttt{water\_monitor\_start}$$

$$WATER\_MON \triangleq \texttt{water\_mon\_start} \odot \ldots$$

Water monitoring involves two action attributes.

state *water_mon* of
water_temp : $\mathbb{N}$
water_working_temp : $\mathbb{N}$
end

The development of the pex proceeds similarly to the motor velocity monitor, giving

MON_WATER_TEMP ≙
  get_water_temp();
  if *water_mon.water_temp* < *water_mon.water_working_temp*
  then *idle*(5) ⊙ *MON_WATER_TEMP*
  else **water_ready** ⊙ *WATER_MON*

The action **water_ready** synchronizes the activation of *PHASE*3.

*PHASE*3 ≙ **water_ready** ⊙ **open_water_valve** ...

The water circulation valve is controlled identically as the motor control giving

WATER_CONTROL ≙
  (**open_water_valve** ⊙ *water_valve.status*: = OPEN; *WATER_CONTROL*)
  ⊕ (**close_water_valve** ⊙ *water_valve.status*: = CLOSED; *WATER_CONTROL*)

The full specification is not given, it is structurally equivalent with specification 5.1. The main switch is modelled by

HOME_HEATING_SYSTEM ≙
  start_mon_temperature ⊙ *idle*(10) ⊙
  stop_mon_temperature ⊙ *idle*(10) ⊙ *HOME_HEATING_SYSTEM*

Each apropiate agent behaviour is made circular, as the system should perform the same behaviour repeatedly. E.g. *PHASE*1 becomes

*PHASE*1 ≙ **start_heater** ⊙
  start_motor ⊙
  motor_mon_start ⊙ *PHASE*1

which indicates that after the last action has been executed the agent is willing to start all over again, ready to perform it actions in the next system phase.

Continuing in this way the whole set of agents is easily completed. Each (composite) A/E action associated with a sex is transformed into an MOSCA agent, each pex is transformed into a specialised agent. All events that initialise sex's are transformed into synchronization between the agents. State components are distributed over the involved agents. Finally all agents are connected through composition, together forming the home heating system.

6 Conclusions and Future Research

The stepwise development of MOSCA specifications from action/event models is far from trivial. The proposed series of steps act as a basic transformation scheme, giving general guidelines to setup explicit thread of controls that together behave in accordance to the action/event model. In the example it shown how to transform pex's and sex's in a general manner into MOSCA agents. However, state control tranformation and distribution of state information over the agents may involve complicated analysis, and will in general be difficult to formalise, as is examplified by the control in *THERMO*.

In the example we have abstracted away many design decisions and omitted many details. It highlights just one particular real-time behaviour, namely cooperative work based on synchronization. It lacks e.g. strong emphasis on hard real-time constraints. It must be shown elsewhere how action/event systems with such constraints can be transformed.

The transformation could be partially supported by tools. Static analysis of action/event systems will result in basic information concerning the different relations used within the transformation. But before engaging in designing these tools much more work remains on fixing formalised descriptions of the transformation process itself.

The notation MOSCA offers is specially dedicated to record data abstraction through VDM domain definitions, functions and operations; process abstractions using the CCS operators and time abstractions using some highly experimental constructions. As such it offers a total range of constructions to model real-time systems on a level far more appropiate to further software development than action/event languages offer. Although the transformation is complicated, we think it is effective by producing a more comprehensive view of the system, aimed at implementation of the system, and not only aimed at simulating the behaviour of the system.

A A A/E representation of the heater system

A.1 Primitive Actions

Primitive Actions	
IMV	Input Motor Velocity, sets state-predicate *MVRNV* to TRUE if normal speed value of motor is reached
IWT	Input Water Temperature, sets state-predicate *WTRWV* to TRUE if water working temperature is reached
IHT	Input Home Temperature, sets state-predicate *HTWDS* to TRUE if home temperature reaches desired setting, sets state-predicate *HTWDS* to FALSE if temperture drops below acceptable setting
SMO	Start MOtor
AFI	Activate Furnace Ignition, sets state predicate *WCMIR* to TRUE
AOV	if state-predicate *FOVIO* is FALSE, Activate Oil Valve
OCV	if state-predicate *WCVIO* is FALSE, Open Circulation Valve
AMM	Activate motor monitoring by setting state-predicate *MVRNV* to FALSE
AWM	Activate water temperature monitoring by setting state-predicate *WTRWV* to FALSE
COV	if state-predicate *FOVIO* is TRUE, Close Oil Valve
*SHT*5	Start Hardware Timer for a delay of 5 seconds
STM	if state-predicate *WCMIR* is TRUE, STop Motor
CCV	if state-predicate *WCVIO* is TRUE, Close Circulation Valve

A.2 Composite Actions

Composite Actions	
*PHASE*1	*SMO* ; *AMM*
*PHASE*2	(*AFI* ‖ *AOV*) ; *AWM*
*PHASE*3	*OCV*
*PHASE*4	*COV* ; *SHT*5
*PHASE*5	*STM* ; *CCV*

A.3 State Predicates

State Predicates	
HHSAC	Home Heating System Activated, initital DISABLED
MVRNV	Motor Velocity Reached Normal Value, initital DISABLED
WTRWV	Water Temperature Reached Working Value, initital DISABLED
HTWDS	Home Temperature Within Desired Setting, initital TRUE
WCVIO	Water Circulation Valve is Open, initial FALSE
FOVIO	Furnace Oil Valve is Open, initial FALSE
WCMIR	Water Circulation Motor is Running, initial FALSE

A.4 External Events

External Events	
HT5I	Hardware Timer 5 seconds delay Interrupt, should occur at least 5 seconds after *SHT*5 action
FFE	Fuel Flow Exception : something's wrong with the fuel flow
FCE	Furnace Combustion Exception
MSTH	Master Switch to Heat position, sets state-predicate *HHSAC* to TRUE
MSTO	Master Switch to Off position, resets all state-predicates to initial value

A.5 Periodic Execute Statements

Periodic Execute Statements			
While	Do	Each #'sec	Delay #'sec
MVRNV = FALSE	IMV	1	
WTRWV = FALSE	*IWT*	5	
HHSAC = TRUE	*IHT*	30	

A.6 Sporadic Execute Statements

Sporadic Execute Statements			
When	Do	Within #'sec	Separation #'sec
HTWDS = FALSE	*PHASE*1	1	
MVRNV = TRUE	*PHASE*2	1	
WTRWV = TRUE	*PHASE*3	1	
HTWDS = TRUE	*PHASE*4	1	
HT5I	*PHASE*5	1	
MSTO	*PHASE*4	1	{*MSTH*,*MSTO*},10
FFE ∨ *FCE*	*PHASE*4	0.1	

References

[1] Problem set for the Fourth International Workshop on Specification and Design, 1987. nr. 2: Heating System.

[2] G. Berry and L. Cosserat. The Esterel Synchronous Programming Language and its Mathematical Semantics. In S.D. Brookes, A.W. Roscoe, and G. Winskel, editors, *Seminar on Concurrency*, volume 197 of *LNCS*, pages 389–448, Carnegie Mellon University, 1984. Springer Verlag.

[3] D. Bjørner. Towards a Meaning of 'M' in VDM. In J. Diaz and F. Oregas, editors, *Tapsoft-89*, volume 352 of *LNCS*, pages 1–35. Springer Verlag, 1989.

[4] D. Bjørner and C.B. Jones. *Formal Specification & Software Development.* PHI. Prentice Hall, 1982.

[5] W.W. Bledsoe and L.M. Hines. Variabale elimination and chaining in a resolution based prover for inequalities. In W. Bibel and Kowalski R., editors, *5th Conference Automated Deduction*, LNCS, pages 70–87, New York, 1980. Springer Verlag.

[6] T. Bolognesi and Brinksma E. Introduction to the ISO Specification Language LOTOS. In P.H.J. van Eijk, C.A. Vissers, and M. Diaz, editors, *The Formal description Technique LOTOS*, pages 23–77. North Holland, 1989.

[7] L. Chen, S. Anderson, and F. Moller. A Timed Calculus of Communicating Systems. Technical Report LFCS-90-127, University of Edinburgh, 1990.

[8] B. Dasarathy. Timing Constraints of Real Time Systems: Constructs for Expressing them, Methods of Validating them. *IEEE Transactions on Software Engineering*, SE-11(1):80–86, January 1985.

[9] M.S. Deutsch. Focussing on Real-Time Systems Analysis on User Operations. *IEEE Software*, pages 39–51, September 1988.

[10] J. Dias-Gonzales and J.E. Urban. ENVISAGER: A Visual Object-Oriented Specification Environment for Real-Time Systems. In *Fourth International Workshop on Software Specification and Design*, pages 13–20, Monterey, California, USA, 1987. IEEE Computer Society Press.

[11] Jahanian F. and A.K-L. Mok. Safety Analysis of Timing Properties in Real-Time Systems. *IEEE TSE*, se-12(9):890–904, September 1986.

[12] ISO SC22/WG19. *VDM Specification Language — Proto-Standard*, 1991. Draft dated 9th March.

[13] C.B. Jones. *Systematic Software Development Using VDM, 2-nd edition.* PHI. Prentice Hall, 1990.

[14] Jarvinen H-M. and R. Kurki-Suonio. The Disco Language. Technical Report, Tampere University, Software Systems Laboratory, Report 8, March 1990.

[15] P.G. Larsen, A. Tarlecki, W. Pawlowski, and M. Borzyszkowski, Wieth. The Dynamic Semantics of the BSI/VDM Specification Language. Technical report, IFAD, The institute of Applied Computer Science, Munkebjergsvaenget 17, DK-5230 Odense M, Denmark, August 1990.

[16] L.Y. Liu and R.K. Shyamasundar. Programming for Real-Time Reliable Reactive Systems. Technical report, Penn. State University, University Park PA 16802, 1988.

[17] L.Y.H. Liu and R.K. Shyamasundar. An Operational Semantics of Real Time Design Language RT-CDL. *ACM SIGSOFT Engineering Notes*, 14(3):75–82, May 1989. (proceedings of the Fifth International Workshop on Software Specification and Design).

[18] Z. Manna and A. Pnueli. Verification of Concurrent Programs: The Temporal Framework. In R.S. Boyer and S. Strother Moore, editors, *The Correctness Problem in Computer Science*, pages 215–273. Academic Press, 1981.

[19] R. Milner. Calculi for Synchrony and Asynchrony. *TCS*, 25:267–310, 1983.

[20] R. Milner. *Communication and Concurrency*. PHI. Prentice Hall, 1989.

[21] N. Plat, van Katwijk J., and Pronk K. A Case For Structured Analysis / Formal Design. accepted for 4th VDM'91 Symposium Noordwijkerhout, The Netherlands, 1991.

[22] N. Plat, J van Katwijk, and W.J. Toetenel. Application and Benefits of Formal Methods in Software Development. Technical Report 91-ZZ, Faculty of Technical Mathematics and Informatics, Delft University of Technology, 1991.

[23] G. Plotkin. A Structural Approach to Operational Semantics. Technical Report DAIMI FN-19, Aarhus University, 1981.

[24] A. Pnueli. Applications of Temporal Logic to the specification and verification of reactive systems. In J.W. de Bakker, W.-P. de Roever, and G. Rozenberg, editors, *Current Trends in Concurrency*, volume 224 of *LNCS*, pages 510–585. Springer Verlag, 1986.

[25] W.J. Toetenel. Model-Oriented Specification of Communicating Agents. In *proceedings of CSN'91*. SION, 1991.

[26] W.J. Toetenel. MOSCA-SL Language Reference Manual. Technical Report 91-YY, Faculty of Technical Mathematics & Informatics, Delft University of Technology, 1991.

[27] W.J. Toetenel. *Model Oriented Specification of Communicating Agents*. PhD thesis, Delft University of Technology, Faculty of Mathematics & Informatics, 1992. (in preparation).

[28] W.J. Toetenel, J. van Katwijk, and N. Plat. Structured Analysis - Formal Design, using Object and Stream Oriented Formal Specification. In M. Moriconi, editor, *ACM SIGSOFT International Workshop on Formal Methods in Software Development.* ACM SEN, 1990.

[29] US Department of Defence, US Printing office. *The Ada Programming Language Reference Manual*, 1983.

[30] Y. Wang. An Interleaving Model for Real Time Systems. In K.G. Larsen and A. Skou, editors, *2nd Nordic Workshop on Program Correctness.* The University of Aalborg, October 1990.

[31] Y. Wang. Real-Time Behaviour of Asynchronous Agents. In J.C.M. Baeten and J.W. Klop, editors, *CONCUR'90 Theories of Concurrency: Unification and Extension*, volume 458 of *LNCS*, pages 502–520. Springer Verlag, 1990.

[32] W.G. Wood. Software Design. In *Fourth International Workshop on Software Specification and Design*, pages 201–204. IEEE Computer Society Press, 1987.

Formal Specification of Fault Tolerant Real Time Systems Using Minimal 3-Sorted Modal Logic

Peter Coesmans
Martin J. Wieczorek

Real Time Systems Group, Department of Informatics, University of Nijmegen
Toernooiveld 1, 6525 ED Nijmegen, The Netherlands
e-mail: mjw@cs.kun.nl

*"I miss the good old days
when all we had to worry about
was bits and bytes."*

Abstract

Fault tolerance is the property of a system to provide a specified service despite the occurrence of faults, i. e. to prevent a system from failing even in the presence of faults. In this paper, we will contribute to the area of formal specification of fault tolerant real time systems to make fault tolerance and real time formally treatable in a unified approach.

According to the paradigm of *separation of concerns* we get separation in two directions: In real time systems, a distinction can be made between *functional*, *locational*, and *temporal* properties. To explicitly state such properties in a formal specification we will use a *three-sorted modal logic.*

In fault tolerant systems, two kinds of behaviour can be distinguished from each other: *normal behaviour*, which takes place if no fault occurs during system execution, and *exceptional behaviour*, which takes place just in the case of a fault occurrence. To separate system properties according to that a logical connective *C (Combine)* will be defined. This connective allows to state predicates about normal behaviour as well as exceptional behaviour and it also provides the possibility to specify the conditions under which the one or the other behaviour will be reached. To ensure that a fault tolerant real time system has precisely the properties stated in its formal specification minimal model interpretation is applied to the logical formulae.

1 Introduction

In formal specification of distributed real time systems, precise *locations*, *times*, and *values* are most often regarded as incidental details. What is considered to be important are the significant events and the relationships between such events. For example, sending a message from one process to another process using an unreliable transmission medium would depend on the right state of the participating components, e. g. ready to receipt a new message, and on the fact that a new message is provided to such a system. Here, the reception of a message would depend on previous occurences of events for reaching the right state, e. g. having finished the transmission of the last message according to some protocol.

Recent research has shown, e. g. Manna and Pnueli [11], Winskel [17] and Hooman [7], that one or more of the above mentioned incidental details must be available for formal reasoning about distributed real time systems: whether this is reached using the details implicitly in the underlying model of the specification formalism or explicitly in the formalism itself is not relevant to the basic argument.

Also for fault tolerant systems a classification into functional (value), locational (location label), and temporal (time point) properties is relevant: faults due to hardware component failures are often classified [13] by *duration* (transient or permanent fault), *extent* (local or distributed effect), and *value* (fixed or varying erroneous logical values). According to the paradigm of *separation of concerns* we will use in this paper a three-sorted modal logic for the specification of fault tolerant real time systems allowing for formal reasoning about such systems.

Another important aspect in fault tolerant computing, contrasted to the non fault tolerant case, is the more refined notion of behaviour: behaviour which is in accordance with the specification—*correct behaviour*—and behaviour which is not in accordance with the specification—*catastrophic behaviour*. Catastrophic behaviour resulting from unanticipated input [4] is not describable; therefore, it cannot be subject to specifications. Correct behaviour resulting from anticipated input [4] can be classified further: behaviour which is intended (required) by the designer of a system—*normal behaviour*—and behaviour which is due to system (hardware) faults—*exceptional behaviour*. Exceptional behaviour as well as normal behaviour must be subject to formal specifications of systems to make fault tolerance formally treatable.

For formal specification of fault tolerant real time systems we will use a *minimal three-sorted modal logic (minimal S_3ML)*. Logic and especially temporal logic has been shown valuable in formal specification and verification of real time systems (e. g. by Pnueli in [12] and by Hooman in [7]). Temporal logic can be regarded as a particular positional logic (cf. Rescher [14]) where the positions denote points in some dating scheme, which can be used to model (flow of) time in a real time system. The propositions there are used in a temporally indefinite manner, i. e. their truth or falsity is not independent of their assertion time.

In (distributed) real time systems not only correct functionality and correct timings but also correct locations are important for correct behaviour. Therefore, our modal logic is three-sorted where the positions not only denote time points but also allow for locations and values. What we get is a propositional logic augmented by positional operators ranging over the three-dimensional position space to model functional, locational, and temporal properties. The propositions are now functionally and locationally and temporally indefinite, i. e. their truth or falsity is now dependent on the position characterized by a value, a location, and a time.

In classical logics and also in S_3ML, the rule of monotonicity is valid: whatever can be obtained from assumption φ can still be obtained if one adds more assumptions to φ [5]. With respect to fault tolerant systems this means that if $\varphi \rightarrow \psi$ is a valid specification of a system and if we add assumption χ about faults to the antecedent we can logically derive the validity of the extended formula, i. e.

$$\frac{\varphi \rightarrow \psi}{\varphi \wedge \chi \rightarrow \psi}$$

But, if we allow for occurrences of (hardware) faults in a system we must take care of the fact that those faults can seriously influence the behaviour of that system and eventually make ψ no longer hold. Only adding propositions about faults in a logical formula still using an implication for specifying dependencies between events seems, therefore, not sufficient.

Several solutions to overcome the problem of formal specification of fault tolerant real time systems seem possible: for example, in [3] Cristian proposed a rigorous approach to fault tolerance using a particular class of faults, namely exceptions. Exceptions raised in the case that faults have occurred during system execution lead to exceptional behaviour in the system. This is specified in [3] by using an extended version of Hoare logic [6]. Although useful in the context there this approach is neither concerned with real time nor with distribution of system components.

Another possibility to look at formal specification of fault tolerant real time systems is from the point of view of temporal logic. In [11] it is argued by Manna and Pnueli that "no specification of a system can be complete without containing some safety properties and some liveness properties". Due to Lamport [9] safety properties describe "what a program is allowed to do or, dually, what a program may not do" and liveness properties describe "what a program must do". Although neither Manna and Pnueli in [11] nor Lamport in [9] are concerned with fault tolerance this leads to an interesting observation: liveness porperties define some "lower bound" on the set of events occurring in a system whereas safety properties define some "upper bound" on the set of events. This view is applied in [10] by Larsen though not for temporal logic but for Hennessy-Milner logic: minimal interpretations are used for liveness properties and maximal interpretations are used to specify safety properties.

As we have seen above correct behaviour can be divided into normal and exceptional behaviour. When we now use minimal interpretation of our logic for the required behaviour (liveness) and introduce a new connective $\mathcal{C}$ (Combine) to separate logical formulae about normal behaviour from those about exceptional behaviour we will get means to reason about fault tolerance and real time in a unified approach. This will be followed in the remaining sections of this paper: In section 2 we will define a three-sorted modal logic with minimal model interpretation. Section 3 contains the definition and discussion of our logical connective $\mathcal{C}$ (Combine) as an extension of the logic defined in section 2. This connective will then be used in section 4 in the formal specification of a Stable Storage. Finally we will summarize our contribution and draw some conclusions.

2 Minimal 3-Sorted Modal Logic

In this section we define a three-sorted modal logic S_3ML. The three sorts are introduced to give more structure to logical formulae and for separation of property classes:

functional, locational, and temporal properties. Logical formulae can then directly be associated with parts of structured systems, as argued in [16], and, therefore, allows for a structured development. Fault tolerant systems especially have this property of structuredness or layering.

To get such a logic we have augmented classical propositional logic with functional, locational, and temporal operators similar to those known from temporal logic. The semantics of our logic is defined in terms of a Kripke model, i. e. we provide a 'possible worlds' semantics based on frames and models [1]. Validity then is restricted to our purposes of minimal interpretation of logical formulae by defining the notion of 'minimal validity'.

2.1 Syntax and Semantics

In table 1 the formation rules of our logic are provided. We assume that $\mathcal{A}_p$ is an alphabet of propositional constants, that φ (possibly indexed) denotes a logical formula. The superscripts of the modal connectives indicate to which of the three sorts the operator belongs.

Propositional Constants:
$\alpha \quad (\in \mathcal{A}_p)$
Propositional Connectives:
$\bot, \neg\varphi, \varphi_1 \rightarrow \varphi_2$
Modal Connectives:
$\bigcirc^L\varphi, \bigcirc^T\varphi, \Box^V\varphi, \Box^L\varphi, \varphi_1\, \mathcal{U}^T \varphi_2$

Table 1: S_3ML formulae

Informally the above defined modal operators have the following meaning:

$\bigcirc^L\varphi$ is satisfied at the current position ('actual world') if φ is satisfied at the next (locationally) reachable position where the value and time component of the new position is the same as of the current position

$\bigcirc^T\varphi$ is satisfied at the current position ('actual world') if φ is satisfied at the next time point where the value and location component of the new position is the same as of the current position

$\Box^V\varphi$ is satisfied at the current position ('actual world') if φ is satisfied at all (functionally) computable positions where the location and time component of the new positions is the same as of the current position

$\Box^L \varphi$ is satisfied at the current position ('actual world') if φ is satisfied at all (locationally) reachable positions where the value and time component of the new positions is the same as of the current position

$\varphi_1 \mathcal{U}^T \varphi_2$ is satisfied at the current position ('actual world') if we can find a future time point (exclusive) where condition φ_2 is satisfied and for all time point between the current time point and that time point φ_1 is satisfied

Remark: In the sequel, we will use meta-logical operators for *conjunction* ' & ', *disjunction* ' | ', *implication* ' $\Rightarrow$ ', and *negation* ' $\sim$ ' with the usual interpretations. Square brackets '[ ]' are used to enclose meta-expressions.

Definition 2.1 (Frame) Let V, L, and T be non-empty sets of values, locations (location labels), and time points, respectively. Let $\prec_V$, $\prec_L$, and $\prec_T$ be relations ('accessibility' relations) on V, L, and T, respectively. Then, a *frame* is a structure (W, R) where $W := (V \times L \times T)$ and $(v_1, l_1, t_1)R(v_2, l_2, t_2) :\Longleftrightarrow v_1 \prec_V v_2 \;\&\; l_1 \prec_L l_2 \;\&\; t_1 \prec_T t_2$.

In table 2, we provide a rudimentary set of conditions on the accessibility relations. A proof system for our logic must be provided in the future elsewhere.

L-Symmetry:	$\forall x, y[x \prec_L y \;\Rightarrow\; y \prec_L x]$
L-Transitivity:	$\forall x, y, z[[x \prec_L y \;\&\; y \prec_L z] \;\Rightarrow\; x \prec_L z]$
L-BackLinearity:	$\forall x, y, z[[y \prec_L x \;\&\; z \prec_L x] \;\Rightarrow\; [y \prec_L z \;\mid\; z =_L y \;\mid\; z \prec_L y]]$
T-Transitivity:	$\forall x, y, z[[x \prec_T y \;\&\; y \prec_T z] \;\Rightarrow\; x \prec_T z]$
T-Linearity:	$\forall x, y[[x \prec_T y \;\mid\; x =_T y \;\mid\; y \prec_T x]$

Table 2: Conditions on Accessibility Relations

Definition 2.2 (Model) Let be given non-empty sets V, L, and T and relations $\prec_V$, $\prec_L$, and $\prec_T$ on V, L, and T, respectively. Let $F := (W, R)$ be a frame where W and R are defined as above. Let $\mathcal{A}_p$ be a set of propositional constants and let $\mathcal{I}$ be an interpretation function on $\mathcal{A}_p$, i. e. $\mathcal{I} : \mathcal{A}_p \longrightarrow \wp(W)$. Then, a *model* is a structure $(F, \mathcal{I})$.

Definition 2.3 (Satisfiability) Let be given non-empty sets V, L, and T and relations $\prec_V$, $\prec_L$, and $\prec_T$ on V, L, and T, respectively. Let $F := (W, R)$ be a frame where W and R are defined as above. Given a model $\mathcal{M} = (F, \mathcal{I})$ and a point of reference ('actual world') $(v, l, t) \in (V \times L \times T)$, a formula φ *is satisfied in a model $\mathcal{M}$ at position* (v, l, t), denoted $\mathcal{M}, (v, l, t) \models \varphi$, is defined recursively in table 3.

Definition 2.4 (Validity) Let $F := (W, R)$ be a frame where W and R are defined

$\mathcal{M},(v,l,t) \models \bot$		for no $\mathcal{M}$ and no (v,l,t)
$\mathcal{M},(v,l,t) \models \alpha$	$\Longleftrightarrow$	$(v,l,t) \in V(\alpha)$
$\mathcal{M},(v,l,t) \models \neg\varphi$	$\Longleftrightarrow$	$\sim [\mathcal{M},(v,l,t) \models \varphi]$
$\mathcal{M},(v,l,t) \models \varphi_1 \rightarrow \varphi_2$	$\Longleftrightarrow$	$[\mathcal{M},(v,l,t) \models \varphi_1] \Rightarrow [\mathcal{M},(v,l,t) \models \varphi_2]$
$\mathcal{M},(v,l,t) \models \bigcirc^T\varphi$	$\Longleftrightarrow$	$\exists t' \in T[t \prec_T t'$ &
		$\sim \exists t'' \in T[t \prec_T t'' \prec_T t'$ & $\mathcal{M},(v,l,t') \models \varphi]]$
$\mathcal{M},(v,l,t) \models \bigcirc^L\varphi$	$\Longleftrightarrow$	$\exists l' \in L[l \prec_L l'$ &
		$\sim \exists l'' \in L[l \prec_L l'' \prec_L l'$ & $\mathcal{M},(v,l',t) \models \varphi]$
$\mathcal{M},(v,l,t) \models \Box^L\varphi$	$\Longleftrightarrow$	$\forall t' \in T[t \prec_T t' \Rightarrow \mathcal{M},(v,l,t') \models \varphi]$
$\mathcal{M},(v,l,t) \models \Box^V\varphi$	$\Longleftrightarrow$	$\forall v' \in V[v \prec_V v' \Rightarrow \mathcal{M},(v,l,t') \models \varphi]$
$\mathcal{M},(v,l,t) \models \varphi_1\, \mathcal{U}^T \varphi_2$	$\Longleftrightarrow$	$\exists t_1 \in T[t \prec_T t_1$ & $\mathcal{M},(v,l,t_1) \models \varphi_2$ &
		$\forall t' \in T[t \prec_T t' \prec_T t_1 \Rightarrow \mathcal{M},(v,l,t') \models \varphi]]$

Table 3: Formal Semantics of S_3ML Formulae

as above. Given a model $\mathcal{M} = (F, \mathcal{I})$ a formula φ is *valid in a model*, denoted $\mathcal{M} \models \varphi$, if $\forall w \in W[\mathcal{M}, w \models \varphi]$.

Additional S_3ML operators can be defined as in table 4.

$\top$	$=$	$\neg\bot$	$\Diamond^V\varphi$	$=$	$\neg\Box^V\neg\varphi$
$\varphi_1 \vee \varphi_2$	$=$	$\neg\varphi_1 \rightarrow \varphi_2$	$\Diamond^L\varphi$	$=$	$\neg\Box^L\neg\varphi$
$\varphi_1 \wedge \varphi_2$	$=$	$\neg(\varphi_1 \rightarrow \neg\varphi_2)$	$\Diamond^T\varphi$	$=$	$\top\, \mathcal{U}\varphi$
			$\Box^T\varphi$	$=$	$\varphi\, \mathcal{U}^T \top$

Table 4: Additional S_3ML Operators

Remark: The bindings of the binary operators are assumed to have priorities in the following order: (highest) $\mathcal{U}, \wedge, \vee, \rightarrow$ (lowest). The priorities of the monadic operators are higher than those of the binary ones. To change this order parentheses can be used.

2.2 Minimality

Minimal model interpretation of the logical formulae of S_3ML will be used in the specification examples in the sequel. Therefore, we define minimal validity of an S_3ML formula based on the notion of a 'submodel' of a given model:

Definition 2.5 (Submodel) Let $F_1 := (W_1, R_1)$ and $F_2 := (W_2, R_2)$ be two frames and let $\mathcal{A}_p$ be an alphabet of propositional constants and $\alpha \in \mathcal{A}_p$. Given two models $\mathcal{M}_1 = (F_1, \mathcal{I}_1)$ and $\mathcal{M}_2 = (F_2, \mathcal{I}_2)$ then $\mathcal{M}_1$ *is a submodel of* $\mathcal{M}_2$, denoted $\mathcal{M}_1 \sqsubseteq \mathcal{M}_2$, if

1. $W_1 \subset W_2$
2. $R_1 = R_2$ (restricted to W_1)
3. $\mathcal{I}_1(\alpha) = \mathcal{I}_2(\alpha) \cap W_1$ (for all $\alpha \in \mathcal{A}_p$)

Definition 2.6 (Minimal Validity) Let $F := (W, R)$ be a frame where W and R are defined as above. Given a model $\mathcal{M} = (F, \mathcal{I})$ a formula φ is *minimally valid*, denoted $\mathcal{M} \models_{min} \varphi$, if

1. $\mathcal{M} \models \varphi$
2. $\sim \exists \mathcal{M}' \sqsubset \mathcal{M}[\mathcal{M}' \models \varphi]$

3 The Connective $\mathcal{C}$ (Combine)

As argued previously in this paper, in fault tolerant systems behaviour can be divided into normal and exceptional behaviour. Classical predicate logic and Hoare Logic does not allow or force one to make such distinctions explicit in the formal specifications. Therefore, we augment our logic S_3ML as defined above with a new logical connective $\mathcal{C}$: in table 5, let φ_1, φ_2, χ_1, and χ_2 denote logical formulae; the superscript of the operator again indicates that it is a connective belonging to the temporal dimension of our model.

$\varphi_1 \, \mathcal{C}^T_{\chi_1;\chi_2} \varphi_2$

Table 5: Connective $\mathcal{C}$

Informally connective $\mathcal{C}$ has the following meaning:

$\varphi_1 \, \mathcal{C}^T_{\chi_1;\chi_2} \varphi_2$ is satisfied at the current position ('actual world') if we can find a subdivision of the whole set of time points into two disjoint subsets such that: the first subset is determined by the current time point and a future time point (exclusive) where condition χ_1 becomes satisfied; the second subset is determined by the future time point given as upper for the first subset and a future time point where condition χ_2 becomes satisfied; φ_1 is satisfied on the first subset and φ_2 on the second subset. This is illustrated in figure 1

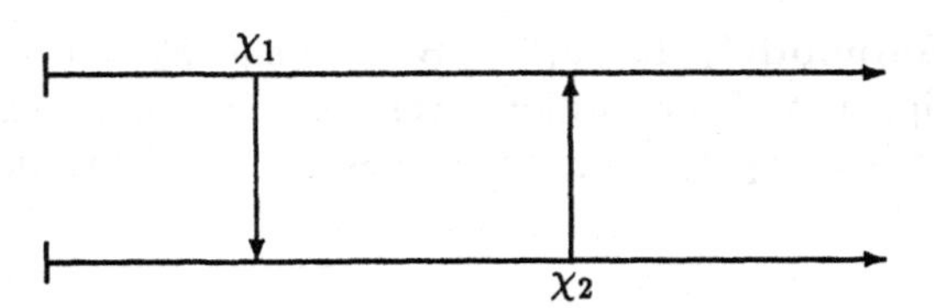

Figure 1: System Behaviour Modelled by $\mathcal{C}$

When using minimal model interpretation the properties given by the two predicates φ_1 and φ_2 are completely separated from each other, i. e. system behaviour first satisfies predicate φ_1 upto the point where χ_1 holds; from this point on upto the point where χ_2 holds the system behaviour satisfies predicate φ_2.

For the definition of the semantics of $\mathcal{C}$ in table 6 let us assume that $\mathcal{A}_p$ is some alphabet of propositional constants, that φ_1, φ_2, χ_1, and χ_2 again denote logical formulae. Furthermore, let $\mathcal{M}$ denote a model as defined above and let $(v, l, t) \in V \times L \times T$.

$$\begin{array}{lcl} \mathcal{M},(v,l,t) \models \varphi_1\, \mathcal{C}^T_{\chi_1;\chi_2}\varphi_2 & \iff & \exists t_1 \in T[t \prec_T t_1 \;\&\; \mathcal{M},(v,l,t_1) \models \chi_1 \;\&\; \\ & & \forall t' \in T[t \prec_T t' \prec_T t_1 \Rightarrow \mathcal{M},(v,l,t') \models \varphi_1 \;\&\; \\ & & \exists t_2 \in T[t_1 \prec_T t_2 \;\&\; \mathcal{M},(v,l,t_2) \models \chi_2 \;\&\; \\ & & \forall t'' \in T[t_1 \prec_T t'' \prec_T t_3 \Rightarrow \mathcal{M},(v,l,t'') \models \varphi_2]]]] \end{array}$$

Table 6: Formal Semantics of $\mathcal{C}$

The $\mathcal{C}$ connective can now be used to provide definitions of previously mentioned modal operators; as an example, we give for connective $\mathcal{U}$ a new definition in table 7.

$$\varphi\, \mathcal{U}^T \chi \quad = \quad \varphi\, \mathcal{C}^T_{\chi;\top} \top$$

Table 7: Until as Abbreviated Combine

Let us now briefly discuss some simple specification examples to clarify the intended meaning of the $\mathcal{C}$-connective. The specifications presented in the sequel define system properties only on a very high level of abstraction.

3.1 A Fault Tolerant System

In general, a fault tolerant system is a system whose behaviour can be divided into two parts: normal behaviour, which is required by the user if no fault has occurred, and exceptional behaviour, which is required to overcome fault situations. To specify such a system let nb describe normal behaviour of a system and let eb describe exceptional behaviour of that system. Conditions leading to exceptional behaviour, e. g. a particular exception raised in case of a hardware fault, are described by exc, conditions which represent the fact that a system has successfully been recovered from exceptional behaviour will be described by rec. Then, a specification of a fault tolerant system would, in general, look like:

$$nb\, C^T_{exc;rec} eb$$

Table 8: Fault Folerant System

The formula in table 8 means that beginning with normal behaviour nb an exception described by exc will lead to exceptional behaviour eb. Until the recovery of the system has been reached, which will be indicated by rec, the system stays in eb and then turns back to nb or whatever.

3.2 A Fault Free System

Suppose we want to specify a system of which no faults are known or at least assumed not to occur—this is usually done in formal specification and verification of systems outside fault tolerance—and where, therefore, no exceptional behaviour has to be provided. Let the identifiers appearing in the following table denote the same as above. Then, a specification of a fault free system without any chance to enter exceptional behaviour would look like:

$$nb\, C^T_{\bot;\top} \top$$

Table 9: Fault Free System

The formula in table 9 means that only normal behaviour, which, in this case, will be the same as correct behaviour will occur during system execution and that no exception is allowed to occur because of $\bot$.

3.3 A Fail Stop System

A fail stop system is a system which stops execution after the occurrence of a fault. Thus, no behaviour at all is allowed to be observable after that event. This can be modelled by the $\perp$ as exceptional behaviour. Let the identifiers appearing in table 10 denote the same as in the above examples. Then, a specification of a fail stop system would look like:

$$nb\, C^{T}_{exc;\perp} \perp$$

Table 10: Fail Stop System

Opposed to the formula in table 9 we now admit occurrences of exceptions, denoted by exc. This leads beginning with normal behaviour nb to no behaviour at all (second $\perp$). A recovery condition is not given ($\perp$ in the superscript) which would lead back to normal behaviour of the system.

3.4 A System with Transient Faults

A transient fault is a fault which may occur at some time during system execution and which is present not longer than some bounded period of time; this period is called in [15] threshold. So, we can be sure that after this threshold interval a transient fault disappears. This can be specified as follows:

$$nb\, C^{T}_{exc;rec}(eb \land \Diamond^{T} rec)$$

Table 11: System with Transient Faults

3.5 A System with Permanent Faults

As opposed to a transient fault a permanent fault is a fault which remains present after the threshold interval (cf. [15]) has expired. Therefore, a specification of a system with permanent faults will be similar to the transient case but must take care that there is no point during the threshold interval that recovery from the exceptional behaviour is possible:

4 A Specification Example: Stable Storage

A disk used for storage and retrieval of data must be as reliable as possible. Unfortunately existing physical media used to store data are not always ideal worlds. Faults may occur which, for example, corrupt data stored on the disk or which store data

$$nb\, C^T_{exc;rec}(eb \wedge \neg\Diamond^T rec)$$

Table 12: System with Permanent Faults

on another location as it was intended. To overcome such fault situations a so called stable storage is used (cf. [3] and [15]). A stable storage in general can be characterized informally by the following requirement:

REQ "If data is stored on the stable storage this same data can be retrieved from the stable storage as long as it is necessary despite the occurrence of faults."

The usual way to design a system which fulfils the requirement **REQ** is by using two or more physical disks where the data is written always to all the available disks if there is a store request (*mirrored disk concept*). If data is not retrievable from one disk it can still be retrieved from another one. The architecture of such a system comprises, in general, three layers: the *physical disk layer*, which is made up from the physical disks possibly suffering from faults, the *logical disk layer*, which ensures depending on some CRC mechanism [15] a certain correctness of the data retrieved from a physical disk, and the *stable storage layer*, which ensures by the mirrored disk concept a certain correctness of the data provided to the environment of the stable storage. This is illustrated in figure 2.

One way of defining the properties of a stable storage is by giving the relation of read and write operations initiated on the several layers of the stable storage. We will undertake this in the subsequent sections where the underlying model of the corresponding layer, according to the previously mentioned three-dimensional position space, will be discussed and informal and formal specification will be provided.

4.1 Top Level

The way to retrieve data from a stable storage on the top level is by using a read operation initiated by the environment of the disk. Data is stored on a stable storage by using a write operation initiated by the environment. The relation between these two operations is given informally by the following list:

- a write must precede all subsequent operations on a disk
- a read must deliver the same value as the last write had stored
- two successive reads deliver the same value

On this level of abstraction, i. e. the top level, we do not need to consider any addressing mechanism. What is important here is that a read request can be distinguished

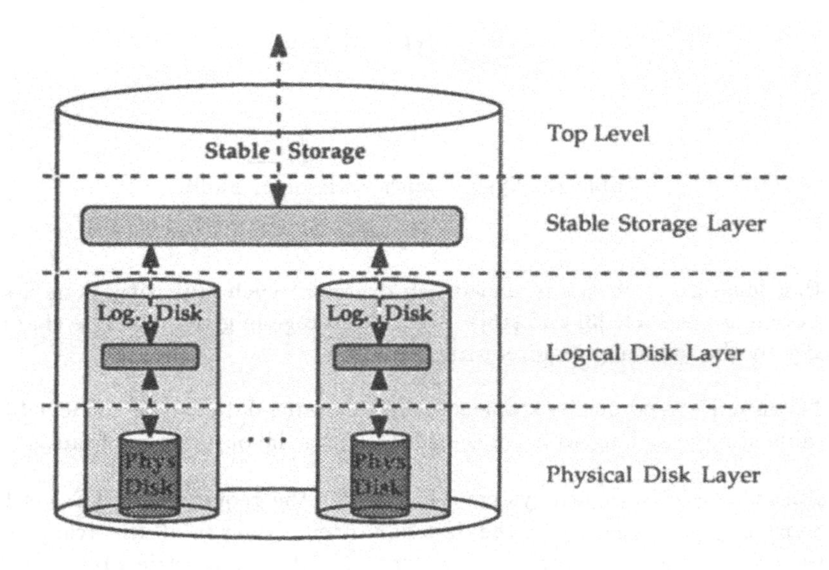

Figure 2: General Structure of a Stable Storage

from a write request and if a write request is given then we need to refer to a particular value.

As concrete model of this layer according to the three dimensions of our abstract model previously defined we get:

- Value Model: some set Val of distinct values
- Location Model: a finte graph where the set of location labels L and the relation $\prec_L$ are given as follows:
  - $N = \{input, output, disk\}$
  - $E = \{(input, disk), (disk, output)\}$
- Time Model: linear discrete model such as the natural numbers

The formal specification of the top level is now given in table 14 where the following propositional constants will be used:

The first three equivalences of table 14 state the properties of read, write, and memorize operations in this order: if a read operation has been initiated ($Rinit$) and if at the same time a particular value ($\in Val$) is available from the disk (indicated by $\bigcirc^L Vgiven$) then this same value will eventually ($\diamond^T(\ldots)$) be output (indicated by $\bigcirc^L \bigcirc^L Vgiven$); eoR indicates that the read operation has been finished.

Similarly, if a write operation has been initiated ($Winit$) and if at the same time a particular value (indicated by $Vgiven$) is input then this value will eventually ($\diamond^T(\ldots)$)

PROPOSITION

Vgiven : "some element from a value domain is given"
Rinit : "a read operation is initiated"
Winit : "a write operation is initiated"
eoR : "a read operation has been finished"
eoW : "a write operation has been finished"

Table 13: General Definitions for Top Level

$$Read \leftrightarrow (Rinit \wedge \bigcirc^L Vgiven \rightarrow \Diamond^T(\bigcirc^L \bigcirc^L Vgiven \wedge eoR)),$$
$$Write \leftrightarrow (Winit \wedge Vgiven \rightarrow \Diamond^T(\bigcirc^L Vgiven \wedge eoW)),$$
$$Memo \leftrightarrow (\bigcirc^L(Vgiven \rightarrow \bigcirc^T Vgiven)),$$
$$\Box^V((\Diamond^T Write)\ \mathcal{C}^T_{eoW;Winit}(Memo\ \mathcal{C}^T_{Rinit;eoR}(Read \wedge Memo)))$$

Table 14: Top Level Specification

be stored on the disk (indicated by $\bigcirc^L \bigcirc^L Vgiven$); *eoW* indicates that the write operation has been finished.

Because of minimal interpretation of the logical formulae a third operation must be defined to save the data on disk, i. e. if a particular value is available on disk (indicated by the antecedent $\bigcirc^L Vgiven$) then it will still be available at the next following time point (indicated by the succedent $\bigcirc^L \bigcirc^T Vgiven$).

The last formula in table 14 now describes the relationships between read, write, and memorize operations on the top level of a stable storage: first of all there will eventually be some write operation; after having finished this write (*eoW* in the subscript of $\mathcal{C}$) memorize and read operations will be possible depending on initiating a read *Rinit* or not.

4.2 Physical Disk

For the specification of a physical disk the environment of the disk also has to be modelled. As in [3] we assume additionally to the input/output interaction with the environment some adverse environment which can damage the physical disk resulting in a transient fault, modelled by f_t, or in a permanent fault, modelled by f_p. Hence, input to the physical disk will either be generated by a write operation from the logical disk or from the adverse environment.

If a fault occurs, the value on disk is changed to a fault value different from the values in the value domain of the top level. The physical read operation is able to read the correct values as well as the fault value. The physical write operation can never write the fault value, since this value can only be caused by the adverse environment. Storage of a value now becomes vulnerable to occurrences of faults. This leads to the following concrete model of the physical disk layer:

- Value Model: the same set Val of distinct values as on the top level and an additional value ft present in case of a fault $(Val \cup ft)$
- Location Model: a finte graph where the set of location labels L and the relation $\prec_L$ are given as follows:
  - $N = \{pdinput, pdoutput, pdisk, adenv\}$
  - $E = \{(pdinput, pdisk), (pdisk, pdoutput), (adenv, pdisk)\}$
- Time Model: linear discrete model such as the natural numbers

The formal specification is provided in table 16 where the following propositional constants will be used:

PROPOSITION
$pdVgiven$: "some element from a value domain is given"
$pdRinit$: "a physical disk read operation is initiated"
$pdWinit$: "a physical disk write operation is initiated"
$pdeoR$: "a physical disk read operation has been finished"
$pdeoW$: "a physical disk write operation has been finished"
$fault$: "a fault has occurred"

Table 15: General Definitions for Physical Disk

$$pdRead \leftrightarrow (pdRinit \wedge \bigcirc^L(pdVgiven \vee fault) \rightarrow \Diamond^T(\bigcirc^L \bigcirc^L (pdVgiven \vee fault) \wedge pdeoR)),$$
$$pdWrite \leftrightarrow (pdWinit \wedge pdVgiven \rightarrow \Diamond^T(\bigcirc^L pdVgiven \wedge pdeoW)),$$
$$pdMemo \leftrightarrow (\bigcirc^L((pdVgiven \rightarrow \bigcirc^T pdVgiven) \wedge (fault \rightarrow \bigcirc^T fault)),$$
$$Fault \leftrightarrow (\Diamond^T(\Diamond^L(fault \rightarrow \bigcirc^T pdVgiven)),$$
$$\Box^V((\Diamond^T pdWrite)\, C^T_{pdeoW;pdWinit}(NB \vee Fault\, C^T_{pdRinit;pdeoR}(pdRead \wedge pdMemo)))$$

Table 16: Physical Disk Specification

A fault can be caused by an adverse environment during normal behaviour if we would have no minimality assumption. Here, minimality on the one hand ensures that normal behaviour, i. e. read, write, and memorize operations are considered atomic and on the other hand that normal behaviour NB and exceptional behaviour EB cannot occur at the same time. Furthermore, no other faults than those mentioned above may occur and read and write operations may not interfere with each other.

4.3 Logical Disk

At this level, the adverse environment no longer is observable. If we want to use the physical disk described in the previous section, we have to make sure that we can recognize whether a value is correct or not. If a value is faulty, we have to signal an error otherwise we have to provide the correct value. Whether or not a certain value is correct can be detected using, for example, a CRC mechanism [15]; in the following specification we will abstract from such a mechanism.

As concrete model of this layer according to the three dimensions of our abstract model previously defined we get:

- Value Model: some set Val of distinct values and additionally some value ft different from all values in Val
- Location Model: a finte graph where the set of location labels L and the relation $\prec_L$ are given as follows:
  - $N = \{ldinput, ldoutput, ldisk\}$
  - $E = \{(ldinput, ldisk), (ldisk, ldoutput)\}$
- Time Model: linear discrete model such as the natural numbers

The formal specification of the top level is now given in table 18 where the following propositional constants will be used:

The task performed by one logical disk is to initiate physical disk read and write operations depending on what was initiated on the logical disk and to provide the results to next level. Therefore, is similar to the stable storage layer (cf. section 4.4) with the difference that on the logical disk we have to take care only of one physical disk where the stable storage layer is concerned with n logical disks.

4.4 Stable Storage Layer

The requirement for the stable storage is given by the top-level requirement. In the sequel we provide a specification which describes how to bridge the gap between the top level specification and the logical disk specification, by using the mirrored disk concept. The logical disk layer consists of n logical disks. Information is stored on and retrieved from these n disks, and therefore the chance that data is damaged is

PROPOSITION

$ldVgiven$: "some element from a value domain is given"
$pdVgiven$: "some element from a value domain is given"
$ldRinit$: "a logical disk read operation is initiated"
$pdRinit$: "a physical disk read operation is initiated"
$ldWinit$: "a logical disk write operation is initiated"
$pdWinit$: "a physical disk write operation is initiated"
$ldeoR$: "a logical disk read operation has been finished"
$pdeoR$: "a physical disk read operation has been finished"
$ldeoW$: "a logical disk write operation has been finished"
$pdeoW$: "a physical disk write operation has been finished"

Table 17: General Definitions for Logical Disk

$ldRead \leftrightarrow (ldRinit \rightarrow \Diamond^T \Diamond^L pdRinit),$
$ldOutput \leftrightarrow (\bigcirc^L(pdeoR \wedge pdVgiven) \rightarrow \Diamond^T(NEXT^L ldVgiven),$
$ldWrite \leftrightarrow (ldWinit \wedge ldVgiven \rightarrow \Diamond^T(\bigcirc^L pdVgiven \wedge pdWinit)),$
$ldackR \leftrightarrow (\bigcirc^L \bigcirc^L pdeoR \rightarrow \Diamond^T ldeoR),$
$ldackW \leftrightarrow (\Box^L pdeoW \rightarrow \Diamond^T ldeoW)$

Table 18: Logical Disk Specification

significantly reduced, depending on the number of disks functioning correctly. The adverse environment has on the stable storage layer.

As concrete model of this layer according to the three dimensions of our abstract model previously defined we get:

- Value Model: some set Val of distinct values
- Location Model: a finte graph where the set of location labels L and the relation $\prec_L$ are given as follows:
  - $N = \{input, output, ldisk_1, \ldots, ldisk_n\}$
  - $E = \{(input, disk_1), \ldots, (input, disk_n), (disk_1, output), \ldots, (disk_n, output)\}$
- Time Model: linear discrete model such as the natural numbers

The formal specification of the top level is now given in table 14 where the following propositional constants will be used:

PROPOSITION

$Vgiven$: "some element from a value domain is given"
$Rinit$: "a stable storage read operation is initiated"
$Winit$: "a stable storage write operation is initiated"
eoR : "a stable storage read operation has been finished"
$ldeoR$: "a stable storage read operation has been finished"
eoW : "a stable storage write operation has been finished"
$ldeoW$: "a stable storage write operation has been finished"

Table 19: General Definitions for Stable Storage Layer

$ssRead \leftrightarrow (Rinit \rightarrow \Diamond^T \Diamond^L ldRinit)$,
$ssOutput \leftrightarrow (\Diamond^L (ldeoR \wedge Vgiven) \rightarrow \Diamond^T (NEXT^L Vgiven)$,
$ssWrite \leftrightarrow (Winit \wedge Vgiven \rightarrow \Diamond^T (\Box^L ldVgiven \wedge ldWinit))$,
$ssackR \leftrightarrow (\Diamond^L ldeoR \rightarrow \Diamond^T eoR)$,
$ssackW \leftrightarrow (\Box^L ldeoW \rightarrow \Diamond^T eoW)$

Table 20: Stable Storage Layer Specification

A stable storage write operation initiates a logical disk write operation on all logical disks. A stable storage read operation initiates a logical disk read operation which delivers a value eventually from one logical disk. If there are acknowledgements from the logical disk an acknowledgement to the top level is given. If there is output from the logical disk layer, output is provided to the top level, and if there are only erroneous outputs, all disks are corrupted and the system can no longer behave as required

4.5 Conclusion of the example

In this section, specifications have been provided for the top level of a stable storage, a physical disk, a logical disk, and the stable storage layer. Compared to other specifications, e.g. [3] and [15], our specifications describe the requirements for the different levels of a stable storage on a very high level of abstraction. Implementation details, for example the CRC mechanism and the addressing mechanism as given in [15] and assumptions about the implementation language as indicated in [3] are omitted.

Because of minimality, our definition of the adverse environment explicitly states that no other causes for faulty behaviour may happen. If other faults do occur, these fall outside the applicability of our specifications. Under a maximal interpretation, this would not always be the case. Minimality also ensures that the interaction between reads and writes is as we would expect it to be: simultaneous read and write cannot occur.

For every specification we have given a separate concrete model for the corresponding level; this is indeed some lack of our specifications given in this paper. But, this can be overcome in the future if we define a suitable graph model for the structure of the overall system with all level incorporated in this graph description.

5 Summary and Conclusion

In this paper, we addressed the problem of formal specification of fault tolerant real time systems. We have defined a three-sorted modal logic which is given a minimal interpretation based on a Kripke model of possible worlds. Minimality has been used to overcome the problem of non-monotonic reasoning which is valid in classical logics but which makes problems in a certain sense in the context of fault tolerant systems.

When we are still interested in monotonic reasoning in the context of fault tolerance other means to reason about such systems are needed. Therefore, we introduced a connective $\mathcal{C}$ to combine specifications. This allows for example for sepparation of normal and exceptional behaviour in a fault tolerant system. Viewed in the context of liveness and safety properties the connective $\mathcal{C}$ allows for specification of some safety properties.

From the specification in the above example of a stable storage the lack of non unified models, e. g. for the location models, becomes obvious. There is work neccessary in the future to get rid of these deficiency and to incorporate all submodel into one model for the whole stable storage according to the three-dimensional model space.

A further point of investigations in the future will be the minimality: several notions of minimality are known from the literature (cf. [2]) so that the question must be considered: is there another kind of minimality will be more adequate in our application area? A third item in future work lies in the specification example of the stable storage: more details, e. g. CRC mechanism, should be considered as it has been done in other papers (cf. [3] and [15]).

Acknowledgements

We are grateful to Wim Koole for many discussions about the logic used here and his competent comments on earlier versions of the paper. We thank Hanno Wupper for careful reading of draft versions of this paper.

The work has been funded partially by SION, the Dutch Research Organisation, under grant number 612-317-016, by ESPRIT BRA 3096 (SPEC), and by STW (NWI 88.1517).

References

[1] J. van Benthem, *Modal and Classical Logic*, Bibliopolis, Naples, 1985

[2] J. van Benthem, *Semantic Parallels in Natural Language and Computation*, in: Logic Colloquium, Granada, M. Garrido (ed.), 1988

[3] F. Cristian, *A Rigorous Approach to Fault-tolerant Programming*, in: IEEE Transactions on Software Engineering, Vol. SE-11, No. 1, January 1985

[4] F. Cristian, *Exception Handling*, in: "Dependability of Resilient Computers", T. Anderson (ed.), Blackwell Scientific Publications, 1989

[5] D. Gabbay, *Intuitionistic Basis for Non-Monotonic Logic*, in: Lecture Notes in Computer Science 138, "Proceedings of the 6th Conference on Automated Deduction", D. W. Loveland (ed.), pp. 260-273, 1982

[6] C. A. R. Hoare, *An Axiomatic Basis for Computer Programming*, in: Communications of the ACM, Vol. 12, pp. 576-580, 1969

[7] J. Hooman, *Specification and Compositional Verification of Real-Time Systems*, Ph.D. Thesis, Eindhoven University of Technology, 1991

[8] R. Koymans, *Specifying Message Passing and Time-Critical Systems with Temporal Logic*, Ph.D. Thesis, Eindhoven University of Technology, 1989

[9] Lamport, *Specifying Concurrent Program Modules*, ACM

[10] K. G. Larsen, *Proof Systems for Hennessy-Milner Logic with Recursion*, Aalborg University Center, Institute for Electronic Systems, Department of Mathematics and Computer Science, Denmark, April 1987

[11] Z. Manna, A. Pnueli, *The Anchored Version of the Temporal Framework*, in: Lecture Notes in Computer Science 354, "Linear Time, Branching Time and Partial Order in Logics and Models for Concurrency", de Bakker, de Roever, Rozenberg (eds.), Springer, 1989

[12] A. Pnueli, E. Harel, *Applications of Temporal Logic to the Specification of Real Time Systems*, in: Lecture Notes in Computer Science 331, "Proceedings of a Symposium on Formal Techniques in Real-Time and Fault-Tolerant Systems", M. Joseph (ed.), Springer, 1989

[13] B. Randell, P. A. Lee, P. C. Treleaven, *Reliability Issues in Computing System Design*, in: ACM Computing Surveys, Vol. 10, No. 2, June 1978

[14] N. Rescher, A. Urquhart, *Temporal Logic*, Springer, 1971

[15] H. Schepers, *Terminology and Paradigms for Fault-tolerance*, Department of Mathematics and Computing Science, Eindhoven University of Technology, Computing Science Notes 91/08, 1991

[16] M. J. Wieczorek, J. Vytopil, *Specification and Verification of Distributed Real-Time Systems*, in: "Proceedings of the Second International Conference on Reliability and Robustness of Engineering Software II", Brebbia/Ferrante (eds.), Wessex Institute of Technology, pp. 99-113, 1991

[17] G. Winskel, *An introduction to event structures*, in: Lecture Notes in Computer Science 354, "Linear Time, Branching Time and Partial Order in Logics and Models for Concurrency", de Bakker, de Roever, Rozenberg (eds.), Springer, 1989

Timed and Hybrid Statecharts and their Textual Representation *

Y. Kesten A. Pnueli
Weizmann Institute of Science †

Abstract.

A structured operational semantics is presented for *Timed* and *Hybrid Statecharts*, which are generalizations of the visual specification language of Statecharts intended to model real-time and hybrid systems. In order to study some of the basic features of Statecharts and the extensions necessary to treat real-time and continuous behaviors without being distracted by the graphical representation, we introduce a concurrent real-time language that can be viewed as a textual representation of Statecharts.

The language contains statements for delays, preemption, and timeouts. A structured operational semantics of the language and an illustrative example of its use for specification are presented. Extensions to the specification of hybrid systems are obtained by allowing a differential equation as a statement of the extended language. Structured operational semantics is also given for the hybrid version.

The same extensions are then applied to the visual Statechart language, and similar compositional semantics are defined.

Keywords: Real-time, timed transitions system, hybrid systems, Statecharts, structured operational semantics.

Contents

*This research was supported in part by the France-Israel project for cooperation in Computer Science, and by the European Community ESPRIT Basic Research Action Project 3096 (SPEC).

†Department of Applied Mathematics, Weizmann Institute, Rehovot, Israel

1 Introduction

The visual formalism of *Statecharts* has been proposed by Harel in [Har87]. Since then it has been used for the specification of complex reactive systems ([HLN+90], [HP85]), and extensively studied from semantic ([HPSS86], [HGdR88], [PS89], [HK89], [Mar91], [Hui91]), and expressive complexity ([DH89], [Har89], [HH90]) points of view.

One of the most attractive and innovative features of the *Statecharts* formalism, is its visual syntax. However, there are some other elements of *Statecharts* as a specification (or a programming) language for reactive systems, which are independent of the graphical syntax and should be studied on their own.

To promote the independent study of the syntax-independent features of *Statecharts* and to make them accessible to people who are not sworn believers in visual syntax and feel more comfortable with text representations of programming and specifications languages, we propose in this note a textual representation of *Statecharts*, to which we refer in this paper as *Statext*.

Statecharts use a *discrete events* approach to model a reactive system. This means that the behavior of a reactive system is described as a sequence of discrete events that cause abrupt changes (taking no time) in the state of the system, separated by intervals in which the system's state remains unchanged. This approach has proven effective for describing the behavior of programs and other digital systems.

The discrete event approach is justified by an assumption that the environment, similar to the system itself, can be faithfully modeled as a digital (discrete) process. This assumption is very beneficial, since it allows a completely symmetrical treatment of the system and its environment, and encourages modular analysis of systems, where what is considered an environment in one stage of the analysis may be considered a component of the system in the next stage.

While this assumption is justified for many systems such as communication networks, where all members of the net are computers, there are certainly many important contexts in which modeling the environment as a discrete process greatly distorts reality, and may lead to unreliable conclusions. For example, a control program driving a robot within a maze or controlling a fast train must take into account that the environment with which it interacts follows continuous rules of change.

This paper suggests an extension to *Statecharts* that will enable it to deal with continuous processes. This extension leads to an integrated approach to *hybrid* systems, i.e., systems consisting of a non-trivial mixture of discrete and continuous components, such as a digital controller that controls a continuous environment, control of process and manufacturing plants, guidance of transport systems, robot planning, and many similar applications.

As a first step towards a specification language that allows a more realistic modeling of programs that interact with a continuous environment, we augment *Statecharts* to deal with *real time*. A *Statecharts* transition is annotated by a *time interval* $\{l, u\}$, denoting the lower and upper time bounds of that transition. The resulting language, to which we refer in this paper as *Timed Statecharts*, is given an *asynchronous* semantics, as opposed to the traditional *synchronous* semantics given to *Statecharts*.

We then extend Statecharts by allowing basic (unstructured) states to be labeled by a set of differential equations describing a continuous change that occurs as long as the system is in that state. *Statext* is extended similarly, by the addition of a *continuous assignment* statement.

Hybrid systems are discussed extensively in [MMP92]. In the above paper, a global semantics is presented for hybrid transition systems, which are referred to as *phased transition systems*.

In this paper, we develop a compositional semantics for both *Timed* and *Hybrid Statecharts* and their respective textual representations *Statext* and *Hybrid Statext*.

2 The *Statext* Language

As any other textual language, *Statext* is constructed out of *statements*, which manipulate *data* and send and receive messages. Below, we specify first the data and expressions of the *Statext* language, then we list its statements.

Data, Expressions and Interval Specifications

- *Variables* and *Channels*
  Similar to *Statecharts*, the *Statext* language is based on *shared variables* and *communication channels*. Both variables and channels are typed.

  It is required that the domain of each channel contains the special value $\perp$, implying no current message (an empty channel).

- *Expressions*
  Expressions are formed by applying operations to constants, variables and channels.

- *Interval specifications*
  An *interval specification* I, is used to specify real time constraints. It is one of

  $$[l, u],\ [l, u),\ (l, u],\ (l, u)$$

  where l and u are real numbers, $0 \leq l \leq u \leq \infty$, called *lower* and *upper* bounds respectively. We write $\{l, u\}$ to specify any of the above four cases. We say that t belongs to the interval I, denoted by $t \in I$, according to the following cases:

  - $t \in [l, u]$ iff $l \leq t \leq u$
  - $t \in (l, u]$ iff $l < t \leq u$
  - $t \in [l, u)$ iff $l \leq t < u$

- $t \in (l,u)$ iff $l < t < u$

We say that t *does not exceed* I, denoted by $t \preceq I$, if there exists a $t' \in I$ such that $t \leq t'$.

Simple Statements

We begin by listing the *simple* statements, that is, statements that do not have substatements.

- **skip**

This statement serves as a filler. It does nothing and terminates in a single execution step.

- **idle**

Like the **skip** statement, the **idle** statement does not change the data state. However, unlike the **skip** statement, the **idle** statement never terminates. The only way to get out of an **idle** statement is by *preemption* which is a major feature of *Statext* (and of *Statecharts*).

- **ascertain**

An *ascertain* statement has the general form

ascertain c **for** I,

where c is a boolean expression (*condition*) and I is an *interval specification.*

Like **skip** and **idle** the *ascertain* statement does not change the data state. It waits until the condition c is *continuously true* for time $t \in I$, at which point it terminates. We write

ascertain c **for** d	for	**ascertain** c **for** $[d,d]$,
await c	for	**ascertain** c **for** 0,
delay I	for	**ascertain** *true* **for** I

- *Assignment*

For $\overline{y}$ a list of variables and $\overline{e}$ a list of expressions of the same length and corresponding types,

$\overline{y} := \overline{e}$

is an *assignment* statement. Its execution evaluates the list of expressions $\overline{e} = e_1, \ldots, e_m$ and *then* assigns them to the variables $\overline{y} = y_1, \ldots, y_m$, respectively. An assignment is enabled whenever it is *ready*, i.e., control is in front of the assignment, and it terminates in a single step.

- *Send*

For a channel α and expression e,

$\alpha!e$

is a *send* (also called a *signal*) statement. Its execution broadcasts the message consisting the value of e over channel α, to be received by whoever is listening at this point. This statement can be viewed as an assignment of e to the channel α.

A special type of channel is the *single message* channel, i.e., a channel that ranges over the domain $\langle\perp,1\rangle$. For a single message channel α, we may write $\alpha!$ as an abbreviation for $\alpha!1$. This represents a frequent case in which we are not interseted in the value of the message but merely in the fact that a message (or a signal) has been broadcasted on channel α.

• *Receive*

For a channel α, and a variable y of compatible type,

$\alpha?y$

is a *receive* statement. The intended meaning of this statement is that it can be taken from a state in which the statement is ready, and at time t, such that some message $v \neq \perp$ has been broadcasted on α at time t, i.e., channel α is nonempty. When executed, the statement places the value v in variable y.

In the case that channel α is single message and we do not intend to examine variable y, we may write $\alpha?$ as an abbreviation for $\alpha?y$. This statement can be viewed as an equivalent to **await** $\alpha \neq \perp$.

Reference to a single message channel α within a boolean expression (*condition*), is equivalent to the atomic formula $(\alpha \neq \perp)$.

Compound Statements

Compound statements consist of a *controlling frame* applied to one or more simpler statements, to which we refer as the *body* of the compound statement. Typically, execution of a compound statement requires several computational steps which are often non-consecutive.

• *Conditional*

For S_1 and S_2 statements and c a boolean expression,

if c **then** S_1 **else** S_2

is a *conditional* statement. Its intended meaning is that the boolean expression c is evaluated and tested. If the condition evaluates to T, statement S_1 is selected for subsequent execution; otherwise, S_2 is selected. Thus, the first step in an execution of the *conditional* statement is the evaluation of c and the selection of S_1 or S_2 for further execution. Subsequent steps continue to execute the selected substatement.

When control reaches a *conditional* statement, the first step in its execution can always be taken. This is because c always evaluates to either T or F and the first step in an execution of the statement is therefore defined for both values of the condition c.

We abbreviate **if** c **then** S_1 **else skip** to

if c **then** S_1.

• **while**

For c a boolean expression and S a statement,

while c **do** S

is a *while* statement. Its execution begins by evaluating c. If c is found to be false, execution of the statement terminates. If c is found to be true, subsequent steps proceed to execute S. If S terminates, c is tested again. Thus, the first step in the execution of a *while* statement is the evaluation of guard c and either finding it true and committing to execution of at least one more repetition of the body S at subsequent steps, or finding the guard c false and terminating execution of the *while* statement. Statement S is called the *body* of the *while* statement.

We write **loop forever do** S as an abbreviation for **while** T **do** S.

- *Concatenation*

For S_1 and S_2 statements,

$$S_1;\ S_2$$

is a *concatenation* statement. Its intended meaning is sequential composition. First S_1 is executed, and when it terminates S_2 is executed. Thus, the first step in an execution of $S_1; S_2$ is the first step in an execution of S_1. Subsequent steps continue to execute the rest of S_1, and when S_1 terminates, proceed to execute S_2.

More than two statements can be combined by concatenation to form a *multiple concatenation* statement S:

$$S_1;\ S_2;\ \ldots;\ S_n.$$

- *Selection*

For S_1 and S_2 statements,

$$S_1 \sqcup S_2$$

is a *selection* statement. Its intended meaning is that, as a first step, one of S_1 and S_2, which is currently enabled, is selected and the first step in the selected statement is executed. Subsequent steps proceed to execute the rest of the selected substatement. If both S_1 and S_2 are enabled, the selection is non-deterministic. If both S_1 and S_2 are currently disabled, then so is the *selection* statement.

Note that all steps in the execution of the *selection* statement are attributed to its body. This is because the first step in the execution of $S_1 \sqcup S_2$ corresponds to a step in the execution of either S_1 or S_2.

More than two statements can be grouped into a *multiple selection* statement

$$S_1 \sqcup S_2 \sqcup \cdots \sqcup S_n,$$

which may be abbreviated to

$$\bigsqcup_{i=1}^{n} S_i.$$

The *selection* statement is often applied to concatenations whose first statement is an *await* or *receive* statement. This combination leads to ***conditional selection***. For example, the general conditional command of the ***guarded command*** language (proposed by Dijkstra), of the form

$$\textbf{if } c_1 \rightarrow S_1 \ \Box\ c_2 \rightarrow S_2 \ \Box\ \ldots \ \Box\ c_n \rightarrow S_n \ \textbf{ fi}$$

can be represented in our language by a multiple *selection* statement formed out of *await* statements:

$$[\textbf{await } c_1 ;\ S_1] \ \sqcup\ [\textbf{await } c_2 ;\ S_2] \ \sqcup\ \cdots\ \sqcup\ [\textbf{await } c_n ;\ S_n],$$

or in shorter form:

$$\bigsqcup_{i=1}^{n} [\textbf{await } c_i ;\ S_i].$$

The first step in the execution of this multiple *selection* statement consists of arbitrarily choosing an i such that c_i is currently true, and ***passing the guard*** c_i. This implies commitment to execute the selected S_i in subsequent steps. The order in which the alternatives are listed is immaterial and does not imply a higher priority to the alternatives appearing earlier in the list.

Note that we do not require that the c_i's be exclusive, i.e., that $c_i \rightarrow (\neg c_j)$ for every $j \neq i$. Nor do we require that the c_i's be exhaustive, i.e., that $\bigvee_{i=1}^{n} c_i$ is true. Non-exclusivity allows non-determinism, while non-exhaustiveness allows the possibility of deadlock, e.g., being at a *selection* statement with all conditions false (which requires waiting until one becomes true).

The *receive* statement can be used similarly for conditional selection:

$$\bigsqcup_{i=1}^{n} [\alpha_i ? v_i;\, S_i]$$

where each $\alpha_i ? v_i$ acts as a guard.

- *Cooperation*

For S_1 and S_2 statements,

$$S_1 \parallel S_2$$

is a *cooperation* statement. It calls for parallel execution of S_1 and S_2. Parallel execution is modeled by *interleaving*. Steps in $[S_1 \parallel S_2]$ are interleaved steps from S_1 and S_2. The cooperation statement terminates when both S_1 and S_2 have terminated. Thus, in the combination $[S_1 \| S_2]; S_3$, execution of S_3 will not start until both S_1 and S_2 are terminated.

Similar to the *selection* statement, more than two statements can be grouped into a *multiple cooperation* statement

$$S_1 \parallel S_2 \parallel \ldots \parallel S_n \qquad \text{or, equivalently,} \qquad \mathop{\parallel}_{i=1}^{n} S_i.$$

- *Preemption*

For statements S_1 and S_2,

$$S_1 \;\mathcal{U}\; S_2,$$

is a *preemption* statement. Steps in the execution of this statement are either steps in the execution of S_1 or any *first* step in the execution of S_2. Any step leading to the termination of S_1 also causes the termination of $S_1 \;\mathcal{U}\; S_2$. Execution of a step of S_2 causes the rest of S_1 to be discarded, and proceeds to execute steps of S_2 until it terminates. Thus, the intended meaning of the *preemption* statement is

Execute S_1 until a first step of S_2 can be taken.

Consider, for example, the statement

$$\left[\begin{array}{l} \textbf{while } \mathrm{T} \textbf{ do} \\ \left[\begin{array}{l} \textbf{delay } I \\ y := y + 1 \end{array}\right] \end{array}\right] \;\mathcal{U}\; \textbf{await } y > 2$$

This statement terminates as soon as y grows above 2, even though the *while* statement, when standing alone, does not terminate.

- *Block*

The *block* statement is intended to allow local declarations and hiding of local variables and channels. In this version of the paper we will omit the presentation of the *block* statement.

3 An Example

Before presenting the formal semantics of the *Statext* language and *Timed Statecharts*, we consider an example.

In Fig. 1 we present a *Timed Statecharts* specification of two machines which communicate by a buffer that can hold only one item at a time. This buffer can represent a receptacle that travels back and forth between the two machines.

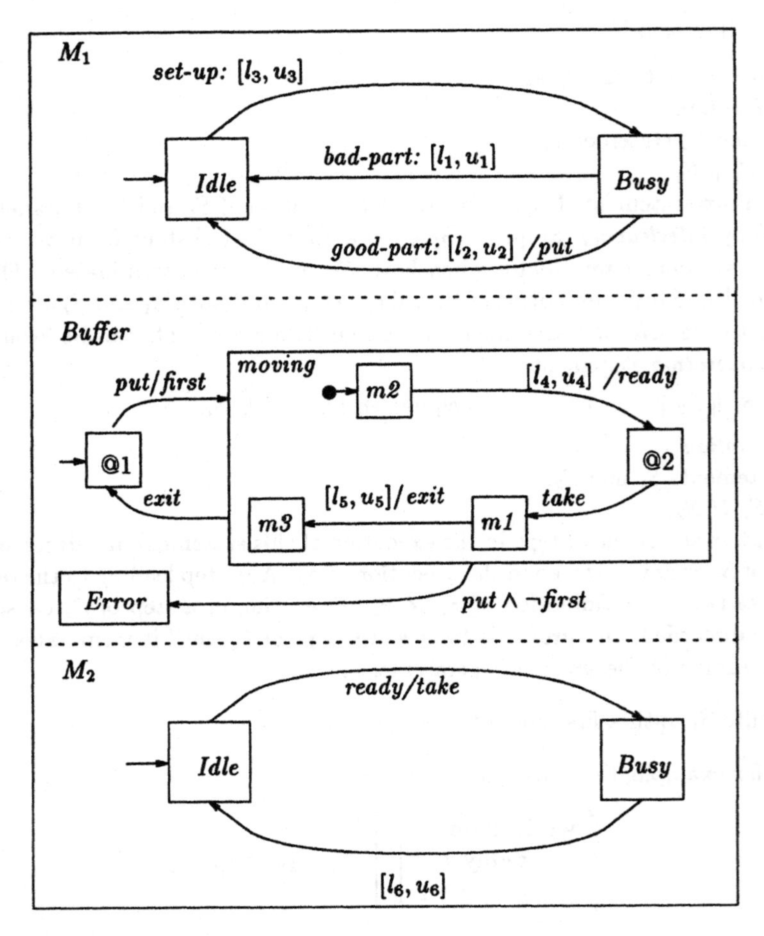

Figure 1: Specification of Two Machines with a Buffer.

The specification consists of three automata: M_1, M_2, and *Buffer*, which act concurrently. These components may represent a machine that does the initial processing of a part, a second machine which applies more advanced processing to the part, and a conveyer device which moves the part from one to another.

A general label of a transition in a *Timed Statecharts* specification has one of the following forms:

- $name : e[c]/a$.
- $name : ([c] \ for \ I)/a$.

In these labels, *name* stands for the transition name (has no semantic meaning), e is an *event expression*, (conjunction of events or negated events), $[c]$ is a *condition* (boolean expression over shared variables), a is an *action* which can generate events or assign values to shared variables when the transition is taken, and I is a *time interval.* All the transition label components are optional.

For the case that $c = true$, we abbreviate the label

$$name : ([c] \text{ for } I)/a$$

to

$$name : I/a$$

When a transition has no triggering event or condition to be satisfied, such as transition *good-part* in the diagram, the transition is enabled whenever the state from which it departs (state *Busy* in the diagram) is active.

The *time interval* $I = \{l, u\}$ specifies the lower and upper bounds on the continuous enabledness interval of the transition, before it is taken. Transitions with time bounds [0,0] are called *immediate.* We require that all transitions which have a triggering event be immediate. Transitions which are not explicitly labeled are considered to have the time bounds $[0, 0]$, and are therefore immediate. A transition is called *relevant* if the state from which it departs is currently active. Transitions that have no triggering event and no condition, are enabled whenever they are relevant. A transition with a triggering event e (condition $[c]$) is enabled if it is relevant and the event e has just occurred (the condition $[c]$ is satisfied).

Some states are *compound.* For example, the state encompassing basic states *m2*, @2, *m1* and *m3* in the automaton *Buffer* is compound. It is considered active whenever one of these basic states is active. A transition departing from a compound state, such as the transition leading into *Error*, is relevant whenever the compound state is active and, when taken, it makes the compound state and all of its substates inactive and activates the state to which the transition leads (*Error* in the example above). A transition entering a compound state, enters the *default state* within that compound state, denoted by small arrow (see *m2* in *moving*).

When a transition which generates an event e is taken, one or more immediate transitions that have e as a triggering event and are currently relevant are taken. This is the mechanism by which the concurrent automata synchronize. For example, if M_1 takes transition *good-part* while *Buffer* is in state @1, then the transition labeled by *put* can be taken next. Note that, since this transition is immediate and ready, it must be taken before time can progress.

Consider for example the transition connecting state *m2* to state @2. Assume that it is taken at position j while, at the same time, machine M_2 is in state *Idle*. On being taken, this transition generates event *ready.* Machine M_2 responds to this generation by taking the transition leading into state *Busy*, generating the event *take.* Since at this point *Buffer* is at state @2, it responds immediately by moving to state *m1*¡. These three transitions are taken before the clock advances.

Another important element of the behavior of Statecharts is that events that are generated at a certain step persist until time progresses. This allows more than one transition triggered by an event e to respond to the generation of e before time progresses.

The given specification describes the following possible scenarios. Both machines start at an idle state. After some time ranging between l_3 and u_3, machine M_1 concludes its set-up procedure and moves to the busy state. While being busy, M_1 may either take time between l_1 and u_1 to produce a bad part, or take time between l_2 and u_2 to produce a good part. In both cases it moves to state *Idle* where it completes another set-up procedure. If a good part is produced the event *put* is generated which triggers the transition departing from state @1 in *Buffer*, if it is ready. This transition represents the initiation of movement of the conveyer between M_1 and M_2. The movement itself may take time between l_4 and u_4 to reach M_2. Reaching M_2 is represented by the transition connecting to state @2 which also generates the event *ready*. This event is sensed by M_2 which removes the part from the conveyer and starts processing it in its *Busy* state.

If all timing are right, M_1 should never issue the *put* signal when the conveyer is at state *moving* (i.e. not at state @1). The diagram represents this expectation by having a transition that moves to state *Error* if that erroneous situation arises.

Note the reason for the conjunct $\neg$*first* in the triggering expression of the transition leading from *moving* to *Error*. State *moving* can be entered from state @1 only when signal *put* is emitted. According to the semantics defined in this paper, several transitions in the same component can be taken at the same time instant. Consequently, if the transition from *moving* to *Error* were labeled only by *put*, it would have been taken immediately following the transition entering *moving*. To prevent this, the transition entering *moving* emits the signal *first*, and the error transition requires that *put* is on, but also that *first* is off.

An interesting analysis question is what should be the relation between the various time constants to ensure that this never happens. Various verification tools, based on algorithms similar to the one proposed in [ACD90], can answer such questions algorithmically.

Now in Text form

The top level of the text form of the specification is

$$Spec :: \quad \left[M_1 \parallel Buffer \parallel M_2\right]$$

M_1, *Buffer*, and M_2 are processes defined as follows:

$$M_1 :: \left[\begin{array}{l} \textbf{loop forever do} \\ \left[\begin{array}{ll} Idle: & \textbf{skip} \\ \text{set-up}: & \textbf{delay}\ [l_3, u_3] \\ Busy: & \left[\begin{array}{lll} \textbf{skip}; & \textit{bad-part}: & \textbf{delay}\ [l_1, u_1] \\ & & \sqcup \\ \textbf{skip}; & \textit{good-part}: & \textbf{delay}\ [l_2, u_2];\quad put! \end{array}\right] \end{array}\right] \end{array}\right]$$

Process *Buffer* is given by

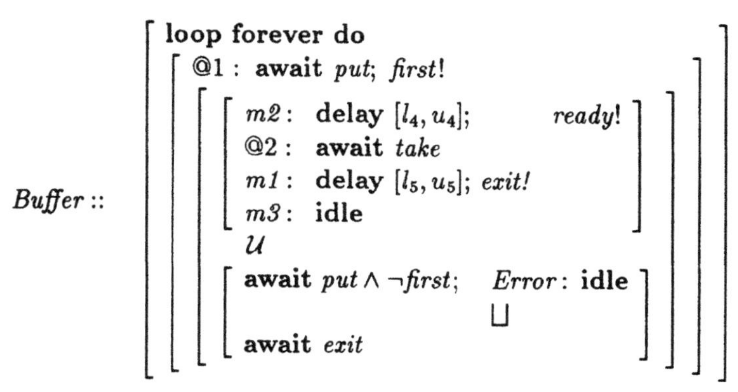

Process M_2 is given by

$$M_2 :: \left[\begin{array}{l} \textbf{loop forever do} \\ \left[\begin{array}{lll} \mathit{Idle}: & \textbf{await}\ \mathit{ready}\ ; & \mathit{take}! \\ \mathit{Busy}: & \textbf{delay}\ [l_6, u_6] & \end{array}\right] \end{array}\right]$$

One may complain that when process *Buffer* detects an error situation, it is the only process affected while M_1 and M_2 continue their activity as though nothing has happened. An alternative specification may require the whole system to stop its activity in such a case. In Fig. 2 we present the Statechart representation of this specification.

The *Statext* representation of this specification is given by

$$\mathit{Spec}_1 :: \left[\begin{array}{c} M_1 \parallel \mathit{Buffer}_1 \parallel M_2 \\ \mathcal{U} \\ \textbf{await}\ \mathit{err};\ \mathit{Error}: \textbf{idle} \end{array}\right]$$

where M_1 and M_2 are as before, and Buffer_1 is given by

$$\mathit{Buffer}_1 :: \left[\begin{array}{l} \textbf{loop forever do} \\ \left[\begin{array}{l} @1: \textbf{await}\ \mathit{put};\ \mathit{first}! \\ \left[\begin{array}{l} \left[\begin{array}{lll} \mathit{m2}: & \textbf{delay}\ [l_4, u_4]; & \mathit{ready}! \\ @2: & \textbf{await}\ \mathit{take} & \\ \mathit{m1}: & \textbf{delay}\ [l_5, u_5];\ \mathit{exit!} & \\ \mathit{m3}: & \textbf{idle} & \end{array}\right] \\ \mathcal{U} \\ \left[\begin{array}{lll} \textbf{await} & \mathit{put} \wedge \neg \mathit{first}; & \mathit{err}!;\ \textbf{idle} \\ & \sqcup & \\ \textbf{await}\ \mathit{exit} & & \end{array}\right] \end{array}\right] \end{array}\right] \end{array}\right]$$

This specification uses the auxiliary signal *err* which is emitted by process Buffer_1 on detecting an error situation. This signal preempts the parallel composition of M_1, Buffer_1, and M_2 and leads to the external *Error* state.

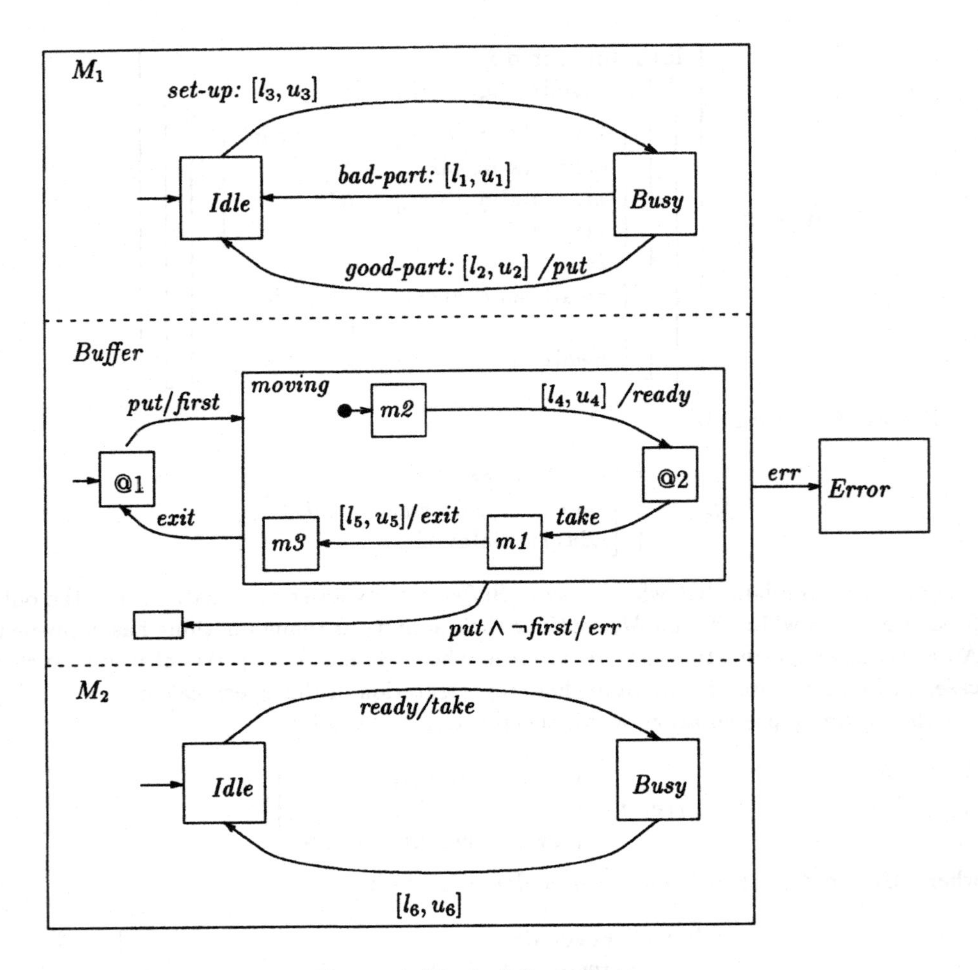

Figure 2: Revised Specification.

4 Timed Transition System

The semantics of *Timed Statecharts* and *Statext* specifications is defined as sets of *timed behaviors.* This is done in two steps. First, we introduce the computation model of *timed transition system* and identify the possible timed behaviors (computations) of any such system. Then, we show how a *Timed Statecharts* and *Statext* specification can be mapped into this generic model.

The notions of *timed behavior* and *timed transition system* are based on [HMP90, HMP91], to which we have added *explicit* timed transitions and partitioned the set of state variables into *persistent* and *volatile* variables.

As in [HMP90, HMP91], we assume a global clock. A *timed transition* (clock tick) advances the global clock by a positive amount. Timed transitions are synchronous. They require the cooperation of all the concurrent processes, which have to agree on the increment by which the clock is advanced, and perform the transition simultanously. Timed transitions are interleaved with other (untimed) systems transitions.

The partition of the state variables into persistent and volatile variables enables a uniform treatment of communication through both shared variables and channels. Channels are represented by *volatile variables* which are reset by every *clock tick*. Shared variables are represented by *persistent variables*, whose values are unaffected by clock ticks, and can only be modified by an explicit assignment.

To model metric time, we assume a totally ordered time domain $\mathbf{T}$ which contains a zero element $0 \in \mathbf{T}$, and a commutative, associative operation $+$, for which 0 is a unique identity element, and such that for every $t_1, t_2 \in \mathbf{T}$,

$t_1 < t_2$ iff There exists a unique $t > 0$ (denoted by $t_2 - t_1$), such that $t_1 + t = t_2$.

We refer to the elements of $\mathbf{T}$ as *time elements* or sometimes as *moments*. In most cases we will take $\mathbf{T}$ to be either the natural numbers N, or the nonnegative real numbers $\mathsf{R}_{\geq 0}$. We define $\mathbf{T}_{>} = \mathbf{T} - \{0\}$.

With a system to be analyzed we associate

- V : A finite set of *state variables*.
- Σ : A set of *states*. Each state $s \in \Sigma$ is an interpretation of V; that is, it assigns to every variable $x \in V$ a value $s[x]$ in its domain.
- $V_t = V \cup \{t\}$: A finite set of *situation variables*. They are obtained by augmenting V, the set of *state variables*, with t a variable representing the current time in each situation.
- Σ_t : A set of situations. Every situation $s \in \Sigma_t$ is an interpretation of V_t. In particular, $s[t] \in \mathbf{T}$ is the value of the real-time clock at situation s. We denote by $s[V]$ the *state* corresponding to the situation s. It is obtained by restricting the interpretation s to the state variables V.

Timed Traces

A *progressive time sequence* is an infinite sequence of time elements

$$\theta : t_0, t_1, \ldots,$$

where $t_i \in \mathbf{T}$, for each $i = 0, 1, \ldots$, satisfying

- $t_0 = 0$.
- Time does not decrease. That is, for every $i \geq 0$, $t_i \leq t_{i+1}$.
- Time eventually increases beyond any bound. That is, for every time element $t \in \mathbf{T}$, $t_i > t$ for some $i \geq 0$. This is called the *Non Zeno* requirement in [AL91].

A *timed trace* is an infinite sequence of situations

$$\sigma: \; s_0, s_1, \ldots,$$

where $s_i \in \Sigma_t$, for each $i = 0, 1, \ldots$. We denote by $t_i = s_i[t]$ the moment at which situation s_i was observed (sampled).

It is required that

- The sequence $t_0, t_1, \ldots$ is a progressive time seqeunce.
- For every $i \geq 0$, state and time do not change at the same time, i.e., either $s_i[V] = s_{i+1}[V]$ or $t_i = t_{i+1}$. This requirement ensures that each state change is precisely timed.

Timed Transition Systems

A *timed transition system* $S = \langle V, \Theta, \mathcal{L}, \mathcal{T} \rangle$ consists of the following components:

- *State variables* V, partitioned into $V = V_p \cup V_v$, where
  - V_p is a set of *persistent variables.* These variables represent data variables. They are not affected by time transitions.
  - V_v is a set of *volatile variables.* These variables represent the communication channels. They are reset by every time transition.

  We denote by Σ the set of all interprations of V.
- The *initial condition* Θ. An assertion that characterizes the states that can appear as initial states in a computation.
- A set of *transition labels* $\mathcal{L}$, partitioned into two disjoint sets $\mathcal{L} = \mathbf{T}_{>} \cup L$, where a label from $\mathbf{T}_{>}$ represents a timed step, while a label from L represents an untimed step.
- A *transition relation* $\mathcal{T} \subseteq (\Sigma \times \mathcal{L} \times \Sigma)$, partitioned into $\mathcal{T} = \mathcal{T}_T \cup \mathcal{T}_L$, where
  - $\mathcal{T}_T$ is the *timed transition relation.* For every $(s_i, \mu_i, s_{i+1}) \in \mathcal{T}_T$,

$$\begin{array}{l} \mu_i \in \mathbf{T}_{>} \\ s_i[V_p] = s_{i+1}[V_p] \\ s_{i+1}[V_v] = \bot \\ s_{i+1}[t] = s_i[t] + \mu_i \end{array}$$

  - $\mathcal{T}_L$ is the *untimed transition relation.* For every $(s_i, \mu_i, s_{i+1}) \in \mathcal{T}_L$,

$$\begin{array}{l} \mu_i \in L \\ s_{i+1}[t] = s_i[t] \end{array}$$

Given the state variables V, we can obtain the corresponding set of situation variables $V_t = V \cup \{t\}$, and the set of situations Σ_t interpreting V_t.
A *computation* of a timed transition system $S = \langle V, \Theta, \mathcal{L}, \mathcal{T} \rangle$ is an infinite sequence of alternating situations and transition labels

$$\sigma: \; s_0, \mu_0, s_1, \mu_1 \ldots,$$

where $s_i \in \Sigma_t$ and $\mu_i \in \mathcal{L}$ for each $i = 0, 1, \ldots$.
It is required that

- [*Time Progress*] The sequence $s_0, s_1, \ldots$ is a timed trace.
- [*Initiality*] $s_0 \models \Theta$.
- [*Consecution*] For all $i \geq 0, \quad (s_i, \mu_i, s_{i+1}) \in \mathcal{T}$.

As shown in [HMP91], the model of timed transition systems is expressive enough to express most of the features specific to real-time programs such as delays, timeouts, preemption, interrupts and multi-programming scheduling.

5 *Statext* Semantics

With every *Statext* specification P, we associate a timed transition system $\langle V, \Theta, \mathcal{L}, \mathcal{T} \rangle$ as follows:

- $V = V_p \cup V_v$, where V_p contains the set of all *shared variables* in P, and the control variable π ranging over statements, and V_v is the set of *communication channels.*
- $\Theta : (\pi = P) \wedge \bigwedge_{\alpha \in V_v} (\alpha = \bot)$
- $\mathcal{L} = \mathbf{T}_{>} \cup L$, where $L = \{\alpha!e \mid \alpha \in V_v$, and $e \neq \bot$ is an expression over the domain of $\alpha\} \cup \{\tau\}$. τ represents non communicating internal operations.
- $\mathcal{T} = \mathcal{T}_T \cup \mathcal{T}_L$, where $\mathcal{T}_T$ and $\mathcal{T}_L$ are defined below.

Statext Transitions

The transition relation associated with a *Statext* specification, is defined by structural induction [Plo81].

A state $s \in \Sigma$ is represented by the tuple $\langle S, D, E \rangle$, where S is a statement interpreting π, and D, E are interpretations of $V_p - \{\pi\}$ and V_v respectively. We use Λ to denote the empty statement.

We use derivations of the form $s_i \xrightarrow{\xi} s_j$, where $s_i, s_j \in \Sigma$, $\xi \in \{\alpha!e, \Delta, \lambda, \mu\}$, and
$\Delta \in \mathbf{T}_{>}$
$\lambda \in L$
$\mu \in \mathcal{L}$

The transition relation is defined by axioms that contain a derivation (possibly with some side conditions) and rules of the form $d_1, d_2, \ldots, d_k \vdash d$, sometimes represented as $\frac{d_1,\ldots,d_k}{d}$.

Untimed Transition Relation

- **skip**

  $\langle \mathbf{skip}, D, E\rangle \xrightarrow{\tau} \langle \Lambda, D, E\rangle$

- **idle**

  The **idle** statement has only associated timed transitions.

- **ascertain**

  To define the semantics of the *ascertain* statement, we introduce a more general form of the statement

  ascertain c **from** t_0 **to** I

  This form specifies an initial value $t_0 \in \mathbf{T}_>$ at which time counting begins. The standard form **ascertain** c **for** I can be viewed as an abbreviation of **ascertain** c **from** 0 **to** I. We define transition semantics for the more general form.

  $\langle \mathbf{ascertain}\ c\ \mathbf{from}\ t_0\ \mathbf{to}\ I, D, E\rangle \xrightarrow{\tau} \langle \Lambda, D, E\rangle$
  provided $t_0 \in I$ and $(D \cup E) \models c$

- *Assignment*

  $\langle \overline{y} := \overline{e}, D, E\rangle \xrightarrow{\tau} \langle \Lambda, D[\overline{y} \leftarrow \overline{e}], E\rangle$

- *Send*

  $\langle \alpha!e, D, E\rangle \xrightarrow{\alpha!e} \langle \Lambda, D, E[\alpha \leftarrow e]\rangle$

- *Receive*

  $\langle \alpha?y, D, E[\alpha \leftarrow v]\rangle \xrightarrow{\tau} \langle \Lambda, D[y \leftarrow v], E[\alpha \leftarrow v]\rangle$, for $v \neq \bot$

  The notation $E[\alpha \leftarrow v]$ denotes any interpretation that can be obtained by sending v on α, i.e. an interpretation in which the value of α is $v \neq \bot$.

- *Conditional*

  $\langle \mathbf{if}\ c\ \mathbf{then}\ S_1\ \mathbf{else}\ S_2, D, E\rangle \xrightarrow{\tau} \langle S_1, D, E\rangle$
  provided $(D \cup E) \models c$

  $\langle \mathbf{if}\ c\ \mathbf{then}\ S_1\ \mathbf{else}\ S_2, D, E\rangle \xrightarrow{\tau} \langle S_2, D, E\rangle$
  provided $(D \cup E) \not\models c$

- **while**

  $\langle \mathbf{while}\ c\ \mathbf{do}\ S, D, E\rangle \xrightarrow{\tau} \langle S; \mathbf{while}\ c\ \mathbf{do}\ S, D, E\rangle$
  provided $(D \cup E) \models c$

$\langle \textbf{while } c \textbf{ do } S, D, E\rangle \xrightarrow{\tau} \langle \Lambda, D, E\rangle$
provided $(D \cup E) \not\models c$

- *Concatenation*

$$\frac{\langle S_1, D, E\rangle \xrightarrow{\mu} \langle \Lambda, D', E'\rangle}{\langle S_1; S_2, D, E\rangle \xrightarrow{\mu} \langle S_2, D', E'\rangle}$$

$$\frac{\langle S_1, D, E\rangle \xrightarrow{\mu} \langle S_1', D', E'\rangle}{\langle S_1; S_2, D, E\rangle \xrightarrow{\mu} \langle S_1'; S_2, D', E'\rangle}, \text{ where } S_1' \neq \Lambda$$

- *Selection*

$$\frac{\langle S_1, D, E\rangle \xrightarrow{\lambda} \langle S_1', D', E'\rangle}{\langle S_1 \bigsqcup S_2, D, E\rangle \xrightarrow{\lambda} \langle S_1', D', E'\rangle}$$

$$\frac{\langle S_2, D, E\rangle \xrightarrow{\lambda} \langle S_2', D', E'\rangle}{\langle S_1 \bigsqcup S_2, D, E\rangle \xrightarrow{\lambda} \langle S_2', D', E'\rangle}$$

- *Cooperation*

$$\frac{\langle S_1, D, E\rangle \xrightarrow{\lambda} \langle S_1', D', E'\rangle}{\langle S_1 \| S_2, D, E\rangle \xrightarrow{\lambda} \langle S_1' \| S_2, D', E'\rangle}, \text{ where } S_1' \neq \Lambda$$

$$\frac{\langle S_2, D, E\rangle \xrightarrow{\lambda} \langle S_2', D', E'\rangle}{\langle S_1 \| S_2, D, E\rangle \xrightarrow{\lambda} \langle S_1 \| S_2', D', E'\rangle}, \text{ where } S_2' \neq \Lambda$$

$$\frac{\langle S_i, D, E\rangle \xrightarrow{\lambda} \langle \Lambda, D', E'\rangle}{\langle S_1 \| S_2, D, E\rangle \xrightarrow{\lambda} \langle S_j, D', E'\rangle}, \text{ for } 1 \leq i, j \leq 2 \text{ and } i \neq j.$$

- *Preemption*

$$\frac{\langle S_1, D, E\rangle \xrightarrow{\lambda} \langle S_1', D', E'\rangle}{\langle S_1 \mathcal{U}\ S_2, D, E\rangle \xrightarrow{\lambda} \langle S_1' \mathcal{U}\ S_2, D', E'\rangle}, \text{ where } S_1' \neq \Lambda$$

$$\frac{\langle S_2, D, E\rangle \xrightarrow{\lambda} \langle S_2', D', E'\rangle}{\langle S_1 \mathcal{U}\ S_2, D, E\rangle \xrightarrow{\lambda} \langle S_2', D', E'\rangle}$$

$$\frac{\langle S_1, D, E\rangle \xrightarrow{\lambda} \langle \Lambda, D', E'\rangle}{\langle S_1 \mathcal{U}\ S_2, D, E\rangle \xrightarrow{\lambda} \langle \Lambda, D', E'\rangle}$$

Timed Transition Relation (clock ticks)

- *Ascertain*

  $\langle \textbf{ascertain } c \textbf{ from } t_0 \textbf{ to } I, D, E\rangle \xrightarrow{\Delta} \langle \textbf{ascertain } c \textbf{ from } (t_0 + \Delta) \textbf{ to } I, D, \bot\rangle$
  provided $(D \cup E) \models c$, $\Delta > 0$ and $\Delta + t_0 \preceq I$.

  $\langle \textbf{ascertain } c \textbf{ from } t_0 \textbf{ to } I, D, E\rangle \xrightarrow{\Delta} \langle \textbf{ascertain } c \textbf{ from } 0 \textbf{ to } I, D, \bot\rangle$,
  for every $\Delta > 0$, provided $(D \cup E) \not\models c$

  In these axioms and the following rules, we use $\bot$ to denote the empty channel interpretation that interprets every channel $\alpha \in V_v$ as $\bot$. All timed transitions reset E to $\bot$.

- *Receive*

  $\langle \alpha?y, D, E[\alpha \leftarrow \bot]\rangle \xrightarrow{\Delta} \langle \alpha?y, D, \bot\rangle$, for every $\Delta > 0$

- *Idle*

  $\langle \textbf{idle}, D, E\rangle \xrightarrow{\Delta} \langle \textbf{idle}, D, \bot\rangle$, for every $\Delta > 0$

- *Compound*

  For $\oplus \in \{\sqcup, \|, \mathcal{U}\}$

  $$\frac{\langle S_i, D, E\rangle \xrightarrow{\Delta} \langle S_i', D, \bot\rangle}{\langle S_1 \oplus S_2, D, E\rangle \xrightarrow{\Delta} \langle S_1' \oplus S_2', D, \bot\rangle}, \text{ for } 1 \leq i \leq 2$$

The following observations can be made concerning a *Statext* computation:

- Computations alternate between transitions in $\mathcal{T}_{\mathbf{T}}$ that advance the global clock and transitions in $\mathcal{T}_L$ that take zero time.
- Transitions in $\mathcal{T}_{\mathbf{T}}$ are executed *synchronously*, while transitions in $\mathcal{T}_L$ are *asynchronous.*
- A transition in $\mathcal{T}_{\mathbf{T}}$ can advance the global clock only by an amount on which all the enabled transitions agree.

6 Timed Statecharts

Timed Statecharts is an extension to the visual specification language *Statecharts* [Har87], in which each transition is annotated by a time interval $\{l, u\}$, denoting the lower and upper time bounds of that transition.

In the definition of Timed Statecharts, we have adopted the approach taken by *Argos* [Mar90], namely, disallowing:

- Inter-level transitions.
- References to state names.

These modifications are intended to improve the modality of the language.

Different than both Statecharts and Argos, Timed Statecharts are *asynchronous*, with concurrency being modelled by interleaving. The only synchronous steps are those in which the time maintained by a global clock, is incremented.

6.1 Abstract syntax

A *rooted automaton* of degree n is a labeled directed graph (r, V, E, ℓ), where

- V is a set of vertices
- E is a set of directed edges
- ℓ is a labeling function, assigning to each edge $e \in E$ a label $\ell(e)$ in some labeling set $\mathcal{L}$.
- $r \in V$ is the *root* vertex. It is required that no edge enters r, and that r has n edges departing from it.

A *single entry automaton*, denoted *SE-automaton* , is a rooted automaton of degree $n \leq 1$.

Two rooted automata (r_1, V_1, E_1, ℓ_1) and (r_2, V_2, E_2, ℓ_2) can be combined into a single rooted automaton

$$(r, V, E, \ell) = (r_1, V_1, E_1, \ell_1) \bigsqcup (r_2, V_2, E_2, \ell_2)$$

by the operation of *fusion*. The automaton (r, V, E, ℓ) is obtained by identifying r_1 with r_2. It is required that $E_1 \cap E_2 = \emptyset$, $r_1 \notin V_2$ and $r_2 \notin V_1$, but $V_1 - \{r_1\}$ and $V_2 - \{r_2\}$ may have a nonempty intersection.

For every labeled directed graph $G : (V, E, \ell)$ (not necessarily rooted) and a vertex $r \in V$, we can construct a rooted automaton $A : (r', V \cup \{r'\}, E', \ell')$ called the *r-unwinding* of G. We write $A = unwind(r, G)$.

Automata G and A are related by the fact that $\ell_1, \ell_2, \ldots, \ell_k$ is a sequenc of labels appearing on a path departing from $r \in G$ *iff* it is a labeling sequence of a path departing from $r' \in A$.

The operation of unwinding is necessary in order to represent graphically the transformation associated with the **while** statement in *Statext* .

It is not difficult to see that every rooted automaton M of degree n can be obtained as a fusion of n *SE-automata*, i.e., $M = \bigsqcup_{i=1}^{n} B_i$, for some *SE-automata* $B_1, \ldots, B_n$.

We can now give a recursive definition for *Timed Statecharts*.

A Timed Statechart S is one of the following:

- A *rooted automaton* .
- A refined *rooted automaton* .
- A parallel composition of *Timed Statecharts* .

A refined *rooted automaton* is defined as follows. Let $A : (r, V, E, \ell)$ be a *rooted automaton* and $v_1, \ldots, v_k \in V$ a list of vertices. Let $S_1, \ldots, S_k$ be a list of timed statecharts. The refinement of A, encapsulating $S_1, \ldots, S_k$ within $v_1, \ldots, v_k$ is obtained by placing for each $i = 1, \ldots, k$, S_i within the box (circle) that graphically represents v_i.

For the presentation of the *Timed Statecharts* semantics, we use a set of icons as shown in figure 3. We use the icon to represent a (not necessarily rooted) labeled directed graph.

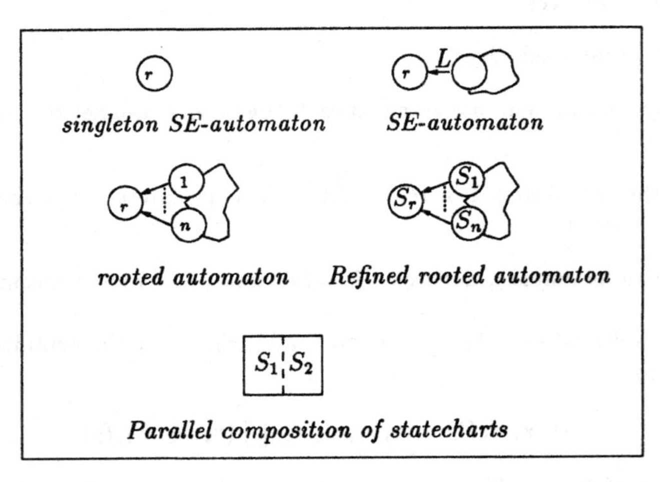

Figure 3: Iconic representation of Statecharts.

6.2 Semantics

With every *Timed Statecharts* specification S, we associate a timed transition system $\langle V, \Theta, \mathcal{L}, \mathcal{T} \rangle$, similar to that of *Statext* programs except for the transition relation that will be defined below, and the fact that π ranges over *Timed Statecharts*.

Timed Statecharts Transitions

For an action a, data-state D and event set E, we denote by D'_a, E'_a the data-state and event set resulting by executing a.

- For a of the form $y := e$, we have

$$D'_a = D[y \leftarrow e] \qquad \text{and} \qquad E'_a = E$$

- For a of the form $\alpha!e$, we have

$$D'_a = D \qquad \text{and} \qquad E'_a = E[\alpha \leftarrow e]$$

Untimed Transition Relation

- ***SE-automaton***

  (r) is equivalent to **idle** in *Statext* , and has no associated *untimed* transitions.

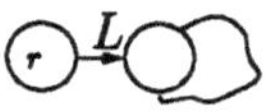

  $\langle$ (r) $\xrightarrow{L}$ $, D, E\rangle \xrightarrow{\lambda} \langle unwind(\quad), D'_a, E'_a\rangle$

  where $L = e[c]/a$, provided $D \cup E \models (e \cup c)$.

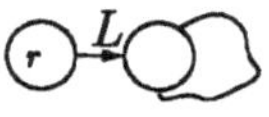

  $\langle$ (r) $\xrightarrow{L}$ $, D, E\rangle \xrightarrow{\lambda} \langle unwind(\quad), D'_a, E'_a\rangle$

  where $L = ([c]$ *from* t_0 to $I)/a$, provided $D \cup E \models c$ and $t_0 \in I$.

- ***Rooted automaton***

  For every i, $1 \leq i \leq n$

  $\langle$ (r) $\xrightarrow{L}$ (i) $, D, E\rangle \xrightarrow{\lambda} \langle unwind($ (i) $), D', E'\rangle$

  $\langle$ (r) → (1) … (n) $, D, E\rangle \xrightarrow{\lambda} \langle unwind($ (i) $), D', E'\rangle$

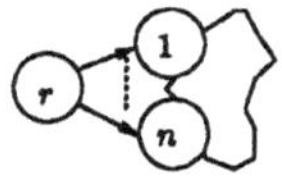

- ***Refined rooted automaton***

  For a statechart S_r,

  $\langle S_r, D, E\rangle \xrightarrow{\lambda} \langle S'_r, D', E'\rangle$

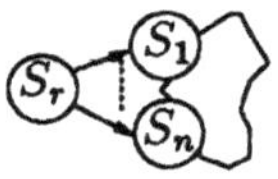

  $\langle$ (S_r) → (S_1) … (S_n) $, D, E\rangle \xrightarrow{\lambda} \langle$ (S'_r) → (S_1) … (S_n) $, D', E'\rangle$

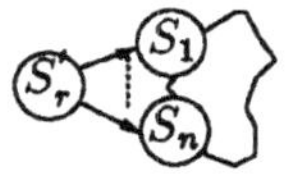

  For basic states $r, 1, \ldots, n$, statecharts $S_r, S_1, \ldots, S_n$ and every i, $0 \leq i \leq n$,

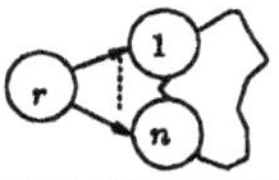

  $\langle$ (r) → (1) … (n) $, D, E\rangle \xrightarrow{\lambda} \langle unwind($ (i) $), D', E'\rangle$

  $\langle$ (S_r) → (S_1) … (S_n) $, D, E\rangle \xrightarrow{\lambda} \langle unwind($ (S_i) $), D', E'\rangle$

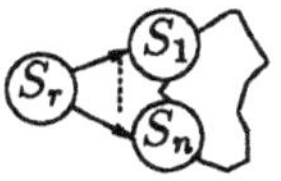

- ***Parallel composition***

$$\frac{\langle S_1, D, E\rangle \xrightarrow{\lambda} \langle S_1', D', E'\rangle}{\langle\ \boxed{S_1 \mid S_2}\ , D, E\rangle \xrightarrow{\lambda} \langle\ \boxed{S_1' \mid S_2}\ , D', E'\rangle}$$

$$\frac{\langle S_2, D, E\rangle \xrightarrow{\lambda} \langle S_2', D', E'\rangle}{\langle\ \boxed{S_1 \mid S_2}\ , D, E\rangle \xrightarrow{\lambda} \langle\ \boxed{S_1 \mid S_2'}\ , D', E'\rangle}$$

Timed Transition Relation

- *SE-automaton*

For every $\Delta > 0$

$\langle\ (r), D, E\rangle \xrightarrow{\Delta} \langle\ (r)\ , D, \perp\rangle$

$\langle\ (r) \xrightarrow{L} \ldots\ , D, E\rangle \xrightarrow{\Delta} \langle\ (r) \xrightarrow{L} \ldots\ , D, \perp\rangle$ where $L = e[c]/a$

provided $D \cup E \not\models (e \cup c)$

$\langle\ (r) \xrightarrow{L} \ldots\ , D, E\rangle \xrightarrow{\Delta} \langle\ (r) \xrightarrow{L'} \ldots\ , D, \perp\rangle$

where $L = ([c]$ *from* t_0 to $I)/a$ and $L' = ([c]\ I)/a$,
provided $D \cup E \not\models c$

$\langle\ (r) \xrightarrow{L} \ldots, D, E\rangle \xrightarrow{\Delta} \langle\ (r) \xrightarrow{L'} \ldots\ , D, \perp\rangle$

where $L = ([c]$ *from* t_0 to $I)/a$ and $L' = ([c]$ *from* $(t_0 + \Delta)$ to $I)/a$,
provided $D \cup E \models c$, $\Delta > 0$ and $\Delta + t_0 \preceq I$

- *Rooted automaton*

$$\frac{\langle\ (r) \xrightarrow{L} (i) \ldots\ , D, E\rangle \xrightarrow{\Delta} \langle\ (r) \xrightarrow{L'} (i) \ldots\ , D, \perp\rangle \quad \text{for all } i \in \{1..n\}}{\langle\ (r) \xrightarrow{L_1} (1) \ldots (r) \xrightarrow{L_n} (n) \ldots\ , D, E\rangle \xrightarrow{\Delta} \langle\ (r) \xrightarrow{L_1'} (1) \ldots (r) \xrightarrow{L_n'} (n) \ldots\ , D, \perp\rangle}$$

- *Refined rooted automaton*

1. $$\langle S_r, D, E\rangle \xrightarrow{\Delta} \langle S'_r, D, E\rangle$$

2. 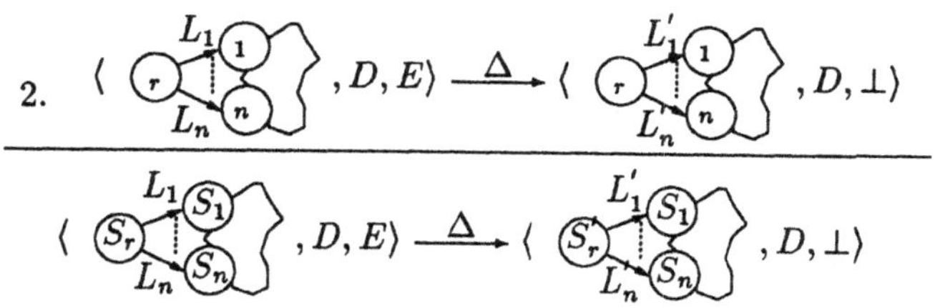

- *Parallel composition*

$$\frac{\langle S_i, D, E\rangle \xrightarrow{\Delta} \langle S'_i, D, \bot\rangle}{\langle \boxed{S_1 \mid S_2}\ D, E\rangle \xrightarrow{\Delta} \langle \boxed{S'_1 \mid S'_2}\ , D, \bot\rangle}, \text{ for } 1 \leq i \leq 2$$

7 Hybrid Systems

We can now extend both the *Statext* and *Timed Statecharts* specification languages, such that they can both be used for the specification of hybrid systems. We first define a *hybrid transition system.* Then, we describe the extended specification languages, and define their compositional semantics.

We conclude with a small example, specified in both *Hybrid Statext* and *Hybrid Statecharts.*

7.1 Hybrid Transition System

A *Hybrid transition system* $S_{\mathcal{H}} = \langle V, \Theta, \mathcal{L}, \mathcal{T}\rangle$ consists of the following components:

- *State variables* V, partitioned into $V = V_p \cup V_v$, where
  - V_p is a set of *persistent variables.* These variables represent data variables. They are not reset by time transitions.
  - V_v is a set of *volatile variables.* These variables represent the communication channels. They are reset by every *time transition.*

  The set of *persistent variables* is further partitioned into $V_p = V_c \cup V_d$, where
  - V_c is the set of *continuous variables.* These variables are modified by continuous activities in the behavior of a hybrid system.
  - V_d is the set of *discrete variables.* These variables are changed by discrete steps.

We denote by Σ the set of all interprations of V.

- The *initial condition* Θ. An assertion that characterizes the states that can appear as initial states in a computation.
- A set of *transition labels* $\mathcal{L}$, partitioned into $\mathcal{L} = \mathcal{A} \cup L$, where
  - Every $\mu \in \mathcal{A}$ is of the form Δ, A where
    - $\Delta \in \mathbf{T}_{>}$.
    - A is an assignment of differentiable functions over $[0, \Delta]$ to a set of continuous variables $Y_A \subseteq V_c$. Thus, A assigns to each $y \in Y_A$ a function $f_y(t)$ differentiable over $t \in [0, \Delta]$.
  - A label $\mu \in L$ represents an *untimed* step.
- A transition relation $\mathcal{T} \subseteq (\Sigma \times \mathcal{L} \times \Sigma)$, partitioned into $\mathcal{T} = \mathcal{T}_{\mathcal{A}} \cup \mathcal{T}_L$, where
  - $\mathcal{T}_{\mathcal{A}}$ is the *timed transition relation.* For every $(s_i, \mu_i, s_{i+1}) \in \mathcal{T}_{\mathcal{A}}$,

$$\begin{array}{ll} \mu_i = (\Delta, A) \in \mathcal{A} & \\ s_i[V_d] = s_{i+1}[V_d] & \\ s_{i+1}[V_v] = \bot & \\ s_{i+1}[t] = s_i[t] + \mu_i[\Delta] & \\ s_{i+1}[y] = s_i[y] & \text{for every } y \in (V_c - Y_A) \\ s_i[y] = f_y[0] \text{ and } s_{i+1}[y] = f_y(\Delta) & \text{for every } y \in Y_A \end{array}$$

    Timed transitions are allowed to change the values of continuous variables.
  - $\mathcal{T}_L$ is the *untimed transition relation.* For every $(s_i, \mu_i, s_{i+1}) \in \mathcal{T}_L$,

$$\begin{array}{l} \mu_i \in L \\ s_{i+1}[t] = s_i[t] \end{array}$$

Given the state variables V, we can obtain the corresponding set of situation variables $V_t = V \cup \{t\}$, and the set of situations Σ_t interpreting V_t.

A *hybrid trace* is an infinite sequence of situations

$$\sigma: s_0, s_1, \ldots,$$

where $s_i \in \Sigma_t$, for each $i = 0, 1, \ldots$. We denote by $t_i = s_i[t]$ the moment at which situation s_i was observed (sampled).

It is required that

- The sequence $t_0, t_1, \ldots$ is a progressive time seqeunce.
- For every $i \geq 0$, discrete state variables and time do not change at the same time, i.e., either $s_i[V_d] = s_{i+1}[V_d]$ or $t_i = t_{i+1}$.

A *computation* of a hybrid transition system is defined in a similar way to the computation of a timed transition system, except for the *Time Progress* requirement which will require the sequence $s_0, s_1, \ldots$ to be a *hybrid trace* instead of a *time trace.*

7.2 Hybrid *Statext*

Syntax

To the set of *Statext* statements, we add the *continuous assignment* statement,

$$\dot{y} = R$$

This statement represents a differential equation, where y is a continuous variable, and R is a term over V_p. We denote by $vars_c(R)$ the set of continuous variables appearing in R.

To the *ascertain* statement, **ascertain** c **for** I, we add the following restriction

If $u_I > 0$, the boolean expression c can only range over V_d.

This restriction may be removed, and is imposed here for the sake of simplicity.

Semantics

The semantics of *Hybrid Statext* is defined by associating with each hybrid specification, a hybrid transition system. Following we present the transition relation corresponding to each of the statements.

Transitions

Untimed Transition Relation

The untimed transition relation remains unchanged, since the *continuous assignment* statement has no associated untimed transitions.

Timed Transition Relation

The set of rules defining the timed transition relation associated with a *Statext* specification, is replaced by the following rules:

- *Continuous Assignment*

  for every $\Delta > 0$
  $\langle \dot{y} = R, D, E \rangle \xrightarrow{\Delta, A} \langle \dot{y} = R, D[Y_A \leftarrow F_A(\Delta)], \perp \rangle$,
  where $Y_A = \{y\} \cup vars_c(R)$.

  The notation $[Y_A \leftarrow F_A(\Delta)]$ stands for the substitution $[\ldots y \leftarrow f_y(\Delta) \ldots]$, updating each $y \in Y_A$ to the value $f_y(\Delta)$.

- *Ascertain*

for $Y_A = vars_c(c)$, and every $\Delta > 0$
$\langle \mathbf{await}\ c, D, E\rangle \xrightarrow{\Delta, A} \langle \mathbf{await}\ c, D[Y_A \leftarrow F_A(\Delta)], \perp\rangle$
provided $D \cup E \not\models c$ and $(D[Y_A \leftarrow F_A(\Delta')] \cup \perp) \not\models c$ for every $\Delta', 0 < \Delta' < \Delta$.

The two other timed transitions for *ascertain*, remain as in *Statext*

- *Receive*

  As in *Statext*

- *Idle*

  As in *Statext*

- *Compound*
  Two function assignments A_1 and A_2 are said to be consistent, if for every $y \in (Y_{A_1} \cap Y_{A_2}$ $f_y^{A_1}(t) = f_y^{A_2}(t)$, where $f_y^{A_1}$ and $f_y^{A_2}$ are the functions assigned to y by A_1 and A_2 respectively.

  For $\oplus \in \{\sqcup, \|, \mathcal{U}\}$

$$\frac{\langle S_i, D, E\rangle \xrightarrow{\Delta, A_i} \langle S_i', D[Y_{A_i} \leftarrow F_{A_i}(\Delta)], \perp\rangle \qquad \text{for } 1 \le i \le 2}{\langle S_1 \oplus S_2, D, E\rangle \xrightarrow{\Delta, A_1 \cup A_2} \langle S_1' \oplus S_2', D[Y_{A_1 \cup A_2} \leftarrow F_{A_1 \cup A_2}], \perp\rangle}$$

  provided A_1 and A_2 are consistent.
  When A_1 and A_2 are consistent, we can define the assignment $A = A_1 \cup A_2$ by taking $Y_A = Y_{A_1} \cup Y_{A_2}$ and assigning to each $y \in Y_{A_i}$ the function $f_y^{A_i}$.

7.3 Hybrid Statecharts

Timed Statecharts are augmented with a notation that allows to annotate a basic state by a differential equation. The implied meaning is that whenever the state is active, the associated differential equation is operational.

The semantics of *Hybrid Statecharts* is defined by associating with each Hybrid Statecharts specification, a hybrid transition system. The associated transition relation is partitioned into $\mathcal{T}_L$, the untimed transition relation and $\mathcal{T}_\mathcal{A}$, the timed relation. As for *Hybrid Statext* , $\mathcal{T}_L$ remains as defined for *Timed Statecharts* , and $\mathcal{T}_\mathcal{A}$ has to be redefined, in a way very similar to the definition given for Hybrid *Statext*.

7.4 An Example

To summarize the discussion on Hybrid systems, we give a small example, specified by both a *Hybrid Statecharts* and a *Hybrid Statext.*

The example can be described as follows: at time $t = 0$, a mouse starts running from a certain position on the floor in a straight line towards a hole in the wall, which is at a distance X_0 from the initial position. The mouse runs at a constant velocity V_m. After Δ

time units, a cat is released at the same initial position and chases the mouse at velocity V_c along the same path. Will the cat catch the mouse, or will the mouse find sanctuary while the cat crushes against the wall?

The Statechart in Fig. 4 describes the possible scenarios.

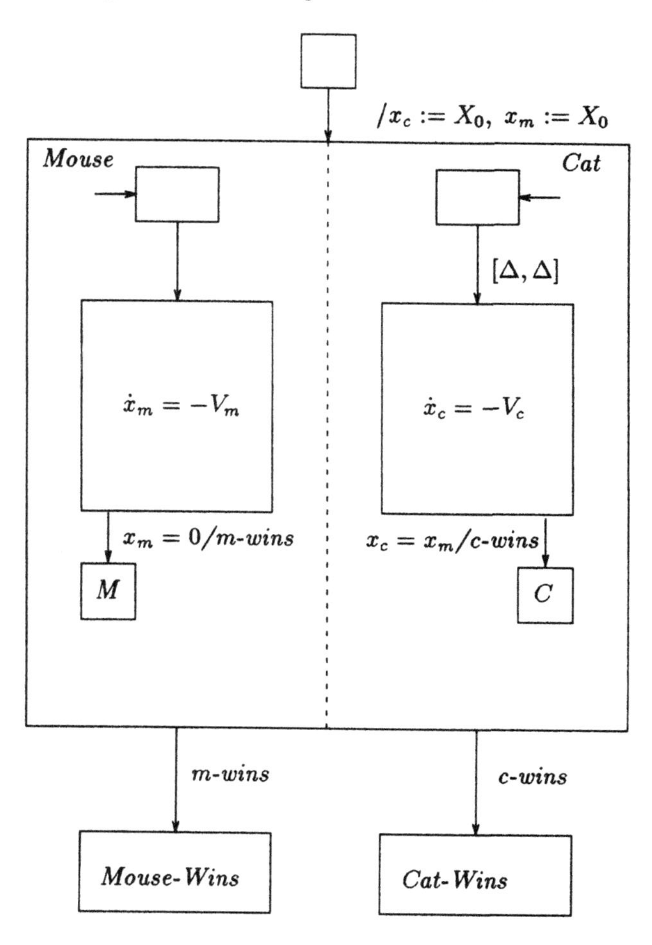

Figure 4: A Hybrid Specification.

The specification (and underlying hybrid transition system) has as continuous variables x_m and x_c, measuring the distance of the mouse and the cat, respectively, from the wall. It refers to the constants X_0, V_m, V_c, and Δ.

The systems starts by the distance variables x_m and x_c being reset to their initial value X_0. Then each of the players proceed in its local behavior. The mouse proceed immediately to the state of running, in which his variable x_m changes continuously according to the equation $\dot{x}_m = -V_m$. The cat waits for a delay of Δ before it enters its running state. Then there are several exits possible. If the event $x_m = 0$ happens first, the mouse wins and moves to its intermediate stage M. At this point, the system moves to state

Mouse-Wins. The other possibility is that the event $x_c = x_m$ occurs first which means that the cat overtook the mouse before the mouse reached sanctuary. In this case they both move to state *Cat-Wins.*

This diagram illustrates the typical interleaving between continuous activities and discrete state changes, which in this example only involved change of control.

In Text form

The top level of the ***hybrid Statext*** specification is

$$
Spec :: \left[\begin{array}{c}
\textbf{local } x_c, x_m \textbf{ where } x_c = x_m = X_0 \\
\\
Mouse \parallel Cat \\
\mathcal{U} \\
\left[\begin{array}{lll}
\textbf{await } c\text{-}wins \ ; & Cat\text{-}Wins: & \textbf{idle} \\
 & \sqcup & \\
\textbf{await } m\text{-}wins \ ; & Mouse\text{-}Wins: & \textbf{idle}
\end{array}\right]
\end{array}\right]
$$

Mouse and *Cat* are processes defined as follows:

$$
Mouse :: \left[\begin{array}{l}
\textbf{skip} \\
\left[\begin{array}{l}
\dot{x}_m = -V_m \\
\mathcal{U} \\
\textbf{await } x_m = 0; m\text{-}wins! \ ; M: \textbf{idle}
\end{array}\right]
\end{array}\right]
$$

$$
Cat :: \left[\begin{array}{l}
\textbf{delay}[\Delta, \Delta] \\
\left[\begin{array}{l}
\dot{x}_c = -V_c \\
\mathcal{U} \\
\textbf{await } x_c = x_m; c\text{-}wins! \ ; C: \textbf{idle}
\end{array}\right]
\end{array}\right]
$$

Note that if $x_m = x_c = 0$, namely, both the cat and the mouse get to the hole simultaneously, the answer to the question who is the winner is nondeterministic.

References

[ACD90] R. Alur, C. Courcoubetis, and D.L. Dill. Model-checking for real-time systems. In *Proc. 5th IEEE Symp. Logic in Comp. Sci.*, 1990.

[AL91] M. Abadi and L. Lamport. An old-fashioned recipe for real time. In *Real-Time: Theory in Practice.* Lec. Notes in Comp. Sci., Springer-Verlag, 1991.

[DH89] D. Drusinsky and D. Harel. On the power of bounded concurrency I: The finite automata level. submitted, 1989. (Preliminary version appeared as "On the Power of Cooperative Concurrency", in *Proc. Intl. Conf. on Concurrency, Concurrency 88*, Lec. Notes in Comp. Sci. 335, Springer, 1988, pp. 74–103).

[Har87] D. Harel. Statecharts: A visual formalism for complex systems. *Sci. Comp. Prog.*, 8:231–274, 1987.

[Har89] D. Harel. A thesis for bounded concurrency. In *Proc. 14th Symp. Math. Found. Comput. Sci.*, pages 35–48. Lec. Notes in Comp. Sci. 379, Springer-Verlag, 1989.

[HGdR88] C. Huizing, R. Gerth, and W.P. de Roever. Modelling statecharts behaviour in a fully abstract way. In *Proc. 13th CAAP*, pages 271–294. Lecture Notes in Comp. Sci. 299, Springer-Verlag, 1988.

[HH90] T. Hirst and D. Harel. On the power of bounded concurrency II: The pushdown automata level. In *Proc. 15th Coll. Trees in Algebra and Programming*. Lec. Notes in Comp. Sci., Springer-Verlag, 1990.

[HK89] D. Harel and H.A. Kahana. On statecharts with overlapping. Technical report, The Weizmann Institute, 1989.

[HLN+90] D. Harel, H. Lachover, A. Naamad, A. Pnueli, M. Politi, R. Sherman, A. Shtull-Trauring, and M. Trakhtenbrot. Statemate: A working environment for the development of complex reactive systems. *IEEE Trans. Software Engin.*, 16:403–414, 1990.

[HMP90] T. Henzinger, Z. Manna, and A. Pnueli. An interleaving model for real time. In *5th Jerusalem Conference on Information Technology*, pages 717–730, 1990.

[HMP91] T. Henzinger, Z. Manna, and A. Pnueli. Temporal proof methodologies for real-time systems. In *Proc. 18th ACM Symp. Princ. of Prog. Lang.*, pages 353–366, 1991.

[HP85] D. Harel and A. Pnueli. On the development of reactive systems. In *Logics and Models of Concurrent Systems*, pages 477–498. Springer-Verlag, 1985.

[HPSS86] D. Harel, A. Pnueli, J.P. Schmidt, and R. Sherman. On the formal semantics of statecharts. In *Proc. First IEEE Symp. Logic in Comp. Sci.*, pages 54–64, 1986.

[Hui91] C. Huizing. *Semantics of reactive systems: comparison and full abstraction.* PhD thesis, Technical University Eindhoven, 1991.

[Mar90] F. Maraninchi. *Argos: un langage graphique pour la conception, la description et la validation des systèmes réactifs.* PhD thesis, University of Grenoble, 1990.

[Mar91] F. Maraninchi. Languages for reactive systems: a common framework for comparing statecharts and argos. Technical report, LGI-IMAG Grenoble, 1991.

[MMP92] O. Maler, Z. Manna, and A. Pnueli. A formal approach to hybrid systems. In *Proceedings of the REX workshop "Real-Time: Theory in Practice"*, LNCS. Springer Verlag, New York, 1992.

[Plo81] G. D. Plotkin. A structural approach to operational semantics. Technical report, Dept. of Comp. Sci., Arhus University, 1981.

[PS89] A. Pnueli and M. Shalev. What is in a step? In J.W. Klop, J.-J.Ch. Meijer, and J.J.M.M. Rutten, editors, *J.W. De Bakker, Liber Amicorum*, pages 373–400. CWI, AMsterdam, 1989.

Lecture Notes in Computer Science

For information about Vols. 1–481
please contact your bookseller or Springer-Verlag

Vol. 482: Y. Kodratoff (Ed.), Machine Learning – EWSL-91. Proceedings, 1991. XI, 537 pages. 1991. (Subseries LNAI).

Vol. 483: G. Rozenberg (Ed.), Advances in Petri Nets 1990. VI, 515 pages. 1991.

Vol. 484: R. H. Möhring (Ed.), Graph-Theoretic Concepts in Computer Science. Proceedings, 1990. IX, 360 pages. 1991.

Vol. 485: K. Furukawa, H. Tanaka, T. Fuijsaki (Eds.), Logic Programming '89. Proceedings, 1989. IX, 183 pages. 1991. (Subseries LNAI).

Vol. 486: J. van Leeuwen, N. Santoro (Eds.), Distributed Algorithms. Proceedings, 1990. VI, 433 pages. 1991.

Vol. 487: A. Bode (Ed.), Distributed Memory Computing. Proceedings, 1991. XI, 506 pages. 1991.

Vol. 488: R. V. Book (Ed.), Rewriting Techniques and Applications. Proceedings, 1991. VII, 458 pages. 1991.

Vol. 489: J. W. de Bakker, W. P. de Roever, G. Rozenberg (Eds.), Foundations of Object-Oriented Languages. Proceedings, 1990. VIII, 442 pages. 1991.

Vol. 490: J. A. Bergstra, L. M. G. Feijs (Eds.), Algebraic Methods II: Theory, Tools and Applicatlons. VI, 434 pages. 1991.

Vol. 491: A. Yonezawa, T. Ito (Eds.), Concurrency: Theory, Language, and Architecture. Proceedings, 1989. VIII, 339 pages. 1991.

Vol. 492: D. Sriram, R. Logcher, S. Fukuda (Eds.), Computer-Aided Cooperative Product Development. Proceedings, 1989 VII, 630 pages. 1991.

Vol. 493: S. Abramsky, T. S. E. Maibaum (Eds.), TAPSOFT '91. Volume 1. Proceedings, 1991. VIII, 455 pages. 1991.

Vol. 494: S. Abramsky, T. S. E. Maibaum (Eds.), TAPSOFT '91. Volume 2. Proceedings, 1991. VIII, 482 pages. 1991.

Vol. 495: 9. Thalheim, J. Demetrovics, H.-D. Gerhardt (Eds.), MFDBS '91. Proceedings, 1991. VI, 395 pages. 1991.

Vol. 496: H.-P. Schwefel, R. Männer (Eds.), Parallel Problem Solving from Nature. Proceedings, 1990. XI, 485 pages. 1991.

Vol. 497: F. Dehne, F. Fiala. W.W. Koczkodaj (Eds.), Advances in Computing and Information - ICCI '91. Proceedings, 1991. VIII, 745 pages. 1991.

Vol. 498: R. Andersen, J. A. Bubenko jr., A. Sølvberg (Eds.), Advanced Information Systems Engineering. Proceedings, 1991. VI, 579 pages. 1991.

Vol. 499: D. Christodoulakis (Ed.), Ada: The Choice for '92. Proceedings, 1991. VI, 411 pages. 1991.

Vol. 500: M. Held, On the Computational Geometry of Pocket Machining. XII, 179 pages. 1991.

Vol. 501: M. Bidoit, H.-J. Kreowski, P. Lescanne, F. Orejas, D. Sannella (Eds.), Algebraic System Specification and Development. VIII, 98 pages. 1991.

Vol. 502: J. Bārzdiņž, D. Bjørner (Eds.), Baltic Computer Science. X, 619 pages. 1991.

Vol. 503: P. America (Ed.), Parallel Database Systems. Proceedings, 1990. VIII, 433 pages. 1991.

Vol. 504: J. W. Schmidt, A. A. Stogny (Eds.), Next Generation Information System Technology. Proceedings, 1990. IX, 450 pages. 1991.

Vol. 505: E. H. L. Aarts, J. van Leeuwen, M. Rem (Eds.), PARLE '91. Parallel Architectures and Languages Europe, Volume I. Proceedings, 1991. XV, 423 pages. 1991.

Vol. 506: E. H. L. Aarts, J. van Leeuwen, M. Rem (Eds.), PARLE '91. Parallel Architectures and Languages Europe, Volume II. Proceedings, 1991. XV, 489 pages. 1991.

Vol. 507: N. A. Sherwani, E. de Doncker, J. A. Kapenga (Eds.), Computing in the 90's. Proceedings, 1989. XIII, 441 pages. 1991.

Vol. 508: S. Sakata (Ed.), Applied Algebra, Algebraic Algorithms and Error-Correcting Codes. Proceedings, 1990. IX, 390 pages. 1991.

Vol. 509: A. Endres, H. Weber (Eds.), Software Development Environments and CASE Technology. Proceedings, 1991. VIII, 286 pages. 1991.

Vol. 510: J. Leach Albert, B. Monien, M. Rodríguez (Eds.), Automata, Languages and Programming. Proceedings, 1991. XII, 763 pages. 1991.

Vol. 511: A. C. F. Colchester, D.J. Hawkes (Eds.), Information Processing in Medical Imaging. Proceedings, 1991. XI, 512 pages. 1991.

Vol. 512: P. America (Ed.), ECOOP '91. European Conference on Object-Oriented Programming. Proceedings, 1991. X, 396 pages. 1991.

Vol. 513: N. M. Mattos, An Approach to Knowledge Base Management. IX, 247 pages. 1991. (Subseries LNAI).

Vol. 514: G. Cohen, P. Charpin (Eds.), EUROCODE '90. Proceedings, 1990. XI, 392 pages. 1991.

Vol. 515: J. P. Martins, M. Reinfrank (Eds.), Truth Maintenance Systems. Proceedings, 1990. VII, 177 pages. 1991. (Subseries LNAI).

Vol. 516: S. Kaplan, M. Okada (Eds.), Conditional and Typed Rewriting Systems. Proceedings, 1990. IX, 461 pages. 1991.

Vol. 517: K. Nökel, Temporally Distributed Symptoms in Technical Diagnosis. IX, 164 pages. 1991. (Subseries LNAI).

Vol. 518: J. G. Williams, Instantiation Theory. VIII, 133 pages. 1991. (Subseries LNAI).

Vol. 519: F. Dehne, J.-R. Sack, N. Santoro (Eds.), Algorithms and Data Structures. Proceedings, 1991. X, 496 pages. 1991.

Vol. 520: A. Tarlecki (Ed.), Mathematical Foundations of Computer Science 1991. Proceedings, 1991. XI, 435 pages. 1991.

Vol. 521: B. Bouchon-Meunier, R. R. Yager, L. A. Zadek (Eds.), Uncertainty in Knowledge-Bases. Proceedings, 1990. X, 609 pages. 1991.

Vol. 522: J. Hertzberg (Ed.), European Workshop on Planning. Proceedings, 1991. VII, 121 pages. 1991. (Subseries LNAI).

Vol. 523: J. Hughes (Ed.), Functional Programming Languages and Computer Architecture. Proceedings, 1991. VIII, 666 pages.

Lecture Notes in Computer Science

This series reports new developments in computer science research and teaching, quickly, informally, and at a high level. The timeliness of a manuscript is more important than its form, which may be unfinished or tentative. The type of material considered for publication includes

- drafts of original papers or monographs,
- technical reports of high quality and broad interest,
- advanced-level lectures,
- reports of meetings, provided they are of exceptional interest and focused on a single topic.

Publication of Lecture Notes is intended as a service to the computer science community in that the publisher Springer-Verlag offers global distribution of documents which would otherwise have a restricted readership. Once published and copyrighted they can be cited in the scientific literature.

Manuscripts

Lecture Notes are printed by photo-offset from the master copy delivered in camera-ready form. Manuscripts should be no less than 100 and preferably no more than 500 pages of text. Authors of monographs receive 50 and editors of proceedings volumes 75 free copies. Authors of contributions to proceedings volumes are free to use the material in other publications upon notification to the publisher. Manuscripts prepared using text processing systems should be printed with a laser or other high-resolution printer onto white paper of reasonable quality. To ensure that the final photo-reduced pages are easily readable, please use one of the following formats:

Font size	Printing area		Final size
(points)	(cm)	(inches)	(%)
10	13.5 x 20.0	5.3 x 7.9	100
12	16.0 x 23.5	6.3 x 9.2	85
14	18.0 x 26.5	7.0 x 10.5	75

On request the publisher will supply a leaflet with more detailed technical instructions or a TEX macro package for the preparation of manuscripts.

Manuscripts should be sent to one of the series editors or directly to:

Springer-Verlag, Computer Science Editorial I, Tiergartenstr. 17, W-6900 Heidelberg 1, FRG

ISBN 3-540-55092-5
ISBN 0-387-55092-5